Thermal and Catalytic Processes in Petroleum Refining

Thermal and Catalytic Processes in Petroleum Refining

Serge Raseev

*Consultant for UNESCO, Paris, France and
former Professor, Institute of Petroleum and Gases,
Bucharest-Ploiesti, Romania*

Technical editor for the
English-language version
G. Dan Suciu

CRC Press
Taylor & Francis Group
Boca Raton London New York

CRC Press is an imprint of the
Taylor & Francis Group, an **informa** business

CRC Press
Taylor & Francis Group
6000 Broken Sound Parkway NW, Suite 300
Boca Raton, FL 33487-2742

First issued in paperback 2019

© 2003 by Taylor & Francis Group, LLC
CRC Press is an imprint of Taylor & Francis Group, an Informa business

No claim to original U.S. Government works

ISBN-13: 978-0-8247-0952-5 (hbk)
ISBN-13: 978-0-367-39544-5 (pbk)

Library of Congress Cataloging-in-Publication Data
A catalog record for this book is available from the Library of Congress.

Visit the Taylor & Francis Web site at
http://www.taylorandfrancis.com

and the CRC Press Web site at
http://www.crcpress.com

To my dear wife Irena

Preface

This book is considered to be a completely new version of the original book published in 3 volumes in Romania, in 1996–1997 under the title *Conversia Hidrocarburilor* ("the conversion of hydrocarbons").

Recent developments in petroleum processing required the complete revision of some of the chapters, the elimination of outdated material and bringing up to date the processes in which the technology was significantly improved. Furthermore, the presentation of theoretical aspects has been somewhat expanded and deepened.

The processes discussed in this book involve the conversion of hydrocarbons by methods that do not introduce other elements (heteroatoms) into hydrocarbon molecules. The first part is devoted to thermal conversion processes (pyrolysis, visbreaking, coking). The second part studies catalytic processes on acidic catalysts (catalytic cracking, alkylation of isoalkanes, oligomerization). The third and fourth parts analyze catalytic processes on metal oxides (hydrofining, hydrotreating) and on bifunctional catalysts (hydroisomerization, hydrocracking, catalytic reforming), respectively.

The importance of all these processes resides in the fact that, when required, they allow large variations in the proportion of the finished products as well as improvement of their quality, as required by increasingly stringent market demands. The products of primary distillation are further processed by means of secondary operations, some fractions being subjected to several processing steps in series. Consequently, the total capacity of the conversion processes is larger than that of the primary distillation.

The development of petroleum refining processes has made it possible to produce products, especially gasoline, of improved quality and also to produce synthetic chemical feedstocks for the industry. The petrochemical branch of the refining industry generates products of much higher value than does the original refining industry from which the feedstocks were derived.

One should not overlook the fact that the two branches are of quite different volume. A few percentage points of the crude oil processed in the refineries are sufficient to cover the needs for feeds of the whole petrochemical and synthetic organic industry and of a large portion of the needs of the inorganic chemicals industry. The continuous development of new products will result in a larger fraction of the crude oil than the approximately 10% used presently being consumed as feedstocks for the chemical industry.

Hydrocarbons conversion processes supply hydrocarbons to the petrochemical industry, but mainly they produce fuels, especially motor fuels and quality lubricating oils. The same basic processes are used in all these different applications. The specific properties of the feedstocks and the operating parameters are controlled in order to regulate the properties of the product for each application. In this book, the processes are grouped by these properties, in order to simplify the presentation and to avoid repetitions.

The presentation of each group of processes begins with the fundamentals common to all the processes: thermodynamics, reaction mechanisms (including catalysis when applicable), and, finally, process kinetics. In this manner, operating parameters practiced in commercial units result as a logical consequence of earlier theoretical discussion. This gives the reader a well-founded understanding of each type of process and supplies the basis on which improvements of the process may be achieved.

The presentation of commercial implementation is followed by a discussion of specific issues pertaining to the design of the reaction equipment, which results in the unity of the theoretical bases with the design solutions adopted for commercial equipment and the quantitative aspects of implementation.

My warmest thanks to Prof. Sarina Feyer-Ionescu, to my son Prof. George Raseev, and especially to my technical editor Dr. G. Dan Suciu, for their support in preparing the English-language version of this book.

Serge Raseev

Preface to the Romanian Edition

This book is the fruit of many years of work in the petrochemical industry, and in research, and of university teaching. It sums up my technical and scientific background and reflects the concepts that I developed over the years, of the manner in which the existing knowledge on chemical process technology—and especially on the processing of hydrocarbons and petroleum fractions—should be treated and conveyed to others.

While initially the discipline of process technology was taught mainly by describing the empirical information, it soon changed to a quantitative discipline that considers the totality of phenomena that occur in the processes of chemical conversion of industrial interest.

The objective of process technology as a discipline is to find methods for the continual improvement of commercial processes. To this purpose it uses the latest advances in chemistry, including catalysis, and applies the tools of thermodynamics and kinetics toward the quantitative description of the processes. In this manner it became possible to progress from the quantitative description provided by the reaction mechanisms to the mathematic formulation for the evolution in time of the processes.

In order to implement the chemical process on a commercial scale, a series of additional issues need to be addressed: the effect of the operating parameters and the selection of the optimal operating conditions, selection of the reactor type, the design of the reaction equipment and of the other processing steps, the limitations due to the heat and mass transfer, and the limitations imposed by the materials of construction.

Process technology thus becomes the convergence point of several theoretical and applicative disciplines called upon to solve in an optimal manner the complex interrelations among quite different sciences and phenomena (chemistry, hydraulics, heat transfer, etc.). This situation requires a multifaceted competence and the full understanding and control of the entire complex phenomenon that is the implementation of chemical conversions in the conditions of the commercial units. Without it, one cannot address the two basic questions about process technology: first, why the commercial processes have been developed in the manner they are presently implemented and second, how they can be continually improved.

In this manner, by mastering the complex phenomena involved, the process engineer is fully equipped to answer the "why" and "how" questions, and will be able to become one of the important driving forces of technical progress. This is the concept that has guided me during my entire professional activity.

This book treats the conversion of hydrocarbons and petroleum fractions by thermal and catalytic methods, while attempting to answer the "why" and "how" questions at the level of the current technical knowledge. In this manner, I hope to contribute to the education of specialists who will advance continuing developments in processing methods.

I am thankful to Mr. Gavril Musca and Dr. Grigore Pop for their help in creating this book. My special gratitude goes to Prof. Sarina Feyer-Ionescu, for her special contributions.

Serge Raseev

Contents

1

Thermodynamic Analysis of Technological Processes

The thermodynamic study of technological processes has two objectives:

Determination of the overall thermal effect of chemical transformations that take place in the industrial process

Determination of the equilibrium composition for a broad range of temperatures and pressures in order to deduce optimum working conditions and performances

The manner in which the two objectives are approached within the conditions of chemical technology is different from the classical approach and requires the use of the specific methodology outlined in this chapter.

1.1 CALCULATION OF THE OVERALL THERMAL EFFECT

In practical conditions under which technological processes operate, the main reaction may be accompanied by secondary reactions. In many cases the transformation is of such complexity that it cannot be expressed by a reasonable number of chemical reactions.

When calculating the heat of reaction in such situations, in order to avoid the difficulties resulting from taking into account all reactions many times in the calculation, simplified approaches are taken. Thus, one may resort to the approximation of limiting the number of the reactions taken into consideration, or to take account only the main reaction. Such approximations may lead to significant errors.

Actually, the exact value of the thermal effect can be calculated without having to resort to such approximations. Since the thermal effect depends only on the initial and the final state of the system (the independence of path, as stipulated by the second principle of thermodynamics), it may be calculated based on the initial and final compositions of the system, without having to take in account the reactions that take place.

Accordingly, the classic equations, which give the thermal effect of a chemical reaction:

$$\Delta H_{rT}^0 = \sum v_p \Delta H_{fT}^0 - \sum v_r \Delta H_{fT}^0 \tag{1.1}$$

$$\Delta H_{rT}^0 = \sum v_r \Delta H_{cT}^0 - \sum v_p \Delta H_{cT}^0 \tag{1.2}$$

may be written under the form:

$$\Delta H_{rT} = \sum n_e \Delta H_{fT} - \sum n_i \Delta H_{fT} \tag{1.3}$$

$$\Delta H_{rT} = \sum n_i \Delta H_{cT} - \sum n_e \Delta H_{cT} \tag{1.4}$$

The heats of formation ΔH_f and of combustion ΔH_c for hydrocarbons and organic compounds, which are of interest in studying petrochemical processes, are given in thermodynamic data books [1,2]. The values are usually given for temperature intervals of 100 K, within which linear interpolation is accurate. Thus, the calculations that use the heat capacities may be avoided.

Example 1.1 shows how to perform the calculations by means of relations (1.3) and (1.4).

Example 1.1. Compute the overall thermal effect of an industrial dehydrogenation process of isopentane to isoprene at 600°C.

The composition of the streams at the inlet and outlet of the reactor is given in Table 1.1. The coke composition by weight, is 95% carbon and 5% hydrogen.

The calculations of the heat of formation at the inlet and the outlet of the reactor at 600°C are collected in Table 1.2.

Table 1.1

Component	Reactor inlet feed + recycle (wt %)	Reactor Outlet (wt %)
H_2	-	1.0
CH_4	-	0.6
C_2H_6	-	0.7
C_2H_4	-	0.7
C_3H_8	-	0.7
C_3H_6	-	1.4
C_4H_{10}	0.3	1.2
C_4H_8	-	2.2
C_4H_6	-	0.2
$i\text{-}C_5H_{12}$	79.3	55.8
$i\text{-}C_5H_{10}$	16.6	17.1
C_5H_8	0.8	12.1
$n\text{-}C_5H_{12}$	1.8	0.8
$n\text{-}C_5H_{10}$	1.7	1.7
$1,3\text{-}C_5H_8$	-	2.0
coke	-	1.8

Table 1.2

Component	Heat of formation ΔH_f^0 (kcal/mol) [2]			Inlet		Outlet	
	800 (K)	900 (K)	873 = 600 (K) (°C)	n_i (mol/kg)	$n_i \Delta H_{f873}^0$ (kcal/kg)	n_e (mol/kg)	$n_e \Delta H_{f873}^0$ (kcal/kg)
H_2	0	0	0	-	-	9.92	0
CH_4	−20.82	−21.15	−21.05	-	-	0.37	−7.79
C_2H_6	−24.54	−24.97	−24.85	-	-	0.23	−5.72
C_2H_4	9.77	9.45	9.54	-	-	0.25	2.39
C_3H_8	−30.11	−30.58	−30.45	-	-	0.16	−4.87
C_3H_6	0.77	0.35	0.46	-	-	0.33	0.15
C_4H_{10}	−36.41	−36.93	−36.79	0.05	−1.84	0.21	−7.73
C_4H_8	−6.32	−6.84	−6.70	-	-	0.39	−2.61
C_4H_6	23.25	22.95	23.03	-	-	0.04	0.92
$i\text{-}C_5H_{12}$	−44.13	−44.65	−44.61	10.99	−489.16	7.73	−344.06
$i\text{-}C_5H_{10}$	−13.45	−13.93	−13.80	2.37	−32.71	2.44	−33.67
C_5H_8	14.16	13.82	13.91	0.12	1.67	1.78	24.76
$n\text{-}C_5H_{12}$	−42.28	−42.85	−42.70	0.25	−10.68	0.11	−4.70
$n\text{-}C_5H_{10}$	−12.23	−12.78	−12.63	0.24	−3.03	0.24	−3.03
$1,3\text{-}C_5H_8$	14.17	13.73	13.85	-	-	0.29	4.02
C	0	0	0	-	-	-	-
Total { kcal/kg				-	−535.75	-	−381.94
{ kJ/kg				-	−2243.1	-	−1599.1

According to Eq. (1.3), the overall thermal effect per unit mass (kg) of feed will be:

$$\Delta H_{r,873} = \sum n_e \Delta H_{f873} - \sum n_i \Delta H_{f873} = -1599 - (-2243.1) = 644 \text{ kJ/kg}$$

Since the process is performed at a temperature much above the critical point and at low pressure, no deviations from the ideal state have to be considered.

In many cases it is convenient to express the thermal effect on the basis of the reacted isopentane or of the formed isoprene.

For this example, according to Table 1.1, $793 - 558 = 235g$, isopentane reacts and $121 - 8 = 113g$, isoprene is formed. In these conditions, the thermal effect expressed per mole of reacted isopentane is:

$$\Delta H_r = \frac{644}{235} \times 72.15 = 197.7 \text{ kJ/mole}$$

and per mole of produced isoprene:

$$H_r = \frac{644}{113} \times 68.11 = 388.2 \text{ kJ/mole}$$

If only the main reaction:

$$i - C_5H_{12} = i - C_5H_8 + 2H_2$$

is taken into account, then according to the Eq. (1.1) one obtains:

$$\Delta H_r = (\Delta H_f)_{C_5H_8} - \Delta H_f)_{C_5H_{12}} = 13.91 - (-44.51) = 58.42 \text{ kcal/mol}$$
$$= 244.59 \text{ kJ/mol}$$

the value being the same whether expressed per mole of isopentane or of isoprene.

This example shows that large errors may result if the computation of the overall thermal effect is not based on the real compositions of the inlet and outlet streams of the reactor.

Eq. (1.4) makes it possible to compute the thermal effects by using the heats of combustion. This is useful for the conversion of petroleum fractions of other feedstocks consisting of unknown components. In such cases it is usually more convenient to perform the calculation in weight units, by modifying the terms n and ΔH accordingly.

For liquid petroleum fractions, the heats of combustion may be determined by using the graph of Figure 1.1 [3], from the known values of the specific gravity and the characterization factor.

The characterization factor of residues may be determined graphically from the viscosity, by means of Figure 1.2 [3].

The heat of combustion of coke is determined experimentally or less precisely on the basis of the elementary composition.

The heats of combustion of gaseous components may be found in data books [1,2], or may be calculated from the heats of formation [2], by applying Eq. (1.1). For hydrocarbons, this equation takes the form:

$$(\Delta H_a)_{C_nH_m} = n(\Delta H_f)_{CO_2} + \frac{m}{2}(\Delta H_f)_{H_2O} - (\Delta H_f)_{C_nH_m} \tag{1.5}$$

This heat of combustion of gases must be brought to the same reference state as that of liquid fractions, i.e. 15°C and liquid water. For these conditions, Eq. (1.5) becomes:

$$(\Delta H_a)_{C_nH_m} = -393.77n - 143.02m - (\Delta H_f)_{C_nH_m} \tag{1.6}$$

It must be noted that Eq. (1.6) gives the heat of combustion in thermodynamic notation, expressed in kJ/mole. Figure 1.1 gives the heat of combustion in technical notation, expressed in kJ/kg.

An illustration of these calculations is given in Example 1.2.

Example 1.2. Calculate the thermal effect of the processing of a vacuum residue by visbreaking. The composition of the produced gases is given in Table 1.3. The yields and the characterization factors, K_{UOP} for the feed and the fuel oil were obtained from Table 1.4.

The characterization factor and the specific gravities were used to determine the heats of combustion for all the liquid fraction from Figure 1.1.

SOLUTION. By introducing the values of the heats of combustion from Tables 1.3 and 1.4 into Eq. (1.4), one obtains:

$$Q_r = 43,645 - (0.0244 \times 51,319 + 0.1166 \times 46,827 + 0.859 \times 43,233)$$
$$= -204 \text{ kJ/kg}$$

Calculation of the thermal effects for a specific reaction, usually a small number obtained as the difference of heats of combustion, usually larger numbers, is

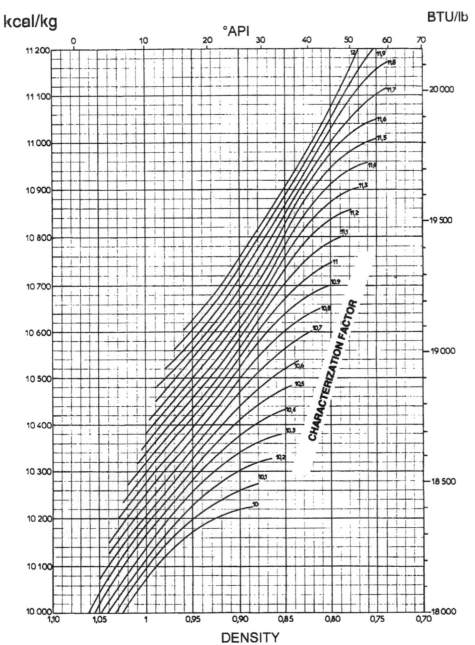

Figure 1.1   Heat of combustion of petroleum fractions. Final state: gaseous CO_2 and liquid water at 15°C.

associated with large errors, unless the determination of the values of the heats of combustion was made with high accuracy. This fact is especially valid for liquid fractions, for which the graphical determination of the combustion heats may give errors. In order to obtain exact results, the determination of the heats of combustion of the liquid fractions by direct calorimetric methods is recommended.

cST at 50° C

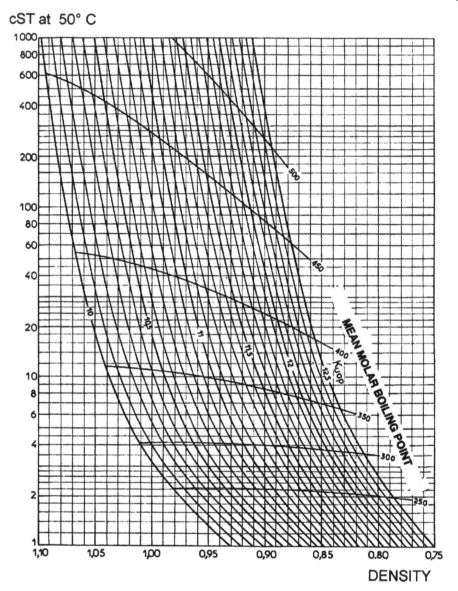

DENSITY

Figure 1.2 K_{UOP} as function of the kinematic viscosity and density.

Graphs and empirical relations are given [4–7] for the calculation of the thermal effect in the petroleum refining processes. The values calculated by their means and the numerical values given in the literature must be critically analyzed, taking into account the characteristics of the feed, the operating conditions, and the conversion. Only values that refer to comparable feeds and conditions should be used in computations.

For the process of thermal cracking, the use of equation [8] is recommended:

$$\Delta H = 117{,}230 \frac{M_a - M_p}{M_a \times M_p} \tag{1.7}$$

Table 1.3

Component	Composition (wt %)	$(\Delta H^0_{288})_C$ (kJ/kg)	$(\Delta H^0_{283})_C$ fraction (kJ/kg)
CH_4	22.32	−55,540	−12,396
C_2H_6	18.84	−51,910	−9,780
C_3H_8	4.57	−50,330	−2,300
C_3H_6	20.56	−50,380	−10,358
i-C_4H_{12}	7.97	−48,950	−3,901
n-C_4H_{12}	2.20	−49,390	−1,093
i-C_4H_{10}	9.20	−49,540	−4,558
n-C_4H_{10}	1.85	−48,170	−891
1-C_4H_8	3.50	−48,470	−1,696
cis-2-C_4H_8	0.55	−48,340	−266
$trans$-2-C_4H_8	2.37	−48,270	−1,144
C_4H_6	2.01	−47,020	−945
C_5+	4.06	−49,050	−1,991

Total $(\Delta H^0_{288})_C$ fraction −51,319 kJ/kg

The calculated ΔH is expressed in kJ/kg of feed. The sign is that used in the thermodynamic notation.

Using the data from example 1.2 (see the Table 1.4), this equation gives:

$$\Delta H = 117,230 \frac{440 - 253}{440 \times 253} = 197 \text{ kJ/kg}$$

which gives the same result as the heats of combustion method.

In the literature, the thermal effect of reactions is often expressed per unit mass of main product and not per unit mass of feed. In some cases, this way of expression is useful, since the thermal effect thus becomes actually independent of conversion [5].

1.2 EQUILIBRIUM CALCULATIONS FOR A WIDE RANGE OF PROCESS CONDITIONS

The computation of the equilibrium compositions for a wide range of process conditions (temperatures and pressures) has the purpose of identifying practical operating conditions that will optimize the performance of the process. Depending on the specifics of the process, the problem may be limited to the calculation of the equilibrium of the main reaction, or may be extended also to the secondary reactions.

In all cases, the composition at equilibrium, calculated on basis of thermodynamic principles, represents the maximum conversion that is possible to achieve in the given conditions. There is however no certainty that such performance will be actually obtained. Nonthermodynamic factors, such as the reaction rate and the residence time within the reactor will determine how close the actual performance will approach the theoretical one.

The use of classical methods for computing equilibrium compositions for the large number of temperature–pressure values needed for thermodynamic analysis of a broad range of process conditions necessitates a large number of calculations. A

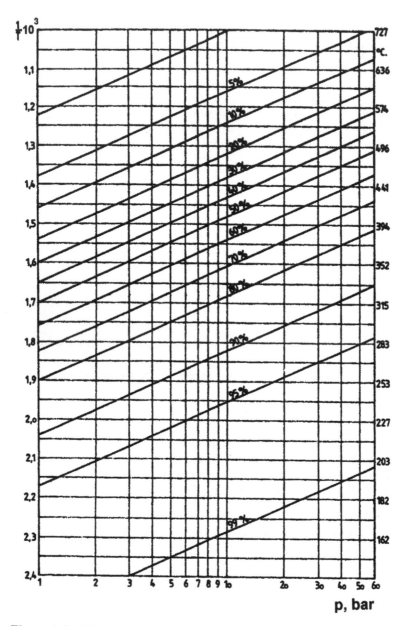

Figure 1.3 Thermodynamic equilibrium of propene dimerization. Parameter: conversion x as %.

method elaborated by the author many years ago [9] provides a simple method for the calculation and graphical representation of the equilibrium. The method is outlined below.

For any chemical reaction, the standard free energy is expressed by the relation:

$$\Delta G_T^0 = \Delta H_T^0 - T\Delta S_T^0 \tag{1.8}$$

Table 1.4

	Yields (wt %)	Density	Viscosity (cSt)	Characterization factor (K_{UOP})	Thermal effect (kJ/Kg)
Feed	1.0000	0.989	1,000	11.38	43,645
Products					
gases	0.0244	-	-	-	51,319
gasoline	0.1166	0.760	-	11.9	46,827
fuel oil	0.8590	1.000	630	11.1	43,233

and as function of the equilibrium constant, by the expression:

At equilibrium, $\Delta G_T = 0$ and $\Delta G_T^0 = RT \ln K_a$ $\qquad$ (1.9)

Assuming that the substances participating in the reaction do not deviate from the behaviour of ideal gas, the equilibrium constant may be expressed by the relation:

$$K_a = K_p = \frac{\varphi_i(x)}{p^{\Delta n}} \qquad (1.10)$$

Here, $\varphi_i(x)$ is a function of the conversion at equilibrium x. The form of this function depends on the stoechiometry of the reaction but is independent on the nature of the substances that participate in the reaction.

Equating Eqs. (1.8) and (1.9), and replacing K_a with the expression (1.10), dividing the right and left sides by $T\Delta H_T^0$, and effecting some elementary transformations, one obtains:

$$\frac{1}{T} = \frac{R\Delta n}{\Delta H_T^0} \ln p + \frac{\Delta S_T^0 - R \ln [\varphi_i(x)]}{\Delta H_T^0} \qquad (1.11)$$

For a given chemical reaction and a temperature range of 200–300°C, which is sufficient for a process analysis, $\Delta H^0 T$ and $\Delta S^0 T$ can be considered constants. In these conditions, using as coordinates $\log p$ and $1/T$, the Eq. (1.11) corresponds to a family of parallel straight lines with the equilibrium conversion x as parameter.

Simple plots are obtained, by writing:

$2.3\, R\Delta n = b$

and

$R \ln[\varphi i(x)] = d$

Equation (1.11) becomes:

$$\frac{1}{T} = \frac{b}{\Delta H_T^0} \log p + \frac{\Delta S_T^0 - d}{\Delta H_T^0} \qquad (1.12)$$

The parameter b depends only on the stoechiometric form of the chemical reaction. Parameter d depends both on the stoechiometry and on x, the conversion at equilibrium. Both b and d are independent of the nature of the chemical substances that take part in the reaction and have been calculated [9] for chemical reactions of various stoechiometric forms (Table 1.5).

For reactions proceeding in the opposite direction, the sign of the constants b and d must be changed, and the meaning of the conversion x reversed (for example $x = 0.95$ from the table will have the meaning $x = 0.05$ for the reverse reaction).

Since in plots of log p versus $1/T$ the straight lines of constant conversion are parallel, it is enough to calculate one point of each line and to determine the slope of all the straight lines by calculating just one point for any other pressure. Thus, the whole family of lines may be obtained by selecting a pressure of 1 bar for the determining one point on each straight line and a pressure of either 10 bar or 0.1 bar for which one calculates the one point needed to determine the slope of all lines.

For these values of the pressure, the relation (1.12) becomes:

$$p = 1\,\text{bar} \qquad \frac{1}{T} = \frac{\Delta S_T^0 - d}{\Delta H_T^0}$$

$$p = 10\,\text{bar} \qquad \frac{1}{T} = \frac{\Delta S_T^0 - d + b}{\Delta H_T^0} \qquad\qquad (1.13)$$

$$p = 0.1\,\text{bar} \qquad \frac{1}{T} = \frac{\Delta S_T^0 - d - b}{\Delta H_T^0}$$

The calculation is illustrated by the Example 1.3.

Example 1.3. For the reaction:

$$2C_3H_6 \rightleftharpoons C_6H_{12}$$

determine the equilibrium graph for pressures comprised between 1 and 100 bar and temperatures between 450–800°C.

SOLUTION. The reaction corresponds to the form 2A ↔ B, in Table 1.5.

Using the thermodynamic constants for 900 K [2] and taking mean values for *i*-hexenes, it results:

$$\Delta H_{900}^0 = -20020 - 2 \times 350 = -20{,}720 \text{ cal/mol}$$

$$\Delta S_{900}^0 = 143.65 - 2 \times 89.75 = -35.85 \text{ cal/mol K}$$

Using the Eq. (1.13) corresponding to the pressure of 1 bar and the values of the constant d from the Table 1.5, following pairs of values are obtained:

X	0.01	0.05	0.10	0.20	0.3	0.40	0.50	0.60	0.70	0.80	0.90	0.95	0.99
$(1/T) \times 10^3$	1.22	1.38	1.46	1.54	1.60	1.65	1.70	1.76	1.82	1.90	2.04	2.17	2.50

For the pressure of 10 bar and $x = 0.5$ and using the constant b from the Table 1.5, one obtains, according to the Eq. (1.13):

$$1/T = 1.48 \times 10^{-3}$$

By using the obtained values, the equilibrium is represented in Figure 1.3.

Note that for temperature ranges of not more than 200–300°C that intervene in the analysis of industrial processes, the variations with the temperature of ΔH^0 and ΔS^0 may be neglected, without consequently introducing any practical errors.

Deviations from ideal conditions are important near the critical state and do not affect the results at temperatures much higher than the critical, as used in the

TABLE 1.5 Values of the Constants b and d for Various Reaction Stoichiometric Types

Reaction form	b	x = equilibrium conversion												
		0.99	0.95	0.90	0.80	0.70	0.60	0.50	0.40	0.30	0.20	0.10	0.05	0.01
A↔B	0	9.13	5.85	4.37	2.75	1.68	0.81	0	−0.81	−1.70	−2.75	−4.37	−5.85	−9.13
2A↔B	4.57	15.85	9.14	6.37	3.56	1.84	0.54	−0.57	−1.61	−2.67	−3.90	−5.66	−7.18	−10.48
3A↔B	9.14	16.38	11.41	7.69	3.94	1.79	0.23	−1.04	−2.19	−3.34	−4.62	−6.40	−7.91	−11.29
A+B↔C	4.57	18.30	11.90	9.13	6.31	4.60	3.29	2.18	1.14	0.08	−1.15	−2.89	−4.42	−7.74
A+B↔2C	0	21.01	14.45	11.48	8.26	6.12	4.36	2.75	1.14	−0.61	−2.75	−5.98	−9.13	−15.54
A+2B↔C	9.14	25.07	15.20	11.48	7.73	5.58	4.03	2.75	1.60	0.46	−0.83	−2.62	−4.17	−7.53
A+3B↔C	13.72	30.65	18.00	13.10	8.61	6.14	4.42	3.04	1.83	0.63	−0.58	−2.49	−4.05	−7.44
A+B↔C+D	0	18.25	11.70	8.73	5.51	3.37	1.61	0	−1.61	−3.37	−5.51	−8.73	−11.70	−18.25
A+2B↔2C	2.29	33.42	19.07	14.76	10.20	7.41	4.77	3.20	0.18	−0.69	−3.03	−6.38	−9.45	−16.10
A+3B↔2C	4.57	31.01	22.71	17.21	11.60	8.16	5.54	3.32	1.22	−1.74	−3.55	−6.80	−9.88	−16.53
A+2B↔C+D	4.57	26.04	16.33	12.03	7.52	4.66	2.42	0.443	−1.45	−3.44	−5.76	−9.15	−12.25	−18.87
A+3B↔C+D	9.14	32.81	20.01	14.47	8.84	5.40	2.80	0.572	−1.51	−3.64	−6.08	−9.56	−12.64	−19.28
A+4B↔C+D	13.72	38.92	23.10	16.40	9.78	5.83	2.99	0.581	−1.63	−3.85	−6.37	−9.90	−13.00	−19.66
A+5B↔C+D	18.28	44.54	25.79	18.00	10.50	6.19	3.09	0.538	−1.77	−4.06	−6.62	−10.19	−13.31	−19.98
A↔B+4C	−18.28	7.34	3.99	2.418	0.594	−0.744	−1.96	−3.22	−4.66	−6.50	−9.20	−14.32	−20.09	−35.03
2C↔A+5B	−18.28	17.22	10.56	7.44	3.87	1.30	−0.99	−3.29	−5.84	−8.95	−13.25	−20.76	−28.54	−47.3

thermal and catalytic processes in petroleum refining. If corrections as such are however needed, they can be accomplished by using the methods elaborated in the original work [9].

This method of equilibrium representation will be widely used in the following chapters for the analysis of practical process conditions.

REFERENCES

1. FD Rossini, KS Pitzer, RL Arnett, RM Braun, GC Pimentel. Selected Values of Physical and Thermodynamical Properties of Hydrocarbons and Related Compounds, Pittsburgh: Carnegie Press, 1953.
2. DR Stull, EF Westrum Jr., GC Sinke. The Chemical Thermodynamics of Organic Compounds, New York: John Wiley, 1969.
3. P Wuithier. Le petrole raffinage et genie chimique, Vol. 1, 2nd Edition, Technip, Paris, 1972.
4. OA Hougen, KM Watson. Chemical Process Principles, vol. 1. New York: John Wiley, 1947.
5. S Raseev. Procese distructive de prelucrate a titeiului, Editura tehnica, Bucuresti, 1964.
6. G Suciu, R Tunescu. Editors. Ingineria prelucrdrii hidrocarburilor, Editura tehnica, Bucuresti, 1973.
7. WL Nelson. Petroleum Refinery Engineering, New York: McGraw-Hill Book Co., 1958.
8. IH Hirsch, EK Ficher. The Chemistry of Petroleum Hydrocarbons, Vol. 2, Chap. 23, New York: Reinhold Publishing Co., 1955.
9. S Raseev, Stud Cercet Chim 5 (2): 267, 285, 1957.

2

Theoretical Background of Thermal Processes

Thermal processes are chemical transformations of pure hydrocarbons or petroleum fractions under the influence of high temperatures. Most of the transformations are cracking by a radicalic mechanism.

The thermal processes comprise the following types of industrial processes:

PYROLYSIS (STEAM CRACKING). Main purpose: the production of ethene and s propene for the chemical industry. The pyrolysis of liquid feed stocks, leads also to butadiene, isoprene, and C_6-C_8 aromatics.

Characteristic for the pyrolysis process are temperatures of about 900–950°C and low pressures (less than 5 bar).

At the present, pyrolysis is the most important thermal process.

VISBREAKING. Used for producing fuel oils from heavy residues.

The process is characterized by relatively mild temperatures (around 500°C) and pressures, generally of 15–20 bar. Recently, processes at much lower pressures, sometimes atmospheric, were also developed (Section 4.2.1).

Of similar type was the old-time cracking process for gasoline production. It was realized at relatively low temperatures (495–510°C) and high pressure (20–40 bar).

COKING. Used for producing petroleum coke from heavy residues.

There are two types of coking processes: the delayed coking realized at about 490°C, and a 5–15 bar in coke drums, and fluid coking realized at about 570°C and 2–3 bar, in a fluidized bed.

Of some importance is the production of needle coke, which is used for the production of electrodes especially for electrometallurgy processes (e.g. aluminum).

2.1 THERMODYNAMICS OF THERMAL PROCESSES

Thermodynamic calculations show that the thermal decomposition of alkanes of higher molecular weight may take place with high conversions even at relatively low temperatures. Thus, *n*-decane may convert to over 90% to form pentene and pentane at 350°C and 1 atmospheric pressure.

The great number of parallel–successive reactions that may take place results in the final product distribution being controlled by the relative rates of the reactions that take place and not by the thermodynamic equilibrium.

The situation is different for the lower alkanes. Thus, in order to achieve a conversion of 90% in the decomposition of butane to ethene and ethane at a pressure of 2 bar, a temperature of near 500°C is required (Figure 2.1). In these conditions the dehydrogenation reaction reaches a conversion at equilibrium of only about 15% (Figure 2.2). This makes possible a comparison of the two possible reaction pathways.

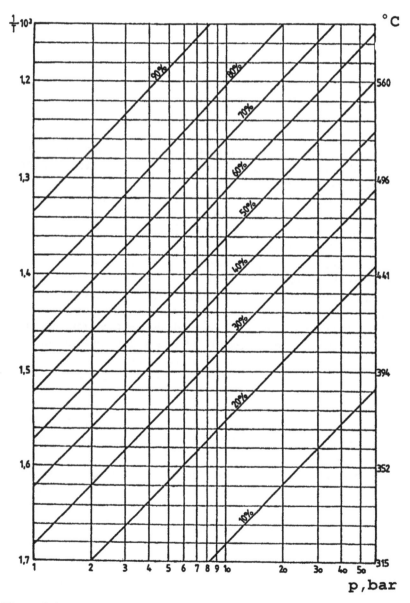

Figure 2.1   The thermodynamic equilibrium for reaction $C_4H_{10} \rightleftharpoons C_2H_6 + C_2H_4$.

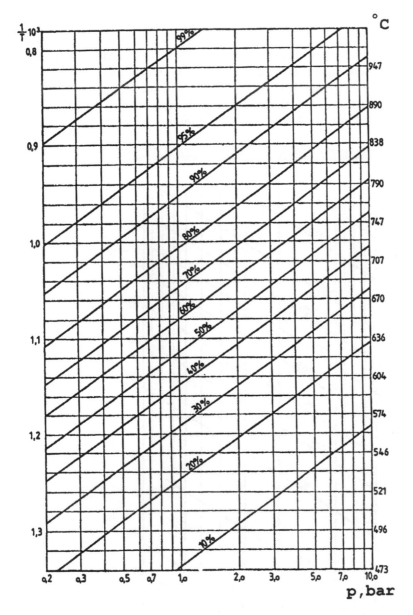

Figure 2.2 The thermodynamic equilibrium for reaction $C_4H_{10} \rightleftharpoons C_4H_8 + H_2$.

The products obtained from the thermal decomposition of ethane, propane, and ethene, are those one would expect from dehydrogenation reactions:

$$C_2H_6 \rightleftharpoons C_2H_4 + H_2 \tag{a}$$
$$C_3H_8 \rightleftharpoons C_3H_6 + H_2 \tag{b}$$
$$C_2H_4 \rightleftharpoons C_2H_2 + H_2 \tag{c}$$

Many studies [1,2,109–111] reached the conclusion that in the pyrolysis process, irrespective of the feedstock used, the values of the concentration ratios ethene/ethane, propene/propane and acetylene/ethene in the reactor effluent are close to those corresponding to the equilibrium of the reactions (a)–(c), Figures (2.3), (2.4), and (2.5).

Even if such assertions are only approximately accurate, the equilibrium of these reactions is of high interest for determining optimum operating conditions in pyrolysis.

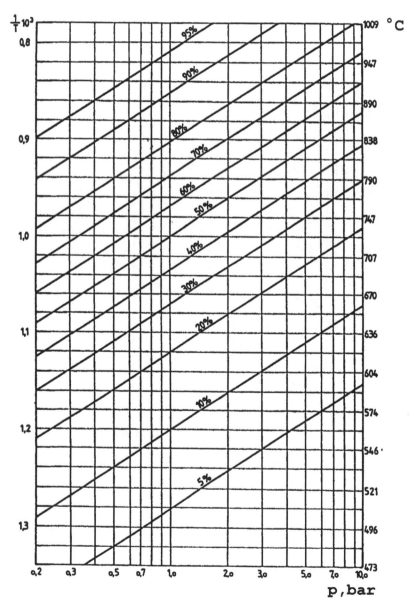

Figure 2.3 The thermodynamic equilibrium for reaction $C_2H_6 \rightleftharpoons C_2H_4 + H_2$.

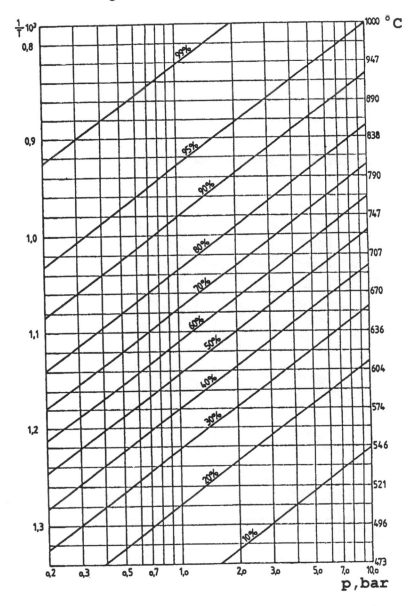

Figure 2.4   The thermodynamic equilibrium for reaction $C_3H_8 \rightleftharpoons C_3H_6 + H_2$.

Figure 2.3 illustrates the significant increase of the conversion of ethane to ethene as the temperature increases from 800 to 950°C. This justifies the use of tubes of special heat-resisting alloys, which make possible the continuous increase of the temperatures towards the coil outlet, as practiced in modern high-conversion pyrolysis furnaces.

The formation of propene (Figure 2.4) is favored by the equilibrium, but the competitive reaction by which propane is cracked to ethene and methane, influences

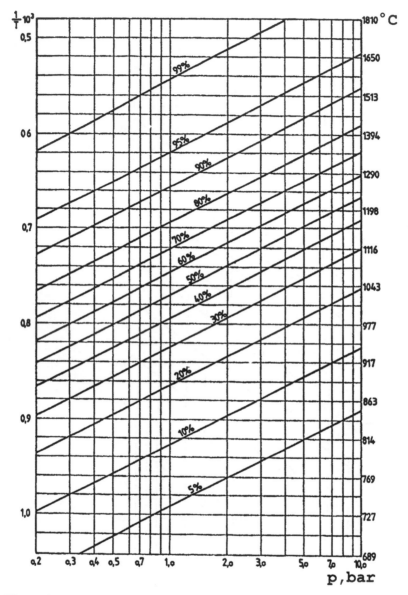

Figure 2.5 The thermodynamic equilibrium for reaction $C_2H_4 \rightleftharpoons C_2H_2 + H_2$.

the final product distribution. The final product composition is determined by the relative rate of the two reactions.

The equilibrium conversion to acetylene is much lower than that to alkenes (Figure 2.5). Still, the presence of acetylene in the reaction products makes their purification necessary. The continuous increase in coil outlet temperatures has made the acetylene removing sections a standard feature of the pyrolysis unit.

Pressure has a strong effect on the conversions at equilibrium of these three reactions. Increased conversions are obtained as the operating pressure decreases. Thus, a 50% lowering of pressure causes a supplementary amount of about 10%

ethane to be transformed to ethene (Figure 2.3). The same effect is obtained by
reducing the partial pressures of the hydrocarbon, e.g. by increasing the proportion
of dilution steam introduced in the reactor.

The equilibrium graph for the dehydrogenation of butene to butadiene (Figure
2.6) shows that in pyrolysis it is possible with high conversions. The parallel reactions of decomposition of the C_4 hydrocarbons take place with high velocity and
reduced the amounts of the butadiene produced. An analogous situation occurs in
the dehydrogenation of isopentene to isoprene.

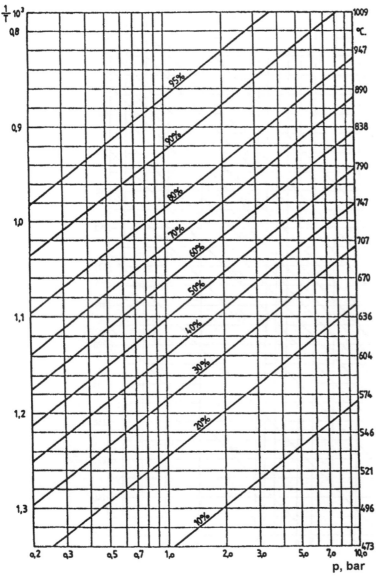

Figure 2.6   The thermodynamic equilibrium for reaction $C_4H_8 \rightleftharpoons C_4H_6 + H_2$.

In visbreaking and delayed coking, the relatively mild operating temperatures lead to a product containing no acetylene and only minor quantities of butadiene.

The equilibrium for the polymerization of alkanes may be illustrated by the graph for the dimerization of propene (Figure 1.3). It results that polymerization cannot take place in the conditions of pyrolysis, but it may be intense in visbreaking, delayed coking, and the older high pressure cracking processes.

The dehydrogenation of alkylcyclohexanes to aromatics is exemplified in Figure 2.7, by the conversion of metilcyclohexane to toluene. While these reactions

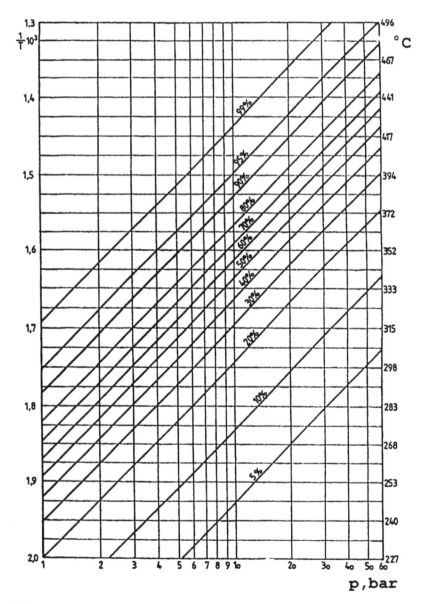

Figure 2.7   Reaction $C_5H_{11} \cdot CH_3 \rightleftharpoons C_6H_5 \cdot CH_3 + 3H_2$.

are thermodynamically favored in all thermal processes, the low reaction rates at the operating temperatures limit the conversions achieved.

2.2 REACTION MECHANISMS

It is well-known that chemical transformations, which are expressed usually by global chemical reactions, actually take place through a large number of parallel and successive elementary reactions. Their knowledge is necessary both for understanding of the chemical aspect of the process and for correctly formulating the kinetic equations.

The initial chemical phenomenon, which take place at high temperatures, is the breaking of the hydrocarbon molecule in two free radicals:

$$RR_1 \overset{k_1}{\underset{k_{-1}}{\rightleftharpoons}} \dot{R} + \dot{R}_1 \tag{a}$$

The reaction energy, which represents in this case the dissociation energy, is given by the equation:

$$\Delta H_r = D_a = E_1 - E_{-1}$$

Since the activation energy for the reverse equation, the recombination of radicals, is equal to zero, the equation becomes:

$$\Delta H_r = D_a = E_1$$

The rate of reaction (a), may be expressed by the relation:

$$r_a = A_a e^{-D_a/RT}[RR_1] \tag{2.1}$$

The molecule RR_1 may also crack to form other two free radicals:

$$RR_1 \overset{k_2}{\underset{k_{-2}}{\rightleftharpoons}} \dot{R}_2 + \dot{R}_3 \tag{b}$$

For this reaction also, one may write:

$$r_b = A_b e^{-D_b/RT}[RR_1] \tag{2.2}$$

The ratio between the relations (2.1) and (2.2) is:

$$\frac{r_a}{r_b} = \frac{A_a}{A_b} e^{(D_b - D_a)/RT} \tag{2.3}$$

For reactions for which data for the values of the pre-exponential factors A are unavailable, as their values are of the same order, it may be assumed that their value is approximately equal to 10^{16} [4–6], and relation (2.3) becomes:

$$\frac{r_a}{r_b} \approx e^{(D_b - D_a)/RT} \tag{2.4}$$

By means of relations (2.3) and (2.4) respectively it is possible to calculate the relative rates of breaking of different bonds within the same molecule and in this way to determine which is the initial step of the thermal decomposition process.

The dissociation energies from Table 2.1 make possible such calculations for hydrocarbons, sulfur compounds, and nitrogen compounds.

Table 2.1 Dissociation Energies of Some Hydrocarbons and Related Substances

Bond	Dissociation energy (kJ/mole)	Ref.
H−H	435	7
$\overset{\bullet}{C}H_2$−H	356	8
CH_3−H	431	8
C_2H_5−H	410	7
n-C_3H_7−H	398	9
i-C_3H_7−H	394	10
n-C_4H_9−H	394	11
i-C_4H_9−H	390	12
$tert$-C_4H_9−H	373	10
neo-C_5H_{11}−H	396	10
CH_2=CH−H	435	7
CH_2=CHCH$_2$−H	322	13
CH_2=C(CH$_3$)CH$_2$−H	322	14
CH_3CH=CHCH$_2$−H	335	14
C_6H_5−H	427	11
$C_6H_5CH_2$−H	348	15
o-$CH_3C_6H_4CH_2$−H	314	10
m-$CH_3C_6H_4CH_2$−H	322	10
p-$CH_3C_6H_4CH_2$−H	318	10
$(C_6H_5)_2$CH−H	310	16
(cyclohexyl)−H or (cyclopentyl)−H	389	16
CH_3−CH_3	360	10
C_2H_5−CH_3	348	17
C_2H_5−C_2H_5	335	17
n-C_3H_7−CH_3	339	16
sec-C_3H_7−CH_3	335	16
n-C_4H_9−CH_3	335	16
n-C_3H_7−C_2H_5	327	16
sec-$C_3H_7C_2H_5$	322	16
n-C_3H_7−n-C_3H_7	318	16
sec-C_3H_7−sec-C_3H_7	320	16
n-C_3H_7−sec-C_3H_7	314	16
n-C_4H_9−C_2H_5	322	16
CH_3CH_2(CH$_3$)−n-C_3H_7	314	18
CH_3C(CH$_3$)$_2$−n-C_3H_7	301	18
CH_3C(CH$_3$)$_2$−CH(CH$_3$)CH$_3$	295	18
n-C_4H_9−n-C_3H_7	314	16
n-C_4H_9−n-C_4H_9	310	16
$tert$-C_4H_9−$tert$-C_4H_9	264	18
CH_2=CH_2	502	11
CH_2=CH−CH_3	394	11
CH_2=CHCH$_2$−CH_3	260	10
CH_2=CHCH$_2$−CH=CH_2	310	11
CH_2=CH−CH=CH_2	435	11
CH_2=CHCH$_2$−CH$_2$CH=CH_2	176	11
CH_2=C(CH$_3$)CH$_2$−CH_3	268	11
CH_3CH=CHCH$_2$−CH_3	275	11
C_6H_5−CH_3	381	16
C_6H_5−C_2H_5	368	16
C_6H_5−n-C_3H_7	360	16

22

Table 2.1 Continued

$C_6H_5CH_2-CH_3$	264	10
$C_6H_5CH_2-C_2H_5$	240	10
$C_6H_5CH_2-C_3H_7$	272	10
$C_6H_5-C_6H_5$	415	16
$C_6H_5CH_2-CH_2C_6H_5$	198	16
$(C_6H_5)_2CH-CH_2C_6H_5$	159	16
$(C_6H_5)_2CH-CH(C_6H_5)_2$	105	16
$(C_6H_5)_3C-C(C_6H_5)_3$	46	16

310	16

293	16

423	16

318	16

364	16

406	16

$H-SH$	377	10
CH_3-SH	293	10
C_2H_5-SH	289	10
$n\text{-}C_3H_7-SH$	295	10
$tert\text{-}C_4H_9-SH$	272	10
$C_6H_5CH_2-SH$	222	19
CH_3-SCH_3	301	10
$C_2H_5-SCH_3$	297	10, 20
$C_2H_5-SC_2H_5$	289	10, 20
$C_6H_5CH_2-SCH_3$	214	21
CH_3S-SCH_3	306	10
$C_2H_5S-SCH_3$	301	10
$C_2H_5S-SC_2H_5$	293	10
CH_3-NH_2	335	22
CH_3-NHCH_3	364	22
$C_6H_5CH_2-NH_2$	394	22

429	16

417	16

For the lower hydrocarbons, more complete data are available, including the values of the pre-exponential factors (Table 2.2).

For instance, the relative rate of breaking of the molecule of ethane at a temperature of 900°C, according to the reactions:

$$C_2H_6 \rightleftharpoons 2\,\overset{\bullet}{C}H_3$$

$$C_2H_6 \rightleftharpoons \overset{\bullet}{C}_2H_5 + \overset{\bullet}{H}$$

may be calculated by using relation (2.4) with R = 8.319 kJ/mol·degree and the data from Table 2.1:

$$\frac{r_a}{r_b} = e^{(410,000-360,000)/(8.319\times1,173)} = 168$$

By using the data of Table 2.2, a more accurate result is obtained:

$$r_a/r_b = 58$$

Both calculations lead to the conclusion that the reaction of breaking a C–C bond, will be prevalent over breaking a C–H bond.

In the same way, the initial prevalent reaction may be determined for other hydrocarbons or hetero-atomic compounds.

Examination of the dissociation data of Table 2.1, shows that for alkanes in general, the energy of the C–C bonds is much lower than the energy of the C–H bonds. Therefore, in the pyrolysis not only for ethane but also of other alkanes, a C–C bond will be preferentially broken. The cracking of the C–H bonds may be neglected, since their rate is approximately two orders of magnitude lower than for C–C bonds.

Table 2.2 Kinetic Constants of Some Dissociation Reactions

Reaction	A (s^{-1})	E (kJ/mol)	Ref.
$C_2H_6 \rightarrow 2\overset{\bullet}{C}H_3$	5.248×10^{16}	372.0	23
	5.185×10^{16}	380.0	24
$C_2H_6 \rightarrow \overset{\bullet}{C}_2H_5 + \overset{\bullet}{H}$	1.3×10^{15}	360.1	25
$C_3H_8 \rightarrow \overset{\bullet}{C}_2H_5 + \overset{\bullet}{C}H_3$	1.995×10^{16}	353.8	26
	2.074×10^{16}	366.9	24
$n - C_4H_{10} \rightarrow 2\overset{\bullet}{C}_2H_5$	1.5×10^{16}	343.7	27
	5.0×10^{15}	339.1	26
$n - C_4H_{10} \rightarrow 1 - \overset{\bullet}{C}_3H_7 + \overset{\bullet}{C}H_3$	9.0×10^{16}	357.6	27
$i - C_4H_{10} \rightarrow 2 - \overset{\bullet}{C}_3H_7 + \overset{\bullet}{C}H_3$	2.0×10^{16}	343.3	27
$C_3H_6 \rightarrow \overset{\bullet}{C}_2H_3 + \overset{\bullet}{C}H_3$	8.0×10^{17}	397.7	27
$C_3H_6 \rightarrow \overset{\bullet}{C}_3H_5 + \overset{\bullet}{H}$	3.5×10^{16}	360.1	27
$1 - C_4H_8 \rightarrow \overset{\bullet}{C}_3H_5 + \overset{\bullet}{C}H_3$	8.0×10^{16}	309.8	28
$2 - C_4H_8 \rightarrow \overset{\bullet}{C}_3H_5 + \overset{\bullet}{C}H_3$	2.0×10^{16}	298.5	27

From the same table one may note that the energy of the $C-C$ bonds in n-alkanes decreases as one moves towards the center of the molecule. Thus, for n-hexane, the energy is about 318 kJ/mole for cracking in two propyl radicals, and approximately 322 kJ/mole for cracking in ethyl and butyl radicals. In the same way, the energy needed for the cracking of n-pentane to ethyl and propyl radicals is lower by 8 kJ/mole than for cracking to methyl and butyl radicals.

According to relation (2.4), such small differences of energy lead to ratios of the respective reaction rates of about 1.5-2.5. Thus, one may not neglect the various ways in which the $C-C$ bonds of a molecule may crack.

The energy of the $C-C$ bonds of tertiary and especially quaternary carbon atoms is lower than the others; thus they will be preferentially cracked.

The double and triple bonds have bonding energies much higher that the single $C-C$ bond. The energy of $C-C$ bonds in the α position to a double bond is higher than that of bonds in the β position, and significantly lower than the energy of the $C-C$ bonds in alkanes.

The energy of the $C_6H_5-C_6H_5$ bond is 415 kJ/mole, but it decreases greatly if the rings are bound by means of an alkylic bridge, or if the bridge is bound to more aromatic rings. Thus, for the molecule $(C_6H_5)_3C-C(C_6H_5)_3$, the dissociation energy decreases to as low as 46 kJ/mole.

For sulphur compounds, the energy of the $C-S$ bonds is of the same order as that of the $C-C$ bonds in n-alkanes.

From the above data, it may be concluded that in all cases, the initial step in the thermal decomposition of hydrocarbons is the breaking of a $C-C$ bond. The cracking of various $C-C$ bonds takes place with comparable rates, with the exception of those in the α position to double bonds, to triple bonds, or to aromatic rings, the rate of which can be neglected.

The formation of free radicals by the cracking of $C-C$ bonds was confirmed experimentally. The presence of methyl and ethyl radicals in the pyrolysis of ethane in a tubular reactor was first identified by mass spectroscopy [29–31].

The concentration of the various species of radicals formed in the pyrolysis of ethane and propane at 850°C, using dilution with steam in conditions similar to those of industrial processes, was determined by Sundaram and Froment in 1977. At a concentration of the feed of 10^{-2} mole/liter, the concentrations of the radicals in the effluent ranged between $10^{-6.6}$ and $10^{-8.3}$ in the pyrolysis of ethane and between $10^{-7.0}$ and $10^{-9.3}$ in the pyrolysis of propane.

Semenov [18] distinguishes three types of transformations undergone by the formed radicals:

isomerization:

$$CH_3-CH_2-\overset{\bullet}{C}H_2 \rightarrow CH_3-\overset{\bullet}{C}H-CH_3 \qquad\qquad (a)$$

decomposition:

$$R-CH_2-\overset{\bullet}{C}H_2 \rightarrow \overset{\bullet}{R}+C_2H_4 \qquad\qquad (b_1)$$

the reverse of which is the addition to the double bond:

$$\overset{\bullet}{R}+C_2H_4 \rightarrow R-CH_2-\overset{\bullet}{C}H_2 \qquad\qquad (b_2)$$

substitution:

$$\dot{R}_1 + R_2H \rightarrow R_1H + \dot{R}_2 \tag{c}$$

Reactions (a), isomerization by odd electron migration, are understood to take place via the formation of a cyclic activated complex, followed by the transfer of an atom of hydrogen [35,38].

For higher radicals (C_5^+), a cyclic-activated complex of 5 or 6 atoms is formed. For the lower radicals (C_5-C_3), the formation of an activated cyclic complex of 3 or 4 atoms is necessary, which would require a higher activation energy. This energy is much higher than that for decomposition or substitution reactions. For these reasons, the isomerization reaction with migration of an odd electron may be neglected for alkanes, which have chains shorter than that of 5 atoms of carbon (Table 2.3).

In conclusion, isomerization by odd electron migration must be taken into account only for molecules which have longer chains [35,40,41].

Reactions such as (b_1) are endothermic, with thermal effects of about 80–160 kJ/mole. Accordingly, at the high temperatures characteristic for thermal processes, the favored transformation is decomposition. Reactions of type (b_2) are exothermic and characteristic for polymerization processes at low temperatures.

The cracking reactions (b_1) of the radicals occurs by the breaking of the $C-C$ bond in β position to carbon with the odd electron. This fact is explained by the excess of bonding energy possessed by the carbon at which the odd electron is located. The excess will be distributed over the three bonds of this carbon atom. The electron-attracting effect of the carbon with the odd electron strengthens the bonds with the carbon atoms in α position. As result of the polarizing of the electrons of the carbon atoms in α toward the carbon with the odd electron, the bonds in β position will be weakened. This explains the preferred cracking in β position.

The electrons of the $C-H$ bonds are less prone to polarization and play no role in the cracking.

The thermal effects for the cracking of some radicals are listed in Table 2.4, while the kinetic constants of such reactions are presented in the Table 2.5.

The decomposition of alkyl radicals or of alkyl chains takes place by successive cracking in β position, occurring at time intervals corresponding to the life of the radicals.

Table 2.3 Kinetic Constants for the Isomerization of Selected Radicals

Reaction	A (s^{-1})	E (kJ/mole)	Ref.
$1 - \dot{C}_4H_9 \rightarrow 2 - \dot{C}_4H_9$	5.2×10^{14}	171.7	42
$1 - \dot{C}_6H_{13} \rightarrow 2 - \dot{C}_6H_{13}$	3.2×10^{10}	49.0	35
$1 - \dot{C}_6H_{13} \rightarrow 3 - \dot{C}_6H_{13}$	2.0×10^{11}	78.3	35
$2 - \dot{C}_6H_{13} \rightarrow 1 - \dot{C}_6H_{13}$	5.0×10^{10}	67.0	35
$3 - \dot{C}_6H_{13} \rightarrow 1 - \dot{C}_6H_{13}$	3.2×10^{11}	96.3	35

For the methyl and ethyl radicals, at temperatures of 600–900°C, the life is of the order of 10^{-3} s. It decreases as the temperature and the molecular mass of the radicals increase.

Since the ethyl and sec-propyl radicals possess only C–H bonds in β position, these will be the ones cracked.

Indeed, according to the Table 2.4, the thermal effect for the reaction:

$$\dot{C}_2H_5 \rightarrow C_2H_4 + \dot{H} \qquad \text{is 165 kJ/mole,}$$

while for the reaction:

$$\dot{C}_2H_5 \rightarrow \dot{C}H_3 + \bullet CH_2 \bullet \qquad \text{is 310 kJ/mole}$$

which shows that the second reaction is unlikely.

The radicals, before or after β cracking, may give substitution reactions of type (c) with molecules of the feed. A new radical, of higher molecular mass, is thereby formed. It also could undergo β cracking, and a new radical with a lower molecular mass is formed. Such reactions may continue and the cracking reactions acquire the character of a chain reaction.*

The kinetic constants of some substitution reactions are given in Table 2.6.

Besides the substitution reactions, additions of the free radicals to double bonds also may take place, leading to the formation of radicals with higher molecular mass. The kinetic constants of such interactions are given in Table 2.7.

The data of Table 2.6 suggest that reactivity diminishes with increasing molecular mass, which is easy to explain by steric hindrances. Thus, if the rates of the reactions which follow are compared, by using relation (2.5):

$$\dot{H} + n\text{-}C_4H_{10} \rightarrow H_2 + 1 - \dot{C}_4H_9 \qquad \text{(a)}$$

$$\dot{C}H_3 + n\text{-}C_4H_{10} \rightarrow H_2 + 1 - \dot{C}_4H_9 \qquad \text{(b)}$$

it follows that for a temperature of 1100 K, the ratio of the reaction rates is:

$$\frac{r_a}{r_b} = \frac{1.5 \times 10^{11}}{3.5 \times 10^{10}} e^{(48,600 \, 40,600)/(1100 \times 8.319)} \times \frac{[\dot{H}]}{[\dot{C}H_3]} = 10.27 \frac{[\dot{H}]}{[\dot{C}H_3]}$$

For similar reactions which lead to the formation of the 2-$\dot{C}_4H_9$ radicals it follows:

$$\frac{r_a}{r_b} = 42.5 \frac{[\dot{H}]}{[\dot{C}H_3]}$$

In both cases, the rates of the reactions initiated by the atomic hydrogen are significantly higher than those initiated by the methyl radical. This may explain the increase of the reaction rates in pyrolysis caused by the introduction of hydrogen into the reactor that was measured experimentally [39].

---

* The formulation of this mechanism for the thermal decomposition of hydrocarbons was first developed by F.C. Rice and K.F. Herzfeld [32].

Table 2.4 Thermal Effects for the Decomposition of Selected Radicals

Reaction	ΔH (kJ/mole)
$\overset{\bullet}{C_2H_5} \rightarrow C_2H_4 + \overset{\bullet}{H}$	165
$\overset{\bullet}{C_2H_5} \rightarrow \overset{\bullet}{CH_3} + \bullet CH_2 \bullet$	310
$CH_3 - CH_2 - \overset{\bullet}{CH_2} \rightarrow C_2H_4 + CH_3$	115
$\qquad\qquad C_3H_6 + \overset{\bullet}{H}$	163
$CH_3 - \overset{\bullet}{CH} - CH_3 \rightarrow C_3H_6 + \overset{\bullet}{H}$	167
$CH_3 - CH_2 - CH_2 - \overset{\bullet}{CH_2} \rightarrow C_2H_4 + \overset{\bullet}{C_2H_5}$	107
$CH_3 - CH_2 - \overset{\bullet}{CH} - CH_3 \rightarrow C_3H_6 + \overset{\bullet}{CH_3}$	113
$CH_3 - \overset{\bullet}{C} - CH_3 \rightarrow i - C_4H_8 + \overset{\bullet}{H}$ $\qquad \mid$ $\qquad CH_3$	180
$n - \overset{\bullet}{C_5H_{11}} \rightarrow C_2H_4 + \overset{\bullet}{C_3H_7}$	98
$n - \overset{\bullet}{C_6H_{13}} \rightarrow C_2H_4 + \overset{\bullet}{C_4H_9}$	96
$n - \overset{\bullet}{C_8H_{17}} \rightarrow C_2H_4 + n - \overset{\bullet}{C_6H_{13}}$	94
$C - \overset{\bullet}{C} - C - C - C - C - C - C \rightarrow C_3H_6 + n - \overset{\bullet}{C_5H_{11}}$	103
$C - C - \overset{\bullet}{C} - C - C - C - C - C \rightarrow C_4H_8 + n - \overset{\bullet}{C_4H_9}$	113
$\qquad\qquad 1 - C_7H_{14} + \overset{\bullet}{CH_3}$	138
$C - C - C - \overset{\bullet}{C} - C - C - C - C \rightarrow C_5H_{10} + \overset{\bullet}{C_3H_7}$	121
$\qquad\qquad C_6H_{12} + \overset{\bullet}{C_2H_5}$	130
$\overset{\bullet}{C} - \overset{\overset{\textstyle C}{\mid}}{C} - C - C \rightarrow C = \overset{\overset{\textstyle C}{\mid}}{C} - \overset{\overset{\textstyle C}{\mid}}{C} - C + \overset{\bullet}{CH_3}$ $\quad \overset{\mid}{C} \quad \overset{\mid}{C}$	90
$\qquad\qquad C = \overset{\overset{\textstyle C}{\mid}}{C} - C + i - \overset{\bullet}{C_3H_7}$	73
$C - \overset{\overset{\textstyle C}{\mid}}{C} - \overset{\bullet}{C} - C \rightarrow C - \overset{\overset{\textstyle C}{\mid}}{C} = \overset{\overset{\textstyle C}{\mid}}{C} - C - + CH_3$ $\quad \overset{\mid}{C} \quad \overset{\mid}{C}$	130
$C_6H_5CH_2\overset{\bullet}{CH_2} \rightarrow C_2H_4 + \overset{\bullet}{C_6H_5}$	77
$\qquad\qquad C_6H_5CH = CH_2 + \overset{\bullet}{H}$	96
(cyclohexyl radical) $\rightarrow$ (cyclohexenyl radical) $+ \overset{\bullet}{H}$	151
$\qquad\qquad C = C - C - C - C - \overset{\bullet}{C}$	75
(methylcyclohexyl radical) $\rightarrow$ (cyclohexenyl) $+ \overset{\bullet}{CH_3}$	88

Table 2.5 Kinetic Constants for the Decomposition of Selected Radicals

Reaction	A (s^{-1})	E (kJ/mole)	Ref.
$\dot{C}_2H_5 \rightarrow C_2H_4 + \dot{H}$	3.2×10^{13}	167.5	33
$1 - \dot{C}_3H_7 \rightarrow C_2H_4 + \dot{C}H_3$	4.0×10^{13}	136.5	33
$1 - \dot{C}_3H_7 \rightarrow C_3H_6 + \dot{H}$	2.0×10^{13}	160.8	33
$2 - \dot{C}_3H_7 \rightarrow C_3H_6 + \dot{H}$	2.0×10^{13}	162.0	27
	2.5×10^{13}	170.8	34
$1 - \dot{C}_4H_9 \rightarrow C_2H_4 + \dot{C}_2H_5$	4.0×10^{13}	121.4	35
$1 - \dot{C}_4H_9 \rightarrow 1 - C_4H_8 + \dot{H}$	1.3×10^{14}	163.3	35
$2 - \dot{C}_4H_9 \rightarrow C_3H_6 + \dot{C}H_3$	1.0×10^{14}	138.1	35
$2 - \dot{C}_4H_9 \rightarrow C_4H_8 + \dot{H}$	1.6×10^{14}	171.6	35
$i - \dot{C}_4H_9 \rightarrow i - C_4H_8 + \dot{H}$	3.3×10^{14}	150.7	27
$i - \dot{C}_4H_9 \rightarrow C_3H_6 + \dot{C}H_3$	2.0×10^{14}	142.3	35
$i - \dot{C}_4H_9 \rightarrow 2 - C_4H_8 + \dot{H}$	4.0×10^{13}	153.2	27
$\dot{C}_5H_{11} \rightarrow C_5H_{10} + \dot{H}$	5.0×10^{13}	153.2	33
$\dot{C}_5H_{11} \rightarrow 1 - C_4H_8 + \dot{C}H_3$	3.2×10^{13}	131.9	27
$\dot{C}_5H_{11} \rightarrow C_2H_4 + 1 - \dot{C}_3H_7$	4.0×10^{12}	120.2	33
$1 - \dot{C}_6H_{13} \rightarrow C_2H_4 + 1 - \dot{C}_4H_9$	1.0×10^{14}	125.6	35
$2 - \dot{C}_6H_{13} \rightarrow C_3H_6 + 1 - \dot{C}_3H_7$	1.0×10^{14}	129.8	35
$3 - \dot{C}_6H_{13} \rightarrow 1 - C_4H_8 + \dot{C}_2H_5$	1.0×10^{14}	129.8	35
$3 - \dot{C}_6H_{13} \rightarrow 1 - C_5H_{10} + \dot{C}H_5$	1.0×10^{14}	136.1	35
$\dot{C}_2H_3 \rightarrow C_2H_2 + \dot{H}$	2.0×10^9	131.9	27
$\dot{C}_3H_5 \rightarrow C_2H_2 + \dot{C}H_3$	3.0×10^{10}	151.6	27
$\dot{C}_4H_7 \rightarrow C_4H_6 + \dot{H}$	1.2×10^{14}	206.4	33
$\dot{C}_4H_7 \rightarrow C_2H_4 + \dot{C}_2H_3$	1.0×10^{11}	154.9	27
Methyl $-$ allyl $\rightarrow C_3H_4 + \dot{C}H_3$	1.0×10^{13}	136.5	27
Methyl $-$ allyl $\rightarrow C_2H_4 + \dot{C}_2H_3$	1.0×10^{12}	117.2	27

The above information on the reactions generated by the radicals formed in the initial breaking of the C–C bonds, make it possible to analyze the succession of transformations that take place during the thermal decomposition of hydrocarbons. These transformations consist of successive breaking of C–C bonds in β position to the odd electron, of the radicals produced following the initial cleavage (reactions of the type b_1). The radicals of lower molecular mass produced will react with hydrocarbon molecules by way of substitution reactions of the type (c) and produce new radicals with a higher molecular mass. The decomposition and substitution reactions

Table 2.6 Kinetic Constants for Selected Substitution Reactions

Reaction	A $(L \cdot mol^{-1}s^{-1})$	E (kJ/mole)	Ref.
$\dot{H} + H_2 \rightarrow H_2 + \dot{H}$	4.6×10^{10}	28.8	6
$\dot{H} + CH_4 \rightarrow H_2 + \dot{C}H_3$	7.24×10^{11}	63.1	36
$\dot{H} + C_2H_6 \rightarrow H_2 + \dot{C}_2H_5$	1.0×10^{11}	40.6	27
$\dot{H} + C_3H_8 \rightarrow H_2 + 1 - \dot{C}_3H_7$	1.0×10^{11}	40.6	27
$\dot{H} + C_3H_8 \rightarrow H_2 + 2 - \dot{C}_3H_7$	9.0×10^{10}	34.8	27
	4.0×10^{10}	34.3	37
$\dot{H} + n - C_4H_{10} \rightarrow H_2 + 1 - \dot{C}_4H_9$	1.5×10^{11}	40.6	27
$\dot{H} + n - C_4H_{10} \rightarrow H_2 + 2 - \dot{C}_4H_9$	9.0×10^{10}	35.2	27
$\dot{H} + i - C_4H_{10} \rightarrow H_2 + i - \dot{C}_4H_9$	1.0×10^{11}	35.2	27
$\dot{C}H_3 + H_2 \rightarrow CH_4 + \dot{H}$	1.55×10^{10}	64.9	36
$\dot{C}H_3 + C_2H_6 \rightarrow CH_4 + \dot{C}_2H_5$	3.8×10^{11}	69.1	27
$\dot{C}H_3 + C_3H_8 \rightarrow CH_4 + 1 - \dot{C}_3H_7$	3.4×10^{10}	48.1	27
$\dot{C}H_3 + C_3H_8 \rightarrow CH_4 + 2 - \dot{C}_3H_7$	4.0×10^9	42.3	27
$\dot{C}H_3 + n - C_4H_{10} \rightarrow CH_4 + 1 - \dot{C}_4H_9$	3.5×10^{10}	48.6	27
$\dot{C}H_3 + n - C_4H_{10} \rightarrow CH_4 + 2 - \dot{C}_4H_9$	3.5×10^9	39.8	27
$\dot{C}H_3 + i - C_4H_{10} \rightarrow CH_4 + i - \dot{C}_4H_9$	9.5×10^9	37.7	27
$\dot{C}H_3 + C_6H_{14} \rightarrow CH_4 + \dot{C}_6H_{13}$	1.0×10^8	33.9	38
$\dot{C}_2H_5 + H_2 \rightarrow C_2H_6 + \dot{H}$	3.0×10^8	45.2	6
$\dot{C}_2H_5 + CH_4 \rightarrow C_2H_6 + \dot{C}H_3$	2.3×10^8	47.7	36
$\dot{C}_2H_5 + C_3H_8 \rightarrow C_2H_6 + 1 - \dot{C}_3H_7$	1.2×10^9	52.8	27
$\dot{C}_2H_5 + C_3H_8 \rightarrow C_2H_6 + 2 - \dot{C}_3H_7$	8.0×10^8	43.5	27
$\dot{C}_2H_5 + n - C_4H_{10} \rightarrow C_2H_6 + 1 - \dot{C}_4H_9$	2.0×10^9	52.8	27
$\dot{C}_2H_5 + n - C_4H_{10} \rightarrow C_2H_6 + 2 - \dot{C}_4H_9$	4.5×10^8	43.5	27
$\dot{C}_2H_5 + i - C_4H_{10} \rightarrow C_2H_6 + i - \dot{C}_4H_9$	1.5×10^9	43.5	27
$\dot{C}_2H_5 + n - C_7H_{16} \rightarrow C_2H_6 + n - \dot{C}_7H_{15}$	9.8×10^9	44.4	6
$1 - \dot{C}_3H_7 + n - C_4H_{10} \rightarrow C_3H_8 + 2 - \dot{C}_4H_9$	2.0×10^8	43.5	27
$2 - \dot{C}_3H_7 + n - C_4H_{10} \rightarrow C_3H_8 + 2 - \dot{C}_4H_9$	2.0×10^8	52.8	27
$2 - \dot{C}_3H_7 + i - C_4H_{10} \rightarrow C_3H_8 + i - \dot{C}_4H_9$	1.0×10^8	56.1	27
$\dot{H} + C_2H_4 \rightarrow H_2 + \dot{C}_2H_3$	8.0×10^8	16.7	27
$\dot{H} + C_3H_6 \rightarrow H_2 + \dot{C}_3H_5$	2.5×10^9	17.2	28
$\dot{H} + 1 - C_4H_8 \rightarrow H_2 + \dot{C}_4H_7$	5.0×10^{10}	16.3	33
$\dot{H} + 2 - C_4H_8 \rightarrow H_2 + \dot{C}_4H_7$	5.0×10^{10}	15.9	27
$\dot{H} + i - C_4H_8 \rightarrow H_2 + Methyl - allyl''$	3.0×10^{10}	15.9	27
$\dot{C}H_3 + C_2H_4 \rightarrow CH_4 + \dot{C}_2H_3$	1.0×10^{10}	54.4	27
$\dot{C}H_3 + C_3H_6 \rightarrow CH_4 + \dot{C}_3H_5$	2.0×10^9	51.1	27
$\dot{C}H_3 + 1 - C_4H_8 \rightarrow CH_4 + \dot{C}_4H_7$	1.0×10^8	30.6	33
$\dot{C}H_3 + 2 - C_4H_8 \rightarrow CH_4 + \dot{C}_4H_7$	1.0×10^8	34.3	27
$\dot{C}H_3 + i - C_4H_8 \rightarrow CH_4 + Methyl - allyl''$	3.0×10^8	30.6	27
$\dot{C}_2H_5 + C_3H_6 \rightarrow C_2H_6 + \dot{C}_3H_5$	1.0×10^8	38.5	33

Table 2.6 Continued

	A	E	Ref
$\overset{\bullet}{C_2H_5} + 1-C_4H_8 \rightarrow C_2H_6 + \overset{\bullet}{C_4H_7}$	2.0×10^8	34.8	27
$\overset{\bullet}{C_2H_5} + i-C_4H_8 \rightarrow C_2H_6 + \text{Methyl} - \text{allyl}''$	6.0×10^7	34.8	27
$\overset{\bullet}{C_2H_5} + 1-C_7H_{14} \rightarrow C_2H_6 + 1-\overset{\bullet}{C_7H_{13}}$	3.1×10^8	34.8	6
$1-\overset{\bullet}{C_3H_7} + C_3H_6 \rightarrow C_3H_8 + \overset{\bullet}{C_3H_5}$	1.0×10^8	38.5	33
$2-\overset{\bullet}{C_3H_7} + C_3H_6 \rightarrow C_3H_8 + \overset{\bullet}{C_3H_5}$	1.0×10^8	42.7	33
$\overset{\bullet}{C_2H_3} + C_3H_8 \rightarrow C_2H_4 + 1-\overset{\bullet}{C_3H_7}$	3.0×10^9	78.7	27
$\overset{\bullet}{C_2H_3} + C_3H_8 \rightarrow C_2H_4 + 2-\overset{\bullet}{C_3H_7}$	1.0×10^9	67.8	27
$\overset{\bullet}{C_2H_3} + n-C_4H_{10} \rightarrow C_2H_4 + 1-\overset{\bullet}{C_4H_9}$	1.0×10^9	75.4	27
$\overset{\bullet}{C_2H_3} + n-C_4H_{10} \rightarrow C_2H_4 + 2-\overset{\bullet}{C_4H_9}$	8.0×10^8	70.3	27
$\overset{\bullet}{C_2H_3} + i-C_4H_{10} \rightarrow C_2H_4 + i-\overset{\bullet}{C_4H_9}$	1.0×10^9	70.3	27
$\overset{\bullet}{C_3H_5} + C_3H_8 \rightarrow C_3H_6 + 1-\overset{\bullet}{C_3H_7}$	1.0×10^9	78.7	33
$\overset{\bullet}{C_3H_5} + C_3H_8 \rightarrow C_3H_6 + 2-\overset{\bullet}{C_3H_7}$	8.0×10^8	67.8	33
$\overset{\bullet}{C_3H_5} + n-C_4H_{10} \rightarrow C_3H_6 + 1-\overset{\bullet}{C_4H_9}$	4.0×10^8	78.7	33
$\overset{\bullet}{C_3H_5} + n-C_4H_{10} \rightarrow C_3H_6 + 2-\overset{\bullet}{C_4H_9}$	8.0×10^8	70.3	33
$\overset{\bullet}{C_3H_5} + i-C_4H_{10} \rightarrow C_3H_6 + i-\overset{\bullet}{C_4H_9}$	1.0×10^9	79.5	27

Table 2.7 Kinetic Constants for Selected Addition
Reactions

Reaction	A $(L \cdot mol^{-1}s^{-1})$	E (kJ/mol)	Ref.
$C_2H_2 + \overset{\bullet}{H} \rightarrow \overset{\bullet}{C_2H_3}$	4.0×10^{10}	5.4	27
$C_2H_4 + \overset{\bullet}{H} \rightarrow \overset{\bullet}{C_2H_5}$	1.0×10^{10}	6.3	33
$C_3H_6 + \overset{\bullet}{H} \rightarrow 1-\overset{\bullet}{C_3H_7}$	1.0×10^{10}	12.1	33
$C_3H_6 + \overset{\bullet}{H} \rightarrow 2-\overset{\bullet}{C_3H_7}$	1.0×10^{10}	6.3	27
$1-C_4H_8 + \overset{\bullet}{H} \rightarrow 2-\overset{\bullet}{C_4H_9}$	1.0×10^{10}	5.0	33
$2-C_4H_8 + \overset{\bullet}{H} \rightarrow 2-\overset{\bullet}{C_4H_9}$	6.3×10^9	5.0	33
$i-C_4H_8 + \overset{\bullet}{H} \rightarrow i-\overset{\bullet}{C_4H_9}$	1.0×10^{10}	5.0	27
$C_2H_4 + \overset{\bullet}{CH_3} \rightarrow 1-\overset{\bullet}{C_3H_7}$	2.0×10^8	33.0	27
$C_3H_6 + \overset{\bullet}{CH_3} \rightarrow 2-\overset{\bullet}{C_4H_9}$	3.2×10^8	31.0	27
$C_3H_6 + \overset{\bullet}{CH_3} \rightarrow i-\overset{\bullet}{C_4H_9}$	3.2×10^8	38.1	33
$i-C_4H_8 + \overset{\bullet}{CH_3} \rightarrow \overset{\bullet}{C_5H_{11}}$	1.0×10^8	30.1	27
$C_2H_4 + \overset{\bullet}{C_2H_5} \rightarrow 1-\overset{\bullet}{C_4H_9}$	1.5×10^7	31.8	33
$C_3H_6 + \overset{\bullet}{C_2H_5} \rightarrow \overset{\bullet}{C_5H_{11}}$	1.3×10^7	31.4	33
$C_2H_4 + 1-\overset{\bullet}{C_3H_7} \rightarrow \overset{\bullet}{C_5H_{11}}$	2.0×10^7	31.0	33
$C_2H_4 + 2-\overset{\bullet}{C_3H_7} \rightarrow \overset{\bullet}{C_5H_{11}}$	1.3×10^7	28.9	33

will be repeated and will lead to the decomposition of a great number of feed molecules, thereby conferring to the reaction the characteristics of a nonramified chain.

The length of the chain is defined as the number of molecules cracked by reactions of types b_1 and c by one radical formed by the initial breaking of a C–C bond. In thermal processes, the length of the chain is 100–200 and above.

The successive interactions of types b_1 and c will be interrupted by reactions that lead to the disappearance of the radical. This may take place by a heterogeneous mechanism (adsorption on the wall) or by a homogenous mechanism such as the interaction between two radicals.

The existence of the heterogeneous chain-interruption reaction may be observed by increasing the ratio: walls surface/volume of the reactor. The resulting decrease of the reaction rate indicates the existence of heterogeneous chain-interruption reactions.

The surface to volume ratio was shown to influence the chain length, in laboratory studies performed in glass reactors, at subatmospheric pressures and at moderate temperatures [43–45]. However, no effect was detected in metallic reactors and at the pressures used in industrial processes [45].

The nature of the reactor walls and the pretreatment of the metallic ones (by oxidation, reduction, sulfidation, etc.) has a strong influence on the formation of coke [46], but has very small effect upon the kinetics of the overall pyrolysis process [45].

For these reasons, in studies of the reaction mechanism and the kinetics of thermal decomposition reactions, only the chain interruptions by homogeneous mechanisms are taken into account.

The interaction between two radicals may be of two types: recombination and disproportionation.

The recombination is the reverse of the initiation reaction and may be illustrated by the reaction:

$$\overset{\bullet}{C}H_3 + \overset{\bullet}{C}_2H_5 \rightarrow C_3H_8$$

The disproportionation can be illustrated by:

$$CH_3\overset{\bullet}{C}H_2 + CH_3\overset{\bullet}{C}H_2 \rightarrow C_2H_4 + C_2H_6$$

The rates of the disproportionation reactions is significantly lower than those of recombination reactions.

Among recombination reactions, the recombination of two hydrogen atoms will take place at a very low rate, since a triple collision of the type:

$$2\overset{\bullet}{H} + M \rightarrow H_2 + M^*$$

is required, with molecule M accepting the energy resulted from the recombination. Since the frequency of triple collisions is very low, the reaction takes place with a very low rate. A similar situation exists in the recombination of a methyl radical with atomic hydrogen.

In the case of radicals with higher molecular mass, the energy developed by the recombination of the radicals is dissipated over a larger number of interatomic bonds. This will favor the recombination of radicals.

The activation energy for the recombination and disproportionation of the radicals is practically equal to zero. Thus, the rates of these reactions is independent of temperature, and the rate constant is equal to the pre-exponential factor.

Pre-exponential factors for the recombination and disproportioning of some radicals are listed in Table 2.8.

Since in thermal processes the length of the kinetic chain exceeds 100–200, it follows that the composition of the products will be determined mainly by the substitution reaction and the decomposition of the radicals, that is, by the reactions of developing the kinetic chain and not by those of initiation or interruption of the chain. This understanding made it possible to calculate the composition of the products resulting from the thermal decomposition of hydrocarbons.

Table 2.8 Pre-exponential Factors for Some Radicalic Recombinations and Disproportionations

Reaction	A (L · mol^{-1}s^{-1})	Ref.
$\dot{H} + \dot{C}H_3 \rightarrow CH_4$	1.3×10^{11}	6
$\dot{H} + \dot{C}_2H_5 \rightarrow C_2H_6$	8.5×10^{10}	25
$\dot{H}$ + any of the radicals: $\quad 1 - \dot{C}_3H_7; 2 - \dot{C}_3H_7;$ $\quad 1 - \dot{C}_4H_9; 2 - \dot{C}_4H_9;$ $\quad i - \dot{C}_4H_9; \dot{C}_5H_{11}$	1.0×10^{10}	33
$\dot{C}H_3 + \dot{C}H_3 \rightarrow C_2H_6$	1.3×10^{10}	25
$\dot{C}H_3 + \dot{C}_2H_5 \rightarrow C_3H_8$	3.2×10^{9}	27
	1.0×10^{11}	44
$\dot{C}H_3 + \dot{C}_3H_7 \rightarrow C_4H_{10}$	3.2×10^{9}	27
$\dot{C}_2H_5 + \dot{C}_2H_5 \rightarrow C_4H_{10}$	7.9×10^{9}	36
$\dot{C}_2H_5 + \dot{C}_3H_7(1;2) \rightarrow C_5H_{12}$	8.0×10^{8}	33
$1 - \dot{C}_3H_7 + 1 - \dot{C}_3H_7 \rightarrow C_6H_{14}$	4.0×10^{8}	6
$i - \dot{C}_4H_9 + i - \dot{C}_4H_9 \rightarrow C_8H_{18}$	3.2×10^{9}	48
$\dot{C}_3H_3 + C_6H_5\dot{C}H_2 \rightarrow C_6H_5CH_2CH_3$	1.6×10^{8}	49
$\dot{H} + \dot{C}_2H_5 \rightarrow H_2 + C_2H_4$	2.1×10^{9}	6
$\dot{H} + 1 - \dot{C}_3H_7 \rightarrow H_2 + C_3H_6$	1.3×10^{9}	6
$\dot{C}H_3 + \dot{C}_2H_5 \rightarrow CH_4 + C_2H_4$	2.3×10^{8}	6
$\dot{C}H_3 + 1 - \dot{C}_3H_7 \rightarrow CH_4 + C_3H_6$	1.9×10^{8}	6
$\dot{C}_2H_5 + \dot{C}_2H_5 \rightarrow C_2H_6 + C_2H_4$	1.2×10^{8}	6
$\dot{C}_2H_5 + 1 - \dot{C}_3H_7 \rightarrow C_2H_6 + C_3H_6$	1.3×10^{8}	6
$1 - \dot{C}_3H_7 + 1 - \dot{C}_3H_7 \rightarrow C_3H_8 + C_3H_6$	1.1×10^{8}	6

Such a calculation is based on the rates of the substitution reactions, corresponding to the extraction of different hydrogen atoms from the initial feed. The rate depends on the number of hydrogen atoms that have the same position in the molecule. It will also depend on the dissociation energy of the $C-H$ bonds of the H atoms in different positions (relation 2.4).

The data of Table 2.1 illustrate the significant differences that might be found among these bond energies.

From the analysis of the data available at that time, F.C. Rice [32] accepted average relative rates of extraction by free radicals, of the hydrogen atoms linked to primary, secondary and tertiary carbon atoms.

The average activation energies accepted by F.C. Rice have the values:

Extraction of Hydrogen from	kcal/mole
Primary C	92.0
Secondary C	90.8
Tertiary C	88.0

which, according to Eq. (2.4) lead to the following expressions for the relative rates:

$$r_{sec}/r_{prim} = e^{604/T}$$

$$r_{tert}/r_{prim} = e^{2013/T}$$

which makes possible their calculation for different temperatures.

According to F.O. Rice [31], the relative rates at 600°C, have the values:

1   for primary
2   for secondary
10   for tertiary carbon atoms

The calculation of the product composition after F.O. Rice is illustrated by means of Example 2.1.

Example 2.1. Calculate the composition of the products resulting from the thermal decomposition of *i*-pentane at 600°C. The assumptions corresponding to different possible substitution reactions are:

ASSUMPTION A.   Hydrogen abstraction takes place from the first carbon atom or equivalently, from the carbon atom of the side chain.

$$CH_3-\overset{\displaystyle |}{\underset{\displaystyle CH_3}{CH}}-CH_2-CH_3 + \dot{R} \rightarrow RH + \overset{\displaystyle |}{\underset{\displaystyle CH_3}{\dot{C}H_2}}-CH-CH_2-CH_3$$

$$
\begin{array}{l}
\overset{(a_1)}{\nearrow} RH + 1-C_4H_8 + \dot{C}H_3 \\
\underset{(a_2)}{\searrow} RH + C_3H_6 + \dot{C}_2H_5
\end{array}
\begin{array}{l}
\overset{(a_1')}{\nearrow} RH + C_3H_6 + C_2H_4 + \dot{H} \\
\underset{(a_2')}{\searrow} RH + C_3H_6 + \dot{C}_2H_5
\end{array}
$$

Assumption (a_1) corresponds to the breaking of the $C-C$ bond in the side chain and (a_2) to the breaking of the $C-C$ bond from the main chain. The chances of breaking these bonds are identical. The assumption (a_2') corresponds to the decomposition of the radicals ethyl to ethene and atomic hydrogen, while assumption hypothesis (a_2'') corresponds to the absence of this decomposition.

The probability of the decomposition (a_2'), which is not taken into account by F.C. Rice, was estimated by A.J. Dintes, to account for 20% of the ethyl radicals formed at 600°C. Taking into account the radical which is reproduced, the overall reactions may be written as follows:

$$i\text{-}C_5H_{12} \rightarrow CH_4 + 1\text{-}C_4H_8 \tag{a_1}$$

$$i\text{-}C_5H_{12} \rightarrow H_2 + C_3H_6 + C_2H_4 \tag{a_1'}$$

$$i\text{-}C_5H_{12} \rightarrow C_2H_6 + C_3H_6 \tag{a_2''}$$

ASSUMPTION B. Abstraction of hydrogen from the tertiary carbon:

$$CH_3-\underset{\underset{CH_3}{|}}{CH}-CH_2-CH_3 + \overset{\bullet}{R} \rightarrow RH + CH_3-\underset{\underset{CH_3}{|}}{\overset{\bullet}{C}}-CH_2-CH_3 \rightarrow$$

$$\rightarrow RH + i\text{-}C_4H_8 + \overset{\bullet}{C}H_3$$

Taking into account the radical which is reproduced, the overall reaction will be:

$$i\text{-}C_5H_{12} \rightarrow CH_4 + i\text{-}C_4H_8 \tag{b}$$

ASSUMPTION C. Abstraction of hydrogen from the third carbon atom:

$$CH_3-\underset{\underset{CH_3}{|}}{CH}-CH_2-CH_3 + \overset{\bullet}{R} \rightarrow RH + CH_3-\underset{\underset{CH_3}{|}}{CH}-\overset{\bullet}{C}H-CH_3 \rightarrow$$

$$\rightarrow RH + 2\text{-}C_4H_8 + \overset{\bullet}{C}H_3$$

The overall reaction will be :

$$i\text{-}C_5H_{12} \rightarrow CH_4 + 2\text{-}C_4H_8 \tag{c}$$

ASSUMPTION D. Abstraction of hydrogen from the 4th carbon atom:

$$CH_3-\underset{\underset{CH_3}{|}}{CH}-CH_2-CH_3 + \overset{\bullet}{R} \rightarrow RH + CH_3-\underset{\underset{CH_3}{|}}{CH}-CH_2-\overset{\bullet}{C}_2 \rightarrow$$

$$\rightarrow RH + C_2H_4 + CH_3-\overset{\bullet}{C}H-CH_3 \rightarrow RH + C_2H_4 + C_3H_6 + \overset{\bullet}{H}$$

The overall reaction will be:

$$i\text{-}C_5H_{12} \rightarrow H_2 + C_2H_4 + C_3H_6 \tag{d}$$

By using the relative rates at 600°C given by F.C. Rice and taking into account the number of carbon atoms that have the same position in the molecule, the relative reaction rates given in Table 2.9 were obtained.

The overall stoechiometric equation for the thermal decomposition of i-pentane, at 600°C may be written:

$$115 \; i\text{-}C_5H_{12} = 18 \; H_2 + 85 \; CH_4 + 18 \; C_2H_4 + 12 \; C_2H_6 + 30 \; C_3H_6 +$$
$$15 \; (1) \; C_4H_8 + 50 \; i\text{-}C_4H_8 + 20 \; (2) \; C_4H_8$$

Table 2.9

Reactions	Hydrogen atoms occupying identical position	Relative rate of hydrogen extraction	Overall relative rate
a	6	1	$6 \times 1 = 6$
$a_1 = 0.5\ a$			$0.5 \times 6 = 3.0$
$a_2' = 0.2\ a_2$			$0.2 \times 3 = 0.6$
$a_2'' = 0.8\ a_2$			$0.8 \times 3 = 2.4$
b	1	10	$1 \times 10 = 10$
c	2	2	$2 \times 2 = 4$
d	3	1	$3 \times 1 = 3$

More detailed studies [50] have shown that, owing to steric hindrances, the extraction rates of the hydrogen atoms linked to secondary and primary carbons depends also upon the nature of the radical that effectuates the extraction. At 545°C, the ratio of the extraction rates is 1.75 for extraction by $\overset{\bullet}{C_2H_5}$; 2.6 for $\overset{\bullet}{C}H_3$ and 4 for $\overset{\bullet}{H}$, as compared with the value of 1, for radicals of higher molecular mass.

The comparison of the theoretical calculations with a great number of experimental data [35] leads to the conclusion that it is necessary to take into account not only the nature of the radical that acts in the substitution reactions but also the relative rates of decomposition of the radicals formed in these reactions.

For radicals of higher molecular mass the isomerization reaction must also be taken into account [40].

Table 2.10 lists the kinetic parameters, recommended for calculations on the basis of these studies [35].

In commercial pyrolysis, it is of great interest to predict the ratio of ethene to propene produced, as a function of the ratio n-/i-alkanes in the feed.

To this purpose, the ratio ethene/propene, resulting from pyrolysis at 1100 K of n-octane, 4-methyl-heptane and 2,5-dimethyl-hexane, were calculated in Appendix 1:

	C_2H_4/C_3H_6
n-octane	8.23
4-methyl-heptane	1.77
2,5-dimethyl-hexane	0.35

This calculation proves that the ethene/propene ratio diminishes strongly with the degree of branching of the alkane feedstock.

The pyrolysis of heavy petroleum fractions (atmospheric and vacuum gas oils) has prompted studies [107,108], aiming at the understanding of the mechanism of the thermal decomposition of the $C_{13} - C_{14}$ n- and i-alkanes.

The decomposition reactions of other classes of hydrocarbons may be examined analogously, by using data collected in Tables 2.1–2.8.

Alkenes. The data of Table 2.1 show, that alkenes may undergo breaking of the C−C bonds, except those in α position to the double bond.

Table 2.10   Kinetic Data Recommended by E. Ranzi, M.Dente, S. Pierucci, and G. Bardi

Reaction	A (s^{-1}) or (L·mol^{-1}s^{-1})	E (kJ/mol)
Hydrogen Abstraction Reactions (Primary H-Atoms)		
by a primary radical	2.0×10^8	58.6
by a secondary radical	2.0×10^8	67.0
by a tertiary radical	2.0×10^8	71.2
Ratio between secondary and primary H-abstraction	–	9.63
Ratio between tertiary and primary H-abstraction	–	18.84
Radicals Isomerization (primary to primary hydrogen transfer)		
1-5 transfer	2×10^{10}	58.6
1-4 transfer	1×10^{11}	87.9
Radicals Decomposition Reactions		
C_2–C_4 Radicals		
$\overset{\bullet}{C_2H_5} \rightarrow C_2H_4 + \overset{\bullet}{H}$	3.2×10^{13}	170.4
$1-\overset{\bullet}{C_3H_7} \rightarrow C_2H_4 + \overset{\bullet}{CH_3}$	5.0×10^{13}	134.0
$1-\overset{\bullet}{C_3H_7} \rightarrow C_3H_6 + \overset{\bullet}{H}$	7.9×10^{13}	159.1
$2-\overset{\bullet}{C_3H_7} \rightarrow C_3H_6 + \overset{\bullet}{H}$	1.1×10^{15}	175.8
$1-\overset{\bullet}{C_4H_9} \rightarrow C_2H_4 + \overset{\bullet}{C_2H_5}$	4.0×10^{13}	121.4
$1-\overset{\bullet}{C_4H_9} \rightarrow 1-C_4H_8 + \overset{\bullet}{H}$	1.3×10^{14}	163.3
$2-\overset{\bullet}{C_4H_9} \rightarrow C_3H_6 + \overset{\bullet}{CH_3}$	1.0×10^{14}	138.2
$2-\overset{\bullet}{C_4H_9} \rightarrow C_4H_8 + \overset{\bullet}{H}$	1.6×10^{14}	171.7
$1,i-\overset{\bullet}{C_4H_9} \rightarrow C_3H_6 + \overset{\bullet}{CH_3}$	2.0×10^{14}	142.4
$1,i-\overset{\bullet}{C_4H_9} \rightarrow i-C_4H_8 + H$	1.0×10^{14}	159.1
$2,i-\overset{\bullet}{C_4H_9} \rightarrow i-C_4H_8 + \overset{\bullet}{H}$	2.0×10^{14}	175.8
Decomposition of Higher Radicals to Primary Radicals		
primary radical	1.0×10^{14}	125.6
secondary radical	1.0×10^{14}	129.8
tertiary radical	1.0×10^{14}	134.0
Correction to the Activation Energy (Related to Primary Radical)		
secondary radical	–	8.4
tertiary radical	–	16.7
methyl radical	–	6.3

Source: Ref. 35.

   The rate of such initiation reactions is comparable with that for breaking the C–C bonds in alkanes.

   It should be observed that the bonding energy of the hydrogen linked to carbon atoms in α position to the double bond is relatively low. Thus, for example in the pyrolysis of propene and isobutene, the initiation takes place by breaking these weaker C–H bonds. In both cases, the activation energy is of 322 kJ/mole and a pre-exponential factor of 10^{13} sec^{-1}, which is comparable with the values of breaking of a bond a C–C at alkanes.

By the breaking of a C–C bond in alkenes, an allyl radical is formed. Similarly, an allyl radical will be formed by the breaking of the C–H bond at the carbon in the α position to the double bond, as well as in the interaction between a radical and an alkene.

Thus, alkenes will preferably give allyl radicals of the formula:

$$H_2C{=}CH{-}\overset{\bullet}{C}H_2 \longrightarrow H_2C{=\!=\!=}CH{=\!=\!=}CH_2 \qquad (a)$$

$$H_2C{=}CH{-}\overset{\bullet}{C}H_2 \longrightarrow H_2C{=\!=\!=}C{=\!=\!=}CH_2 \qquad (b)$$
$$\underset{CH_3}{|} \qquad\qquad \underset{CH_3}{|}$$

or, give dienes:

$$H_2C{=}CH{-}\overset{\bullet}{C}H{-}CH_2R \longrightarrow H_2C{=}CH{-}CH{=}CH_2 + \overset{\bullet}{R} \qquad (c)$$

$$H_2C{=}CH{-}\overset{\bullet}{C}{-}CH_2{-}R \longrightarrow H_2C{=}CH{-}C{=}CH_2 + \overset{\bullet}{R} \qquad (d)$$
$$\underset{CH_3}{|} \qquad\qquad\qquad \underset{CH_3}{|}$$

To a lesser extent, higher allyl radicals may be formed. They may have the odd electron in a position different than α, as for example:

$$H_2C{=}CH{-}CH_2{-}CH_2{-}\overset{\bullet}{C}H_2 \rightarrow C_2H_4 + \overset{\bullet}{C}_3H_5$$

$$H_3C{-}CH{=}CH{-}CH_2{-}\overset{\bullet}{C}H_2 \rightarrow H_3C{-}CH{=}CH{-}CH{=}CH_2 + \overset{\bullet}{H}$$

$$H_2C{=}CH{-}CH_2{-}\overset{\bullet}{C}H{-}CH_2{-}CH_3 \rightarrow H_2C{=}CH{-}CH_2{-}CH{=}CH_2 + \overset{\bullet}{C}H_3$$

Such radicals may appear also as result of breaking of a C–C bond of a higher alkene.

The main types of radicals and the specific reactions that may be produced in the case of alkene are those indicated by the letters (a)–(d). Further, the additional reactions of the radicals to the double bond given in the Table 2.7 must be considered.

Radicals (a) and (b) are stabilized by the conjugation of the odd electron with the π electrons of the double bond, producing three conjugated electrons. The distances between the carbon atoms become equal, intermediary to that corresponding to the single C–C bond and double bond. The available energy will be distributed equally between the two marginal and the central carbon atoms. These radicals are inactive, since they do not posses sufficient energy for the extraction of a hydrogen atom from a feed molecule and for continuation of the reaction chain. They will be adsorbed on the wall or combined with another radical. If the radicals (a) or (b) were formed as a result of the reaction of the alkene with an alkyl radical, the interruption of the reaction chain will occur, leading to a slowdown of the process of thermal decomposition.

Reactions (c) and (d) explain why at the high conversions, which are specific to the pyrolysis process, liquid feed stocks produce large amounts of butadiene and isoprene in comparison with much lower amounts of other dienes.

The experimental studies of the pyrolysis of alkenes and the interpretation of the obtained results [38] are in full agreement with the above considerations.

Cycloalkanes. The bond energies between the carbon atoms of the 5 and 6 atoms cycles (Table 2.1) were estimated by indirect calculations. Their values are comparable to those between the carbon atoms in alkanes.

The breaking of the cycles leads to the formation of biradicals, which can react according to one of the following schemes:

$$\text{(cyclohexane)} \longrightarrow \overset{\bullet}{C}H_2-CH_2-CH_2-CH_2-CH_2-\overset{\bullet}{C}H_2 \overset{(a)}{\longrightarrow} 3C_2H_4$$

$$\swarrow_{(c)} \qquad \searrow_{(b)}$$

$$C_2H_4 + C_4H_8 \qquad n\text{-}C_6H_{12}$$

$$\text{(cyclopentane)} \longrightarrow \overset{\bullet}{C}H_2-CH_2-CH_2-CH_2-\overset{\bullet}{C}H_2 \overset{(a)}{\longrightarrow} C_2H_4 + C_3H_6$$

$$\downarrow_{(b)}$$

$$n\text{-}C_5H_{10}$$

Among these, the formation of alkenes with the same number of carbon atoms—reactions (b)—was not be confirmed with certainty.

It accepted that in case of alkylcycloalkanes, the break will occur more easily between the carbon atoms of the side chains, than in the cycle. The formed alkyl radical will follow the specific reaction paths of these radicals. The radical containing the cycle will decompose according by cleavages in β position, until finally, a cyclic radical is produced. A cyclic radical will also result following the action of a free radical upon a cycloalkane.

The decomposition of cyclic radicals takes place as follows:

$$\text{(cyclic radical)} \longrightarrow CH_2=CH-CH_2-CH_2-\overset{\bullet}{C}H_2 \longrightarrow C_2H_4 + \overset{\bullet}{CH_2=CH=CH_2}$$

$$\downarrow$$

$$CH_2=CH-\overset{\bullet}{C}H-CH_2-CH_3 \longrightarrow C_4H_6 + \overset{\bullet}{C}H_3$$

The methyl radicals may continue the reaction chain.

$$\text{(cyclic radical)} \longrightarrow CH_2=CH-CH_2-CH_2-CH_2-\overset{\bullet}{C}H_2 \longrightarrow CH_2=CH-CH_2-\overset{\bullet}{C}H_2 + C_2H_4$$

$$\downarrow$$

$$CH_2=CH-CH_2-CH_2-\overset{\bullet}{C}H-CH_3 \longrightarrow C_3H_6 + \overset{\bullet}{C}_3H_5$$

$$\downarrow$$

$$CH_2=CH-CH_2-\overset{\bullet}{C}H-CH_2-CH_3 \longrightarrow C_5H_8 + \overset{\bullet}{C}H_3$$

$$\downarrow$$

$$CH_2=CH-\overset{\bullet}{C}H-CH_2-CH_2-CH_3 \longrightarrow C_4H_6 + C_2H_4 + \overset{\bullet}{H}$$

These reactions show the competition between the isomerization reaction, by migration of the odd electron, and the reaction of decomposition of the radical. There are no reliable methods for estimating their relative rates at this writing. However, the high proportion of butadiene, which is formed in the pyrolysis of cyclohexane, may be an indication of a higher rate of the isomerization compared to the decomposition reaction.

The reactions of the radicals formed following the initial decomposition of cyclohexane deserve special attention. Two reaction paths are observed:

$$CH_2{=}CH{-}CH_2{-}\overset{\bullet}{C}H_2 \longrightarrow CH_2{=}CH{-}CH{=}CH_2 + \overset{\bullet}{H} \tag{a}$$

$$\searrow CH_2{=}\overset{\bullet}{C}H + C_2H_4 \tag{b}$$

Both decompositions can be justified by the effect of the odd electron on the bonding energies of the carbon situated in β position.

The dissociation energy of the bonds broken according to reactions (a) and (b) are not known. It is however possible to estimate them by a calculation based upon the use of relation (1.1) for the dissociation reaction [11,16].

For reaction (a) one obtains:

$$(D_{C_4H_7})_a = (\Delta H^0_{fT})_{C_4H_6} + (\Delta H^0_{fT})_{\overset{\bullet}{H}} - (\Delta H^0_{fT})_{\overset{\bullet}{C_4H_7}}$$

and for equation (b):

$$(D_{C_4H_7})_b = (\Delta H^0_{fT})_{\overset{\bullet}{C_2H_3}} + (\Delta H^0_{fT})_{C_2H_4} - (\Delta H^0_{fT})_{\overset{\bullet}{C_4H_7}}$$

By substitution in Eq. (2.4) it results:

$$\frac{r_a}{r_b} = e^{\frac{1}{RT}\left[(\Delta H^0_{fT})_{\overset{\bullet}{C_2H_3}} + (\Delta H^0_{fT})_{C_2H_4} - (\Delta H^0_{fT})_{C_4H_6} - (\Delta H^0_{fT})_{\overset{\bullet}{H}}\right]}$$

Using for radicals the data from Table 2.11 recalculated for the temperature of 1000 K, and for hydrocarbons the heats of formation published by Stull [52], it results:

$$\frac{r_a}{r_b} = e^{\frac{(330,637+38,560-95,208-232,749)}{8.319\times1,000}} = e^{4.954} = 142$$

This means that during pyrolysis, the reaction (a) of formation of butadiene will be predominant, while (b) is negligible.

This method of calculation could be applied at the analysis of the results of other reactions of thermal decomposition, for which the data on the dissociation energy are not available.

The data of Table 2.11 make possible to compare on thermodynamic grounds the expressions for rate constants for reversible reactions in which radicals participate, such as the reactions of dissociation of the molecules versus those of recombination of the resulting free radicals.

Table 2.11 Thermodynamic Constants of Free Radicals

Radical	$\Delta H^0_{f,298}$ (kJ/mol)	$\Delta S^o_{f,298}$ (J/mol · grd)	C_p (J/mol · grd) 300 K	500 K	1000 K	Source
•H	218.14	114.72	20.81	20.81	20.81	6
•CH$_3$	142.36	193.02	28.89	29.30	30.14	6
•C$_2$H$_3$	288.90	235.73	40.61	54.84	78.30	6
•C$_2$H$_5$	108.86	250.38	46.47	68.25	107.61	6
•C$_3$H$_5$	169.99	265.46	59.04	83.74	125.61	6
n-C$_3$H$_7$	87.93	286.81	71.60	105.51	159.52	6
•i-C$_3$H$_7$	73.69	279.27	71.18	104.26	158.69	6
•n-C$_4$H$_9$	66.32	336.64	97.05	143.15	213.20	6
•sec-C$_4$H$_9$	51.67	339.06	96.47	141.81	212.26	6
•$tert$-C$_4$H$_9$	28.05	312.35	79.97	127.70	205.16	6
•n-C$_5$H$_{11}$	45.60	376.29	120.08	177.70	264.87	6
•sec-C$_5$H$_{11}$	30.94	378.50	119.50	176.36	264.03	6
•C$_6$H$_5$	293.0					11
C$_6$H$_5$CH$_2$ •	180 ± 12					51
o-CH$_3$C$_6$H$_4$CH$_2$•	113 ± 4					11
m-CH$_3$C$_6$H$_4$CH$_2$•	121 ± 8					11
p-CH$_3$C$_6$H$_4$CH$_2$•	117 ± 6					11

For other temperatures, the following equations may be used:

$$\Delta H^0_{fT} = \Delta H^0_{f298} + \overline{\Delta C}_p(T - 298) \qquad S^0_{fT} = S^0_{298} + \overline{\Delta C}_p \ln\frac{T}{298}$$

$$\text{where: } \overline{\Delta C}_p = \frac{C_{pT} + C_{p298}}{2}$$

In favorable thermodynamic conditions, the cyclohexane and its alkyl derivatives may undergo dehydrogenation reactions with the formation of aromatic hydrocarbons.

The graph of Figure 2.7, which indicates the equilibrium for the dehydrogenation of methyl-cyclohexane to toluene, proves that such reactions may take place during pyrolysis, as well as in other thermal processes.

Experimental studies on the pyrolysis of unsubstituted poly-cycloalkanes [53] confirmed the above mechanisms for cyclohexane and lead to the reaction schemes for other hydrocarbons.

For decahydronaphthalene the mechanism is [53,54]:

1. Radical formation

2. Decomposition
2.1 Cleavage in β position

$+ 3/2H_2$

$+$

$2C_2H_4$

$+ 2H_2$

$+$

C_2H_4

2.2 Isomerization followed by cleavage in β position

$+ CH_4 + C_2H_4$

$+ CH_4 + C_2H_4$

Similarly, for perhydrophenanthrene [54]:

1. Radical formation

$+ R^{\bullet} \longrightarrow$

2. Decomposition
2.1

$+ H_2$

$+$

$+ H^{\bullet}$

$+ 2H_2$

32%

$+$

C_2H_4

2.2a,b

a)

$$+ C_2H_4 \qquad + C_2H_4$$

$$+ C_2H_4 \qquad + H^\bullet \qquad + H_2 \qquad\qquad 12\%$$

b)

$$+$$
$$C_3H_7^\bullet \longrightarrow C_2H_4 + CH_3^\bullet \qquad\qquad 30\%$$

2.3

$$\qquad\qquad 26\%$$

Aromatics The aromatic ring is remarkably stable, the average binding energy between the carbon atoms within the ring being 407 kJ/mole [10]. The binding energy of the hydrogen atoms bound to the carbons of the ring is also higher than in other hydrocarbons (see Table 2.1). For these reasons, the decomposition reactions affect especially the side chains attached to the rings.

The interaction of the free radicals with alkylbenzenes is illustrated by n-propylbenzene, i-propylbenzene and $tert$-isobutylbenzene.

For n-propylbenzene the following reactions will take place:

$$C_6H_5-CH_2-CH_2-CH_3 + \overset{\bullet}{R} \rightarrow RH + C_6H_5-\overset{\bullet}{C}H-CH_2-CH_3$$

$$\rightarrow RH + C_6H_5-CH=CH_2 + \overset{\bullet}{C}H_3 \qquad\qquad (a)$$

$$C_6H_5-CH_2-CH_2-CH_3 + \overset{\bullet}{R} \rightarrow RH + C_6H_5-CH_2-\overset{\bullet}{C}H-CH_3$$

$$\rightarrow RH + \overset{\bullet}{C}_6H_5 + C_3H_6 \qquad\qquad (b)$$

$$C_6H_5-CH_2-CH_2-CH_3 + \overset{\bullet}{R} \rightarrow RH + C_6H_5-CH_2-CH_2-\overset{\bullet}{C}H_2$$

$$\rightarrow RH + C_6H_5-\overset{\bullet}{C}H_2 + C_2H_4 \qquad\qquad (c)$$

Overall reactions are:

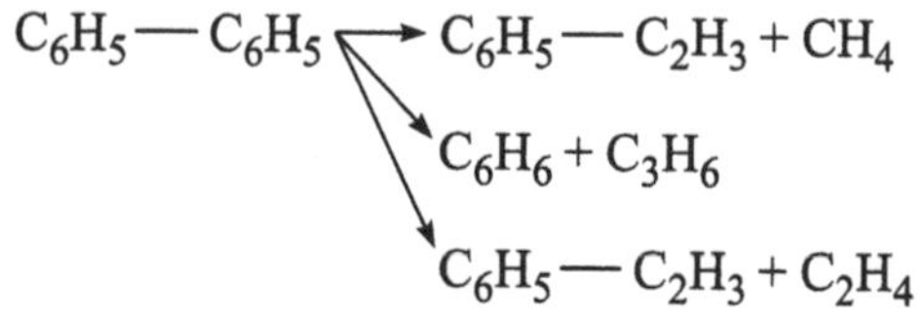

$$C_6H_5-C_6H_5 \longrightarrow \begin{cases} C_6H_5-C_2H_3+CH_4 \\ C_6H_6+C_3H_6 \\ C_6H_5-C_2H_3+C_2H_4 \end{cases}$$

For *i*-propyl-benzene:

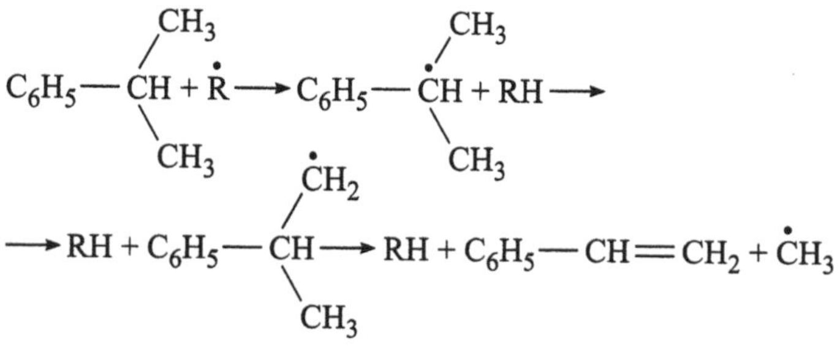

(a)

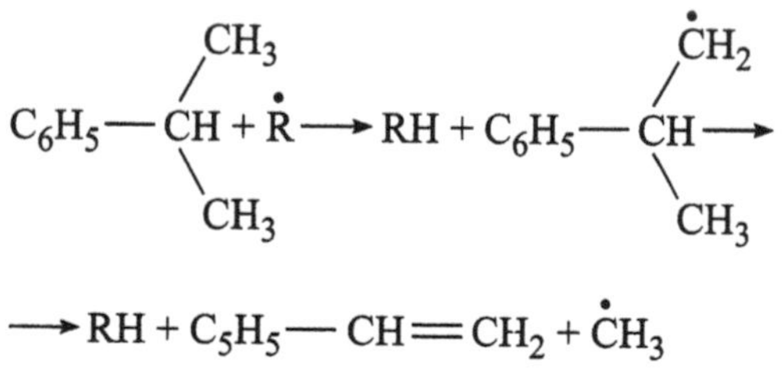

(b)

Both reactions lead finally to the formation of styrene and methane, which is in accordance with the experimental data:

For *tert*-isobutylbenzene:

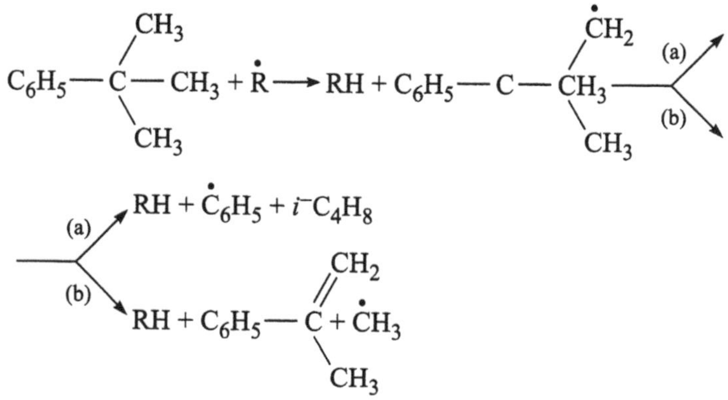

The occurrence of reaction (b) is doubtful. If and when it however takes place, the formed hydrocarbon will isomerize to α-methyl-styrene, similar to the dehydrogenation of isopropylbenzene.

Thus, the overall reaction may be written:

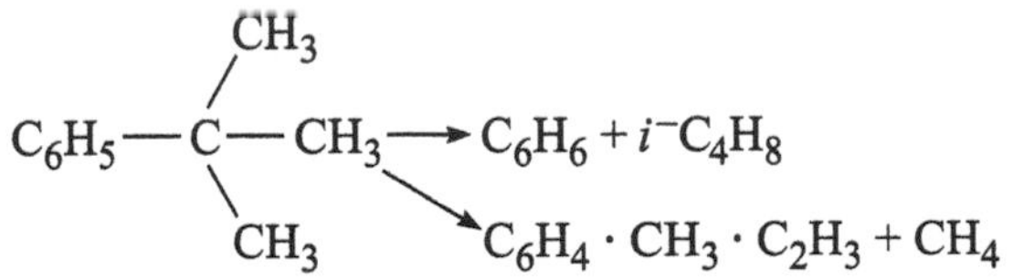

$$C_6H_5 - \overset{\overset{\displaystyle CH_3}{|}}{\underset{\underset{\displaystyle CH_3}{|}}{C}} - CH_3 \longrightarrow \begin{matrix} C_6H_6 + i\text{-}C_4H_8 \\ C_6H_4 \cdot CH_3 \cdot C_2H_3 + CH_4 \end{matrix}$$

The radicals formed by breaking the alkyl chains off the benzene ring will decompose further, following a similar pattern.

The use of atmospheric and vacuum gas oils as pyrolysis feed stocks has induced studies of the pyrolysis of model compounds with naphthenic-aromatic, aromatic and substituted-aromatic structures [54].

The thermal cracking of substituted aromatics having a long aliphatic chain was examined using dodecylbenzene as a model compound. It was confirmed that the radicals formed by the initial breaking of the side chain extract one of the hydrogen atoms from the alkyl chain. The resulting alkyl-aromatic radical further decomposes by successive β cleavages. The final product distribution is dependent on the position of the carbon atom from which the hydrogen atom was extracted. The final result of the decomposition was formulated by the authors [54] by means of six overall reactions. They are:

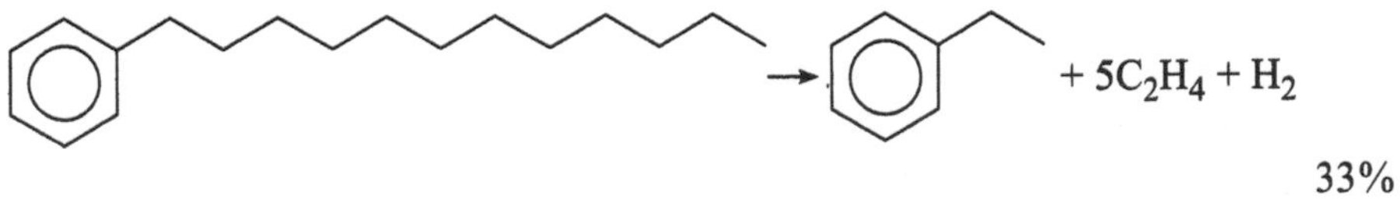

$+ 5C_2H_4 + H_2$

33%

Odd electron at carbon atom in odd position (1,3,...) from cycle

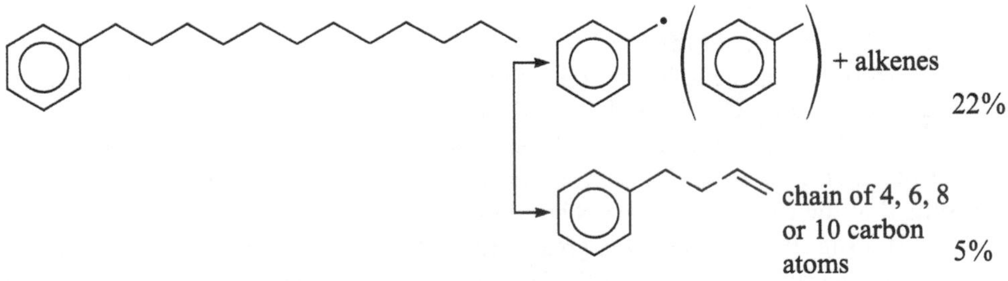

+ alkenes

22%

chain of 4, 6, 8 or 10 carbon atoms

5%

Odd electron at carbon atom in even position (2,4,6...) from cycle

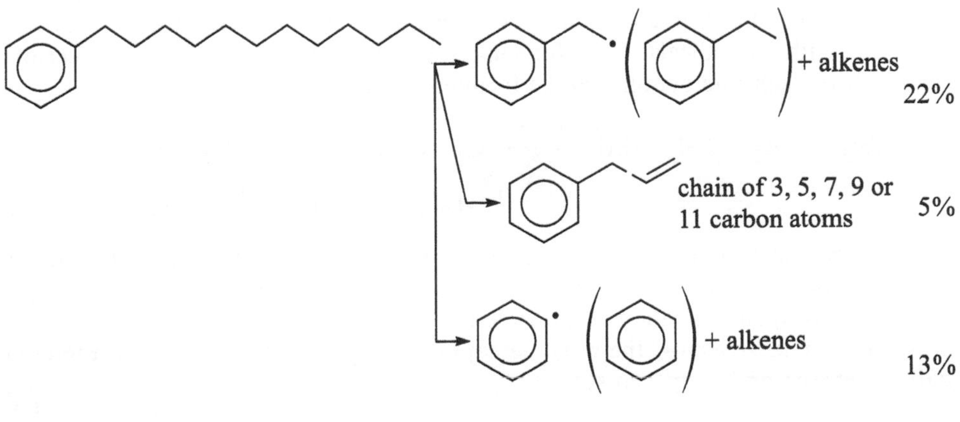

+ alkenes

22%

chain of 3, 5, 7, 9 or 11 carbon atoms

5%

+ alkenes

13%

In this study the following breaking energies of the side chain bonds were used:

Position of the bond relative to ring	α	β	γ	δ and following
Dissociation energy, kJ/mole	407	287	340	342

The decomposition of the naphthenic-aromatic hydrocarbons was examined taking as a model tetrahydro-naphthalene and octahydro-phenanthrene [53–56]. The results of the studies indicated that the presence of the aromatic rings favors the dehydrogenation of the naphthenic rings. This leads to the conclusion that for these types of hydrocarbons the dehydrogenation reactions are in competition with the thermal cracking reactions. The latter lead to the formation of monocyclic aromatic hydrocarbons with short chains, which are refractory to subsequent decompositions.

The nonsubstituted condensed aromatic hydrocarbons are completely refractory against the decompositions; but it seems that they have also a stabilizing effect, of hindering the decomposition of other hydrocarbons which might be present. This effect was observed at the pyrolysis of n-decane in the presence of naphthalene [54].

The condensed aromatics with methyl groups attached have a different behavior. Thus, the methyl-naphthalene gives methyl and naphthyl radicals, the latter leading also to condensation reactions.

Heteroatomic compounds The data contained in Table 2.1 show that in compounds with similar structures, the energy of C−S bonds is lower than that of the C−C bonds in hydrocarbons. It results that the initiation of thermal reactions in sulphur compounds will probably take place by breaking of the bonds C−S or S−S, with formation of H_2S and mercaptans respectively, which agrees with the experimental findings.

The lower energy of the C-S bonds leads to a greater energy of the C−H bonds for the carbon atoms neighboring the sulphur atom, as compared with the energies of the corresponding C−H bonds in hydrocarbons.

Also, taking into account that the energy of the H−S in hydrogen sulfide is of 377 kJ/mole, and the energy of the C−S bond in mercaptans is of about 293 kJ/mole, it results that the energy of the H−S bond in mercaptans is of the order of $2 \times 377 - 293 = 461$ kJ/mole. This energy is greater than that of the C−H bond in hydrocarbons. Thus, the hydrogen atom in the H−S of mercaptans can not be extracted by a radical.

As in the case of hydrocarbons, the composition of the products of thermal decomposition of sulfur compounds is determined by the interaction of the feed molecules with the radicals. Thus, for some sulphur compounds, schemes concerning the result of such interactions can be formulated as:

$$CH_3—CH_2—CH_2—SH + \overset{\bullet}{R} \overset{(a)}{\underset{(b)}{\rightarrow}} \begin{matrix} RH + \overset{\bullet}{C}H_2—CH_2 \dashrightarrow CH_2—SH \\ \\ RH + CH_3—\overset{\bullet}{C}H—CH_2 \dashrightarrow SH \end{matrix}$$

From this reaction one can anticipate that the products obtained from the decomposition of propylmercaptan are methylmercaptan and ethene or hydrogen sulfide and propene, for the two possible pathways shown.

For similar reasons, the ethylmercaptan is expected to decompose preferentially to ethene and hydrogen sulfide.

The decomposition of *n*-butyl-mercaptan can take place according to the following reactions:

$$CH_3-CH_2-CH_2-CH_2-SH + \overset{\bullet}{R} \overset{(a)}{\longrightarrow} RH + \overset{\bullet}{C}H_2-CH_2\overset{}{\rightsquigarrow}CH_2-CH_2-SH$$

$$\overset{(b)}{\searrow} RH + CH_2-\overset{\bullet}{C}H_2-CH_2\overset{}{\rightsquigarrow}CH_2-SH$$

$$\overset{(c)}{\searrow} RH + CH_2\overset{}{\rightsquigarrow}CH_2-\overset{\bullet}{C}H_2-CH_2\overset{}{\rightsquigarrow}SH$$
$$\qquad\qquad (c_1) \qquad\qquad\qquad (c_2)$$

It results that the decomposition of *n*-butyl-mercaptan and of higher mercaptans will produce besides hydrogen sulfide, methyl-mercaptan and hydrocarbons, also mercaptans with a carbon-carbon double bond, according to pathway c_1.

The decomposition of the sulfides may be illustrated by:

$$CH_3-CH_2-S-CH_2-CH_3 + \overset{\bullet}{R} \rightarrow \overset{\bullet}{C}H_2-CH_2-S-CH_2-CH_3$$

$$\rightarrow C_2H_4 + CH_3-CH_2-\overset{\bullet}{S}$$

with formation of ethene and mercaptan, as final products.

The experimental results confirm the decomposition of sulfides to alkenes and mercaptans and the decomposition of mercaptans to alkenes and hydrogen sulfide. Also, the industrial practice confirms the important amounts of methyl-mercaptan and hydrogen sulfide expected in the cracking gases.

For disulfides and cyclic sulfur compounds (thiophenes, thiophanes, etc.) the available data do not allow the formulation of even hypothetical cracking mechanisms. It is known that the cyclic compounds are much more stable than the acyclic ones.

The presence of elementary sulfur in the heavy products resulting from pyrolysis may be explained as result of the decomposition of disulfides. Thus, it may be assumed that the interaction of the disulfides with the radicals, followed by the cleavage in β position, may lead, among others, to the formation of two types of radicals:

$$R-\overset{\bullet}{S} \quad \text{and} \quad R-S-\overset{\bullet}{S}$$

The first one, may produce a mercaptan:

$$R-\overset{\bullet}{S} + R_1H \rightarrow RSH + \overset{\bullet}{R}_1$$

while the second, by a decomposition in β, may give elemental sulfur:

$$R-S-\overset{\bullet}{S} \rightarrow S_2 + \overset{\bullet}{R}$$

The thermal decomposition, of oxygen compounds produces mainly phenolic compounds. They (especially *m*-cresol) can be recovered from the gasoline obtained in thermal cracking.

In a similar manner, the nitrogen compounds give pyridyne bases. These can be recovered from gasoline fractions, and may be sold as useful byproducts

Coke formation   Thermal processes in the presence of a liquid phase, may lead to the formation of coke. The coke is constituted of carbenes, which are insoluble in benzene, but are soluble in carbon disulfide, and carboids, which are not soluble in any solvent. The atomic C/H ratio in coke is of 2–4. In coke produced at low temperatures this ratio is much lower, of only 1.1–1.25.

The carbenes content of coke is low (about 2%), going lower as the temperature at which they are formed goes higher. The molecular mass of carbenes is about 100,000–135,000.

The carboids are condensed, cross-linked polymers, in which the greatest part of the carbon atoms belong to aromatic condensed structures.

The carboids are not formed directly as a result of the thermal decomposition of alkanes or cyclo-alkanes, but only as a subsequent result of the very advanced conversion of the products obtained from their decomposition. The mechanism is different for feeds containing aromatic bi- and polycyclic hydrocarbons, which directly undergo condensation reactions resulting in the final formation of carboids. The rate of such reactions depends very much upon the structure of the aromatic hydrocarbons used as feed.

Experimental tests performed in autoclave, at 450°C, show that dibenzyl produces 1% carboids, after about 40 minutes, while α-methylnaphthalene requires 400 minutes. The conversion of condensed aromatics without side chains, to advanced condensation products is even slower. For example, naphthalene requires 670,000 minutes.

These findings may be explained by the fact that hydrocarbons that possesses alkyl chains or bridges much more easily form radicals, which are required for generating the condensation reactions, than those lacking side chains. The presence of important concentrations of free radicals in the reactions of thermal decomposition of higher hydrocarbons in liquid phase was proven experimentally by means of paramagnetic electronic resonance [57].

Aromatic hydrocarbons lacking side chains or alkyl bridges have a very high thermal stability. Actually, they do not generate free radicals. Aromatic hydrocarbons bound to alkyl groups easily generate free radicals of the form $Ar-\overset{\bullet}{C}H_2$ or $Ar-CH-\overset{\bullet}{C}H$.

Thus, in the case of dibenzyl, one of the possible sequences of reactions is the following:

(a)

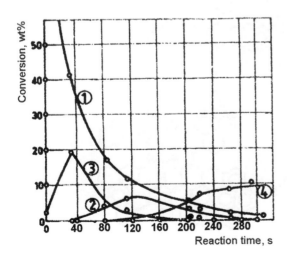

As shown, in this scheme, the reaction (a) regenerates the free radical, which ensures the continuation of the reaction chain. In the same time, higher molecular weight aromatics are produced by condensation reactions of type (b).

The importance of the free radicals in these reactions was demonstrated experimentally. Addition of high molecular weight alkanes (paraffin wax) to polycyclic aromatic hydrocarbons submitted to thermal processing, greatly increases the rate of carboids formation. This fact is explained by the high rate of radical formation as a result of the thermal decomposition of alkanes.

The formation of coke doesn't take place directly, but takes place through several steps. The scheme which is generally accepted for this succession of transformations is:

Hydrocarbons → Resins → Asphalthenes →Coke

The experimental data represented in Figure 2.8 [58] illustrate clearly the succession of the transformations which take place.

In this succession, the most important step is the flocculation of the asphaltenes. The condensation reactions that take place lead to an increasingly higher degree of aromatization with the result that the condensation products gradually

Figure 2.8 Products formed by the coking of a deasphalted oil. 1-oil, 2-asphaltenes, 3-resins, 4-carboids. (From Ref. 58.)

become insoluble in the liquid reaction mixture. At a certain point, the liquid phase becomes unstable and the asphaltenes begin to flocculate [59].

The formation of carboids can take place only after the flocculation of the asphaltenes has occurred. Thus, in processes such as visbreaking, which have its target liquid products, the reaction conditions are selected such that the flocculation of the asphaltenes is prevented. The reaction is stopped before reaching the conversion corresponding to the flocculation of asphaltenes. To this purpose, several analytical methods were suggested, which make it possible to determine the state in which the asphaltenes are peptized or flocculated [60–62].

Studies by means of small angle X-ray scattering (SAXS) and small angle neutron scattering (SANS), ascertained that the resins and the asphaltenes have a bidimensional structure, more or less open, constituted of naphtheno-aromatic precondensed rings connected by alkyl, chains or bridges, which may contain sulphur atoms [63].

The structure for the asphaltenes resulted from an Arab crude oil is depicted in Figure 2.9 [64].

The mechanism of the structural transformations that were observed under the effect of severe thermal treatment is depicted in Figure 2.10 [62,65]. The dealkylation and condensation of asphaltenes was observed, by X-ray scattering [2,3]. The initially peptized molecules are destabilized and begin to organize in asphaltenic micelles with superposed layers [4]. These structures grow and the distances between the layers diminish, approaching progressively the interlayer distance within graphite [5].

The conversion of asphaltenes to carboids was described as a polycondensation reaction with a chain as in the scheme:

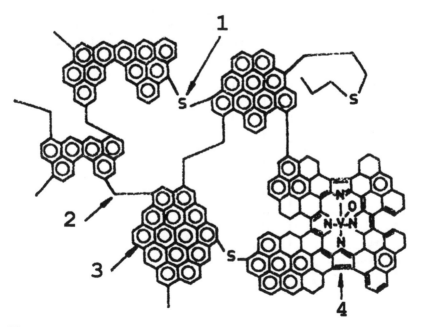

Figure 2.9 Typical asphaltene structure. 1-sulphur bridge, 2-aliphatic bridge, 3-bidimensional aromatic structure, 4-porphyrine. (From Ref. 64.)

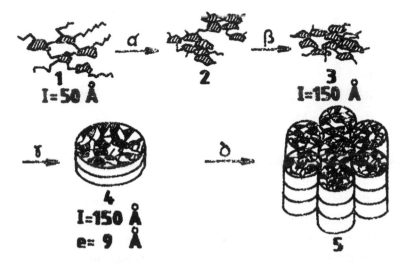

Figure 2.10 Condensation of resins and asphaltenes in steam cracking: α dealkylation, β-asphaltenes and resins condensation, γ-stratification, δ-grouping in multilayered structures. ⬡—Naphthenic-aromatic polycondensates; ⬿—aliphatic chains. (From Ref 62, 65.)

$$A \rightarrow A \bullet + R \bullet \qquad \text{initiation}$$
$$R \bullet + A \rightarrow A' \bullet + RH$$
$$A' \bullet + A \rightarrow A'A \bullet$$
$$A'A \bullet \rightarrow \Sigma M + A'A' \bullet$$
$$A'A' \bullet + A \rightarrow A'A'A \bullet \qquad \text{propagation}$$
$$- - - - - - - - - - - - - -$$
$$(A' \bullet)_X + A \rightarrow (A'')_X A \bullet \qquad \text{interruption}$$

wherein A is the asphaltene molecule, ΣM are light products which go into the vapor phase and $(A'')_X A \bullet$ represents inactive radicals.

From the average molecular masses of the asphaltenes and of the produced carboids, it was deduced that the length of the kinetic chain is 120–150 reactions.

2.3 KINETICS OF THERMAL PROCESSES

2.3.1 Reaction order

The kinetic equations can be deduced on the basis of the reaction mechanism. For the thermal decomposition of a hydrocarbon, the chain mechanism may be represented by the generalized scheme:

Initiation	$M \xrightarrow{k_1} \dot{R}_1 + \dot{R}_2$	(a)
Propagation	$\dot{R}_1 + M \xrightarrow{k_2} R_1 H + \dot{R}_3$	(b$_1$)
	$\dot{R}_2 + M \xrightarrow{k_2} R_1 H + \dot{R}_3$	(b$_2$)
	$\dot{R}_3 \xrightarrow{k_3} \Sigma M_f + \dot{R}_1$	(b$_3$)

Interruption $\overset{\bullet}{R}_1 + \overset{\bullet}{R}_3 \overset{k_4}{\to} M_2$ (b$_4$)

$\overset{\bullet}{R}_2 + \overset{\bullet}{R}_3 \overset{k_5}{\to} M_3$ (b$_5$)

$\overset{\bullet}{R}_1 + \overset{\bullet}{R}_1 \overset{k_6}{\to} M_4$ (c$_3$)

$\overset{\bullet}{R}_2 + \overset{\bullet}{R}_2 \overset{k_7}{\to} M_5$ (c$_4$)

$\overset{\bullet}{R}_3 + \overset{\bullet}{R}_3 \overset{k_8}{\to} M_6$ (c$_5$)

$\overset{\bullet}{R}_1 + \overset{\bullet}{R}_2 \overset{k_9}{\to} M_7$ (c$_6$)

A semiquantitative analysis of these equations is of interest.

The initiation reaction results in the formation of two types of radicals: $\overset{\bullet}{R}_1$— which can propagate the reaction chain and $\overset{\bullet}{R}_2$—which cannot. Since the length of the kinetic chain is about 200 reactions, it results that the concentration of the $\overset{\bullet}{R}_2$ radicals is about 200 times lower than of the radicals $\overset{\bullet}{R}_1$ and $\overset{\bullet}{R}_3$. Considering that the rate constants are of the same order of magnitude, it results that the interruption reactions that involve the $\overset{\bullet}{R}_2$ radicals, i.e. (c$_2$), (c$_4$), and (c$_6$) may be neglected.

From among the remaining reactions, reaction (c$_1$) takes place between two different radicals, whereas the reactions (c$_3$) and (c$_5$) take place between two identical radicals.

Since again the rate constants may be considered to be equal, it is possible to deduce the condition required in order that the preferred (most probable) interaction be that among identical radicals.

If in an arbitrary volume, the number of radicals of different species is $n_1, n_2,...,n_i,..., n_k$, the probability of interaction between two identical radicals is proportional to n_i^2 and that for all the possible reactions are proportional to:

$$\left(\sum_1^k n_i\right)^2$$

The probability of interaction between different radicals will be proportional to the difference:

$$\left(\sum_1^k n_i\right)^2 - \sum_1^k n_i^2$$

Thus, the condition that the interaction between two identical radicals, for example of species 1, should be more probable than the interactions between different radicals will be given by the relation:

$$\left(\sum_1^k n_i\right)^2 - \sum_1^k n_i^2 < n_i^2 \qquad (2.5)$$

Using this relation one can compute that the interaction between two identical radicals (n_i) is more probable only if their concentration is about 3 times larger than the sum of concentrations of the rest of the radicals.

It is obvious that this condition can not be fulfilled during the thermal reactions of higher hydrocarbon or petroleum fractions. Consequently, the only interruption reaction which needs to be taken into account is reaction (c$_1$).

By applying the theorem of stationary states to the reactions of scheme (a)–(c$_1$), one obtains:

$$\frac{d[\dot{R}_1]}{d\tau} = k_1[M] - k_2[\dot{R}_1][M] + k_3[\dot{R}_3] - k_4[\dot{R}_1][R_3] = 0 \tag{2.6}$$

$$\frac{d[\dot{R}_2]}{d\tau} = k_1[M] - k_2[\dot{R}_2][M] = 0 \tag{2.7}$$

$$\frac{d[\dot{R}_3]}{d\tau} = k_2[\dot{R}_1][M] + k_2[\dot{R}_2][M] - k_3[\dot{R}_3] - k_4[\dot{R}_1][\dot{R}_3] = 0 \tag{2.8}$$

From Eq. (2.7) it results:

$$[\dot{R}_2] = \frac{k_1}{k_2} \tag{2.9}$$

By substitution, reaction (2.8) becomes:

$$k_2[\dot{R}_1][M] + k_1[M] - k_3[\dot{R}_3] - k_2[\dot{R}_1][\dot{R}_3] = 0$$

By adding this to Eq. (2.6), it results:

$$[\dot{R}_3] = k_1 \frac{[M]}{k_4[\dot{R}_1]}$$

and by substitution in (2.6), after simplifications one obtains:

$$[\dot{R}_1] = \left(\frac{k_1 k_3}{k_2 k_4}\right)^{1/2} \tag{2.10}$$

The initial decomposition rate of the molecules M is given by the equation:

$$\frac{-d[M]}{d\tau} = k_1[M] + k_2[\dot{R}_1][M] + k_2[\dot{R}_2][M]$$

By substituting for $[\dot{R}_1]$ and $[\dot{R}_2]$ the expressions (2.9) and (2.10), it results:

$$\frac{-d[M]}{d\tau} = \left\{ 2k_1 + \left(\frac{k_1 k_2 k_3}{k_4}\right)^{1/2} \right\}[M]$$

which shows that the reaction is of first order in accordance with the experimental data.

This expression can be written as:

$$\frac{d[M]}{d\tau} = k[M] \tag{2.11}$$

where the overall rate constant is a function of the rate constants of the elementary reactions according to the relation:

$$k = 2k_1 + \left(\frac{k_1 k_2 k_3}{k_4}\right)^{1/2} = \left(\frac{k_1 k_2 k_3}{k_4}\right)^{1/2} \tag{2.12}$$

The simplification of neglecting $2k_1$ is justified by the fact that:

$$2k_1 << \left(\frac{k_1 k_2 k_3}{k_4}\right)^{1/2}$$

i.e., the rate constant for the initiation step is much smaller than the other rate constants.

If in the simplified expression (2.12), one replaces the overall rate constant and the rate constants of the elementary reactions by expressions of the Arrhenius type, it results:

$$Ae^{-E/RT} = \left(\frac{A_1 A_2 A_3}{A_4}\right)^{1/2} e^{-\frac{1}{2RT}(E_1+E_2+E_3-E_4)}$$

from where:

$$E = \frac{1}{2}(E_1 + E_2 + E_3 - E_4)$$

By substituting approximate values for the activation energies of the elementary reactions, it results:

$$E = \frac{1}{2}(335 + 33 + 126 - 0) = 247 \text{ kJ/mole}$$

From the above discussion it results that the overall activation energy for the chain reaction is much lower than that of the initiation reaction, or of the decomposition following a molecular mechanism. Usually, this is expressed by: *in thermal decompositions, the mechanism of chain reactions is "energetically more advantageous" than the molecular mechanism.*

The general scheme discussed above represents a gross simplification of the actual mechanism of the thermal decomposition of hydrocarbons with high molecular mass, or of fractions of crude oil. It does not take into account the formation during the initiation reactions of different types of radicals depending on the position in the molecule of the $C-C$ bond which is broken of the further decomposition of the radicals initially formed, or of the fact that the R_3 radicals are of various types, depending on the position of the extracted hydrogen within the various molecules M.

This larger diversity of the species of free radicals involved in a more elaborated mechanism does not change the results of the previous evaluation. If one replaces M, $\dot{R}_1$, $\dot{R}_2$, and $\dot{R}_3$ by ΣM, $\Sigma\dot{R}_1$, and $\Sigma\dot{R}_3$, the results obtained by the application of the theorem of the stationary states will be exactly the same.

It is to be noticed that the reaction order is changed only by modifying the nature of the initiation reaction (bimolecular instead of monomolecular) or of the interruption reaction (between the identical radicals $\dot{R}_1$ or $\dot{R}_3$ instead of between different radicals). In the first case (bimolecular initiation), by applying the theorem of stationary states, the order 2 is obtained. In the second case (interruption by the collision of two radicals R_1), the order 1/2 is obtained and in the third case (interruption by the collision of the $\dot{R}_3$ radicals) the order 3/2 is obtained [66].

Of great interest is the application of the general scheme for ethane and propane, which are important feeds for commercial pyrolysis.

For *ethane*, the mechanism may be expressed by the reactions:

Initiation $\quad C_2H_6 \xrightarrow{k_1} 2\overset{\bullet}{C}H_3$ (a)

$\qquad\qquad C_2H_6 + \overset{\bullet}{C}H_3 \xrightarrow{k_2} CH_4 + \overset{\bullet}{C}_2H_5$ (b)

Propagation $\quad \overset{\bullet}{C}_2H_5 \xrightarrow{k_3} C_2H_4 + \overset{\bullet}{H}$ (c)

$\qquad\qquad C_2H_6 + \overset{\bullet}{H} \xrightarrow{k_4} H_2 + \overset{\bullet}{C}_2H_5$ (d)

Interruption $\quad \overset{\bullet}{C}_2H_5 + \overset{\bullet}{H} \xrightarrow{k_5} C_2H_6$ (e)

$\qquad\qquad 2\overset{\bullet}{C}_2H_5 \xrightarrow{k_6} C_4H_{10}$ (f)

$\qquad\qquad \overset{\bullet}{C}H_3 + \overset{\bullet}{C}_2H_5 \xrightarrow{k_6} C_3H_8$ (g)

$\qquad\qquad 2\overset{\bullet}{C}H_3 \xrightarrow{k_7} C_2H_6$ (h)

$\qquad\qquad \overset{\bullet}{C}H_3 + \overset{\bullet}{H} \xrightarrow{k_8} CH_4$ (i)

$\qquad\qquad 2\overset{\bullet}{H} \xrightarrow{k_9} H_2$ (j)

The last two reactions (i) and (j) are very slow, since they imply triple collisions and may be ignored.

Because the length of the chain is of 100–200, it results that the concentration of the methyl radicals must be by 2 orders of magnitude less than that of the ethyl radicals and of the atomic hydrogen [16,66]. For this reason reactions (g) and (h) may be also neglected.

Concerning the reactions (e) and (f), the number of only 3 different radicals involved is too small to conclude—as in the previous case, on the basis of the relation (2.5)—whether the interaction between different radicals is the most probable. In order to determine which of these reactions should be taken into account, their rates must be compared.

The rate constants for these reactions are given in Table 2.8:

$$k_5 = 8.5 \times 10^{10}$$
$$k_6 = 0.79 \times 10^{10}$$

Since they are recombinations of radicals, the activation energy of these reactions is zero.

The concentrations of the ethyl radical and of the atomic hydrogen at 1100 K and 1 bar are the same [16]. At lower temperatures, the concentration of the atomic hydrogen is lower than that of the ethyl radicals. At 900 K and 0.1 bar, they are correlated by: $[\overset{\bullet}{H}] = 0.05[\overset{\bullet}{C}_2H_5][69]$.

Taking into account the values of the rate constants, it results that at the pyrolysis temperature of about 1100 K and at pressures only slightly larger than atmospheric, the first interruption reaction (e) is dominant. At lower temperatures and at pressures of a few bar, the reaction (f) appears to become prevalent.

By taking into account the interruption reaction (e), and by applying the theorem of stationary states, to the system of reactions (a)–(e), one obtains:

$$\frac{d[\overset{\bullet}{C}H_3]}{d\tau} = 2k_1[C_2H_6] - k_2[\overset{\bullet}{C}H_3][C_2H_6] = 0 \qquad (2.18)$$

$$\frac{d[\dot{C}_2H_5]}{d\tau} = k_2[\dot{C}H_3][C_2H_6] - k_3[\dot{C}_2H_5] + k_4[C_2H_6][\dot{H}]-$$

$$- k_5[\dot{C}_2H_5][\dot{H}] = 0 \tag{2.19}$$

$$\frac{d[\dot{H}]}{d\tau} = k_3[\dot{C}_2H_5] - k_4[C_2H_6][\dot{H}] - k_5[\dot{C}_2H_5][\dot{H}] = 0 \tag{2.20}$$

From the relation (2.18) it results:

$$[\dot{C}H_3] = \frac{2k_1}{k_2} \tag{2.21}$$

Replacing this value for $[\dot{C}H_3]$ into equation (2.19) and adding it to equation (2.20), one obtains:

$$2k_1[C_2H_6] - 2k_5[\dot{C}_2H_5][\dot{H}] = 0$$

from which

$$[\dot{C}_2H_5] = \frac{k_1[C_2H_6]}{k_5[\dot{H}]} \tag{2.22}$$

By substituting this result in (2.20), multiplying by $[\dot{H}]$, and making a series of simplifications, one obtains:

$$k_4[\dot{H}]^2 + k_1[\dot{H}] - \frac{k_1k_3}{k_5} = 0$$

from which

$$[\dot{H}] = \frac{-k_1 \pm \sqrt{k_1^2 + \frac{4k_1k_3k_4}{k_5}}}{2k_4} \tag{2.23}$$

Only the positive values are valid, since $[\dot{H}] > 0$.

The rate of ethane decomposition is given by the differential equation:

$$-\frac{d[C_2H_6]}{d\tau} = k_1[C_2H_6] + k_2[C_2H_6][\dot{C}H_3] + k_4[C_2H_6][\dot{H}]$$

$$- k_5[\dot{C}_2H_5][\dot{H}] \tag{2.24}$$

Substituting in this equation $[\dot{C}H_3]$, $[\dot{C}_2H_5]$, and $[\dot{H}]$ by their corresponding values given by the expressions (2.21)–(2.23), after a series of simplifications one obtains:

$$-\frac{d[C_2H_6]}{d\tau} = \left\{\frac{3}{2}k_1 + \frac{1}{2}\left(k_1^2 + 4\frac{k_1k_3k_4}{k_5}\right)^{1/2}\right\}[C_2H_6] \tag{2.25}$$

This relation may be written also as:

$$-\frac{d[C_2H_6]}{d\tau} = k[C_2H_6]$$

where,

$$k = \left\{ \frac{3}{2}k_1 + \left(\frac{k_1^2}{4} + \frac{k_1 k_3 k_4}{k_5} \right)^{1/2} \right\} \approx \left(\frac{k_1 k_3 k_4}{k_5} \right)^{1/2}$$

The latter approximation is admissible, since the k_1 value is lower than the other constants.

If instead of the reaction (e), one assumes that the chain interruption follows equation (f), the relations (2.18) and (2.21) remain valid, while relations (2.19) and (2.20) become:

$$\frac{d[\overset{\bullet}{C}_2H_5]}{d\tau} = k_2[\overset{\bullet}{C}H_3][C_2H_6] - k_3[\overset{\bullet}{C}_2H_5] + k_4[C_2H_6][\overset{\bullet}{H}] \tag{2.26}$$

$$- k_6[\overset{\bullet}{C}_2H_5]^2 = 0$$

$$\frac{d[\overset{\bullet}{H}]}{d\tau} = k_3[\overset{\bullet}{C}_2H_5] - k_4[C_2H_6][\overset{\bullet}{H}] = 0 \tag{2.27}$$

Substituting $[\overset{\bullet}{C}H_3]$ given relation (2.21), in relation (2.26), and adding to relation (2.27), one obtains:

$$2k_1[C_2H_6] - k_6[\overset{\bullet}{C}_2H_5]^2 = 0$$

from which,

$$[\overset{\bullet}{C}_2H_5] = \left(\frac{2k_1}{k_6}[C_2H_6] \right)^{1/2} \tag{2.28}$$

Substituting this value in relation (2.27) it results:

$$[\overset{\bullet}{H}] = \frac{k_3}{k_4} \left(\frac{2k_1}{k_6[C_2H_6]} \right)^{1/2} \tag{2.29}$$

The overall decomposition rate of the ethane for this case is given by the relation:

$$- \frac{d[C_2H_6]}{d\tau} = k_1[C_2H_6] + k_2[C_2H_6][\overset{\bullet}{C}H_3] + k_4[C_2H_6][\overset{\bullet}{H}]$$

After performing the substitutions and simplifications, one obtains:

$$- \frac{d[C_2H_6]}{d\tau} = 3k_1[C_2H_6] + k_3 \left(\frac{2k_1}{k_6} \right)^{1/2} [C_2H_6]^{1/2} \tag{2.30}$$

from which it results that the order of reaction is intermediary, between 1 and $\frac{1}{2}$, closer to the order 1/2, since the constant k_3 has a larger value than the other rate constants.

On basis of the final relations (2.25) and (2.30), the thermal decomposition of ethane is of order 1, if the interruption of the reaction chains is controlled by equation (e), the collision of the ethyl radicals with atomic hydrogen; it is of order 1/2 if the interruption is controlled by equation (f), the collision of two ethyl radicals.

Experimental results in flow system, as well as measurements in industrial furnaces, have proven that in the case of ethane the first order equation is fully satisfactory and it is not necessary to resort to the equation of order 1/2.

This confirms the earlier statement that according to the facts shown previously, in the conditions of the pyrolysis the prevalent interruption occurs by reaction (e).

The proposed model and the deduced equations allow one to obtain not only the equation for the overall conversion, but also the kinetic equations of the formation of different components.

On the basis of the reaction system described by equations (a)–(e), one may write:

$$\frac{d[C_2H_4]}{d\tau} = k_3[\dot{C}_2H_5]$$

Using the expressions (2.22) and (2.23) it results:

$$\frac{d[C_2H_4]}{d\tau} = \frac{2k_1k_3k_4}{k_5\left(-k_1 + \sqrt{k_1^2 + 4\frac{k_1k_3k_4}{k_5}}\right)}[C_2H_6]$$

Multiplying the numerator and the denominator by $k_1 + \sqrt{k_1^2 + 4\frac{k_1k_3k_4}{k_5}}$ and performing the simplifications, one obtains finally:

$$\frac{d[C_2H_4]}{d\tau} = \frac{1}{2}\left(k_1 + \sqrt{k_1^2 + 4\frac{k_1k_3k_4}{k_5}}\right)[C_2H_6]$$

In a similar manner,

$$\frac{d[CH_4]}{d\tau} = k_2[C_2H_6][\dot{C}H_3] = 2k_1[C_2H_6]$$

$$\frac{d[H_2]}{d\tau} = k_4[C_2H_6][\dot{H}] = \frac{1}{2}\left(-k_1 + \sqrt{k_1^2 + 4\frac{k_1k_3k_4}{k_5}}\right)[C_2H_6]$$

These relations correspond to the stoichiometric equations:

$$C_2H_6 = C_2H_4 + H_2$$
$$2C_2H_6 = C_2H_4 + 2CH_4$$

The pyrolysis of *propane* may be represented by the following scheme:

Initiation $\quad C_3H_8 \xrightarrow{k_1} \dot{C}H_3 + \dot{C}_2H_5$ $\qquad\qquad$ (a)

$\qquad\qquad\quad \dot{C}_2H_5 \xrightarrow{k_2} C_2H_4 + \dot{H}$ $\qquad\qquad$ (b)

Propagation $\quad C_3H_8 + \overset{\bullet}{H} \overset{k_3}{\rightarrow} H_2 + n{-}\overset{\bullet}{C}_3H_7$ $\qquad\qquad$ (c)

$\qquad\qquad\qquad\qquad \searrow^{k_4} H_2 + i - \overset{\bullet}{C}_3H_7$ $\qquad\qquad$ (d)

$\qquad\quad C_3H_8 + \overset{\bullet}{C}H_3 \overset{k_5}{\rightarrow} CH_4 + n - \overset{\bullet}{C}_3H_7$ $\qquad\qquad$ (e)

$\qquad\qquad\qquad\qquad \searrow^{k_6} CH_4 + i - \overset{\bullet}{C}_3H_7$ $\qquad\qquad$ (f)

$\qquad\quad n - \overset{\bullet}{C}_3H_7 \overset{k_7}{\rightarrow} C_2H_4 + \overset{\bullet}{C}H_3$ $\qquad\qquad$ (g)

$\qquad\qquad\quad i - \overset{\bullet}{C}_3H_7 \overset{k_8}{\rightarrow} C_3H_6 + \overset{\bullet}{H}$ $\qquad\qquad$ (h)

Interruption $\quad n - \overset{\bullet}{C}_3H_7 + \overset{\bullet}{H} \overset{k_9}{\rightarrow} C_3H_8$ $\qquad\qquad$ (i)

$\qquad\qquad\quad i - \overset{\bullet}{C}_3H_7 + \overset{\bullet}{H} \overset{k_{10}}{\rightarrow} C_3H_8$ $\qquad\qquad$ (j)

$\qquad\quad n{-}\overset{\bullet}{C}_3H_7 + \overset{\bullet}{C}H_3 \overset{k_{11}}{\rightarrow} C_4H_{10}$ $\qquad\qquad$ (k)

$\qquad\quad i - \overset{\bullet}{C}_3H_7 + \overset{\bullet}{C}H_3 \overset{k_{12}}{\rightarrow} C_4H_{10}$ $\qquad\qquad$ (l)

$\qquad\quad 2\overset{\bullet}{C}H_3 \overset{k_{12}}{\rightarrow} \overset{\bullet}{C}_2H_6$ $\qquad\qquad$ (m)

In this scheme the interruption reactions:

$$\overset{\bullet}{H} + \overset{\bullet}{H} \rightarrow H_2 \text{ and } \overset{\bullet}{H} + \overset{\bullet}{C}H_3 \rightarrow CH_4,$$

which involve three-body collisions, were neglected.

The reaction between two identical or different propyl radicals, was also neglected, as justified by the steric factor, which is very unfavorable for such inter-actions.

The analysis of this system of reactions leads to the conclusion that the kinetic order will be one, if the interruption of the reaction chain takes place by any of the reactions (i) to (l). It will be different from one, if the interruption is due to the collision between two identical radicals, as in reaction (m).

The relative values for the concentrations of the various radicals at 1100 K and a pressure of 3 bar, are [16]:

$$[i\overset{\bullet}{C}_3H_7] = 1.7[\overset{\bullet}{C}H_3];$$

$$[\overset{\bullet}{H}] = 2.5[\overset{\bullet}{C}H_3];$$

$$[n{-}\overset{\bullet}{C}_3H_7] = 0.012[\overset{\bullet}{C}H_3].$$

The low concentration of n-propyl radicals allows one to neglect the reactions (i) and (k).

Since the activation energy of the recombination reactions of the radicals is actually zero, their rate will be determined by the pre-exponential factors. Using the values given in Table 2.8 for the pre-exponential factors, and neglecting the reactions involving the n-propyl radical, the rates of the latter three interruption reactions are:

$$r_j = 1.0 \times 10^{10} \times 1.7[\overset{\bullet}{C}H_3] \times 2.5[\overset{\bullet}{C}H_3] = 4.25 \times 10^{10}[\overset{\bullet}{C}H_3]^2,$$

$$r_l = 0.32 \times 10^{10} \times 1.7[\overset{\bullet}{C}H_3]^2 = 0.54 \times 10^{10}[\overset{\bullet}{C}H_3]^2,$$

$$r_m = 1.3 \times 10^{10}[\overset{\bullet}{C}H_3]^2.$$

It follows that in the conditions of pyrolysis, reaction (j) is the most important chain interruption reaction.

Neglecting the other interruption reactions and applying the theorem of the stationary states, one obtains:

$$\frac{d[\dot{C}H_3]}{d\tau} = k_1[C_3H_8] - (k_5 + k_6)[C_3H_8][\dot{C}H_3] + k_7[n - \dot{C}_3H_7] = 0 \qquad (2.31)$$

$$\frac{d[\dot{H}]}{d\tau} = k_2[\dot{C}_2H_5] - (k_3 + k_4)[C_3H_8][\dot{H}] + k_8[i - \dot{C}_3H_7]$$
$$- k_{10}[i - \dot{C}_3H_7][\dot{H}] = 0 \qquad (2.32)$$

$$\frac{d[\dot{C}_2H_5]}{d\tau} = k_1[C_3H_8] - k_2[\dot{C}_2H_5] = 0 \qquad (2.33)$$

$$\frac{d[n - \dot{C}_3H_7]}{d\tau} = k_3[C_3H_8][\dot{H}] + k_5[C_3H_8][\dot{C}H_3] - k_7[n - \dot{C}_3H_7] = 0 \qquad (2.34)$$

$$\frac{d[i - \dot{C}_3H_7]}{d\tau} = k_4[C_3H_8][\dot{H}] + k_6[C_3H_8][\dot{C}H_3] - k_8[i - \dot{C}_3H_7]$$
$$- k_{10}[i - \dot{C}_3H_7][\dot{H}] = 0 \qquad (2.35)$$

From the relation (2.33) it results:

$$[\dot{C}_2H_5] = \frac{k_1}{k_2}[C_3H_8] \qquad (2.36)$$

Substituting this in equation (2.32), one obtains:

$$k_1[C_3H_8] - (k_3 + k_4)[C_3H_8][\dot{H}] + k_8[i - \dot{C}_3H_7]$$
$$- k_{10}[i - \dot{C}_3H_7][\dot{H}] = 0 \qquad (2.37)$$

Adding with (2.35) it results:

$$k_1[C_3H_8] - k_3[C_3H_8][\dot{H}] + k_6[C_3H_8][\dot{C}H_3] - 2k_{10}[i - \dot{C}_3H_7][\dot{H}] = 0$$

By the addition of the relations (2.31) and (2.34) it results:

$$k_1[C_3H_8] + k_3[C_3H_8][\dot{H}] - k_6[C_3H_8][\dot{C}H_3] = 0$$

from which:

$$[\dot{C}H_3] = \frac{k_1 + k_3[\dot{H}]}{k_6} \qquad (2.38)$$

Adding the equations (2.37) and (2.35) gives:

$$k_1[C_3H_8] - k_3[C_3H_8][\dot{H}] + k_6[C_3H_8][\dot{C}H_3] - 2k_{10}[i - \dot{C}_3H_7][\dot{H}] = 0$$

from which, after the replacing the concentration of $[\overset{\bullet}{C}H_3]$ with (2.38) and simplifying, one gets:

$$k_1[C_3H_8] = k_{10}[i - \overset{\bullet}{C}_3H_7][\overset{\bullet}{H}] \qquad (2.39)$$

Subtracting Eq. (2.35) from (2.37), it results:

$$k_1[C_3H_8] - (k_1 + 2k_4)[C_3H_8][\overset{\bullet}{H}] - k_6[C_3H_8][\overset{\bullet}{C}H_3] + 2k_8[i - \overset{\bullet}{C}_3H_7] = 0$$

Replacing the $[\overset{\bullet}{C}H_3]$ concentration by (2.38) and after simplifications, it results:

$$(k_3 + k_4)[C_3H_8][\overset{\bullet}{H}] = k_8[i - \overset{\bullet}{C}_3H_7] \qquad (2.40)$$

From the relation (2.40) it results:

$$[\overset{\bullet}{H}] = \left(\frac{k_1k_8}{k_{10}(k_3 + k_4)}\right)^{1/2} \qquad (2.41)$$

The decomposition rate of propane will be given by the expression:

$$-\frac{d[C_3H_8]}{d\tau} = k_1[C_3H_8] + (k_3 + k_4)[C_3H_8][\overset{\bullet}{H}] + (k_5 + k_6)[C_3H_8][\overset{\bullet}{C}H_3]$$

$$- k_{10}[i - \overset{\bullet}{C}_3H_7][\overset{\bullet}{H}]$$

By taking into account relation (2.39), the first and the last terms of the right side cancel out and by replacing $[CH_3]$ with (2.38), it results:

$$-\frac{d[C_3H_8]}{d\tau} = \left\{(k_3 + k_4)[\overset{\bullet}{H}] + \frac{(k_5 + k_6)k_1}{k_6} + \frac{(k_5 + k_6)k_3}{k_6}[\overset{\bullet}{H}]\right\}[C_3H_8]$$

Because k_1 has a very low value, the second term in the parenthesis can be neglected, and after replacing $[\overset{\bullet}{H}]$ with the expression (2.41), one obtains:

$$-\frac{d[C_3H_8]}{d\tau} = \left((k_3 + k_4) + \frac{(k_5 + k_6)k_3}{k_6}\right)\left(\frac{k_1k_8}{k_{10}(k_3 + k_4)}\right)^{1/2}[C_3H_8] \qquad (2.42)$$

which shows that the pyrolysis proceeds according to first order kinetics.

Referring to the data of Table 2.9, the activation energy, which corresponds to the ratio between the rates of extraction of a hydrogen atom bound to a secondary carbon atom and one bound to a primary carbon atom, is 9630 J/mole [35]. Taking into account the number of hydrogen atoms that have the same position in the propane molecule, it results that at the temperature of 1100 K:

$$k_3/k_4 = k_5/k_6 = (6/2)e^{-9630/1100 \cdot R} = 1.04$$

Using this approximation, the relation (2.42) becomes:

$$-\frac{d[C_3H_8]}{d\tau} = 2.9\left(\frac{k_1k_8k_4}{k_{10}}\right)^{1/2}[C_3H_8]$$

Replacing in this expression the values of rate constants, one obtains the following approximate expression for the overall rate constant:

$$k_{C_3H_8} \approx 11.3 \times 10^{15}e^{-279,500/RT} s^{-1} \qquad (2.43)$$

The length of the reaction chain v may be obtained by dividing the overall rate expression (2.43) by the initiation reaction rate (a), which is equal to $k_1[C_3H_8]$, resulting in:

$$v \approx 2.9 \left(\frac{k_8 k_4}{k_1 k_{10}} \right)^{1/2}$$

By using the values for rate constants for the elementary reactions, one obtains:

$$v \approx 0.546 \times e^{74350/RT}$$

Thus, at the temperature of 1100 K the length of the kinetic chain is 1800. Actually, it may be shorter, since some interruption reactions were neglected.

The thermal decomposition of the propane proceeds following two main directions, according to the overall reactions:

$$C_3H_8 \rightarrow C_3H_6 + H_2$$
$$C_3H_8 \rightarrow C_2H_6 + CH_4$$

It is important to know not only the overall decomposition reaction rate given by relation (2.43), but also the rates for the formation of the compounds resulting from the decomposition. According to the considered reaction scheme and by using the expressions for the concentrations of the radicals, it results:

$$\frac{d[C_3H_6]}{d\tau} = \frac{d[H_2]}{d\tau} = k_8[i - \dot{C}_3H_7] = (k_3 + k_4)[\dot{H}][C_3H_8]$$

$$\frac{d[CH_4]}{d\tau} = \frac{d[C_2H_4]}{d\tau} = (k_5 + k_6)[\dot{C}H_3][C_3H_8]$$
$$- \left(\frac{(k_5 + k_6)k_1}{k_6} + \frac{(k_5 + k_6)k_3}{k_6}[\dot{H}] \right)[C_3H_8]$$

Comparing these expressions with the relation (2.42), one observes that the first term in the parenthesis of equation (2.42) corresponds to the formation of propene and of hydrogen. The other two terms correspond to the formation of ethene and of methane.

By substituting in the above relations the concentration of the atomic hydrogen $[\dot{H}]$ by the expression (2.41), after simplification, it results:

$$\frac{d[C_3H_6]}{d\tau} = \frac{d[H_2]}{d\tau} = \left(\frac{k_1 k_8 (k_3 + k_4)}{k_{10}} \right)^{1/2} [C_3H_8] \qquad (2.44)$$

$$\frac{d[CH_4]}{d\tau} = \frac{d[C_2H_4]}{d\tau} = \frac{(k_5 + k_6)k_3}{k_6} \left(\frac{k_1 k_8}{k_{10}(k_3 + k_4)} \right)[C_3H_8] \qquad (2.45)$$

Similar considerations were used for normal- and iso-alkanes of higher molecular mass [16,73]. They proved that in all these cases, the decomposition reactions are also of first order. The same kinetic order results for the decomposition of alkyl chains connected to naphthenic or to aromatics rings.

Other experimental studies confirmed these conclusions [44,74,112], at times with little differences of practical importance. Thus, for temperatures of 600-720°C, the kinetic reaction order was 1.07–1.18 for n-tetracosane, 1.06–1.16 for 6 methyl-arcosane and 1.11–1.18 for dodecil-benzene [75].

In propane pyrolysis at 525–600°C and a pressure of 10–150 mm Hg, the measured reaction order was of 1.2–1.3 [16]. An order of 1.5 was found in the pyrolysis of n-butane at 530–600°C and 1 atmospheric pressure [76].

All these considerations refer to conditions very far from those practiced in industrial plants. The general opinion is that in industrial plants at 2–3 bar, above 750–800°C and at high conversion, the chemical transformations follow of first order kinetics. The kinetics is affected only by the product inhibition (retardation) effect, which will be discussed in the next section.

2.3.2 Retardation Effect

The interactions between free radicals and molecules of propene or i-butene, formed during decomposition reactions, lead to the formations of allylic radicals. These lack the energy necessary for the continuation of the reaction chain. As result the kinetic chain is shortened; respectively the extent of the decomposition is reduced. This retardation effect, acting during the cracking of hydrocarbons was discovered in 1934 and named "braking effect" by A.V. Dintses and A.V. Frost [84]. Nitrogen oxide (NO), has a similar inhibition effect, well-known for a long time.

It must be mentioned that the inactive allylic radicals may be formed not only by the extraction of a hydrogen atom from a molecule of propene or of isobutene, but also by the decomposition of radicals of higher molecular mass produced by the decomposition of some higher alkenes.

Since alkenes are generally not present in the feed stocks submitted to pyrolysis or cracking process, the inhibition phenomenon becomes apparent only at advanced conversion, when the decomposition of other hydrocarbons leads to the formation of significant amounts of alkenes.

The influence of the inhibition effect upon the kinetics of thermal decomposition can be analyzed by considering the reactions between the propene produced during the decomposition and the free radicals:

$$C_3H_6 + \overset{\bullet}{R} \rightarrow RH + \overset{\bullet}{C_3H_5}$$

Reactions of this type will result in the reduction of the concentration of the active radicals and of the reaction rate.

The allylic radical formed is not able to extract a hydrogen atom and to continue the reaction chain. It will interact with another active radical:

$$\overset{\bullet}{C_3H_5} + \overset{\bullet}{R} \rightarrow C_3H_5R$$

or it will dimerize to produce a diene:

$$2\overset{\bullet}{C_3H_5} \rightarrow C_6H_{10}$$

The adsorption of the allylic radicals to the wall may be considered as prevalent only at pressures that are less than those practiced in the industrial processes.

If the destruction of the active radicals were the only effect produced by the propene, the reaction rate should continuously diminish as more propene is produced in the process. However, one finds experimentally that a residual reaction rate is reached, below which addition of propene does not further diminish the feed conversion.

This effect is explained by the fact that the propene does not only interact with the active radicals, but also it may be decomposed according to the reaction:

$$C_3H_6 \rightarrow \overset{\bullet}{C}_3H_5 + \overset{\bullet}{H} \qquad (p)$$

In addition to the allyl radical, an active radical of atomic hydrogen- also is produced.

According to the data of Table 2.2, the rate constant of this reaction is given by the expression:

$$k_p = 3.5 \times 10^{16} e^{-306.100/RT}$$

For comparison, for the dissociation of propane:

$$C_3H_8 \rightarrow \overset{\bullet}{C}_2H_5 + \overset{\bullet}{C}H_3 \qquad (a)$$

the rate constant will be:

$$k_a = 2.074 \times 10^{16} e^{-366.900/RT}$$

At the temperature of 1100 K, the ratio of the two reaction rates will be:

$$\frac{r_p}{r_a} = 3.54 \frac{[C_3H_6]}{[C_3H_8]}$$

In order to compare the extent to which the two reactions promote the decomposition by chain reaction, one must take also into account the relative abstraction rates of the hydrogen atoms, by the radicals formed in reactions (p) and (a).

Using the data of Table 2.5, the abstraction rates of a primary hydrogen atom from a molecule of propane at 1100 K will be:

$$r_{\overset{\bullet}{H}} = 1.2 \times 10^9 [\overset{\bullet}{H}][C_3H_8]$$

$$r_{\overset{\bullet}{C}H_3} = 1.7 \times 10^8 [\overset{\bullet}{C}H_3][C_3H_8]$$

$$r_{\overset{\bullet}{C}_2H_5} = 4.7 \times 10^6 [\overset{\bullet}{C}_2H_5][C_3H_8]$$

$$r_{\overset{\bullet}{C}_3H_5} = 1.8 \times 10^5 [\overset{\bullet}{C}_3H_5][C_3H_8]$$

It is obvious that only the first two reactions have significant rates, while the last two may be neglected. The ratio of these rates will be:

$$\frac{r_{\overset{\bullet}{H}}}{r_{\overset{\bullet}{C}H_3}} = 7.06 \frac{[\overset{\bullet}{H}]}{[\overset{\bullet}{C}H_3]}$$

Taking into account the fact that the atomic hydrogen is generated by reaction (p) and the methyl radicals by reaction (a), and considering the relative rates of the two reactions deduced earlier, it results that propene will initiate chain decomposition reactions of the hydrocarbons present in the system $3.54 \times 7.06 \approx 25$-times faster than propane.

After a certain propene concentration is reached, this effect will counterbalance the inhibition effect.

In the analysis of the inhibition phenomenon, one has to consider both the reactions (a) to (h) (see p. 58,59) for the initiation and propagation of the thermal decomposition of the propane, but also those undergone by the propene produced:

1. The interaction of propene with active radicals, which produces the allyl radical, which is inactive
2. The cleavage of propene to form an allyl radical and atomic hydrogen

Concerning the termination reaction, it must be remembered that the allyl radicals are practically inert towards the feed and therefore their concentration in the reaction mixture will be much greater than of other radicals. In this situation, the most probable termination reaction is the reaction between two allyl radicals with the formation of dienes.

In these conditions the reactions to be considered are:

$$C_3H_8 \xrightarrow{k_1} \overset{\bullet}{C}H_3 + \overset{\bullet}{C}_2H_5 \tag{a}$$

$$\overset{\bullet}{C}_2H_5 \xrightarrow{k_2} C_2H_4 + \overset{\bullet}{H} \tag{b}$$

$$C_3H_8 + \overset{\bullet}{H} \xrightarrow{k_3} H_2 + n-\overset{\bullet}{C}_3H_7 \tag{c}$$

$$\xrightarrow{k_4} H_2 + i-\overset{\bullet}{C}_3H_7 \tag{d}$$

$$C_3H_8 + \overset{\bullet}{H} \xrightarrow{k_5} CH_4 + n-\overset{\bullet}{C}_3H_7 \tag{c}$$

$$\xrightarrow{k_6} CH_4 + i-\overset{\bullet}{C}_3H_7 \tag{d}$$

$$n-\overset{\bullet}{C}_3H_7 \xrightarrow{k_7} C_2H_4 + \overset{\bullet}{C}H_3 \tag{g}$$

$$i-\overset{\bullet}{C}_3H_7 \xrightarrow{k_8} C_3H_6 + \overset{\bullet}{H} \tag{h}$$

$$C_3H_6 + \overset{\bullet}{H} \xrightarrow{k_{14}} H_2 + \overset{\bullet}{C}_3H_5 \tag{n}$$

$$C_3H_6 + \overset{\bullet}{C}H_3 \xrightarrow{k_{15}} CH_4 + \overset{\bullet}{C}_3H_5 \tag{o}$$

$$C_3H_6 \xrightarrow{k_{16}} \overset{\bullet}{C}_3H_5 + \overset{\bullet}{H} \tag{p}$$

$$2\overset{\bullet}{C}_3H_5 \xrightarrow{k_{17}} C_6H_{10} \tag{q}$$

Applying the theorem of stationary state, it results:

$$\frac{d[\overset{\bullet}{C}_3]}{d\tau} = k_1[C_3H_8] - (k_5 + k_6)[C_3H_8][\overset{\bullet}{C}H_3] + k_7[n-\overset{\bullet}{C}_3H_7] \tag{2.46}$$

$$- k_{15}[C_3H_6][\overset{\bullet}{C}H_3] = 0$$

$$\frac{d[\overset{\bullet}{H}]}{d\tau} = k_1[C_3H_8] - (k_3 + k_4)[C_3H_8][\overset{\bullet}{H}] + k_8[i-\overset{\bullet}{C}_3H_7] \tag{2.47)*}$$

$$- k_{14}[C_3H_6][\overset{\bullet}{H}] + k_{16}[C_3H_6] = 0$$

---

* The first term in relation (2.47) was a result of substituting for the ethyl ($\overset{\bullet}{C}_2H_5$) radicals the relation (2.36).

$$\frac{d[\overset{\bullet}{C_3H_5}]}{d\tau} = k_{14}[C_3H_6][\overset{\bullet}{H}] + k_{15}[C_3H_6][\overset{\bullet}{CH_3}] + k_{16}[C_3H_6]$$

$$- k_{17}[\overset{\bullet}{C_3H_5}]^2 = 0 \tag{2.48}$$

$$\frac{d[n-\overset{\bullet}{C_3H_7}]}{d\tau} = k_3[C_3H_8][\overset{\bullet}{H}] + k_5[C_3H_8][\overset{\bullet}{CH_3}] - k_7[n-\overset{\bullet}{C_3H_7}] = 0 \tag{2.49}$$

$$\frac{d[\overset{\bullet}{C_3H_7}]}{d\tau} = k_4[C_3H_8][\overset{\bullet}{H}] + k_6[C_3H_8][\overset{\bullet}{CH_3}] - k_8[i-\overset{\bullet}{C_3H_7}] = 0 \tag{2.50}$$

Adding the expressions (2.46), (2.47), (2.49) and (2.50) together, it results:

$$2k_1[C_3H_8] - k_{14}[C_3H_6][\overset{\bullet}{H}] - k_{15}[C_3H_6][\overset{\bullet}{CH_3}] + k_{16}[C_3H_6] = 0$$

The concentration of the methyl radical results in:

$$[\overset{\bullet}{CH_3}] = \frac{2k_1[C_3H_8]}{k_{15}[C_3H_6]} - \frac{k_{14}}{k_{15}}[\overset{\bullet}{H}] + \frac{k_{16}}{k_{15}} \tag{2.51}$$

The rate of decomposition of propane is therefore given by the relation:

$$-\frac{d[C_3H_8]}{d\tau} = k_1[C_3H_8] + (k_3 + k_4)[C_3H_8][\overset{\bullet}{H}] + (k_5 + k_6)[C_3H_8][\overset{\bullet}{CH_3}]$$

Substituting $[\overset{\bullet}{CH_3}]$ from (2.51), one obtains:

$$-\frac{d[C_3H_8]}{d\tau} = k_1[C_3H_8] + (k_3 + k_4)[C_3H_8][\overset{\bullet}{H}] + \frac{2k_1(k_5 + k_6)[C_3H_8]^2}{k_{15}[C_3H_6]} -$$

$$- \frac{k_{14}}{k_{15}}(k_5 + k_6)[C_3H_8][\overset{\bullet}{H}] + \frac{k_{16}}{k_{15}}(k_5 + k_6)[C_3H_8] = 0 \tag{2.52}$$

It is reasonable to accept that the relative rate of extraction of a hydrogen atom from the molecule of propene by the radicals $[\overset{\bullet}{H}]$ and $[\overset{\bullet}{CH_3}]$ is the same as from propane, which was deduced previously as 7.06. It results:

$$\frac{k_{14}}{k_{15}}(k_5 + k_6) = k_3 + k_4$$

and the two terms that contain the radicals $[\overset{\bullet}{H}]$ cancel out. The relation (2.52) becomes:

$$-\frac{d[C_3H_8]}{d\tau} = k_1[C_3H_8] + \frac{k_5 + k_6}{k_{15}}[C_3H_8]\left\{\frac{2k_1[C_3H_8]}{[C_3H_6]} + k_{16}\right\}$$

The first term on the right hand side may be neglected, and one obtains finally:

$$-\frac{d[C_3H_8]}{d\tau} = \frac{k_5 + k_6}{k_{15}}\left\{\frac{2k_1[C_3H_8]}{[C_3H_6]} + k_{16}\right\}[C_3H_8] \tag{2.53}$$

By substituting the values of the kinetic constants for the elementary reactions from Tables 2.2 and 2.5 at a temperature of 1100 K, the following values of the rate constants are calculated:

$$k_5 + k_6 = 1.8 \times 10^8 + 0.39 \times 10^8 = 2.19 \times 10^8 \, \text{L/mol} \cdot \text{s}$$

$$k_{15} = 0.07 \times 10^8 \, \text{L/mol} \cdot \text{s}$$

$$k_1 = 0.033 \, \text{s}^{-1}$$

$$k_{16} = 0.28 \, \text{s}^{-1}$$

By substituting them in the relation (2.53), it results:

$$-\frac{d[C_3H_8]}{d\tau} = 31.3 \left\{ \frac{0.066[C_3H_8]}{[C_3H_6]} + 0.28 \right\} [C_3H_8]$$

On the other hand, by using Eq (2.43), which does not take into account the retardation effect, at a temperature of 1100 K one obtains :

$$-\frac{d[C_3H_8]}{d\tau} = 614[C_3H_8]$$

Since equation (2.53), was deduced on basis of the theorem of steady states, it is not valid for the conditions existing at the start of the reaction or while the concentration of the propene is still very low. Also, taking into account the approximations used in its deduction, the value of the equation is only qualitative, not suitable for calculations. However, it shows clearly the retardation effect produced by the propene and the existence of a residual rate for a fully inhibited process.

The retardation phenomenon may be analyzed also on basis of the general mechanism for the thermal decomposition of hydrocarbons or of fractions of crude oil, which was presented previously (see Section 2.4.1). To this purpose, the mechanism must be completed with the reaction for the formation of propene and/or of isobutene and with the reactions showing their interactions with the active radicals, by which inactive radicals are being formed.

By noting with A the propene or the isobutene, this mechanism becomes:

$$M \xrightarrow{k_1} \dot{R}_1 + \dot{R}_2 \qquad \qquad \text{(a)}$$

$$M + \dot{R}_1 \xrightarrow{k_2} R_1H + \dot{R}_3 \qquad \qquad \text{(b}_1\text{)}$$

$$M + \dot{R}_2 \xrightarrow{k_2} R_2H + \dot{R}_3 \qquad \qquad \text{(b}_2\text{)}$$

$$\dot{R}_3 \xrightarrow{k_3} \sum M_f + \dot{R}_1 + A \qquad \qquad \text{(b}_3\text{)}$$

$$\dot{R}_1 + \dot{R}_3 \xrightarrow{k_4} M_2 \qquad \qquad \text{(c}_1\text{)}$$

$$\dot{R}_1 + A \xrightarrow{k_5} R_1H + \dot{R}_0 \qquad \qquad \text{(d)}$$

where $\dot{R}_0$ are the inactive radicals.

In this scheme, it was considered that the radicals $\overset{\bullet}{R}_3$ are decomposed very quickly. Therefore, their interaction with the propene A may be neglected.

By applying the steady state theorem to the concentration of the radicals $\overset{\bullet}{R}_2$ and $\overset{\bullet}{R}_3$, equations identical to (2.7) and (2.8) are obtained. For the radicals $\overset{\bullet}{R}_1{}^*$ and the retarding molecules A, the same theorem gives:

$$\frac{d[\overset{\bullet}{R}_1]_i}{d\tau} = k_1[M] - k_2[\overset{\bullet}{R}_1]_i[M] + k_3[\overset{\bullet}{R}_3] - k_4[\overset{\bullet}{R}_1]_i[\overset{\bullet}{R}_3] - k_5[\overset{\bullet}{R}_1]_i[A] = 0 \qquad (2.55)$$

$$\frac{d[A]}{d\tau} = k_3[\overset{\bullet}{R}_3] - k_5[\overset{\bullet}{R}_1]_i[A] = 0 \qquad (2.56)$$

For $[\overset{\bullet}{R}_2]$ Eq. (2.9) remains unchanged:
From (2.56) it results:

$$[A] = \frac{k_3[\overset{\bullet}{R}_3]}{k_5[\overset{\bullet}{R}_1]_i} \qquad (2.57)$$

Subtracting Eq. (2.55) from (2.8), in which $[R_2\}$ was substituted by (2.9), it results:

$$2k_2[\overset{\bullet}{R}_1]_i[M] - 2k_3[\overset{\bullet}{R}_3] + k_5[\overset{\bullet}{R}_1]_i[A] = 0$$

By substituting [A] with (2.57) and after performing simplifications, one obtains

$$2k_2[\overset{\bullet}{R}_1]_i[M] - k_3[\overset{\bullet}{R}_3] = 0$$

from which:

$$[\overset{\bullet}{R}_3] = \frac{2k_2[\overset{\bullet}{R}_1]_i[M]}{k_3} \qquad (2.58)$$

Substituting (2.58) in (2.8), it results after simplifying:

$$k_1 - k_2[\overset{\bullet}{R}_1]_i - \frac{2k_1k_4}{k_3}[\overset{\bullet}{R}_1]_i^2 = 0$$

or

$$[\overset{\bullet}{R}_1]_i = \frac{-k_2k_3 \pm \sqrt{k_2^2k_3^2 + 8k_1k_2k_3k_4}}{4k_2k_4}$$

the sign before the square root being obviously plus.
This equation may be written also as:

$$[\overset{\bullet}{R}]_i = -\frac{k_3}{4k_4} + \sqrt{\left(\frac{k_3}{4k_4}\right)^2 + \frac{k_1k_3}{2k_2k_4}} \qquad (2.59)$$

---

* In this deduction, the concentration of the radicals $\overset{\bullet}{R}_1$ is indicated by $[\overset{\bullet}{R}_1]_i$ in order to distinguish it from that corresponding to the non-inhibited reaction, as given by equations (2.6) and (2.10).

Taking into account Eq. (2.10), which gives the concentration of the radicals R_1 of the nonretarded reaction, Eq. (2.59) may be written also as:

$$[\dot{R}]_i = -\frac{k_3}{4k_4} + \sqrt{\left(\frac{k_3}{4k_4}\right)^2 + \frac{1}{2}[\dot{R}_1]^2}$$

Writing:

$$\frac{k_3}{4k_4} = k_A$$

one obtains:

$$([\dot{R}_1]_i + k_A)^2 = k_A^2 + \frac{1}{2}[\dot{R}_1]^2$$

from which:

$$\frac{[\dot{R}_1]}{[\dot{R}_1]_i} = \sqrt{2 + \frac{4k_A}{[\dot{R}_1]_i}}$$

This ratio is obviously greater than one (>1), which means that the retardation (inhibition) of the reaction by propene leads to a reduction of the free radicals concentration $[\dot{R}_1]$.

As shown earlier, the decomposition rate of the feed molecules M, will be given, by the relation:

$$\frac{d[M]}{d\tau} = \left(2k_1 + k_2[\dot{R}_1]\right)[M] \tag{2.60}$$

The lowering of the radicals concentration as result of the retardation caused by propene, leads to the reduction of the decomposition rate of the feed molecules, i.e. of the overall rate constant. The expression for the overall rate constant, as a function of the constants of the individual reactions is obtained by substituting $[R_1]$ from equation (2.60) with (2.59). The order of reaction is not changed thereby.

In this derivation, the concentration of the propene was considered to be at steady state. Actually, the rate constant will decrease gradually, until it reaches the value corresponding to Eqs. (2.59) and (2.60).

An opposite effect, that of acceleration of the thermal decomposition, is caused by the presence of oxygen and oxygenated compounds as well as by the hydrogen sulfide and sulfur compounds.

The kinetics of the propane pyrolysis in the presence of acetone and of acetaldehyde was examined by Layokun and Slater [85]. They developed a kinetic model and performed calculations using the kinetic constants of the elementary reactions. In both cases, good agreement was obtained with the experimental results obtained in a continuous flow operation.

In both cases, the addition of oxygenated compounds increases the conversion without however modifying the reaction order. The enhancement of the reaction rate decreases with the temperature, becoming insignificant at temperatures above 750°C.

Concerning the products obtained, the conversion increase caused by these additions leads to higher yields mainly for methane and ethane, making this approach of no practical interest.

The effect of addition of oxygen in proportion of 2–3% in the pyrolysis of propane, was studied by Layokun [86], with much more interesting results. In this case also, a kinetic model was elaborated and kinetic constants were computed for the elementary reactions. The effect of the added oxygen is to increase of the conversion to ethylene and propylene and is accentuated at higher temperature. The temperature range up to 700°C was investigated. The results confirmed those reported previously by other researchers, for the temperatures of 500–600°C.

Research effected with additions of oxygen during the pyrolysis of *n*-butane [76] did not confirm these results. This reaction was investigated at temperatures ranging between 530 and 600°C, at 1 atmospheric pressure, conversion of butane of only 1% and 5–500 ppm of oxygen added. In these conditions the overall conversion decreased, compared to the absence of oxygen, while the distribution of the products remained basically the same, with the exception of the increased yields of butenes and 1-3 butadiene.

All these results, to some extent contradictory, indicate the need for supplementary research and suggest exercising prudence in the analysis of experimental results and in the formulation of final conclusions.

2.3.3 The kinetics of coke formation

The formation of coke during thermal processes evidences two different aspects:

1. The formation of coke deposits inside the pyrolysis tubes, where the feed and the reaction products are all in vapor phase
2. The formation of coke as a consequence of the reactions taking place in the liquid phase, even if a vapor phase is also present—as in the processes of visbreaking, delayed coking, and in all classical processes of thermal cracking

These two phenomena will be treated separately, taking into account their fundamental differences.

2.3.3.1 Coke formation in pyrolysis processes

Coke forms gradually on the inner wall of the furnace tubes during pyrolysis. The more intense the coke formation, the more frequent the process of decoking has to be performed. The rate of coke deposition increases as the molecular weight of the feed and its aromatic character increase. However, coke formation is exhibited even by very light raw materials such as ethane.

Detailed studies using electron microscopy and X-ray dispersion [87] revealed that the formation of coke in the furnace tubes is a stagewise process.

In the first stage, coke filaments are formed due to reactions on the metal surface catalyzed by iron and nickel. This catalytic effect is stronger at the beginning of the cycle when the tubes are clean.

Once the coke filaments have appeared, coke formation is amplified in subsequent stages, by two mechanisms: a) very small droplets of heavy liquid molecules resulting from reactions of dehydrogenation and condensation accumulate in the filaments and b) the trapping of methyl, ethyl, phenyl radicals and acetylene by free radicals existing on the coke surface.

The entire process is determined by the first of these stages, which takes place as result of catalytic reactions on the metal surface. The elimination or reduction of this stage leads to significant reduction in coke formation and consequently to longer working cycles for the furnace. The reduction of the catalytic effect of iron and nickel is accomplished in two ways:

1. The use of furnace tubes made of steel with high nickel content, which have the inner surface aluminated by a diffusion process at high temperatures
2. Passivation of the furnace tubes by treatment with certain substances (passivators) after the decoking operation

The first method has the advantage of diminishing the formation of metal carbides (carburation of the steel), which is the cause of extensive wear of the furnace tubes. This method is described in Section 4.4.2, where criteria for the selection of tubes for the pyrolysis furnace are presented. It should be mentioned that the carbide formation phenomenon is accompanied by the incorporation of metal particles into the coke.

The second method used in industry consists of introducing into the feed, after decoking, hydrogen sulfide or organic sulfur compounds [45]. Studies in a specially designed laboratory plant [42] made it possible to compare the effect of different passivators. The results are presented in Figure 2.11, for a reactor made of 1Cr18Ni9Ti steel. It is important to note the significant reduction in the formation of coke due to treatment of the reactor by hydrogen sulfide.

An interesting result of this research [42] was the observed influence of the reactor wall and of the passivation treatment on the overall conversion and especially on the distribution of the products obtained from propane pyrolysis at 800°C, at a contact time of 0.75 sec. The results are presented in Table 2.12.

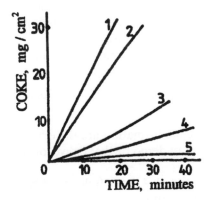

Figure 2.11   Coke formation on the walls of a reactor made of 1Cr18Ni9Ti stainless steel for various passivation treatments   1-no treatment; 2-mechanical erosion; 3-reduction with hydrogen; 4-pretreatment by H₂S; 5-pretreatment by diluted sulphuric acid. Experimental conditions: $t = 850°C$ atmospheric pressure, duration of contact 1 s., feed: propane, conversion approx. 25%. (From Ref. 42.)

Table 2.12 The Effect of the Properties of the Reactor Wall on Selectivity, Conversion, and Coke Formation in Propane Pyrolysis

Reactor and treatment	Selectivity (moles/mole)					Molar fraction of unreacted C_3H_8	Coke formation after 40 min (mg/cm²)
	H_2	CH_4	C_2H_6	C_2H_4	C_3H_6		
Quartz	0.65	0.75	0.033	0.54	0.070	0.016	3.2
Stainless steel treated by H_2S	0.87	0.72	0.032	0.51	0.064	0.036	5.3
Stainless steel treated by H_2	0.89	0.68	0.031	0.41	0.060	0.051	11.6
Stainless steel treated by mechanical erosion	0.99	0.66	0.030	0.35	0.050	0.053	26.1
Oxidized stainless steel	1.51	0.62	0.030	0.29	0.048	0.054	31.5
Nickel	3.28	0.60	0.031	0.21	0.035	0.039	39.3

Conditions: 800°C, contact time of 0.75 s. Stainless steel 1Cr18Ni9Ti walls treated with 3% H_2SO_4 solution before each run.
Source: Ref. 42.

The results were compared with those obtained in a reactor made of quartz, a material which is usually considered to have no catalytic effect. It is obvious from the results presented that passivating stainless steel by means of hydrogen sulfide ensures selectivities close to these obtained in the quartz reactor excepting a higher rate of coke formation. At the same time, untreated stainless steel leads to a reduced selectivity to ethylene while the rate of coke formation is much higher (up to 10 times) than in the quartz reactor.

The results of this study explain the attention paid to the materials used in building laboratory plants and their treatment before experimentation. Furthermore, this research draws attention to the problems encountered when comparing literature results and the importance of taking into account a particular selection of materials and procedures. Such differences may explain the significant variation in the values obtained by different authors for the kinetic constants.

It is generally accepted that the mechanism of coke formation during the pyrolysis of ethane and propane is represented by the following sequence:

ethylene, propylene → cyclic olefins → benzene, alkyl-benzenes → condensed aromatics → coke

According to this sequence, the precursors of coke formation are the ethylene and propylene adsorbed onto the reactor walls, which undergo further dehydrogenation and condensation until coke is produced.

After 21 models based on this mechanism were tested [42], the following kinetic model seems to best represent the process of coke formation:

$$C_2H_4 \xrightarrow{k_1} coke \tag{a}$$

$$1/3 C_3H_6 \xrightarrow{k_2} coke \tag{b}$$

with the rate of coke formation given by the equation:

$$r_{coke} = r_a + r_b = k_1[C_2H_4] + k_2[C_3H_6]^{1/3} \qquad (2.61)$$

where the two rate constants are given by:

$$k_1 = 5.89 \times 10^{10} e^{-230,290/RT}$$

$$k_2 = 2.21 \times 10^8 e^{-165,220/RT}$$

2.3.3.2 Coke formation during delayed coking and visbraking

In these processes, the initial stages of the thermal decomposition reactions have a specific character since they take place in liquid phase.

Indeed, one liter of gas at 500°C and 1 atmosphere pressure contains approx. 10^{22} molecules while one liter of liquid contains approx. 10^{24} molecules. Therefore, the concentrations in the liquid phase are equivalent to the concentrations in the gas phase at pressures of about 100 bar.

Due to this effect, there will be an increase in the liquid phase of bimolecular reactions such as polymerization, condensation and the bimolecular interactions of thermal decomposition, as compared to the gas phase.

As a result, the cracking reactions in the liquid phase convert the feed to products of a higher average molecular mass such as asphaltenes, which remain mostly in the liquid phase and undergo increasingly advanced condensation and dehydrogenation, eventually becoming coke.

The kinetic analysis of coke formation in these circumstances should take into account several specific effects.

The cellular effect. In contrast to the gas phase, the radicals in the liquid phase are surrounded by molecules, especially of the polynuclear aromatic type. The delocalized electrons of such large molecules have a "stabilizing" effect upon the free radicals. The result is a significant concentration of radicals in the liquid phase, which was measured experimentally by electronic paramagnetic resonance [57]. Together with the surrounding molecules, the radicals form so-called "cells." In order to become free, independent radicals and to be able to react, the radicals need to migrate out of the cell, for which energy must be spent.

The decomposition of a molecule AB in the liquid phase could be described by:

$$BC \underset{k_{-1}}{\overset{k_1}{\rightleftharpoons}} (\overset{\bullet}{B} \cdots \overset{\bullet}{C}) \overset{k_2}{\longrightarrow} \overset{\bullet}{B} + \overset{\bullet}{C}$$

where $(\overset{\bullet}{B} \cdots \overset{\bullet}{C})$ is the cellular complex of the two radicals.

Applying the steady-state theorem to the cellular complex leads to:

$$[(\overset{\bullet}{B} \cdots \overset{\bullet}{C})] = \frac{k_1[BC]}{k_{-1} + k_2}$$

Since the rate constant k_{-1} for the recombination of radicals is much higher than the rate constant k_2 for diffusion out of cell, the equation above may be written as:

$$[(\overset{\bullet}{B} \cdots \overset{\bullet}{C})] = \frac{k_1}{k_{-1}}[BC]$$

The rate of formation of out-of-cell, free radicals, i.e. of kinetically independent radicals is:

$$-\frac{d[BC]}{d\tau} = k_2[(\overset{\bullet}{B}\cdots\overset{\bullet}{C})] = \frac{k_1 k_2}{k_{-1}}[BC] \tag{2.62}$$

Since the activation energy for recombination of the radicals is equal to zero, the overall activation energy for this reaction becomes:

$$E = E_R + E_D$$

where E_R is the energy of breaking the B–C bond (corresponding to the k_1 constant) and E_D is the energy of diffusion of the radicals out of the cell (corresponding to the k_2 constant). The latter will depend on the size of the radicals and on the viscosity of the fluid.

As a first approximation, the pre-exponential factors for the reaction constants k_2 and k_{-1} may be considered to be equal (having a value of approx. 10^{-10} cm^3 sec) in which case the rate of formation of independent radicals in the liquid phase, according to Eq. 2.62 becomes:

$$-\frac{d[BC]}{d\tau} = A_1 e^{-(E_1 + E_2)/RT}[BC] \tag{2.63}$$

while if the same reaction would take place in the gas phase, the rate of formation would be:

$$-\frac{d[BC]}{d\tau} = A_1 e^{-E_1/RT}[BC] \tag{2.64}$$

On the other hand, taking into account that $E_{-1} = 0$ and considering a steric coefficient p equal to 1, the rate of radical recombination in the gas phase is:

$$10^{-10}[\overset{\bullet}{B}][\overset{\bullet}{C}] \tag{2.65}$$

The rate equation for radical recombination in the liquid phase accounts for the energy barrier E_2, corresponding to the diffusion of radicals in the cell:

$$10^{-10} e^{-E_2/RT}[\overset{\bullet}{B}][\overset{\bullet}{C}] \tag{2.66}$$

At the rate of formation and recombination steady state of the radicals is the same in both phases. By equaling the equation (2.64) with (2.65), or (2.63) with (2.66), it results:

$$[\overset{\bullet}{B}] = [\overset{\bullet}{C}] = \left(\frac{A_1 e^{-E_1/RT}}{10^{-10}}[BC]\right)^{1/2} \tag{2.67}$$

The solvation effect. The solvation interaction between adjacent particles may influence the rate of reaction in the liquid phase, especially when polar molecules are involved. In the case of oil fractions, the kinetics of thermal reactions is influenced by the formation of π complexes between the free radicals with aromatic hydrocarbons.

However, this effect is not strong and it decreases significantly with increasing temperature. Hence, its influence on the rate of radical recombination in the cracking process may be neglected.

Coke formation. The following chain mechanism was suggested for the polycondensation reaction, which transforms asphaltenes into coke precursors such as carboids [38]:

$$A \xrightarrow{k_a} \overset{\bullet}{A}' + \overset{\bullet}{R} \tag{a}$$

$$\overset{\bullet}{R} + A \xrightarrow{k_b} \overset{\bullet}{A}' + RH \tag{b}$$

$$A' + A \xrightarrow{k_c} A' \overset{\bullet}{A} \tag{c}$$

$$A' \overset{\bullet}{A} \xrightarrow{k_d} \Sigma M + A' \overset{\bullet}{A}' \tag{d}$$

$$A' \overset{\bullet}{A} + A \xrightarrow{k_c} A' A' \overset{\bullet}{A} \tag{c'}$$

$$\vdots \qquad \vdots \qquad \vdots$$

$$(\overset{\bullet}{A}')_x + A \xrightarrow{k_n} (A'')_x \overset{\bullet}{A} \tag{n}$$

where A is a molecule of asphaltene and ΣM represents the light molecules that go into the gas phase. Reaction (a) is the chain initiation; reactions (b),...,(d) are the chain propagation and reaction (n) is the chain termination. $(A'')_x \overset{\bullet}{A}$ is an inactive radical.

The reaction of chain termination (n) corresponds to the behavior of asphaltenes within crude oil. In the case of asphaltenes with a more aromatic character, such as those that result from cracking processes, the chain termination reaction that is in better agreement with the experimental data is:

$$2 (\overset{\bullet}{A})_x \xrightarrow{k_m} (A')_{2x} \tag{m}$$

For a sufficiently long chain, the rate of asphaltenes consumption depends only on the reactions of chain propagation, i.e. the (c) reactions, while the reactions of chain initiation and termination may be neglected. Therefore:

$$-\frac{d[A]}{d\tau} = k_c [(\overset{\bullet}{A})_i][A] \tag{2.68}$$

Considering *(n)* as the reaction of chain termination and applying the steady-state theorem for all formed radicals, one obtains the equation:

$$-\frac{d[(\overset{\bullet}{A}')_i]}{d\tau} = k_a[A] - k_n[(\overset{\bullet}{A}')_n][A] = 0$$

which leads to:

$$[(\overset{\bullet}{A}')_i] = \frac{k_a}{k_n}$$

After substituting in Eq. (2.68), an equation for a first-order reaction rate is obtained:

$$-\frac{d[A]}{d\tau} = k_c \frac{k_a}{k_n}[A] \tag{2.69}$$

Alternately, considering (m) to be the chain termination reaction corresponding to the polycondensation of asphaltenes of a secondary origin, i.e., produced in the cracking process one obtains:

$$-\frac{d[(\dot{A})_i]}{d\tau} = k_a[A] - k_m[(\dot{A}')_i]^2 = 0$$

which leads to:

$$[(\dot{A}')_i] = \pm\sqrt{\frac{k_a}{k_m}[A]}$$

Considering only the positive solution and substituting in Eq. (2.66), a rate equation for a 3/2 reaction order is obtained:

$$-\frac{d[A]}{d\tau} = k_c\left(\frac{k_a}{k_m}\right)^{1/2}[A]^{3/2} \tag{2.70}$$

Eqs. (2.69) and (2.70) agree satisfactorily with the experimental data. The experimental value for the overall rate constant for asphaltenes that were separated from crude oil, becomes:

$$k = k_c\frac{k_a}{k_n} = 6 \times 10^{19}e^{-34425/T}, \text{s}^{-1} \tag{2.71}$$

and Eq. (2.69) becomes:

$$-\frac{dA}{d\tau} = 6 \times 10^{19}e^{-34425/T}[A], \text{s}^{-1} \tag{2.72}$$

Using the overall rate constant calculated from experimental data for asphaltenes obtained from thermal cracking residue:

$$k = k_c\left(\frac{k_a}{k_m}\right)^{1/2} = 4 \times 10^{-3}e^{-17865/T}\text{cm}^{2/3}\cdot\text{moles}^{-1/2}\cdot\text{s}^{-1/2} \tag{2.73}$$

Eq. (2.70) becomes:

$$-\frac{d[A]}{d\tau} = 4 \times 10^{-3}e^{-17865/T}[A]^{3/2} \tag{2.74}$$

In oil fractions, the asphaltenes occur as mixtures with other substances that may have the role of solvent. The presence of such solvents may prevent the polycondensation reaction by way of interaction with the free radicals. Such interactions have a higher probability than an uninterrupted succession of 110–150 reactions of type (c) and (d).

Thus, the interaction of a radical with a molecule of asphaltene has the same probability as the destruction of the radical by way of interacting with a molecule of solvent. Hence, the probability of completing a succession of 120 reactions required for the formation of a molecule of carbene would be $0.5^{120} = 7.5 \times 10^{-37}$, i.e., negligible. Therefore, the formation of carbenes, (and consequently of coke) is very unlikely unless the asphaltenes will separate from the solution forming a separate phase, in which their concentration is much higher than in the original solution.

This separation process of asphaltenes from the solution could be represented by the following:

$$
\begin{array}{ccccc}
\text{Asphaltenes dispersed} & k_1 & \text{Associated} & k_2 & \text{Precipitated} \\
\text{in solution} & \rightleftharpoons & \text{asphaltenes} & \rightarrow & \text{asphaltenes} \\
\text{(A)} & k_{-1} & \text{(A}_x) & & \text{(A}_p)
\end{array}
$$

Applying the steady-state theorem to the associated asphaltenes, gives the following:

$$
\frac{d[A_x]}{d\tau} = k_1[A]^2 - k_{-1}[A_x] - k_2[A_x] = 0
$$

from which

$$
[A_x] = \frac{k_1[A]^2}{k_{-1} + k_2}
$$

Assuming that the separation of asphaltenes from solution is the rate-controlling step for coke formation, its rate is given by:

$$
\frac{d[A_p]}{d\tau} = k_2[A_x] = \frac{k_1 k_2[A]^2}{k_{-1} + k_2} \tag{2.75}
$$

This corresponds to a second-order reaction.

The apparent activation energy for the precipitation of asphaltenes described by the rate equation 2.75 is very low. Indeed, the constant k_1 refers to a diffusion process. Therefore $E_1 \cong 20$ kJ/mol. Moreover, the constant k_2 of precipitation of the asphaltenes from solution depends very little on temperature (which means that $E_2 \cong 0$). Also, the activation energy E_{-1}, which represents the difference between the energy of interaction of asphaltenes molecules with each other and the energy of interaction of molecules of asphaltenes with the molecules of solvent, is also very small (8–20 kJ/mole).

The process of precipitation of asphaltenes from the solution will therefore be influenced very little by temperature, less so than the reaction of transformation of asphaltenes into carboids (coke) described by the Eqs. (2.72) and (2.74). Coke formation starts taking place only after the asphaltenes have precipitated. Hence two domains are evident:

At lower temperatures, the rate-controlling step of the coking process will be the reaction of transformation of asphaltenes into carboids according to Eq. (2.72) or according to Eq. (2.74), depending on the nature of the asphaltenes.

At higher temperatures, the rate-controlling step will be the precipitation of asphaltenes.

The temperature level that borders the domains of validity of the two rate-controlling steps depends on the nature and on the concentration of asphaltenes, but especially on the type of solvent, i.e., the type of molecules surrounding the asphaltenes. When these molecules have a structure close to that of the asphaltenes such as aromatic hydrocarbons, condensed hydroaromatics, or malthenes as resulting from the cracking process, the precipitation of asphaltenes is slower and starts at relatively higher temperatures. On the other hand, asphaltenes will precipitate at relatively lower temperatures, from a solution with a less aromatic, i.e., a more "alkane" character*. Without giving precise ranges of temperature, it may be said that below 350°C the rate-controlling step is the chemical reaction of the asphaltenes [38].

In the case of "good solvents," i.e. with aromatic character, the asphaltenes separate from solution as three-dimensional structures with gel consistency. The "holes" within these gels are occupied by the solvent, which induces a high porosity in the resulting coke. This is usually the case for coking taking place at moderate pressures. Under these conditions the light products formed during the decomposition are eliminated, the residue becoming concentrated in asphaltenes that remain in solution until they precipitate as three-dimensional structures. This takes place only after a certain degree of conversion (decomposition) is reached, corresponding to a high enough concentration of asphaltenes (Figure 2.12).

During the mixed phase vapor-liquid cracking processes carried out at high pressures, the liquid phase is enriched in light products that decrease the solubility of asphaltenes and accelerate their precipitation as compact structures.

2.3.4 Overall kinetic equations

From Section 2.3.1 it results that first-order kinetics can be accepted for all the thermal decomposition reactions of the hydrocarbons and petroleum fractions.

Accordingly, for conditions of constant volume, with x being the overall conversion expressed in weight fractions, the differential rate equation will be:

$$\frac{dx}{d\tau} = k(1 - x) \tag{2.76}$$

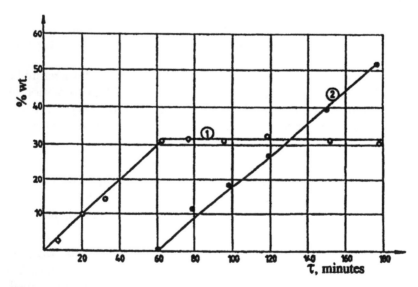

Figure 2.12   The decomposition of resins at 400°C 1-Formation of asphaltenes, 2-coke formation.

---

* In this context the term "solution" is used for colloidal solutions which in certain conditions may flocculate.

and the integral form:

$$k = \frac{1}{\tau} \ln \frac{1}{1-x} \qquad (2.77)$$

Eqs. (2.76) and (2.77) are valid for irreversible reactions, as are those of thermal cracking.

An exception is the pyrolysis of ethane, which is generally carried out in conditions for which thermodynamic considerations indicate that the transformation of ethane to ethene is reversible (Figure 2.3). For this case, the rate equations will have the form:

$$\frac{dx}{d\tau} = k(x_\infty - x) \qquad (2.78)$$

$$\text{and } k = \frac{1}{\tau} \ln \frac{x_\infty}{x_\infty - x} \qquad (2.79)$$

where x_∞ is the conversion at equilibrium.

For processes in which the retardation effect must be taken into account, several ways were suggested for expressing the gradual reduction of the rate constant with increasing conversion [84,88].

Dintes and Frost [84] express this decrease of the rate constant by the relation:

$$k = \frac{k_D}{1 - \beta(1-x)} \qquad (2.80)$$

where β is the retardation constant that may take values between 0 (lack of retardation) and 1 (maximum retardation), k_D = the rate constant in absence of a retardation effect, k = the rate constant for the retardation process.

Buekens and Froment [88] expressed the decrease of the rate constant by the relation:

$$k = \frac{k_0}{1 + \alpha x} \qquad (2.81)$$

where k_0 is the rate constant extrapolated to zero conversion (for $x = 0$, $k = k_0$). The constant α may vary between zero (lack of retardation) and infinity (at full inhibition).

By equating Eqs. (2.80) and (2.81) and by using for x the two limit values, one obtains:

$$\begin{aligned} \text{for } x = 0 \qquad k_0 &= \frac{k_D}{1 - \beta} \\ \text{for } x = 1 \qquad k_D &= \frac{k_0}{1 + \alpha} \end{aligned} \qquad (2.82)$$

from which it results: $\alpha = \beta/(1 - \beta)$ and, $\beta = \alpha/(1 + \alpha)$.

These correlations allow one to convert the results obtained from Eq. (2.80) to those resulted from Eq. (2.81) and vice versa.

Substituting the expression (2.80) in (2.76) one obtains:

$$\frac{dx}{d\tau} = \frac{k_D(1-x)}{1 - \beta(1-x)} \qquad (2.83)$$

Separating the variables it results:

$$\frac{dx}{1-x} - \beta dx = k_D d\tau$$

which, by integration, and using the integration limits $x = 0$ for $\tau = 0$, gives:

$$k_D = \frac{1}{\tau}\left[\ln\frac{1}{1-x} - \beta x\right] \tag{2.84}$$

Substituting (2.81) in (2.76), it results:

$$\frac{dx}{d\tau} = \frac{k_0(1-x)}{1+\alpha x} \tag{2.85}$$

Separating the variables, one obtains:

$$\frac{1+\alpha}{1-x}dx - \alpha dx = k_0 d\tau \tag{2.86}$$

which, by integration, and by using the same integration limits gives:

$$k_0 = \frac{1}{\tau}\left[(1+\alpha)\ln\frac{1}{1-x} - \alpha x\right] \tag{2.87}$$

For high conversions, one has to use equations which take into account the retardation effect, as illustrated by the data for the thermal cracking of n-octane at 570°C presented in Table 2.13 [89].

The results given in the table show that, whereas the application of Eq. (2.81) leads to a decrease of the rate constant that increasing conversion, the relations (2.86) and (2.87) give values of the rate constant that are independent of conversion. Some observed deviation in the obtained values, which do not have a regular character, are probably the result of experimental errors.

In the pyrolysis of fractions of crude oil, besides the retardation effect produced by the propene and isobutene produced in the decomposition reactions, the

Table 2.13 n-Octane Pyrolysis at 570°C

	Experimental data	Rate constants by indicated equation		
τ (s)	x (wt. fraction)	(2.81)	(2.86)	(2.87)
3.65	0.079	0.0227	0.0020	0.065
8.85	0.131	0.0160	0.0018	0.056
32.8	0.271	0.0100	0.0018	0.061
80.1	0.414	0.0067	0.0018	0.059
109.5	0.525	0.0068	0.0022	0.075
157.0	0.595	0.0058	0.0022	0.071
175.5	0.593	0.0051	0.0019	0.063
215.0	0.611	0.0044	0.0017	0.055

Rate constants were calculated using various equations.
$\beta = 0.95$ $a = 19$.

Source: Ref. 89.

lowering of the rate constant with the conversion is also the result of the differences between the reaction rates of the various components present in the original feed. The more reactive components are consumed first, so that the unreacted portion of the feed stock is gradually concentrated in more stable components.

The reduction of the cracking rate with the conversion produced by the increasing concentration of the more stable components, may be illustrated by the successive cracking* of a paraffinic distillate (Table 2.14).

In the next-to-the-last column of the table, the rates of cracking are given, expressed in percents of gasoline formed per minute; they are decreasing as the conversion of the feed increases. Thus, upon cracking the 36.5% of the initial feedstock which was left unconverted after the first operation, the rate of formation of the gasoline was half the value measured during the first operation. In the sixth operation, when the 5.5% unconverted portion of the initial feedstock was pyrolyzed, the rate of the formation of gasoline was almost 8 times lower.

The use of kinetic relations for describing the pyrolysis of crude oil fractions raises two questions of the manner that x must be actually defined in the conversion. Rigorously, x represents the molar fraction or the weight fraction of the feed stock converted.

The first difficulty is that a portion of the products obtained in the pyrolysis has the same distillation range as the feedstock. The corresponding decomposition products cannot be separated and determined quantitatively. In this situation, the conventional solution is to consider the fraction with the same distillation limits as the feedstock as being nonreacted feedstock. The error introduced by this convention increases as the distillation range of the feedstock is larger. The error is quite large in

Table 2.14 Successive Cracking at 400°C of a Paraffinic Distillate from Grozny Crude

Operation	Duration (min)	% Gasoline (EP 200°C)	Density Gasoline	Density Cut range: 200–300°C	Cracking rate: Gasoline yield in %/min	Feed to each successive cracking (% of initial feed)
1	16.2	19.7	0.737	0.838	1.22	100.0
2	41.0	23.5	0.743	0.852	0.57	36.5
4	57.0	16.7	0.764	0.910	0.27	13.1
6	60.0	9.9	0.812	0.963	0.16	5.5
9	80.0	6.9	0.854	0.982	0.086	2.2

Source: Ref. 90.

---

* The term "successive cracking" refers to a series of cracking operations in which the cut submitted to the operation is the unreacted portion, recovered from the previous operation. In this way the influence of the decomposition products upon the rate of the following cracking operation is excluded or to a great extent limited.

the case of residues, where one cannot distinguish the heavy condensation products resulted from the reaction from those present in the initial feed stock.

In the case of cracking processes performed for producing gasoline or gasoline and gases, the conversion x is expressed conventionally in terms of the desired product, i.e., as the weight fraction of gasoline plus gases, weight fraction gasoline, or volume fraction of gasoline. Note that the numerical value of the reaction rate constant is dependent on the manner in which the conversion of x is expressed. This fact must be taken into account at whenever using or comparing values of the rate constants taken from the literature.

2.3.5 The modeling of thermal processes

Studies performed since 1970 led to the development of models by means of which it is now possible to compute not only the overall kinetics of the process, but also the compositions of many of the products obtained in the process. The driving force for the modeling studies was not the scientific interest by itself, but the need to improve the design of the tubular furnaces and of the separation sections of the pyrolysis units.

The early models started from a system of stoichiometric equations, to which the theorem of steady state was applied. In order to improve concordance with experiments, the initially accepted simplifications of a single termination mechanism for the chain reaction had to be abandoned. This led to a system of nonlinear differential equations, that could not be solved by analytical methods.

Initially, the integration of such systems was performed by means of computers, using the Gear method (1971) [77,78], the integration program of Kershenbaum (1973), or other similar methods.

While the application of the steady state theorem is very useful for establishing overall kinetics (the reaction order), it gives errors if applied for computing the rates of formation of the reaction products in a broad range of feed conversions. [27]. Therefore, the steady-state method was abandoned [24], when improvements in computing techniques allowed the development of models in which the stoichiometric equations were replaced by reactions involving large numbers of the radicals participating in the pyrolysis process. [79,80].

The studies from the school of G. F. Froment, proved that the results obtained by modeling the pyrolysis of ethane, propane, n-butane, ethane, propene and i-butene predicted accurately the results obtained in the plug flow pyrolysis pilot plant [27]. Similar studies were performed by numerous researchers [74,81]. This method has become the established technique for the quantitative evaluation of the pyrolysis of lower paraffins.

Complications were encountered with reference to the modeling of the pyrolysis of mixtures of hydrocarbons, such as ethane-propane [25,27,82,83]. Technological reasons—the optimal reaction time is different for the two hydrocarbons—require that in practice the pyrolysis of ethane and propane have to be performed in separate furnaces. Therefore, the modeling of the pyrolysis process for this mixture is of low interest.

The described approach made it possible to include in the model radical reactions the methyl-acetylene and dienes, which are the precursors of acetylene. Some of the models include reactions for molecular decomposition (not involving radicals).

The models based on elementary radicalic reactions were extended also to the pyrolysis of heavier feedstocks, such as gasoline and gas oils, by including simplifying assumptions (see Section 3.1.).

The products from the "classic" cracking processes (visbreaking, delayed coking and the old processes of residue cracking) are complex mixtures, which are used as motor fuels (gasoline, gas oil) and fuel oils. These products are characterized mainly by distillation limits. Knowledge of their detailed chemical composition is of no practical interest.

For these processes, the modeling refers to a process involving only successive reactions or parallel and successive reactions. Such a treatment is sufficient for producing satisfactory information on the effluent composition, and also for the analysis of the influence of the various process parameters and for selecting the operating conditions.

The most simple model is that consisting of two successive reaction steps:

$$A \xrightarrow{k_1} v_1 B \xrightarrow{k_2} v_2 C$$

This model may be used to simulate a cracking process under elevated pressures or in some cases a pyrolysis process. In these reactions, A is the feedstock, B is the main desired product (a light distillate in the case of the residue cracking, or ethene and propene in the case of pyrolysis). C is the final product resulting from the decomposition of the desired main product.

In industrial conditions, the definition of the product B and of its separation from A and C is to a large extent arbitrary, depending on the practical objective of the kinetic analysis. This fact doesn't diminish the utility of such calculations and the correctness of the obtained results, on condition that correct values, corresponding to the operating conditions were selected for the rate constants k_1 and k_2.

Noting the time parameter with τ, and with y, z' and u' the number of moles of the substances A, B, C, the following equations may be written for the two successive transformations at constant volume, for first-order kinetics:

$$-\frac{dy}{d\tau} = k_1 y \tag{2.88}$$

$$\frac{dz'}{d\tau} = k_1 v_1 y - k_2 z' \tag{2.89}$$

$$\frac{du'}{d\tau} = k_2 \frac{v_2}{v_1} z' \tag{2.90}$$

By noting with z the number of moles of substance A transformed to B and with u the number of moles of the same substance transformed to C, one has:

$$z' = v_1 z \qquad u' = v_2 u$$

Making these substitutions in Eqs. (2.89) and (2.90), they become:

$$\frac{dz}{d\tau} = k_1 y - k_2 z \tag{2.91}$$

$$\frac{du}{d\tau} = k_2 z \tag{2.92}$$

Since y, z and u represent the number of moles of the same substance of reference A, which was transformed or not in the substances B and C, these amounts will be identical with the weight fractions of substance A not converted and respectively, of substances B and C which were formed.

Using the Laplace transform and taking into account the initial conditions, i.e. $\tau = 0$, $y = 1$, $z = 0$, $u = 0$, the Eqs. (2.88), (2.91), and (2.92) become:

$$- Py + P = k_1 y$$
$$Pz = k_1 y - k_2 z$$
$$Pu = k_2 z$$

from which, by means of elementary algebraic operations, one obtains the images:

$$y = \frac{P}{P + k_1}$$

$$z = k_1 \frac{P}{(P + k_1)(P + k_2)}$$

$$u = k_1 k_2 \frac{1}{(P + k_1)(P + k_2)}$$

For these images, the integrated forms of the equations (2.88), (2.91), and (2.92) are:

$$z = \frac{k_1}{k_2 - k_1}\left(e^{-k_1 \tau} - e^{-k_2 \tau}\right) \tag{2.94}$$

$$u = 1 - \frac{k_2}{k_2 - k_1}e^{-k_1 \tau} - \frac{k_1}{k_2 - k_1}e^{-k_2 \tau} \tag{2.95}$$

During chemical process, the yield of the intermediary product B will pass through a maximum, because its formation rate decreases as feed A is consumed, while its decomposition increased with its accumulation.

The time corresponding to the maximum yield, $\tau_{z_{max}}$, is obtained from equation (2.94) by differentiating and equating to zero, with the result:

$$\frac{dz}{d\tau} = \frac{k_1}{k_2 - k_1}\left(-k_1 e^{-k_1 \tau_{z_{max}}} + k_2 e^{-k_2 \tau_{z_{max}}}\right) = 0$$

By equating the expression in parenthesis to zero, it results:

$$\tau_{z_{max}} = \frac{\ln \dfrac{k_1}{k_2}}{k_1 - k_2} \tag{2.96}$$

This expression represents the reciprocal of the logarithmic mean of the two rate constants and thus $\tau_{z_{max}}$ decreases if any of them increases.

The value of z_{max} is obtained by replacing (2.96) in (2.94):

$$z_{max} = \frac{k_1}{k_2 - k_1}\left(e^{\frac{k_1}{k_2 - k_1}\ln\frac{k_1}{k_2}} - e^{\frac{k_2}{k_2 - k_1}\ln\frac{k_1}{k_2}}\right)$$

This may be written as:

$$z_{max} = \frac{k_1}{k_2 - k_1}\left[\left(\frac{k_1}{k_2}\right)^{\frac{k_1}{k_2-k_1}} - \left(\frac{k_1}{k_2}\right)^{\frac{k_2}{(k_2-k_1)}}\right]$$

Multiplying and dividing the first term in the parenthesis by k_1/k_2, it results:

$$z_{max} = \frac{k_1}{k_2 - k_1}\left[\left(\frac{k_2}{k_1}\right)\left(\frac{k_1}{k_2}\right)^{\frac{k_1}{k_2-k_1}+1} - \left(\frac{k_1}{k_2}\right)^{\frac{k_2}{k_2-k_1}}\right]$$

from which: $z_{max} = \frac{k_1}{k_2 - k_1}\left(\frac{k_2}{k_1} - 1\right)\left(\frac{k_1}{k_2}\right)^{\frac{k_2}{k_2-k_1}}$

and finally:

$$z_{max} = \left(\frac{k_1}{k_2}\right)^{\frac{k_2}{k_2-k_1}} \tag{2.97}$$

Writing $k_1/k_2 = r$ and dividing the numerator and the denominator of the exponent by k_2, it results:

$$z_{max} = r^{1/(1-r)} \tag{2.98}$$

This result shows that the maximum yield of the intermediary product B does not depend on the absolute values of the reaction rate constants, but only upon their ratio.

In Eq. (2.98), r may vary from very low positive values up to positive values which tend to infinity. Therefore, z_{max} will have very low values for low values of r; it increases with increasing values of r, passing through the value e^{-1} for $z = 1$ and tends to 1 as r increases towards infinity.

Indeed, for $r = 0$, $z_{max} = 0$. For $z = 1$ a nondetermination is reached, namely $z_{max} = 1^\infty$. Taking the logarithm of the expression and applying the l'Hospital's rule one gets:

$$\lim_{r \to 1} \ln z_{max} = \lim_{r \to 1} \frac{\ln r}{1 - r} = \lim_{r \to 1} -\frac{1}{r} = -1$$

Therefore:

$$\lim_{r \to 1} z_{max} = e^{-1}$$

For $r = \infty$, the expression is undetermined of the form $z_{max} = \infty^0$, which is treated analogously to give:

$$\lim_{r \to \infty} \ln z_{max} = \lim_{r \to \infty} -\frac{1}{r} = 0$$

Thus :

$$\lim_{r \to \infty} z_{max} = 1$$

It results that the maximum of the yield of the intermediary product B will increase as the ratio of the rate constants increases. This result may be deduced qualitatively also by the logical analysis of the formation and decomposition phenomena of the intermediary product B.

To obtain the variation of the yield of the final product C, the relation (2.95) is differentiated:

$$\frac{du}{d\tau} = \frac{k_1 k_2}{k_2 - k_1} e^{-k_1 \tau} - \frac{k_1 k_2}{k_2 - k_1} e^{-k_2 \tau}$$

By equating the derivative to zero:

$$\frac{k_1 k_2}{k_1 - k_2} \left(e^{-k_1 \tau} - e^{-k_2 \tau} \right) = 0$$

which is satisfied for $\tau = 0$.

From here, it results that in the coordinates τ, u, the curve which represents the yield of the final product C is tangent to the abscissa in the origin. Therefore, the final product is not formed at the beginning of the process, as long as the concentration of B is low.

By equating to zero the second derivative of relation (2.95), one gets:

$$\frac{k_1 k_2}{k_1 - k_2} \left(k_2 e^{-k_2 \tau} - k_1 e^{-k_1 \tau} \right) = 0$$

As the fraction multiplying the parenthesis cannot be zero, it follows:

$$k_2 e^{-k_2 \tau} - k_1 e^{-k_1 \tau} = 0$$

and therefore

$$\tau = \frac{\ln \dfrac{k_1}{k_2}}{k_1 - k_2}$$

It follows that the curve that represents the time variation of the yield of the final product C, will have an inflexion point, corresponding to the maximum yield of intermediary product B.

For a more accurate modeling of the process, the direct formation of the final product also has to be considered. In this case, the kinetic scheme becomes:

$$\underbrace{A \xrightarrow{k_1} v_1 B \xrightarrow{k_2} v_2 C}_{k_3}$$

(b)

Using the same notation, the following system of differential equations is obtained.

$$-\frac{dy}{d\tau} = (k_1 + k_3) y$$

$$\frac{dz}{d\tau} = k_1 y - k_2 z$$

(2.99)

$$\frac{du}{d\tau} = k_2 z + k_3 y$$

Solving these equations similarly to the previous case, it results finally:

$$y = e^{-(k_1 + k_3)\tau}$$

(2.100)

$$z = \frac{k_1}{k_2 - k_1 - k_3}\left(e^{-(k_1+k_3)\tau} - e^{-k_2\tau}\right) \qquad (2.101)$$

$$u = 1 - \left[\frac{k_1 k_2}{(k_1 + k_3)(k_2 - k_1 - k_3)} + \frac{k_3}{k_1 + k_3}\right]e^{-(k_1+k_3)\tau}$$
$$- \frac{k_1}{k_1 + k_3 - k_2}e^{-k_2\tau} \qquad (2.102)$$

$$\tau_{z_{max}} = \frac{\ln\dfrac{k_1 + k_3}{k_2}}{(k_1 + k_3) - k_2} \qquad (2.103)$$

and by using the notation $(k_1 + k_2)/k_2 = p$:

$$z_{max} = \frac{k_1}{k_1 + k_3}p^{\frac{1}{1-p}} \qquad (2.104)$$

The form of these two expressions is similar to Eqs. (2.96) and (2.98) and lets us draw similar conclusions concerning the variation of the intermediary and final products.

These results are confirmed by experimental data. Figure 2.13 plots the gasoline yields obtained by the cracking of a heavier feedstock. Figure 2.14 shows the formation of ethene and of propene during the pyrolysis of a gasoline. In both cases it may be observed that the intermediary product (in the first case the gasoline, in the second, the propene and ethene respectively), pass through a maximum.

The position of the maximum makes it possible to use Eq. (2.98) in order to obtain the ratio of the rate constants k_1 and k_2. By using also relation (2.96), the numerical values of the two constants may be deduced.

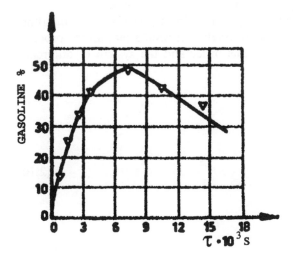

Figure 2.13 Gasoline yield by high pressure cracking of a wax of Rangoon crude. (From Ref. 91.)

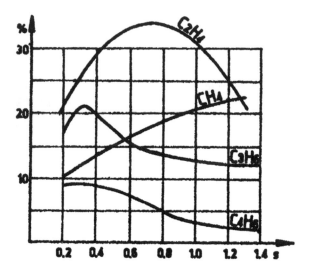

Figure 2.14 Main products from pyrolysis at 810°C and atmospheric pressure, of a 85–120°C gasoline cut.

In a number of cases, for example, in those where besides the decomposition reactions, condensation reactions also occur with the formation of heavier products, schemes (a) or (b) are not sufficient for performing a satisfactory analysis of the process. In such cases, more complex schemes must be developed in order to answer the needs of the analysis.

An example of such a more complex scheme, is represented as (c), below:.

$$\nu_4 E \xrightarrow{k_4} A \xrightarrow{k_1} \nu_1 B \xrightarrow{k_2} \nu_2 C \xrightarrow{k_3} \nu_3 D \qquad \text{(c)}$$

It corresponds to a coking process where: A = raw material, E = products of condensation, B = gas oil, C = gasoline and D = produced gases.

Since the formation of the condensed products (resins, asphaltenes, and other) that lead eventually to coke obey first order kinetics, it results that all the processes represented in the reaction model are of this order and the system of differential equations may be written as:

$$-\frac{dy}{d\tau} = (k_1 + k_4)y$$

$$\frac{dz'}{d\tau} = k_1 \nu_1 y - k_2 z'$$

$$\frac{du'}{d\tau} = k_2 \frac{\nu_2}{\nu_1} z' - k_3 u' \qquad (2.105)$$

$$\frac{dv'}{d\tau} = k_2 \frac{\nu_3}{\nu_2} u'$$

$$\frac{dw'}{d\tau} = k_4 \nu_4 y$$

If the formation of the coke occurs by the precipitation of the asphaltenes from the solution, then Eq. (2.75) should be used for the formation of the coke. The process becomes controlled by reactions of second order and the solution of the differential equations system becomes more difficult.

Using as in the previous case, z, u, v, w, for the weight ratios of the substances B, C, D and E that are formed, one may write:

$$z' = v_1 z$$
$$u' = v_2 u$$
$$v' = v_3 v$$
$$w' = v_4 w$$

Replacing these expressions in the last four differential equations, taking into account that v_1, v_2, v_3 and v_4 may be considered as constants, and simplifying, one obtains:

$$\frac{dz}{d\tau} = k_1 y - k_2 z \tag{2.106}$$

$$\frac{du}{d\tau} = k_2 z - k_3 u \tag{2.107}$$

$$\frac{dv}{d\tau} = k_3 u \tag{2.108}$$

$$\frac{dw}{d\tau} = k_4 y \tag{2.109}$$

Using the Laplace transform and taking into account the initial condition $y = 1$, the Eqs. (2.105–2.109) become:

$$-Py + P = (k_1 + k_4)y$$
$$Pz = k_1 y - k_2 z$$
$$Pu = k_2 z - k_3 u \tag{2.110}$$
$$Pv = k_3 u$$
$$Pw = k_4 y$$

The corresponding images are:

$$y = \frac{P}{P + (k_1 + k_4)}$$
$$z = k_1 \frac{P}{(P + k_1 + k_4)(P + k_2)}$$
$$u = k_1 k_2 \frac{P}{(P + k_1 + k_4)(P + k_2)(P + k_3)} \tag{2.111}$$
$$v = k_1 k_2 k_3 \frac{1}{(P + k_1 + k_4)(P + k_2)(P + k_3)}$$
$$w = k_4 \frac{1}{P + (k_1 + k_4)}$$

The integrated forms of this system of equations are:

$$y = e^{-(k_1+k_4)\tau} \tag{2.112}$$

$$z = \frac{k_1}{k_2 - (k_1 + k_4)}(e^{-(k_1+k_4)\tau} - e^{-k_2\tau}) \tag{2.113}$$

$$u = k_1 k_2 \left[\frac{1}{(k_2 - k_1 - k_4)(k_3 - k_1 - k_4)} e^{-(k_1+k_4)\tau} \right.$$
$$\left. + \frac{1}{(k_1 + k_4 - k_2)(k_3 - k_2)} e^{-k_2\tau} + \frac{1}{(k_1 + k_4 - k_3)(k_2 - k_3)} e^{-k_3\tau} \right] \tag{2.114}$$

$$v = \frac{k_1}{k_1 k_4} - \frac{k_1 k_2 k_3}{(k_1 + k_4)(k_3 - k_1 - k_4)(k_2 - k_1 - k_4)} e^{-(k_1+k_4)\tau} -$$
$$- \frac{k_1 k_3}{(k_2 + k_4 - k_2)(k_3 - k_2)} e^{-k_2\tau} - \frac{k_1 k_2}{(k_1 + k_4 - k_3)(k_2 - k_3)} e^{-k_3\tau} \tag{2.115}$$

$$w = \frac{k_4}{k_1 + k_4}(1 - e^{-(k_1+k_4)\tau}) \tag{2.116}$$

For residue fractions, the feed A contains a certain fraction of resins and asphaltenes (compounds E) that can be expressed as weight fractions by "a." The initial conditions that must be taken into account for such feedstocks will be:

$$y_0 = 1 - a \qquad w_0' = a \qquad z_0 = 0 \qquad u_0 = 0 \qquad v_0 = 0$$

In this case, in the system of equations (2.110) the first and the last equation are modified and become:

$$- Py + P(1 - a) = (k_1 + k_4)y$$
$$Pw - Pa = k_4 y$$

while the other equations remain unchanged.

The equations system (2.111) acquires the form:

$$y = \frac{P(1 - a)}{P + (k_1 + k_4)}$$

$$z = k_1 \frac{P(1 - a)}{(P + k_1 + k_4)(P + k_2)}$$

$$u = k_1 k_2 \frac{P(1 - a)}{(P + k_1 + k_4)(P + k_2)(P + k_3)} \tag{2.117}$$

$$v = k_1 k_2 k_3 \frac{1 - a}{(P + k_1 + k_4)(P + k_2)(P + k_3)}$$

$$w = k_4 \frac{1 - a}{P + (k_1 + k_4)} + a$$

In a similar manner, the integrated forms corresponding to the transforms (2.117) are obtained. The equations for y, z, u and v will be different from equations (2.111–2.115) only by the factor $(1 - a)$, which multiplies the right-hand side of these equations. The Eq. (2.116) will have the form:

$$w = \frac{k_4(1-a)}{k_1+k_4}(1 - e^{-(k_1+k_4)\tau}) - a$$

By examining the Eqs. (2.113) and (2.101) one notes that they are identical if the constant k_3, of Eq. (2.101) is replaced by k_4 in Eq. (2.113). Accordingly, Eq. (2.113) will lead to the following expressions of the maximum of product B:

$$z_{max} = \frac{\ln\dfrac{k_1+k_4}{k_2}}{(k_1+k_4) - k_2} \tag{2.118}$$

$$z_{max} = \frac{k_1}{k_1+k_4} \times p^{1/(1-p)} \tag{2.119}$$

wherein:

$$p = (k_1 + k_4)/k_2$$

The analysis of Eqs. (2.118) and (2.119) leads to similar conclusions as the analysis of Eqs. (2.103) and (2.104).

In order to determine the maximum of the product u, Eq. (2.114) is differentiated and set equal to zero, obtaining:

$$
\begin{aligned}
&k_2(k_3 - k_1 - k_4)e^{-k_2\tau} - (k_1 + k_4)(k_3 - k_2)e^{-(k_1+k_4)\tau} \\
&- k_3(k_2 - k_1 - k_4)e^{-k_3\tau} = 0
\end{aligned} \tag{2.120}
$$

This relation is satisfied for the values $\tau = 0$, $\tau = \infty$ and for a value u_{max} corresponding to the maximum of the intermediary product. However, the value $\tau_{u_{max}}$ cannot be obtained explicitly from Eq. (2.120) and therefore an analytical expression for u_{max} may not be obtained. The values of $\tau_{u_{max}}$ and u_{max} for the specific values of the kinetic constants may be obtained by trial and error.

Within the range of realistic values for the rate constants, the variation of $\tau_{u_{max}}$ and u_{max} may be deduced, as well as their values relative to $\tau_{z_{max}}$ and z_{max}.

Thus, if k_2 is very large, the reaction step corresponding to the conversion of B to C actually disappears. In this case, the maximum of the weight fraction (or yield) of C will be confounded with the maximum for B, which was determined previously and will be expressed by relations similar to Eqs. (2.103) and (2.104), but where the constant k_2 and k_3 are replaced by k_3 and k_4, respectively.

When the value of the constant k_2 is very small, the product C will be formed very slowly and $\tau_{u_{max}}$ will be very far, relative to $\tau_{z_{max}}$.

If the value of the constant k_3 is small when compared to k_2, the decomposition of the product C and the formation of the final product D take place slowly, and u_{max} will have a high value. The contrary occurs when the value of the constant k_3 is large compared to that of k_2.

Accordingly, the curve that represents the variation in time of the weight fraction of the second intermediary product (in coordinates τ versus weight fraction of products) will be tangent to the abscissa in the origin, will pass through a maximum situated at times beyond those for corresponding to the maximum of the first intermediary product, and at longer times will again approach the abscissa.

The variation with time of the weight fraction of the final product D will be analogous to the case in which only one intermediary product is formed.

The time variation of the weight fractions of the feed and products for the case presented are sketched in Fig. 2.15.

Coke formation is represented in Figure 2.15 by a dotted curve, corresponding to the kinetics of coke formation described in Section 2.3.3. The coke formation process passes through the step of precipitation of asphaltenes from the colloidal solution, followed by their subsequent conversion to coke. Thus, coke formation does not start at the beginning of the thermal treatment but after a delay that depends on the nature of the "solvent" in which the asphaltenes are dispersed. The variation of the weight fraction of coke is shown in Figure 2.15 as a dotted curve. The delay in the formation of coke was also illustrated in Figures 2.8 and 2.12.

2.3.6 Kinetics of continuous processes

So far in this chapter, the integration of differential rate equations was performed for conditions of constant volume. Since in practice the thermal cracking processes are performed in continuous flow, it is necessary to examine the form taken by the kinetic equations for this kind of operation.

The general definition of the reaction rate, in moles of a reactant converted per unit time and unit volume, is expressed by the equation:

$$\tau_A = -\frac{1}{V}\frac{dn_A}{d\tau} \tag{2.121}$$

If A is a product the plus sign will be used.

For reactions at constant volume, by incorporating V in the variable which is differentiated, it follows:

$$\tau_A = -\frac{dC_A}{d\tau} \tag{2.122}$$

the form of which was implicitly used in the previous deductions.

Thus, when Eqs. (2.88–2.90) were written directly in terms of moles instead of concentrations, it was assumed implicitly that the volume is constant.

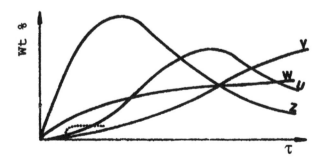

Figure 2.15 Products formation according to kinetic scheme (c) on p. 88. The dotted line depicts coke formation. (From Ref. 92.)

For a continuous process, $Vd\tau$ represents the differential element of volume of the ideal plug flow reactor dV_R, and Eq. (2.76) may be written as:

$$r_A = -\frac{n_{0A}d(1-x)}{dV_R} = \frac{n_{0A}dx}{dV_R} \tag{2.123}$$

$$r_A = \frac{dx}{d\left(\dfrac{V_R}{n_{0A}}\right)} \tag{2.124}$$

where, n_{0A} is the molar feedrate, V_R, is the volume of the reaction zone and x is the fractional conversion.

In Eq. (2.124) the denominator $\dfrac{V_R}{n_{0A}}$ is the reciprocal of the feedrate, in mole units, i.e., moles per second per unit volume of the reaction zone: $\dfrac{n_{0A}}{V_R}$, also called *molar feedrate*.

The feedrate may be expressed also in weight units, *mass feedrate*, or in volume units, *volume feedrate*.

In catalytic processes, the feedrate is often referred to as the mass of catalyst present in the reaction zone and not to the volume of the reaction zone.

Expressing the reaction rate of the thermal cracking of hydrocarbons by a first order kinetic equation:

$$r_A = kC_A \tag{2.125}$$

and setting this expression equal to (2.123) one obtains:

$$\frac{n_{0A}dx}{dV_R} = kC_A \tag{2.126}$$

By expressing the concentration of reactant C_A as:

$$C_A = \frac{n_{0A}(1-x)}{V} \tag{2.127}$$

and the volume V by means of the gas law:

$$V = \frac{\sum nzRT}{p}$$

it follows:

$$C_A = \frac{n_{0A}(1-x)p}{\sum nzRT} \tag{2.128}$$

The sum of moles $\sum n$ that correspond to a conversion x, may be easily expressed in terms of the increase of the volume (or number of moles) during the reaction δ, using the relation:

$$\sum n = n_{0A}(1 + \delta x) \tag{2.129}$$

Replacing this in (2.128) and the result in (2.126), it results:

$$\frac{n_{0A}dx}{dV_R} = k\frac{(1-x)p}{(1+\delta x)zRT} \tag{2.130}$$

which can be written as:

$$n_{0A} \frac{1 + \delta x}{1 - x} dx = \frac{kp}{zRT} dV_R$$

Integration of this expression in conditions of constant temperature and pressure, leads to:

$$k = \frac{n_{0A} \bar{z} RT}{pV_R} \left[(1 + \delta x) \ln \frac{1}{1 - x} - \delta x \right] \qquad (2.131)$$

This expression corresponds to the conditions of constant pressure and is different from the one obtained previously for constant volume (2.77).

For processes that take place without a variation of the number of moles, $\delta x = 0$ the relations (2.77) and (2.131) become identical.

The expansion coefficient δ may be obtained on basis of the overall stoichiometric equation. Thus, for the equation:

$$A \rightarrow v_1 A_1 + v_2 A_2 + \cdots + v_n A_n$$

the expansion coefficient has the value:

$$\delta = \sum_{i=1}^{n} v_i - 1 \qquad (2.132)$$

In situations where the overall stoichiometric equation cannot be formulated, such as in the case of the cracking of the petroleum fractions, the expansion coefficient δ can be obtained on the basis of the molecular mass of the feed and of the average molecular mass of the products, by using the relation:

$$\delta = \frac{M_{0A}}{\overline{M}} - 1 \qquad (2.133)$$

Similarly, other kinetic expressions may also be integrated for the conditions of the continuous flow reaction at constant pressure and temperature.

Thus, if the pyrolysis of ethane, is assumed to be a reversible reaction, the reaction rate will be expressed by:

$$r_{C_2H_6} = k_1 \times C_{C_2H_6} - k_{-1} \times C_{C_2H_4} \times C_{H_2} \qquad (2.134)$$

The concentrations of the participating species will have the expressions:

$$C_{C_2H_6} = \frac{p(1 - x)}{(1 + x)RT} \qquad C_{C_2H_4} = C_{H_2} = \frac{px}{(1 + x)RT} \qquad (2.135)$$

while also $\delta = 1$ and $z = 1$.

Expressing the reaction rate by Eq. (2.123) and replacing the concentrations by the relations (2.135), Eq. (2.134) becomes:

$$\frac{n_0 dx}{dV_R} = k_1 \frac{p(1 - x)}{(1 + x)RT} - k_{-1} \frac{p^2 x^2}{(1 + x)^2 R^2 T^2} \qquad (2.136)$$

The rate constant for the reverse reaction, is given by:

$$k_{-1} = \frac{k_1}{K_c} = \frac{k_1 RT}{K_p}$$

Substitution in (2.136), gives:

$$\frac{n_0 dx}{dV_R} = k_1 \frac{p}{RT}\left[\frac{1-x}{1+x} - \frac{p}{K_p}\frac{x^2}{(1+x)^2}\right] = \frac{k_1 p}{RT}\frac{1-x^2-\left(\frac{p}{K_p}\right)x^2}{(1+x)^2}$$

By separating the variables, one obtains:

$$\frac{k_1 p}{n_0 RT} V_R = \int_0^x -\frac{(1+x)^2}{1-\left(1+\frac{p}{K_p}\right)x^2}\,dx \tag{2.137}$$

For ethane pyrolysis, the equilibrium constant K_p may be expressed by:

$$K_p = \frac{x_\infty^2}{(1-x_\infty)}\cdot\frac{p}{(1+x_\infty)} = \frac{x_\infty^2}{1-x_\infty^2}p \tag{2.138}$$

from which

$$1 + \frac{p}{K_p} = \frac{1}{x_\infty^2}$$

Replacing in (2.137) and performing the integration yields:

$$\frac{k_1 p}{n_0 RT} V_R = \int_0^x \frac{(1+x)^2}{1-x^2/x_\infty^2}dx = \int_0^x \frac{dx}{1-x^2/x_\infty^2} + 2\int_0^x \frac{xdx}{1-x^2/x_\infty^2}$$

$$+ \int_0^x \frac{x^2 dx}{1-x^2/x_\infty^2} = \frac{x_\infty}{2}\ln\left[\frac{1+x/x_\infty}{1-x/x_\infty}\right] - x_\infty^2\ln(1-x^2/x_\infty^2) - x\cdot x_\infty^2$$

$$+ \frac{x_\infty^3}{2}\ln\left[\frac{1+x/x_\infty}{1-x/x_\infty}\right]$$

and finally:

$$\frac{k_1 p}{n_0 RT} V_R = x_\infty\left[\frac{1}{2}(1+x_\infty^2)\ln\left(\frac{x_\infty+x}{x_\infty-x}\right) - x_\infty\ln\left(1-\frac{x^2}{x_\infty^2}\right) - x_\infty x\right] \tag{2.139}$$

The integration of differential rate equations for continuous flow systems operating at constant pressure and temperature is of interest for processing laboratory experiments generated in such reaction systems. In industrial tubular reactors used for thermal processes, the pressure decreases significantly along the reactor. Therefore, the integrated equations at constant pressure are not adequate.

In calculations pertaining to such systems, besides the exact models set forth in Chapter 4, approximate results may be obtained by a shortcut method that uses the concept of average volume of product mixture $\overline{V}$, that passes through the plug flow reactor or through the portion of the reactor considered in unit time.

Replacing in Eq. (2.127) V with $\overline{V}$ and substituting in (2.126), after simplifying and rearranging one obtains:

$$\frac{dx}{d\left(\frac{V_R}{\overline{V}}\right)} = k(1-x)$$

which, after integration becomes:

$$k = \frac{\overline{V}}{V_R} \ln \frac{1}{1-x}$$

which is of the same form as Eq. (2.77) integrated at constant volume, in which the time τ was replaced by $\frac{V_R}{\overline{V}}$. Actually, the equality:

$$\tau = \frac{V_R}{\overline{V}} \tag{2.140}$$

defines the residence time in terms of the variables used.

In this way, if the residence time is defined by means of relation (2.140), all the relations obtained by integration at conditions of constant volume are valid for the flow processes.

The average volume $\overline{V}$ may be calculated as the arithmetic mean of the volumes at the inlet and outlet of the flow reactor or of the considered portion of the reactor. This simplifies the calculations also in the case when the pressure varies along the reactor.

For reactions taking place in the gas phase, replacing $\overline{V}$ in relation (2.140) with its expression obtained from the gas law:

$$\overline{V} = \frac{\overline{n}\overline{z}RT}{p}$$

yields: $\tau = \frac{V_R p}{\overline{n}\overline{z}RT}$ (2.141)

where: $\overline{n}$ is the average number of moles that cross the reaction zone in unit time. The number $\overline{n}$ may be expressed using relation (2.129) and, as for the volume $\overline{V}$, calculating the average conditions by the arithmetic mean between the entrance and the exit, obtaining:

$$\overline{n} = n_{0A}\left(1 + \frac{\delta x}{2}\right)$$

Accordingly, Eq. (2.141) may be written also as:

$$\tau = \frac{V_R p}{n_{0A}\left(1 + \frac{\delta x}{2}\right)\overline{z}RT} \tag{2.142}$$

from which, it results that the reaction time in continuous systems is a function of the reciprocal of the feedrate $\frac{n_{0A}}{V_R}$.

This correlation depends on temperature, pressure and conversion.

2.4 INFLUENCE OF OPERATING CONDITIONS

In the previous sections, the thermodynamics, mechanisms, and kinetics of the thermal processes were examined. On this basis, it is possible to analyze the influence of the operating conditions on the product yields of these processes. The influence of the temperature, pressure, feedstock properties, and steam introduced into the reactor will be examined in the following sections. Since reaction time is a direct consequence of the process kinetics discussed in Chapter 1, its effect will not be examined here on itself, but only in connection with other factors.

2.4.1 Temperature

The influence of temperature on a complex process such as thermal cracking must be examined from the point of view of both the modification of the thermodynamic equilibriums and of the relative rates of the decomposition and other reactions that take place within the process.

The examination in Section 2.1 of the thermodynamic factors of thermal cracking, made possible identifying the reactions whose equilibrium influences the composition of the products obtained from the process.

Practical results show that in pyrolysis, the ratios ethene/ethane and propene/propane are close to those for equilibrium. The graphs of Figures 2.3 and 2.4 allow one to estimate that increase in temperature has a favorable effect on the conversion to alkenes. Therefore, considerable efforts are made to increase the coil outlet temperatures (COT), in pyrolysis furnaces.

Higher COT favor also the formation of acetylene from ethene, as represented in Figure 2.5 and make it necessary to increase the capacity of the ethene purification section.

Increased temperature has a favorable effect also on the formation of butene from butane and of butadiene from butene (Figures 2.2 and 2.6), which however, enter in competition with the cracking of butane to ethene and ethane (Figure 2.1)

Despite the fact that the increase of temperature favors the production of ethene, propene, and butadiene, it is not possible to reach a maximum production of all these three hydrocarbons in the same time. The reason is in the difference among the rates of the reactions that produce them and in the fact that the desired products suffer further decomposition. Accordingly, the production of each of the three hydrocarbons passes through a maximum, which for propene and butadiene, occurs earlier than for ethene (Figure 2.14). For this reason, a pyrolysis furnace may be operated either for maximizing the production of ethene, or that of propene. The two objectives can not be reached in the same time.

Higher operating temperatures also favor the equilibrium for the formation of aromatic hydrocarbons from alkylcyclohexanes (Figure 2.7), which leads to an increased production of aromatic hydrocarbons with increased temperature in the pyrolysis of naphta.

Concerning the influence of temperature on the kinetics of the chain decomposition involving free radicals, it is interesting to note that with increasing temperature the average molecular mass of the formed products decreases.

Since in pyrolysis, the length of the kinetic chain is of a few hundred interactions, the composition of the products will be determined by the propagation reactions, which are repeated over and over again. The initiation and termination reactions can be neglected.

The propagation reactions are:

(a) β cleavage of the radicals
(b) interaction of the radicals with feed molecules

It is obvious that if reactions (a) may proceed without the chain being interrupted by reactions of type (b), products of low molecular mass are prevalently formed. In the opposite situation, products of medium molecular mass are formed.

The presented facts may be expressed by the following kinetic scheme:

$$\left.\begin{array}{l} \dot{R} \xrightarrow{k_2} M_1 + \dot{R}_1 \\ \dot{R}_1 \xrightarrow{k_2} M_2 + \dot{R}_2 \\ \quad \cdots\cdots\cdots \\ \dot{R}_{n-1} \xrightarrow{k_2} M_n + \dot{R}_n \end{array}\right\} \quad \dot{R}_i \xrightarrow{k_2} M_i + \dot{R}_n \tag{a}$$

$$\left.\begin{array}{l} \dot{R} + M \xrightarrow{k_3} RH + \dot{R} \\ \dot{R}_1 + M \xrightarrow{k_3} R_1H + \dot{R} \\ \quad \cdots\cdots\cdots \\ \dot{R}_{n-1} \xrightarrow{k_3} R_{n-1}H + \dot{R} \end{array}\right\} \quad \dot{R}_i + M \xrightarrow{k_3} R_iM + \dot{R} \tag{b}$$

where, the initiation and termination reactions were omitted. In the right-hand side of the scheme, all reactions of type (a) and (b) were added together, in the form of overall reactions.

Using x for the total number of moles resulting from the reactions of type (a) and with y the moles resulted from the reactions of type (b) one may write:

$$\frac{1}{V}\frac{dx}{d\tau} = k_2[\dot{R}_i] = A_2 e^{-E_2/RT} \cdot \frac{p_{\dot{R}}}{RT} \tag{2.143}$$

$$\frac{1}{V}\frac{dy}{d\tau} = k_3[\dot{R}_i][M] = A_3 e^{-E_3/RT} \cdot \frac{p_{\dot{R}} p_M}{(RT)^2} \tag{2.144}$$

The generalization of the reactions of each type implied by the formulation of these two equations is correct, since the competition between the β cleavage of a radical and its interaction with a feed molecule takes place after each cleavage step.

Division of these equations yields:

$$\frac{dx}{dy} = \frac{A_2 R}{A_3} e^{(E_3 - E_2)/RT} \cdot \frac{T}{p_M} \tag{2.145}$$

For a unit mass of feedstock, p_M may be expressed by the relation:

$$p_M = p(1 - x - y)$$

By substitution in (2.145) it results:

$$\frac{dx}{dy} = \frac{A_2 R}{A_3} e^{(E_3 - E_2)/RT} \cdot \frac{T}{p(1 - x - y)} \tag{2.146}$$

According to Tables 2.3 and 2.5, the two activation energies range between the values:

$$E_2 = 105 - 165 \text{ kJ/mole}$$
$$E_3 = 30 - 45 \text{ kJ/mole}$$

Using as average values $E_2 = 135$ kJ/mole and $E_3 = 35$ kJ/mole, Eq. (2.146) becomes:

$$\frac{dx}{dy} = \frac{A_2 R}{A_3} e^{-100,000/RT} \times \frac{T}{p(1-x-y)}$$

It is of interest to compare the products obtained in pyrolysis, which is operated typically at a characteristic temperature of 850°C (1123 K), to those obtained in a milder cracking process (such as visbreaking), which is operated at 500°C (773 K).

Thus, for the same pressure and overall conversion, since the pre-exponential factors vary very little with temperature, one may write:

$$\frac{\left(\dfrac{dx}{dy}\right)_{1123\ K}}{\left(\dfrac{dx}{dy}\right)_{773\ K}} = 1.45 \frac{e^{-100,000/R \times 1123}}{e^{-100,000/R \times 773}} = 185$$

Despite the approximations implied in this calculation, the result proves that at the temperatures used in pyrolysis, a very advanced decomposition of the feed is favored for products of small molecular masses. For processes that are carried out at moderate temperatures, the formation of products with intermediate molecular mass is predominant. These conclusions are in agreement with the results obtained in industrial units.

Another conclusion may be obtained from the kinetic analysis by modeling the thermal decomposition as a process consisting of successive steps, wherein the intermediary product is the one desired.

In the cracking processes of crude oil fractions, the reactions of the intermediary products resulting from the decomposition, have higher activation energies than the reactions of the feed. Higher operating temperatures will therefore reduce the ratio of the rate constants and, according to the Eq. (2.98), also reduce the maximum concentration of the intermediary product. The time for obtaining the maximum—as per Eq. (2.96)—will also be shorter (Figure 2.16) [92].

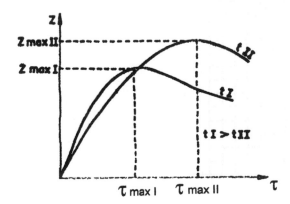

Figure 2.16 Effect of temperature on the formation of intermediate product in thermal cracking. (From Ref. 92.)

For pyrolysis, the situation is somewhat more complex. Thus, assuming that the alkenes are the intermediate product in the pyrolysis of alkanes, the activation energy for their further conversion is larger than that for the conversion of the feed alkanes. A temperature increase has the same. effect as in the previous case. The thermodynamic equilibrium is displaced towards the formation of alkenes and produces a larger maximum of intermediary product. This has two opposed effects. First, situations may occur when the increase in temperature will increase the maximum of the alkenes produced as seen from the experimental data plotted in Figure 2.17.

In processes with two or more successive steps, the change in temperature may also modify the rate determining step. This is the case of the formation of coke from asphaltenes that are present as colloidal solution in heavy and especially in the residual fractions, discussed earlier. The first step, that of precipitation of the asphaltenes, has a low activation energy as compared to the activation energy of the reactions by which coke is formed from the precipitated asphaltenes. Therefore, at lower temperatures (generally under 350°C), the transformation of the asphaltenes to coke is controlling the overall rate. At higher temperatures, rate controlling is the precipitation of the asphaltenes (Figure 2.18). The difference between the structure of the coke obtained at high temperatures (pyrolysis) and at moderate temperatures (delayed coking), is related to the difference in the activation energies.

The effect of temperature on the rate constants of thermal processes is generally expressed by the Arrhenius equation:

$$k = Ae^{-E/RT}$$

(2.147)

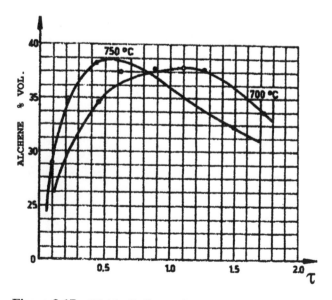

Figure 2.17 Yield of alkenes from the pyrolysis at 700 and 750°C, of butane with 10% propane. (From Ref. 3.)

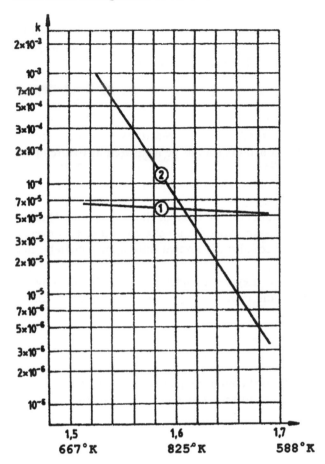

Figure 2.18 Rate constants for the conversion of asphaltenes to coke: 1-asphaltenes precipitation; 2-coke formation.

Usually, the logarithm of Eq. (2.147) is plotted as a straight line in coordinates: $1/T$ log k. If the activation energy does not change in these coordinates*, Eq. (2.147) is represented by a straight line, with the slope $-(E/R)$.

In cases where changes of the activation energy take place as result of changes of the direction of the chemical transformation or of the rate controlling step in a process with successive steps, the slope will also vary and the plot of the rate constant will present a curvature.

By measuring the activation energies in a wide range of temperatures, modifications in the mechanism and in the chemistry of the transformations that take place may be identified.

---

* Sometimes the gradation of the abscissa is reversed in order to have increasing temperatures along it.

In order to compute the activation energy from experimental data, the Arrhenius equation is integrated between the limits T_1 and T_2. In this way, the pre-exponential factor A falls out. This equation has the form:

$$\ln\frac{k_{T_1}}{k_{T_2}} = \frac{E}{R}\left(\frac{1}{T_2} - \frac{1}{T_1}\right) \qquad (2.148)$$

If the two rate constants are measured at two temperatures, the only unknown is the activation energy, E. The rate constants should be measured at two temperatures T_1 and T_2, which do not differ from each other by more than 10–20°C. In this way, the danger of incorporating a domain where modifications of the value of the activation energy take place is avoided.

The calculation of the rate constants on the basis of the conversions obtained at the temperatures T_1 and T_2 requires knowing the kinetic equation. However, in many practical cases the form of the equation is known only approximately, if at all.

As shown below, this difficulty may be avoided without diminishing the correctness of the obtained activation energy. The general form of a kinetic equation is:

$$k = \frac{1}{\tau}f(x)$$

where $f(x)$ depends on the form of the kinetic equation. For two temperatures at which the same feedstock was reacted to the same conversion one may write:

$$\frac{k_{T_1}}{k_{T_2}} = \left(\frac{\tau_{T_2}}{\tau_{T_1}}\right)_{x=\text{const}} \qquad (2.149)$$

By substitution in relation (2.148) it follows:

$$\ln\left(\frac{\tau_{T_2}}{\tau_{T_1}}\right)_{x=\text{const}} = \frac{E}{R}\left(\frac{1}{T_2} - \frac{1}{T_1}\right) \qquad (2.150)$$

The application if this equation eliminates the need to know the form of the kinetic equation.

The experimental difficulty of obtaining the same conversion at two different temperatures may be avoided by plotting a series of experimental values of the conversion versus the corresponding reaction times, at the two temperatures. Such a plot is given in Figure 2.19, which shows that it is possible to obtain the correct ratio

$$\left(\frac{\tau_{T_2}}{\tau_{T_1}}\right)_{x=\text{constant}}$$

by graphical interpolation, without having to operate experimentally at exactly the same values of the conversion.

The plotting of the experimental results at two temperatures as in Figure 2.19, makes it possible to verify whether in the range of the investigated parameters the characteristics of the chemical process do not change with the conversion increase.

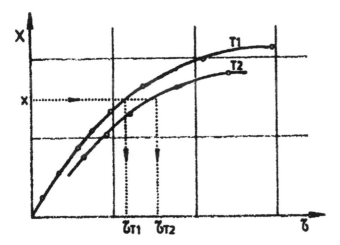

Figure 2.19 The use of conversion versus time plots for obtaining the same conversion at two different temperatures T_1 and T_2.

Such a change in the direction of the reactions, if present, will result in different values of the ratio

$$\left(\frac{\tau_{T_2}}{\tau_{T_1}}\right)_{x=\text{constant}}$$

for different values of the conversion x.

The experimental determination of the rate constants and of the activation energies for thermal processes presents some specific difficulties. Indeed, in the case of catalytic processes, the reaction begins only when the feed is in contact with the catalyst and it is interrupted as soon as this contact is stopped. Therefore, it is possible to determine without difficulties the duration of the reaction. On the other hand, in thermal processes the reaction begins progressively, as the temperature increases and the measured time of reaction corresponds to a variable temperature. To diminish the inherent errors that derive from this situation, a number of procedures were invented in order to perform extremely rapid heating and cooling. For heating, the feed may be introduced in a vessel with molten metal, in a fluidized bed maintained at the desired temperature, or contacted with a super-heated inert gas. For fast cooling, "quenching" by contacting the reactor effluent with a cold liquid (usually water) is routinely practiced.

Whatever the adopted system is, the temperature profile will always be a curve with a zone of increasing temperature, one in which the temperature is at the desired level and a third zone of decreasing temperature. The various systems used are different only by the degree to which they succeed in reducing the zones for heating and cooling, thus reducing, but not eliminating completely, the errors.

In this situation two methods were developed for computing the isothermal equivalent for the reaction, which actually proceeds at variable temperature.

The first one, is applied to continuous reaction systems [93], and determines the equivalent reaction volume (V_{RE}) i.e., the volume of an isothermal hypothetical reaction zone that produces the same conversion as the real (nonisothermal) reaction zone.

On the basis of Eq. (2.124), the equivalent temperature T_e, for which we wish to express V_{RE} is given by the relation:

$$\frac{dx}{d\left(\dfrac{V_{RE}}{n_{0A}}\right)} = k_{T_e} f(x)$$

which can be integrated for isothermal conditions and equivalent temperature:

$$\int_0^x \frac{dx}{f(x)} = \int_0^{V_{RE}} k_{T_e} d\left(\frac{V_{RE}}{n_{0A}}\right)$$

or for the real conditions of variable temperature:

$$\int_0^x \frac{dx}{f(x)} = \int_0^{V_R} k_T d\left(\frac{V_R}{n_{0A}}\right) \tag{2.151}$$

For equal conversions, the integrals on the left-hand side are equal. Since the first equation corresponds to the isothermal process, k_{T_e} is constant over the whole length of the reactor. Accordingly, one may write:

$$\frac{V_{RE}}{n_{0A}} = \int_0^{V_R} \frac{k_T}{k_{T_e}} d\left(\frac{V_R}{n_{0A}}\right) \tag{2.152}$$

By using Eq. (2.148) and the fact that the feed is the same in both cases, Eq. (2.152) also may be written as:

$$V_{RE} = \int_0^{V_R} e^{-E/R\left(\frac{1}{T}-\frac{1}{T_e}\right)} dV_R \tag{2.153}$$

The equivalent volume V_{RE} is determined by numerical or graphical integration of the curve that represents the variation of temperature with the length of the reaction zone, in coordinates V_R and $e^{-E/R\left(\frac{1}{T}-\frac{1}{T_e}\right)}$ (Figure 2.20).

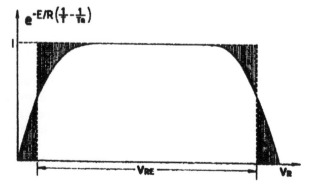

Figure 2.20 Determination of the equivalent reaction volume – V_{RE} by graphical integration.

Since, according to Eq. (2.142), the time is a function of $\frac{V_R}{V_{0A}}$, Eq. (2.152) may be written as:

$$\tau_e = \int_0^x e^{-E/R\left(\frac{1}{T}-\frac{1}{T_e}\right)} d\tau \tag{2.154}$$

This equation may be used in a similar manner as Eq. (2.153), for computing the isothermal equivalent residence time of a reaction that takes place at constant volume.

In the industrial practice, two notions are used in place of the activation energy as a measure of the effect of temperature on the reaction rate. They let us directly estimate the effect of temperature on the process. These notions are:

the temperature coefficient of the reaction rate, k_t which represents the ratio of the rate constants for two temperatures that differ by 10°C

the temperature gradient of the reaction rate, α, which expresses how many degrees the temperature must increase in order to double the reaction rate

The relationship between the temperature coefficient k_t and the activation energy E may be deduced by substituting the definition of k_t in relation (2.148). It follows:

$$\ln k_t = \frac{E}{R}\left(\frac{1}{T_2} - \frac{1}{T_2 + 10}\right) \tag{2.155}$$

from which:

$$\ln k_t = 10\frac{E}{RT_2(T_2 + 10)}$$

By taking into account that T_2 is much larger than 10, converting to common logarithms, and replacing the value of R, one obtains:

$$\log k_t \approx 0.52\frac{E}{T^2} \tag{2.156}$$

In a similar manner, by making the corresponding substitutions in relation (2.148), one obtains the correlation of the temperature gradient of the reaction rate α with the activation energy:

$$\ln 2 = \frac{E}{R}\left(\frac{1}{T_2} + \frac{1}{T_2 + \alpha}\right) \tag{2.157}$$

After making simplifications and substitutions, analogous to the previous equations obtains:

$$\alpha \approx \frac{5.78T^2}{E} \tag{2.158}$$

The correlation between k_t and α is obtained by eliminating E/T^2 from Eqs. (2.156) and (2.158). It follows:

$$\alpha = \frac{3.01}{\log k_t} \tag{2.159}$$

From Eqs. (2.156) and (2.158) it follows that, different from the activation energy, the k_t and α depend also of the temperature. Therefore, when using them one also must bear in mind the temperature range within which these values are valid.

The knowledge of k_t or α for a given process and for the temperature range in which it takes place is useful not only for operating the industrial plant, but also for performing a number of simple but extremely useful calculations.

Thus, by dividing Eq. (2.148) by (2.157), after performing the simplifications, one gets:

$$\frac{\ln \dfrac{k_{T_1}}{k_{T_2}}}{\ln 2} = \frac{(T_1 - T_2)(T_2 + \alpha)}{T_1 \alpha}$$

As $T_2 + \alpha = T_1$, it follows after simplifying:

$$k_{T_1} = k_{T_2} \times 2^{\frac{t_1 - t_2}{\alpha}} \tag{2.160}$$

Dividing Eq. (2.148) by (2.155), an equation similar to that for k_t is obtained:

$$k_{T_1} = k_{T_2} k_t^{0.1(t_1 - t_2)} \tag{2.161}$$

The Eqs. (2.160) and (2.161) make it possible to compute the rate constant for the same feedstock but at a different temperature.

In a similar way, by taking into account Eq. (2.149) an expression may be obtained on basis of the relationships (2.160) and (2.161) for the reaction time required to get the desired conversion at a different temperature. These expressions have the form:

$$\begin{aligned} \tau_2 &= \tau_1 \times 2^{\frac{t_1 - t_2}{\alpha}} \\ \tau_2 &= \tau_1 k_t^{0.1(t_1 - t_2)} \end{aligned} \tag{2.162}$$

The relationships (2.160–2.162) allow one to compute the influence of the temperature on a process that takes place in isothermal conditions.

Most processes usually take place in nonisothermal conditions. It is therefore of interest to determine the equivalent isothermal processes, that would lead to the same conversions in the same reaction time. The use of the arithmetic mean for this purpose may lead to errors since the variation of the reaction rate with temperature is an exponential function.

For processes for which the variation of temperature with time is linear, such an equivalence may be obtained by using the following equations [94]:

For adiabatic processes:

$$t_{evma} = t_f - \frac{10}{\log k_t} \log \frac{k_t^{0.1(t_f - t_i)} - 1}{0.23(t_f - t_i) \log k_t} \tag{2.163}$$

or

$$t_{evma} = t_f - \frac{\alpha}{0.301} \log \frac{\alpha \left(2^{(t_f - t_i)/\alpha} - 1 \right)}{0.696(t_f - t_i)} \tag{2.164}$$

For nonadiabatic processes:

$$t_{\text{evmp}} = t_f - \frac{10}{\log k_t} \times \log \frac{0.23(t_f - t_i)\log k_t}{1 - k_t^{-0.1(t_f - t_i)}} \tag{2.165}$$

or

$$t_{\text{evmp}} = t_f - \frac{\alpha}{0.301} \log \frac{0.696(t_f - t_i)}{\alpha\left(1 - 2^{\frac{t_f - t_i}{\alpha}}\right)} \tag{2.166}$$

In these equations, the notations t_{evma} and t_{evmp} refer to the temperature for which an isothermal process would result in the same final conversion as a adiabatic or polytropic process, t_i and t_f being the respective initial and final temperatures.

In the case when the variation of the temperature with time is not linear, it is replaced by a number of linear intervals.

2.4.2 Pressure

The effect of pressure on the equilibrium of the reactions that take place in pyrolysis is traced in the graphs of Figures 1.3, and 2.1–2.7 and was discussed in Section 2.2.

Recall the important increase of the equilibrium conversion that accompanies the decrease of the operating pressure for the conversion of ethane to ethene (Figure 2.3), of propane to propene (Figure 2.2) and, especially of 1-butane to butadiene (Figure 2.6). The conversion at equilibrium increases even stronger with the decrease of pressure in the dehydrogenations of the 6-carbon atoms ring cyclo-alkanes to aromatic hydrocarbons (Figure 2.7). Concerning the equilibrium for the formation of acetylene from ethene (Figure 2.5), the reduction of pressure favors the increase in conversion. This effect however, seems to be somewhat less important than for the conversion of the ethane to ethene.

The polymerization of the alkenes, produced during the processes of cracking under high pressure, is favored by the increase in pressure (Figure 1.3).

The effect of pressure upon the kinetics of the radical chain decomposition may be analyzed also by examining its influence on the mechanisms of the following reactions:

1. β cleavage of free radicals
2. substitution reactions, i.e. the reactions of radicals with feed molecules

As in the analysis of the effect of temperature, here also the effect of the initiation and termination reactions on the products of the decomposition may be neglected.

The decomposition of the radicals is a succession of cracking steps that occur in β position to the odd electron and take place at time intervals corresponding to the medium life duration of the radicals. Since the reaction kinetics is of first order, the time during which the transformation took place to some extent does not depend on the concentration; thus, the half-time is given by the relation: $\tau_{1/2} = 0.693/k$.

The substitution relations are of second order. Therefore, the time to reach a certain degree of transformation is inversely proportional to the concentration. The equation for the half-time is given by:

$$\tau_{1/2} = \frac{1}{k[M]}$$

The increase of pressure, causes an increase of the concentration. The result is that the frequency of the substitution reactions is higher than that of the decomposition reactions. This means that the substitution reactions may take place before all the cleavages in β position took place. For this reason the average molecular mass resulting from the decomposition is higher at higher pressure.

The effect of pressure on the reaction mechanism may be expressed quantitatively. The notation r_b is used for the rate of the decomposition reactions and r_c for the rate of the substitution reactions.

Expressing the reaction rates by the number of molecules M reacted per second in a cm^3 and accepting for the activation energies average values from the Tables 2.5 and 2.6, one obtains:

$$r_b = 10^{13} \times e^{-134,000/RT}[\overset{\bullet}{R}]$$

$$r_c = 10^{-26} \times e^{-40,000/RT}[\overset{\bullet}{R}][M]$$

Dividing the second equation by the first, at 800 K one gets:

$$\frac{r_c}{r_b} = 10^{-26} \times e^{\frac{134,000-40,000}{RT}} \times [M] = 1.36 \times 10^{-20}[M]$$

The number of molecules per cm^3 M can be expressed by using Avogadro's number (N_A) and the gas law, by means of the relation:

$$[M] = \frac{N_A}{V} = \frac{N_A \cdot p}{RT} = \frac{6.02 \cdot 10^{23}p}{82.06T} = 7.34 \cdot 10^{21} \cdot \frac{p}{T}$$

For a temperature of 800 K, substituting in the previous relation we obtain:

$$\frac{r_c}{r_b} = 0.12p \tag{2.167}$$

In the thermal decomposition of alkenes, increasing pressure produces a higher proportion of alkanes with medium molecular mass. These are formed by substitution reactions at the expense of the alkenes and alkanes of lower molecular mass, that are formed by the continuation of the β-cleavage. Similar effects will be also apparent during the thermal decomposition of other classes of hydrocarbons.

The pressure also influences the precipitation of asphaltenes, which is a determining stage in the formation of coke at the temperatures of thermal cracking (see Section 2.4.4). By increasing the pressure, the liquid phase will become enriched in lighter hydrocarbons, in which asphaltenes are less soluble. This facilitates their precipitation and increases the rate of coke formation. This effect is the usual cause of coking in thermal cracking.

However, if the system pressure exceeds the critical pressure of the hydrocarbons mixture in the vapor phase, an opposite effect may come into play. In supercritical conditions, the solubility of the heavy hydrocarbons is increased and their precipitation as well as coking will be retarded. As an end effect, the formation of coke becomes disfavored by the increase of pressure. With respect to coke formation, the processes of cracking at high pressure requires a thorough analysis in order to take into account the nature of the hydrocarbons present in the system and the pressure in the reactor.

The system pressure might influence the overall kinetics of the process. Since the thermal decomposition reactions are of the first order, the rate of feedstock conversion is independent of pressure. The polymerization reactions follow second order kinetics and increasing the pressure will increase the rate of conversion.

It must be remarked that for reactions in vapor phase in continuous systems at a constant molar flowrate n_{0A}/V_R, increasing the operating pressure also increases the reaction time. Therefore, in such cases there appear two types of overlapping effects: some are due to the influence of the pressure on the reaction chemistry while others are due to the increase of the reaction time. The interdependence between the flowrate, pressure, and contact time is given by Eq. (2.142)*.

In order to study correctly the effect of pressure on chemical reactions in a continuous system, the experiments must be carried out in conditions that maintain a constant reaction time. This can be achieved by correspondingly adjusting the molar flowrate when the system pressure is changed. According to Eq. (2.142) the following condition must be respected:

$$\frac{(n_{0A}/V_R)_1}{p_1} = \frac{(n_{0A}/V_R)_2}{p_2} = \text{const.} \tag{2.168}$$

The effect of the pressure on product distribution may be analyzed by modeling the system as a process made of successive or successively parallel reactions.

Thus, when using a model consisting of two successive steps, as in scheme (a), the maximum of the intermediary product is given by:

$$z_{max} = r^{1/(1-r)} \tag{2.98}$$

where r is the ratio of the rate constants of the first and second reactions:

$$r = k_1/k_2$$

The time corresponding to this maximum is:

$$\tau_{z_{max}} = \frac{\ln \dfrac{k_1}{k_2}}{k_1 - k_2} \tag{2.96}$$

If the effect of pressure on the two rate constants is known, it is possible in the graph $x = f(\tau)$, to estimate the displacement of the maximum of the intermediary product, as function of the system pressure.

The increase of pressure favors the reactions of alkene polymerization, leading to a decrease of both rate constants. Since, the final product usually contains more alkenes than the intermediate product, k_2 will decrease more than k_1 and r will

---

* Some works ignore this interdependence and arrive in this way to erroneous interpretations concerning the influence of the pressure.

increase. These effects will be apparent only for the cracking under high pressures. In pyrolysis, the polymerization reactions are not possible.

The effect of pressure upon the mechanism of the thermal decomposition discussed before is equivalent to a decrease of the rate constant k_2.

According to Eq. (2.98), the increase in pressure leads to an increase of the yield of intermediary product, and according to the relation (2.96) to an increase of the time until this maximum is obtained (Figure 2.21).

Similar conclusions are obtained by the analysis of the schemes that are modeling the cracking process as successive parallel reactions, with two or three successive decomposition reactions.

With respect to the composition of the products resulting from the cracking process, the increased importance of polymerization and influence of pressure on the decomposition reactions determines the decrease of the unsaturated character of the products, especially of the gases.

In the presence of other favorable process parameters such as high temperatures, a reduced pressure in the process may lead to a strong increase of the aromatization reactions especially by the dehydrogenation of cycloalkanes.

The mentioned effects of the temperature and pressure on the thermal processes, resulted in the selection of two extreme regimes, determined by the objective of the process: a) very high temperatures (limited only by the possibilities of technical implementation) and partial pressures as close as possible to the atmospheric, for pyrolysis processes. Here, the main objective is the production of ethene and propene. b) moderate temperatures (as low as possible, but sufficient for achieving reaction rates conducive to reasonable reactor sizes) and high pressures (limited by economic criteria), when the production of liquid fractions is targeted.

The consequences of these two extreme regimes on the yields and the chemical character of the products, may be illustrated by the following data:

In the pyrolysis of liquid fractions, the gases represent approximately 73% for the pyrolysis of gasoline and about 50% for the pyrolysis of gas oil, whereas for the cracking at higher pressures, the maximal yield is about 12%.

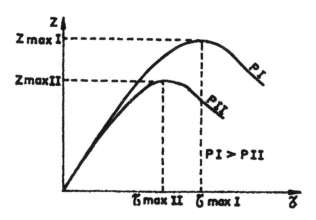

Figure 2.21   The effect of pressure on the maximum of intermediary product – z. (From Ref. 92.)

In the gases from pyrolysis (the fraction C_1-C_4), ethene represents 30–35% by weight, propene 14–18%, butadiene about 5%, and unsaturated hydrocarbons represent around 75%. In the gases produced in the processes operating at high pressures, the alkenes generally represent about 25%.

Comparative data for liquid fractions are gathered in Table 2.15.

2.4.3 Feed composition

The molecular mass and the chemical composition of the feedstock determine the conversion as well as the distribution of the products obtained from thermal processes. This connection between the characteristics of the feedstock and the result of the process is expressed quantitatively by the kinetic equations and the rate constants.

The kinetic constants of some reactions of thermal decomposition are presented in the Table 2.16, for several hydrocarbons.

The table shows that for the same reaction, large differences may exist between the results of various authors for the values of the pre-exponential factors and of the activation energies. This fact was noted by many authors [1]. However, despite these differences, for the temperatures practiced in the pyrolysis processes the Arrhenius-type equation leads to values of the rate constants, close to each other.

This fact results with clarity from the values of the rate constant for the temperature of 1100 K, computed by us and recorded in the last column of the table.

The rate constants of the decomposition of the C_1-C_5- hydrocarbons are presented in Figure 2.22. For higher hydrocarbons they may be determined by means of Figure 2.23, which gives the rate constants for different classes of hydrocarbons and different number of carbon atoms in the molecule, with reference to n-pentane [99].

The following relation is recommended for n-alkanes, beginning with n-hexane [38]:

$$k = (n-2) \times 10^{14} \times e^{-272,000/RT} \; s^{-1}$$

Table 2.15 Chemical Composition (weight percent) of Gasoline Cuts Produced in Thermal Processes

	Cracking	Pyrolysis
Light gasoline EP 60°C		
Alkanes	64.0	30.0
Alkenes	36.0	70.0
Naphtha IBP 120°C		
Alkanes	9.0	17.0
Alkenes	31.0	43.0
Cycloalkanes	55.0	15.0
Aromatics	5.0	25.0

EP: end point;
IBP: initial boiling point.
Source: Ref. 95.

Table 2.16 Kinetic Constants for the Thermal Decomposition of Selected Hydrocarbons

Reaction	A (s^{-1}) or $(L/mol \cdot s)$	E (kJ/mol)	Source	k at 1100 K (s^{-1}) or $(L/mol \cdot s)$
$C_2H_6 \rightarrow C_2H_4 + H_2$	6.04×10^{16}	343.0	96	3.07
	1.46×10^{15}	304.8	98	5.025
	4.65×10^{13}	273.0	97	5.16
$C_3H_8 \rightarrow C_2H_4 + CH_3$	3.2×10^{13}	264.0	96	9.46
	4.69×10^{10}	212.0	97	4.10
	1.6×10^{9}	184.2	44	2.90
$C_3H_8 \rightarrow C_3H_6 + H_2$	2.9×10^{13}	265.0	96	7.60
	5.89×10^{10}	214.7	97	3.80
	1.4×10^{9}	184.2	44	2.54
$C_2H_4 \rightarrow C_2H_2 + H_2$	1.8×10^{13}	318.0	96	1.43×10^{-2}
	6.0×10^{13}	318.2	44	4.75×10^{-2}
$C_2H_6 \rightarrow CH_4 + 1/2C_2H_4$	1.6×10^{12}	280.0	96	8.2×10^{-2}
	4.0×10^{12}	280.5	44	1.9×10^{-1}
$2C_2H_2 \rightarrow C_4'$ and higher	3.3×10^{11}	188.0	96	9.3×10^{2}

The value of the retarding factor β, required for the use of the Eq. (2.84), is 0.90–1.000 for *n*-alkanes and 0.82–0.90 for *i*-alkanes, decreasing with increased degree of branching of the molecule.

Important differences exist between the values of the rate constants that are obtained by using the various equations, graphs and other methods of calculation. Some of the important reasons for these differences are the following:

1. The experimental conditions: the pressure, temperatures, steam dilution, and especially the conversion for which the rate constant was calculated. It is known that in many cases the rate constant decreases with conversion.
2. The material from which the reactor was manufactured (see Table 2.11 and Figure 2.11). For metallic reactors, this may significantly modify the conversion and the selectivity. For a long time, this influence was not considered. One has to assume that in many older studies, the reported results were strongly influenced by the composition of the used steels.
3. The method used to determine the correct reaction time in continuous, nonisothermal reactors.

The errors that may occur as result of the scatter within the values of the kinetic constants may lead to undesired consequences for the design and operation of commercial plants. The consequences may be of two types: first, differences between the computed value of the overall conversion and that actually obtained; second, differences in the distribution of the reaction products.

Because the thermal cracking and the pyrolysis processes are strongly influenced by temperature, the overall conversion may be modified without difficulties

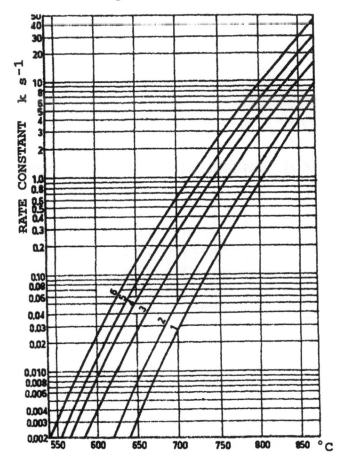

Figure 2.22   Rate constants for the pyrolysis of light hydrocarbons: 1-ethane, 2-propene, 3-propane, 4-*i*-butane, 5-*n*-butane, 6-*n*-pentane. (From Ref. 99.)

by a small change in the temperature at the reactor exit. It must be taken into account that, according to relation (2.158) in visbreaking and delayed coking, which are performed at temperatures around 500°C, the reaction rate doubles for a temperature increase of 12–15°C. In pyrolysis, the doubling of the rate requires an increase of about 25°C. A more important change of the temperature may be limited by the highest value tolerated by the tubes of the furnace. This is, among others, the reason for the special care paid to the selection of the value of the kinetic constants used in the design of the pyrolysis furnaces. This is of lesser importance for processes that use lower temperatures (thermal cracking, visbreaking, delayed coking, etc.).

    The problem of obtaining by calculation the correct distribution of reaction products from the pyrolysis process is very important. This distribution is the starting point for the sizing of the whole downstream system of separation and purification of the products. In addition to the careful selection of the constants, two recommendations should contribute to obtaining correct results. The first: preference

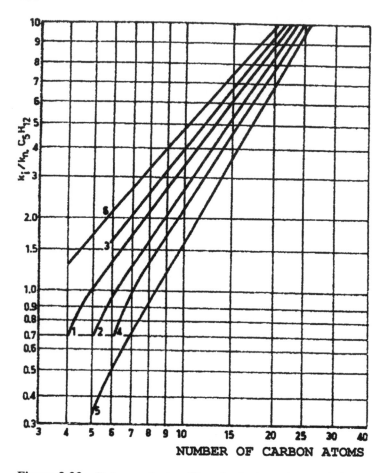

Figure 2.23 Rate constants of heavier hydrocarbons relative to that of *n*-pentane: 1-*n*-alkanes, 2-2-methylalkanes, 3-dimethylalkanes, 4-alkylcyclohexanes, 5-alkylcyclopentanes, 6-*n*-alkenes. (From Ref. 99.)

must be given to the models based on systems of elementary reactions, which cover the steps of initiation, propagation, and termination of the chain over the models based on systems of molecular reactions. In the recommended models, the interdependence of the phenomena of formation of the various products is inherent, and lower errors occur.

The second recommendation is that if possible, for all the reactions considered in the model, the kinetic constants should be by the same author. In this way, all the constants are obtained by using the same procedure of testing and of processing the data. Therefore, they will be influenced to the same extent by the possible errors or assumptions.

At any rate, it is necessary to compare the final product distribution obtained by calculation with the experimental data or with data from the operation of commercial plants.

For the petroleum fractions, the graph of Figure 2.24, depicts values of the reaction rate constants as function of temperature for the overall thermal reactions of some petroleum fractions and of some selected hydrocarbons.

When using this graph, the conversion x must be expressed as follows:

For pure hydrocarbons—in molar fraction of decomposed hydrocarbon
For feedstock used in the thermal cracking for gasoline production—in volume
    fraction of gasoline with the final distillation temperature of 204°C
For residues—in volume fraction of liquid obtained by vacuum distillation

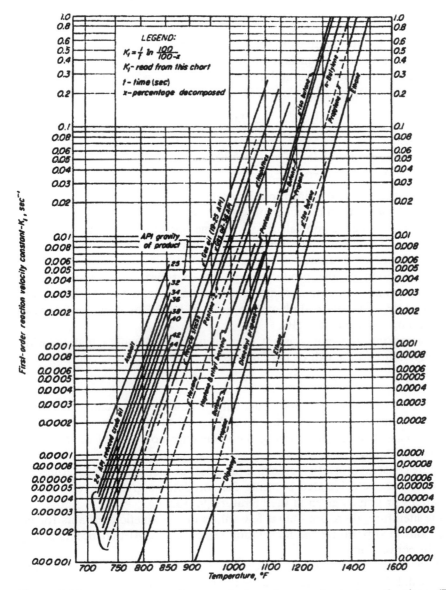

Figure 2.24 Rate constants for some hydrocarbons and petroleum fractions. (From Ref. 100.)

The rate constants determined with this graph are valid, in the pressure range of 0–3.5 bar for ethane, propane, and butane. For other feedstocks, the rate constants correspond to the pressures usually used in the respective processes.

The reported values of the rate constants correspond to average conversions as practiced in the respective industrial processes. The pyrolysis of pure hydrocarbons, is usually carried out at high conversions where the retardation effect is considerable. The rate constants for these conditions are lower than the constants deduced for the incipient phases of the reaction when the retardation does not occur, and higher than those corresponding to the thermal decomposition of the last portions of the feedstock.

For the pyrolysis of gasolines and gas oil, the overall reaction rate constants may be estimated by taking into account their composition in terms of classes of hydrocarbons by means of the graphs of Figures 2.22 and 2.23, which were correlated on the basis of data obtained in industrial furnaces.

Of use may also be the constants in the Arrhenius equation obtained for the following pure hydrocarbons, the boiling temperatures of which allow them to be illustrative for the pyrolysis of gas oils [75].

	$A(sec^{-1})$	$E(kJ)$
n-tetracosan	1.6×10^{14}	243,700
6-methyleicosan	1.2×10^{14}	235,700
dodecylbenzene	1.2×10^{11}	188,400

These constants were obtained in a laboratory unit, using a plug flow reactor passivated by chromium-aluminum plating. The data processing was made for a volume of the reaction zone equivalent to a volume operating at the reference temperature, using a kinetic equation of first order in which the retardation effect and the expansion factor were ignored.

The distribution of the products from the pyrolysis of gasoline may be obtained by assuming data obtained in commercial units, or by way of semi-empirical correlations derived from real plant data, or even better, by using models based on the elementary reaction steps [10].

For visbreaking, the rate constant is given by the following equation, which uses the temperature at the exit of the furnace:

$$k = 1.456 \times 10^{12} \times e^{-213,000/RT}$$

The rate constant is for a first order kinetic equation, the conversion being expressed as volume fraction of the cut with end b.p. of 484°C. The residence time in the furnace is calculated on the basis of the volumetric flow rate of the feedstock, considered to be liquid.

The rate constants for other cracking processes at moderate temperatures and high pressures may be calculated by using the correlations given in Table 2.17 or other equations or graphs given in the literature [92,102].

In the literature, there are only a few values of the kinetic constants for first order reactions incorporating the retardation effect (Eq. 2.84) that occurs at high values of the conversion.

Such data may be obtained by processing by means of the regression method the experimental data. Of practical value is the linearization of Eq. (2.84):

$$k_D = Y - \beta X$$

Table 2.17 Equations for the Rate Constants for Pyrolysis of Selected Petroleum Cuts

Straight run gas oil	$\log k = 13.26 - \dfrac{12{,}470}{T}$	Conversion given as % volume gasoline	[103]
Thermal cracking gas oil	$\log k = 12.88 - \dfrac{12{,}470}{T}$	up to max. 20% conversion	[103]
Atmospheric residue	$\log k = 13.35 - \dfrac{12{,}470}{T}$		[103]
Atmospheric residue	$\log k = 11.822 - \dfrac{11{,}450}{T}$	Conversion expressed as	[104]
Cracking gas oil IBP 350°C	$\log k = 13.768 - \dfrac{12{,}680}{T}$	weight % of products with EP = IBP of feed	[104]

where

$$X = \frac{x}{\tau} \quad \text{and} \quad Y = \frac{1}{\tau}\ln\frac{1}{1-x}$$

For preliminary evaluations, the values of the constants may be obtained by plotting the experimental data in X and Y coordinates.

Another possible way is to obtain the values of the constants k_D for use in Eq. (2.84) from the values of the constants k used in the classical Eq. (2.77). To this purpose, one makes use of Eq. (2.80), that correlates the two constants. The value of the constant β is estimated and to the conversion x, a value is given contained within the range of validity of the equations with and without retardation (for example $x = 0.1$). The value of the constants k_D may be now computed. The constant β may be also estimated on the basis of the values given at the beginning of this section.

A similar method may be used with respect to the constants k_0 and α required for using Eq. (2.37).

For the kinetic calculations, where the cracking process is represented by successive or successive-parallel reactions, the rate constants are determined in the same manner, separately for each reaction step. When selecting the method for computing the constants one must pay attention that the units used for expressing the conversion, for which the selected rate constant is valid, should be compatible with the kinetic reactions used. One must also take into account whether the product for which the rate constant is determined, has undergone or not a previous cracking. Thus, the constant k_1 in the kinetic scheme (c) of Section 2.3.5 for the cracking of residue (or coking) refers to a feedstock that has not undergone a previous cracking. It must be valid for the expression of the conversion as the totality of decomposition products, that are lighter than the residue. The constant k_2 from the same kinetic scheme, refers to a cracked gas-oil fraction and is valid when the conversion is expressed as a fraction by weight of produced gasoline plus gases. Analogously, the constant k_3 must be valid for the conversion of cracked gasoline to gases.

Since it is difficult to find the values of kinetic constants when the conversion is expressed in these units, even the few published data concerning the ratio of the rate constants may be useful. Thus, the ratio between the rate constants for the formation and decomposition of gasoline was determined [94], finding the value of 2.35 in the case when the feedstock is a straight run residue and 1.8 for paraffin wax. The ratio of the constants can be established also on the basis of the maximum of intermediary product, as determined experimentally, using the Eq. (2.98) or (A2.15). In the latter case it is necessary to determine the ratio between the constants k_1 and k_2, from experimental data.

The given graphs and equations allow one to draw some conclusions concerning the variation of the kinetic constants with the composition of the petroleum fractions. The rate constant increases with the distillation limits and with the characterization factor of the feedstock. Therefore it will be larger for the heavier fractions and for those, which have a more pronounced paraffinic character. For the cracking products, the constant will have a lower value for cracked products than for the uncracked fractions having the same distillation limits.

Even from the scarce available experimental data it may be concluded that the value of the activation energy decreases with the increase of the boiling temperature and with the characterization factor of the fraction submitted to cracking. Thus, for three straight run cuts, separated from the Surahani crude oil, the obtained values are [90]:

for the fraction 172–215°C E = 261.5 kJ/mole

for the fraction 266–288°C E = 246.0 kJ/mole

for the fraction 380–415°C E = 232.0 kJ/mole

For products that were previously submitted to cracking, the activation energy is higher than for the fractions with the same boiling limits that were not cracked.

The values of the activation energies that may be recommended for calculation are:

for straight run   E = 232 kJ/mole

for naphtha   E = 270–290 kJ/mole

Knowledge of the variation of the values of the rate constants and of the activation energy as a function of the chemical character of the feedstock allow one to analyze the influence of its quality on the maximum weight fraction of intermediary product.

For the successive process with two reaction steps:

$$A \xrightarrow{k_1} v_1 B \xrightarrow{k_2} v_2 C$$
$$\quad y \qquad z \qquad u$$

the value of the constant k_1 depends on the characteristics of the feedstock, whereas the constant k_2 for the decomposition of the cracked product will not depend on the nature of the feedstock.

Therefore it follows that for a more paraffinic feed stock, the ratio k_1/k_2 will be higher and, according to relation (2.98) the maximum of the intermediary product will be higher. The time interval necessary for reaching this maximum will be shorter according to Eq. (2.96), since one of the constants has a higher value than the other one.

On basis of these conclusions, the curves of formation of an intermediary product (gasoline or gas oil) are plotted in Figure 2.25 for two feedstocks having the same distillation limits but different chemical character.

For feedstocks consisting of residue, the content of resins and asphaltenes is high and an intense formation of carboids begins much before the maximum for the first intermediary product—gas oil—is reached. The formation of insoluble material is generally dependent on the content of asphaltenes, resins, and polycyclic hydro-carbons in the feedstock (Figure 2.25). For this reason, in the visbreaking process and generally during the cracking of the residues in tubular furnaces, where the intense formation of carboids leads to the coking of the tubes, the weight fraction of the obtained gas oil depends on the position of the curve of carboids. Figure 2.25 shows the curves of the formation of carboides and gas oil, indicating the highest weight fractions of gas oil that can be obtained during cracking in a tubular furnace of a feedstock paraffinic-slightly resinous character (1) in comparison with a feed-stock with naphthene-aromatic character and a high content of resins and asphaltenes (2).

In industrial practice, the highest conversion that may be obtained in the furnace during the cracking of a residue is expressed usually as a percentage of gasoline produced. Typical conversions are of: 6–10% from straight run residue and 2–3% from heavy residues.

Recall that the position of the curve for the formation of carboids does not depend only on the constituents of the feedstock and on its general chemical character, but also on the operating conditions of the process. Thus, a higher pressure in the reaction system may sometimes delay the formation of the carboids (Section 2.4.2). A similar result is caused by the introduction of steam (Section 2.4.4).

The formation of carboids does not limit the conversion for cracking performed in systems with a solid heat carrier in moving or fluidized beds.

2.4.4 The influence of steam introduced in the reactor

Water vapor (steam) introduced into the reactor has a complex effect on thermal processes. Steam acts firstly as an inert diluent, decreasing the partial pressures of all components of the reaction. The effect is analogous to decreased system pressure

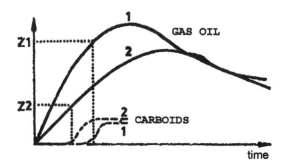

Figure 2.25   The variation of the conversion to gas-oil for two feeds having different chemical character: 1-paraffinic, low in resins and asphaltenes; 2-naphthenic—aromatic, with more resins and asphaltenes. (From Ref. 92.)

with all the consequences on the displacements of the thermodynamic equilibriums discussed in Section 2.4.2.

Second, steam is usually being introduced into the last sector of the furnace tubes, where it causes a significant increase of turbulence. This prevents the deposit of asphaltenes on tubes walls that possibly begin to be separated from the colloidal solutions. In this manner phenomena leading to coking of the tubes are prevented.

This effect is especially important in the processes of residue cracking—visbreaking and delayed coking. In this process, steam increases the conversion to the first intermediary product (see the Figure 2.25) and increases the length of the cycle between two decoking operations. Water was introduced to this purpose in the cracking furnaces of straight run residue for the first time by G.C. Suciu in the Standard Oil refinery of Teleajen-Romania in 1939, much before this method was used elsewhere in the world or mentioned in the literature. The effect is very important in delayed coking because it significantly improves process performances.

In the pyrolysis process the action of the steam is more complex. The main effect is the fact that for the same final pressure at the exit from the coil, by decreasing the partial pressures of the hydrocarbons the steam displaces the equilibrium towards the formation of supplementary amounts of ethene, propene, butadiene, isoprene, and aromatic monocyclic hydrocarbons, by means of the very products that constitute the objective of the process.

Figure 2.26 depicts a typical example for the yield increase for ethene, or ethene and propene, as the ratio steam/feedstock, is increased.

Note that the exit pressure in the pyrolisis furnaces cannot be decreased below some limits. These limits are determined by the aspiration pressure of the compressors and by the pressure drop in the succession of cooling and separation equipment situated between the exit from the furnaces and the compressor group. The introduction of steam leads to the decrease of partial pressures, therefore having an effect similar to that which could be obtained if it were possible to reduce the pressure below the limits imposed by the system.

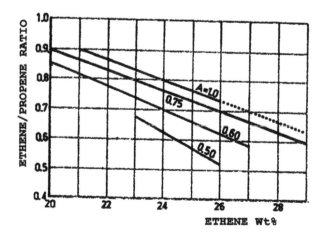

Figure 2.26   The influence of the steam/feedstock weight ratio (A) on the yield of ethene and the ethene/propene ratio.

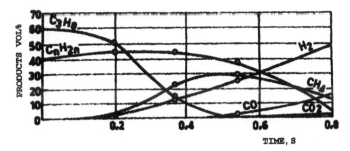

Figure 2.27 Formation of carbon monoxide and dioxide during propane pyrolysis at 850°C, at molar ratio $H_2O/C_3H_8 = 4$. (From Ref. 106.)

Concerning the coking of the tubes of the furnaces, the favorable effect of introducing steam in the tubes is similar to that described for cracking residual fractions.

An additional effect appears at temperatures above 650–700°C, namely the formation of carbon monoxide and dioxide, proving that above these temperatures, water participates in chemical reactions.

Studies on cracking of gas oil, ethane, propane, and other hydrocarbons in the presence of steam, led to the assumption [105] that water vapors do not react directly with hydrocarbons contained in the feedstock, but with some products of their decomposition. This assumption was confirmed by the fact that carbon oxide and dioxide are formed only after an advanced degree of decomposition of the feedstock is achieved (Figure 2.27) [106].

The fact that carbon monoxide and dioxide appear only when reactions take place at high temperatures led to the assumption that the steam reacts only with some of the hydrocarbons formed at these high temperatures, such as acetylene.

Despite all these findings, the mathematic modeling of the pyrolysis processes usually leads to a more simple solution, which is to consider that carbon monoxide and dioxide are the result of the interaction of steam with the deposits of carbon, according to the reactions:

$$C + H_2O \rightarrow CO + H_2$$
$$C + 2H_2O \rightarrow CO_2 + 2H_2$$

Such a modeling, although contrary to some findings that support the assumption that steam may act during the formation of the deposits but before their actual production, constitutes nevertheless a satisfactory solution from the practical point of view and for this reason it is generally accepted.

REFERENCES

1. V Vintu, R Mihail, V Macris, G Ivanus. Piroliza hidrocarburilor, Editura Tehnica, Bucuresti, 1981.
2. IM Pauschin, TP Visniakova. Proizvodstvo olefinosoderjascih i goriutchih gazov, Ed. AN SSSR, Moskva. 1960.
3. S Raseev, K Clemens. Rev Chim 14:123, 1963.
4. W Tsang, J Phys Chem 76:143, 1972.

5.   W Tsang, Int J Chem, Kinet. 5:929, 1973.
6.   AD Stepuhovici, Ulitchi VA, Kinetica i termodinamica radicalnyh reactii crekinga, Himia, Moskva, 1975.
7.   BE Knox, HE Palmer, Chem. Rev. 61:247, 1961.
8.   LV Gurvici et al., Termodinamiceskie svoistva individualnih vescestv, Ed AN SSSR, Moscva, 1962.
9.   JA Kerr, JG Colvert, J Amer Chem Soc 83:3391, 1961.
10. TL Cottrel, The Strengths of Chemical Bonds, London: Butterworths Sci. Publ., 1954.
11. VI Vendeneev, LV Gurvici, VN Kondratiev, VA Medvedev, EL Frankevici. Energhiia razryva khimiceskih sviazei, potentsialy ionizatsii i srodstvo k electronu, Ed. AN SSSR, Moskva, 1962.
12. CC Fettis, AF Trotman-Dickenson, J Amer Soc 81:5260, 1959.
13. M Szwerc, Chem Rev 47:75, 1950.
14. CA McBowel, Can J Chem 34:345, 1956.
15. AH Schon, M Szwarc. Am Rev Phys Chem 8:439, 1957.
16. RZ Magaril, Mehanizm i Kinetica gomaghennyh termiceskih prevrashcenii uglevodorodov, Himia, Moskva, 1970.
17. B Steiner, CB Giese, MG Inghram, J. Chem Phys 34:189, 1961.
18. NN Semenov, 0 nektoryh prohlemah himiceskoi kinetiki i reactionnoi sposobnosti, Ed. AN SSSR, Moskva, 1958.
19. AH Sehon, BJ Danvent, J Amer Chem Soc 76:4806, 1954.
20. JL Franklin, HE Lumpkin, J Amer Chem Soc 74:1023, 1952.
21. EH Braye et al, J Amer Soc 77:5282, 1955.
22. VH Dibeler et al., J Amer Chem Soc 81:68, 1959.
23. AB Trenwith, J Chem Soc Faraday Trans. I 75:614, 1979.
24. TC Tsai, Z Renjun, Z Sin, Oil Gas J 85, Dec. 21:38, 1987.
25. GF Froment, BO Van de Stoone. Oil Gas J 77, (nr. 10):87, 1979.
26. G Pratt, D Rogers, J. Chem Soc Faraday Trans. I. 75:1089, 2688, 1979.
27. KM Sundaram, GF Froment, Ind Eng Chem Fundam 17(3):174, 1978.
28. T Kunugi, T Sakai, K Soma, Y Sasaki, Ind Eng Chem Fundam 8:374, 1969.
29. GGJ Eltenton, J Chew Phvs 15:445, 1947.
30. FP Lossing et al., Faraday Discuss Chem Soc 14:34, 1953.
31. AJB Robertson, Proc Roy Soc, London A199:394, 1949.
32. FO Rice, KP Herzfeld, J Amer Chem Soc 56:284, 1934.
33. DL Altara, D Edelson, Int J Chem Kinet 7:479, 1975.
34. P Camileri, KM Marshall, JH Purnell, J Chem Soc Farday Trans I 71:1491, 1975.
35. E Ranzi, M Dente, S Pierucci, G Bardi, Ind Eng Chem Fundam 22:132, 1983.
36. F Christonher, F McConnell, BD Nead. In: Pyrolysis Theory and Industrial Practice, LF Albright, ed. New York: Academic Press, 1983, p. 25.
37. Bradley JN, J Chem Soc Faraday Trans 1, 75:2819, 1979.
38. RZ Magaril, Teoreticheskie ositovy himiceskoi pererabotki nefti, Himiia, Moskva, 1976.
39. J Shabtai, R Ramakrishnan, AG Oblad. In: Thermal Chemistry, Oblad AG, Davis HG, Eddinger RT Eds, p. 293. Advances in Chemistry Series 183, American Chemical Society, Washington, 1979.
40. C Rebick. In: Pyrolysis Theory and Industrial Practice, LF Albright, Ed. Academic Press, New York, 1983, p. 70.
41. A Kossiakoff, FO Rice, J Amer Chem Soc 65:590, 1943.
42. R Zou, Q Lou, H Lin, F Niu, Ind Eng Chem Res 26:2528–2532, 1987.
43. HG Davis, VD Williamson, Proc. 5th World Pet. Congr. New York, 1959, sect. IV, p. 37.
44. AG Volkan, GC April, Ind Eng Chem Process Des Dev 16:429, 1977.

45. LF Albright, Ind Eng Chem Process Des Dev 17:377, 1978.
46. MC Lin, NH Back, Can J Chem 44:2369, 1968.
47. J Grotewold, JA Kerr, J Chem Soc:4337, 4342, 1963.
48. EL Metcalfe, J Chem Soc:3560, 1963.
49. RJ Kominar et al, Can J Chem 45:575, 1967.
50. RD Powers, WH Corcoran, Ind Eng Chem Fundam 13:351, 1974.
51. WK Bustfeld, KJ Irvin, H Mackie, PAG O'Hare, Trans Farad Soc 57:106, 1961.
52. DR Stull, EF Westrum Jr, GC Sinke, The Chemical Thermodynamics of Organic Cnmpoimils, John Wiley, New York, 1969.
53. PS Virk, A Korosi, HN Woebcke, In: Thermal Hydrocarbon Chemistry, Advances in Chemistry Series 183, AG Oblad, HG Davis, RT Eddingerr eds, Amer Chem Soc, Washington, 1979, p. 67.
54. P Chaverot, M Berthelin, E Freund, Rev Inst Fr Petrol 41, nr. 4, 529, nr. 5:649 (1986).
55. B Bettens, F Ninauve, Valoisation chimique et physique du charbon, Commission of The European Community, Brussels, Nov. 1980, p. 125.
56. H Juntgen, Fuel 63:731, 1984.
57. IC Lewis, LS Singer. In: Chemistry and Physics of Carbon, vol. 17, PL Walker, PA Thrower, eds. New York: Marcel Dekker, Inc., 1981, p. 1.
58. SA Caganov, MH Levinter, MJ Medvedeva, Chim Tekhn Top Masel, 7(7):38, 1962.
59. JPh Pfeiffer, RNJ Saal, J Phys Chem 44:129, 1940.
60. CPG Lewis et al., Oil Gas J 83, Apr 8:77, 1985.
61. Thermal Stability of U.S. Navy Special Fuel Oil, ASTM D 1661.
62. JF Le Page, SG Chatila, M Davidson, Raffinage et conversion des produifs lourds du petrole, Technip, Paris, 1990.
63. WL Nelson, JD McKinney, DE Blasser, In: Encyclopedia of Chemical Processing and Design, vol. 10, McKetta JJ, Cunningham WA, eds, Marcel Dekker Inc., New York, 1979, pp. 1–41.
64. FM Dautzenberg, JC De Dekken, Prepr Am Chem Soc Div Pet Chem 30(1):8, 1985.
65. JF Le Page, M Davidson, Rev Inst Fr Ptrol, 41(1):131, 1986.
66. EWR Steacie, S Bywater. The Chemistry of Petroleum Hydrocarbons, Chapter 2, Broocks et al., eds., Chap. 22, New York: Reinhold Publishing Co., 1955.
67. N Emmanuel, D Knerre, Cinetique Chimique, Mir, Moscow, 1975.
68. T Erdey-Gruz, G Schay, Chimie fizica teoretica, vol. 1, Ed. Tehnica, Bucuresti, 1957.
69. SW Benson, The Foundations of Chemical Kinetics, New York: McGraw-Hill Book Co., 1960.
70. AD Stepuhovici, VA Ulitschi, J Fiz Chim SSSR 37:689, 1963, Usp Chim 35:487, 1966.
71. S Schepp, KO Kutschke, J Chem Phys, 26:1020, 1957.
72. GB Kistiakowski, WW Ranson, J Chem Phys, 7:725,1939.
73. WH Corcoran, In: Pyrolisis Theory and Industrial Practice, LF Albricht ed, New York: Academic Press, 1983.
74. SK Layokun, DH Slater, Ind. Eng. Chem. Process. Des. Dev, 18:232, 1979.
75. B Biouri, J Giraud, S Nouri, D Herault, Ind Eng Chem Process Des., Dev., 29:307, 1981.
76. JE Blakermore, JR Barker, WH Corcoran, Ind Eng Chem Fundam 12:147, 1973.
77. CW Gear, Commun., ACM 14(3):176, 1971.
78. CW Gear, Numerical Initial Value Problems, In: Ordinary Differential Equations, Engelwood Cliffs, New Jersey: Prentice Hall, Inc., 1971.
79. KM Sundaram, GF Froment, Chem Eng Sci 32:609, 1977.
80. KM Sundaram, GF Froment, Chem Eng Sci 32:601, 1977.
81. MC Lin, MH Back, Can J Chem. 44:505, 1966.
82. R Zou, J Zou, Ind Eng Chem Process Des Dev, 25(3):828, 1986.
83. G Froment, B Van der Steene, P Berghe, A Goossens, AJChE J 21(1):93, 1977.

84. AJ Dimes, AV Frost, Dokl Akad Nauk SSSR 3:510 (1934).
85. SK Layokun, DH Slater, Ind Eng Chem Process Des Dev 18(2):236, 1979.
86. SK Layokun, Ind Eng Chem Process Des Dev 18(2):241, 1979.
87. LF Albricht, JC Marek, Ind Eng Chem Res 27:751, 1988; 27:755, 1988.
88. AG Buekens, GF Froment, Ind Eng Chem Process Des Dev. 7:435, 1968.
89. AJ Dintes, AV Frost, J Obshch chim SSSR 6(1):68, 1936.
90. SN Obreadcicov, Tehnologia nefti vol. 2, Gostoptehizdat, Moskva, 1952.
91. SA Kiss, Ind Eng Chem 23:315, 1931.
92. S Raseev, Procese distructive de prelucrare a titeiului, Ed. Tehnica Bucuresti, 1964.
93. OA Hougen, KM Watson, Chemical Process Principles, vol. 3, New York: John Wiley, 1947.
94. DJ Orocico, Teoreticeskie osnovy vedeniia sintezov jidkih topliv, Gostoplehizdat, Moscow-Leningrad, 1951.
95. EV Smidovici, Sovremenye melody termicheskogo crekinga, Gostoptehizdat, Moscow-Leningrad, 1948.
96. MJ Shah, Ind Eng Chem 59(5):71 (1967).
97. GF Froment, BO Van de Steene, PS Van Damme, Ind Eng Chem Process Des Dev 15,(4):495, 1976.
98. V Illes, L Pleszkats, L Szepsy, Arta Chim Hung 80;267, 1974.
99. SB Zdonik, EJ Geen, LP Hallee, Manufacturing Ethylene, Tulsa, Okla: The Petroleum Publishing Co., 1971.
100. WL Nelson, Petroleum Refinery Engineering, New York: McGraw-Hill Book Co., 1958.
101. GC Suciu, Progrese in procesele de prelucrare a hidrocarburilor, Editura Tehnica, Bucuresti, 1977.
102. JH Hirsh, EK Fischer, The Chemistry of Pletroleum Hydrocarbons, BT Brooks, ed., vol. 2, Reinhold Publ. Co, New York, 1955.
103. WL Nelson, Oil Gas J, 38, Oct., 1939.
104. NL Barabanov et al., Chim Tehn Top Masel, 4(8):19, 1959.
105. AS Gordon J Amer Chem Soc 70:395, 1948; Ind Eng Chem 44:1587, 1952.
106. MF Guyomard, J. Usines Gas (6):210, 1954.
107. Y Jian, E Semith, IECh., 36:574–584, 1997.
108. Y Jian, E Semith, IECh., 36:585–591, 1997.
109. HR Linder, RE Peck, Ind Eng Chem 47(12):2470, 1955.
110. G Froment, BO Van de Steene, PS Damme, S Narayanan, AG Crossens, Ind Eng Chem Process Des Dev 15, nr. 4:495–504, 1976.
111. RH Snow, HC Schmitt, Chem Eng Prog, 53(3):133-M, 1957.
112. S Raseev, C Iorgulescu, D Besnea, Revista de Chimie, 31, nr. 4, 1980.
113. HG Davis, KD Williamson, In: Thermal Hydrocarbon Chemistry, Oblad AG, Davis HG, Eddinger RT, eds. Advanced in Chemistry Series 183, American Chemical Soc., Washington 1979.
114. Ch Rebic, Pyrolysis of Heavy Hydrocarbons, In: Pyrolysis Theory and Industrial Practice, 87 (75), Academic Press, 1983, p. 67–87.

3

Reaction Systems

The reaction systems used in all cracking processes, i.e. thermal, catalytic, and hydrocracking, have specific features. These features follow from the fact that in all these processes the desired product is an intermediary in a number of successive and parallel decomposition reactions.

In order to maximize the yield of the desired product, special attention is given to

The type of reactor selected

The reaction systems i.e., the equipment that directly influences the operation of the reactor or reactors

3.1 SELECTION OF REACTOR TYPE

The flow conditions in all the reactors for continuous processes are situated between the limiting cases: the ideal plug flow reactor and the ideal perfectly mixed reactor.

In order to examine the problem of selecting the best suited reactor type, we will refer to a process consisting of two successive steps:

$$A \xrightarrow{k_1} v_1 B \xrightarrow{k_2} v_2 C$$
$$y \quad z \quad u$$

For the ideal plug flow reactor, the equation for the maximum of the intermediary product was derived in section 2.3.5:

$$z_{max} = r^{1/(1-r)} \tag{2.98}$$

where $r = k_1/k_2$.

For the perfectly mixed reactor, since the flow leaving the reactor has the same composition as the reactor content, we will have for the above kinetic scheme the relations [1]:

$$\frac{1-y}{V_R/V_e} = k_1 y \tag{3.1}$$

$$\frac{z}{V_R/V_e} = k_1 y - k_2 z \tag{3.2}$$

$$\frac{u}{V_R/V_e} = k_2 z \tag{3.3}$$

From (3.1) it follows:

$$y = \frac{1}{k_1 \dfrac{V_R}{V_e} + 1} \tag{3.4}$$

which by substitution in (3.2), and following a series of simplifications and transformations, gives the expression:

$$z = \frac{\dfrac{V_R}{V_e}}{\left(\dfrac{V_R}{V_e} + \dfrac{1}{k_1}\right)\left(1 + k_2 \dfrac{V_R}{V_e}\right)}$$

or:

$$z = \frac{\dfrac{V_R}{V_e}}{k_2 \left(\dfrac{V_R}{V_e}\right)^2 + \left(\dfrac{k_2}{k_1} + 1\right)\left(\dfrac{V_R}{V_e}\right) + \dfrac{1}{k_1}} \tag{3.5}$$

By differentiation of this expression and equating to zero, it results:

$$\frac{dz}{d\left(\dfrac{V_R}{V_e}\right)} = \frac{k_2 \left(\dfrac{V_R}{V_e}\right)^2 + \left(\dfrac{k_2}{k_1} + 1\right)\left(\dfrac{V_R}{V_e}\right) + \dfrac{1}{k_1} - 2k_2 \left(\dfrac{V_R}{V_e}\right)^2 - \left(\dfrac{k_2}{k_1} + 1\right)\left(\dfrac{V_R}{V_e}\right)}{\left[k_2 \left(\dfrac{V_R}{V_e}\right)^2 + \left(\dfrac{k_2}{k_1} + 1\right)\left(\dfrac{V_R}{V_e}\right) + \dfrac{1}{k_1}\right]^2} = 0$$

The denominator of this equation cannot be equal to zero. By setting the numerator equal to zero and making simplifications, the condition that corresponds to the maximum of the intermediary product z_{max} is:

$$\frac{1}{k_1} - k_2 \left(\frac{V_R}{V_e}\right)^2 = 0,$$

or

$$\frac{V_R}{V_e} = \sqrt{\frac{1}{k_1 k_2}} \tag{3.6}$$

Replacing this value in (3.5), after performing simplifications it follows:

$$z_{max} = \frac{1}{\left(1 + \sqrt{\frac{k_2}{k_1}}\right)^2} \qquad (3.7)$$

This equation gives the maximum yield of intermediary product that can be obtained in a perfectly mixed reactor.

The values of z_{max} for the two types of reactors for different values of the ratio k_1/k_2 are given in Table 3.1.

The values in this table lead to the conclusion that when the target is to maximize the yield of the intermediary product in a process consisting of successive steps, the reactor should be as close as possible to the ideal plug flow reactor.

The tubular furnaces used in thermal processes satisfy this requirement.

In the fluid catalytic cracking (FCC), the flow conditions in the riser approach the conditions within an ideal plug flow reactor.

If the reactor is a cylindrical vessel, a shape with the highest practical ratio height/diameter should be selected. Also, internal baffles may be used in order to decrease internal mixing.

The data of Table 3.1 show that, even in the case when the decomposition of the intermediate product seems at first glance to be unimportant—since the value of the rate constant k_2 is much smaller than k_1—the effect upon the maximum yield of the intermediary product is significant.

Table 3.1 z_{max} in Ideal Plug Flow and Perfectly Mixed Reactors

k_1/k_z	Ideal plug flow reactor $z_{max}D$	Ideal perfectly mixed reactor $z_{max}A$	$\dfrac{z_{max}D}{z_{max}A}$
0.05	0.043	0.033	1.30
0.1	0.077	0.058	1.33
0.2	0.134	0.095	1.41
0.3	0.179	0.125	1.43
0.5	0.250	0.172	1.45
1.0	0.368	0.250	1.47
2.0	0.500	0.343	1.46
4.0	0.630	0.444	1.42
7.0	0.723	0.527	1.37
10.0	0.774	0.577	1.34
15.0	0.824	0.632	1.30
20.0	0.854	0.668	1.29
30.0	0.889	0.715	1.24
50.0	0.923	0.768	1.20
80.0	0.946	0.809	1.17
100.0	0.955	0.826	1.16

A good example is the pyrolysis of ethane for the production of ethene. By noting with k_1 the rate constant for the formation of ethene from ethane and with k_2 that for the decomposition of ethane to acetylene, the maximum conversions to ethene were computed in Table 3.2.[*] for temperatures of 850°C and 950°C.

The values in the table clearly show the significant increase of the theoretical yield for ethene if the ideal plug flow reactor is used. This explains the exclusive use of tubular furnaces for pyrolysis at a commercial scale and the difficulty to replace them with other types of reactors.

In catalytic processes such as catalytic cracking or hydrocracking, the reaction begins when the feed comes in contact with the catalyst and ends when the products separate from the catalyst.

In thermal processes, the chemical transformation begins when the feed temperature exceeds a certain level and the reaction rate increases with the increase of temperature. Owing to high values of the activation energy, the value of the reaction rate doubles for a temperature increase of only 12–25°C. The following consequences result:

1. The reaction rate is insignificant during the heating of the feed, until the temperature has risen to 100–150°C below that corresponding to the exit from the furnace (the maximum value of 150°C corresponds to pyrolysis);
2. The final reaction temperature should be controlled with high precision in order to maintain the conversion in the desired range.
3. In order to stop the reaction, it is necessary to resort to a sudden cooling (quenching) of the reactor effluent. Its temperature is reduced suddenly by about 80–150°C below that at the exit from the reactor.

The thermal processes are strongly endothermic, with reaction heats situated in the range 250–1500 kJ/(kg feed). The lower values are for processes at high pressures, where the strongly exothermic secondary polymerization reactions partially compensate for the endothermal effect of the decomposition reactions. The largest indicated values are for the pyrolysis process.

The endothermic character of these reactions requires a continuous heat input to the reaction zones. The tubular furnace ensures the necessary heat flux,

Table 3.2 Maximum of Intermediary Product in Ethane Pyrolysis

	850°C	900°C
k_1	9.42 s^{-1}	32.92 s^{-1}
k_2	0.0968 s^{-1}	0.4135 s^{-1}
k_1/k_2	97.3	79.6
z_{max} for plug flow reactor	0.954	0.946
z_{max} for perfectly mixed reactor	0.824	0.809

---

[*] For the calculation of the constants, the most recent data from Table 2.16 were used.

that may reach high values in pyrolysis, since a maximum for the temperature at the exit from the reaction zone ensures a maximum for the production of ethene and propene.

3.2 REACTION SYSTEMS

The term "reaction system" stands for all the equipment items that serve the transformation of the feed to products. The reaction system is preceded by equipment for the operations for feed preparation and it is followed by equipment for the separation of the obtained products.

The reaction system may comprise one or more reactors and the equipment for interconnecting them.

The most simple continuous system is that in which the feed enters the reactor, passes through it, and leaves it as a mixture of reaction products with unreacted feed. Such a system is called a "once-through" system.

In most cases such a system is not completely satisfactory since a portion of feed remains unreacted. The reasons for the limited conversion may be the decomposition of the intermediary product or some chemical phenomena that prevent the increase of the conversion, such as the formation of coke.

For these reasons, in order to increase the overall yield one submits the unreacted portion of the feed to repeated cracking operations while removing before the repetition of those reaction products, that hindered the continuation of the process.

The repetition of the cracking operation may be made each time in a new zone of reaction, in which case the cracking system in its totality is called a "successive" system, or in the same reactor, in which case it is named a "recycling" system.

The solution of recycling in the same reactor is usually more advantageous from an economic point of view, since the investments are lower than in successive system, and the transformation of the feedstock to final products is complete. The recycling may not be used when the reaction conditions required for processing the recycled product are different from those for processing the feedstock. In such cases, the solution is to use the successive system or a combination of the successive and recycling systems.

Such a situation, where the conversion of the unreacted portion requires a more severe regime, is depicted by the data of Table 2.14, which shows a significant decrease of the reaction rate as the cracking operation is repeated. Since each successive repetition of cracking requires for this case a higher temperature, the recycling in the same reactor would be inefficient.

In general, when the feedstock has a homogeneous composition, the stability of the nondecomposed portion is similar to that of the feedstock and may be recycled in the same reactor. When the feedstock is not homogeneous the more reactive components are converted faster, the nonreacted portion is more stable than the feed, and its recycling in the same reactor is inefficient. Such a situation occurs as a rule when the product or a portion of it has lower distillation limits than the feedstock, and is in fact a decomposition product. In such cases, the high-conversion cracking requires another reactor with a more severe regime. This leads to the necessity to use successive or combined systems. Such situations appear not only in thermal cracking, but also in catalytic cracking and hydrocracking.

The equations of the material balance of a successive system with n reaction zones is given in Table 3.3.

The total capacity of the reaction zones per unit mass of feedstock introduced in the process may be expressed, according to the notations of Table 3.3, by the equation:

$$\sum_{i=0}^{n-1} A_i = 1 + y_{\mathrm{I}} + y_{\mathrm{I}}y_{\mathrm{II}} + y_{\mathrm{I}}y_{\mathrm{II}}y_{\mathrm{III}} + \cdots + \prod_{i=0}^{n-1} y_i \tag{3.8}$$

The mass fraction of feed left untransformed at the end to the final products is given by the equation:

$$A_n = \prod_{i=0}^{n} y_i \tag{3.9}$$

The total quantities of products per unit mass of feed will be:

$$\sum_{i=1}^{n} B_i = z_{\mathrm{I}} + y_{\mathrm{I}}z_{\mathrm{II}} + y_{\mathrm{I}}y_{\mathrm{II}}z_{\mathrm{III}} + \cdots + z_n \prod_{i=0}^{n-1} y_i$$

$$\sum_{i=1}^{n} C_i = u_{\mathrm{I}} + y_{\mathrm{I}}u_{\mathrm{II}} + y_{\mathrm{I}}y_{\mathrm{II}}u_{\mathrm{III}} + \cdots + u_n \prod_{i=0}^{n-1} y_i$$

$$\sum_{i=1}^{n} D_i = v_{\mathrm{I}} + y_{\mathrm{I}}v_{\mathrm{II}} + y_{\mathrm{I}}y_{\mathrm{II}}v_{\mathrm{III}} + \cdots + v_n \prod_{i=0}^{n-1} y_i \tag{3.10}$$

$$\sum_{i=1}^{n} E_i = w_{\mathrm{I}} + y_{\mathrm{I}}w_{\mathrm{II}} + y_{\mathrm{I}}y_{\mathrm{II}}w_{\mathrm{III}} + \cdots + w_n \prod_{i=0}^{n-1} y_i$$

In order to use Eqs. (3.8–3.10) one needs to know the conversions in each reactor step. These may be calculated by using the kinetic constants corresponding to each step.

The difficulties of the calculation consist in the correct estimation of the rate constants for each reaction step. The problem is simplified by the fact that the products B and C are cracked products, irrespective of the steps from which they resulted. Accordingly, the values of the constants k_2 and k_3 may be considered, for simplification, identical for all the steps. The situation is different for the rate constants k_1 and k_4. The constant k_1 will decrease significantly from step to step. This decrease may be evaluated on the basis of some experimental determinations or published data.

Concerning the constant k_4, if the feedstock A is a residue, and E the coke, the variation of this constant is more difficult to specify.

Economic reasons impose that in the commercial practice of successive cracking of petroleum fractions, not more than two successive steps are used. For the first step, the kinetic constants are those corresponding to the feedstock. The values of the constants for the second step may be determined on the basis of the characteristics of the nonconverted portion of the feed that resulted from the first step.

If the feedstock is a distillate, the characteristics and the rate constants for the portion of the feed not transformed in the first step may be determined by means of

Table 3.3 Material Balance for a System of Successive Reactions

$$E \xrightarrow{k_4} A \xrightarrow{k_1} B \xrightarrow{k_2} C \xrightarrow{k_3} D$$

E	A	B	C	D
$E_0 = w_0 = 0$	$A_0 = y_0 = 1$	$B_0 = z_0 = 0$	$C_0 = u_0 = 0$	$D_0 = v_0 = 0$
$E_I = w_I$	$A_I = y_I$	$B_I = z_I$	$C_I = u_I$	$D_I = v_I$
$E_{II} = y_I w_{II}$	$A_{II} = y_I y_{II}$	$B_{II} = y_I z_{II}$	$C_{II} = y_I u_{II}$	$D_{II} = y_I v_{II}$
$E_{III} = y_{II} w_{III}$	$A_{III} = y_I y_{II} y_{III}$	$B_{III} = y_I y_{II} z_{III}$	$C_{III} = y_I y_{II} u_{III}$	$D_{III} = y_I y_{II} v_{III}$

$$A_{n-1} = y_n \prod_{i=0}^{n-1} y_i$$

E	A	B	C	D
$E_n = w_n \displaystyle\prod_{i=0}^{n-1} y_i$	$A_n = \displaystyle\prod_{i=0}^{a} y_i$	$B_n = z_0 \displaystyle\prod_{i=0}^{n-1} y_i$	$C_n = u_n \displaystyle\prod_{i=0}^{n-1} y_i$	$D_n = v_n \displaystyle\prod_{i=0}^{n-1} y_i$

graphic correlations [2]. For residual feedstocks, the estimation is more difficult because the published experimental data are scarce.

The recycling system (Figure 3.1) represents in fact the gathering in a single step of an infinite number of successive reaction steps, with identical reaction time and temperature. On this basis one may deduce the material balance equations for recycling systems by using those for the successive system.

In the case of feeds of homogenous compositions, the equality of the temperatures and of the reaction times in all the steps lead to the equality of the conversions obtained, which means:

$$y_I = y_{II} = y_i = \cdots y_n$$

$$z_I = z_{II} = z_i = \cdots z_n$$

The Eq. (3.8), which gives the total capacity of the n reactors per unit mass of feed introduced in process, becomes in this case:

$$\sum_{i=0}^{n-1} A_i = 1 + y + y^2 + y^3 + \cdots + y^{n-1} \tag{3.11}$$

Since $y < 1$, the equation represents a decreasing geometric progression; therefore:

$$\sum_{i=0}^{n-1} A_i = \frac{1 - y^n}{1 - y} \tag{3.12}$$

For the recycling system, which corresponds to a successive system with an infinite number of reactors, the capacity A of the reactor per unit mass of processed feed will be:

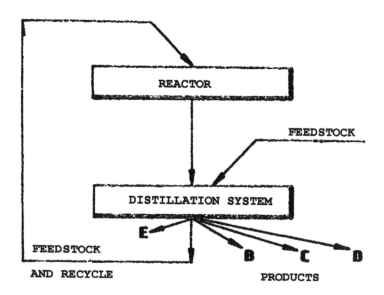

Figure 3.1   The recycling system.

$$A = \lim_{n \to \infty} \frac{1 - y^n}{1 - y} = \frac{1}{1 - y} \tag{3.13}$$

Similarly, Eq. (3.9) becomes:

$$A_n = \lim_{n \to \infty} y^n = 0$$

Thus, the amount of feed left unconverted in a recycle system is zero, which is in fact obvious.

The amount F of recycled material will be:

$$F = A - A_0 = \frac{1}{1 - y} - 1 = \frac{y}{1 - y} \tag{3.14}$$

If equal conversions are obtained in each step, the Eqs. (3.10) become:

$$\sum_{i=1}^{n} B_i = z(1 + y + y^2 + y^3 + \cdots + y^{n-1})$$

$$\sum_{i=1}^{n} C_i = u(1 + y + y^2 + y^3 + \cdots + y^{n-1})$$

$$\sum_{i=1}^{n} D_i = v(1 + y + y^2 + y^3 + \cdots + y^{n-1})$$

$$\sum_{i=1}^{n} E_i = w(1 + y + y^2 + y^3 + \cdots + y^{n-1})$$

The progressions within the brackets are identical to Eq. (3.11), which for the recycling system gives at the end expression (3.13). Thus, for the system with recycling one obtains:

$$\begin{aligned} B &= \frac{z}{1 - y} \\ C &= \frac{u}{1 - y} \\ D &= \frac{v}{1 - y} \\ E &= \frac{w}{1 - y} \end{aligned} \tag{3.15}$$

The systems with recycling are characterized by the recycling coefficient K, which represents the ratio between the flowrate to the reaction zone (therefore the sum of the feed and the recycled material) and the net feed rate of raw material to be processed. Thus, for $A_0 = 1$, according to the Eqs. (3.13) and (3.14):

$$K = \frac{A}{A_0} = \frac{A_0 + F}{A_0} = \frac{1}{1 - y} \tag{3.16}$$

which according to the Eqs. (3.15) may be written also as:

$$K = \frac{B}{z} = \frac{C}{u} = \frac{D}{v} = \frac{E}{w}$$

The use of the recycling coefficient for the characterization of such processes is very useful for solving various practical problems, especially those concerning the influence of the conversion upon the processing capacity of the plant.

The analysis of such problems may be simplified, by the combination of the Eqs. (3.15) and (3.16), and by taking into account that for a given plant the processing capacity of the reactor $A = A_0 + F$ is constant and the yields of products B, C, D, and E depend on the nature of the feedstock. From here, one may obtain the interdependence between the conversions z, u, v, or w and the amount of the raw material processed, A_0.

It is at times convenient [5], to express Eq. (3.16) in terms of volume flowrates. The recycling is frequently expressed as:

$$\text{Recycle ratio} = K^* = F/A_o \tag{3.17}$$

in terms of volume flowrates.

For processes in which the feedstock has a very broad composition and/or contain components with quite different reactivities, as well as when it is desired to obtain a second intermediary product not only from the feedstock but also from the first intermediary product, one has to use mixed systems containing two successive reaction zones, both with recycling streams. This is the case of old plants for the thermal cracking of atmospheric residue for the production of gasoline, today completely outdated, but also for modern plants for fluid catalytic cracking, that have two reaction zones, often placed in the same reactor. Hydrocracking often applies similar solutions.

The scheme of such a process is given in Figure 3.2.

It is customary to perform the preheating of the feedstock to the final temperature, by introducing it in an adequate point into the products separation system.

The scheme of Figure 3.2 is different from that of Figure 3.1, by the introduction of a second reaction zone, which has the role of cracking the recycle II, resulted

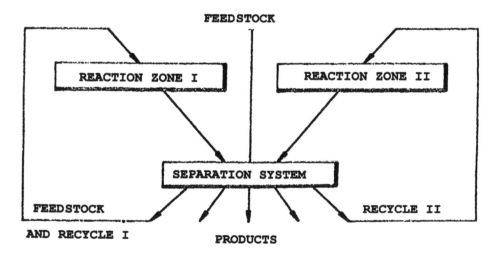

Figure 3.2 System with two successive reactor zones, each of them with recycling. Second zone at higher severity.

from the cracking. Recycle II has distillation limits lower than those of the feedstock, and for this reason it is much more stable. Such a situation may occur in both thermal and catalytic cracking when a heavy feedstock is cracked for maximizing the gasoline output while minimizing the gas oil production.

Some details concerning the specifics of selecting the recycling regime for the thermal cracking at high pressures for the purpose of gasoline production are discussed in a previous work of Raseev [3,4]. They are not repeated here, since these processes find little interest among modern petroleum processing technologies.

REFERENCES

1. S Raseev. Postgraduat lectures at Zulia Uniyersity, Maracaibo, Venezuela 1980.
2. IH Hirsch, EK Fischer. In: BT Brooks, ed, The Chemistry of Petroleum Hydrocarbons vol. II, Reinhold Publ. Co., New York 1955, p. 27.
3. S Raseev. Procese distinctive de prelucrare a titeiului, Editura Tehnica, Bucurest, 1964.
4. S Raseev. Procesy razkladowe w przerobce ropy naftowey, Wydawnictwa Naukowo Techniczne, Warsaw, 1967.
5. AW Gruse, RD Stevens. Chemical Technology of Petroleum, 3rd Edition, McGraw-Hill Inc., New York, 1960.

4

Industrial Implementation of Thermal Processes

The tubular furnace is the most suitable reactor for performing thermal processes. It has the residence time distribution closest to the ideal plug flow reactor and it can achieve heat transfer rates required for reaching reaction temperatures and for balancing endothermal effects. The maximum temperature at the exit from the furnace, which in pyrolysis is known as *coil outlet temperature* (COT), is limited by the coil metallurgy.

The thermal processing of residues poses special problems when they are processed for the production of either coke or liquid fuel. In these processes a portion of the reactor content is in liquid phase and contain asphaltenes and/or their precursors (resins, polycyclic aromatic hydrocarbons, etc.), which are transformed or may be converted to coke. The generation of coke, whether desired as in the coking processes or undesired, as in pyrolysis, etc., profoundly influences the operating conditions and design of the reactors used in each of the affected processes. The specifics will be described in the following chapters devoted to these processes.

The only processes that require a supplementary reaction zone after the furnace are those for residue cracking (soaker visbreaking and coking). The equipment for the supplementary reaction is specifically bound to the phenomena resulting in coke formation.

Attempts to perform the cracking processes in other types of reactors, such as those involving moving or fluidized beds of solid particles, heat carriers, or other systems of ensuring an intensive heat transfer, were not successful. Although initially they looked very promising and despite the large efforts made, most of these reaction systems did not result in commercial processes. The only exception is fluid coking.

4.1 THERMAL CRACKING AT HIGH PRESSURES AND MODERATE TEMPERATURES

The objective of thermal cracking at high pressure is the production of liquid fractions that should be lighter or have a lower viscosity than the feedstock [13].

In the past, the main purpose of these processes was the production of gasoline. The cracking at high pressures was the first process that produced supplementary gasoline, after that from straight-run distillation.

The first plant of this type, a Dubbs patent, was put into operation in 1919. This was a semicommercial demo unit with a capacity of 64 m³/day and a cycle length of 10 days [4]. Beginning in 1928, when the pumps were first developed that allowed feeding the furnace with hot raw material, the process underwent rapid development, amplified by the introduction in 1932 of the "selective" cracking plants equipped with two furnaces. Concomitantly, the duration of the cycle between two decokings of the furnaces increased to 120 days, reaching in some cases 210 days.

Units for the thermal cracking of naphtha (thermal reforming) were developed to increase the gasoline octane number. Also, two-furnace units for the thermal cracking of straight run residue were developed to increase the total amount of gasoline obtained from crude oil. The latter plants reached a great extension. In the U.S., during the period 1938–1940, the whole amount of primary residue was cracked, with the exception of that which used in the fabrication of oils. In parallel, other cracking processes were developed for converting liquid residues to coke.

The appearance in 1940 of catalytic cracking produced a fundamental change in the petroleum processing industry. Catalytic cracking made it possible to obtain gasoline with an octane number sensibly higher and more stable than that obtained from thermal cracking. The construction of thermal cracking units gradually ceased and the existent plants were transformed or liquidated.

In parallel with the expansion of catalytic cracking, the preferred feedstock became the overhead product from the vacuum distillation of the atmospheric residue. A vacuum residue was a result of this operation. Its use as fuel presented difficulties due to its high viscosity.

This fact determined the appearance and the expansion of the visbreaking process, by means of which the vacuum residue is converted to fuel oil, gas-oil fractions, and small amounts of gasoline and gases. Visbreaking represents today the last important process of thermal cracking that produces a liquid residue. This process is widely used in Western Europe, where the residue obtained from the preparation of feedstock for catalytic cracking is not submitted to coking.

A thermal cracking process which has limited extension but is of great importance for the preparation of the strongly aromatic residue necessary for the fabrication of needle coke for electrodes used in electrochemical industries and especially in the production of aluminum.

4.1.1 Visbreaking

The main purpose of the visbreaking process is to produce a fuel with lower viscosity than that of the feed, a vacuum residue.

In the past, the viscosity of the visbreaking residue was controlled by means of the operating conditions of the flasher. The amount of cracked lighter material left in the residue (indigenous diluent) controlled the viscosity and stability of the fuel oil.

This way of operation was almost completely abandoned. Presently, in order to reach a desired viscosity, several other diluents might be selected. In this manner, the quantity of diluents required for reaching a specified viscosity is smaller than when indigenous diluent is used, and also, the stability of the fuel oil is improved.

In addition to fuel oil, the visbreaking unit produces one or two distillates fractions (light and heavy gas oil or only light gas oil) and small amounts of gasoline and gases.

Overall, by converting some of the vacuum residue to distillates, the presence of a visbreaking unit reduces by about 20% the amount of fuel oil produced in a refinery. To this amount about 10% distillates should be added, when visbreaking is missing, for diluting the vacuum residue in order to bring it to the fuel oil specifications.

Visbreaking units are classified in those without soaker (Figure 4.1), and those with soaker (Figure 4.2). In each of these units, a vacuum column may be present for producing a heavy gas oil, generally used as feed for catalytic cracking (Figure 4.3).

Other less important differences may exist among various visbreaking units. Thus, the flasher and the fractionator may be two independent units (Figure 4.1 and Figure 4.2), or may be combined by superposition, in a single vessel (Figure 4.3). Also, several heat exchange schemes are being practiced.

The presence or the absence of the soaker is the main distinguishing feature of visbreaking units. In units without soaker (coil visbreaking) the cracking reaction takes place only in the furnace. The desired conversion is achieved by maintaining the feed for 1–2 minutes in the reaction zone of the coil. The pressure in the furnace is approximately 15 bar, and the coil exit temperature is about 480–490°C.

At the exit, the reactions are stopped by a sudden decrease of temperature (to about 350°C) performed by adiabatic flashing by means of a pressure relief valve and is completed by the injection of cold liquid. Usually, the light gas oil fraction obtained in the unit is used as quench liquid.

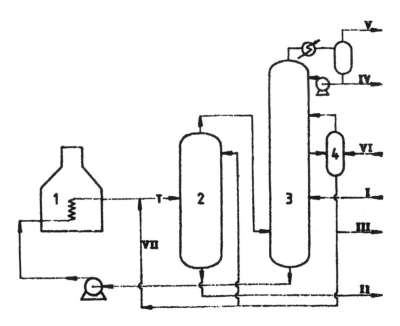

Figure 4.1 Visbreaking without soaker. 1—furnace, 2—flasher, 3—fractionation column; I—feedstock, II—residue; III—gas oil, IV—gasoline, V—gases, VI—steam, VII—cold gas oil.

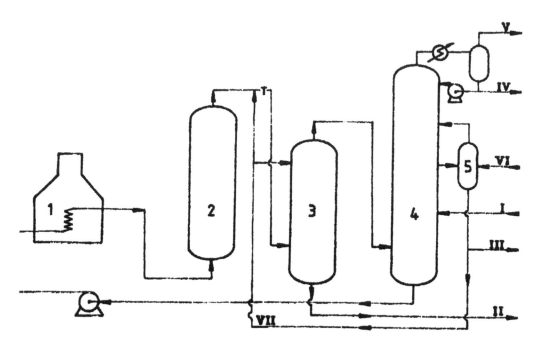

Figure 4.2 Visbreaking with soaker. 1—furnace, 2—soaker, 3—flasher, 4—fractionation column; 5—stripper; I—feedstock, II—residue; III—gas oil, IV—gasoline, V—gases, VI—steam, VII—cold gas oil.

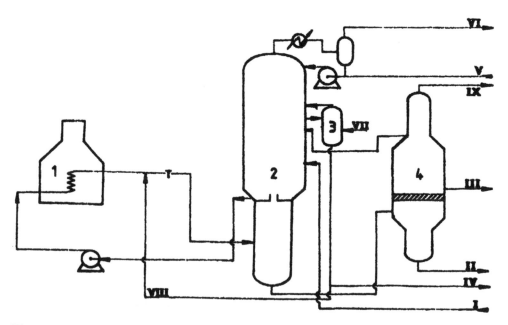

Figure 4.3 Visbreaking with vacuum column. 1—furnace, 2—flasher, 3—fractionation column; 4—vacuum column; I—feedstock, II—residue; III—heavy vacuum gas oil, IV—gas oil, V—gasoline, VI—gases, VII—steam, VIII—cold gas oil, IX—to vacuum system.

In the soaker visbreaker, the greatest part of the reactions takes place in this vessel. The conversion at the exit from the furnace is only about 20–25% of the final conversion, and the soaking temperature is approximately 440–460°C. For the same feedstock, the coil outlet temperature is 30–40°C lower than in coil visbreaking units. Since in the soaker, the reactions take place practically adiabatically, there is a temperature decrease of 15–25°C, to about 420–430°C. After the soaker, the pressure reduction valve and the quenching with cold liquid cause the temperature to drop to 350°C.

In either unit, with or without soaker, the reaction system is followed by flasher, fractionator with stripper for the light gas oil, and sometimes, by a vacuum column for the recovery of a heavy gas oil.

Typical profiles of the temperature, pressure, and conversion along the reaction system are given in Figure 4.4 for units with and without soakers. The conversion in these graphs is expressed in weight percent of gases plus gasoline with an end point (EP) of 165°C [5].

The soaker is a cylindrical, vertical vessel with a height/diameter ratio of about 6 and a volume that must ensure about 10–20 minutes residence time for the liquid effluent from the furnace [6]. The need for a longer time in comparison with that practiced in coil visbreaking (1–2 minutes) can be easily understood by taking into account the difference between the corresponding reaction temperatures. Depending on the design, the pressure in the soaker varies between 5 and 15 bars.

Studies with radioactive tracers [5] demonstrated the benefits of upflow circulation and of horizontal perforated plates for reducing the backmixing of the liquid phase. Backmixing has a negative effect on the conversion to light gas oil and especially on the stability of the residue.

The inclusion of the soaker in the viscosity breaking units has many advantages that can be summarized as follows:

1. The lower temperature at the exit from the furnace has a fuel savings of 30–35% as a direct result [5,7]. Taking into account the lower amounts of steam produced in the auxiliary coil of the furnace, the net economy of fuel is still 10–15%.

2. As result of the lower temperature in the coil, the coking tendency is reduced and therefore a significant extension of the working cycle between two decoking operations of about 3 months to 12 months results. Figure 4.5 presents a typical example of the increase with time of the temperature of the furnace tube wall in units with and without soaker, as a consequence of the coke formation. From this figure, the increase of the cycle length is obvious.

The graph is provided for tubes with 5% Cr and 0.5% Mo. For tubes made of higher alloyed steels, the duration of the cycles will increase accordingly.

Concerning the duration of the decoking operation, it is about a week for the soaker and is performed using hydraulic, pneumatic, or mechanical methods. This coincides with the time necessary for decoking the coil and the maintenance operations for the furnace [7].

3. A larger gas oil yield is the result of two phenomena: a) The activation energy for the formation of the gas oil from the residue is lower than that for its further decomposition to form gasoline and gases (about 170 kJ/mole in compar-

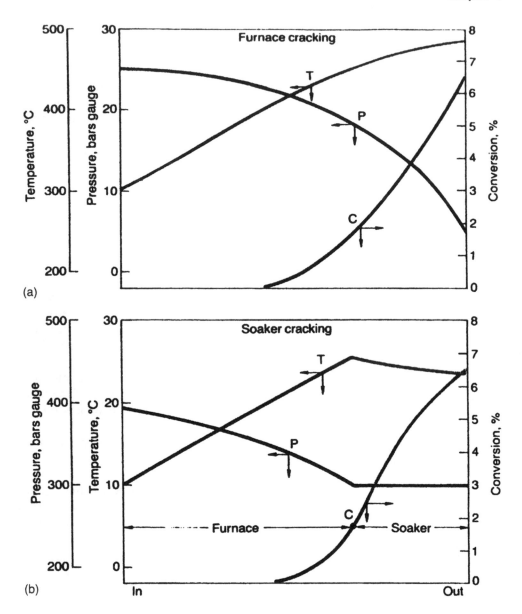

Figure 4.4 Typical process conditions in visbreaking plants (a) without and (b) with soaker [5].

ison with 210 kJ/mole). Therefore, a lower temperature in the reaction zone (as in the soaker) will lead to a higher yield of intermediary product—the gas oil (Figure 2.16). b) In both systems with and without soaker, the effluent from the reaction zone is in a mixed vapor-liquid phase. Actually, the feedstock and the final residue are found wholly in the liquid phase, while the gasoline and the gases are mostly in the vapor phase, while the gas oil cut is distributed between the two phases with

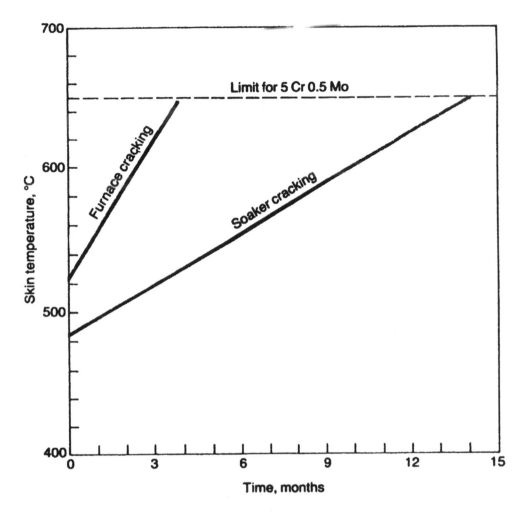

Figure 4.5   Influence of coke deposits on the furnace tube temperature.

the light gas oil, mainly in the vapor phase. Also, the residence times of the two phases in the reaction zone are not identical. It was estimated [5] that the ratio of the residence times of the vapor and liquid phases ranges between 1/2 and 1/5 in the furnace coil and is only 1/10 within the soaker. This difference leads to a shorter reaction time for the light gas oil in soaker visbreakers (as compared to coil visbreakers) and accordingly, to a decrease of its decomposition to gasoline and gases.

4.   The stability of the residue, a factor which is of special importance, seems in general to be better for well-designed soaker units [5,7], i.e., for units with a proper height/diameter ratio and provided with perforated plates to prevent back-mixing.

5.   The visbreaking units are sensitive to fluctuations of the operating conditions. The operation of the soaker units is more easier to control [7].

A very important requirement for the visbreaking units is to produce fuel oil that has good stability. By good stability, the absence of flocculation phenomena of the asphaltenes is understood. In stable fuel oils, the asphaltenes must remain in a stable colloidal state, even if the storage time is long, (see the Section 2.3.3.).

A fuel which has a reduced stability will generate deposits in the storage tanks and cause the blocking of filters and create deposits on the tubes of the heat exchangers*. Since the flocculation and the formation of deposits are very slow processes removing them does not ensure good stability for the rest of the product. From a technological point of view, it is important to know the causes of the flocculation phenomena and the possible measures for their elimination.

The flocculation tendency of asphaltenes (Section 2.3.3) depends on the properties of the liquid surrounding the medium in which they are found. A medium with a chemical structure similar to that of the asphaltenes, i.e., composed of malthenes and polycyclic aromatic hydrocarbons, ensures their existence in peptized form, that is as a stable colloidal solution, which does not present a danger of flocculation. On the other hand, a medium with alkane or cycloalkane character, and especially one with low molecular mass, will produce a rapid flocculation of the asphaltenes.

From here, it follows that attention which must be given to the nature of the fractions used for diluting of the visbreaking residue. From this point of view, the most adequate are the gas oils from catalytic cracking, which have a strong aromatic character. Other recommended diluents are the extracts from the solvent extraction of lube oils and gas oils, and also the gas oils from the fluidized bed coking units.

Awareness of these problems makes possible the analysis of why the residue obtained in the soaker visbreaking units are generally more stable than those obtained in coil visbreaking units.

The overall activation energy of the condensation reactions that occur during visbreaking is lower than 93 kJ/mole while that for the thermal decomposition reactions is 170–230 kJ/mole [6]. In this situation, the fact that in the soaker systems the average temperature is lower than in coil visbreaking cannot explain the greater stability of the produced residue. On the contrary, one would expect that at lower temperatures the condensation reactions will be favored at the expense of the decompositions.

To a certain extent, a low stability of the residue is the result of the phenomena of overcracking. Such phenomena are inherent high temperature heating in coil furnaces and occur mainly in the laminar boundary layer formed inside the tubes. Overcracking negatively influences the stability of the obtained residue [5].

Similar phenomena take place and were confirmed in soakers without perforated plates, where the backmixing leads to overcracking.

Still, overcracking cannot explain by itself the difference in stability between the residues obtained in the two types of units. One must suppose that the concentration of the asphaltenes and the chemical nature of the medium in which they are present also play an important role in their stability.

---

* The stability could be determined by the method ASTM D 1661, used by the U.S. Navy or using other described methods [6,8–10].

Before the flocculation phenomenon takes place, a sufficient concentration of asphaltenes must be formed as a result of condensation reactions. Without it, neither the flocculation, nor the subsequent coking can take place. Table 4.1.

However, according to other opinions [7], decomposition of asphaltenes contained in crude oil takes place simultaneously with their recombination to form other structures that usually tend to flocculate. All these facts prove that the flocculation is the result of a series of previous phenomena and reactions.

More important is the fact that flocculation of asphaltenes takes place exclusively when the structure of the medium in which they are dispersed is much different from their own. For example, it has a low molecular weight and alkane-, alkene- or cycloalkane-character. In the soaker visbreaking units, the vapor phase is rapidly eliminated from the system, which causes additional portions of light fractions to be stripped out from the liquid phase. As result, the asphaltenes are left in a state of stable emulsion. It seems to us that this phenomenon probably contributes to the greatest extent to the more stable character of the residues produced by the soaker visbreakers.

The lower pressure that prevails in soaker visbreakers explains the increased amounts of light fractions that are flashed out compared to coil visbreakers. Maintaining such low pressures in the furnace coil visbreakers would lead however to units of large physical sizes and investment costs [5].

Although the soaker plants were not introduced before the year 1978 [2], they were so rapidly accepted that today the inclusion of a soaker in the process flow sheet is almost universal. It is worth mentioning that in a monograph published in 1964 [3], when discussing the problem of the conversion of the thermal crackers for straight-run residue to visbreaking units, the author suggested maintaining the reaction chamber in order to achieve a more advanced conversion of the feedstock. This anticipated by many years the actual commercial implementation of soaker visbreakers.

There are also other possible methods for improving the visbreaking process. An interesting proposal is that the exit from the furnace is at essentially atmospheric pressure, and the effluent is introduced directly into the flasher-fractionator without passing through a pressure-reduction valve [13]. This system makes possible collect-

Table 4.1 Coke Formation by Cracking at 410°C

Asphaltenes in feedstock wt%	% Asphaltenes converted to coke	
	Propane deasphalting residue dissolved in transformer oil	Thermal cracking residue dissolved in anthracene oil
10	0	traces
20	0	traces
30	0	traces
40	0	traces
50	0	traces
55	34.8	traces
60	–	6.1
70	–	41.2

One hour in autoclave.
Source: Ref. 11.

ing also from the fractionator a heavy gas oil cut, that otherwise would require a vacuum column. In this way the investment for the soaker and vacuum column are eliminated, which [13] should be economically advantageous.

The flow sheet of such a unit is given in Figure 4.6.

In this system, when using a vacuum residue feed the exit temperature from the furnace is 488–500°C, and that at the entry point in the flasher-column, is 480°C. The cooling liquid is introduced in the flasher below the feedpoint of the furnace effluent in order to stop cracking reactions, that would worsen the stability of the residue.

The higher than usual exit temperatures from the furnace, are the consequence of the relatively low pressure at which the effluent leaves the furnace. If the temperature were not increased, the longer residence time required in the reaction zone would have led to exaggerated dimensions of the furnace and to higher investment costs.

No information is available on the application of this system or on the obtained results. Important operating difficulties are foreseeable, related to coking is expected to occur in the flasher. Indeed, it is difficult to implement an efficient contacting between the cooling liquid and the residue in the flasher so that coking is prevented, while ensuring in the same time a correct vapor-liquid separation.

In a later paper (1991), published by the same author [14], a new unit is described the implementation of which (1986) he actively participated. This unit has no soaker and was named HIRI Visbreaker. After it was first started, the unit performed a cycle of 843 days without decoking of the furnace. Although not explicitly stated in the paper, the unit seems to incorporate the ideas contained in the previous publication [13] of the same author. It is however mentioned that minimal modifications will allow obtaining a heavy gas oil cut directly from the fractionator. The process flow sheet of this unit is given in Figure 4.7.

The process flow diagram and the paper [14], allow one to deduce that several measures were taken for avoiding coking in the flasher zone, and for the simultaneous, direct introduction into the flasher of the furnace effluent and the cooling

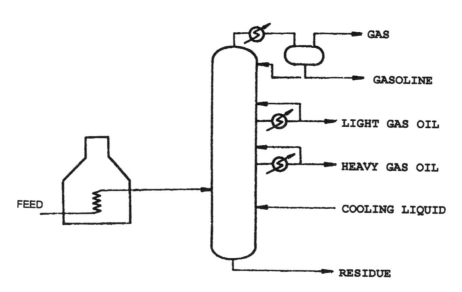

Figure 4.6   Low pressure visbreaking [13].

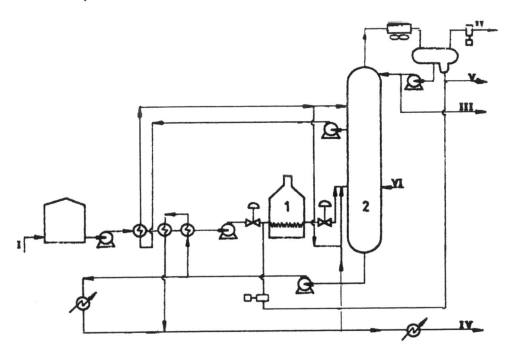

Figure 4.7 HIRI visbreaking. 1—furnace, 2—fractionator; I—feed, II—gases, III—gasoline, IV—residue, V—water, VI—steam. Produced gas oils not shown.

liquid. One may also assume that the new unit maintained the concepts of working at reduced but rigorously controlled pressure in the furnace and of operating at a higher temperature at the entrance in the flasher.

The results obtained in visbreaking depend to a great extent on the feedstock, more exactly on the maximum severity it tolerates, which imposes a corresponding control of the conversion. Special attention must be given to changing to a different feedstock. The mixing of feedstocks of different natures must be completely avoided since this may lead to disastrous coking. Thus, a feedstock with a high content of asphaltenes, but with a pronounced aromatic character may have acceptable behavior in the process and allow normal severities. Also, a feedstock of paraffinic character, but with a reduced content in asphaltenes, may behave well. The mixing of these two feedstocks will lead to rapid flocculation of the asphaltenes, which even at reduced conversions will have negative consequences for the operability of the unit.

Several methods were developed for determining the stability of the residue obtained from visbreaking. This stability limits the severity at which the unit may be operated [8–10]. In the following, a method is described [6] that besides allowing to perform a complete study, is sufficiently rapid for being of practical value for regulating the severity of the operation of the unit.

For each given residue, the method consists of the construction of the flocculation curve C_f of the asphaltenes, similar to that given in the graph of Figure 4.8.

The appearance of the flocculated state is determined by means of light dispersion measurement or, more simply, by establishing if an insoluble spot is left by

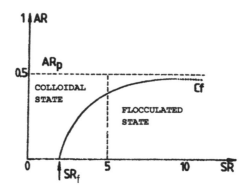

Figure 4.8 Asphaltenes flocculation. SR-isooctane + xylene/sample, AR-xylene/isooctane + xylene, SR_f-minimal quantity of isooctane that produces the flocculation, AR_p-minimal quantity of xylene in solvent, that eliminates the flocculation even at infinite dilution.

the product on a filter paper. A drop of the residue is placed on a filter paper. After this spot is kept in prescribed conditions in a heated chamber, the oils are washed with a solvent mixture of known composition. The flocculated state appears visible as a dark central zone that will not be washed and remains on the place of the spot.

Since the construction of the complete curve and the determination of the characteristic points (as indicated in the caption of the figure) requires a longer time than available for the operation control of the unit, a simplified test is performed. The test consists in changing the composition of the solvent (isooctane-xylene), while maintaining the solvent/residue ratio at the constant value of 5 (Figure 4.8). The fraction of volume of xylene, in the solvent mixture that corresponds to the last trace left on the place of the spot, multiplied by 10, gives the characteristic value of the severity SF^* [6].

SF can vary in the limits 1–10. The residue is qualified as stable if $SF \leq 7$. The operating conditions of the unit are controlled such that the obtained residue satisfies this condition for SF.

The construction of the flocculation curve is useful also for the preliminary comparative study of feedstocks.

An example for the yields obtained and the economics of a typical unit without soaker is given in Table 4.2 [95].

The results obtained in a typical soaker visbreaker are given in Table 4.3 [96].

The operation of a similar soaker unit at three severities (SF = 4.5, SF = 7, and SF = 10) for the same crude is given in Table 4.4 [6]. The severity SF = 7, corresponds to normal working conditions.

The octane number F_1 of the visbreaking gasoline is usually in the range 60–70, while the cetane number of gas oil, is in the range 40–50, being lower than for the straight-run gas oil obtained from the same crude oil.

The distribution of sulphur among various distillation fractions, is completely analogous to that obtained by the atmospheric distillation of the same crude oil and is essentially independent of the conversion. The situation is different for

---

* SF is sometimes expressed in % xylene : 10, which gives the same value.

Table 4.2 Foster Wheeler/UOP "Coil"-type Visbreaking Process

Yields			
Feed source type	Light arabian atm. residue	Light arabian Vacuum residue	
Gravity, °API	15.9	7.1	
Density, d_{15}^{15}	0.9600	1.0209	
Sulfur, wt %	3.0	4.0	
Concarbon, wt %	8.5	20.3	
Viscosity			
cSt 130°F (54.4°C)	150	30,000	
cSt 210°F (98.9°C)	25	900	
Products, wt. %			
Gas	3.1	2.4	
Naphtha C_5–330°F (165°C)	7.9	6.0	
Gas oil 330–600°F (165–315°C)	14.5	–	
Gas oil 330–662°F (165–350°C)	–	15.5	
Visbreaking residue 600°F+ (315°C+)	74.5	–	
Visbreaking residue 669°F+ (350°C+)	–	76.1	

Economics
Investment (basis 40,000–10,000 bpsd, 4th Q 1998, US Gulf); $ per bpsd 785–1,650 ($ per m^3 psd 125–262)

Utilities per bbl feed					
Fuel,	MMBTU	0.1195	10^3 Joules	126.1	
Power,	kW/bpsd	0.0358	kW/bpsd	0.0358	
MP steam,	lb	6.4	kg	2.9	
Water cooling	gal	71.0	m^3	0.27	

Installations: over 50 units worldwide (1998).
Operating conditions: furnace outlet temperature = 850–910°F (454–488°C); quench temperature = 710–800°F (377–427°C).
Source: Ref. 95.

nitrogen. Its concentration in the distillates increases markedly with the visbreaking conversion.

Concerning the residue, the important characteristics are those related to its stability. The data of Table 4.5 give a qualitative comparison between various methods for determining stability. It must be also mentioned that complex additives may be used for delaying the flocculation phenomena.

Often, a very high sulphur content of the residue can constitute a problem, since its desulfurization is particularly difficult. In such situations, it is preferred to perform the desulfurization of the distillate cuts used as diluents. The diluted fuel oil will have acceptable sulfur content.

Graphs for estimating the yields and quality of the products obtained in visbreaking have been reported [3,15,69]. Their practical value is limited since they refer to units without soaker and therefore will not be reproduced here.

Table 4.3   Shell/Lummus Soaker Type Visbreaking Process

Yields				
Feed				Middle East
Vacuum residue				
Viscosity, cSt at 212°F (100°C)				770
Products, wt %				
Gas				2.3
Gasoline, 330°F (165°C) EP				4.7
Gas oil, 662°F (350°C) EP				14.0
Waxy distillate, 968°F (520°C) EP				20.0
Residue, 968°F+ (520°C +)				59.0
Viscosity, 330°F (165°C) plus cSt at 212°F (100°C)				97
Economics				
Investment basis 1998			1000–1400 \$/bbl	
Utilities per bbl feed				
Fuel	10^3 BTU	63.5	10^3 Joules	67
Electricity	kWh	0.5	kWh	0.5
Net steam production	lb	39.7	kg	18
Cooling water	gal	26.4	m^3	0.1

Installations: 70 units have been built or are under construction (1998).
Source: Ref. 96.

4.1.2 Cracking of Straight Run Residue

The thermal cracking of the straight run residue was for many years the most important process supplementing the straight run gasoline, and many such units were built in various countries.

This position was lost as FCC, which produces a much higher octane number gasoline, became widespread. However, many such units still exist in less developed countries, for instance in Eastern Europe, and their conversion and improvement is an important problem for these countries.

In such units, in order to obtain only gasoline and gases it is necessary to use a parallel-successive reaction system, as shown in Figure 3.2. The first reactor (furnace) converts the feedstock at 15–20 bar, with exit temperature of 490–495°C. The second furnace converts the cracking gas oil produced in the first reactor at a higher pressure and at 510–525°C exit temperature.

There are two main types of thermal cracking units: with and without reaction chamber. Figure 4.9 gives the scheme of a plant with reaction chamber. In the plants without reaction chamber, the products from the two furnaces pass through pressure-reduction valves, are injected with cold gas oil, and enter the flasher.

The reaction chamber is an empty vessel, shaped as a vertical cylinder. The mixing of the two flows increases the temperature of the heavy feed to the cracking furnace (2 in Figure 4.9). Therefore, the conversion is increased and also the amount of gas oil produced. The gas oil is converted to gasoline and gases in the other furnace.

Table 4.4 Yields as Severity Function: Soaker Type Visbreaking Plant

Feed

Source	Middle East
Type	Vacuum residue
API	8.05
d_4^{20}	1.014
Viscosity, cSt at 212°F (100°C)	736
S, wt %	5.35
N, wt %	0.37
Concarbon, wt %	17.5
AsC$_5$ (wt %)	13.5
AsC$_7$ (wt %)	6.5

Yields, wt%

Severity (SF*)	4.5	7	10
Temperature	824°F (440°C)	842°F(450°C)	851°F (455°C)
Residence time	(t)	(t)	(1.5t)
H$_2$S	0.20	0.44	0.68
C$_1$–C$_2$	0.26	0.44	0.68
C$_3$–C$_4$	0.71	1.18	1.33
C$_5$- 302°F (150°C)EP	1.80	4.30	7.28
302–482°F (150–250°C)	1.81	4.44	7.51
482–707°F (250–375°C)	5.86	9.24	13.94
Vacuum gas oil	12.02	16.00	18.54
Residue	77.54	64.00	50.3

* Flocculation limit.
Source: Ref. 6.

Table 4.5 Correlation Between Methods for Stability Determination

	Residue			
	Atmospheric		Vacuum	
Conversion IBB 350°C, wt %	11.03	22.10	11.40	19.60
Severity	4.50	6.50	4.00	6.50
ASTM D 2781	2	5	1	5
Shell, wt % after warm filtration	0.08	0.13	0.17	0.17
Mobil, % vol. precipitate	0.12	6	0.12	5.2
Charact. Points in the graph, Fig. 4.10:				
P = 1 + SR$_f$—peptization state of asphaltenes	2.00	1.59	2.12	1.89
P$_a$ = 1 AP$_p$—asphaltenes tendency to remain in colloidal solution	0.53	0.41	0.55	0.38

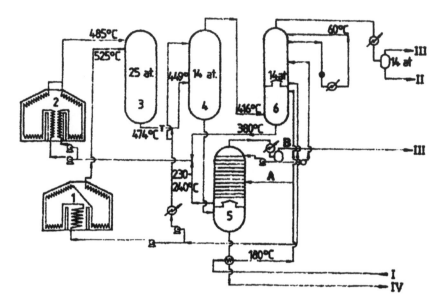

Figure 4.9 Thermal cracking of straight run residue. 1—gas oil cracking furnace, 2—feedstock cracking furnace; 3—reaction chamber, 4—evaporator, 5—low pressure flasher, 6—fractionation column; I—feedstock, II—gasoline, III—gases, IV—residue. A and B—two feeding possibilities.

The limitations of the concentration of carboids in the cracking residue led to adopting a descending flow in the reaction chamber. Such a flow pattern reduces the residence time of the residue within the reaction chamber.

The situation is exactly the opposite to that for visbreaking, where it is desired to have a more rapid elimination of the vapors and not of the liquid. For this reason, the circulation in the soaker is ascendant.

The pressure in the reaction chamber is essentially identical to that of the exit from the furnaces, while the temperature is lower (the chamber works actually adiabatically). The polymerization reactions of alkenes are favored, and the gasoline yield increases as well as its octane number.

More details about the thermal cracking process for the straight run residue are given in previous publications [1,3,97] of the author.

The typical yields that are obtained in the thermal cracking units with a reaction chamber for straight run residue are given in Table 4.6. In the table, besides the classic method of exclusively producing gasoline, gases, and residue, it is given also the regime for the simultaneous production of gasoline and distillate (gas oil).

4.1.3 Conversion of Thermal Cracking Units to Visbreaking

As discussed before, a significant number of units for the thermal cracking of straight run residue still exists today, especially in countries where catalytic cracking has not taken over. Also, the decrease in the demand for residual fuels determines an increased need for the visbreaking process to the extent that the conversion of the available units for straight run distillation to visbreaking units is actively considered [20].

Table 4.6 Straight Run Residue Thermal Cracking Including Gas Polymerization

	Gasoline production	Gasoline + gas oil production
Yields		
Cracking gasoline, % vol.	48.5	34.5
Polymerization gasoline, % vol.	5.0	3.5
Gas oil, % vol.	–	23.0
Residue, % vol.	37.5	34.0
Gases Nm^3/m^3 feedstock	8.9	5.15
Feedstock		
Straight run residue		
d_4^{20}		0.904
Viscosity, °E at 50°C		5.0
Products		
Total gasoline F_2ON	71–72	70–71
Vapor pressure, mm Hg	517	517
End point, °C	205	205
Residue density d_4^{20}	1.022	1.014
Residue viscosity °E at 50°C	55	55

In these conditions, a problem of great interest for many countries is the conversion to visbreakers of the straight run residue cracking units. In fact, from the technological point of view, such a transformation is quite natural.

Similarly, the conversion of straight run residue cracking units so as to maximize the production of gas oil may become an attractive issue. This conversion was examined by this author in detail in previous works [3,97] and will not be repeated here.

The main problem in the conversion of units for the thermal cracking of fuel oil to visbreakers is the conversion of the reaction chamber to soaker. For units having the rest of the equipment of similar size, the volume of the reaction chamber is almost twice that of the soaker. Therefore, the reaction chamber can accommodate the use of both furnaces in parallel when converting to visbreaking. The reaction chamber is built to operate at temperatures of about 480°C and pressures of about 25 bar, thus at much more severe conditions than those of the soaker, and fully corresponds to this purpose. The only problem related to this new utilization is that of the overall direction of the flow, which must be upwards and not downwards, and of the backmixing, which in the soaker must be prevented as much as possible by means of perforated plates and other similar devices. Therefore, besides modifying the connections for reversing the flow, the inner part of the reaction chamber must be supplied with devices for preventing backmixing, designed on basis of broad practical experience that exists in the design of soakers. Special attention must be paid to this problem, especially since the diameter of the reaction chamber (about 3 m) is generally larger that of soaker (about 2 m), for volumes of 100 m^3 and 50 m^3, respectively [5].

The product recovery system must be supplemented with means for the collection (and stripping) of a light gas oil from the corresponding tray, and possibly also of a heavy gas oil cut.

4.1.4 Visbreaking with Additional Heater

This is a process which was first introduced in the early 1960s, initially for the processing of paraffinic crude oils of North African origin. It uses an additional heater for the thermal cracking of the heavy gas oil cut obtained in the vacuum column of a visbreaker. Such an approach may be of interest to the refineries that do not have catalytic cracking.

The available information does not indicate whether the unit incorporates a soaker.

The flow diagram of such a unit is presented in Figure 4.10.

Results obtained by the visbreaking of the same feedstock in units with or without an additional heating furnace are compared in Table 4.7 [4].

Insights gained from visbreaking with an additional heater can help in finding the optimal configuration when the conversion of old units for the thermal cracking of straight run residue is contemplated.

4.1.5 Preparation of Needle Coke Feed

Besides the fractions used directly in the production of needle coke (see Section 4.3.2), a feed of very good quality is obtained by the thermal cracking of certain distillates in a system with two heaters and a reaction chamber, similar to the one

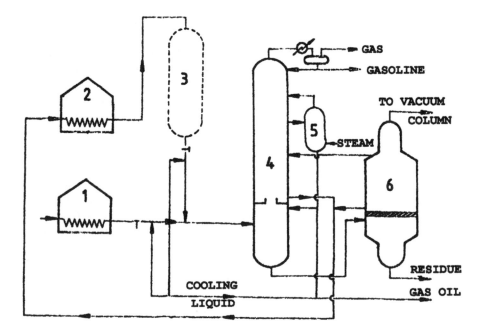

Figure 4.10 Visbreaking with additional furnace. 1—visbreaking furnace, 2—additional furnace, 3—reaction chamber, 4—flasher-fractionator, 5—stripper, 6—vacuum column.

Table 4.7 Visbreaking with Additional Furnace

Feedstock	Arabian light atmospheric residue: d_4^{20} = 0.954, K_{UOP} = 11.7, S = 3.0%, N = 0.16%, insoluble in C_7H_{16} = 3.5, Conradson carbon = 7.6, freezing point = +15°C, Ni = 8 ppm, V = 26 ppm, ash = 0.12, viscosity = 480 cSt at 50°C

Products	wt (%)	d_4^{20}	S (%)	$N_{(ppm)}$	Bromine index	Freezing point (°C)	Viscosity (cSt at 50°C)	Cetane index
				Visbreaking without additional furnace				
H_2S	0.2							
C_1–C_4	2.1							
C_5–C_6	1.4	0.663	0.8					
C_7–185°C	4.7	0.774	0.9	30	60			
185–370°C	10.7	0.859	1.3	600	26	−1	2.6	49
Residue	80.9	0.968	3.2				300	
				Visbreaking with additional furnace				
H_2S	0.3							
C_1–C_4	6.2							
C_5–C_6	3.5	0.663	0.7					
C_7–185°C	11.8	0.768	0.8	30	60			
185–370°C	31.8	0.868	1.3	600	26	−1	2.8	44
Residue	46.4	1.056	4.5				6000	

used in the thermal cracking of straight-run residue depicted in Figure 4.11. The flow diagram, and the operating conditions [4] indicate that a residue cracking unit that incorporates a reaction chamber may be used directly or with minimal modifications for this operation.

The feedstocks have distillation range similar to the gas oil and must have a low content of asphaltenes and contaminants. Often, the feedstock is a mixture of gas oils produced in catalytic cracking, coking, pyrolysis, and straight run distillation.

The cracking severity (conversion) is rigorously controlled; a too high conversion would lead to a residue (tar) of low stability and to coking in the heaters, the reaction chamber, and the flasher.

This tar, which constitutes the feed for needle coke production, is a product of a highly aromatic character, having high density and reduced viscosity. It is the result of advanced aromatization reactions and synthesis of certain molecules of high molecular mass.

An example of such operation, in which the feed was a mixture of gas oils from catalytic cracking, coking and straight run distillation, is given in Table 4.8.

The gasoline produced in the process may be submitted to catalytic reforming after a deep hydrofining, or may be used directly as component of mixtures following stabilization.

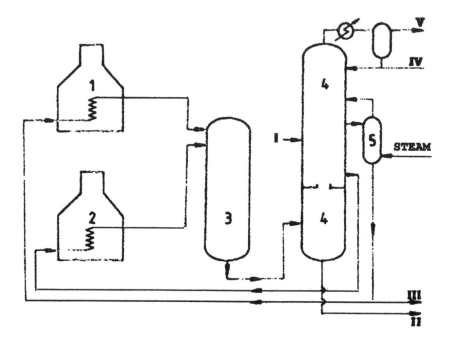

Figure 4.11 Preparation of needle coke feed. 1,2—furnaces, 3—reaction chamber, 4—flasher-fractionator, 5—stripper; I—feed, II—feed for needle coke, III—gas oil, IV—gasoline, V—gases, VI—steam.

4.1.6 Cracking of Heavy Residues and of Bitumen

These are thermal processes that have features similar to both thermal cracking and coking. Their product is a very viscous residue that becomes brittle at room temperature. Although they are not widely practiced, they are useful especially for processing natural bitumens that resemble the very heavy crude oils, such as those from Orinoco (Venezuela). There are enormous reserves of these natural deposits, but their exploitation is only borderline rentable. In favorable economic conditions, they could become extremely important.

In the following, two processes will be examined, that belong to this category: the HSC process (High-conversion Soaker Cracking), the first unit of this type with a capacity of 14,000 bpsd (1260 m^3/day) was first started in September 1988 at the Schwedt refinery, in Germany [102].The second process EURECA, was the object of demonstration tests and of limited industrialization [6] as the Shell Deep thermal conversion process [103].

The HSC Process [102]. This process makes possible the processing of very heavy crude oils (Orinoco) of the bitumen derived from the bituminous shales and of the visbreaking residues.

The process flow diagram of the unit is depicted in Figure 4.12.

The process is characterized by a very short residence time in the heater, for reducing cracking reactions to a minimum. This is achieved by using a large circula-

Table 4.8   Cracking for Needle Coke Feed Preparation

	Yields (% wt)	d_{15}^{15}	S (% wt)	N (ppm)	Viscosity (cSt at 50°C)
Products					
H$_2$S	0.1	–	–	–	
C$_1$–C$_4$	18.4	–	–	–	
C$_5$–180°C	37.6	0.758	0.10	20	
180–232°C	12.6	0.806	0.15	150	1.1
Tar	31.0	1.076	0.48	–	900

Feed   Gasoline C$_5$—180°C composition
d_{15}^{15} = 0.895   PONA   P = 40
K = 11.89   O = 34   40% cyclooleffins from this
S wt% = 0.312   N = 14
Ash wt% = 0.007   A = 12
                                                  Bromine index—70

tion rate through the tubes and by injecting steam into the coil. The coil exit temperature is 440–460°C. The soaker that follows is much larger than the usual ones, and a massive amount of steam is injected into it. Thus, the soaker serves both the cracking and the stripping of the volatiles from the residue. Atmospheric pressure prevails at the exit of the furnace and of the soaker.

The soaker has perforated plates in order to minimize the backmixing. The plates and the steam injected at the top and bottom of the soaker produce a uniform dispersion of the coke precursors. This makes it possible to operate at much higher severity than in classic visbreaking.

The stability of the residue is improved by stripping out the lighter products produced by cracking, as soon as they are formed. This measure prevents the

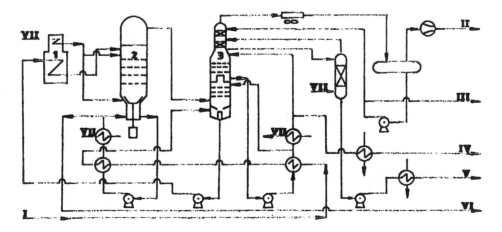

Figure 4.12   HSC Plant scheme. 1—furnace, 2—soaker, 3—fractionation column; I—feed, II—gases, III—gasoline, IV—light gas oil, V—heavy gas oil, VI—residue, VII—steam.

accumulation in the liquid phase of aliphatic products, that would flocculate the asphaltenes.

Operating the soaker at atmospheric pressure makes it possible to recover from the distillation column not only light gas oil, but also heavy gas oil, without the need to operate under vacuum.

If the residue obtained in the process will be used after dilution as liquid fuel, the authors of the process estimate that it should have a softening point (by the ring and ball method) not higher than 100°C. If the residue will be used as solid fuel, the severity of the process may be increased so that softening may reach 150°C.

At the Schwedt refinery the HSC unit processes the residue from a classic visbreaking unit. The residue obtained in the HSC plant can be used directly as liquid fuel in a power plant by maintaining its temperature at 250°C so that its viscosity stays below 40 cSt. The results obtained in that unit are presented in Table 4.9.

It is remarkable that the distilled fractions have very low metals content. This is explained by the very advanced conversion achieved in the unit.

The gas oils obtained in the unit are hydrofined and are included in the feed for the catalytic cracking unit.

Table 4.9 Typical HSC Plant Performances

Feed:	Sp. gr. 1.027, CCR 21 wt. %, S = 3.94%, Iranian Heavy blends 550°C				
Products	Cracked gas	Naphtha	Light gas oil	Heavy gas oil	Residue
Cut point, °C	C_7^-	$C_5/200$	200/350	350/520	520^+
Yield, wt%	2.1	4.4	12.0	27.5	54.0
LV%	–	5.8	14.0	29.5	50.8
Sp. gr.	–	0.775	0.878	0.956	1.087
S, wt%	–	0.94	2.30	2.95	4.70
Br. no.	–	55	30	10	–
n-C_7 insols., wt%	–	–	–	< 0.05	–
Visc., cSt, at 260°C	–	–	–	–	40
Soft, pt, R&B, °C	–	–	–	–	100

Economics

Investment (basis: 20,000 bpsd, US Gulf Coast 1991), $ per bpsd	1,600
Utilities, typical per bbl feed	
Fuel fired, 10^6 cal	40
Electricity, kWh	1.9
Steam (net produced), ton	0 (balance)
Water, boiler feed, m^3	0.02
Water, cooling, m^3	0.11

Installation: A 14,000 bpsd plant in Petrolchemie und Kraftstoffe AG, Schwedt, Germany; licenser: Toyo Engineering Corp. and Mitsui Kozan Chemicals Ltd.
Source: Ref. 102.

The EURECA process. This process closely resembles the coking process in that the soaker is replaced by two reactors operating alternatively. The working cycle consists of 2 hours of filling and 1 hour soaking. During the whole duration the product is being continuously stripped. A final stripping is performed before emptying the reactor. The complete elimination of the lighter cracking products leads to a very aromatic tar, that is completely devoid of coke particles. The tar is maintained at temperatures of 250–350°C, when it is in liquid state and can be pumped. It is then cooled and crushed into pieces that can be easily packed and transported. It is claimed that no dust is produced during the crushing operation.

The vacuum residue, used as feedstock is heated in the tubular heater until the temperature reaches almost 500°C and is fed to the reactors. Steam, superheated to 600–700°C is used for stripping and the temperature in the reactor is maintained at 400–450°C [23–25].

This tar is heavier than the one produced in the previous process. Main specifications for two typical tars are given in Table 4.10.

Tar is a preferred feed for the production of coke briquets. The product properties are quite equivalent to those using cokery tar.

Two units are mentioned in a published paper. The unit in the Fuji Oil refinery in Japan was started in 1976 and has a capacity of 2,800 m³/day. The other one, in the petrochemical complex in Nanking, China has a capacity of 3,200 m³/day, and was started in 1986. A unit of this type was proposed for the conversion of very heavy crude from Orinoco-Venezuela.

Deep Thermal Conversion

This process is very similar to EURECA. It produces a "liquid coke" (tar) which is not suitable for blending with commercial fuels. It is used for gasification or as solid fuel. Typical yields and process economics data are given in Table 4.11 [103].

4.1.7 Hydrovisbreaking

Three visbreaking processes belong to this category:

> Under hydrogen pressure
> With hydrogen donor
> Under hydrogen pressure, with catalyst in suspension (slurry)

Table 4.10 Tar Characteristics of EURECA Process

Flow point, °C	220	160
d_4^{15}	1.25	1.17
H/C ratio	0.85	1.05
Insolubles, wt% in:		
n-C_7H_{16}	80	65
C_6H_6	55	32
Quinoline	18	4

Table 4.11   Typical Example of the "Deep Thermal
Conversion" Process

Feed, vacuum residue	Middle East
Viscosity, cSt at 100°C	770
Products, wt %	
Gas	4.0
Gasoline ECP 165°C	8.0
Gas oil ECP 350°C	18.1
Waxy distillate ECP 520°C	22.5
Residue ECP 520°C+	47.4
Economics	
Investment: 1,300–1,600 US $/bbl (basis 1998)	
Utilities per bbl	
Fuel, Mcal	26
Electricity, kWh	0.5
Net steam production	20
Cooling water, m^3	0.15

Installation: Until 1998, four units have been licensed; licenser: Shell
International Oil Products B.V.
Source: Ref. 103.

All three processes have as their purpose the fixation of hydrogen during decomposition reactions via free-radicals. The third process uses catalysts with a weak hydrogenating activity (iron oxides). Although the process is exothermic, the products are olefinic in character. This process is in fact borderline between the thermal and catalytic processes. It has many common points with the hydrocracking processes of residues. For this reason it will be described in Section 10.8.

Visbreaking under hydrogen pressure.   The operation uses operating conditions similar to those of classical visbreaking, but under a hydrogen pressure between 80 and 180 bar.

The flow diagram of this unit differs from the classic one by the fact that the products leaving the soaker enter a high temperature separator. The vapor phase leaving the separator is condensed and cooled. The liquid is knocked out and the gases rich in hydrogen are compressed and recycled. The system is identical to other processes that involve hydrogen treating of liquid fractions.

The reaction order and the activation energy are identical to those for classic visbreaking. The distribution of the distillation cuts, their olefinic character, and the distribution of sulfur and nitrogen are also the same as in visbraking. The main differences are in the quality of the residues [6]. The one from hydrovisbreaking is more stable, has a lower viscosity and a lower content of n-pentane and n-heptane insolubles (see Table 4.12).

The consumption of hydrogen is 0.2–0.4% by weight, but its action has not been yet completely elucidated. It is certain that, thermodynamically, it limits the dehydrogenation of the polycondensed naphthene-aromatics, thus reducing the formation of precursors for polycondensation. From the chemical point of view, hydro-

Table 4.12 Residues ibp 375°C of Visbreaking and
Hydrovisbreaking IFP Data

Yield of distillate > 375°C	Feed 17%	Visbreaking 31.76%	Hydrovisbreaking 35.94%
d_4^{20}	0.998	1.046	1.040
Viscosity, cSt at 100°C	198	842	511
S, wt %	5.25	5.09	5.24
N, wt %	0.7	0.95	1.06
Conradson carbon	15.2	24.3	24.2
Insolubles in:			
C_5H_{12}	19.5	29.6	24.2
C_7H_{16}	12.3	26.7	20.0
H/C ratio	0.127	0.112	0.116
Economics IFP data [104]			
Investment US $/bpsd		1,500	2,050
Utilities per bbl			
Fuel, kg		2.4	2.4
Electricity, kWh		0.3	1.9
Steam consumed, kg		–	4.8
Steam produced, kg		2.4	–
Water, m^3		–	–
Hydrogen, $S \cdot m^3$		–	4.8

Source: Ref. 6.

gen is taken up by acceptors. The mechanism of this action is not completely understood. Several hypotheses have been proposed:

Direct activation of the H−H bonds by collision with radicals in the liquid phase [26]. It is however difficult to accept that the radicals in the liquid phase possess enough energy to dissociate the hydrogen molecule.
Hydrogenation of the very reactive pericondensed aromatics that could act as hydrogen donor solvents [27].
The catalytic action of the vanadium and nickel sulfides present in the feed [6].

Our more nuanced interpretation is that in the first step, interactions take place between hydrogen and the free radicals, especially methyl in the vapor phase, with the formation of atomic hydrogen. This is similar to the process that takes place in hydropyrolysis (see Section 4.4.4). The atomic hydrogen formed is much more reactive than the molecular one or the methyl radicals. Atomic hydrogen interacts with the radicals in the liquid phase and interrupts the polycondensation reactions. This hypothesis does not contradict but can accommodate the participation of the pericondensed aromatics and of the vanadium and nickel sulfides.

In view of the high working pressures and of the weak participation of the hydrogen to the reaction, it is doubtful that this process will have an economic justification (Table 4.12).

Visbreaking with hydrogen donor. These processes are much more efficient with respect to the participation of hydrogen to the reaction, than those under pressure of hydrogen discussed above. As a consequence, they lead to larger improvements in the yields and product quality.

The use of hydrogen donating solvents was first mentioned in 1933 in the hydrogenation of coal. For the treatment of petroleum residues, it was patented for the first time in 1947, then developed due to the work of Varga and realized as an industrial process by ESSO as an HDDC cracking process [28] and an HDDV visbreaking [29]. The process continued to be improved thanks to other contributions [30–32].

The process uses as hydrogen donors fractions rich in polycyclic aromatic hydrocarbons, such as: tetraline, dihydroanthracene, dihydrophenantrene, dihydropyrene etc., by themselves or in the mixtures. While in the heater and soaker, these hydrocarbons give up the hydrogen that participates in the reactions. After that, they are resubmitted to hydrogenation, which proceeds without difficulties using the classic methods.

As donor, one may use the easily available, strongly aromatic fractions, such as: the bottom product from the fractionator or the recycle gas oil from the catalytic cracking unit, the tar from the pyrolysis plants, and the tar from the coals coking [32].

Concerning the pressure in the furnace and soaker, it is determined by the used donor and must be enough to maintain it integrally in a liquid phase. In this way the pressure in a unit that uses a hydrogen donor does not exceed the pressure used in classic visbreaking. The retrofitting of a classical visbreaking unit to hydrogen donor operation becomes easy.

The typical flow diagram of a unit for hydrogen donor visbreaking is given in Figure 4.13. The comparison of operating conditions and yields with those of a classic plant (both with soaker) is given in Table 4.13.

The measure for donor efficiency is the change in its composition before and after participation in visbreaking presented in Table 4.14. The data refer to the use as donor of a light cycle oil obtained from a catalytic cracking unit operating under severe operating condition [31].

The usage efficiency of hydrogen is very high. Only very small amounts are eliminated together with the gases of the process (about 1.4% by weight for the C_1-C_3 fraction) [31].

Besides the significant increase of the distillates yield in the detriment of the residue, its increased stability is especially important.

This increase is explained by the interaction of the donor with the radicals present in the liquid phase, which sensibly reduces the condensation reaction. Besides, the asphaltenes contained in the feed are hydrogenated and their concentration in the final product becomes lower than in the feedstock; in some cited cases [32], the decrease is 25%.

The difference between the stability of the residue from a classic visbreaking operation and from hydrovisbreaking with hydrogen donor as measured by the MNI test (percent of insolubles in a paraffinic gasoline of a specified composition) is similar to that for n-C_5H_{12} insolubles), the result of which is given in Figure 4.14 [32].

The figure shows that the yield of the C_4–205°C fraction can reach 30% in hydrovisbreaking, while it is only 10% in the process without hydrogen donor. This explains the differences between the final yields obtained by the two processes.

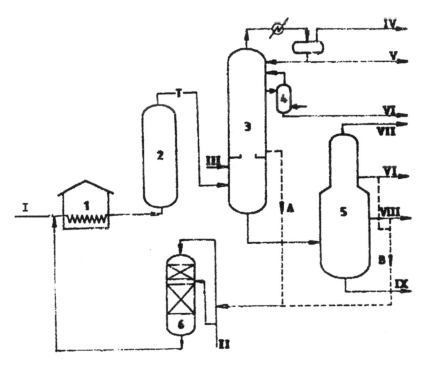

Figure 4.13 Hydrovisbreaking with hydrogen donor. 1—furnace, 2—soaker, 3—flasher-fractionator, 4—stripper, 5—vacuum column, 6—donor rehydrogenation; I—vacuum residue feed, II—hydrogen feed, III—donor, IV—gases, V—gasoline, VI—gas oil, VII—vacuum system, VIII—heavy gas oil, IX—residue; A—recycling with aromatic donors, B—recycling with polyaromatic donors.

The increase of residue stability depends to a great measure on the residue/donor ratio and on the amount of hydrogen used (Figure 4.15).

In selecting optimal operating conditions one must take into account not only the quality of the obtained residue, but also the economics of the process.

Here are some orientation figures for the costs associated with such a process: for a unit processing 7200 m^3/day Canada tar, the investment was of USD73 mill. The cost of production, including labor, maintenance, amortization of the investment, utilities, etc. amounts to USD26/m^3 (all in 1981 dollars).

Hydrogen donor hydrovisbreaking produces high yields of valuable products, operates at moderate pressure (generally below 30 bar) and uses inexpensive and readily available donors. It is expected that in the future this process will further expand and find new applications.

4.2 COKING

Between 1900–1910 some countries made coke by adding petroleum residues to coal. After the emergence of thermal cracking units to produce coke, the pressure-reducing valve on the outlet of the reaction chamber was taken out of use, heater outlet temperature was increased, and fractions boiling above 300°C were integrally

Table 4.13 Comparison of Visbreaking and Hydrogen Donor Visbreaking

Feedstock properties

Sulfur	4.27	Softening point, °F	130
Specific gravity	1.023	Modified naphtha insolubles	21.2
Conradson carbon	23.6	Characterization factor	10.2

Operating parameters	Normal visbreaker	Hydrogen donor visbreaker
Unit type	soaking drum	soaking drum
H_2 added scf/bbl donor		400
Make-up donor		FCCU fract. bottoms
Residue/donor ratio		2
Pressure, psig	350	350
Furnace outlet temperature, °F	865	860
Residue to furnace, °F	500	500
Donor to furnace, °F		630
Heat flux, same residue and same furnace		11% increase

Yields	vol. (%)	wt. (%)	vol. (%)	pc > wt. (%)
C_2 and lighter		0.7		1.5
C_3		0.5		0.9
C_4	0.8	0.5	1.6	0.9
$C_5 - 430°F$	15.0	11.3	11.0	8.3
$430° - 1,000°F$	32.0	28.8	55.8	50.6
$1,000°F$ + residue		58.2		38.1
Hydrogen added				0.3
MNI increase		37		12

Table 4.14 Typical Hydrocarbon Composition of Donor Before and After Processing[a]

	Fresh	Spent[b]
Paraffins	7.2	8.3
Cycloparaffins	2.1	3.5
Condensed cycloparaffins	0.5	2.0
Benzenes	10.9	6.5
Tetralins	48.7	17.3
Dicycloalkylbenzenes	7.4	2.3
Naphthalenes	19.4	50.2
Other aromatics	3.8	7.4
Aromatic sulfur	–	2.4
Σtetralins + naphthalenes	68.1	67.5

[a] Mass %.
[b] Recovered donor after bitumen conversion.

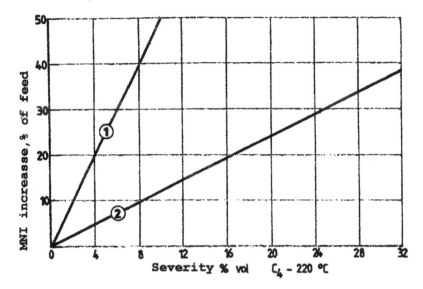

Figure 4.14 Residue stability. 1—visbreaking with soaker, 2—hydrovisbreaking. Residue/ donor ratio = 2; H_2/feed ratio: 70 $STPm^3$ H_2/m^3 feed.

recycled. The coke deposited in the reaction chamber, which became thus a coking chamber, and the operation of the unit became intermittent.

Later on, in order to extend the working cycle the number of the chambers was increased. From 1933 on, owing to improvements in the methods for removing coke from the chambers, the operation of the furnace and the separation equipment

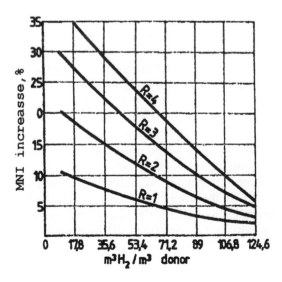

Figure 4.15 Stability increase when using a polycyclic donor. R = feed/donor ratio. (From Ref. 32.)

became continuous, while the operation of the coke chambers alone remained cyclical. In 1934, by feeding the unit with gas oil instead of residue, needle coke was obtained for the first time.

As with other thermal cracking processes, the interest in coking diminished following the appearance of catalytic cracking. However, beginning in the 1950s a new generation of coking process named "delayed coking" emerges. In these units, light gas oil and heavy gas oil are not recycled but serve as supplementary feed for the catalytic cracking unit. More recently, light gas oil is preferably used as a component for diesel fuel and only the heavy gas oil is submitted to catalytic cracking. Concomitantly, general use of electrical desalting of crude allowed one to obtain coke with a very low ash content, used for the fabrication of electrodes for production of aluminum.

Through the 60s and 70s, the production of petroleum coke increased continuously, especially after 1980. During the years 1983–1984 alone, the additional needle coke capacity put into operation in the U.S. was 5 million tons [33]. From 1960–1990, needle coke world production increased fourfold.

In 1990 world production of petroleum coke was about 33 million tons, distributed among the various regions of the world, as follows:

%	Region
71	North America
9	Western Europe
6	Eastern Europe
3	Middle East
7	Asia and Oceania
4	South America

New capacities built, especially in the U.S., contributed to world production of 40 million tons/year in 2000.

Concerning the relative importance of various types of coking processes, delayed coking leads with 88% of the world capacity (1990). Coking in a fluidized bed was first commercialized in 1954 and initially showed high promise. However, in 1990 it represented only 11% of production capacity, and its more recent variation, "flexicoking," which first appeared in 1976, represented only 1%. Other processes are only prototypes of experimental units, with negligible impact on world capacity.

The uses of petroleum coke in various regions of the world are quite varied. Some of the main uses: solid fuel for heating (Japan—46%), production of cement (Europe—49%), of steel, in the ceramic industry, gasification, etc.

Besides the coke, this process produces important amounts of distillate and smaller amounts of gasoline and of gases. In all the coking processes, the final boiling temperature of the distillate may be controlled as desired, the heavy fractions being recycled. In most cases, the process produces two fractions of distillate: a light fraction with the distillation range of 200–340°C and is used as a component for Diesel fuel after hydrofining, and a heavy fraction that is sent as feed to the catalytic cracking.

The conversion to coke and the decomposition products resulting from the process are presented in the graph of Figure 4.16.

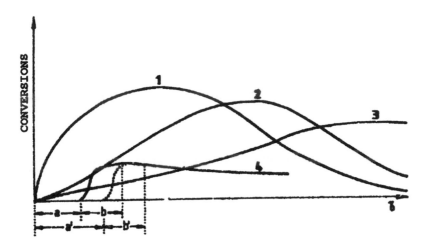

Figure 4.16 Evolution of conversions in coking. 1—heavy gas oil, 2—light gas oil, 3—gasoline and gases, 4—coke; a, b—conversions in delayed coking (a—in furnace, b—in coke drum); a′, b′—change of conversions by steam injection.

After the formation of coke has leveled off, the condensation processes inside the coke mass continues. This results in the decrease of the content of hydrogen and volatile substances within the coke mass. Coke with a low content of volatile substances is obtained from processes in which the conversion is not limited by technology considerations, such as in the moving bed or fluidized bed processes. The low content of volatile substances represents the essential advantage of these processes. Thus, for coking in a moving bed the need for additional calcinations of the coke in special units is eliminated. It is much limited for the product of coking in a fluidized bed.

The differences that exist between the operating conditions of different industrial coking processes (temperature, pressure, and residence time) determine important differences also in the quality of other products, especially of the midrange fractions. The high temperatures in the fluid coking reactor lead to a pronounced aromatic character of the midrange fractions. In absence of a deep hydrotreatment, these fractions are difficult or impossible to use, either as components for Diesel fuel or as feed for catalytic cracking. Actually, this is the reason for the stagnation of the expansion of the fluidized bed coking process. It is also to be mentioned [6], that the coke obtained in the fluidized bed process, despite its reduced content of volatile substances, is not suitable for the fabrication of electrodes for the electrochemical or metallurgical industries.

4.2.1 Delayed Coking

Delayed coking uses the tubular furnace and coking chambers as reaction equipment. It was developed on the basis of industrial experience with thermal cracking but has specific features.

The main characteristics of the process follow from the time evolution of the products formation, as a result of the decomposition and condensation reactions.

Thus, due to the fact that the feed is vacuum residue, the intense formation of carboids (coke) occurs before the maximum of the first intermediary product is reached (Figure 4.16).

This fact limits the function of the tubular furnace to heating of the feed until reaching a conversion that is lower than that at which the intense formation of the coke begins (point "a" on the abscissa in the graph in Figure 4.16). The injection of water or steam in the furnace delays the formation of coke deposits inside the tubes, and makes it possible to increase conversion (from point "a" to point "a'" on the abscissa of the graph).

After reaching this conversion, the heater effluent enters one of the coking chambers (the chambers work alternately) where the reactions continue in adiabatic mode and the conversion exceeds the plateau on the coke conversion curve (Figure 4.16).

As result, coke is deposited on the walls of the chamber and the vapors pass on to the fractionation system.

Figure 4.17 depicts the flow diagram of the delayed coking unit.

It is noteworthy that polycondensation reactions continue in the formed coke mass even after reaching the upper plateau on its formation curve. For this reason, the volatiles content of the coke decreases with increasing total conversion. Since the process in the coking chamber is adiabatic, the final conversion is in principle limited. The higher the furnace outlet temperature (generally it varies between 900–950°F, or 482–510°C), the higher will be the final conversion, the lower the content of volatiles in the coke, and the larger the amount of collected distillate.

The requirement to increase the heater outlet temperature without exceeding the outlet conversion means increased thermal stress on the heater tubes and increased flowrate within the tubes. That may be opposed to the present trend of striving to lengthen the operation cycle of the furnace between two decoking operations [34,35]. In order to increase the operation cycle while achieving the desired conversion, beginning in 1980 changes were brought to delayed coking heaters

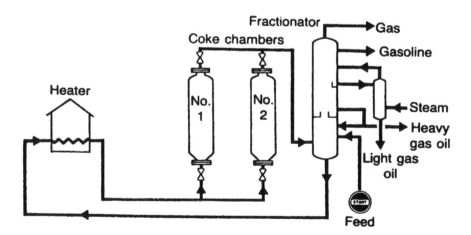

Figure 4.17   Delayed coking flow diagram.

(Figure 4.18). The combustion chamber was increased and the mean thermal stress in the radiation section was fixed at 10^5 kJ/m^{2}h, while those used previously were of 1.13×10^5–1.36×10^5 kJ/m^{2}h. During the same period linear velocity in the tubes was set at 1.8 m/s, based on cold feed [35].

The result of these measures was the increase of the length of the operation cycle of the furnace to 9–12 months, reaching in some cases 18 months.

In order to increase the final conversion, occasionally one introduces into the flow entering the coking chamber, a heating agent preheated to a convenient temperature in a separate coil. Relatively light fractions are used for this purpose,

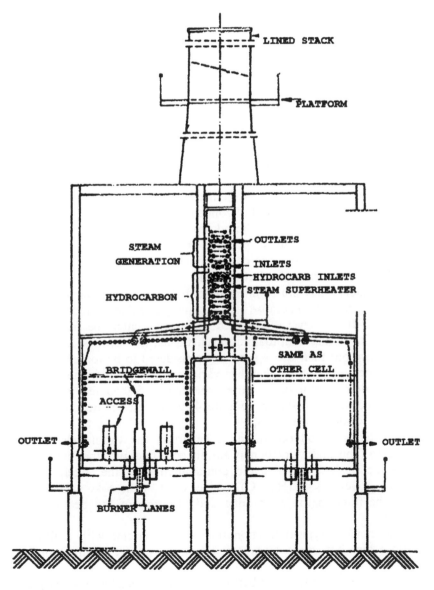

Figure 4.18 Section and elevation of typical coker furnace.

since they may be heated at higher temperatures, without undergoing decomposition reactions.

However, even when using these possibilities to the maximum, the conversion in the delayed coking units is limited by the adiabatic character of the formation of coke in the chambers.

As discussed, the process consists of displacement of the phenomena of coke formation from the furnace to the chambers and it appears as a delay in coke formation, thus the name of the process: "delayed coking."

The deposition of coke in the chambers determines the cyclic character of their operation; the chamber filled with coke is taken out of the circuit for removing the coke and for cleaning. In its place another chamber is connected to ensure the continuous operation of the rest of the plant.

Accordingly, a coking unit comprises 2–3 chambers, their number being determined by the ratio of the length of time for emptying the deposited coke from the chamber and the duration of filling the chamber with coke formed during the process.

The coke is formed from the liquid products after they leave the furnace. It is therefore necessary to maximize the duration it spends at high temperature.

For this reason, the circulation of the products through the chamber in delayed coking is upflow, which is the opposite to that through the reaction chambers of the cracking units for straight run residue and is analogous to the circulation in the soakers of the visbreaking.

The upflow circulation of vapors through the coking chamber and the continuation of decomposition reactions and of condensation in the coke mass causes it to have a porous structure with included liquid products. Upon taking the chamber out of the filling stage, intense steam stripping is practiced for removing the liquid included in the pores. Nevertheless, the coke produced by this process still has a content of 6–11% volatiles (determined at 620°C).

The coking chambers may not be filled more than 2/3–3/4 of their volume (20 feet below the outlet nozzle [105]) because of the foaming phenomena that take place above the coke level. If this level is exceeded, foam may enter the fractionation column and lead to its coking. Antifoaming agents, such as silicones, are efficient in controlling the foam but their use is generally considered to be too expensive.

The vapor velocity in a coke chamber varies upon feed characteristics and pressure. Without adding antifoam it varies between 0.35–0.55 feet per second and with antifoam between 0.4–0.7 fps. Figure 4.19 indicates more precisely the maximum allowable vapor velocity [105].

If quench water is introduced in the coke drum, the vapor velocity may increase to over 2 fps. This high velocity may entrain coke fines into the fractionator. These fines must be washed out to avoid contamination of the coker distillate streams. The heavy coker gas oil (HCGO) receives most of the coke fines that are not removed in the fractionator wash zone and filtration of the heavy gas oil becomes necessary.

Figure 4.20 shows a typical fractionator bottoms circuit with a filtration system loop and a high-hat screen to protect the heater charge pump from large coke pieces [105].

The removal of coke from the chambers is presently done exclusively by hydraulic cutting. After the chamber is taken off-stream and following stripping, cooling,

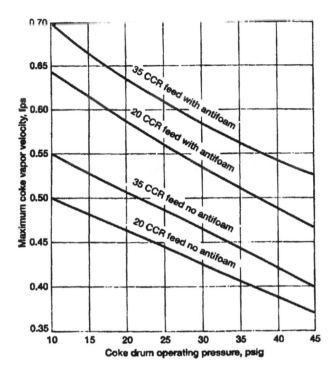

Figure 4.19   Coke drum maximum vapor velocity. (From Ref. 105).

and opening of the upper and lower ports, a vertical hole is drilled through the mass of coke by means of a special drilling installation placed on top of the coking chambers. A hydraulic cutting tool is then fastened to the drill stem. Water under a pressure of 80–120 bar feeds the cutting tool and forms high-velocity, horizontal jets that impact

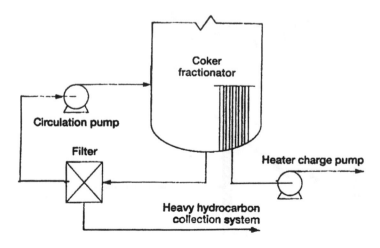

Figure 4.20   A typical fractionator bottoms circuit with a filtration system and high-hat screen.

the hot coke, shattering it. The cutting tool rotates continuously and advances from the bottom to the top, and in a second pass, from the top to the bottom of the chamber. In this manner, the coke mass is cut into chunks that fall and are evacuated together with the water through the bottom port of the chamber.

There are several systems for the collection and separation of the mixture of water and coke that leave through the lower opening of the chamber [34]. The most representative systems are described below.

A quite simple system is the direct collection of the mixture in railcars. The water and the fine particles of coke flow to a separator, while the chunks of coke are retained in the railcar. The system has the lowest investment, but has the disadvantage of requiring a relatively longer time for decoking the chamber.

In another system, the water and coke are drained into a concreted tank from which the water is separated by overflowing a system of weirs and the coke is loaded to trucks by means of a crane (Figure 4.21).

The most advanced systems have water separation vessels. These systems have the advantage of eliminating the pollution of the environment that occurs inherently in the previous systems that handle the coke in open systems.

There are two types of such separators (see Figures 4.22 and 4.23 [34]). Their operation is self-explanatory. The second system (Figure 4.23) has the advantage of eliminating the operation of crushing the coke to the sizes suitable for pumping the coke in water slurry. The investments for this system are somewhat higher because of the necessity to place the coking chambers at a higher elevation.

In all systems, coke fines remain in the cutting water. Their separation can be performed in settling tanks, or more efficiently, by means of hydrocyclones [107].

Process automation of delayed coking [108] will significantly improve the safety and efficiency of coke extraction operation.

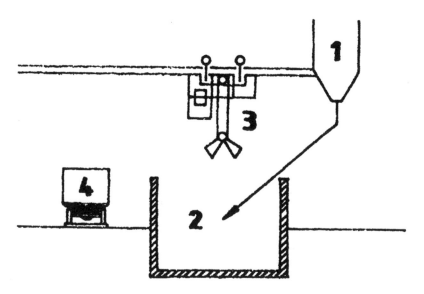

Figure 4.21 Coke discharge system. 1—coke drum, 2—concrete tank, 3—handling system, 4—railcar.

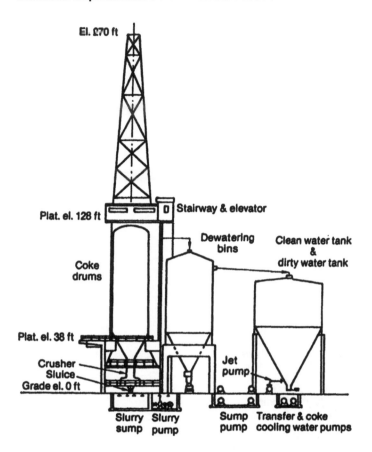

Figure 4.22   Coke discharge system. Type 1.

The vapors leaving the coking chamber at a temperature of about 450°C enter the fractioning column that recovers overhead gasoline and the gases and usually two side fractions of distillate.

The feedstock is preheated in heat exchangers and introduced into the fractionation column, above the feedpoint for the vapors. In contact with the vapors, the heavy components are condensed back into the feedstock and are recycled to the heater. The recycle ratio is regulated (Figure 4.17) by returning to the column a portion of the heavy gas oil (cooled or not) and/or by having a portion of the feed bypass the contact with the hot vapors. Other methods than that in Figure 4.17 may be used in order to regulate the recycle ratio, and concomitantly, the distillation end point of the heavy gas oil. It must be stressed that the recycle ratio has a strong effect on the yields of final products, as illustrated by the data of Table 4.15.

The table shows that the increase of the recycle ratio leads to the increase of the yields of coke and light distillate while reducing the yield of heavy distillate. Due to their high stability, heavy distillates may not be completely converted to other products even if high recycle ratios are practiced. The recycle ratio varies in the range 1.05–3. The present trend is to reduce it to values that produce the desired distillate quality.

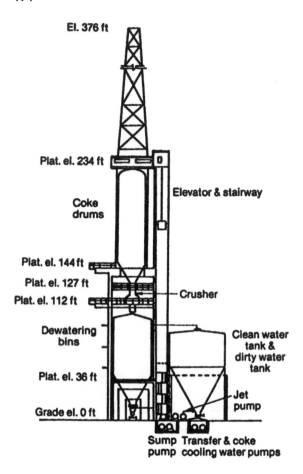

Figure 4.23 Coke discharge system. Type 2.

Delayed coking is a very flexible process that allows the processing of various feeds. They may range between residues from the vacuum distillation of the crude with densities of the order of 0.941 and 6% coke, and the tar sands bitumen or tar from the coal calcining with density of 1.28 and 48% coke.

Table 4.15 Influence of the Recycling Coefficient on Final Yields

Products	Recycling coefficient		
	1.0	1.2	1.4
Gases C_1–C_3	4.0	5.5	6.0
Gasoline + C_4 fraction, fbp 204°C	13.0	15.0	15.5
Gas-oil fbp 343°C	21.5	26.0	28.5
Heavy distillate	46.0	36.0	29.5
Coke	15.5	18.0	20.0

Typical yields and economics for the delayed coking of various feedstocks are given in Table 4.16 [37,38].

The yield of coke and of distillates varies as a function of the pressure in the coking chambers. The effect of pressure at various recycle ratios is illustrated in Fig. 4.24.

The present trend is to operate in conditions yielding a maximum of distillates [39,99], that requires the lowest possible pressure in the coking chambers. If one takes into account the pressure drops and also the lowest acceptable overpressure in the separator located at the top of the fractionation column (0.14 bar), the following values are required in various points of the system:

Location	Bar
Top of the fractionation column	1.46
Bottom of the fractionation column	1.70
Top of the coking chamber	2.05

It must be mentioned that the operation of the unit at such low pressures, i.e., below 2.75 bar in the coking chambers, requires the inclusion in the flow sheet, and a compressor for introducing the gases produced by the delayed coking unit into the gas circuit of the refinery [39].

The apparently slight modification of the pressure (from 2.75–2.05 bar), together with the decrease of the recycle ratio from 1.15 to its low limit of 1.05, influence strongly the yield and characteristics of the heavy gas oil (see Table 4.17).

The requirement to maximize the yield of distillates determines also the operation at high limit of the range for the heater outlet temperature, which produces a coke with 8–10 wt.% volatiles.

If the temperature is too high, the volatiles content decreases below this value. The coke becomes very hard and difficulties appear in the operation of the hydraulic cutting devices [39].

Approximate relationships, that can be used for obtaining approximate values for the yields are [35,40]:

Coke % by weight $= 1.6 \times C$

Gases $(< C_4)$, % by weight $= 7.8 + 0.144\ C$

Gasoline, % by weight $= 11.29 + 0.343\ C$

Gas oil, % by weight $=$ by difference

where C is Conradson carbon in weight %.

A much more exact graphical estimation, which takes into account the operating condition of the unit, is given by B.P. Castiglioni [41] and reproduced in the previous work by this author [3,97].

An estimate of the yields of the products of delayed coking of heavy feedstocks using measured asphaltenes content and carbon residue was published by J.F. Schabron and J.G. Speight [100].

The bromine number of the distilled fractions is about 60 for gasoline, 30 for light gas oil and 15 for heavy gas oil. The cetane number of the gas oil is about 40.

Table 4.16 Typical Delayed Coking Data by Licenser

	ABB Lummus Global Inc.				Foster Wheeler/UOP	
Operating conditions						
Furnace outlet °F	900–950				900–950	
Coke drum pressure, psig	15–90				15–100	
Recycle ratio vol/vol feed	0.0–1.0				0.05–1.0	
Yields						
Feed						
source	Middle East vacuum residue	Hydrotreated vacuum residue	Coal tar pitch	Venezuela vac. res.	N, Africa vac. res.	Decant. oil
Type						
Gravity °API	7.4	1.3	11.0	2.6	15.2	−0.7
Sulfur, wt %	4.2	2.3	0.5	4.4	0.7	0.5
Concarbon, wt %	20.0	27.6	–	23.3	16.7	–
Products, wt %						
Gas + LPG	7.9	9.0	3.9	8.7	7.7	9.8
Naphtha	12.6	11.1	–	10.0	16.9	8.8
Gas oil	50.8	44.0	31.0	50.3	46.0	41.6
Coke	28.7	35.9	65.1	31.0	26.4	40.2
Operation	Fuel grade coke			Max. dis.	Anode coke	Needle coke
Economics						
Investment						
Basis U.S. Gulf 1998	20,000 bpsd vacuum residue				65,000–100,000 bpsd	
US $ per bpsd	4,000				2,500 – 4,000	
Utilities, per bbl feed						
Fuel, 10^3 Btu	145				120	
Electricity, kWh	3.9				3.6	
Steam (exported), lb	20				40	
Water cooling, gal	180				36	
Installation						
Until 1998	More than 55 units				More than 58,000 tpd of fuel, anode and needle coke	

Source: Ref. 37, 38.

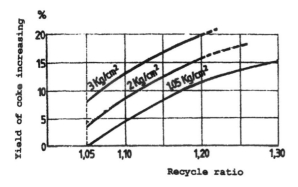

Figure 4.24 Effect of recycle ratio and of pressure on coke yield.

The distribution of sulfur and nitrogen in weight% is given in Table 4.18 [40]. Useful economic data for delayed coking units are given in Table 4.19.

4.2.2 Needle Coke Production

Needle coke is produced in delayed coking units, using selected feedstocks and suitable operating conditions.

A high quality needle coke must have a low coefficient of thermal expansion, high density, good electrical conductivity, and a low tendency to swell [42].

The essential condition for the production of needle coke of good quality is to use feedstocks having a very strong aromatic character and a low content of asphaltenes, sulfur, and heavy metals, especially of vanadium. For this purpose the feeds of choice are: the distillate obtained at the bottom of the fractionation columns of catalytic cracking units (after the removal of catalyst traces); the tar obtained in the process of pyrolysis; the extracts from the selective solvent refining of oils; the coal tars; feedstock prepared intentionally by the thermal cracking of distillates (Section 4.1.5). Despite not being mentioned in the literature, we think that gas oils obtained from fluid cocking, which have a very strong aromatic character, could be also used.

Table 4.17 Heavy Gas Oil Characteristics

	Operating conditions	
	$p = 2.75$ bar $K = 1.15$	$p = 2.05$ bar $K = 1.05$
Yield, wt %	25.7	35.2
End point °C	493	570
d_4^{20}	0.9365	0.9574
Conradson carbon	0.35	0.8–1.0
Ni + V ppm	0.5	1.0

Table 4.18 Sulfur and
Nitrogen Distribution in
Delayed Cooking in wt %

	Sulfur	Nitrogen
Gases	30	–
Gasoline	5	1
Kerosene	35	2
Gas oil	35	22
Coke	30	75

Operating conditions depend on the used feedstock being situated near to the upper limits of the operating range of delayed coking. Since the feedstock for needle coke is much more stable than the vacuum residue, which is the usual feedstock for delayed coking units, the inlet temperature in the coking chambers must be higher, usually from 500–505°C. The operating temperature depends on the used feedstock and on the danger of forming coke deposits inside the furnace tubes. In some cases, the heater outlet temperature can reach 524°C [35].

The operating pressure is generally limited only by the design of the unit. As confirmed by research studies [42] operation at higher pressure is beneficial. Operation at a pressure of 10–11 bar was recommended [35].

The entire amount of heavy gas oil produced during the process is recycled in order to increase the yield of the coke produced. Thus, in some cases, a recycling ratio of about 2 is used, which is much higher than when the unit processes vacuum residue. For this reason, when producing needle coke the actual capacity of the unit will be lower than the nominal one. Two examples on the conditions used in a delayed coking unit for producing needle coke are given in Table 4.20.

The coke obtained in the process is generally submitted to calcination in order to lower the content of volatiles.

Table 4.19 Economics of Some Delayed Coking Units

	Lummus	Foster Wheler
Investment $/m^3 feed	25,000	15,000–25,000
Plant capacity, m^3/day	3,200	1,600–10,000
Reference year	2000	1998
Consumptions per m^3 feed		
Fuel, 10^3 kJ	960	795
Electricity, kWh	24.5	22.6
Cooling water, m^3	4.3	0.85
Produced steam	57	115

Table 4.20 Examples of Needle Coke Production

	Decanted Catalytic Cracking Column Bottom	Tar
Feed		
d_4^{20}	1.082	1.085
S, %	0.5	0.56
Conradson carbon	–	8.6
Yields, wt %		
Gases	9.8	18.1
Gasoline	8.4	0.9
Gas oil	41.6	21.1
Coke	40.2	59.9
Operating conditions		
Furnace outlet, °C	510	502
Pressure in drum, bar	–	3.5
Recycle coefficient	2.0	2.08

4.2.3 Coking on a Heat Carrier

The use of a solid heat carrier has the advantage that it makes possible reaching any conversion of the feedstock to coke and lighter products. Indeed, contrary to delayed coking, the conversion is not limited any more by formation of coke deposits inside the tubes of the furnace and on the adiabatic operation of the cracking chambers.

The reaction temperature is reached by the contact of the feedstock with the heat carrier that was preheated to the necessary temperature. The desired conversion is obtained by controlling the residence time in the reactor. The coke is deposited on the heat carrier, which may be made if a material that is inert towards the reaction or coke particles. The products in the vapor state are separated from the carrier. The carrier particles are sent back to be reheated by the partial or total combustion of the coke deposited on them.

Depending on the type of heat carrier used in the reactor, there are processes with a moving bed of heat carrier and units with a fluidized bed carrier. The latter may be implemented as "flexicoking" where the coke formed in the process is gasified.

Coking in the fluidized bed. The first industrial unit for coking in a fluidized bed had a capacity of 600 t/day, and was put in operation in December 1954. Presently, this process is second after delayed coking, and accounted for 11% of world coke production in 1990.

The basic process flow diagram is given in Figure 4.25.

Variations to this scheme occur mainly in the transport system and with reference to the place where the preheated feedstock is fed: directly in the reactor as in Figure 4.25, or in the lower portion of the column situated above the reactor, from which it goes in the reactor together with the recycling material. The heat carrier is formed by coke particles.

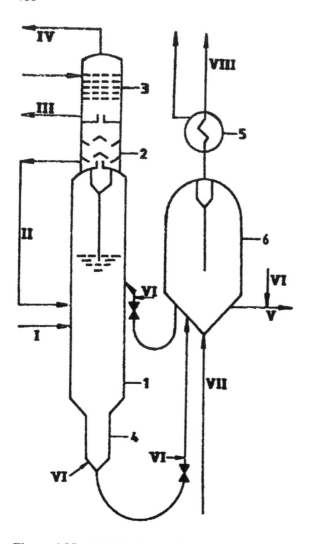

Figure 4.25   Fluidized bed coking. 1—reactor, 2,3—fractionation column, 4—stripper, 5—
steam production, 6—heat carrier reheating; I—feed, II—recycle, III—heavy gas oil, IV—to
fractionation, V—coke, VI—steam, VII—air for coke burning, VIII—flue gases.

In comparison with other units, using fluidization, coking units have the particular trait that the feedstock, being introduced in the reactor in liquid state, cannot
ensure the fluidized state. The fluidization is ensured by the steam introduced in the
lower part of the reactor. Its speed in the reaction zone must range between 0.3–0.9
m/sec, depending on the operation conditions, especially reaction temperature [36].

In the upper portion, the reactor has a larger diameter, because through the
upper section flows not only the steam but also the vapors of the products formed in
the process.

This portion serves as the reaction zone, whereas in the lower portion of the
reactor only the steam circulates, performing concomitantly the stripping of volatiles
from coke particles.

The cyclones, traditionally used for capturing particles from a gas stream, cannot be used here due to the danger of coking them. The coke particles entrained from the fluidized bed are partly carried in the column that is placed directly above the reactor.

This placement avoids the transfer lines and the danger of their becoming filled with coke. The entrained coke particles are returned back to the reactor together with the recycled material; a minimum flowrate of recycling is necessary for good operation of the unit.

The column is provided in the lower section with baffles and in the upper section with fractionating trays.

In order to obtain coke with a reduced content of volatiles, the residence time of the feed in the reactor is 15–20 sec. However, even with this residence time the coke remains with a residual content of 1–2% of volatiles, which requires an additional calcination. This occurrence is due to the backmixing present in the fluidized bed and leads to some particles leaving the reactor before the coke formation process is completed.

The rate of feed supplied to the reactor, expressed as kg per hour and per kg of heat carrier circulating through the reactor, depends on the coke content of the feedstock, on the temperature, on the recycle rate of the heavy products from the column, and on the velocity of the steam through the reactor.

An improper control of the parameters leads to an increasing amount of sticky products, which are intermediaries in the formation of coke, the agglomeration of the particles of coke, and finally to the blocking of the reactor.

The estimation of the feedrate, as a function of the reactor temperature and the content of coke in the feedstock, may be performed by means of the graph of Figure 4.26 [45].

In this graph, the feedrate is expressed by the ratio of the hourly circulation rates of the feed and of the heat carrier (coke).

A high steam velocity in the reactor may excessively shorten the contact time, and make it become insufficient for achieving the desired conversion. For this reason, steam velocity of the order of 0.9 m/sec, which is needed in order to maintain the reactor diameter at reasonable values, will require quite high reaction temperatures. For operating at 510–525°C, the vapors velocity must be maintained within the limits 0.15–0.30 m/sec.

It seems that best results are obtained at temperatures of 540–550°C and at a velocity of 0.3 m/sec.

At reaction temperatures over 565°C, gas oil becomes excessively aromatized, and its use is difficult without a deep hydrotreatment.

The results obtained in coking depend to a great extent on the recycling of the heavy products formed in the process.

In normal operating conditions the fractions boiling above 545°C are recycled and the coke formed in the process represents 110–130% of the Conradson carbon of the feedstock. This percent decreases sensibly if recycling is zero [36]. The improvement of the quality of the heavy distillate, in view of its use in good conditions in the catalytic cracking process (d_4^{20} between 0.876 and 0.904, coke under 0.3%), imposes a high recycle ratio. This fact however diminishes the capacity of the unit. The recycled substance, being more stable as the feedstock, needs a somewhat longer residence time in the reactor or a temperature situated towards the highest admissible limit.

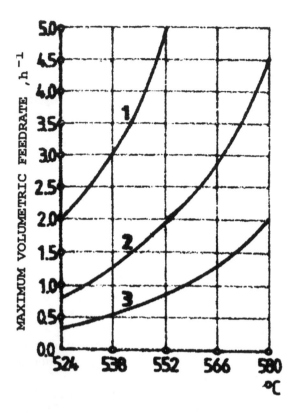

Figure 4.26 The maximum feeding velocity function of the reactor temperature, for three feed qualities. 1—Feed with Conradson carbon = 12%, 2—Same = 22%, 3—Same = 29%.

The coke particles are reheated by burning a portion of the coke in a fluidized bed furnace. Laboratory tests using a fluidized bed with a depth of 0.15–0.25 m, with 90% of the particles having diameters comprised between 0.25–0.67 mm, were reported. Operating the fluidized bed at the temperature of 600°C, with superficial gas velocities of 0.5–0.8 m/s, burning rates of 270–300 kg/hour·m^3 were measured [46].

In commercial units, the height of the fluidized bed is of 3–4 m, and the burning rate of coke is somewhat lower.

The mechanical qualities of the coke do not change if the residence time in the furnace is between 5–10 minutes. At excessively long residence times the particles become brittle and their volume density decrease from 1.030 to 0.805 g/cm^3.

The coke burnt for preheating represents generally about 5% of the feedstock. The rest, preferably the larger particles, is removed from the system.

The yield and the quality of the products obtained by the fluidized bed coking of various feedstocks, with recycling of the fractions with boiling points above 545°C, are given in Table 4.21.

Typical characteristics of the coke are: 0.6% volatiles at 593°C and 5.0% at 950°C; hydrogen content in the coke is 1.6%.

Table 4.21 Fluidized Bed Coking of Various Feedstocks

	Feedstocks				
	I	II	III	IV	V
Density	1.0427	0.9509	0.9484	1.0772	1.1010
Conradson carbon	24.5	11.0	5.0	33.0	41.0
Sulfur, %	4.3	0.7	0.5	2.3	2.1
Nitrogen, %	0.28	0.32	0.2	1.9	2.1
C/H ratio	8.4	6.2	7.0	9.2	9.9
Viscosity at 135°C, cSt	88	55	–	–	–
Distillation at 538°C, % vol.	10	0.0	40	0.0	5.0
Gases C_3, wt %	9.5	7.0	6.0	10.0	11.5
Overall gasoline, wt %	27.5	11.5	8.0	36.0	48.5
C_4, % vol.	3.5	2.0	2.0	3.0	3.0
Debutanized gasoline, fbp 221, % vol.	19.5	21	17	17.5	14.5
Gas oil 221–546°C, % vol.	52	68.5	74	44.5	32.5
Gasoline					
Density	0.7547	0.7425	0.7547	0.7628	0.7587
Octane number F1	77	66	73	82	–
Sulfur, wt %	0.9	0.2	0.1	1.7	0.8
Gas oil					
Density	0.9529	0.8816	0.9100	0.9729	0.9659
Conradson carbon	2.8	1.2	0.8	3.1	1.6
Sulfur, %	3.7	0.4	0.4	2.2	1.7
K_{UOP}	11.17	11.98	11.81	10.97	11.01
Temperature 50% dist.	416	416	453	416	423

The yields of coke and gases for coking in a fluidized bed may be estimated by means of the relations [36]:

$$C = 1.15c$$
$$C + G = 5.0 + 1.30c \tag{4.1}$$

In these relations, C and G represent the percent of coke and the percent of gases by weight, and c, the content of Conradson carbon in the feedstock.

The relations (4.1) are approximate since they do not take into account the recycle ratio.

More exact results are obtained by using Figures 4.27 and 4.28 [48]. Figure 4.27 allows the determination of the yields of coke and gas oil and of their main characteristics.

When the feed is straight run residue, the yield of debutanized gasoline represents 17% (density of gasoline 0.755) and when feeding vacuum residue, it represents 21% (density of gasoline 0.765) of the product.

The yield of the C_1–C_3 fraction is calculated by difference.

Figure 4.28 gives the octane number of the gasoline, the distribution of sulfur among the fractions, and the yield of C_4 fraction.

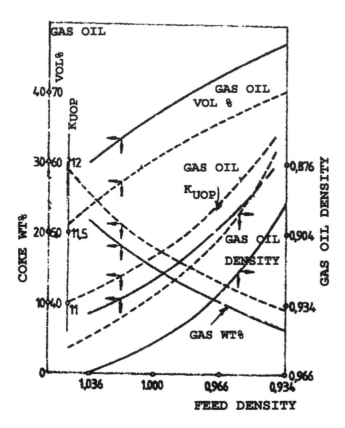

Figure 4.27 Gas oil and coke yields and density in fluid coking. (From Ref. 48). Continuous line: straight run residue, dotted line: vacuum, thermal cracking or deasphalting residue.

The investments for fluid coking are approximately the same as for the delayed coking, but operating costs are higher [40].

Flexicoking. Flexicoking is a more recent variation (1976) of coking in a fluidized bed. It is the combination between the classic fluidized bed coking and the gasification of the produced coke. In 1990 the process accounted for 1% of world coking capacity.

The typical simplified process flow diagram is given in Figure 4.29. This process flow diagram depicts both the classic system, with a single gasifier (without the part delimited by the dotted line) and the Dual system, provided with two gasifiers. In the second gasifier, the gasification is performed with steam and synthesis gas is obtained [49].

In both versions (classic and dual), the reheating of the coke in the dedicated equipment is not done by injection of air and partial burning of coke, but instead by contact with gases emitted from the gasifier. In the gasifier, partial burning of coke is carried out with a mixture of air and steam, and fuel gas is obtained. When a second gasifier is added, the coke from the first gasifier is sent to the second, where the

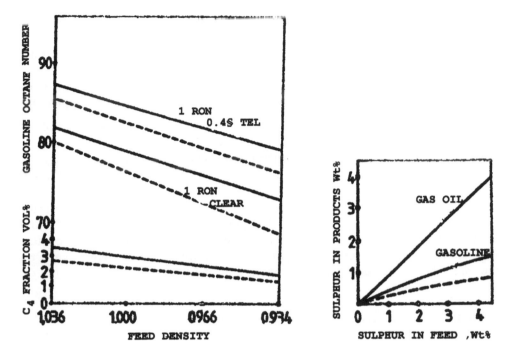

Figure 4.28   Gasoline octane number, C$_4$ yield, sulfur distribution in fluid coking. Continuous line: straight run residue, dotted line: vacuum, thermal cracking or desasphalting residue. (From Ref. 48.)

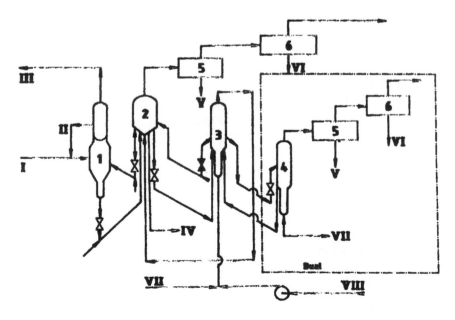

Figure 4.29   Flexicoking process flow diagram. 1—reactor, 2—carrier reheating, 3—gasifier, 4—"dual" gasifier, 5—coke dust removal, 6—desulfurization; I—feed, II—recycle, III—vapors to fractionation, IV—coke, V—coke dust, VI—sulfur, VII—steam, VIII—air.

synthesis gas is obtained. This may be used for the synthesis of methanol or of hydrogen, for the production of ammonia.

Typical yields and economics for the fluid coking and flexicoking processes are given in Table 4.22 [109].

A comparison between the yields obtained by delayed-, fluid- and flexicoking of the same feedstock, is given in J.H. Gary and G.E. Handwerk's monograph [101].

Moving bed systems. The initial attempts to develop systems with moving beds of coke particles, or inert material encountered great difficulties due especially to coke depositing on the walls of the reactor and eventually blocking circulation. In order to eliminate this phenomenon, a very high carrier/feed circulation ratio and other measures were needed, which made the system to be uneconomical [3].

Recently, the solution to these difficulties was obtained by shaping the reactor as a horizontal cylinder. The coke particles enter at a temperature of 600–700°C and are moved by means of a screw conveyor, whereby they are brought into contact with the feedstock. The screw conveyor hinders the formation of coke deposits on the wall. The coking takes place at temperatures of 500–600°C.

Table 4.22 Typical Fluid Coking and Flexicoking Data

Feedstock		
°API		3.2
Conradson carbon		28.5
Sulfur, wt %		5.6
Vacuum distillate fbp 1050°F from heavy Saudi Arabian crude		
Yields		
Gases C_1–C_4, wt %		12.9
C_5–430°F, % vol		14.4
430–650°F, % vol		10.2
650–975°F, % vol		27.3
For fluid coking, net coke yield, tons/bbl		0.05
For flexicoking low Btu gas, 10^3 BTU/bbl		1,320
	Flexicoking	Fluid Coking
Economics		
Investment 96, $/bpsd	2,400–3,100	1,600–2,100
Utilities per bbl feed		
Electricity, kWh	30	30
Steam, 125 psig, lb	150	25
Water, boiler feed, gal	35	35
Water, cooling, gal	700	45
Steam 600 psig (produced), lb	(200)	(160)
Air blower		
Compressor HP-hr	0.6	0.2
Steam 125 psig (produced), lb	(660)	(230)
Steam 600 psig, lb	660	230

Installation: Until 1994, 5 flexicooking units built, over 160 Kbpd capacity; Licenser: Exxon Research & Engineering Co.
Source: Ref. 109.

The flow diagram of the plant, known under the name of *Coking-LR* is presented in Figure 4.30.

The burning of coke and reheating of the coke granules take place in a vertical transport pipe.

The unit converts not only heavy residues obtained in crude oil processing but also natural solid bitumen or other similar feedstocks.

Although the construction of several pilot plants and also commercial scale units was reported [44], no information concerning capacities and plant performance is available.

Another coking process with solid heat carrier which was proposed as early as 1981 and is known as *Dynacracking*. The process uses alumina as heat carrier.

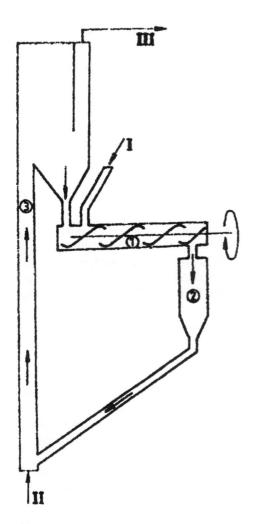

Figure 4.30 Moving bed LR coking unit. 1—reactor, 2—buffer vessel, 3—riser and reheater; I—feed, II—air, III—flue gases.

The burning of coke uses a mixture of oxygen and steam, their proportions being controlled so that the following reaction is favored:

$$C + H_2O \rightarrow CO + H_2$$

and the consumption of oxygen is minimized [51].

The produced gases are rich in hydrogen. As they pass through the reaction zone, coking takes place in the presence of hydrogen at a hydrogen partial pressure of 5–15 bar. In this way, the amount of coke decreases and the quality of the products is favorably influenced.

Other proposed processes that have not yet become popular in industry are described summarily [6,12].

4.2.4 Coke Calcination

Coke calcination is carried out mostly in tilted rotary cylindrical furnaces similar to those used in the cement industry. In the past, such operations were competitive only at very high capacities. For this reason the units were built as separate entities that processed the coke produced by several refineries. The subsequent development of plants of lower capacities (40,000–300,000 t/year) [53] led in the last two decades to the trend to perform coke calcination in the refineries themselves.

Calcination is used for treating the spongy coke obtained in the delayed coking, the needle coke, and the coke obtained in the fluid coking.

The calcination of needle coke produces graphite, which is used for the construction of special equipment (heat exchangers, pumps, valves, chemical reactors, crucibles, pipes a.o.) in the nuclear industry, construction of electrical motors, in the fabrication of electrodes for the caustic and chlorine industry, and for electrical metallurgical furnaces.

The coke obtained by calcining that was produced in delayed coking and in fluid coking is used for the fabrication of anodes for the aluminum industry and in some electrometallurgical and electrochemical industries. In the production of aluminum the consumption of coke is of 0.4–0.5 kg coke/kg aluminum. Since the world production of aluminum is 17–20 mill. t/year (1985), this industry alone consumes yearly 8–10 million tons of calcined coke.

Typical characteristics of raw coke and specifications for the calcined coke are given in Table 4.23.

Tilted rotary furnaces are used in about 95% of plants for calcination of petroleum coke. See Figure 4.31.

The heat required for calcination is supplied by the burning of volatile substances contained in the coke and of gaseous, liquid, or solid additional fuel as well as by the calcination of small amounts of coke. The complete combustion of the volatile substances is ensured by the secondary and tertiary air introduced into the furnace.

The calcined coke is discharged in a rotary cooler where the temperature is reduced to 90–120°C. Various cooling systems are in operation: water pulverization, air preheating, indirect cooling etc.

After being discharged from the cooler, and before it is sent to storage, an oil is usually pulverized over the coke, in order to reduce the dust produced during handling.

Table 4.23   Typical Coke and Calcined Coke Characteristics

Characteristics	From delayed coking		Needle coke	
	Crude	Calcined	Crude	Calcined
Moisture, %	6–10	0.1	6–10	0.1
Volatiles, %	8–14	0.5	4–7	0.5
Sulfur, %	1.0–4.0	1.0–4.0	0.2–2.0	0.5–1.0
SiO_2, %	0.02	0.02	0.02	0.02
Fe, %	0.013	0.02	0.013	0.02
Ni, %	0.02	0.03	0.02	0.03
Ash, %	0.25	0.4	0.25	0.4
V, %	0.015	0.03	0.015	0.02
Bulk density	0.720–0.800	0.673–0.720	0.720–0.800	0.673–0.720
Real density		2.06		2.11
Thermal expansion between 25 and 130°C, per °C				5×10^{-7}

The volatile substances that were not burned in the furnace are burned in an incinerator.

Special attention is paid to the recovery of heat from burnt gases and steam production.

The temperature profiles along the furnace and the identification of the main stages of the process are illustrated in Figure 4.32.

As shown in the figure, the temperature in the final part of the calcination zone is 1200–1300°C. It is important to remark that the real density of the

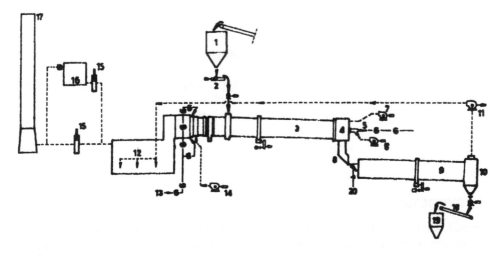

Figure 4.31   Kiln for coke calcination. 1,2—feed, 3—kiln, 4—combustion zone, 5—burner, 6—primary air, 7—secondary air, 8—transfer duct, 9—rotary cooler, 10—receiver, 11—cooling air vent, 12—incinerator, 13—additional burners, 14—incineration air, 15—flue gas gates, 16—steamgenerator, 17—stack, 18—calcined coke conveyor, 19—calcined coke buffer tank, 20—cooling water.

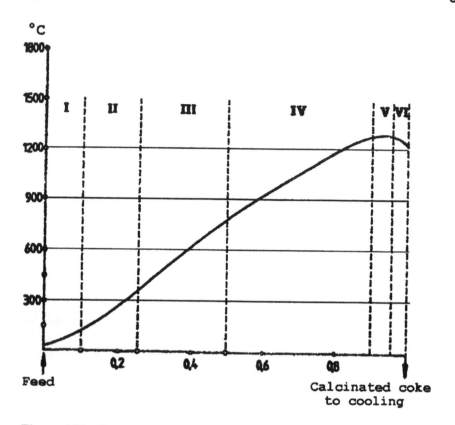

Figure 4.32 Temperature of coke particles along the calcination kiln. I—heating, II—drying, III—heating, IV—calcination, V—heating, VI—cooling.

calcined coke depends on the calcination temperature, a correlation is given in Figure 4.33 [54].

The main process data of a unit of this type, with a capacity of 200,000 t/year of raw delayed coking are given in Table 4.24. [54]

A more recent system based on the same principles is named *rotary-hearth* and is shown schematically in Figure 4.34. The system seems to require a more reduced space than other similar systems.

4.3 PYROLYSIS

Thermal cracking at low pressures—pyrolysis or steam cracking—was used worldwide earlier than cracking at higher pressures. Its commercialization was based on observations made in the years 1850–1860 concerning the decomposition of fractions of crude oil at high temperatures, with the formation of alkenes and of aromatic hydrocarbons. These observations allowed Mendeleev to predict, at the end of the 19th century, that this process would become extremely important for the chemical industry.

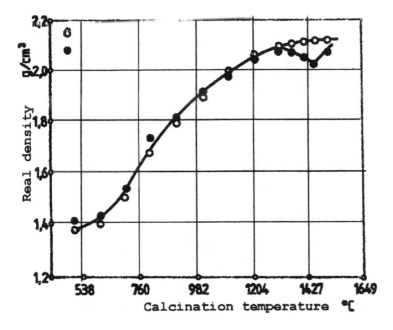

Figure 4.33 Correlation between calcination temperature and coke real density (heating rate = 28°C/min). ○ – low sulfur coke; ● – high sulfur coke.

Under the name of pyrolysis, the process was used before and during the first World War for the production of benzene, toluene, xylenes, and naphthalene using kerosene and gas oil as feedstock.

In those times alkenes production was of no interest. The production of aromatics was the only economic justification for the process. The low yield of aromatics and difficulties in operating at high temperatures around 700°C caused the process to be soon abandoned.

Under the name of "cracking at low pressures" pyrolysis reappeared in the years 1930–1935, with the purpose of producing gasoline with an aromatic character

Table 4.24 Process Data for a Delayed Coking Coke Calcination Plant

Feed	35 t/h
Calcined coke	25 t/h
Fuel	95×10^6 kJ/h
Kiln (furnace) exit gases	64 t/h
Incinerator exit gases	165 t/h
Cooler exit gases	78 °C
Temperature of coke living Kiln	1200–1260 °C
Temperature of coke living cooler	95–120 °C
Temperature of coke living incinerator	1100–1300 °C
Estimated steam production, 60 bar, 440°C	72.5 t/h

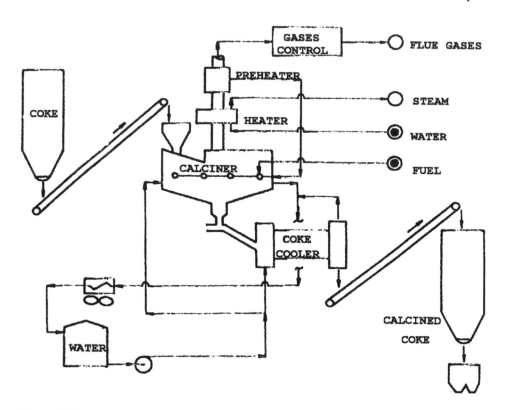

Figure 4.34   Rotary-hearth coke calciner.

and therefore with a high octane number. Under this form also, the process was soon abandoned because it could not resist the competition of other processes for producing gasoline.

   After a pause of nearly two decades, during which pyrolysis was only of historical interest, it reappeared and underwent rapid development as a result of the high demand for lower alkenes and especially for ethene, demands which could not be covered by the recovery of ethene from refinery gases.

   This radical change was a consequence of perfecting the processes for the production of several synthetic products of widespread use and large tonnage: polyethylene, polypropylene, ethyl-benzene, isopropyl-benzene, ethylene-oxide, vinyl chloride, glycols, etc., which in turn are raw materials for chemical products in large demand.

   Thus it is confirmed once again that the development of the petroleum refining industry and the emergence and development of new processes is the result of the development and improvement of the industrial domains that use the products of this industry and especially of the emergence and development of new consumers. The development of petrochemistry is an eloquent example of this situation. Whenever scientific research had perfected the industrial fabrication of a new product that requires a certain hydrocarbon, the petroleum refining industry found the ways to produce it by development of the corresponding production, separation, and purifying processes.

The pyrolysis process (steam cracking), is today the most important cracking process and the most important purveyor of raw materials for the petrochemical industry. Its impact may be characterized by world consumption of the main products supplied by this process, presented in Table 4.25.

More detailed data concerning the demand and the use of the monomers supplied by the petroleum refining industry to the chemical industry were published also in Romania [55,114,115].

The production of chemicals brings the highest added value to petroleum. Therein resides the present importance of pyrolysis as the main supplier of raw materials for the organic synthesis industry.

However, if one takes into account the relative volumes of the organic synthesis industry and of petroleum refining, this role of purveyor of feedstock for chemistry represents but a few percentage points of the total volume of petroleum processed. The main role of petroleum refining remains that of energy supplier, with a share of about 40% of the world energy production.

4.3.1 General Issues of Commercial Pyrolysis

Feedstock selection. The selection of feedstocks for pyrolysis is based on technical and economical conditions, namely:

1. The feed availabilities, taking into account the most reasonable use of various petroleum fractions
2. The products that pyrolysis should deliver for various downstream petrochemical processes
3. The limitations imposed by pyrolysis technology

The last aspect refers to the fact that the pyrolysis in tubular heaters cannot process residual fractions due to excessive coke formation. The pyrolysis of fuel oil and of other residues is possible only in systems with a solid heat carrier where the coke is deposited on the carrier and it is burned during the reheating step. These systems have not reached commercial acceptance. Only experimental units are in existence. For this reason, feedstocks that can be presently used in commercial units range from ethane to vacuum distillates.

Table 4.25 World Production of Main Products for Chemical Industry Provided by the Pyrolysis (thousands metric tons/year)

Year	Ethene	Propene	Butadiene
1970	19,380	8,670	–
1975	23,650	12,060	3,445
1980	34,900	17,790	5,060
1985	41,840	21,900	5,979
1990	52,400	25,200	–
1995	70,200	–	6,200

The use of ethane constitutes the most reasonable solution if ethene is the only product desired. Indeed, besides pyrolysis, the only other use for ethane is as fuel. However, the use of ethane in pyrolysis presupposes the existence of crude oil or of rich fields of natural gas. Such conditions exist in the U.S., where the ethane has been used as pyrolysis feed for quite a long time, representing in 1975 39% (and together with the liquefied gases 75%) of pyrolysis feedstock. In Western Europe, until the discovery of North Sea gases and petroleum fields, there were no sources of ethane and naphtha was the feedstock of choice. In other countries with petroleum resources, the use of ethane for pyrolysis is more recent, since its separation from oil field gases makes use of technology which became available only recently.

Propane and butane are also used as feed for pyrolysis. There is strong competition for using these hydrocarbons as liquefied gases. In some regions of the world, this use completely eliminates especially butane as feed for pyrolysis. The data mentioned previously for the U.S. indicate that it is very important in some situations to submit to pyrolysis propane together with ethane. Concerning *n*-butane, it must be mentioned that besides its use as liquefied gas, important amounts are used as an additive to gasoline for increasing vapor tension, especially as a replacement for isopentane. Smaller amounts of *n*-butane are used in dehydrogenation.

The use of naphtha as pyrolysis feedstock was developed initially in countries lacking natural sources of natural gases, especially in Western Europe. Besides ethene and propene, important quantities of butadiene, isoprene and aromatic C_6-C_8-hydrocarbons are produced. Unlike in the U.S., in Western Europe there is no production of butadiene by the catalytic dehydrogenation of *n*-butane. Also, catalytic reforming was developed especially for producing high octane gasoline and not aromatic C_6-C_8-hydrocarbons.

The use of gas oil and especially of vacuum gas oil as pyrolysis feedstock occurred more recently [56]. The pyrolysis of gas oils in tubular heaters requires the solution of specific problems related to the higher propensity for coking and the higher reaction rate of these feedstocks. Following the solution of these technical difficulties, the selection of the liquid fraction feed became a problem bound to the relative consumption of gasoline or gas oil in the respective country and to the forecast for the evolution of this consumption.

Thus in the U.S., where the consumption of gasoline is very high, the future development of pyrolysis capacity is oriented to the use of gas oil, especially of vacuum gas oil. In other countries, the future direction is unknown. Both options, the pyrolysis of naphtha or of gas oil, are being evaluated. In Romania, it was estimated [57] that new capacities will be designed for the pyrolysis of gas oils, including vacuum gas-oil, and not of naphtha. This problem was addressed again in a 1991 study [58].

It must be mentioned that the trend to expand the sources of feedstocks and the possibilities offered by their preliminary hydrotreating allows considering not only the vacuum gas oils [59], but even the oils extracted from the bituminous shales [60].

Concerning the chemical composition of liquid feedstocks, note the need to limit their content in aromatic hydrocarbons and to practically eliminate alkenes, both conditions leading to the formation of coke. The concentrations of other classes of hydrocarbons determine especially the ethene/propene ratio in the effluent. The *n*-alkanes favor the formation of ethene and the *i*-alkanes favor the formation of propene (see Appendix 1). In present conditions, when pyrolysis aims at maximizing the production of ethene, feedstocks with a prevalent concention of *n*-alkanes are preferred.

The use in pyrolysis of liquid feedstocks as opposed to gases may be determined also by the necessity to obtain, besides ethene, other important hydrocarbons in petrochemistry, especially butadiene, C_6-C_8 aromatics, and isoprene. Other hydrocarbons will be added certainly to the above ones if economical methods will be developed for their recovery and use. From this point of view it may be estimated that the use of gas oil in pyrolysis offers opportunities not yet sufficiently explored and developed.

Table 4.26 presents data on the products derived from the pyrolysis of various feedstocks expressed as tons per 1,000 tons of produced ethene. Table 4.27 [61] shows the relative investments for units specialized for the pyrolysis, of various feedstocks. The data were confirmed also by other publications [57,62] and is useful in selecting pyrolysis feedstock.

Process conditions and reactor types. Ethene currently represents the main product of the pyrolysis process. Accordingly, the reaction system and technological conditions are selected so as to maximize the yield of this product.

These conditions could be set on basis of theoretical, thermodynamic, mechanistic, and kinetic considerations set forth in Chapter 2.

Thus, thermodynamic considerations on the approach to equilibrium in a process condition for the reactions:

$$C_2H_6 \rightarrow C_2H_4 + H_2 \tag{a}$$

$$C_3H_8 \rightarrow C_3H_6 + H_2 \tag{b}$$

show that the maximum yield of ethene is favored by high temperatures and reduced partial pressures.

Table 4.26 Yields for 1000 Tons of Produced Ethene Using Various Feedstocks

				Feedstocks					
				Naphtha		Heavy Naphtha		Gas oil	
Products	C_2H_6	C_3H_8	C_4H_{10}	High severity	Moderate severity	High severity	Moderate severity	High severity	Moderate severity
Fuel gases	159	652	660	521	453	551	505	457	416
C_2H_4	1,000	1,000	1,000	1,000	1,000	1,000	1,000	1,000	1,000
C_3 fraction	19	682	460	420	627	445	650	557	614
C_4 fraction	35	103	440	248	408	275	450	413	458
Gasoline	43	102	326	741	1101	877	1,220	794	868
Fuel		1		195	118	241	137	729	1,006
Acid gases	2	3	3						
Total	1,258	2,543	2,889	3,125	3,707	3,389	3,962	3,950	4,362
C_3H_6		626	410	360	518	390	530		
C_4H_6		65		141	171	152	180	187	203
BTX fraction	10	44		500	412	612	585	409	451

Table 4.27 Relative Investment for Pyrolysis of
Various Feedstocks

Feedstock	Relative investment
Ethane	80–85
Ethane/propane (50/50)	82–87
Propane	85–90
Butane	90–95
Naphtha/liquefied gases (50/50)	90–95
Naphtha	100
Heavy Naphtha	100–105
Naphtha/gas oil (50/50)	115–120
Naphtha/gas oil/C_4H_{10}	120–125
Straight run gas oil	110–115
Vacuum gas oil	120–125

It is to be mentioned that reaction (b) is in competition with the parallel transformation:

$$C_3H_8 \rightarrow C_2H_4 + CH_4 \tag{c}$$

the equilibrium of which, at the temperatures of the pyrolysis, is completely displaced to the right.

The ethene formed by way of reactions (a) and (c) also suffers decompositions, especially following the reaction:

$$C_2H_4 \rightarrow C_2H_2 + H_2 \tag{d}$$

followed by

$$C_2H_2 \rightarrow 2C + H_2 \tag{e}$$

At the usual temperatures practiced in various pyrolysis systems[*], the amount of acetylene present in the effluent is limited by the equilibrium of reaction (d) to a few percentage points. The excess of acetylene is decomposed according to reaction (e).

It follows that the yield of ethene passes through a maximum, that imposes the limitation of the residence time in the reactor.

Analogously, the yield of propene also passes through a maximum. The difference is that, while the maximum of ethene yield shows a strong increase with temperature due to the displacement to the right of the equilibrium of the reaction (a), the maximum of propene increases much slower due to the competition of reactions (b) and (c). The latter orients the transformation towards producing ethene and not propene.

---

[*] According to several authors the formation of acetylene occurs by the reaction:
$$C_3H_6 \rightarrow C_2H_2 + CH_4$$
In the case of ethane, the propene results from secondary reactions [65].

In general terms, the operating conditions that maximize the yield of ethane are:

1. High coil outlet temperatures (COT), limited by the increase of acetylene formation that occurs at temperatures that exceed 1000°C.
2. Low partial pressures of the reactants and of the products at the outlet from the system, corresponding to low pressure and to a possible dilution with inert gases or steam.
3. Short reaction residence time, of the order of seconds, or fractions of a second. The higher the temperature, and molecular mass of the feedstock, the lower should be the reaction time.

It is remarkable that the increase of the maximum value of ethene yield with increasing temperature and lowering of the partial pressure is independent of the nature of the feedstock.

Therefore, the reaction equipment should ensure reaching the same high COT value, irrespective of the nature of the feedstock submitted to pyrolysis. Accordingly, in order to not obtain ethane yields beyond the maximum, the reaction time should be shorter for heavier feedstocks.

Ethene is an intermediary product in a system of parallel-successive reactions and leads to the fact that the actual value of this maximum depends also on the type of reactor used.

For plug flow reactors, this maximum is given by Eq. (2.98), and the time for reaching it by Eq. (2.96). Since the flow in the tubular heater closely approaches the ideal plug flow, these relations can serve to determine the maximum of ethene yield that will be obtained in commercial tubular heaters. For reactors where flow conditions approach perfect mixing, the maximum is given by Eq. (3.7).

In other reaction systems such as fluidized bed, moving bed etc., the presence of backmixing leads to the decrease of the maximum yield of the intermediary product. An exception is the riser system, wherein backmixing is insignificant.

The comparison between the ideal plug flow reactors and the reactors with perfect mixing, and the corresponding relationships, were deduced in Section 3.1.1. Real-life reactors have flow regimes situated between the two extreme limits.

Values of the ratio between the values of the maximum of the yield of the intermediary product obtained in the two types of ideal reactors are given in Table 3.1. They prove that in a very large domain of ratios of velocity constants, the ideal plug flow reactor produces 15–40% more intermediary product than the reactor with perfect mixing (see also Table 3.2, for the pyrolysis of ethane).

Despite the fact that in real reactors the difference will be lower than shown, it follows that those reactors approaching the ideal plug flow must be selected for pyrolysis.

The tubular heater completely satisfies these requirements. Other systems will give comparable results only if they approach the ideal plug flow reactor, as for instance the riser reactor.

Moreover, if other reactor systems are to be competitive with the tubular heater they must satisfy the condition of increasing temperature along the reactor with the outlet temperature between 900–1000°C. Thus, the riser reactor system, which practically satisfies the conditions of an ideal plug flow reactor, cannot satisfy the conditions of increasing temperature because of the strong endothermal character of the cracking reactions that occur within it.

The high overall conversions at which pyrolysis operates lead to situations in which coke formation becomes the limiting factor. The injection of steam diminishes the coking process and makes possible pyrolysis in tubular furnaces of gas oils and of vacuum distillates. However, at this time, pyrolysis in tubular heaters of residual fractions cannot be realized.

Pyrolysis of these products is possible only in systems with a heat carrier, in a moving or fluidized bed, or in other nonconventional systems that are described in the Section 4.3. Although such systems did not lead so far to units beyond the experimental phase, they provide interesting perspectives especially because they allow reaching higher temperatures than those possible in the tubular furnace.

In order to be efficient, the system must ensure, as it was described, a flow that is closer to the ideal plug flow reactor and maximum temperatures at outlet.

Products separation. The problems of the separation of the products from pyrolysis do not depend on the reaction system but is in great measure dependent on the processed feedstock.

The separation is divided conventionally in the hot section and the cold section. The hot section contains the rapid cooling of the effluent as soon as it leaves the reactor (quenching), its cooling, and the separation of the heavy products, used as fuel, of the gases and of the gasoline. A typical flow diagram of the hot sector is given in Figure 4.35.

The cold section contains the processing of the gases, which in the case of pyrolysis of liquid fractions consists of the following main operations: compression, separation of H_2S and CO_2 with diethanolamine followed by washing with a solution of NaOH, cooling, removal of the water vapors by condensation and of the water traces by drying. Then, the demethanation and the separation of the C_2, C_3, and C_4 fractions are carried out, and also of the heavier hydrocarbons. From each of these three fractions, the acetylene and its homologues are removed by selective hydro-

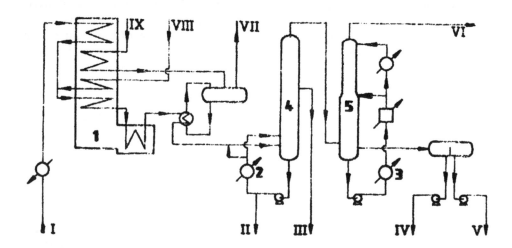

Figure 4.35 Hot section of a pyrolysis unit. 1—furnace, 2, 3—heat exchangers, 4—primary column, 5—water quench column; I—feed, II—fuel, III—gas oil, IV—process water, V—gasoline, VI—gases, VII—high pressure steam, VIII—dilution steam, IX—boiler feed water.

genation, after which the separation by fractional distillation of the ethene from ethane and of the propene from propane is carried out.

The ethane, the propane, and other fractions are pyrolyzed separately in special heaters. After quenching, the effluents are introduced in the effluents deriving from the other heaters.

Butadiene is recovered from the C_4 fraction by selective extraction. Other individual components may be extracted (for instance 1-butene by adsorbtion on zeolites) if required by the downstream petrochemical industry. Also, the separation of isoprene from the C_5 fraction is performed. The purification of isoprene to the specification purity for the fabrication of the stereospecific rubber is quite difficult.

Pyrolysis naphtha can be used as a component for automotive fuel after it is selectively hydrogenated for removing unstable hydrocarbons, which are gum generators. From the fraction 60–135°C, the aromatic C_6-C_8-hydrocarbons could be extracted with selective solvents after the complete hydrogenation of the alkenes.

Following the pyrolysis of naphtha and gas oils, other products for the chemical industry may be recovered from the effluent (naphthalene).

Of course, differences exist in the design of separation systems by different constructors of plants [62], but they don't affect the main principles presented here.

In the case of the gaseous fractions C_2 and C_3, the flow diagram of the separation system is simplified accordingly.

In Figures 4.36 and 4.37 flow diagrams are shown for the separation operations following the pyrolysis of ethane, propane, and of liquid fractions of naphtha and gas oil.

4.3.2 Pyrolysis in Tubular Heaters

Tubular heaters are used in the present nearly exclusively as reactors for pyrolysis. They present the advantage that the process takes place in practically identical conditions as in the ideal plug flow reactor and the temperature is the highest at the outlet. Their limitations concern the temperature that can be reached at the coil outlet and the nature of the feedstock with reference to coke formation.

The first limitation is related to the highest temperature that is acceptable by the steel of the tubes. A function of the thickness of the coke layer and of the thermal stress, it determines the highest temperature at which the reaction products leave the heater.

For a heater with thermal stress φ_t, the heat transfer from the metal to the product that flows inside the tube can be expressed by the relation:

$$\varphi_t = k(t_m - t_p) \tag{4.2}$$

where:

t_m is the temperature of the metal, t_p is the temperature of the product

k is the overall heat transfer coefficient from the outer surface of the metal to the product and is given by the equation:

$$\frac{1}{k} = \frac{1}{\lambda_m/b_m} + \frac{1}{\lambda_c/b_c} + \frac{1}{\alpha_i} \tag{4.3}$$

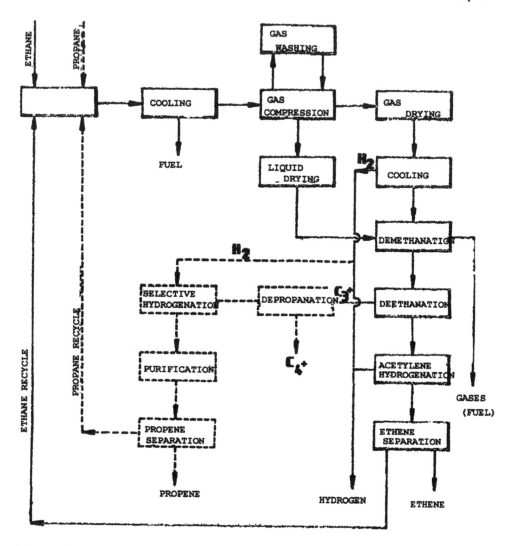

Figure 4.36 Sequence of operations in ethane and propane pyrolysis. If feed contains propane, the dotted part must be added.

In this equation λ_m and λ_c are the respective heat conductivities of the metallic wall, and of the coke; b_m and b_c, the respective thicknesses; and α_i, the partial heat transfer coefficient from the wall to the product, calculated with the equation [5]:

$$\alpha_i = 0.025 \frac{\lambda}{D_i}\left(\frac{v\rho D_i}{\mu}\right)^{0.8} \tag{4.4}$$

where:

λ is the thermal conductivity of the product in J/m·s·degree;
D_i = the inside diameter of the tube in m
ρ = the density of the product at the conditions of the tube in kg/m^3

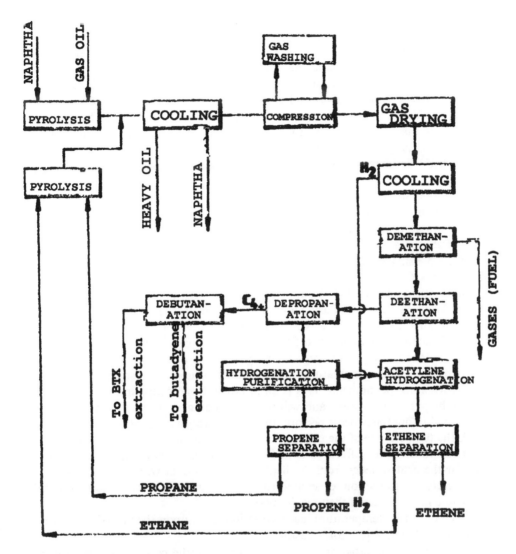

Figure 4.37   Sequence of operations in pyrolysis of liquid fractions.

v = the flowrate in m/s
μ = the dynamic viscosity in kg/m·s.

Using the Eqs. (4.2)–(4.4), each temperature drop through the walls of the tube, the coke layer, and the liquid film can be calculated.

For the coke layer, by using $\lambda_c = 0.81$ J/ms·degree [66], for the thickness of the coke layer of 1 mm, one obtains a partial heat transfer coefficient through the coke $k_c = 810$ J/m²·s·degree (Eq. (4.3)) that for thermal tensions of 35,000–105,000 J/m²·s used in pyrolysis furnaces (35,000 in old plants and over 100,000 in modern heaters), represents a difference of temperature of 40–120°C (Eq. (4.2)). This calculation agrees with some published data [67], that demonstrates that after a period of time it is possible that coke forerunners penetrate through the gas film, the formation of the coke layer takes place, and the temperature of the tube wall increases rapidly, by about 85°C. Other published data [57], based on older furnaces with lower thermal stresses, estimate that after a layer of 1 mm coke had deposited the temperature increase is 30–40°C. Concerning the temperature differences for the heat transfer through the tube metal and through the gas film, measurements estimate these to be 20–40°C, and 100°C respectively 100°C [57].

Among these, the only temperature drop that can be influenced by operating conditions is that corresponding to gas film. According to Eq. (4.4) it could be diminished actually only by increasing the flow velocity through the tubes, which however will also increase the pressure drop along the coil.

Since, as seen, the possibilities of increasing the heat transfer coefficient are limited, the possibility of increasing the coil outlet temperature depends mainly on progress in the development of high temperature resistant tubes. Early tubes for the radiation section of pyrolysis heaters allowed a coil outlet temperature of 830–840°C. Tubes resistant to temperatures higher up to 1100°C were developed around 1975 using steels with 19–23% Cr, 30–35% Ni, 1.5% Mn, about 1% Si, and 0.1–0.5% C [57].

Subsequent progress [47,68], allowed operating at a tube temperature of 1150°C and even 1200°C as a result of using higher alloyed steels such as the steel Supertherm which contains 28% Cr, 48% Ni, 15% Co, 5.5% W, 1.2% Mn, 1.2% Si and 0.5% C. Concomitantly, aluminated tubes were developed[*] [69] that show increased resistance to corrosion. At the present, research aims at finding alloys that cannot only operate at higher temperatures but are stable in time and also resistant towards carbide formation phenomena. This is an important preoccupation in the efforts to improve the pyrolysis process, and will be discussed again in the following section.

In evaluating the process performance of a pyrolysis unit, it is considered that, overall, the temperature of the effluent that leaves the furnace is about 170°C below that of the tubes. This difference is in fact in agreement with the values indicated above for the heat transfer from the heater to the product within the tubes.

Therefore, product outlet temperatures of about 835°C were obtained in older units, while in newer units they reach 1000°C.

The coil outlet temperature has a great influence upon the yields, as shown by the data of Table 4.28 [69] for a gasoline.

The data of this table prove that by increasing the coil outlet temperature a substantial increase in the production of ethene and benzene takes place, while propene production is reduced and that of butadiene remains unchanged.

The increase of the reaction temperature at a given conversion imposes the shortening of the residence time in the reaction coil. This requires increasing the rate of heating of the product, which can be achieved by increasing the thermal stress and/or by decreasing the tube diameter. By diminishing the tube diameter, the heating rate is increased. This is easy to understand: for the same linear velocity through the tubes, the mass flowrate of the product is proportional to the cross section and therefore it decreases with the square of the diameter, whereas the transferred heat is proportional to the circumference of the tube, thus decreasing directly proportional to the diameter.

---

[*] The alumination is a treatment of the steel at high temperatures which leads to the diffusion of the aluminum in the steel [68].

Table 4.28 Influence of the Coil
Outlet Temperature on Pyrolysis
Yields of a Naphtha

Feed stock analysis			
Distillation limits	35–160°C		
50% distillate at	90°C		
Density	0.710		
PONA analysis	P = 80%		
	N = 15%		
	A = 5%		
Steam/gasoline ratio	0.6 in weight		

Outlet composition	Temperature		
	815°C	835°C	855°C
H_2	0.66	0.74	0.81
CH_4	13.82	15.65	17.40
C_2H_4	24.71	27.06	29.17
C_3H_6	17.34	16.28	14.44
C_4H_6	4.18	4.17	3.99
C_6H_6	4.89	5.90	7.08
C_5–200°C	22.64	20.89	20.01

Source: Ref. 69.

The influence of reducing the residence time on the yields for a coil outlet temperature of 835°C, at 0.6 kg/kg steam dilution and for the same feedstock as in Table 4.28, is given in Table 4.29 [69].

The data show that the decrease of the residence time leads to yield increases for ethene and propene, no change for butadiene, and yield decreases for benzene and methane. In connection with this fact, it is to be mentioned that the minimum

Table 4.29 Influence of Furnace
Tubes Diameter (residence time) on
Yields

Inside tube diameter	4 in.	3 in.
Residence time	0.88	0.55
Yields		
H_2	0.86	0.74
CH_4	18.27	15.65
C_2H_4	25.47	27.06
C_3H_6	14.37	16.28
C_4H_6	4.12	4.17
C_6H_6	6.79	5.90
C_5–200°C	21.52	20.89

Same feedstock as in Table 4.28.

amount of methane corresponds to a maximum efficiency of the pyrolysis, since methane is the product with the lowest value among all those produced.

A correlation between the tube diameter, the temperature of the wall, and the ethane yield is shown in Figure 4.38 [69]. Despite the fact that the authors do not specify the feedstock used, on basis of the ethane yield, one may assume that it is the pyrolysis of a naphtha.

The correlation of the analyzed parameters with the temperature of the tube wall can be obtained by equating the transmitted heat with that consumed in the coil:

$$\varphi_t L \pi D_e = G(\Delta i_h + q_r + n \Delta i_a) \tag{4.5}$$

where:

φ_t = the thermal stress in $J/m^2 \cdot s$
L = the length of the coil in m
D_e = the exterior diameter of the tube
G = the feed flow in kg/s
q_r = heat of reaction in J/kg
n = the weight ratio steam/hydrocarbons
Δ_i = the enthalpy difference between the coil outlet and inlet in J/kg

subscripts: h = hydrocarbons, a = steam.

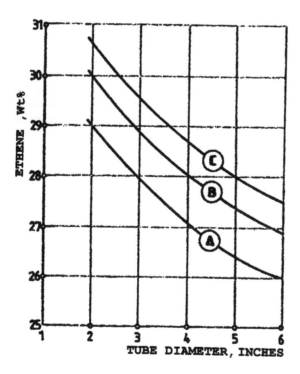

Figure 4.38 Variation of ethene yield with inside tube diameter, for clean tubes at three wall temperatures: A—925°C, B—980°C, C—1040°C.

The feed flowrate (G), can be correlated with the residence time in the coil (τ) by the equation:

$$G/V = \rho_r/\tau \qquad (4.6)$$

where:

V = volume of the coil in m^3
ρ_r = the specific mass of the hydrocarbons (without steam), in the conditions in the coil.

Replacing G given by Eq. (4.6) in (4.5) and since $V = \frac{1}{4}\pi D_i^2 L$, where D_i is the inner diameter of the tube, it follows:

$$\varphi_t = \frac{1}{4}\frac{D_i^2}{D_e}\frac{\rho_r}{\tau}(\Delta i_h + q_r + n\Delta i_a) \qquad (4.7)$$

which correlates the thermal tension, the diameter of the tubes, and the residence time in the coil.

Replacing φ_t with Eq. (4.2), it follows:

$$t_m = \frac{1}{4k}\frac{D_i^2}{D_e}\frac{\rho_r}{\tau}(\Delta i_h + q_r + n\Delta i_a) + t_p \qquad (4.8)$$

which correlates the temperature of the tube metal, the temperature of the product inside the tube, the tube diameter, and the residence time in the coil.

Eqs. (4.7) and (4.8) allow one to analyze the effect upon the process of the modification of some parameters, such as the tube diameter, metal temperature, the thermal stress, and residence time.

Thus, from Eq. (4.8) it follows that for a given feedstock, conversion, and operating conditions, the only efficient measure for decreasing the temperature of the metal (or for increasing the product outlet temperature and therefore the ethene yield at a given metal temperature) is to decrease the tube diameter. This explains why in modern designs of pyrolysis heaters, several tubes of small diameter area connected in parallel, and are placed in the same heater, the number of parallel tubes towards the exit of the coil is increased [47,70].

The data on the influence of the cited parameters on the conversion to ethene, propene, and methane are given in the graphs of Figure 4.39 [69].

The data were obtained in three experimental heaters using the same naphtha, which had the characteristics:

Initial Boiling Point = 35°C $t_{50\%}$ = 80°C end Point = 135°C

The PONA analysis

  n-alkanes = 47%
  i-alkanes = 37%
  cyclanes = 11%
  aromatics = 5%

The graphs show the favorable effect of the decrease the diameters of the tubes, of the increase of thermal tension, and of the shortening of the residence time upon

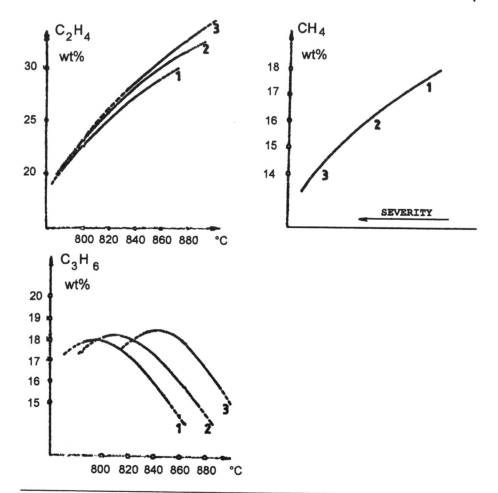

Figure 4.39 Effect of furnace parameters on yields (From Ref. 69.).

	Furnace parameter		
	1	2	3
Inside tube diameter, inches	3.5–4.0	3.0–3.5	2.5–3.0
Thermal tension, kJ/m^2·h	272×10^3	314×10^3	377×10^3
Residence time, s,	0.9	0.65	0.35
Steam/feed ratio, kg/kg	0.6	0.6	0.6

the conversion to ethene. Simultaneously, the conversion to methane decreased. Also, the conditions that allow one to obtain a maximum of propene are easy to identify.

The evolution of pyrolysis furnaces from 1950 on is presented very clearly in Figure 4.40 [47,110]. It correlates the main parameters previously analyzed, namely: the tube diameter, the used alloy, the metal temperature, the effluent temperature, and the residence time. This figure shows the evolution of the parameters over the last 40–50 years, including the switch to vertical reaction tubes around 1965.

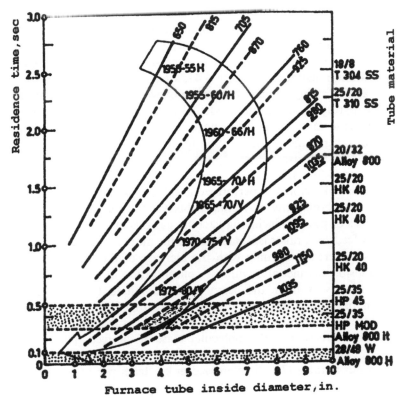

Figure 4.40 Evolution of heaters used in ethane pyrolysis. Numbers on arrow indicate the year of first commercial application, H-horizontal coil, V—vertical coil. — Coil outlet temperature, °C, - - - Tube wall temperature, °C.

The data that correlate the working parameters of a classic naphtha pyrolysis furnace are presented in Table 4.30 [67].

It must be mentioned that, besides the indicated parameters, the conversion is influenced also by the temperature profile along the coil length [71]. This fact allows maintaining a constant conversion during the whole duration of the cycle, using heaters that permit the modification of the temperature profile. By changing the

Table 4.30 Operating Parameters of a Classic Furnace for Naphtha Pyrolysis with 3 in. Tubes

Clean metal temperature (°C)	Coil length (m)	Pressure drop (kg/cm^2)	Thermal tension (kJ/m^{2}h)	Residence time over 650°C (s)
926	53.35	1.18	183,000	0.46
982	39.60	0.95	258,000	0.35
1038	32.90	0.72	315,000	0.27

Source: Ref. 67.

concave temperature profile to a convex one, the total conversion for the same final temperature will increase. Conversion can be maintained constant by gradually changing from the concave profile to the convex one with the concomitant reduction of the outlet temperature. This reduction is made necessary by the coke deposits. This manner of operation, illustrated in Figure 4.41, seems to lead to a decrease in the deposits of coke and to a lengthening of the cycle between decoking operations.

Concerning the pyrolysis of various feedstocks, the reaction rate constant increases with the increase in the molecular mass. Therefore, in order to obtain the maximum yield of ethene, the residence time in the coil must decrease as the molecular mass of the feed increases.

From Eqs. (4.7) and (4.8), it follows that, in order to satisfy this requirement the pyrolysis of heavier feedstocks should be performed in tubes of lower diameter and in heaters capable of higher thermal stresses. Actually this was the manner that allowed industrial scale pyrolysis of gas oils, including vacuum gas oils in 1970–1972 [72].

The use of heavy feedstock raises issues related to the more intensive deposition of coke. These deposits increase not only with the increase of the distillation end point, but in a very sensible manner also with the concentration of aromatic hydrocarbons. For this reason, very often the heavy feeds submitted to pyrolysis are submitted to de-aromatization by means of deep hydrotreating [59].

In order to reduce coke deposits, as a general rule steam is introduced into the coil in proportions that increase with the molecular mass of the feed. The optimal value of these ratios based on economic calculations [73], range between limits recommended by various authors [73,77], expressed in kg steam per kg feedstock:

Ethane	0.25–0.40
Propane	0.30–0.50
Naphtha	0.50–0.80
Gas oil	0.80–1.00

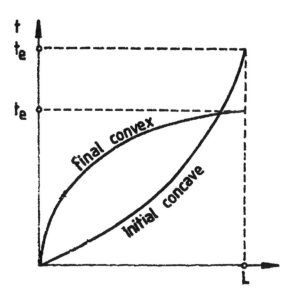

Figure 4.41 Temperature profile for maintaining a constant conversion during the cycle.

In older systems, the decoking of pyrolysis coils was performed by stopping the furnace, stripping, and cooling, after which the coke was burned, introducing a mixture of air and inert gases or air and steam, under such conditions that the temperature of the tubes did not exceed the admitted limits. In newer systems, the hydrocarbon feeding is stopped and is replaced by steam, the coke being transformed to a mixture of CO and CO_2. This decoking system does not need to stop the combustion in the fire box. Therefore, it is performed in modern furnaces that contain in the same fire box a large number of coils, of which at all times one is undergoing decoking. The length of the working cycle between two decokings depends on the feedstock, the steam rate, and other operating conditions and can reach 30–50 days or even longer. The duration of decoking ranges between 0.5 and 3 days.

Beginning with the years 1965–1967, vertical heaters were introduced in the pyrolysis units. In such heaters, the convection section is placed in the upper part of the furnace and has horizontal coils for heating the feedstock, vaporizing boiler feed water, and superheating steam. The cracking coil is made of vertical tubes welded on the return curves and is placed integrally in the radiant section, so as to be submitted to direct radiation on both sides of the tubes, for achieving a high thermal flux.

In Figure 4.42 a schematic of such a furnace is given.

At the outlet from the coil, the effluent must be suddenly cooled (quenched) for interrupting the radicalic reactions. In the pyrolysis of gases and light fractions the quenching is performed in a *transfer line exchanger*, generating high pressure steam. The coil effluent enters from the tube side of the exchanger by a very short pipe. Water under pressure circulates on the shell side and generates steam, which collects in the steam drum (Figure 4.42). The tube sheets of the transfer line exchangers are curved in order to compensate for their thermal expansion. Useful data were published [75] concerning the optimization of the operation of this equipment.

Additional difficulties appear when quenching the effluent resulting from the pyrolysis of gas oils and especially of vacuum gas oils. In these cases, the circulation of the effluent through the transfer line exchanger is downflow and a high circulation rate is applied in order to prevent the formation of a coke-generating film inside its tubes.

Special attention has been given to the formation and structure of coke formed inside the tubes of the coil [112,113] and in the transfer line exchanger during the pyrolysis of various feedstocks. Also, finding methods for avoiding the phenomena of carburization of the tube steel, and corrosion will lead to the most efficient ways of optimizing the quality and service life of the tubes [47,68,76–78].

The problem of coke formation inside the tubes of the pyrolysis coil was examined in Section 2.3.3.1. The problem of the interaction of coke with the metal of the tube will be examined below. The phenomenon that takes place is the carburization of the steel. For highly alloyed steels, carburization is insignificant up to temperatures of about 1050°C but becomes rapid after this temperature is exceeded (Figure 4.43) [77,110]. This dynamic explains why the problem of carburization gained importance as the temperature practiced in the pyrolysis process increased.

The penetration of the carbon into the metal is strongest up to a depth of 1 mm and then decreases gradually with the distance from the surface (Figure 4.44) [111]. The same figure illustrates that the resistance to carburization increases with nickel content.

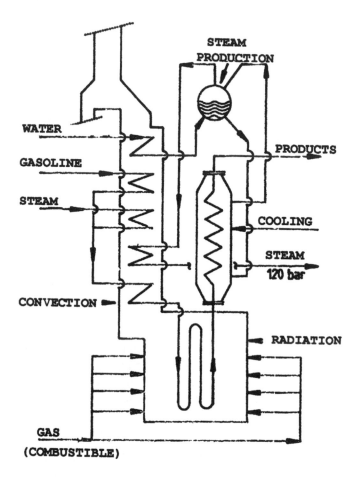

Figure 4.42 Schematics of the streams in a vertical pyrolysis heater.

The curves of this figure were obtained by maintaining the steel samples at 1100°C for 520 hours in granulated coal. Both figures show that the degree of carburization is less for higher alloyed steels.

The carburization phenomenon causes wear of the tubes. For this reason, protection measures are necessary, among which the most important is alumination. The effect of alumination is seen with clarity on the micrographs of cross sections of aluminated and not aluminated tube walls after carburization (Figure 4.45) [68].

A very important and efficient operation is to retrofit old furnaces by replacing the tubes in the radiant section with tubes which can ensure a very short residence time, thus producing a substantial increase in the selectivity to ethene and of the flexibility in the feedstock selection [79].

In order to visualize the extent to which such a modification is possible, Figure 4.46 represents a cross section through a modern heater.

Usually, the convection section can be used without modifications. There are two possibilities to replace the reaction coil situated in the radiant section: *Conventional* operates at a reaction time of 0.2–0.5 seconds and *Millisecond* operates

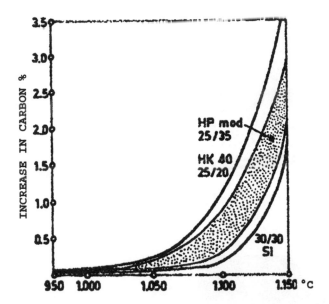

Figure 4.43 Temperature influence on steel alloys carburization.

at a reaction time of 0.05–0.1 seconds. Some information concerning the two systems is given in Table 4.31. To the information of Table 4.31 for the "Millisecond" heater, some particular features should be mentioned: the burners are placed exclusively on the floor to ensure uniform temperature profile along the whole length of the reaction tube. Each tube crosses the fire box only once, which makes the return bends super-fluous. The decoking is performed with steam for 12 hours, without taking the heater

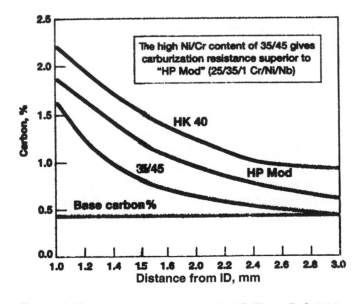

Figure 4.44 Carbon absorption by Cr-Ni (From Ref. 111.)

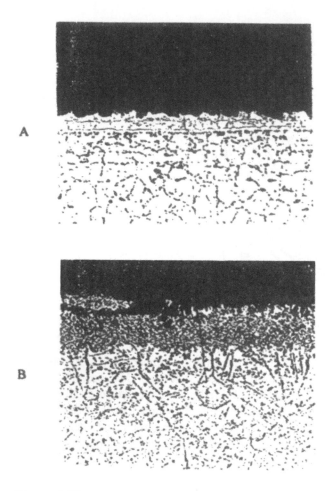

Figure 4.45 Protection of Cr-Ni steel by alumination. A—aluminated tubes, B—nonalumi-
nated tubes (crosssection magnified 100 times). (From Ref. 68.).

off-stream. The cooling system is situated above the radiant section. The pyrolysis
effluent flows downwards, is injected with cooling liquid for 0.015–0.03 seconds, and
the products enter the transfer line exchanger, which produces high pressure steam.

A similar process is licensed by ABB Lummus as "Short Residence Time" or
SRT cracker. More than 1000 SRT crackers were built up to March 2001. The
pyrolysis is carried out in two parallel reaction coils located in the radiant section
of the furnace. The residence time is from 0.130–0.180 seconds. The diameter of the
coil tubes is somewhat larger than those of the Millisecond cracker. While the yields
in the two crackers are quite similar, the main advantage of the SRT system resides
in the longer cycle length (60 days) compared to the Millisecond system (10–20 days).

It is important to know the composition of the effluent and to correctly predict
it for various feedstocks and operating conditions. This is particularly important in
the pyrolysis of liquid fractions, in view of the large number of components produced that must be recovered and purified, and of the components that might affect
the operation of the separation system.

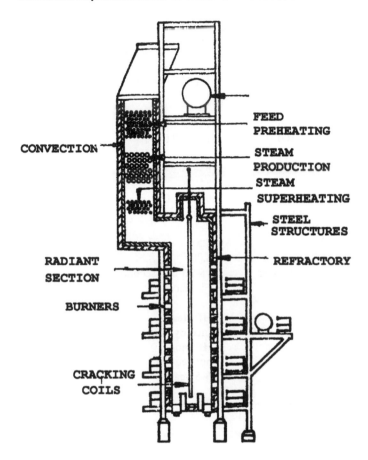

CONVECTION

RADIANT
SECTION

BURNERS

CRACKING
COILS

FEED
PREHEATING

STEAM
PRODUCTION
STEAM
SUPERHEATING

STEEL
STRUCTURES

REFRACTORY

Figure 4.46 Section through a modern pyrolysis furnace.

In Table 4.32, a typical composition is given for the effluent obtained in the pyrolysis of ethane, propane, butane, and naphtha in conventional systems [80] and in the Millisecond and SRT systems [121].

Somewhat simplified data for the pyrolysis of a naphtha, in conventional systems at different operating conditions were given previously in Tables 4.28 and 4.29. The results of the pyrolysis at a low severity and a high severity for a naphtha, a gas oil, and a vacuum distillate are given in Table 4.33.

In the pyrolysis of a naphtha, the number of components of which the concentration in the effluent must be known is estimated at about 16. It is to be foreseen that, with the use of gas oils and of heavy fractions in pyrolysis and with the increase of the number of hydrocarbons used in petrochemistry, the number of components and concentration in the effluent will increase. This will make correct estimation more difficult.

The estimation of the effluent composition on the basis of kinetic equations (which will be treated in Chapter 5) is limited by the fact that the kinetic constants for the less usual components are not available, although their concentration is required.

Other methods for estimating the composition of the effluent may be used. They apply various expressions for the severity factor [62,67,74,57] and graphs that

Table 4.31 High Severity Tubular Furnace Design

	Conventional	Millisecond
Furnace type	2- and 4-pass vertical	1–pass vertical
Tube size (ID, in.)	2–4	1–1.3
Tube length, ft	25–40	30–40
Outlet temperature, °F	1,700–1,800	1,600–1,700
Metal skin temperature, °F	1,950–2,100	1,850–2,000
Range ethylene yield, wt %		
naphtha feedstock	28–32	33–38
vacuum gas oil	19–22	24–28
Residence time, sec	0.2–0.5	0.05–0.10
Decoking	steam/air	steam
Decoking frequency, days	30–50	7–10
Decoking duration, days	2–3	0.5
Operating hours per year	8,000	8,000
Quench system	Heat exchanger	15–30 millisecond quench
Burner type	multilevel side wall	Floor only
Heat recovery	HP steam	HP steam
Tube material	HP mod 25–35 Cb	Incoloy 800H or HP mod
Return bends	HP mod	not required
Approx. tube life (excluding transfer lines)	3–4 years	6–8 years

Typical data, latest technology.

correlate it with the composition of the effluent. Such estimations are useful especially after they were confirmed by comparing them to values obtained by other methods or with data obtained in operating units.

Selected economic data concerning the investment and utilities consumption in the conditions of Western Europe are given in Table 4.34 [61].

4.3.3 Pyrolysis in Nonconventional Systems

The limitations presented by the tubular furnace concerning the maximum possible reaction temperature and the impossibility of using feedstocks that produce much coke (residues) inspired the idea of performing the pyrolysis by using heat carrier systems. The coke formed in the process is deposited on the heat carrier, and is partly or totally burned in the furnace that reheats it. Ceramic materials or coke particles were used as heat carriers.

The circulation systems used in pyrolysis are similar to those applied in coking (Section 4.2):

1. Systems with moving bed, where the granules of the carrier, having sizes of the order of several millimeters, circulate through the reactor and the reheating furnace under the action of their own weight
2. Systems with the heat carrier in a fluidized bed

Table 4.32 Typical Yields of Cracking Furnace, % wt

Products	Conventional [80]				Millisecond [121]		SRT [121]	
	Ethane	Propane	Butane	Naphtha	Ethane	Naphtha	Ethane	Naphtha
H_2	3.6	1.4	1.2	0.95	3.94	1.00	3.93	1.00
CO	0.3	0.3	0.25	0.08	–	–	–	–
CO_2	0.1	0.1	0.08	0.03	–	–	–	–
H_2S	–	–	–	0.02	–	–	–	–
CH_4	3.6	21.5	20.8	14.9	3.70	17.83	3.82	18.00
C_2H_2	0.25	0.5	0.6	0.65	0.48	1.05	0.43	0.95
C_2H_4	48.9	33.0	31.0	28.4	53.15	34.70	53.00	34.30
C_2H_6	38.2	4.0	4.0	3.9	35.00	3.62	35.00	3.80
C_3H_4	0.1	0.35	0.45	0.8	0.06	1.07	0.06	1.02
C_3H_6	1.1	16.0	15.0	12.0	0.86	14.20	0.89	14.10
C_3H_8	0.1	8.5	0.4	0.4	0.17	0.33	0.17	0.35
C_4H_6	1.4	2.3	3.5	4.3	1.17	4.52	1.19	4.45
C_4H_8	0.2	1.2	5.0	4.1	0.18	3.70	0.18	3.70
C_4H_{10}	0.1	0.1	6.5	0.3	0.21	0.20	0.22	0.20
C_5S	0.25	3.7	3.8	4.0				
B	1.4	2.3	2.5	6.8				
T	0.15	0.9	0.9	3.5		$C_5 - 200°C$		
X	0.10			1.8				
ETB	–	0.8	0.7	0.3				
STY	0.05			1.1	1.08	13.77	1.11	13.96
C_6–C_8NA	0.1	2.0	2.02	3.57				
C_9–200°C	–	0.5	0.5	1.8				
Residue	–	0.5	0.7	6.2	0.00	4.01	0.00	4.17
Total	100.0	100.0	100.0	100.0	100.0	100.0	100.0	100.0
Steam dilution ratio	0.4	0.4[a]	0.4[a]	0.6	0.3	0.5	0.3	0.5
		0.8[b]	0.8[b]					

[a] In gas cracking furnaces.
[b] In liquid cracking furnaces.

Despite multiple and repeated attempts, including the construction and operation of experimental plants, these systems did not establish themselves. Pyrolysis in a tubular furnace is the only system generally used.

Besides the systems with solid carrier, units named *autothermal* also were tested, where the heat required for endothermic pyrolysis is supplied by the combustion of a fraction of the feedstock [70].

The majority of these processes uses much higher temperatures than those practiced in coil pyrolysis. Indeed, at temperatures of the order of 2000°C and even more, the main product is a mixture of acetylene and hydrogen. Since these are processes of a different type than pyrolysis, they will not be further analyzed in this work. In fact, none of these processes were developed beyond the stage of pilot or semicommercial units.

Table 4.33 Typical Yields from Liquid Feedstocks at High and Low Severity

	Naphtha Kuwait		Gas oil Kuwait		Vacuum distillate Es Siden	
Density 15/15°C	0.713		0.832		0.876	
Distillation, °C	30–170		230–315		300–540	
K_{UOP}	12.3		11.98		12.21	
Hydrogen, wt %	15.2		13.7		13.0	
Aromatics, wt %	7		24		28	
severity	low	high	low	high	low	high
Products						
CH_4	10.5	15.0	8.0	13.7	6.6	9.4
C_2H_4	25.8	31.3	19.5	26.0	19.4	23.0
C_2H_6	3.3	3.4	3.3	3.0	2.8	3.0
C_3H_6	16.0	12.1	14.0	9.0	13.9	13.7
C_4H_6	4.5	4.2	4.5	4.2	5.0	6.3
C_4H_8	7.9	2.8	6.4	2.0	7.0	4.9
BTX	10.0	13.0	10.7	12.6		
C_5-204°C	17.0	9.0	10.0	8.0		
Fuel	3.0	6.0	21.8	19.0	25.0	21.0
$H_2, C_2H_2, C_3H_4, C_3H_8$	2.0	3.2	1.8	2.5	1.4	1.8
TOTAL	100.0	100.0	100.0	100.0	100.0	100.0

Table 4.34 Economic Data for a Pyrolysis unit Producing 450,000 t/year Ethylene

	Feed			
	Ethane	Ethane/Propane 50/50	Naphtha	Straight run gas oil
Relative investment	1200	1300	1500	1800
Consumptions per ton C_2H_4				
NaOH (100%), kg	2	3	0.5	2
monoethanolamine, kg			0.2	1
Catalysts, FF	2	2	5	5
Utilities per ton C_2H_4				
steam, ton	1	2	−0.15	0.9
fuel (10^6 kJ)	10	4		
electricity, kWh	30	40	80	100
cooling water, m^3	200	220	280	300
process water, m^3	2	2	2	2
Number of operators	8	8	12	12

Source: Ref 61.

The fact that all these attempts did not succeed in competing with the pyrolysis in tubular furnaces deserves a more detailed analysis in order to understand the reasons why they did not succeed and possibly to identify pathways that could lead to the development of viable processes.

From the information presented earlier concerning the theoretical bases of the process (Chapter 2) and the criteria for the selection of the reactor type for cracking processes (Section 3.1), we conclude that a pyrolysis reactor should satisfy the following conditions:

1. Approach as much as possible the ideal plug flow reactor.
2. The temperature should be rising along the reaction zone with the highest temperature in the range of 1000–1100°C at the outlet. The effluent is quenched immediately thereafter.
3. The increase of the temperature in the reactor should be very rapid especially in its last segment of it, so that the duration of the reaction should be of the order of tenth of a second or even less.
4. To be capable of processing various feedstocks, including those producing much coke.

The tubular furnace satisfies these demands with the exception of the last one and of the final temperature level of the effluent. The increase of the latter depends on the progress achieved in producing tubes resistant to high temperatures. Considerable progress was achieved also in the processing of heavy feedstocks. Despite all the improvements that may intervene, pyrolysis in tubular furnaces is not suitable for feeds that generate much coke.

The systems with heat carrier, while satisfying without difficulties the last condition, namely the ability to process feedstocks generating much coke, satisfy only partially the first three.

Thus, in systems with a moving bed, the backmixing of the reactants cannot be completely avoided, even though it can be reduced to satisfactory limits. The ascending temperature profile in the reaction zone, with a strong increase towards the outlet, can be obtained in such systems only if the circulation of the heat carrier is countercurrent to the reactants. In order to achieve a short reaction time, a large rate of circulation of the product vapors through the reaction zone is required. The countercurrent circulation, upflow for the vapors and downflow for the carrier, makes it necessary to use carrier particles of a size sufficient for not being entrained by the reaction mixture or for producing backmixing. At the same time, the particles must be small enough to ensure the necessary surface for the heat exchange.

The main difficulty still remains the fact that the upflow circulation, in the case of a heavy, only partially vaporized feedstock, leads to the coking of the walls of the lower part of the reactor and to the blocking of the circulation system. The large solid/feed circulation ratios necessary to avoid this phenomenon, make this system noneconomical.

Concerning systems with a fluidized bed, the classic one in dense phase is a typical system with intensive backmixing, which does not satisfy the first condition. In Section 3.1, it was shown how the conversion decreases in backmixed reactors. For this reason, the yield of ethene will be always lower than that obtained in the tubular or moving bed systems.

The use of the riser reactor in pyrolysis has the disadvantage that it cannot satisfy either the second or third condition. The heat exchange in the fluidized bed is so intensive such that, after the contact of the feed with the heat carrier, the temperature at the basis of the riser rapidly becomes uniform. The temperature will decrease along the riser due to the reaction endotherm.

These deficiencies of the classic fluidization reactors lead to the development of the "spouted" (or "fountain") bed, which significantly reduces backmixing and ensures a very short contact time.

An additional difficulty when using fluidized bed systems is to rapidly separate and quench the effluent. Since the classic cyclones cannot be used, one resorts to the injection of a cooling liquid into the flow that leaves the system. This creates the problem of efficiently recovering the heat removed by the cooling liquid.

4.3.3.1 The Pyrolysis in Systems With Moving Bed

So far, several experimental pyrolysis units were built using as heat carriers moving beds of spherical particles made of refractory material (or coke particles) [81–84]. Despite the fact that the process makes it possible to obtain higher temperatures than the tubular furnace and to use heavy and even residual feedstocks, all efforts to date have failed to make it economically competitive. The process flow diagram of the reactor-furnace system of such a plant is given in Figure 4.47.

The heat carrier, shaped as ceramic balls, circulates continuously from the bottom of the furnace to the top of the reactor, where it is contacted with the partially converted feedstock. Countercurrent circulation allows the effluent to reach the highest outlet temperature.

From the bottom of the reactor, after stripping the heat carrier is transported pneumatically to the top of the furnace, where it is reheated by the combustion of the deposited coke. The heat of the flue gases is used for the production of steam.

The gases exiting the reactor are quenched by using a system similar to that used in tubular pyrolysis furnaces.

In order to prevent the mixing of the reactor products with the air and with the combustion gases from the furnace, steam is introduced in the duct which connects the two vessels.

Carborundum, quartz, or other inert ceramic materials are used as heat carriers. They must have high heat capacity and a low thermal expansion coefficient in order not to crack at the large temperature differences that appear when the regenerated particles come in contact with the feedstock, and high mechanical resistance in order not to be eroded in the pneumatic transport system. In fact, at the highest point of the transport system, cyclone separators are provided for the elimination of the dust from the mechanical wear of the heat carrier.

The plants with heat carrier in the moving bed perform the cracking of residual feedstocks that produce large coke deposits.

The main difficulty consists of the fact that coke deposits are formed (especially in the case of heavy feedstocks) not only on the surface of the heat carrier particles but also on the walls of the lower zone of the reactor, so that they can agglomerate and even hinder the circulation of the carrier particles. These deposits are more extensive in the case of the countercurrent circulation of the feedstock and the heat carrier. They become larger with increasing molecular mass of the feedstock.

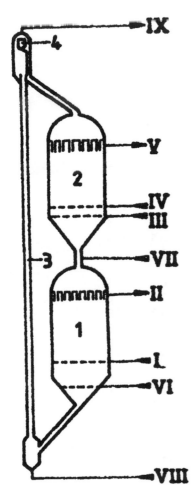

Figure 4.47 Schematics of a moving bed pyrolysis unit for light fractions. 1—reactor, 2—heater, 3—pneumatic-elevator, 4—cyclone; I—feed, II—products, III—air, IV—fuel (gases), V—flue gases, VI—stripping steam, VII—separation steam, VIII,IX—flue gases for conveying.

In order to avoid the formation of coke deposits on the walls of the reactor despite using the countercurrent flow required for achieving high effluent temperatures, high contact ratio carrier/feed is used. When pyrolyzing residues, the value of 65 for the carrier/feed ratio is reached [82].

To improve the situation, the contact ratio was decreased by increasing the rate of circulation of the heat carrier near the reactor walls. In this manner, the feedstock was introduced into a zone limited to the central portion of the reactor.

Published data concerning pyrolysis in such a system of a vacuum distillate are given in Table 4.35.

4.3.3.2 Pyrolysis in Fluidized Bed Systems

Pyrolysis in a fluidized bed, the same as pyrolysis with a moving bed of the heat carrier, was the object of a large number of studies in experimental plants at a semi-

Table 4.35 Pyrolysis of Vacuum Distillate in a Moving Bed System

Feed stock analysis

Density	0.9606
Viscosity, cSt at 99°C	160
Sulfur, wt %	0.59
Ramsbottom, carbon wt. %	8.2

Operating conditions and yields	Circulation		
	Cocurrent	Countercurrent	
Furnace exit, °C	939	1004	1204
Mean reactor temperature, °C	693	677	718
Heat carrier from reactor exit, °C	621	627	652
Steam/feed ratio	1.4	5.5	2.5
Heat carrier/feed ratio	40.1	70.5	96.0
Heat carrier residence time in reactor, min.	10	7	6
Yields, wt. %			
gases C_1–C_4	29.7	36.7	49.5
gasoline	17.5	17.0	17.0
light gas oil	12.2	6.2	7.3
heavy gas oil	7.8	8.5	3.7
heavy oil	24.6	22.5	17.6
coke (on heat carrier)	8.2	9.1	4.0
Gases composition, molar %			
H_2	16.2	14.4	16.2
CH_4	36.7	21.8	22.4
C_2H_2			0.7
C_2H_4	24.5	38.5	36.9
C_2H_6	6.5	3.0	3.6
C_3H_6	10.1	12.3	11.7
C_3H_8	1.1	0.6	0.5
C_4H_6			4.7
C_4H_8	4.5	9.1	3.1
C_4H_{10}	0.4	0.3	0.2
C_2H_4 % of reacted feedstock	12.2	20.0	26.2

industrial scale [62,70, 85–88], The same advantages as for pyrolysis with heat carrier in a moving bed are expected: the possibility of obtaining high reaction temperatures and to process feedstocks that produce much coke. The published reports refer mostly to reactor systems with fluidized bed in the dense phase, which does not eliminate backmixing that diminishes the reactor performance.

The data of Table 4.36 compares 5 of the pyrolysis processes of this type [88]. As indicated in the table, oxygen injection is practiced in some reactors in order to reach the desired reaction temperature.

The process flow diagrams of the units of Lurgi—pyrolysis using sand heat carrier; BASF—on coke particles, and of the KK process are presented in Figures 4.48–4.50.

The UBE process uses a spouted bed reactor Figure 4.51 [62].

Table 4.36 Comparison Between Different Systems of Fluidized Bed Pyrolysis

	Lurgi	BASF	BASF	UBE	KK
Reactor temperature, °C	705–845	700–750	760	830–880	750–800
Steam/feed ratio		1.0			1.06
Reaction time, s	0.3–0.5			0.2–0.3	1.7–2.1
C_2H_4, wt. %	23.1	20.6	25.0	28.1	20.9
C_3H_6/C_2H_4	0.3–0.9	0.56	0.45	0.40	0.52
Fuel/C_2H_4	0.96	0.81	0.70	0.16	1.3
O_2 injection	NOT	YES	NOT	YES	NOT
Heat carrier	sand	coke	coke	oxides	coke
Feedstock preheating °C	350–400			400	400
Licensor	Lurgi	BASF	BASF		Kunugi & Kuni
Experimental plants realized in	Dormagen, Germany	Ludwigshaven, Germany			Japan
C_2H_4 production	13–18 t/year	36 t/year	45 t/year	250 t/day	120 t/day
In operation since	1958	1960	1970	1979	1980

As shown above, none of the suggested processes uses the "riser" reactor. The reason is that, since the reaction is endothermic, the temperature will decrease along the reaction path, which doesn't correspond to the requirements of the pyrolysis process.

It is to be noted that the ethene yield increases as temperature in the reactor is higher and the reaction time is shorter. These trends must be used as guidance in

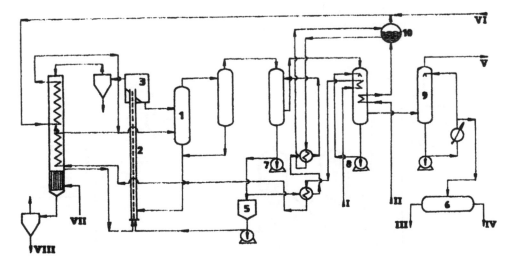

Figure 4.48 Lurgi pyrolysis unit with sand heat carrier. 1—reactor, 2—riser-coke burning, 3—gas-solids separator, 4—furnace, 5—heavy oil tank, 6—separator, 7,8,9—washing coolers, 10—steam production; I—feed, II—water, III—gasoline, IV—process water, V—gases, VI—steam, VII—air, VIII—sand fines, IX—combustion gases.

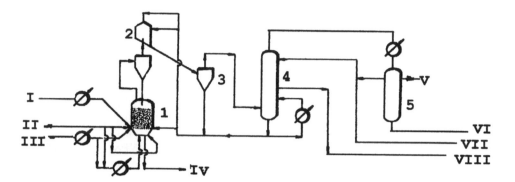

Figure 4.49 BASF pyrolysis. 1—reactor, 2—quenching, 3—separator, 4—column, 5—separator; I—feed, II—steam, III—oxygen, IV—coke, V—gases, VI—water, VII—light oil, VIII—heavy oil.

subsequent developments in order to develop processes that will be able to compete with the tubular furnace.

4.3.3.3 The ACR Process

The process is not different in principle from those described previously. The only difference is that the heat carrier is superheated steam obtained by burning gaseous fuel and oxygen in a stream of steam. Thus, the temperature of the steam heat carrier can reach 2000°C.

The flow diagram of the process is given in Figure 4.52.

The quenching of the effluent is performed in an apparatus that combines the injection of cold liquid with the production of steam.

Some operation data for this process are [89]:

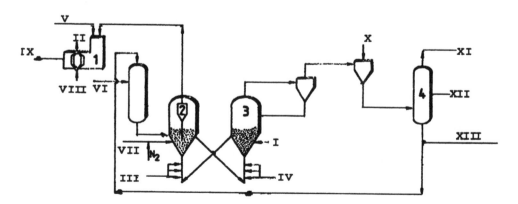

Figure 4.50 K.K. pyrolysis process with coke heat carrier. 1—steam generator, 2—heater, 3—reactor, 4—distillation column; I—feed, II,III,IV—steam, V,VI,VII—air, VIII—water, IX—flue gases, X—cooling oil, XI—gasoline and gases, XII—middle distillate, XIII—heavy oil.

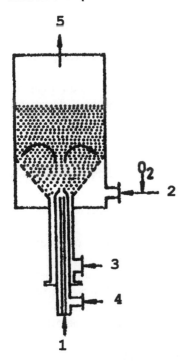

Figure 4.51   UBE spouted bed reactor. 1—feed, 2,3,4—steam, 5—products.

Temperature of the heat carrier steam (°C)	2000
Temperature in the injection point for the feed (°C)	1200
Reaction residence time (s)	0.015–0.03
Effluent temperature at the outlet from the reactor (°C)	900
Temperature after quenching (°C)	335

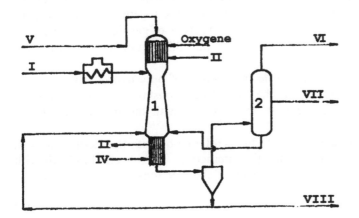

Figure 4.52   ACR pyrolysis process. 1—reactor, 2—column; I—feed, II, III—steam, IV—water, V—flue gases, VI—liquid products, VII—gaseous products, VIII—tar.

The process yields 37.5% ethene for gasoline feed and 26.2–31.6% (depending on the severity) for pyrolysis of vacuum gas oil. These conversions exceed those obtained in tubular furnaces, which is understandable by the higher temperature of reaction.

The propene/ethene ratio is 0.42 for gasoline feed and 0.20 for vacuum gas oil feed. In the latter case, it decreases as the severity of the process decreases. Concomitantly, between 5–10% butadiene and 7–8% BTX fraction is obtained.

4.3.4 Hydropyrolysis

Research done over several years on pyrolysis process under pressure in the presence of hydrogen [26,90–94], led in 1976 to the operation of a pilot plant with a capacity of 1000 t/year [94]. This confirmed the outstanding performances expected from this process.

The process is performed at temperatures of 800–900°C, pressures of 10–30 bar and, contact times of the order of 0.1 sec.

Despite the fact that operation under hydrogen pressure should, from the thermodynamic point of view, disfavor the formation of lower alkenes, the results of the research prove that this is perfectly valid at high pressures and moderate temperatures. The situation is quite different at the pressures and temperatures of classical pyrolysis.

Thus, the hydropyrolysis of *n*-hexadecane and decaline in presence of nitrogen at temperatures not exceeding 600°C and pressures of 70 bar [26], showed a decrease of the yields of ethene and propene when nitrogen was replaced by hydrogen. This result is obvious from the examination of Figures 2.3 and 2.4, which give the equilibrium of the reaction of the formation of ethene from ethane and of propene from propane.

However, the same graphs show that in the conditions of pyrolysis, i.e., at approximately 850°C and at relatively low pressures, the thermodynamically possible conversion is 80–90% and the formation of lower alkenes will depend therefore on the mechanism and relative rates of the elementary decomposition reactions.

The experimental data show that in such conditions, hydrogen promotes decomposition reactions: the conversion to gases increases, together with the amount of ethene. The data of Table 4.37 on the hydropyrolysis of gasoline, kerosene, and gas oil as compared to classical pyrolysis with steam dilution instead of hydrogen, illustrate these findings [11].

Similar results occur also in the decomposition of pure hydrocarbons. Thus, at a molar ratio H_2/1-hexene = 5.5, the overall rate constant for the decomposition at 700°C is 1.5-times higher than for mixtures with the molar ratio He/1-hexene = 1 [11].

This effect of hydrogen is explained by its interaction with the methyl radicals, which proceed with higher velocity than the interactions of the methyl radicals with the higher alkanes. Thus, the relative velocity of the reactions:

$$\overset{\bullet}{C}H_3 + H_2 \rightarrow CH_4 + \overset{\bullet}{H} \tag{a}$$

$$\overset{\bullet}{C}H_3 + C_6H_{14} \rightarrow CH_4 + \overset{\bullet}{C}_6H_{13} \tag{b}$$

Table 1.37 Pyrolysis at 850°C and Atmospheric Pressure with Steam or Hydrogen Dilution

	Light naphtha		Naphtha		Gas oil	
Distillation limits, °C	104–181		148–240		180–375	
Composition, vol %						
alkanes	65		59		83	
cycloalkanes	20		23			
aromatics	15		18		17	
Molecular mass	130		170		220	
Dilution with	H_2O	H_2	H_2O	H_2	H_2O	H_2
Diluent/feed molar ratio	7.00	6.80	6.90	7.30	6.70	5.30
Reaction time, s.	0.13	0.11	0.12	0.16	0.15	0.16
Products, wt %						
H_2	1.0	0.1	0.9	−0.5	0.7	−0.1
CH_4	10.6	16.3	10.9	17.5	9.4	14.3
C_2H_4	27.4	32.0	27.3	33.7	23.3	27.6
C_2H_6	2.0	5.0	3.2	5.5	2.5	4.7
C_3H_6	12.6	10.9	13.0	14.0	11.8	13.1
C_3H_8	0.9	0.8	0.9	1.1	0.7	1.0
C_4H_8	3.8	2.5	4.8	1.8	3.5	3.0
C_4H_6	6.4	3.3	6.0	3.5	4.2	3.9
Total gases	64.7	70.9	67.0	76.6	56.1	67.5
Coke	0.14	0.038	0.31	0.12	0.55	0.52

are, according to the data of Table 2.5:

$$\frac{r_a}{r_b} = \frac{3.2 \times 10^9 e^{-42,700/RT}}{10^8 e^{-33,900/RT}} \times \frac{[H_2][\overset{\bullet}{C}H_3]}{[C_6H_{14}][\overset{\bullet}{C}H_3]}$$

which at 1100°C gives:

$$\frac{r_a}{r_b} = 12 \frac{[H_2]}{[C_6H_{14}]}$$

Since the velocity of hydrogen extraction from a molecule by atomic hydrogen is 1–2 times faster than the extraction of the same atom by the methyl radical, the formation of atomic hydrogen by the reaction (a) will increase the overall decomposition rate.

Observe that, according to the above equation, the ratio of the rates of the reactions (a) and (b) depends not only on the ratio of the two rate constants, but also on the concentration of the molecular hydrogen in the reactor. Therefore the increase of this concentration will lead to the increase of the overall decomposition rate that was confirmed experimentally.

The presence of hydrogen decreases the rate of the secondary reactions that form diolefines, such as among others, butadiene. This is explained by competition between the following reactions:

$$\overset{\bullet}{C}_2H_3 + C_2H_4 \rightarrow CH_2{=}CH{-}CH_2{-}\overset{\bullet}{C}H_2 \rightarrow CH_2{=}CH{-}CH{=}CH_2 + \overset{\bullet}{H} \qquad (c)$$

$$\dot{C}_2H_3 + H_2 \rightarrow C_2H_4 + \dot{H} \tag{d}$$

The rate constants of these two reactions are:

$$k_c = 5.0 \times 10^8 e^{-30,500/RT} \, \text{L/mol} \cdot \text{s}$$

$$k_d = 7.95 \times 10^9 e^{-31,000/RT} \, \text{L/mol} \cdot \text{s}$$

The relative velocity of the two reactions at 1100°C will be:

$$\frac{r_d}{r_c} = 15 \frac{[H_2]}{[C_2H_4]}$$

The presence of hydrogen will thus sensibly reduce the formation of butadiene by way of reaction (c).

The authors of the cited process [94], which led to the construction of the hydropyrolysis pilot plant, show that because of the exothermal character of the hydrogenation reactions, the process could be carried out in such a way that the obtained coil outlet temperature could not be reached in classic pyrolysis furnaces.

In the hydropyrolysis pilot plant, 35% ethene, 18% propene and 35% ethane were obtained from a fraction with a boiling range of 178–375°C.

The authors of the process claim lower investment and lower energy usage than in the case of classic pyrolysis.

It is difficult to foresee to what extent such a process will be expand and will prove its viability at the industrial scale. In any case, it opens a new way, worthy to be taken into account.

A thermal process, with a typical radicalic reaction mechanism is the thermal dealkylation of toluene, which is similar to the dealkylation of alkyl-naphthalenes [95,96].

The reaction mechanism is the following:

$$C_6H_5CH_3 \xrightarrow{k_1} C_6H_5\dot{C}H_2 + \dot{H} \tag{a}$$

$$\dot{H} + C_6H_5CH_3 \xrightarrow{k_2} H_2 + C_6H_5\dot{C}H_2 \tag{b}$$

$$\dot{H} + C_6H_5CH_3 \xrightarrow{k_2'} C_6H_6 + \dot{C}H_3 \tag{c}$$

$$\dot{C}H_3 + H_2 \xrightarrow{k_3} CH_4 + \dot{H} \tag{d}$$

$$C_6H_5\dot{C}H_2 + H_2 \xrightarrow{k_4} C_6H_5CH_3 + \dot{H} \tag{e}$$

$$C_6H_5\dot{C}H_2 + \dot{H} \xrightarrow{k_5} C_6H_5CH_3 \tag{f}$$

Reaction (a) is the chain initiation, (b)–(e) is the propagation and (f) the chain termination.

Applying the steady state theorem it follows:

$$\frac{d[C_6H_5\dot{C}H_2]}{d\tau} = k_1[C_6H_5CH_3] + k_2[C_6H_5CH_3][\dot{H}]$$

$$- k_4[C_6H_5\dot{C}H_2][H_2] - k_5[C_6H_5\dot{C}H_2][\dot{H}] = 0 \tag{4.9}$$

$$\frac{d[\overset{\bullet}{H}]}{d\tau} = k_1[C_6H_5CH_3] - (k_2 + k_2')[C_6H_5CH_3][\overset{\bullet}{H}] + k_3[\overset{\bullet}{CH_3}][H_2]$$

$$+ k_4[C_6H_5\overset{\bullet}{CH_2}][H_2] - k_5[C_6H_5\overset{\bullet}{CH_2}][\overset{\bullet}{H}] = 0$$

(4.10)

$$\frac{d[\overset{\bullet}{CH_3}]}{d\tau} = k_2'[C_6H_5CH_3][\overset{\bullet}{H}] - k_3[\overset{\bullet}{CH_3}][H_2] = 0 \tag{4.11}$$

Replacing $k_3[\overset{\bullet}{CH_3}][H_2]$ from (4.11) in (4.10), it results:

$$\frac{d[\overset{\bullet}{H}]}{d\tau} = k_1[C_6H_5CH_3] - k_2[C_6H_5CH_3][\overset{\bullet}{H}] + k_4[C_6H_5\overset{\bullet}{CH_2}][H_2]$$

$$- k_5[C_6H_5\overset{\bullet}{CH_2}][\overset{\bullet}{H}] = 0 \tag{4.12}$$

Adding (4.9) and (4.12), we get:

$$k_1[C_6H_5CH_3] = k_5[C_6H_5\overset{\bullet}{CH_2}][\overset{\bullet}{H}] \tag{4.13}$$

$$[\overset{\bullet}{H}] = \frac{k_1[C_6H_5CH_3]}{k_5[C_6H_5\overset{\bullet}{CH_2}]} \tag{4.14}$$

Introducing (4.12) after simplifying gives:

$$\frac{k_1k_2[C_6H_5CH_3]^2}{k_5[C_6H_5\overset{\bullet}{CH_2}]} = k_4[C_6H_5\overset{\bullet}{CH_2}][H_2]$$

and:

$$[C_6H_5\overset{\bullet}{CH_2}] = \left(\frac{k_1k_2}{k_4k_5}\right)^{1/2} \times \frac{[C_6H_5CH_3]}{[H_2]^{1/2}} \tag{4.15}$$

Replacing in (4.14), it follows:

$$[\overset{\bullet}{H}] = \left(\frac{k_1k_4}{k_2k_5}\right)^{1/2} \times [H_2]^{1/2} \tag{4.16}$$

The reaction rate of toluene decomposition will be:

$$-\frac{d[C_6H_5CH_3]}{d\tau} = k_1[C_6H_5CH_3] + (k_2 + k_2')[C_6H_5CH_3][\overset{\bullet}{H}] -$$

$$- k_4[C_6H_5\overset{\bullet}{CH_2}][H_2] - k_5[C_6H_5\overset{\bullet}{CH_2}][\overset{\bullet}{H}]$$

Considering (4.13), it gives:

$$-\frac{d[C_6H_5CH_3]}{d\tau} = (k_2 + k_2')[C_6H_5CH_3][\overset{\bullet}{H}] - k_4[C_6H_5\overset{\bullet}{CH_2}][H_2]$$

Replacing the radicals concentrations with (4.15) and (4.16), one obtains finally:

$$-\frac{d[C_6H_5CH_3]}{d\tau} = k_2'\left(\frac{k_1k_4}{k_2k_5}\right)^{1/2}[C_6H_5CH_3][H_2]^{1/2} \tag{4.17}$$

The reaction is of the first order in toluene concentration and of the 1/2 order in hydrogen concentration, orders confirmed by the experimental data.

The activation energy of this reaction is 226–234 kJ/mol and the heat of reaction, 50 kJ/mol.

The process is carried out at temperatures of 650–750°C, pressure of 50 bar, and hydrogen/feedstock ratio, of 4 mol/mol.

The industrial implementation of this process will be presented in Section 13.9.3 together with the catalytic toluene dealkylation and in our research work concerning these two processes.

REFERENCES

1. S Raseev. Postgradued lectures at Zulia University, Maracaibo, Venezuela, 1980.
2. IH Hirsh, EK Fischer. In: BT Brooks, ed. The Chemistry of Petroleum Hydrocarbons, vol. 2, Reinhold Publ. Co., New York, 1955.
3. S Raseev. Procese distructive de prelucrarea titeiului, Editura Tehnica Bucuresti, 1964.
4. F Stolfa. Hydrocarbon Processing 59 (5): 101, 1980.
5. M Akbar, H Geelen. Hydrocarbon Processing 60 (5): 81, 1981.
6. JF Le Page, SG Chatila, M Davidson. Reffinage et conversion des produits lourds du ptrole. Technip, Paris, 1990.
7. VD Singh. Erdol Kohle Erdgas Petrochem. 39 (1): 19, 1986.
8. CPG Lewis et al. Oil Gas J 83, 8 Apr. 77, 1985.
9. S Rossarie, V Devanneau. Petr. Tech. nr. 310: 7, 1984.
10. Van Kerkyoort WJ. et al., IV Congrs International du chauffage industriel, 1952.
11. RZ Magaril. Teoreticeskie osnovy himiceskih protsesov pererabotki nefti, Himiia, Moskva, 1976.
12. Alaund. Oil Gas J 76 Nov. 6: 56, 1978.
13. JR Wood. Oil Gas J 83 22 Apr: 80, 1985.
14. JR Wood, CK Marino. Oil Gas J 89, 25 Febr: 42, 1991.
15. D Nedelcu. Inginerie prelucrarii hidrocarburilor, In: G Suciu, R Tunescu, eds., vol. 2, p. 186, Editura Tehnica, Bucuresti, 1975.
16. Lummus Crest, Inc., Refining Handbook 1990, Hydrocarbon Processing 69 (9): 104, 1990.
17. Institut Francais du Petrole, Refining Handbook 1990, 69 (9): 104, 1990.
18. WL. Nelson. Petroleum Refinery Engineering. New York, London, Tokyo: McGraw-Hill Book Co. 1958.
19. Petroleum Refiner, 39 (8) : 214, 1960.
20. GL Ypsen, JH Jenkins. Hydrocarbon Processing 60 (9): 117, 1981.
21. K Waskimi, H Limmer. Hydrocarbon Processing 68 (9): 69, 1969.
22. Y Ogata, H Fukuyama. Hydrocarbon Technology International, P. Harrison, ed. London: Sterling Publishing International Ltd, 1991, p. 17.
23. Hydrocarbon Processing 61 (9): 161, 1982.
24. K Washimi, K Ozaki et al. The Second Pacific Chemical Engineering Congress (PAChEC77), vol. 1, Library of Congress Catalog nr. 77-82322, New York: AJChE, 1977.

25. R Tukahashl, K Washlml. Hydrocarbon Processing 55 (11): 93, 1976.
26. TY Shabtai, R Ramakrishnan, AG Oblad. In: AG Oblad, HG Davis, RT Eddinger, eds. Thermal Hydrocarbons Chemistry, ACS Advances in Chemistry Series no. 183, Washington, DC: American Chemical Society, 1979, p. 297.
27. P Le Perche et al. Petr. Tech. 311: 23, 1984.
28. AW, Langer, J Stevard, CE Thompson, T White, RM Hil. Ind Eng Chem 50: 1067, 1958; 53: 27, 1961.
29. AW Langer et al. Ind Eng Chem Process Des Dev 1: 309, 1962.
30. D Decrocq, M Thomas, B Fixari, P le Perchec, M Bigots, L Lena, T. des Cairieres, J Rossarie. Rev. Inst. Fr. Petrol 47 (1): 103, 1992.
31. LP Fisher, F Souhrada, HJ Woods. Oil Gas J 89, 22 Nov., 111, 1982.
32. AS Bakshi, IH Lutx. Oil Gas J 85, 13 Iul 84: 1987.
33. RE Dymond. Hydrocarbon Processing 70 (9): 161 D, 1991.
34. R De Blase, JD Elliot. Hydrocarbon Processing 61 (5): 99, 1982.
35. HM Feintuch, JA Bonilla, Godino RL. FW Delayed-Coking Process. In: Handbook of Petroleum Refining Process, RA Meyers, ed., New York: McGraw-Hill Book Co, 1986.
36. SW Martin, LX Wills. Advances in Petroleum Chemistry and Refining, vol. II, New York, London: Interscience Publ. Co. p. 157.
37. Hydrocarbon Processing 79: 11, 98, 2000.
38. Hydrocarbon Processing, 79: 11, 98, 2000.
39. JD Elliot. Oil Gas J 89, 4 Feb: 41 (1991).
40. JH Gary, GE Handwerk. Petroleum Refining, Technology and Economics. 2nd ed. New York, Basel: Marcel Dekker Inc., 1984.
41. BP Castiglioni. Hydrocarbon Processing 62 (9): 77, 1983.
42. M Isao, Y Korai, YQ Fei, J Oyama. Oil Gas J 86, 2 mat: 73, 1988.
43. Hydrocarbon Processing 77: 62, Nov 1998.
44. Hydrocarbon Processing 77: 64, Nov. 1998.
45. TP Forbath. Chem. Eng. 64 (11): 222, 1957.
46. BK Americ, JA Botnicov, KP Lavrovschi, AJ Scoblo, VS Aliev, AM Brodski. Fifth World Petrol Cong. Sect. 3, New York, 1959, p. 373.
47. CM S Schillmoler. Hydrocarbon Processing 64 (9): 101, 1985.
48. P Wuither. Rev. Inst. Fr. Petrol (9): 1156, 1959.
49. DJ Allan, OJ Blaser, MM Lambert. Oil Gas J, 17 May: 93, 1982.
50. Hydrocarbon Processing 61 (9): 163, 1982.
51. FN Dawson Jr., Hydrocarbon Processing, 60 (5): 86, 1981.
52. T Miyauchi, Y Ikeda, T Kikuchi, T Tsutsui. Proc. 12th World. Pet. Congr., vol. 4, 335, 1987.
53. Hydrocarbon Processing 69 (11): 144, 146, 1990.
54. EA Bagdoyan, E Gootzalt, Hydrocarbon Processing 64 (9): 85, 1985.
55. I Velea, G Ivanus. Monomeri de sinteza, vol. 1 (1989), 2 (1990) Tehnica Bucuresti.
56. JJ Roos, D Pearce. Proc. 11th World petroleum Congress, vol. 4, 153, 1983.
57. CG Suciu. Progrese in procesele de prelucrare a hidrocarburitor, Editura Tehnica, Bucuresti, 1977.
58. DG Tanasescu. Oil Gas J 89, Dec. 30: 89, 1991.
59. EF Gallei, HD Dreyer, W Himmel. Hydrocarbon Processing, 61 (11): 97, 1982.
60. VF Yesavage, CF Griswold, PF Dickson. Hydrocarbon Processing 60 (4): 155, 1981.
61. A Chanvel, G Lefebore, L Castex. Procede de Petrochimie, vol. 1, Technip, 1985.
62. V Vintu, R Mihail, V Macris, G Ivanus. Piroliza hidrocarburilor, Editura Tehnica, Bucuresti, 1980.
63. BT Baba, JR Kennedy. Chem Eng 5 Jan: 166, 1976.
64. KM Sundarom, GF Froment. Chem Eng Science, 32: 601, 1977.
65. DD Parcey, JH Prunell. Ind Eng Chem Process Des Dev 7: 435, 1968.

66. AI Scoblo, IA Tregubova, NN Egarov. Preotzesy i aparaty neftepererabotyvaiuscei i neftehmicescoi promyslennosti, Gostoptehizdat, Moskva, 1972.

67. LE Chembers, WS Potter. Hydrocarbon Processing, 53 (1): 121, 1974; (3): 95, 1974; (8): 39, 1974.

68. LF Albright, WA McGill. Oil Gas J. 85, 31 Aug: 46, 1987.

69. GL Jaques, P Dubois. Rev Assoc Fr Techn. Petr, nr. 217, 43, 1973.

70. V Kaiser, M Leroy, R Amonyal. Rev Inst Fr Petr. 37, (5) 651, 1982.

71. JL James, et al. Rev. Assoc. Fr Techn Petr, nr. 217 56, 1973.

72. D Newman, ME Brooks. Proc. 8th World Petroleum Congress, Simposium 18 Moscow, 1971.

73. B Lehr, W Schwab. In: Thermal Hydrocarbon Chemistry, AG Oblad, HG Davis, RT Eddinger, eds., p. 153, Advances in Chemistry Series 183, Washington: American Chemical Society, 1979, p. 153.

74. SB Zdonik, EJ Green, LP Halless. Manufacturing ethylene, Tulsa: The Petroleum Publishing Co., 1971.

75. J Barton. Oil Gas7, 29 Jan, 81: 88, 1990.

76. F Albricht. Oil Gas J, 86, 1 Aug: 35, 1988; 15 Aug: 69, 1988; 29 Aug: 44, 1988; 19 Sept: 90, 1988.

77. CM Schillmoier, VW Van den Bruck. Hydrocarbon Processing 63 (12): 55, 1984.

78. O Szechy, T-C Luan, LF Albricht. Pretreatment of High-Alloy Steels to minimise coking in ethylene furnaces, In: LF Albright, ed., Novel Production methods for ethylene, light hydrocarbons and aromatics, New York: Marcel Dekker, 1992, p. 341.

79. G Merz, H Zimmerman. Modern Furnace design for steam crackers. In: P Harrison ed. Hydrocarbon Technology International 1992, London: Sterling Publ. Intermat. Ltd, 1992.

80. M Picciotti. Hydrocarbon Processing 59 (4): 223, 1980.

81. SC Eastwood, AE Potas. Oil Gas J, 47: 104, 1948.

82. RA Findiay, RR Goins. Advances in Petroleum Chemistry and Refining, vol. 2, p. 126, New York-London: Interscience, 1959.

83. MO Kilpatrick, et al. Petr. Process. 53:3, 903, 1954.

84. MO Kilpatrik, et al. Oil Gas. J. 53, (1): 162, 1954.

85. World Petr. 30 (6): 62, 1959.

86. D Kunii, T Kunugi, et al. Proc. 9th World petroleum Congress, Tokyo 1975, vol. 5, 1975, 137–143.

87. LM Pauschin, TP Visniacova. Proizvodstvo olefinosoderjascih i gorucih gazov, Ed. Acad. SSSR, Moscva, 1960.

88. YC Hu. Hydrocarbon Processing 61, (11): 109, 1982.

89. RL Baldwin, GR Kamm. Hydrocarbon Processing 61, (11): 127, 1982.

90. L Kramer, J Happel. Ind Eng Chem 59: 39, 1967.

91. J Andersen, L Case. Ind Eng Chem Process Des Dev 1: 161, 1962.

92. T Kunugi, H Tominaga, S Abico. Proc. 7th World Petroleum Congress, Mexico City, 1967, vol. 5, 239–245.

93. VM Rabin, IP Iampolachi. Neftehimia 16: 729, 1976.

94. C Barre, E Chahvekilian, R Dumon. Hydrocarbon Processing 55 (11): 176–178, 1976.

95. Hydrocarbon Processing 77: 111, 1998; 79: 142, 2000.

96. Hydrocarbon Processing 77: 112, 1998, 79, 142, 2000.

97. S Raseev. Procesy rozkladove w przerabce ropy naftowy, Wydawnictwo Naukovo Technicne, Warszawa, 1967.

98. A Stefani. Hydrocarbon Processing 74: 61–66, 1995.

99. A Stefani. Hydrocarbon Processing 75: 99, 103, 1996.

100. JF Schabron, JG Speight. Revue de l'Institut Francais du Pétrole, 52, Jan-Feb 1997, 73–85.

101. JII Gary, GE Handwerk. Petroleum Refining Technology and Economics. 3rd ed., Marcel Dekker Inc., New York, Basel, Hong Kong, 1993.
102. Hydrocarbon Processing 71: 207, 1992.
103. Hydrocarbon Processing 77: 69, 1998.
104. Hydrocarbon Processing 77: 111, 1998.
105. A Stefani. Hydrocarbon Processing 76, Aug. 1997: 110–113.
106. KM Sundaram, JV Albano and K Goldman. Erdöl, Erdgas, Kohle, 111 (3): 125, 1995.
107. GP Kelton, DL Torres, A Rawlins. Hydrocarbon processing 77, March 1998: 111–114.
108. B Kerr. Hydrocarbon Processing 78, May 1999: 109–112.
109. B Kerr. Hydrocarbon Processing 77, Nov. 1998: 62.
110. SD Parks, CM Schillmoller. Hydrocarbon Processing 75, March 1996: 53–61.
111. AG Wysiekierski, G Fisher, CM Schillmoller. Hydrocarbon Processing 78, Jan. 1999: 97–100.
112. F Billaud, C Guéret, P Bronin, J Weil. Revue de l'Institut Français du Pétrole, 47, 1992, 537–549: July–Aug.
113. JJ De Saepher, T Deterumerrmon, GF Froment. Revue de l'Institut Francais du Pétrole, 51, March–April 1996.
114. M Florescu. Este eficienta petrochimia, Chiminform Data SA, Bucarest, Romania 1996.
115. M Florescu. Industria comparata a polimerilor sintetici in Romania Chiminform Data SA, Bucarest, Romania.
116. AD Estman, JH Kolts, JB Kimble. Catalytic production of ethylene at Philips, II Ethan oxidative dehydrogenation, In: LF Albricht, ed., Pyrolisis Theory and Industrial Practice, New York: Academic Press, 1983, p. 21–40.
117. GA Detzer, JH Koltz. Catalytic production of ethylene at Philips, III Ethylene from butane cracking catalysts, In: LF Albricht, ed., Pyrolisis theory and Industrial Practice, New York: Academic Press, 1983, p. 41–60.
118. KK Pant, D Kunzrn. Catalytic pyrolisis of n; helptane: Kinetic and modeling, Ind. Eng. Chem. Res., 36, 2059–2064 (1997).
119. SV Adelson, FG Jafarov, OV Masloboyschicova, EV Tourunina. Himia I Tehnologia Topliv I Masel, nr. 5, 1990, 15–17.
120. AD Guseinova, LM Morozaieva, RY Ganbarov, IA Timakov, SI Adigesalov. Himia i Tehnologia Topliv i Masel, nr. 7, 9–10, 1990.
121. KM Sundaram, EF Olszewski, MM Schreehan, Ethylene, in Kirk-Othmer Encyclopedia of Chemical Technology, 4th Ed. Vol. 9, 1994.

5

Elements of Reactor Design

In this chapter specific aspects of reactor design for thermal processes are treated. The treatment is based on the general theory of chemical reactor design .

5.1 DESIGN OF THE REACTION SECTION OF TUBULAR FURNACES

In the following, the design of tubular furnaces for thermal processes is limited to the final part of the coil where the chemical transformations take place, without references to the design of the anterior part where only physical phenomena of heating and vaporizing occur. Also, general principles for the design of the furnace are assumed to be known.

The reaction rate of thermal processes shows a strong variation with temperature. In cracking processes (visbreaking and coking), where the furnace outlet temperature is approximately 500°C, one may consider that chemical reactions take place only in the portion of the coil corresponding to the last 100°C of temperature increase.

For pyrolysis furnaces, where the coil outlet temperature is much higher, one may consider that chemical reactions begin when the temperature of the reactants reaches 150–200°C below the outlet temperature.

In essence, the design of the reactor volume where the reactions take place takes into account the coil length required for reaching the desired conversion. This is achieved by computing the evolution of the temperature, pressure, and chemical composition, along this portion of the coil length. The designers applies mathematical equations describing the interaction of the flow, thermal, and chemical phenomena that take place simultaneously within the reaction coil.

There are two ways of performing the calculations:

1.   The reaction portion of the coil is equally divided in a finite number of elements. The larger their number, the more accurate the calculation. For each element, on basis of the inlet conditions and of the transformations that take place within (using balance, transfer, and kinetic equations) the outlet conditions (temperature, pressure, and chemical composition) are computed. In the calcula-

tions, the integral form of the kinetic equations are used, expressing time on the basis of the mean volume of the feed and of the products flow at the inlet and outlet from the element.

2.   The chemical, thermal, and flow phenomena that take place along the coil are expressed by a system of differential equations with specific boundary conditions. The integration of these equations leads to the outlet conditions from the coil.

Three methods were developed for computing the extent of the chemical transformations:

1.   Use one kinetic equation that expresses the overall conversion. Experimental data, preferably from a commercial unit of the same type using similar feedstock, is used in order to obtain the conversions of individual major components of the feed or of individual fractions.

2.   Use a system of kinetic equations based on molecular (stoichiometric reactions) reaction mechanisms. When petroleum fractions are submitted to pyrolysis, a model written for successive or successive-parallel reactions may be used, as described in Section 2.4.5. The number of kinetic equations used in this case must be at least equal to that of the individual components or of the fractions contained in the feed. The composition of the feed must be known in terms of the individual components or narrow fractions. Some difficulties might appear in relation to the values of the required kinetic constants.

3.   The use of a system of kinetic equations corresponding to elementary radical-molecular interactions, deduced on basis of the reaction mechanism—*mechanistic modeling*. In the case of pyrolysis, all elementary radical reactions corresponding to the accepted reaction mechanism must be taken into account. This computation method was used also for the validation of proposed reaction mechanisms, by comparing the results of the computation with experimental data [1–3]. This method takes into account implicitly a large number of interactions that may take place in the reaction conditions. The method implies the availability of a large number of kinetic constants, and still its application to the pyrolysis of gasoline and gas oils requires simplifying assumptions. Application of this method to the design process of the reaction coil gives results that are more exact than those based on the system of kinetic equations based upon stoichiometric reactions.

The first method of computing, which uses only one overall kinetic equation, is suitable when there is a limited number of components or fractions, the proportion of which must be known, as is the case in coking and visbreaking.

In the case of pyrolysis, dividing the overall conversion into a large number of components and fractions is difficult. In order to be realistic, it must be based upon a large number of experimental data produced in the operation of similar pyrolysis furnaces processing similar feedstocks. In this context, one should note that in classic cracking processes, the errors that might result from the use of existing methods of dividing the overall conversion are not able to sensibly influence the operation of the equipment downstream of the furnace. In the case of pyrolysis, the operation of the equipment for recovering individual components from the light fraction of the effluent makes necessary a more accurate estimate of the concentration in the feed of these individual components. However, one may assume that the established engineering and construction companies that build pyrolysis plants, which have accumulated considerable operating experience, preferably use semi-empirical methods together with some sophisticated methods of kinetic modeling.

A second way of computing, based on stoichiometric modeling, was used to obtain the results of the pyrolysis of lower alkanes [4,3] and earlier, also for the pyrolysis of liquid fractions, including gas oils [5].

The latter method is more recent. Its use requires knowledge of the kinetic constants for elementary radicalic reactions.

Other methods, similar to the two last mentioned, were developed for the optimization of existing pyrolysis units [6].

Usually, when the method that employs one overall kinetic equation is used, the length of the coil is divided into a number of sections or steps. When the modeling of the pyrolysis is made by using a system of kinetic, stoichiometric, or mechanistic equations, a system of differential equations is written that expresses the variation along the coil of the composition, temperature and pressure.

Also, a sufficient precision is obtained by using a calculation in which the division of the coil length in a number of steps is combined with the system of kinetic equations written in integral form. The residence time in each segment is expressed by equations of the form:

$$\tau = \frac{V_R}{V} \tag{2.124}$$

Such methods of tackling the problem may be applied also to the processes of coking and visbreaking by using the system of kinetic equations obtained from the modeling of the process as either successive or as successive-parallel reactions.

In the following, the equations are deduced and the calculation methods are presented for each of these methods.

5.1.1 Stepwise Design Using One Overall Kinetic Equation

The thermal balance for each step (j) is expressed by the equation:

$$G\left[\sum_j g_e i_e - \sum_j g_i i_i - q_r x_j + n\Delta i_a\right] = \varphi_t \pi D L_{t_j} \tag{5.1}*$$

where:

G = flowrate through the furnace in kg/h;
g = weight fractions,
i = enthalpies of products, respectively of steam;
q_r = heat of reaction in kJ/kg formed product;
n = steam/feed mass ratio;
φ_t = thermal tension in kJ/h m^2;
L_t = length of tubes within fire box, in m.
Subscripts: e = exit from the considered segment;
$\quad\quad\quad\quad\quad$ i = entry into segment;
$\quad\quad\quad\quad\quad$ a = steam

The pressure drop Δp_j in step j is calculated with the relations [7,8]:

---

* The thermal cracking being endothermic, φ will be negative and the sign fore φ in the equation (3.1) will be plus.

$$\Delta p_j = f\,\frac{w^2 \cdot L_{e_j}}{2g \cdot d \cdot \overline{\rho}_j} = f\,\frac{v^2 L_{e_j}\overline{\rho}_j}{2g \cdot d} \tag{5.2}$$

where:

Δp = the pressure drop in segment j, in kg/m^2;
f = friction factor, dimensionless;
w = mass velocity (including steam) through the tubes, in kg/m^2 s;
L_{ej} = equivalent hydraulic length of the tubes in segment j, in m;
g = gravitational acceleration in m/s^2;
d = tube inside diameter, in m;
ρ_j = mean density of reaction mixture, including steam in kg/m^3;
v = average linear velocity in tubes, in m/s

The equivalent hydraulic length of the tubes is calculated by the equation:

$$L_{e_j} = m_j L_d + C(m_j - 1)d \tag{5.3}$$

where:

m_j = the number of tubes corresponding to the step j
L_d = the length of the straight part of the tube
C = a correction coefficient that has the value of 30 for welded return bends and 50 for forged return bends.

Eq. (5.2) may be used for calculating the pressure drop in single-phase flow (vapors or liquid). In the case of two-phase vapors/liquid flow, several calculation methods exist [7–9].

For the portion of the coils where the heating, vaporizing, and reaction take place, an empirical approximate equation is recommended:

$$\Delta p = 1.65 \cdot 10^8\,\frac{v^{1.77}}{p_e^{0.73}}L \tag{5.4}$$

where Δp and p_e are expressed in Pascal* and v in cold liquid feed in m/s.

In order to obtain reasonable pressure drops, it is recommended [9] to use the following linear rates in the tubes:

Visbreaking	0.6–1.8 m/s
Delayed coking	2.1 m/s
Thermal cracking:	
Outlet from the convection section	1.5 m/s
Outlet from the radiant section	1.8 m/s

The residence time in each segment needed for the kinetic calculation is obtained from Eq. (2.137), written for segment j as:

$$\tau_j = \frac{V_{R_j}}{\overline{V}_j} = \frac{L_j \pi d^2/4}{\overline{V}_j} \tag{2.137 bis}$$

---

* 1 Pa = 10^{-5} bar.

The mean volume of the products $\overline{V}_j$ is calculated as the arithmetic mean between the volume flowrates at the inlet and at the outlet of the sector.

If the reaction mixture is in vapor phase, the volume flowrate of the hydrocarbons is calculated using the ideal gas law, with the equation:

$$V_h = \frac{G}{3600} \frac{RT}{p} \sum \frac{g}{M} z \qquad (5.5)$$

where the summing extends to the hydrocarbons in the reaction mixture. The volume of the steam must be added to that resulting from Eq. (5.5).

In the case of mixed phase flow, the tracing of the equilibrium vaporization curves at the partial pressure of the coil is required. To this purpose the usual methods of computing the equilibrium vaporization curves may be used [7,10]. More detailed data concerning the calculation of the volume of the mixed phase as applied to cracking furnaces are given in the monograph of W. I. Nelson [11].

The values $\overline{p}_j$, in Eq. (5.2) and $\overline{V}_j$ from Eq. (2.137a) can be correlated easily by using the obvious expression:

$$\overline{V}_j \cdot \overline{p}_j = G(1+n) \qquad (5.6)$$

The calculation of the overall conversion makes use of a first order kinetic equation written for sector j as:

$$k_j = \frac{1}{\tau_j} \ln \frac{1 - \sum\limits^{j-1} x_i}{1 - \sum\limits^{j} x_i} \qquad (5.7)$$

The rate constant k_j is determined for the equivalent mean polytropic temperature calculated with Eqs. (2.161) or (2.163) on the basis of inlet and outlet temperatures of the segment and on considering that temperature variation with the length of the element is linear.

It is to be observed that Eqs. (5.1), (5.2), (2.137 bis), and (5.5), applied to the condition at outlet from the segment, Eq. (5.7), and the equation of the equivalent mean polytropic temperature are interdependent. All these equations contain in explicit or implicit form the conversion and the temperature at the outlet of the segment. In these conditions, the computation is by trial and error until the values of the parameters resulting from the calculation coincide with those assumed (convergence).

The following succession of computations is recommended in order to reduce the number of iterations to convergence:

The overall conversion x_j achieved in the segment j is assumed. By distributing the conversion among the products, the composition at the outlet is obtained.

These compositions and conditions at the inlet of the segment are used in Eq. (5.1) to calculate the enthalpy and from here the outlet temperature.

Using these data and assuming the outlet pressure, the mean density $\overline{p}_j$ is calculated. Then by means of the Eq. (5.2) the pressure drop is determined. The calculation is repeated until the calculated pressure converges to an acceptable degree with the assumed pressure.

(t_{evmp}) is now calculated using the temperatures of the inlet and outlet of the segment and k_t or α corresponding to the activation energy and temperature level.

For t_{evmp} obtained in this way, the the reaction rate constant k_j is calculated. Using the value ρ_j obtained previously and Eq. (5.6), $\overline{V}_j$ is calculated, and by means of the expression (2.137a) the time τ_j is obtained.

Eq. (5.7) is now used to calculate the overall conversion x_j. This value should coincide to an acceptable degree with the value assumed at the onset of the iteration. If needed, the computation is repeated until convergence is obtained.

The calculation begins from the first of the considered segments. The composition at the inlet of this segment corresponds to the nonconverted feedstock. The calculation is continued segment by segment, taking into account the fact that, for each instance, the composition and condition at the inlet of a segment are those at the outlet of the previous segment. For the last segment, the calculation is carried out on basis of the conversion at the outlet of the coil that was established on basis of process considerations. In order to obtain the convergence but not the conversion, the length of the coil corresponding to this last segment is modified.

In performing these calculations one must take into account that the meaning and numerical values for the coil lengths used are different for each of the equations (5.1), (5.2), (5.4) and (2.137a), as indicated above (the length situated in the fire box, the equivalent hydraulic length, the real length).

In place of the kinetic Eq. (5.7), a kinetic equation can be used, that takes into account the inhibition effect (Section 2.3.2). For a certain segment j, such a kinetic equation will have the form:

$$ k_j = \frac{1}{\tau_j} \left[\ln \frac{1 - \sum\limits_{}^{j-1} x_i}{1 - \sum\limits_{}^{j} x_i} - \beta x_j \right] \qquad (5.8) $$

All the above computations are routinely performed by a variety of computer algorithms.

5.1.2 Stepwise Design Using a System of Integral Kinetic Equations

This method uses a system of equations that computes the yields of the various resulting products without the need to use experimental data for dividing the overall conversion, as in the previous method.

The thermal reactions of petroleum fractions can be treated by using the system of kinetic equations written for processes involving successive reactions or successive-parallel reactions. Thus:

The system of Eqs. (2.93)–(2.95) can be used for modeling cracking as a successive process with two reaction steps.

The system of kinetic Eqs. (2.99) takes into account also the conversion of feed directly to the final product.

The Eq. system (2.112)–(2.116) or (2.117) corresponds to the cracking of residual feedstock.

Other kinetic equation systems that adequately models the process are taken
into consideration.

For the pyrolysis of pure hydrocarbons or of mixtures of hydrocarbons, one
has to consider the system of overall chemical reactions completed on the basis of
thermodynamic calculations, with corrections for the reversible character of some of
the reactions. Some secondary reactions, that are important for the process are also
added, such as the syn-gas reaction that produces carbon monoxide and hydrogen by
the reaction between steam and the carbon deposits on the tubes:

$$C + H_2O \rightarrow CO + H_2$$

If such a reaction system is used in the computations, the equation of the
thermal balance for the segment j may be written in the form:

$$G\left[\sum_j g_e i_e - \sum_j g_i i_i - \sum_i q_{r_i} x_{l_j} + n\Delta i_a \right] = \varphi_t \pi DL_{t_j} \qquad (5.9)$$

This equation is analogous to (5.1), the difference being that the thermal effects must
be totalized for all the reactions taken into consideration x_{l_j} is the conversion achieved
by reaction l in segment j and expressed for the reference component to which the
reaction heat q_{r_i} was reported. Often, it is easier to perform the computations in terms
of mols instead of weight units, in which case Eq. (5.9) has to be modified accordingly.

Eqs. (5.2)–(5.6), which serve to compute the pressure drop and the residence
time within the segment, remain unchanged.

Because the residence time within each segment is expressed by Eq. (2.137 bis),
it results that the system of kinetic equations, written for the l reactions, must be
integrated for conditions of constant volume and used in the integrated form instead
of Eq. (5.7).

The computation is carried out by trial and error, as in the previous case.
Taking into account that a system of possibly many kinetic equations is used, the
iterations of the conversion variable until agreement with Eqs. (5.9), (5.2), (5.4), and
(2.137 bis) is obtained, might prove to be a lengthy process. In this case, the modification of the length of the coil of the respective segment is a practical solution.

It must be mentioned that this method allows modeling of the pyrolysis coil by
a system of chemical reactions without resorting to systems of mathematical equations that would require a large computation capacity. The only approximations
required and deemed acceptable are the expression of the residence time in the
coil segment, on the basis of Eq. (2.137 bis) and of the volume $\overline{V}$ as an arithmetic
mean between the volumes at the inlet and at the outlet conditions of the segment. In
a given coil segment the increase of the number of moles, due to the reactions taking
place, and the temperature, vary linearly with the length of the coil.

5.1.3 Design Using Kinetic Differential Equations

In this method, the chemical as well as the thermal and flow processes are expressed
by differential equations.

The method is applied for computations of the pyrolysis of lower alkanes [4].
For the pyrolysis of petroleum fractions, a semi-empirical method was recommended
[5], but no details were given. One may assume that a computation method similar to
that of the preceding paragraph may be applied.

For a system of chemical reactions marked by the subscript i that describes the pyrolysis of a pure hydrocarbon or of a mixture of pure hydrocarbons, the rates of the irreversible decomposition reactions will be expressed by equations of the form:

$$r_i = k_i C_k \tag{5.10}$$

C_k being the concentration of the reactant k. The rates of the reversible reactions are written as:

$$r_i = k_i C_k - k_{-i} C_{k'} C_{k''} \tag{5.11}$$

or similar, depending on the stoichiometry of the reaction. Where $C_{k'}$, $C_{k''}$ are the concentrations of the products k' and k''.

The pyrolysis of ethane may be modeled by two chemical reactions that are the result of the free radicals mechanism, namely:

$$C_2H_6 \rightarrow C_2H_4 + H_2$$
$$C_2H_6 \rightarrow \tfrac{1}{2}C_2H_4 + CH_4$$

where the first reaction is reversible. Completing the model with the reactions for the formation of acetylene, for the formation of carbon deposits, and for their interaction with steam, the following system of reactions results:

$$\left. \begin{array}{l} C_2H_6 \underset{k_{-1}}{\overset{k_1}{\rightleftharpoons}} C_2H_4 + H_2 \\[2mm] C_2H_6 \overset{k_2}{\rightarrow} \tfrac{1}{2}C_2H_4 + H_4 \\[2mm] C_2H_4 \underset{k_{-3}}{\overset{k_3}{\rightleftharpoons}} C_2H_2 + H_2 \\[2mm] C_2H_2 \overset{k_4}{\rightarrow} 2C + H_2 \\[2mm] C + H_2O \overset{k_5}{\rightarrow} CO + H_2 \end{array} \right\} \text{(a)}$$

According to Eq. (5.11), for the first reaction the expression for the reaction rate is:

$$r_1 = k_1 C_{C_2H_6} - k_{-1} C_{C_2H_4} \cdot C_{H_2} \tag{5.12}$$

and for the second reaction:

$$r_2 = k_2 C_{C_2H_6} \tag{5.13}$$

In the same manner, the expressions for the reaction rates for the other reactions that were incorporated into the model will be written.

The concentration of the component k is defined by the expression:

$$C_k = \frac{n_k}{V} \tag{5.14}$$

where n_k is the number of moles of the component k. Expressing the volume by means of the ideal gas law[*]:

$$V = \frac{\sum n_k RT}{p}$$

---

[*] Since the pyrolysis takes place at low pressures and high temperatures, it is not necessary to take into account the compressibility factor—z.

the concentrations may be obtained from the expression:

$$C_k = \frac{n_k}{\sum n_k} \cdot \frac{p}{RT} \tag{5.15}$$

The rate constants are expressed by Arrhenius-type equations:

$$k_1 = A_1 e^{-E_1/RT} \tag{5.16}$$

and the constants of the reverse reactions by means of the equilibrium constants:

$$k_{-1} = \frac{k_1}{K_{C_1}} \tag{5.17}$$

Neglecting deviations from ideality, the equilibrium constant K_{C_1} expressed in terms of concentrations, may be written as:

$$K_{C_1} = \left(\frac{1}{RT}\right)^{\Delta n} K_{p_1} = \left(\frac{1}{RT}\right)^{\Delta n} e^{-\Delta G_T^\circ/RT} \tag{5.18}$$

where Δn is the increase of the number of moles as per the stoichiometric equation.

In the small temperature range for which the calculation is performed, the variation of the free energy of reaction ΔG^0 with temperature is assumed to be linear, using the relation:

$$\Delta G_{T_2}^\circ = \Delta G_{T_1}^\circ + B(T_2 - T_1) \tag{5.19}$$

The value of the constant B is calculated using the values $\Delta G_{T_1}^\circ$ and $\Delta G_{T_2}^\circ$ taken from tables of thermodynamic constants [12,13].

By substituting in Eqs. (5.12) and (5.13) the concentrations and the rate constants with the values that were computed, and after simplifying, one obtains:

$$r_1 = A_1 e^{-E_1/RT} \frac{p}{RT} \left[\frac{n_{C_2H_6}}{\sum n} - \frac{p}{e^{-\Delta G_T^\circ/RT}} \cdot \frac{n_{C_2H_6} \cdot n_{H_2}}{\sum n} \right] \tag{5.20}$$

$$r_2 = A_2 e^{-E_2/RT} \frac{p}{RT} \cdot \frac{n_{C_2H_6}}{\sum n} \tag{5.21}$$

In the same way the expressions may be written for the reaction rates of all the chemical reactions within the system constituting the model.

In order to obtain the change of the chemical composition along the coil, it is necessary to express the rates of formation and/or the rates of disappearance for each component k as a function of the rates of the reactions within the model.

For the pyrolysis of ethane, on the basis of the reaction model (a), which was accepted, one obtains:

$$\left.\begin{array}{l} -r_{C_2H_6} = r_1 + r_2 \\ r_{C_2H_4} = r_1 + 1/2 r_2 - r_3 \\ r_{CH_4} = r_2 \\ r_{C_2H_4} = r_3 - r_4 \\ r_{H_2} = r_1 + r_3 + r_4 + r_5 \\ r_{CO} = r_5 \end{array}\right\} \tag{5.22}$$

For each component k the reaction rate is defined by the expression:

$$r_k = \frac{1}{V}\frac{dn_k}{d\tau} \tag{5.23}$$

For a plug flow reactor one may write:

$$V d\tau = dV_R$$

where dV_R is the volume element which is flown through by the reaction mixture in the time $d\tau$.

Substituting in (5.23) and taking into account that the diameter of the coil is constant, it results:

$$r_k = \frac{dn_k}{\dfrac{\pi D_i^2}{4}dL} \tag{5.24}$$

Replacing r_k in the expression (5.24) written for all the components, with the respective expressions (5.22), and the reaction rates r_1 in the latter with equations of the type (5.20) and (5.21), one obtains a system of differential equations that gives the variation of the number of moles of each component along the coil.

The design of the furnace, besides requiring a system of differential equations that expresses the chemical transformations, an equation that also gives in differential form the heat transfer along the coil. Such an equation may be written as:

$$\sum_k n_k C_{p_k} dT + \sum_1 (\Delta H_r)_1 dn_1 = \frac{\pi D_e \varphi_t dL}{3600} \tag{5.25}$$

The heat capacities are expressed by equations of the form:

$$C_{p_k} = a_k + b_k T + c_k T^2 \tag{5.26}$$

Since in a 100ºC interval the variation of the heat capacities can be considered to be linear with temperature [12], one can extend the linearity without introducing a large error to a 200°C interval (to which this calculation refers), by using a relationship of the form:

$$(C_{p_k})_T = (C_{p_k})_{T_1} + C(T_2 - T_1) \tag{5.27}$$

The calculation of the constant C uses the C_p values for all components as provided in thermodynamic tables [12,13].

The heats of reaction (ΔH_r) can be taken as being approximately equal to those corresponding to the ideal gases. Their variation with temperature is expressed by the equation:

$$(\Delta H_r)_1 \approx (\Delta H_r^{\circ})_1 = (\Delta H_{298r}^{\circ})_1 + \int_{298}^{T} \Delta C_{p_1} dT \tag{5.28}$$

where: ΔC_{p_1} represents the difference between the heat capacities of the products and the reactants for reaction 1.

When the heat capacities of the components are expressed by equations of the form (5.26), for ΔC_{p_1}, one gets:

$$\Delta C_{p_1} = \Delta a_1 + \Delta b_1 T + \Delta c_1 T^2$$

where Δa_1, Δb_1, and Δc_1 are the differences between the corresponding coefficients for products and for the reactants.

Replacing this expression in (5.28), one obtains:

$$(\Delta H_r^0)_1 = (\Delta H_{298r}^0)_1 + \Delta a_1(T - 298) + \frac{\Delta b_1}{2}(T^2 - 298^2) + \frac{\Delta c_1}{3}(T^3 - 298^3)$$

$$(5.29)$$

In the temperature range that intervenes in the calculation of the coil, one can use a simplified relationship for the variation of the heat of reaction similar to those used in computing the specific heats and free energy of the reaction:

$$(\Delta H_{T_2 r})_1 = (\Delta H_{T_1 r})_1 + D(T_2 - T_1) \tag{5.30}$$

where the coefficient D is determined from the heats of reaction that are calculated from the heats of formation for the temperatures T_1 and T_2.

If a more exact calculation is needed, the coefficients B, C, and D in the Eqs. (5.19), (5.27) and (5.30), are given new values calculated for each 100°C range.

The pressure drop in the coil may be computed by Eqs. (5.2), (5.4) or by the following equation [4]:

$$p(L_e) = p_o + B_1 L_e + B_2 L_e^2$$

where L_c is the equivalent hydraulic length calculated with Eq. (5.3).

The system of differential equations can be conveniently integrated for instance by the method of Runge-Kutta and Milne of the fourth order [4].

5.1.4   Design Based on the Mechanistic Modeling

The computation is different from that given previously. Instead of the overall stoichiometric reaction system, which reproduces more or less faithfully the chemical transformations, a model is used that is based on the reaction mechanism and takes directly into account the elementary reactions of the free radicals. Therefore, the term *mechanistic modeling*.

It is obvious that such a modeling applies the knowledge of a large number of kinetic constants and could be developed only after results obtained in many fundamental studies were accumulated and systematized. From this point of view, it is interesting to cite the statements made in 1970 by S.W. Benson [14], a recognized authority in the field of chemical kinetics: "There is now sufficient understanding and both kinetic and thermodynamic data available to describe the behavior of even the most complex pyrolysis in terms of a finite number of elementary step reactions ... On this basis, it can be expected that the pyrolysis of hydrocarbons is a ripe candidate for quantitative modeling."

For the implementation of this task, Benson suggests: " ... the inclusion of too many starting reactants can impose intolerable burdens on even large computers. Under the circumstances, the scheme must be simplified by a combination of methods which include careful fitting of average kinetic parameters, by using the available data. Many of these can be done reasonably well by analyzing products yields."

The first step in the development of the model is the collection, selection, and correlation of kinetic constants for elementary reactions as given by various authors and based on the pyrolysis of various hydrocarbons and feedstocks. The development of a model strives to be as complete as possible although certain simplifications will be used in order to not overload the computer. On the other hand, the simplifications should not be excessive, otherwise they could reduce the model to one that is not significantly better than that based on the stoichiometric equations.

A correctly developed mechanistic model presents the following advantages:

Is flexible and may be applied to various feedstocks and operating conditions and even extrapolated to feedstocks which were not tested.

It is the only model which allows the correlation and the concomitant use of chemical kinetics together with results obtained experimentally.

Following the refinement of the model, it is no longer necessary to resort to pilot plant testing of the feedstock.

The results obtained by the modification of several operating conditions of the industrial installation may be estimated.

The free radical reactions taken into account in the development of the model are those described in Sections 2.3 and 2.4, namely the initiation, propagation, and interruption of the chain, including those of isomerization of the radicals and of addition to the double bond. Some authors [14] found that the modeling will produce results in agreement with the experimental data, especially concerning the formation of olefins and diolefins, if selected molecular reactions are incorporated such as:

$$1\text{-}C_5H_{10} \rightarrow C_2H_4 + C_3H_6$$
$$1,3\text{-}C_6H_{10} \rightarrow 2,4\text{-}C_6H_{10}$$

$$\bigcirc \rightarrow C_2H_4 + C_4H_6$$

The development of the model on the basis of the elementary radicalic reactions may be illustrated by a paper published in 1987 [3] on the pyrolysis of mixtures of ethane and propane.

The study has as its objective the clarification of the mutual influence of the two hydrocarbons and not the design of the reaction system. Therefore, it does not extend the model to the elementary reactions leading to the formation of acetylene, methyl-acetylene and higher hydrocarbons. This makes the methodology of developing the model more simple and accessible.

The elementary radicalic reactions taken into account, the expressions of the reaction rates and the kinetic constants used by the authors are given in Table 5.1.

The data allow one to formulate the equation corresponding to the evolution of the chemical composition along the reactor, which is the "mechanistic" model, written on basis of the elementary reactions taken into account:

$$\left.\begin{aligned}
r_{CH_4} &= r_3 + r_8 + r_9 \\
r_{C_2H_6} &= -r_1 - r_3 - r_4 + r_{10} + r_{11} + \frac{r_{17}}{2} + r_{18} \\
r_{C_2H_4} &= r_5 + r_{12} - r_{16} \\
r_{C_3H_8} &= -r_2 - r_6 - r_7 - r_8 - r_9 - r_{10} - r_{11} \\
r_{C_3H_6} &= r_{13} - r_{14} - r_{15}
\end{aligned}\right\} \qquad (5.31)$$

Table 5.1 Kinetic Constants for the Pyrolysis of Ethane-Propane Mixtures

	Reaction rate (r_i)	A (s^{-1}) or $(L/mol \cdot s)$	E (kJ/mol)
$C_2H_6 \rightarrow 2\overset{\bullet}{C}H_3$	$r_1 = k_1[C_2H_6]$	5.185×10^{16}	380.0
$C_3H_8 \rightarrow \overset{\bullet}{C}H_3 + \overset{\bullet}{C}_2H_5$	$r_2 = k_2[C_3H_8]$	2.074×10^{16}	366.9
$C_2H_6 + \overset{\bullet}{C}H_3 \rightarrow CH_4 + \overset{\bullet}{C}_2H_5$	$r_3 = k_3[C_2H_6][\overset{\bullet}{C}H_3]$	3.941×10^{11}	35.0
$C_2H_6 + \overset{\bullet}{H} \rightarrow H_2 + \overset{\bullet}{C}_2H_5$	$r_4 = k_4[C_2H_6][\overset{\bullet}{H}]$	7.537×10^{10}	82.2
$\overset{\bullet}{C}_2H_5 \rightarrow C_2H_4 + \overset{\bullet}{H}$	$r_5 = k_5[\overset{\bullet}{C}_2H_5]$	1.013×10^{14}	217.1
$C_3H_8 + \overset{\bullet}{H} \rightarrow H_2 + 1 - \overset{\bullet}{C}_3H_7$	$r_6 = k_6[C_3H_8][\overset{\bullet}{H}]$	5.096×10^{10}	31.6
$C_3H_8 + \overset{\bullet}{H} \rightarrow H_2 + 2 - \overset{\bullet}{C}_3H_7$	$r_7 = k_7[C_3H_8][\overset{\bullet}{H}]$	2.401×10^{10}	45.1
$C_3H_8 + \overset{\bullet}{C}H_3 \rightarrow CH_4 + 1 - \overset{\bullet}{C}_3H_7$	$r_8 = k_8[C_3H_8][\overset{\bullet}{C}H_3]$	2.813×10^{10}	28.0
$C_3H_8 + \overset{\bullet}{C}H_3 \rightarrow CH_4 + 2 - \overset{\bullet}{C}_3H_7$	$r_9 = k_9[C_3H_8][\overset{\bullet}{C}H_3]$	2.119×10^9	43.8
$C_3H_8 + \overset{\bullet}{C}_2H_5 \rightarrow C_2H_6 + 1 - \overset{\bullet}{C}_3H_7$	$r_{10} = k_{10}[C_3H_8]\overset{\bullet}{C}_2H_5]$	3.440×10^9	54.7
$C_3H_8 + \overset{\bullet}{C}_2H_5 \rightarrow C_2H_6 + 2 - \overset{\bullet}{C}_3H_7$	$r_{11} = k_{11}[C_3H_8]\overset{\bullet}{C}_2H_5]$	3.973×10^8	56.5
$1 - \overset{\bullet}{C}_3H_7 \rightarrow C_2H_4 + \overset{\bullet}{C}H_3$	$r_{12} = k_{12}[1 - \overset{\bullet}{C}_3H_7]$	2.119×10^{13}	141.6
$2 - \overset{\bullet}{C}_3H_7 \rightarrow C_3H_6 + \overset{\bullet}{H}$	$r_{13} = k_{13}[2 - \overset{\bullet}{C}_3H_7]$	3.195×10^{13}	210.1
$C_3H_6 + \overset{\bullet}{H} \rightarrow 1 - \overset{\bullet}{C}_3H_7$	$r_{14} = k_{14}[C_3H_6][\overset{\bullet}{H}]$	9.705×10^9	19.7
$C_3H_6 + \overset{\bullet}{H} \rightarrow 2 - \overset{\bullet}{C}_3H_7$	$r_{15} = k_{15}[C_3H_6][\overset{\bullet}{H}]$	1.151×10^{11}	5.2
$C_2H_4 + \overset{\bullet}{H} \rightarrow \overset{\bullet}{C}_2H_5$	$r_{16} = k_{16}[C_2H_4][\overset{\bullet}{H}]$	4.559×10^9	8.1
$\overset{\bullet}{C}H_3 + \overset{\bullet}{C}H_3 \rightarrow C_2H_6$	$r_{17} = k_{17}[\overset{\bullet}{C}H_3]^2$	1.349×10^{10}	0.0
$\overset{\bullet}{C}_2H_5 + \overset{\bullet}{H} \rightarrow C_2H_6$	$r_{18} = k_{18}\overset{\bullet}{C}_2H_5][\overset{\bullet}{H}]$	5.189×10^{10}	0.0

The number of the elementary reactions that must be taken into account for the design of a pyrolysis furnace is much larger. The model [3] shown above takes into account only the reactions that lead to the formation of ethene and propene. Thus, the more complete model, SPYRO, for the pyrolysis of ethane, propane, and butane is written on the basis of a number of 91 free radical elementary reactions and 18 molecular reactions [15].

The first step is to collect all available kinetic data (pre-exponential factors and activation energies) for the equations to be incorporated in the model. Then, one has to assess which simplification can be used without negatively influencing the correctness of the final results.

The first problem is approached by comparing the constants calculated by theoretical methods with those obtained experimentally. This also makes it possible to extrapolate and deduce the constants for analogous reactions for which there are no experimental data available.

Following the examination of the results obtained this way, a number of simplifications and analogies are now generally accepted.

The rate constant for the reactions of hydrogen extraction depends only on the radical that performs the extraction and on the type of the extracted hydrogen atom (primary, secondary or tertiary from alkyl groups, primary or secondary, from allyl groups, etc.). More concretely, the pre-exponential factor depends only on the nature of the radical that performs the extraction while the activation energy depends only on the nature (position in the molecule) of the extracted hydrogen atom.

For reactions of additions to the carbon=carbon double or multiple bonds, the rate constant depends only on the radical, on the bond position (α or β), and bond type (double, triple, dienic etc.).

The recombination of radicals has zero activation energy and the pre-exponential factors given in Table 2.8 may also be calculated on a theoretical basis. When the rate constants for the reactions of recombination of the radicals are known, the calculation for the reverse reactions, which is the initiation, can be made on a thermodynamic basis.

In the isomerization of radicals, pre-exponential factors and activation energies may be calculated without difficulties on a theoretical basis [16].

To determine the variation of the rate constant as a function of the molecular mass of the radical is of great interest, because it allows some important simplifications to be performed. Thus, owing to steric hindrances, the rate constant of extraction of the hydrogen atom decreases proportionally to the molecular mass of the radical that performs the extraction. For the same reason, for radicals with large molecular mass the addition reaction to the double bond and the chain interruption reaction by interaction with other radicals become much less probable. Accordingly, it is estimated that for radicals with more than 4 carbon atoms, the addition reactions to the double bond, those of extraction of a hydrogen atom from another molecule, and participation in the chain interruption are negligible. Such radicals undergo exclusively reactions of isomerization and decomposition. Since bimolecular reactions are thus excluded, the distribution of the decomposition products will be independent of pressure. Also, since interactions with other molecules do not occur, the result of the decomposition of such radicals will be independent of the nature of the feedstock from which they originate, and on the hydrocarbons in the presence of which their decomposition takes place. The distribution of products from decomposition of radicals with more than 4 carbon atoms will thus depend exclusively on the temperature and will be determined by the decomposition and isomerization reactions.

For illustration, Figure 5.1 shows a scheme of the reaction that must be taken into consideration at the pyrolysis of n-octane and the distribution of the resulted products [15].

Another factor must be considered in this scheme, i.e. that all intermediary species are isomerized with high rates and, therefore, they must be considered at equilibrium (in Figure 5.1, these transformations are surrounded by a dotted line.). This makes it useless to take into consideration, in molecules with over 4 carbon atoms, what is the position of the hydrogen atom that was extracted in order to produce the radical (Figure 5.1).

Such isomerizations, which are practically instantaneous, will take place also in 1,3- and 2,4- hexadiene or methyl-hexadiene. The cyclohexadienes will be converted to benzene with a high rate with the release of a molecule of hydrogen.

Figure 5.2 illustrates the result of such interactions [15], which lead, among results to the distribution of decomposition products being the same for cyclohexane and for any of the three hexenes.

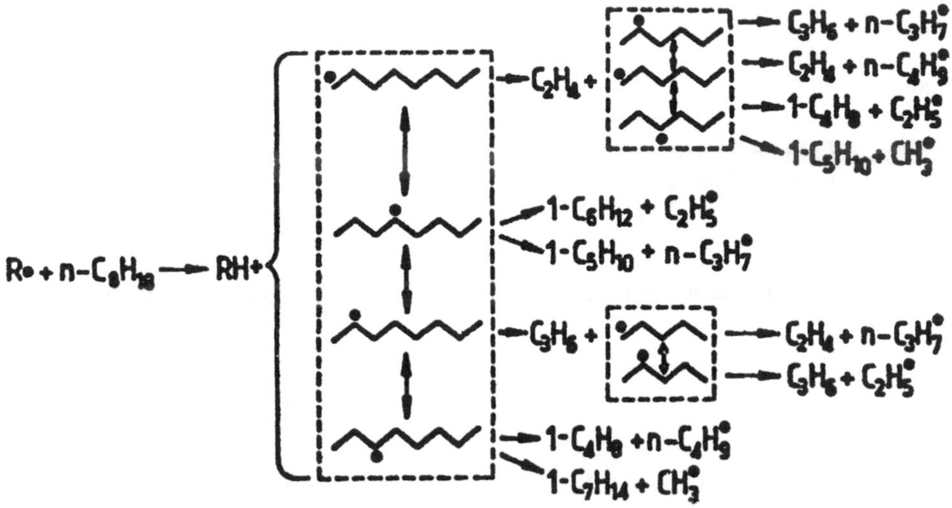

Figure 5.1 Reaction pathways in *n*-octane pyrolysis [15].

A mechanistic model for the pyrolysis of the gases (alkanes and alkenes C_2–C_4– and their mixtures), which gives very good agreement with the experimental data and allows the modeling of the industrial furnaces, was developed by Sundaram and Froment [1].

Elementary reactions that intervene in the calculations and the corresponding kinetic constants selected by the authors are given in Table 5.2. The reactions that must be taken into account in the pyrolysis of ethane, propane, *n*- and *iso*-butane, ethene, and propene, and the radicals and the molecular species that intervene, are given in Table 5.3.

Similar data and kinetic constants for the pyrolysis of gases were published also by other authors, such as those incorporated in the SPYRO model [15]. This model makes use also of a number of 18 molecular reactions and thus produces correct results for the formation of some of the higher hydrocarbons.

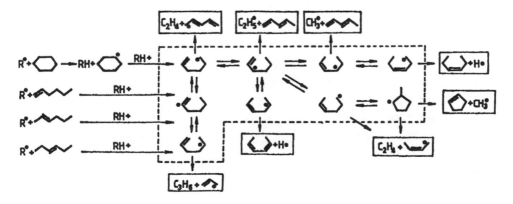

Figure 5.2 Reaction products obtained in the pyrolysis of cyclohexane, 1-, 2-, and 3-hexene. (From Ref. 15)

Table 5.2 Kinetic Constants for the Pyrolysis of Lower Alkanes and Alkenes

Number	Reaction	A (s^{-1}) or (Lmol$^{-1}\cdot$s^{-1})	E (kJ/mol)
1	$C_2H_6 \rightarrow 2\,\dot{C}H_3$	4.0×10^{16}	366.3
2	$C_3H_8 \rightarrow \dot{C}_2H_5 + \dot{C}H_3$	2.0×10^{16}	353.8
3	$n-C_4H_{10} \rightarrow 2\dot{C}_2H_5$	1.5×10^{16}	343.7
4	$n-C_4H_{10} \rightarrow 1-\dot{C}_3H_7 + \dot{C}H_3$	9.0×10^{16}	357.6
5	$i-C_4H_{10} \rightarrow 2-\dot{C}_3H_7 + \dot{C}H_3$	2.0×10^{16}	343.3
6	$2C_2H_4 \rightarrow \dot{C}_2H_3 + \dot{C}_2H_5$	9.0×10^{13}	272.1
7	$C_3H_6 \rightarrow \dot{C}_2H_3 + \dot{C}H_3$	8.0×10^{17}	397.7
8	$C_3H_6 \rightarrow \dot{C}_3H_5 + \dot{H}$	3.5×10^{16}	360.0
9	$2C_3H_6 \rightarrow 1-\dot{C}_3H_7 + \dot{C}_3H_5$	3.5×10^{11}	213.5
10	$1-C_4H_8 \rightarrow \dot{C}_3H_5 + \dot{C}H_3$	8.0×10^{16}	309.8
11	$2-C_4H_8 \rightarrow \dot{C}_3H_5 + \dot{C}H_3$	2.0×10^{16}	298.5
12	$C_2H_4 + \dot{H} \rightarrow \dot{C}_2H_3 + H_2$	8.0×10^{8}	16.7
13	$C_2H_6 + \dot{H} \rightarrow \dot{C}_2H_5 + H_2$	1.0×10^{11}	40.6
14	$C_3H_6 + \dot{H} \rightarrow \dot{C}_3H_5 + H_2$	2.5×10^{9}	4.6
15	$C_3H_8 + \dot{H} \rightarrow 1-\dot{C}_3H_7 + H_2$	1.0×10^{11}	40.6
16	$C_3H_8 + \dot{H} \rightarrow 2-\dot{C}_3H_7 + H_2$	9.0×10^{10}	34.8
17	$1-C_4H_8 + \dot{H} \rightarrow \dot{C}_4H_7 + H_2$	5.0×10^{10}	16.3
18	$2-C_4H_8 + \dot{H} \rightarrow \dot{C}_4H_7 + H_2$	5.0×10^{10}	15.9
19	$i-C_4H_8 + \dot{H} \rightarrow$ Me allyl $+ H_2$	3.0×10^{10}	15.9
20	$n-C_4H_{10} + \dot{H} \rightarrow 1-\dot{C}_4H_9 + H_2$	1.5×10^{11}	40.6
21	$n-C_4H_{10} + \dot{H} \rightarrow 2-\dot{C}_4H_9 + H_2$	9.0×10^{10}	35.2
22	$i-C_4H_{10} + \dot{H} \rightarrow i-\dot{C}_4H_9 + H_2$	1.0×10^{11}	35.2
23	$C_2H_4 + \dot{C}H_3 \rightarrow \dot{C}_2H_3 + CH_4$	1.0×10^{10}	54.4
24	$C_2H_6 + \dot{C}H_3 \rightarrow \dot{C}_2H_5 + CH_4$	3.8×10^{11}	69.1
25	$C_3H_6 + \dot{C}H_3 \rightarrow \dot{C}_3H_5 + CH_4$	2.0×10^{9}	51.1
26	$C_3H_8 + \dot{C}H_3 \rightarrow 1-\dot{C}_3H_7 + CH_4$	3.4×10^{10}	48.1
27	$C_3H_8 + \dot{C}H_3 \rightarrow 2-\dot{C}_3H_7 + CH_4$	4.0×10^{9}	42.3
28	$1-C_4H_8 + \dot{C}H_3 \rightarrow \dot{C}_4H_7 + CH_4$	1.0×10^{8}	30.6
29	$2-C_4H_8 + \dot{C}H_3 \rightarrow \dot{C}_4H_7 + CH_4$	1.0×10^{8}	34.3
30	$i-C_4H_8 + \dot{C}H_3 \rightarrow$ Me allyl $+ CH_4$	3.0×10^{8}	30.6
31	$n-C_4H_{10} + \dot{C}H_3 \rightarrow 1-\dot{C}_4H_9 + CH_4$	3.5×10^{10}	48.6
32	$n-C_4H_{10} + \dot{C}H_3 \rightarrow 2-\dot{C}_4H_9 + CH_4$	3.5×10^{9}	39.8
33	$i-C_4H_{10} + \dot{C}H_3 \rightarrow i-\dot{C}_4H_9 + CH_4$	9.5×10^{9}	37.7
34	$C_3H_6 + \dot{C}_2H_3 \rightarrow \dot{C}_3H_5 + C_2H_4$	3.0×10^{9}	60.7
35	$C_3H_8 + \dot{C}_2H_3 \rightarrow 1-\dot{C}_3H_7 + C_2H_4$	3.0×10^{9}	78.7
36	$C_3H_8 + \dot{C}_2H_3 \rightarrow 2-\dot{C}_3H_7 + C_2H_4$	1.0×10^{9}	67.8
37	$i-C_4H_8 + \dot{C}_2H_3 \rightarrow$ Me allyl $+ C_2H_4$	1.0×10^{9}	54.4
38	$n-C_4H_{10} + \dot{C}_2H_3 \rightarrow 1-\dot{C}_4H_9 + C_2H_4$	1.0×10^{9}	75.4
39	$n-C_4H_{10} + \dot{C}_2H_3 \rightarrow 2-\dot{C}_4H_9 + C_2H_4$	8.0×10^{8}	70.3

40	$i - C_4H_{10} + \overset{\bullet}{C}_2H_3 \rightarrow i - \overset{\bullet}{C}_4H_9 + C_2H_4$	1.0×10^9	70.3
41	$C_2H_4 + \overset{\bullet}{C}_2H_5 \rightarrow \overset{\bullet}{C}H_3 + C_3H_6$	3.0×10^9	79.5
42	$C_3H_6 + \overset{\bullet}{C}_2H_5 \rightarrow \overset{\bullet}{C}_3H_5 + C_2H_6$	1.0×10^8	38.5
43	$C_3H_8 + \overset{\bullet}{C}_2H_5 \rightarrow 1 - \overset{\bullet}{C}_3H_7 + C_2H_6$	1.2×10^9	52.8
44	$C_3H_8 + \overset{\bullet}{C}_2H_5 \rightarrow 2 - \overset{\bullet}{C}_3H_7 + C_2H_6$	8.0×10^8	43.5
45	$1 - C_4H_8 + \overset{\bullet}{C}_2H_5 \rightarrow \overset{\bullet}{C}_4H_7 + C_2H_6$	2.0×10^8	34.8
46	$i - C_4H_8 + \overset{\bullet}{C}_2H_5 \rightarrow \text{Me allyl} + C_2H_6$	6.0×10^7	34.8
47	$n - C_4H_{10} + \overset{\bullet}{C}_2H_5 \rightarrow 1 - \overset{\bullet}{C}_4H_9 + C_2H_6$	2.0×10^9	52.8
48	$n - C_4H_{10} + \overset{\bullet}{C}_2H_5 \rightarrow 2 - \overset{\bullet}{C}_4H_9 + C_2H_6$	4.5×10^8	43.5
49	$i - C_4H_{10} + \overset{\bullet}{C}_2H_5 \rightarrow i - \overset{\bullet}{C}_4H_9 + C_2H_6$	1.5×10^9	43.5
50	$C_3H_8 + \overset{\bullet}{C}_3H_5 \rightarrow 1 - \overset{\bullet}{C}_3H_7 + C_3H_6$	1.0×10^9	78.7
51	$C_3H_8 + \overset{\bullet}{C}_3H_5 \rightarrow 2 - \overset{\bullet}{C}_3H_7 + C_3H_6$	8.0×10^8	67.8
52	$i - C_4H_8 + \overset{\bullet}{C}_3H_5 \rightarrow \text{Me allyl} + C_3H_6$	2.0×10^8	56.5
53	$n - C_4H_{10} + \overset{\bullet}{C}_3H_5 \rightarrow 1 - \overset{\bullet}{C}_4H_9 + C_3H_6$	4.0×10^8	78.7
54	$n - C_4H_{10} + \overset{\bullet}{C}_3H_5 \rightarrow 2 - \overset{\bullet}{C}_4H_9 + C_3H_6$	8.0×10^8	70.3
55	$i - C_4H_{10} + \overset{\bullet}{C}_3H_5 \rightarrow i - \overset{\bullet}{C}_4H_9 + C_3H_6$	1.0×10^9	79.5
56	$C_3H_6 + 1 - \overset{\bullet}{C}_3H_7 \rightarrow \overset{\bullet}{C}_3H_5 + C_3H_8$	1.0×10^8	38.5
57	$C_3H_6 + 2 - \overset{\bullet}{C}_3H_7 \rightarrow \overset{\bullet}{C}_3H_5 + C_3H_8$	1.0×10^8	42.7
58	$n - C_4H_{10} + 1 - \overset{\bullet}{C}_3H_7 \rightarrow 2 - \overset{\bullet}{C}_4H_9 + C_3H_8$	2.0×10^8	43.5
59	$n - C_4H_{10} + 2 - \overset{\bullet}{C}_3H_7 \rightarrow 2 - \overset{\bullet}{C}_4H_9 + C_3H_8$	2.0×10^8	52.8
60	$i - C_4H_{10} + 2 - \overset{\bullet}{C}_3H_7 \rightarrow i - \overset{\bullet}{C}_4H_9 + C_3H_8$	1.0×10^8	56.1
61	$\overset{\bullet}{C}_2H_3 \rightarrow C_2H_2 + \overset{\bullet}{H}$	2.0×10^9	131.9
62	$\overset{\bullet}{C}_2H_5 \rightarrow C_2H_4 + \overset{\bullet}{H}$	3.2×10^{13}	167.5
63	$\overset{\bullet}{C}_3H_5 \rightarrow C_2H_2 + \overset{\bullet}{C}H_3$	3.0×10^{10}	151.6
64	$1 - \overset{\bullet}{C}_3H_7 \rightarrow C_2H_4 + \overset{\bullet}{C}H_3$	4.0×10^{13}	136.5
65	$1 - \overset{\bullet}{C}_3H_7 \rightarrow C_3H_6 + \overset{\bullet}{H}$	2.0×10^{13}	160.8
66	$2 - \overset{\bullet}{C}_3H_7 \rightarrow C_3H_6 + \overset{\bullet}{H}$	2.0×10^{13}	162.0
67	$\overset{\bullet}{C}_4H_7 \rightarrow C_4H_6 + \overset{\bullet}{H}$	1.2×10^{14}	206.4
68	$\overset{\bullet}{C}_4H_7 \rightarrow C_2H_4 + \overset{\bullet}{C}_2H_3$	1.0×10^{11}	154.9
69	$\text{Me allyl} \rightarrow C_3H_4 + \overset{\bullet}{C}H_3$	1.0×10^{13}	136.5
70	$\text{Me allyl} \rightarrow C_2H_4 + \overset{\bullet}{C}_2H_3$	1.0×10^{12}	117.2
71	$1 - \overset{\bullet}{C}_4H_9 \rightarrow C_2H_4 + \overset{\bullet}{C}_2H_5$	1.6×10^{12}	117.2
72	$1 - \overset{\bullet}{C}_4H_9 \rightarrow 1 - C_4H_8 + \overset{\bullet}{H}$	1.0×10^{13}	153.2
73	$2 - \overset{\bullet}{C}_4H_9 \rightarrow C_3H_6 + \overset{\bullet}{C}H_3$	2.5×10^{13}	133.6
74	$2 - \overset{\bullet}{C}_4H_9 \rightarrow 1 - C_4H_8 + \overset{\bullet}{H}$	2.0×10^{13}	166.6
75	$i - \overset{\bullet}{C}_4H_9 \rightarrow i - C_4H_8 + \overset{\bullet}{H}$	3.3×10^{14}	150.7
76	$i - \overset{\bullet}{C}_4H_9 \rightarrow C_3H_6 + \overset{\bullet}{C}H_3$	8.0×10^{13}	138.2
77	$i - \overset{\bullet}{C}_4H_9 \rightarrow 2 - C_4H_8 + \overset{\bullet}{H}$	4.0×10^{13}	153.2
78	$\overset{\bullet}{C}_5H_{11} \rightarrow C_5H_{10} + \overset{\bullet}{H}$	5.0×10^{13}	153.2
79	$\overset{\bullet}{C}_5H_{11} \rightarrow 1 - C_4H_8 + \overset{\bullet}{C}H_3$	3.2×10^{13}	131.9
80	$\overset{\bullet}{C}_5H_{11} \rightarrow C_2H_4 + 1 - \overset{\bullet}{C}_3H_7$	4.0×10^{12}	120.2
81	$C_2H_2 + \overset{\bullet}{H} \rightarrow \overset{\bullet}{C}H_3$	4.0×10^{10}	5.4

Table 5.2 Continued

Number	Reaction	A (s^{-1}) or ($L\,mol^{-1}\cdot s^{-1}$)	E (kJ/mol)
82	$C_2H_4 + \dot{H} \rightarrow \dot{C}_2H_5$	1.0×10^{10}	6.3
83	$C_3H_4 + \dot{H} \rightarrow \dot{C}_3H_5$	1.0×10^{10}	6.3
84	$C_3H_6 + \dot{H} \rightarrow 1 - \dot{C}_3H_7$	1.0×10^{10}	12.1
85	$C_3H_6 + \dot{H} \rightarrow 2 - \dot{C}_3H_7$	1.0×10^{10}	6.3
86	$C_4H_6 + \dot{H} \rightarrow \dot{C}_4H_7$	4.0×10^{10}	5.4
87	$1 - C_4H_8 + \dot{H} \rightarrow 2 - \dot{C}_4H_9$	1.0×10^{10}	5.0
88	$2 - C_4H_8 + \dot{H} \rightarrow 2 - \dot{C}_4H_9$	6.3×10^{9}	5.0
89	$i - C_4H_8 + \dot{H} \rightarrow i - \dot{C}_4H_9$	1.0×10^{10}	5.0
90	$C_2H_4 + \dot{C}H_3 \rightarrow i - \dot{C}_3H_7$	2.0×10^{8}	33.1
91	$C_3H_4 + \dot{C}H_3 \rightarrow$ Me allyl	1.5×10^{8}	31.0
92	$C_3H_6 + \dot{C}H_3 \rightarrow 2 - \dot{C}_4H_9$	3.2×10^{8}	31.0
93	$C_3H_6 + \dot{C}H_3 \rightarrow i - \dot{C}_4H_9$	3.2×10^{8}	38.1
94	$i - C_4H_8 + \dot{C}H_3 \rightarrow \dot{C}_5H_{11}$	1.0×10^{8}	30.1
95	$C_2H_4 + \dot{C}_2H_3 \rightarrow \dot{C}_4H_7$	5.0×10^{7}	29.3
96	$C_2H_4 + \dot{C}_2H_5 \rightarrow 1 - \dot{C}_4H_9$	1.5×10^{7}	31.8
97	$C_3H_6 + \dot{C}_2H_5 \rightarrow \dot{C}_5H_{11}$	1.3×10^{7}	31.4
98	$C_2H_4 + 1 - \dot{C}_3H_7 \rightarrow \dot{C}_5H_{11}$	2.0×10^{7}	31.0
99	$C_2H_4 + 2 - \dot{C}_3H_7 \rightarrow \dot{C}_5H_{11}$	1.3×10^{7}	28.9
100	$1 - \dot{C}_4H_9 \rightarrow 2 - \dot{C}_3H_9$	5.2×10^{14}	171.7
101	$\dot{C}_2H_3 + \dot{H} \rightarrow C_2H_4$	1.0×10^{10}	0
102	$\dot{C}_2H_5 + \dot{H} \rightarrow C_2H_6$	4.0×10^{10}	0
103	$\dot{C}_3H_5 + \dot{H} \rightarrow C_3H_6$	2.0×10^{10}	0
104	$1 - \dot{C}_3H_7 + \dot{H} \rightarrow C_3H_8$	1.0×10^{10}	0
105	$2 - \dot{C}_3H_7 + \dot{H} \rightarrow C_3H_8$	1.0×10^{10}	0
106	$\dot{C}_4H_7 + \dot{H} \rightarrow 1 - C_4H_8$	2.0×10^{10}	0
107	Me allyl $+ \dot{H} \rightarrow i - C_4H_8$	2.0×10^{10}	0
108	$1 - \dot{C}_4H_9 + \dot{H} \rightarrow n - C_4H_{10}$	1.0×10^{10}	0
109	$2 - \dot{C}_4H_9 + \dot{H} \rightarrow n - C_4H_{10}$	1.0×10^{10}	0
110	$i - \dot{C}_4H_9 + \dot{H} \rightarrow i - C_4H_{10}$	1.0×10^{10}	0
111	$\dot{C}_5H_{11} + \dot{H} \rightarrow C_5H_{12}$	1.0×10^{10}	0
112	$\dot{C}H_3 + \dot{C}H_3 \rightarrow C_2H_6$	1.3×10^{10}	0
113	$\dot{C}_2H_5 + \dot{C}H_3 \rightarrow C_3H_8$	3.2×10^{9}	0
114	$\dot{C}_3H_5 + \dot{C}H_3 \rightarrow 1 - C_4H_8$	3.2×10^{9}	0
115	$1 - \dot{C}_3H_7 + \dot{C}H_3 \rightarrow n - C_4H_{10}$	3.2×10^{9}	0
116	$2 - \dot{C}_3H_7 + \dot{C}H_3 \rightarrow C_4H_{10}$	3.2×10^{9}	0
117	$\dot{C}_4H_7 + \dot{C}H_3 \rightarrow C_5 +^{\circ}$	3.2×10^{9}	0
118	Me allyl $+ \dot{C}H_3 \rightarrow C_5^+$	3.2×10^{9}	0

119	$C_2H_3 + \dot{C}_2H_3 \rightarrow C_4H_6$	1.3×10^{10}	0
120	$\dot{C}_4H_7 + \dot{C}_2H_3 \rightarrow C_5^+$	1.3×10^{10}	0
121	$\dot{C}_2H_5 + \dot{C}_2H_5 \rightarrow n-C_4H_{10}$	4.0×10^{8}	0
122	$\dot{C}_2H_5 + \dot{C}_2H_5 \rightarrow C_2H_4 + C_2H_6$	5.0×10^{7}	0
123	$\dot{C}_3H_5 + \dot{C}_2H_5 \rightarrow C_5^+$	3.2×10^{9}	0
124	$1 - \dot{C}_3H_7 + \dot{C}_2H_5 \rightarrow C_5^+$	8.0×10^{8}	0
125	$2 - \dot{C}_3H_7 + \dot{C}_2H_5 \rightarrow C_5^+$	8.0×10^{8}	0
126	$\dot{C}_4H_7 + \dot{C}_2H_5 \rightarrow C_5^+$	3.2×10^{9}	0
127	$\dot{C}_3H_5 + \dot{C}_3H_5 \rightarrow C_5^+$	3.2×10^{9}	0
128	$\dot{C}_4H_7 + \dot{C}_3H_5 \rightarrow C_5^+$	1.3×10^{10}	0
129	Me allyl $+ \dot{C}_3H_5 \rightarrow C_5^+$	1.3×10^{10}	0
130	$\dot{C}_4H_7 + \dot{C}_4H_7 \rightarrow C_5^+$	3.2×10^{9}	0
131	$C_2H_2 \rightarrow 2C + H_2$	5.0×10^{12}	259.6
132	$C_2H_4 + H_2 \rightarrow C_2H_6$	9.2×10^{8}	137.3
133	$C_2H_4 + C_4H_6 \rightarrow C_6H_{10}$	3.0×10^{7}	115.1

Source: Ref. 1.

The detailed, mechanistic modeling of the pyrolysis of liquid fractions, such as gasoline, naphtha and gas oil, requires the concentrations of the components contained in the feedstock. This means to take into consideration of hundreds molecular species, leading to thousands of elementary radicalic reactions. Such modeling, besides the difficulties in obtaining accurate composition data for the feedstock, requires a powerful computer.

Besides the simplifications indicated previously, concerning the behavior of radicals with more than 4 carbon atoms it was found practical to represent large fractions of gasoline or gas oils by a relatively small number of individual hydrocarbons.

This simplification is the result of the detailed analysis of the composition (individual chemical components) of a large number of gasolines and gas oils from various sources. It was surprising to find certain regularities in the distribution of the isomers and the practical absence of some isomers, including those containing a quaternary carbon. Thus, in place of fifteen iso-C_8H_{18}, that are theoretically possible, only eight have to be taken into account: the three mono-methyl-heptanes, 3 ethyl hexane, and the four dimethyl-hexanes that do not contain in their structure quaternary carbon atoms.

For similar reasons, one can ignore a large number of isomers of hydrocarbons with higher molecular mass, as well as of naphthenic and aromatic hydrocarbons.

A second simplification is possible due to the quite similar distribution of some of the isomers in feedstocks from different sources. Thus, the crude oils from three different sources show very similar concentrations of isomers, which must be taken into account according to the above considerations (see Table 5.4) [15]:

This situation makes it possible to derive a typical distribution of isomers. Since the products resulting from decomposition may be predicted on the basis of the theory of Rice-Kassiakoff (Section 2.2), the mixture of isomers can be replaced

Table 5.3 Molecular Species, Radicals, and Reactions from Table 5.2 that Must Be Considered in Pyrolysis of Hydrocarbons.

Hydrocarbon	Number of reactions	Molecular species	Radicals	Reactions to be considered
Ethane	49	H_2, CH_4, C_2H_2, C_2H_4, C_2H_6, C_3H_6, C_3H_8, C_4H_6, 1-C_4H_8, n-C_4H_{10}, C_5^+	$\dot{H}$, $\dot{C}H_3$, $\dot{C}_2H_3$, $\dot{C}_2H_5$, $\dot{C}_3H_5$, 1-$\dot{C}_3H_7$, $\dot{C}_4H_7$, 1-$\dot{C}_4H_9$, $\dot{C}_5H_{11}$	1, 3, 4, 10, 12, 13, 23, 24, 41, 61–65, 67, 68, 71, 72, 78–82, 84, 86, 90, 95–98, 101–104, 106, 108, 111–114, 117, 119–122, 126, 128, 130, 131
Propane	80	H_2, CH_4, C_2H_2, C_2H_4, C_2H_6, C_3H_6, C_3H_8, C_4H_8, C_4H_6, 1-C_4H_8, n-C_4H_{10}, C_5^+	$\dot{H}$, $\dot{C}H_3$, $\dot{C}_2H_3$, $\dot{C}_2H_5$, $\dot{C}_3H_5$, 1-$\dot{C}_3H_7$, 2-$\dot{C}_3H_7$, $\dot{C}_4H_7$, 1-$\dot{C}_4H_9$, 2-$\dot{C}_4H_9$, $\dot{C}_5H_{11}$	1, 2, 10, 12–17, 23–28, 34–36, 41–44, 50, 51, 61–68, 71–74, 78–82, 84–87, 90, 93, 95–106, 108, 109, 111–117, 119–128, 130, 131
n-butane	86	the same as for propane	the same as for propane	1, 3, 4, 10, 12–17, 20, 21, 23–28, 31, 32, 34, 38, 39, 41, 42, 45, 47, 48, 53, 54, 58, 59, 61–68, 71–74, 78–82, 84–87, 90, 92, 95–106, 108, 109, 111–117, 119–123, 126–128, 130, 131
iso-butane	86	H_2, CH_4, C_2H_2, C_2H_4, C_2H_6, C_3H_4, C_3H_6, C_3H_8, C_4H_6, 1-C_4H_8, 2-C_4H_8, i-C_4H_8, i-C_4H_{10}, C_5^+	$\dot{H}$, $\dot{C}H_3$, $\dot{C}_2H_3$, $\dot{C}_2H_5$, 1-$\dot{C}_3H_7$, 2-$\dot{C}_3H_7$, $\dot{C}_4H_7$, Me allyl, 2-$\dot{C}_4H_9$, i-$\dot{C}_4H_9$, $\dot{C}_5H_{11}$	5, 10–14, 17–19, 22–25, 28–30, 33, 34, 37, 40, 42, 46, 49, 52, 55–57, 60–64, 66–70, 73–80, 81–95, 97–99, 101–107, 109–114, 117–120, 125, 127–131
Ethene	66	the same as for propane	the same as for propane	1, 3, 4, 6, 10, 12–14, 23–25, 34, 41, 61–68, 71–74, 78–82, 84–87, 90, 92, 95–97, 101–106, 108, 109, 111–114, 117, 119–128, 130–133
Propene	68	H_2, CH_4, C_2H_2, C_2H_4, C_2H_6, C_3H_6, C_3H_8, C_4H_6, 1-C_4H_8, C_5^+	$\dot{H}$, $\dot{C}H_3$, $\dot{C}_2H_3$, $\dot{C}_2H_5$, $\dot{C}_3H_5$, 1-$\dot{C}_3H_7$, 2-$\dot{C}_3H_7$, 1-$\dot{C}_4H_9$, 2-$\dot{C}_4H_9$, $\dot{C}_5H_{11}$	7–10, 12–14, 23–25, 34, 41, 42, 56, 57, 61–68, 71–74, 78–82, 84–87, 90, 92, 95–106, 108, 109, 111–117, 119, 120, 123–126, 128, 130, 131

Table 5.4 Octane Isomers Proportion in
Weight Percent in Different Crude Oils

Isomers	Crude oils		
	Ponca	Occidental	Texas
2-methyl-heptane	46.3	36.9	42.1
3-methyl-heptane	15.4	28.5	23.4
4-methyl-heptane	10.3	10.2	9.3
2,3-dimethyl-hexane	3.6	5.4	6.3
2,4-dimethyl-hexane	3.1	5.5	4.2
2,5-dimethyl-hexane	3.1	5.7	4.0
3,4-dimethyl-hexane	6.7	2.6	3.7
3-ethyl-hexane	4.6	3.5	3.1
Neglected isomers	6.9	1.7	3.5

by an "equivalent component." The products resulting from its pyrolysis may be calculated as a weighted average of the products resulting from the pyrolysis of the isomers that compose it. The basis for this calculation may be either the actual analysis of the feedstock or the analysis of some similar feedstocks, the similarity being based on the usual analysis (PONA, the H/C ratio, etc.).

Similar simplifications are applied also to the hydrocarbons that result from decomposition, the concentrations of which being calculated as shown previously on the basis of the reaction mechanism. Some of these compounds will be found in very small amounts, so that their subsequent conversion may be neglected. Others will be grouped together with similar products obtained by the decomposition of other hydrocarbons, to which also was applied the method of the equivalent component.

The evaluation of the conversion of light hydrocarbons produced by pyrolysis of liquid fractions makes use of the same kinetic constants as the pyrolysis of gases (Table 5.2).

Despite all the indicated simplifications, a general scheme of pyrolysis of a gas oil needs to take into account about 2,000 reactions to which participate approximately 100 radicals and molecular species, respectively equivalent components.

Programs such as SPYRO, used for modeling pyrolysis of naphthas and gas oils, were submitted to detailed verifications and were checked with experimental data, obtaining very good agreements [15].

Along these lines, one should mention that the development of a kinetic model, or the improvement of a model for which not all the necessary data are available, is a stepwise process. It comprises a succession of operations and comparisons with the experimental data, producing gradual improvements until a satisfactory agreement is obtained. The succession of operations leading to the development, improvement, and finalization of the resulted model is given in Figure 5.3.

Besides the problems related to the development of the kinetic model, mechanistic modeling may present some additional complications concerning the solution of the system of ordinary nonlinear differential equations containing, besides the kinetic equations, also those for heat transfer and pressure drop.

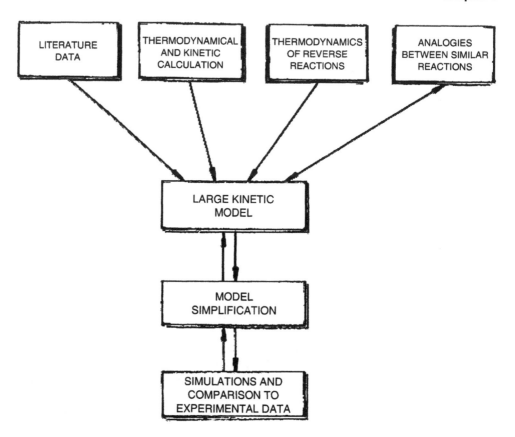

Figure 5.3   Steps for developing a kinetic model.

As in the case of other design methods, the initial available data are those concerning the temperature and composition at the coil inlet and the imposed temperature and pressure at the outlet from the furnace.

The particular feature of mechanistic modeling is the very large range of values for the kinetic constants and the very low concentrations of some of the components, namely of the radicals, that appear and disappear at very high rates. When standard methods of integration are applied, this situation requires the use of a large number of integration steps and therefore substantial computer time.

The difficulty is avoided by the division of the program in two stages. In the first stage, simplified kinetics and the usual numerical methods of integration are used. One obtains the pressure profile and approximate temperature and composition profiles. In the second stage, the exact kinetic equations belonging to the model, are used.

Starting with the temperature profiles calculated in the first stage, the mean arithmetic temperature is calculated for the first step (or incremental coil length). For this temperature, the system of kinetic equations (of material balance) is solved, after which, from the thermal balance, the outlet temperature from this step is determined. In the case that the temperature does not agree with the one assumed, the operation is repeated. In most cases, two iterations are sufficient. The operation is continued for the following increments (steps) until the final conditions imposed for the exit from the furnace are reached. The computing algorithm is therefore similar to the one deduced in Section 5.1.2.

An additional very useful simplification, taking into account the very large number of kinetic equations, is to apply the so called "steady state which varies continuously" or "continuously varying steady-state" assumption (CVSS). In this case, the steady state is applied to each step individually, transforming the differential equations written for the radicals into algebraic equations. The material balance of the stage becomes thus a system of differential equations for the molecules and of algebraic equations for the radicals, that are easy to solve and reduces computer time. The detailed verifications performed [15] proved that this simplification produces only negligible differences for the concentrations of the molecular species and to differences below 5% for the concentrations of the radicals.

Formation of coke on the walls of the tubes is more difficult to incorporate in the model. Vinyl and phenyl radicals are considered to be coke precursors, while the rate of formation of the deposits of coke is controlled by the concentration of the unsaturated radicals and molecules on the superficial layer of polymers formed on the internal surface of the tubes and gradually converted to coke [15].

Besides, one should remark that in the design calculations that use the mechanistic model, and generally in the calculation of the modern pyrolysis furnace provided with tubes with small diameter and short residence times, the radial gradient of temperature inside the tube can not be neglected.

Taking into account this gradient, the effective temperature, which must be introduced in the kinetic calculations for the determination of the rate constants, is given by equation:

$$T_{effect.} = T_f + \frac{R}{E} T_f^2 \ln\left[1 + \frac{4}{N_u}\left(\frac{\exp\sigma - (1+\sigma)}{\sigma}\right)\right] \tag{5.32}$$

where:

$T_{effect.}$ = temperature (in K) which must be used in the kinetic calculations
T_f = temperature of the flow (in K), calculated without taking into account the radial temperature gradient
R = gas constant
E = activation energy
N_u = Nusselt number, and σ is obtained from:

$$\sigma = \frac{E}{RT_b^2}(T_f - T_b) \tag{5.33}$$

wherein T_f is the maximum temperature of the film formed on the walls of the tube (K).

5.1.5   Selection of Reaction Coil Parameters

The selection of reaction coil parameters may be done by using Eq. (5.25) given in the previous section. This equation may be written as:

$$\frac{dT}{dL} = \frac{1}{\sum n_K C_{p_K}} \left[\frac{\pi D_e \varphi_t}{3600} - \sum_1 (\Delta H_r)_1 \frac{dn_1}{dL} \right] \tag{5.34}$$

Introducing the definition of mean heat capacity of the stream flowing through the coil, one can write:

$$\sum n_K C_{p_K} = \overline{C}_p \sum_K n_K$$

where $\sum_K n_K$ can be expressed by means of the gas law $pV = \sum_K n_K RT$ in the form:

$$\sum_K n_K C_{p_K} = \overline{C}_p \frac{p\overline{V}}{RT} \tag{5.35}$$

where $\overline{V}$ is the volume flowrate of fluid (m³/s) through the coil.

Using v for the velocity of the fluid in the tubes, one may write:

$$\overline{V} = \frac{\pi D_i^2}{4} v \tag{5.36}$$

Replacing Eq. (5.35) and (5.36) in Eq. (5.34), one obtains:

$$\frac{dT}{dL} = \frac{4RT}{\overline{C}_p p \pi D_i^2 v} \left[\frac{\pi D_e \varphi_t}{3600} - \sum_1 (\Delta H_r)_1 \frac{dn_1}{dL} \right] \tag{5.37}$$

In terms of the velocity v and the residence time, the differential coil length is given by: $v d\tau = dL$.

Substituting in (5.37), one obtains finally:

$$\frac{dT}{dL} = \frac{4RT}{\overline{C}_p p \pi D_i^2 v} \left[\frac{D_e \varphi_t}{3600} - \frac{1}{\pi v} \sum_1 (\Delta H_r)_1 \frac{dn_1}{d\tau} \right] \tag{5.38}$$

This equation makes it possible to deduce how to modify the furnace parameters when changing from pyrolysis of light feedstocks to pyrolysis of heavy feedstocks. In this case, the reaction rates expressed by:

$$\frac{dn_1}{d\tau}$$

will increase, which will determine the increase of the second term inside the parentheses. Also, since the final temperature at the outlet of the furnace must be the same while the reaction rate is higher, the rate of temperature increases along the coil, as expressed by $\frac{dT}{dL}$ must be higher. In order to accommodate these two opposed requirements within the constraints of Eq. (5.38), the following solutions are possible:

Increase thermal tension φ_t, which must be within the values accepted by the quality of the metal of the tubes.

Decrease the diameter of the tubes, which is correlated with the length of the cycle; the same layer of coke deposited inside thinner tubes will lead to higher pressure drop than in larger tubes.

Decrease of the pressure in the coil (the operation at lowest possible pressure is generally applied in the modern plants).

Increase the flowrate through the coil, but which leads to the increase of the pressure drop in the coil and is limited by the speed of the sound through the fluid inside the coil.

When considering this latter solution, one must bear in mind that the speed of sound varies with the nature of the gas and the temperature. Thus, the speed of sound must be determined for the conditions inside the coil, corresponding to the starting period and reaching the normal operating conditions to which a supplementary safety factor should be applied.

Taking into account that the components of the mixture inside the coil are much above the critical state, the calculation of the speed of sound can be done with the equation which is valid for ideal gases [17]:

$$v_s = \sqrt{\frac{C_p RT}{C_v M}} \tag{5.39}$$

where C_p and C_v are the heat capacities $\dfrac{C_p}{C_v} = \dfrac{4}{3}$ at constant pressure and volume respectively.

To express the speed in m/sec of polyatomic molecules, relation (5.39) becomes:

$$v_s = 105.3\sqrt{\frac{T}{M}} \tag{5.40}$$

As this relation shows, the speed limit inside the tubes increases with the temperature and decreases with the mean molecular mass of the gases flowing inside the tubes. The temperature is approximately the same for all feeds, since it is determined by technological considerations and by ones related to the metal of the tubes. Therefore, the main variable in Eq. (5.40) becomes the mean molecular mass. It follows that especially in the pyrolysis of heavy feedstocks, attention must be given to this problem. For the generally recommended maximum gas velocity in pyrolysis coils of 300–320 m/sec [18], it results, according to Eq. (5.40) that the maximum molecular mass inside the coil for operation in these conditions should be approximately 150 for a temperature of 1200 K or about 110 for a temperature of 1000 K.

It must be mentioned that all the possible solutions that were mentioned above when analyzing Eq. (5.38), are considered when selecting the operating conditions of the pyrolysis furnaces, especially when the feed is heavier fractions such as gas oils, or especially vacuum gas oils.

In 1992, [29] three of Linde's commercially proven concepts for radiant reaction coil layout, covering the range of residence time of the HC/steam mixture from 0.15 to 0.5 seconds were compared and specific requirements for heavy feedstocks were established. The best results came from the LSCC1-1 with small tube diameter and short coil length (residence time 0.16 seconds).

This work confirmed our previously deduced equations and the conclusions presented in this chapter.

5.2 DESIGN OF SOAKERS, COKE DRUMS, AND REACTION CHAMBERS

The design of soakers, coking drums, and reaction chambers is done on basis of practical data obtained from the operation experience of existent plants, since there is uncertainty of some parameters that are derived from theoretical deductions. Still, the equations that express the phenomena that take place are of high interest. Among other factors, they allow a more correct estimate of the effects that would be obtained by the modification of some operation parameters during the exploitation of the unit.

In the case when the feed to the chamber consists of two flows of different compositions and temperatures, the temperature established at the inlet may be determined by means of a thermal balance. Such is the case of the soakers in the two-furnaces visbreaking plants and in the reaction chambers,

$$G_A \sum_A g_A \overline{C}_{p_A}(t_A - t_{ic}) + G_B \sum_B g_B \overline{C}_{p_B}(t_B - t_{ic}) = 0$$

from which

$$t_{ic} = \frac{G_A t_A \sum_B g_A \overline{C}_{p_A} + G_B t_B \sum_B g_B \overline{C}_{p_B}}{G_A \sum_B g_A \overline{C}_{p_A} + G_B \sum_B g_B \overline{C}_{p_B}} \tag{5.41}$$

where the indexes A and B refer to the two flows, and the index $_{ic}$ to the inlet in the soaker or chamber.

In the case of coil or soaker visbreaking or coking drum, there is only a single stream going in and this calculation has no meaning.

The outlet temperature may be determined also on basis of a heat balance:

$$(G_A + G_B)\left[(t_{ic} - t_{ec})\sum_{A+B} g_{A+B} + \overline{C}_{p_{A+B}} + q_r x_c\right] + Q_p = 0 \tag{5.42}$$

where x_c is the conversion achieved in the chamber.

In principle, starting from an assumed value for x_c, this relation allows the calculation of the outlet temperature and, from it, of the temperature equivalent to the adiabatic mean conversion rate. This value is used to check the value of x_c assumed at the beginning.

Such a calculation encounters a series of major difficulties, which at the present cannot be avoided because the chemical transformations taking place in the soaker or chambers are different from those that took place previously, in the furnaces. Thus:

The reaction heats are strongly influenced by the exothermic reactions of condensation and polymerization.

The residence time of the vapors is different from that of the liquid, and the correct assessment of the vapors/liquid ratio and of the respective residence times is very difficult.

It is difficult to estimate the backmixing and the degree to which it is limited in some cases, by the perforated plates used in the soaker.

There are uncertainties concerning which kinetic equations which should be used.

Eq. (5.42) applied to values taken from operating units, could contribute to the gradual elucidation of these problems.

The amount of cooling liquid that must be mixed with the reactor effluent for quenching the reactions can be determined on the basis of a thermal balance. The enthalpies of the reaction stream before and after mixing with the cooling liquid must be calculated taking into account the ratio of vapors/liquid and the influence of pressure on the enthalpies [19].

5.3 SYSTEMS USING SOLID HEAT CARRIER

The presence in the reactor of solid particles makes it necessary to introduce the definition of some specific parameters and also of the space velocity–residence time correlation.

Also, some problems of the heat carrier in moving or fluidized beds are examined.

5.3.1 Definition of Some Specific Parameters

The presence in the reactor of solid particles, either being a catalyst or inert towards the reaction, makes necessary the use of parameters that characterize the fraction of the reactor volume that is not occupied by the solid, which is where the reaction takes place. In addition to this, when the bed is moving or fluidized, it becomes necessary to characterize the circulation rate of the solid.

The free reactor volume may be divided into the volume of the pores and the volume of the spaces between the solid particles—the apparent free volume. Together, they form the total free volume. Hydrodynamic problems such as the pressure drop through the bed or the maintaining of the desired fluidization conditions, must be solved by taking into account the apparent free volume. On the other hand, the determination of the residence time in the reactor, which is necessary for the calculation of the conversion achieved in the thermal process, must take into account the total free volume. Of course, for a solid which is not porous, such as are the majority of heat carriers, the volume of pores being practically equal to zero, the total free volume becomes equal to the apparent free volume.

Noting with V_R the volume inside the bed where the reaction takes place, and with V_{ZR} the volume of the reactor, the total void fraction η_t may be defined by the equation:

$$\eta_t = \frac{V_R}{V_{ZR}} \tag{5.43}$$

The total void fraction can be expressed as a function of the bulk density in the reactor and on the real density of the solid material. By noting with V_S the real volume of the solid material and taking into account that $V_S = V_{ZR} - V_R$, the obvious equation results:

$$V_{ZR}\gamma_v = (V_{ZR} - V_R)\gamma_r$$

Taking into account Eq. (5.43), it results:

$$\eta_t = 1 - \frac{\gamma_v}{\gamma_r} \tag{5.44}$$

Because the bed density of the stationary bed, the moving bed, and especially the fluidized bed are different, the total void fraction will be different in the three cases. In the case of the fluidized bed it will depend on the expansion of the bed.

The apparent void fraction may be defined by an equation similar to Eq. (5.43):

$$\eta_a = \frac{V_{LA}}{V_{ZR}} \tag{5.45}$$

Noting with V_A the apparent volume of the particles and with γ_a their apparent density, one may write:

$$V_{ZR}\gamma_v = (V_{ZR} - V_{LA})\gamma_a \tag{5.46}$$

from which, by taking into account Eq. (5.45), it results:

$$\eta_a = 1 - \frac{\gamma_v}{\gamma_a} \tag{5.47}$$

The volume of the pores can be determined by using the equation:

$$\eta_p = \eta_t - \eta_a = \frac{\gamma_v}{\gamma_a} - \frac{\gamma_v}{\gamma_r} \tag{5.48}$$

The use of Eqs. (5.44), (5.47), and (5.48) needs the determination of the real and apparent densities, respectively of the pore volume. The first two determinations do not pose special problems. The determination of the bed density can be performed by weighing a known volume of solid particles, while the real density is determined by means of the pycnometer. For determining the real density of the porous particles, a liquid is used that penetrates into the pores. In order to ensure the complete penetration of the liquid into the pores the particles shall be boiled in the liquid for 2–3 hours.

The apparent density can be determined by means of the pycnometer, using a liquid that does not penetrate into the pores, or indirectly, by the determination of the volume of the pores by means of the methods currently used for the study of adsorbents [20]. For porous particles of very small size, titration methods can also be used. The particles lose the relative mobility in the moment when their pores are filled with liquid and the moisture extends to the external surface of the particles [30].

A specific parameter of systems with solid recycling is the contacting (or the recycle) ratio, defined by the equation:

$$a = \frac{G_s}{G_{mp}} \tag{5.49}$$

Here, G_s and G_{mp} are the mass flowrates of solids and raw material flowing through the reactor in unit time.

5.3.2 The Residence Time-Space Velocity Correlation

In systems with a solid carrier, the feedrate is expressed usually as mass or volume flowrate. The application of kinetic equations to such systems requires the correlation of these parameters with the contact time.

The volume flowrate is defined by the equation:

$$w = \frac{D}{V_{ZR}} \qquad (5.50)$$

where D is the feed flowrate expressed as the volume in the state corresponding to the ambient temperature.

For flow reactors, the contact time is defined by Eq. (2.137):

$$\tau = \frac{V_R}{V} \qquad (2.137)$$

which, by taking into account (5.43) for reactors containing solid particles becomes:

$$\tau = \frac{V_{ZR}\eta_t}{V} \qquad (5.51)$$

Eliminating V_{ZR} from the Eqs. (5.50) and (5.51), one obtains:

$$\tau = \frac{D\eta_t}{V w} \qquad (5.52)$$

Expressing by $\bar{\rho}_R$ the mean density of the reaction mixture inside the reactor and by ρ_i the density of the feed, one may write:

$$\frac{D}{V} = \frac{\bar{\rho}_R}{\rho_i}$$

After substituting in (5.52) it follows finally:

$$\tau = \frac{\eta_t \bar{\rho}_R}{w \rho_i} \qquad (5.53)$$

an equation that correlates the contact time, τ with the volume flowrate, w. It must be mentioned that, if the volume flowrate is expressed in hours^{-1} and the time in seconds, then Eq. (5.51) should be written as:

$$\tau = \frac{3600 \eta_t \bar{\rho}_R}{w \rho_i} \qquad (5.53a)$$

If in the reaction zone the reaction mixture is totally in the vapor phase, by the application of the ideal gas law the equation (5.53a) becomes:

$$\tau = \frac{3600 \eta_t \bar{M} p}{\bar{z} R T w \rho_i} \qquad (5.54)$$

In a similar manner, the mass flow rate can be expressed by the following equations:

$$\tau = \frac{3600 \eta_t \bar{\rho}_R}{n} \qquad (5.55)$$

and

$$\tau = \frac{3600\eta_t \overline{M}_p}{\overline{z}RTn} \tag{5.56}$$

where n is the mass flowrate, defined in a way similar to expression (5.50) by the equation:

$$n = \frac{G}{V_{ZR}} \tag{5.57}$$

where G is the mass feedrate.

5.3.3 Characteristics of the Moving Bed

The implementation of systems involving beds of moving particles poses some typical circulation and heat exchange problems. They have to be known in order to understand the industrial performance of the processes using such beds.

A continuous circulation of solids, without the blocking of the pipes or of passages, is obtained generally when the minimum diameter of the flow cross section is at least 6 times larger than the maximum diameter of the granules. An exception is dust, which can become electrified by friction to the walls and agglomerates. In this case, the blockage could occur also at larger ratios of the diameters.

In order to ensure uniform circulation through pipes, orifices, and ducts, the industrial practice is to size them in such a way that the flow opening should be 15–20 times larger than the diameter of the largest particle.

The bulk density of the moving bed is less than that of the settled bed. It can be calculated by means of the following equation [21]:

$$\gamma_m = \gamma_v - 0.3\frac{d}{R} + \frac{d^2}{R^2}$$

where d is the diameter of the granules, and R is the radius of the tube or of the apparatus.

This equation shows that the bulk density of the moving bed is different from that of the stationary bed only when the particles move through tubes of a relatively small diameter.

Contrary to the flow of liquids, in the case of granulated solids the flow velocity does not actually depend on the height of the bed situated above the level of the flow area. This fact is understandable if we take into account that the forces resulting from the weight of the bed are transmitted at an angle of about 30° from the vertical and they are thus transmitted to the walls.

Empirical equations for calculating the flowrate of solid particles confirm these findings [22], as shown by the following equations:.

$$V = 0.408d^{0.96}H^{0.004}$$

where:

V = solids flowrate in liters per minute and sq. cm;
d = diameter of the nozzle in cm;
H = bed height in meters

More recent equations distinguish between the flow through orifices and the flow through pipes. The following equations were proposed [23]:

$$\left.\begin{aligned} D_c &= 0.0195 \left(\frac{d_c}{2.54} - n\right)^{7/3} \\ D_o &= 0.0132 \left(\frac{d_o}{2.54} - n\right)^{5/2} \end{aligned}\right\} \tag{5.58}$$

where:

D_c and D_o = flowrates in m^3/min
d_c and d_o = diameters of the pipe of the orifice in cm
n = constant depending on the shape of the solid granules and has the values 0.4 for spheres with a diameter of 2.5 mm and 0.8 for disks with a diameter of 4.3 mm.

Granular material flows freely on a flat horizontal surface and forms a cone with an inclination corresponding to the angle of repose, which is of about 38°. This angle is necessary for designing the shape of the upper part of equipment into which solid material, introduced through a central pipe, must be uniformly distributed on the cross section of the vessel.

The flow profiles of solids inside equipment containing moving beds is very complex (the Figure 5.4) When the solids are drained through the bottom of the vessel, if the apparatus is provided in the lower part only with a central orifice, an important portion of granules, limited by an angle of about 71°, remains in the vessel. In order to avoid important amounts of solids being blocked inside the apparatus, one makes use of several emptying pipes of the solid at the inferior part of the apparatus, which are then unified by means of connecting pipes, a system which is known as *spider*.

A common problem is how to introduce continuously in the upper part of an apparatus, the contents of which are under a pressure above atmospheric, a granulated solid from a vessel at atmospheric pressure. Such cases occur frequently when the solids are lifted by pneumatic transport from a vessel (bunker) at atmospheric pressure into a reactor where the pressure is superior to the atmospheric one (with the purpose of ensuring the circulation of the vapors towards the separation system). In such cases, the bunker and the reactor are connected by a transport line (usually vertical) that overcomes the pressure difference. The length of such a pipe is calculated by the equation:

$$\frac{\Delta p}{\Delta L_{max}} = \gamma_m (\sin\theta - \mu\cos\theta)\sin\theta \tag{5.59}$$

where:

$\dfrac{\Delta p}{\Delta L_{max}}$ = pressure gradient (expressed in kg/m^2) per linear meter of height
γ_m = bulk density of the solid material
μ = coefficient of friction between the granules
θ = the inclination angle of the pipe towards the horizontal.

The transfer of heat between the gases or vapors and the granulated solid is very intense. The overall heat transfer coefficient has values comprised between

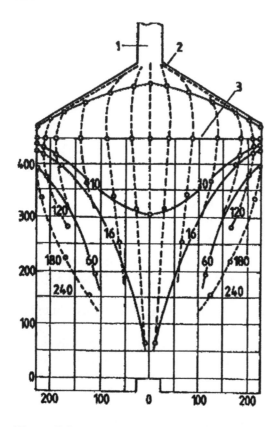

Figure 5.4 Flow pattern for discharging solids from a vessel with one axial discharge point. 1—Feed nozzle; 2—Angle of repose, approx. 38°; 3—initial level of particles. - - - stream lines. —Position of surface of bed, after discharging indicated volumes of particles.

600–3,000 kJ/m^2 · h · °C, which correspond to a heat exchange with the unit volume of moving bed, between $600 \cdot 10^3$ and $3000 \cdot 10^3$ kJ/m^3·h [24]. This intensive heat exchange leads to the almost instantaneous equalization of the temperature when granulated solids are contacted with reaction vapors.

The heat transfer coefficients between the granular solids and the walls of the apparatus are of approximately 350 kJ/ m^2· h · °C [24].

5.3.4 Design of Moving Bed Reactors

The temperature profile along the height of the reactor is required information for designing moving bed reactors. It depends upon the direction of the circulation of the reactants, i.e., upwards or downwards.

Generally, downward circulation is preferred in the case when the reactants are in liquid phase or partially in liquid phase.

In ascendent or upwards circulation the danger exists of blocking circulation by the agglomeration of particles of the carrier.

As will be shown below, the heat brought by the carrier is used more efficiently when the feed and the heat carrier are in cocurrent. Descendent circulation is preferred in all cases where there is no requirement of a maximum temperature at the outlet from the reactor.

In the case of descendent circulation, that means in cocurrent with the heat carrier, which is the typical circulation for the processes of coking and catalytic cracking, the evolution of the temperatures along the height of the reactor is depicted in Figure 5.5.

Owing to the very intense heat exchange between the heat carrier and the feed, the temperature is equalized immediately following the contact of the carrier with the feed. Therefore, one may neglect the chemical transformations that take place during heating. One may consider that the reaction begins at the temperature t_{ir}. In these conditions t_{ir} can be determined from the thermal balance at the inlet of the reactor.

$$G_{mp}\bar{c}_p(t_a - t_{ir}) + G_s c_s(t_p - t_{ir}) = 0$$

from which

$$t_{ir} = \frac{G_{mp}\bar{c}_p t_a + G_s c_s t_p}{G_{mp}\bar{c}_p + G_s c_s} \tag{5.60}$$

The final temperature at the end of the reaction, t_{fr}, can be determined from the thermal balance on the reactor, namely:

$$(G_{mp}\bar{c}_p + G_s c_s)(t_{ir} - t_{fr}) + G_{mp} q_r x - Q_p = 0$$

from which

$$t_{fr} = \frac{(G_{mp}\bar{c}_p + G_s c_s)t_{ir} + G_{mp} q_r x - Q_p}{G_{mp}\bar{c}_p + G_s c_s} \tag{5.61}$$

The upwards circulation of the reactants, which is in countercurrent with the heat carrier, is characteristic for the processes in which it is necessary to reach a maximum temperature at the outlet from the reactor, as in pyrolysis. In this case, the tempera-

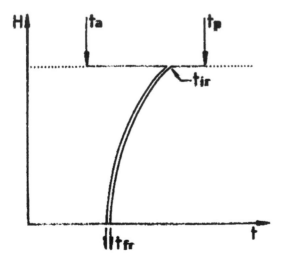

Figure 5.5   Variation with bed height of temperature of carrier (t_p) and of feed (t_a) for downwards, cocurrent flow.

ture profile is obtained by first performing an overall thermal balance on the reactor from which the outlet temperatures of the products and of the heat carrier are calculated.

The equation for the thermal balance of the reactor is:

$$G_{mp}[\bar{c}_p(t_{fr} - t_{ir}) - q_r x] + G_s c_s(t_{fs} - t_{is}) + Q_p = 0 \tag{5.62}$$

Since the countercurrent circulation imposes the restricting conditions $t_{fr} \leq t_{is}$ and $t_{fs} \geq t_{ir}$ and since the heat exchange between the two flows leads to the equalization of the temperatures between the solid and the reactants at any level inside the reactor, the following situations could take place (the heat losses through the walls are considered negligible, i.e. $Q_p = 0$):

a.   The heat contribution of the carrier is higher than the consumption of heat corresponding to the reaction and to the maximum possible heating of the reactants; it follows:

$$t_{fr} = t_{is} \qquad t_{fs} > t_{ir}$$

b.   The two amounts of heat are equal:

$$t_{fr} = t_{is} \qquad t_{fs} = t_{ir}$$

c.   The amount of heat that could be given up by the solid is lower than that necessary for the reaction and heating of the resulting products up to the inlet temperature of the heat carrier, resulting in:

$$t_{fr} < t_{is} \qquad t_{fs} = t_{ir}$$

Since the system in countercurrent is used when it is required to reach a temperature at the outlet t_{fr} imposed by technological considerations, situation "c" must be avoided. Between "a" and "b" situation "b" is preferred. The heat brought into the system by the carrier is used completely in this case.

By imposing condition "b" on Eq. (5.62), the flow of solid carrier G_s will result, respectively the contacting ratio "a" given by Eq. (5.49).

The application of Eqs. (5.61) or (5.62) needs to assume a value for the conversion x at the outlet from the reactor, which must be checked with the value resulting from the kinetic calculation. Often, the final conversion x is imposed by the process conditions. The kinetic calculation is performed after the reaction time, necessary for performing the assumed conversion was established based on the volume of the reactor.

From the combination of the Eqs. (5.55) and (5.57), it results:

$$\tau = \frac{3600 \eta_t \bar{\rho}_R V_{ZR}}{G_{mp}} \tag{5.63}$$

which may be written also as:

$$V_{ZR} = \frac{G_{mp}}{3600 \eta_t \rho_R} \tau \tag{5.63a}$$

As a first approximation, it can be assumed that the variation of the temperature along the height of the reactor is linear. In this case, introducing t_{ir}, t_{if}, respeclively k_t or α in Eqs. (2.158) or (2.159), gives the temperature corresponding to the adiabatic mean rate (t_{evma}) for the conditions in the reactor.

This temperature is used for obtaining the reaction rate for the investigated reaction. Imposing the final conversion x, which was previously established from process considerations, gives the reaction time τ needed for achieving this conversion.

Introducing this value τ in equation (5.63a), gives the volume of the reaction zone V_{ZR}.

Since the variation of the temperature is not linear with the height of the reactor, for a more exact calculation one has to divide the reaction zones into segments and to perform the calculation in a manner similar to that presented for tubular furnaces in Section 5.1.1.

For moving bed reactors, the same systems of kinetic equations as for tubular furnaces may be used. In such cases, a calculation technique similar to that described in Sections 5.1.2–5.1.4 is used.

Consider that the pressure drop through a moving bed is higher than through a stationary bed; the latter may be calculated by means of classic equations [25–27].

The following equation is recommended [27]:

$$\frac{\Delta p}{L} = \frac{2fG^2(1 - \eta_a)^3}{10^5 \cdot g \cdot \overline{p}_R \cdot d_p} \tag{5.64}$$

where f is the friction factor determined by means of the graph shown in Figure 5.6.

The kinetic equations (5.61–5.63) and Eq. (5.64) are interdependent. In many cases the computation is done by trial and error. The iterative method is repeated until convergence is obtained, i.e. the assumed and calculated values coincide.

In processes in mixed phase, vapor-liquid, additional difficulties appear: the estimation of the residence time in the reactor for each phase and the calculation of the pressure drop through the bed.

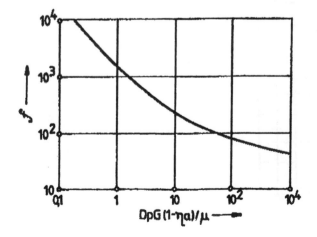

Figure 5.6 Friction factor f plotted by using Eq. (5.64).

In order to determine the heat that the carrier must supply to the reactor, one must calculate the heat losses in the loop of circulating solids and the combustion in the furnace that ensures the reheating of the carrier.

The heat losses in the transport system are not important and can be calculated by classical methods and by using the heat transfer coefficients between the transport gas and the granulated solid, and between the transported solids and the walls of the transport pipe.

The reheating of the solid heat carrier used in coking and the pyrolysis of heavy feeds, such as the vacuum distillates, is based on the partial burning of the coke deposited on the carrier. The coke deposited during the pyrolysis of light feeds may not be sufficient, for reheating the carrier and additional gaseous fuel might be necessary.

The design of the reheating furnace, in addition to the usual combustion calculations [7], also requires knowledge of the burning rate of the coke deposited on the granules. The calculation of this combustion rate [27] can be carried out using the results of Hottel [28], (see Figure 5.7).

This graph shows that up to temperatures of 800°C, the burning takes place in the reaction domain, whereas at higher temperatures, the limitation factor becomes the external diffusion through the boundary layer formed around the granules. In this case the burning rate is dependent not only on the temperature, but also on the rate of the air that feeds the combustion.

Note that the phenomenon of burning coke deposited on the granules is quite different from the case of porous solids, for example the regeneration of catalysts [19] in the process of fluid catalytic cracking.

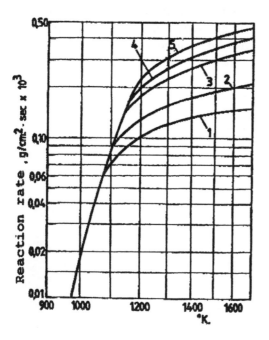

Figure 5.7 Effect of air velocity on rate of coke combustion.   1-3.51 cm/s; 2-7.52 cm/s; 3-27.4 cm/s; 4-39.8 cm/s; 5-50.0 cm/s.

5.3.5   Design Elements for Fluidized Bed Coking Units

Backmixing that occurs in the dense phase fluidized bed is very intensive and the reactor and coke burning bed approach a perfectly mixed reactor. Kinetic calculations must take this into account.

Backmixing also has as a result the homogenization of the temperature inside the dense phase fluidized bed. The difference between different points of the bed does not exceed 1–2°C.

The reactor and the heater in coking and flexicoking plants used dense phase fluidized beds. In these conditions, the reaction temperature inside the bed may be easily correlated with the temperature of the two streams, solid and fluid, entering the reactor, using in this purpose the thermal balance of the reactor[*].

$$G_{mp}\left[\bar{c}_p(t_r - t_{im}) - q_r x\right] + G_s c_s(t_r - t_{is}) + Q_p = 0 \qquad (5.65)$$

When the solid is a catalyst, the reactions will cease altogether at the separation of the solid from the fluid. In the case of thermal (noncatalytic) processes, they will go on as long as the effluent is not quenched when leaving the reactor. This fact must be taken into account in the design of the process.

Two systems are used in thermal cracking for transporting the solids between the reactor and the heater:

The ascendant transport at large dilutions of the solid similar to the pneumatic transport

The transport in fluidized dense phase, the fluidized state being maintained by injections of inert fluid from place to place, along the transport pipe

The first transport system—the pneumatic transport—is a classic, well-known system that does not need any further explanations.

The second system is based on the fact that a homogeneous fluidized bed in a dense phase behaves in some respects as a liquid. The bed flows under the action of differences in the hydrostatic level or of a pressure gradient provided that the fluidization is maintained. Generally, this system is used in combination with the pneumatic transport.

The main advantage of the system that transports solids in a dense fluidized phase is the practical elimination of the erosion of the transport pipes and of the solid particles. These advantages appear especially obvious in the points of the unit where the transport pipes change direction and where strong erosions are observed as a result of operation in the pneumatic transport mode. For this reason, in the combined systems the pneumatic transport is used as much as possible through ascending straight pipes.

The design of the system depends on the particular process and may meet additional difficulties, such as in the case of fluid coking, when a liquid phase is injected into the bed. In this case the theoretical calculation is very difficult and uncertain.

In fluid coking, the amount of coke deposited on the coke particles in the reactor exceeds the amount that must be burnt in order to reheat the coke particles. Accordingly, the thermal balance of the reheating furnace has the form:

---

[*] The thermal balance of the furnace used for reheating is carried out in a similar manner.

$$G_s c_s(t_{ec} - t_{ic}) + G_a \bar{c}_{pa}(t_{ec} - t_{ia}) + Q_p$$
$$= G_{mp} x_c q_a + G_g \bar{c}_{pg} t_{ec} + G_{cb} q_{cb} \tag{5.66}$$

where:

$$x_c = \text{coke burnt for reheating, reported to the feed}$$
$$q = \text{the thermal effects}$$
$$Q_p = \text{the losses of heat through the walls}$$
$$\text{subscript } s = \text{heat carrier}$$
$$a = \text{air,}$$
$$mp = \text{feed}$$
$$g = \text{burnt gases}$$
$$cb = \text{the supplementary fuel}$$
$$ic = \text{the inlet in the furnace}$$
$$ec = \text{outlet from the furnace}$$
$$ia = \text{inlet of air}$$

The fact that the heat losses of the transport system are not important and may be estimated to cause a temperature decrease of 10–20ºC allows the correlation of the inlet and outlet temperatures of the heat carrier in the reactor and in the reheater. Thus, for the correlation of Eqs. (5.65) and (5.66), the following values are accepted:

For the transport of the reheated carrier:

$$t_{ec} - t_{is} = 15 - 20°C \tag{5.67}$$

For the transport of the carrier that left the dense phase reactor:

$$t_{ir} - t_{ic} = 10 - 15°C \tag{5.68}$$

The decrease of temperature is higher for the transport of the reheated carrier. This is justified by its much higher temperature. A final, exact calculation will specify the Δt values.

A last problem concerning the design of the reactor-reheater system is the correlation of the pressures in these two vessels and of their relative height in order to ensure a correct circulation of the heat carrier (coke particles).

The calculation depends on whether the transport is performed in dense phase through semicircular pipes and in diluted phase only in a part of straight ascendant pipes (coking plant, Figure 4.23) or through straight ascendant and descendent pipes (Figure 4.27). In the second case the solids are in diluted phase in the ascendant pipes and in dense phase in the descendent ones.

In both cases, the pressure in the reactor is determined by the hydrostatic and hydrodynamic pressure drops in the fractionation system and by the aspiration pressure of the gas compressor.

The pressure in the reheating furnace is determined by the pressure drop through the heat recovery system and cleaning of the flue gases.

The height and pressure parameters that play a part in the design of the transport of solids through semicircular pipes is given in Figure 5.8.

The gas injections in the semicircular portions of the pipes have the role of maintaining the state of dense phase fluidization. The valves located on the two pipes serve exclusively for the isolation, in case of need, of the two vessels. In normal operation they are in the fully open position.

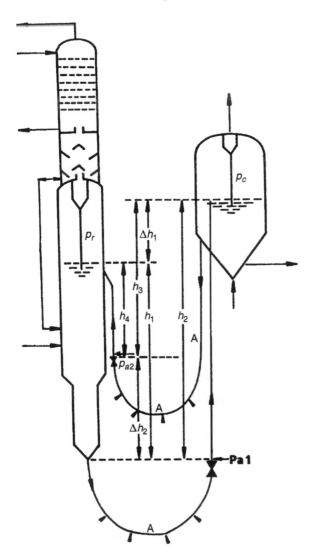

Figure 5.8 Circulation of coke particles in a transport system with curved pipes.

For simplification: the hydrodynamic pressure drop in the pipes for dense phase transport can be neglected, the bulk densities in the pipes for dense phase transport can be considered identical. The same is valid for the pipes for dilute phase transport.

The conditions for accurate circulation may be expressed by the following equations:

$$\left.\begin{array}{l} p_{a_1} \geq h_2\gamma_{\text{dil}} + p_c + \Delta p_2 \\ p_{a_1} < h_1\gamma_{\text{dens}} + p_r \\ p_{a_2} \geq h_4\gamma_{\text{dil}} + p_r + \Delta p_4 \\ p_{a_2} < h_3\gamma_{\text{dens}} + p_c \end{array}\right\} \tag{5.69}$$

where:

p_{a_1} and p_{a_2} = pressures at the feed points of the transport agent, indicated in Figure 5.8,

Δp = hydrodynamic pressure drops

γ_{dens} and γ_{dil} = the densities of the dense and of the diluted phases

The rest of the notations are shown in the same figure.

Equating these equations pairwise, one obtains:

$$\left.\begin{aligned} h_2\gamma_{\text{dil}} + p_c + \Delta p_2 = h_1\gamma_{\text{dens}} + p_r \\ h_4\gamma_{\text{dil}} + p_r + \Delta p_4 = h_3\gamma_{\text{dens}} + p_c \end{aligned}\right\} \tag{5.70}$$

By subtracting the second equation from the first, it results:

$$(h_2 - h_4)\gamma_{\text{dil}} + p_c - p_r + \Delta p_2 - \Delta p_4 = (h_1 - h_3)\gamma_{\text{dens}} + p_r - p_c \tag{5.71}$$

According to the figure,

$$h_2 - h_4 = \Delta h_1 + \Delta h_2$$

and

$$h_1 - h_3 = \Delta h_2 - \Delta h_1$$

By substituting and regrouping the terms of the last expression, one obtains:

$$2(p_r - p_c) = \Delta h_1(\gamma_{\text{dens}} + \gamma_{\text{dil}}) - \Delta h_2(\gamma_{\text{dens}} - \gamma_{\text{dil}}) + \Delta p_2 - \Delta p_4 \tag{5.72}$$

As the pressures in the two vessels and the bulk density in the dense phase are imposed by the process conditions, the correct circulation is ensured by the corresponding selection of the relative height of the two vessels—Δh_1, and of the position of the transport pipes that determines the distance—Δh_2. To a lesser extent it is possible to change the density, γ_{dil}, of the diluted phase. Its selection is determined by considerations related to the erosion of the ascending transport pipes.

Remark: in this transport system, accidental pressure increase in one of the vessels cannot lead to the inversion of circulation; even a beginning of reverse circulation leads to the blocking of the semicircular pipes by changing the transported material from the fluidized state to a settled bed.

In the transport system with straight pipes (Figure 5.9) solids transport is not blocked if a sudden increase of pressure occurs in one of the vessels. To absorb such shocks and to prevent the inversion of circulation, the two valves situated at the lower ends of the descendent transport pipes in dense phase are controlled to ensure an approximately 0.2–0.3 bar pressure drop in normal operation. An accidental increase of the pressure in one of the vessels diminishes this pressure drop, thus preventing the inversion of the circulation.

The equations that express the conditions of proper circulation will have in this case the form:

$$\left.\begin{aligned} p_{a_1} &\geq h_2\gamma_{\text{dil}} + p_c + \Delta p_2 \\ p_{a_1} &< h_1\gamma_{\text{dens}} + p_r - \Delta p_{v_1} \\ p_{a_2} &\geq h_4\gamma_{\text{dil}} + p_r + \Delta p_4 \\ p_{a_2} &< h_3\gamma_{\text{dens}} + p_c - \Delta p_{v_2} \end{aligned}\right\} \tag{5.73}$$

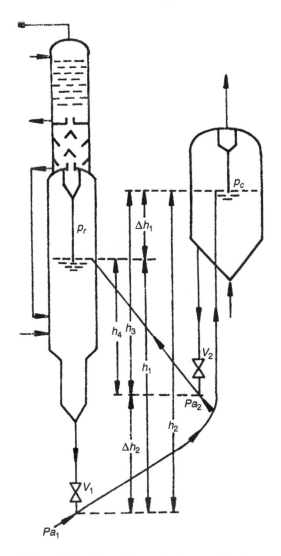

Figure 5.9   Circulation of coke particles in a transport system with straight pipes

By equating the equations pairwise, it follows:

$$\left.\begin{array}{l} h_2\gamma_{\text{dil}} + p_c + \Delta p_2 = h_1\gamma_{\text{dens}} + p_r - \Delta p_{v_1} \\ h_4\gamma_{\text{dil}} + p_r + \Delta p_4 = h_3\gamma_{\text{dens}} + p_c - \Delta p_{v_2} \end{array}\right\} \tag{5.74}$$

Finally, operating as in the previous case, it results:

$$2(p_r - p_c) = \Delta h_1(\gamma_{\text{dens}} + \gamma_{\text{dil}}) - \Delta h_2(\gamma_{\text{dens}} - \gamma_{\text{dil}}) + \Delta p_2 - \Delta p_4 + \\ + \Delta p_{v_1} - \Delta p_{v_2} \tag{5.75}$$

If, as in the case of normal operation the pressure drops through the two valves are equal, Eq. (5.75) becomes identical to (5.72).

REFERENCES

1. KM Sundaram, FG Froment, Ind Eng Chem Fundam 17 (3): 174, 1978.
2. SK Layokun, DH Slater, Ind Eng Chem Process Des Dev 18 (2): 232, 1979.
3. TC Tsai, Z Renjun, Z Jin, Oil Gas J 85, 21 Dec.: 38, 1987.
4. MJ Shah, Ind Eng Chem 59 (5): 70, 1967.
5. GL Jacques, P Dubois, Rev Assoc Fr Tehn Petrol, N 217: 43, 1973.
6. KR Jinkerson, JL Gaddy, Ind Eng Chem Process Des Dev 18: 579, 1979.
7. G Suciu, R Tunescu, eds. Ingineria prelucrarii Hidrocarburilor, vol. 1, ed. Tehnica, Bucuresti, 1973.
8. SV Andelson, Calcul tehnologic si forma constructiva a cuptoarelor din rafmariile de petrol, ed. Tehnica, Bucuresti, 1959.
9. GG Rabinovici, PM Riabih, A Hohriakov, IK Molokanov, EV Sudacov, H Rasacioty osnovnyh protesov i aparatov neftepererabotki. EN Sudakov Editor, Himia Moscow, 1979.
10. R Tunescu, Tehnologia distilarii titewlui Ed. Didactica si Pedagosica Bucuresti, 1970.
11. L Nelson, Petroleum Rafwery Engineering, 4th ed., Tokyo: McGraw Hill Book Company, 1958.
12. FD Rossini, KS Pitzer, RL Amett, RM Braun, GC Pimentel, Selected Values of Physical and Thermodynamical Properties of Hydrocarbons and Related Compounds, Pittsburgh: Carnegie Press, 1953.
13. DR Stull, EF Westrum Jr, GC Sinke, The Chemical Temodinamics of Organic Compounds, New York: John Wiley, 1969.
14. SW Benson, Refining Petroleum for Chemicals, Advances in Chemistry Series vol. 97, Chap. 1, Washington D.C.: Amer. Chem. Soc., 1970.
15. ME Dente, EM Ranzi. In Pyrolysis Theory and Industrial Practice, p. 133–175, LF Albright, eds. 2nd ed. New York: Academic Press Inc., 1983.
16. SW Benson, Thermochemical Kinetics. 2nd ed. New York: John Wiley, 1976.
17. J Rossel, Physique Generale, Paris: Griffon, 1960.
18. V Maoris, In: Piroliza Hidrocarburilor, V Vintu ed., Bucuresti: Editura Tehnica, 1980, p. 161.
19. S Raseev, Procese distinctive de prelucrare a titeiului, Bucuresti: Tehnica, Editura, 1964.
20. JM Thomas, J Thomas, Introduction to the Principles of Heterogeneous Catalysis, Academic Press, New York, 1967.
21. PI Lukianov, Vsesoiuznoie soveshcianie po kataliceskim prozesem pererabotki nefti, Report IV/3 G.N.T.K., Moscow, 1958.
22. Oil and Gas J, Sept. 17: 147, 1939.
23. AW Hage, RF Ashwill, EA White, Petrol Eng, 32, C 37–C 44, 1960.
24. B Bondarenk, Instalatii de cracare catalitica, Ed. Tehnica, Bucuresti, 1860.
25. J Happel, Ind Eng Chem 41: 1161, 1949.
26. G Suciu, R Tunescu, eds. Ingineria prelucrarii hidrocarburilor, Vol. 1, Editura Tehnica, Bucuresti 1973.
27. MW Stanley, Reaction Kinetics for Chemical Engineers, New York: McGraw-Hill Book Co. Inc., 1959.
28. Tu, Davis, Hottel. Ind Eng Chem, 26: 749, 1934.
29. G Merz, H Zimmerman, Hydrocarbon Technology International, p. 133–136, Editor P. Harrison, London: Sterling Publ International Ltd., 1992.

6

Theoretical Basis of Catalytic Cracking

6.1 PROCESS THERMODYNAMICS

The thermodynamic analysis of catalytic cracking, as in any process, involves:

The determination of the heat of reaction in the process

The calculation of the chemical equilibrium of the main and secondary reactions as a means of understanding the chemical transformations taking place

The first issue was dealt with in Section 1.1 for all processes involving the conversion of hydrocarbons.

The method suitable for catalytic cracking determines the heat of reaction as the difference between the heats of combustion of the products and of the feed, using for this purpose the graphs of Figures 1.1 and 1.2.

Experimental studies [232] using a reactor simulating the conditions in a practical isothermal riser reactor found at 500–550°C a reaction heat of 800–1070 kJ/kg. The feed was a gas oil: $d = 0.9292$, $S = 0.72$ wt %, IBP $= 259°C$, temperature at 50 wt % $= 377°C$, EP $= 527°C$, 10.62% paraffins, and 39.76% cycloparaffins. The two catalysts used were Octanat and GX30.

The lower values for the heat of reaction—677 kJ/kg—obtained earlier [233] are attributed to nonisothermal reaction conditions: a 20–40°C temperature drop is caused by the endothermic reaction.

The heat of combustion of the coke deposited on the catalyst depends on its hydrogen content and on the ratio CO/CO_2 in the flue gases. Its calculation is presented in Section 6.6.

The same method used for computing equilibrium concentrations for thermal cracking is valid also for catalytic cracking (see Figures 1.3 and 2.1–2.7), since the presence of the solid catalyst does not influence the equilibrium of the reactions that occur in vapor phase. Note however that knowledge of the equilibria of the reactions of the C_2–C_4 hydrocarbons is not sufficient.

The thermodynamic analysis of catalytic cracking requires information on the behavior of the heavier hydrocarbons contained in the gas oils, vacuum distillates, and even residual fractions. A major difficulty in performing this analysis resides in the limited knowledge of the thermodynamic constants for the hydrocarbons, which are typical for such fractions.

Despite the fact that for the above reason such an analysis is of limited value, it still gives useful information for understanding the thermodynamics of this process.

A problem of special interest is the adsorption equilibrium of the reaction products between the surface of the catalyst and the vapor phase. The adsorbed substances lead finally, following polymerization and condensations, to the formation of coke. The importance of this problem was emphasized in an earlier study by Raseev [1].

It is obvious that the main coke generators present in the feed are the resins and some of the condensed aromatic and hydro-aromatic hydrocarbons, and in the case of residue cracking, the asphaltenes. All these components are usually directly adsorbed on the catalyst and are gradually converted to coke.

But they are not the only coke generators. Coke deposits on the catalyst are produced even during the catalytic cracking of white oils or of paraffins. This proves that coke deposits are formed also as result of the decomposition of saturated hydrocarbons.

In order to clarify these processes, the thermodynamic calculations must be carried out for conditions in which the reaction products remain adsorbed on the catalyst. Since it is difficult to know the concentrations in the adsorbed layer, it was found useful to assume them to be equal to those in liquid phase. Indeed, at 500°C, a given volume of liquid contains approximately the same number of molecules as an equal volume of gas under 100–200 bar. One plots in the same graph the equilibria for the liquid and gaseous phases. At high pressures (100–200 bar) the equilibrium curves for the two phases must intersect. Such plots are used in this book for thermodynamic analysis of catalytic cracking and also for other catalytic processes.

The computation of the equilibrium concentrations for the case presented below was performed by using the method given in the Chapter 1.2 [2], and taking into account that for reactions in the liquid phase the constant b in Table 1.5 is equal to zero.

Published thermodynamic constants were used [3]. For reactions in gaseous phase, the constants for 800 K, which is close to the catalytic cracking, were used. For reactions in liquid phase, the only available constants are for 298 K, and they were used as such.

In order to characterize the behavior of various cuts or of certain classes of compounds, some typical average values were used for the thermodynamic constants $\Delta H°_{800}$ and $\Delta S°_{800}$.

The results of equilibrium calculations were plotted for the following types of reactions:

Alkanes and alkenes cracking
Alkenes polymerization
Cycloalkanes ring opening and cyclization of alkenes
Dealkylation of alkyl-cyclanes
Dehydrogenation of cyclohexanes

Dealkylation of alkyl-aromatics
Dealkylation of polycyclic hydrocarbons
Cracking of sulfur-containing compounds
Cracking of nitrogen-containing compounds

In the final part of this section, conclusions are formulated about the vapor phase reactions that may take place during catalytic cracking, and about the reactions that may lead to the formation of products that remain adsorbed on the surface of the catalyst and are converted to coke.

6.1.1 Alkanes Cracking

The decomposition of butane, which was analyzed in Section 2.1, shows a conversion at equilibrium of about 90% at atmospheric pressure and at a temperature of 500°C, conversion which, at a pressure of 100 bar, decreased to about 20%.

The thermodynamic equilibrium taking into account for reactions of the form:

$$C_{(m+n)}H_{2(m+n)+2} \rightleftharpoons C_mH_{2m+2} + C_nH_{2n}$$

is analyzed in the following, for higher hydrocarbons, such as the normal and iso C_6–C_{20} paraffins. The calculations were performed for the extreme terms of the considered series of hydrocarbons, since their behavior allows one to draw the correct conclusions for the intermediary terms of the series.

The selection of specific iso-alkanes was guided by the structure of the hydrocarbons contained in the straight run gas oil: the monomethyl-derivatives are preponderant, the dimethyl- and ethyl-derivatives are present only in small amounts and compounds with quaternary carbon atoms are absent.

Concerning the products, those resulting from catalytic cracking were the only ones considered: the formation of hydrocarbons with less than 3 carbon atoms was neglected; among alkenes, only those with double bond in position 1 were considered.

The heats of reaction and the variation of the entropies for the selected reactions are:

	$(\Delta H^0{}_{800})_r$ kcal/mol	$(\Delta S^0{}_{800})_r$ cal/mol·degree
$C_{20}H_{42} \rightleftharpoons C_3H_8 + C_{17}H_{34}$	18.93	33.62
$C_{10}H_{22} + C_{10}H_{20}$	18.60	34.23
$C_{17}H_{36} + C_3H_6$	18.61	33.91
$C_6H_{14} \rightleftharpoons C_3H_8 + C_3H_6$	18.92	33.42
2-methyl-nonane $\rightleftharpoons C_6H_{14}$ + i-C_4H_8	16.03	32.66
3-methyl-nonane $\rightleftharpoons C_5H_{12}$ + 2-methyl-butene-1	15.81	33.13
4-methyl-nonane $\rightleftharpoons C_4H_{10}$ + 2-methyl-pentene-1	17.07	34.51
2,3-dimethyl-octane $\rightleftharpoons C_5H_{12}$ + 2-methyl-butene-2	15.59	35.25
$\rightleftharpoons C_4H_{10}$ + 2,3-dimethyl-butene-1	15.30	35.51

These data show the almost identical behavior of the C_6–C_{20} n-alkanes, with very small differences that depend on the products of the reactions (the first 4 reactions in the table).

Concerning the iso-alkanes, somewhat larger differences are observed between the mono- and dimethyl-derivatives.

By using the simplified method developed in Section 1.2 and the thermo-dynamic constants given in the table above, the graph of Figure 6.1 for the equilibriums of these reactions was plotted. In the graph, the straight lines of constant conversion are shown for the values of 99% and 60%, which is enough for the analysis of the results. As in the graphs within Section 1.2, the scale of the temperature in the ordinate is ascending.

Similar calculations were performed for the reactions in liquid phase, for which thermodynamic constants were available, and which can be assimilated with reactions in the adsorbed layer, namely:

$$C_{20}H_{42} \rightleftharpoons C_{10}H_{22} + C_{10}H_{20} \quad (\Delta H_{298}^0)_{rl} = 19.32 \text{ kcal/mol}$$
$$(\Delta S_{298}^0)_{rl} = 24.27 \text{ cal/mol·degree}$$

These data made it possible to plot the equilibrium (see Figure 6.1), correlated with the equilibrium for the same reactions in gas phase.

The data plotted in this graph confirm the well-known fact [1] that in the conditions for the catalytic cracking of distillates (temperatures of about 500°C and pressure a little above atmospheric), the decomposition of alkanes is not limited thermodynamically. The conversions at equilibrium are approximately 99% for n-alkanes and over 99% for i-alkanes.

The equilibrium conversions in liquid phase are identical with those in vapor phase at pressures of about 200–500 bar. They could be considered representative for

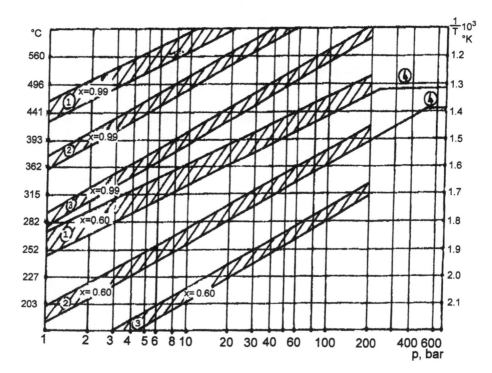

Figure 6.1 Alkanes and alkenes decomposition equilibrium. 1 – n-alkanes and n-alkenes, 2 – methyl-nonane, 3 – dimethyl-octanes, 4 – equilibrium in adsorbed layer equivalent to liquid state, x – conversion in molar fraction.

those in the adsorbed film on the surface of the catalyst. Different from those in the vapor phase, the conversions at equilibrium are in this case about 60% for n-alkanes and a little higher for i-alkanes at temperatures of catalytic cracking.

6.1.2 Alkenes Cracking

As in the previous case, the following table contains the reactions considered and the values of the heats and entropies of the reactions.

	$(\Delta H^0_{800})_r$ kcal/mol	$(\Delta S^0_{800})_r$ cal/mol·degree
$C_{20}H_{40} \rightleftharpoons 2C_{10}H_{20}$	18.60	34.24
$\rightleftharpoons C_5H_{10} + C_{15}H_{30}$	18.63	34.26
$C_6H_{12} \rightleftharpoons 2C_3H_6$	18.61	33.61

The values are practically identical to those of alkanes having the same structure.

The reaction of 1-dodecyl-hexene to 1-octene is taken as typical for reactions in liquid phase.

$$C_{16}H_{32} \rightleftharpoons 2C_8H_{16} \quad \begin{array}{l} (\Delta H^0_{298})_{rl} = 18.30 \text{ kcal/mol} \\ (\Delta S^0_{298})_{rl} = 24.40 \text{ cal/mol·degree} \end{array}$$

Since the thermodynamic values are practically identical with those corresponding to n-alkanes, the graph of Figure 6.1 is valid also for the equilibrium of the decomposition reactions of alkenes.

6.1.3 Alkenes Polymerization

The reactions of polymerization of alkenes are of great importance especially since the produced polymers adsorb on the catalyst and following further condensation and dehydrogenation reactions, leading to the formation of coke.

The available heats of reaction and the variation of the entropies for some representative dimerization reactions are listed in the following table:

	$(\Delta H^0_{800})_r$ kcal/mol	$(\Delta S^0_{800})_r$ cal/mol·degree
$2C_{10}H_{20} \rightleftharpoons n\text{-}C_{20}H_{40}$	−18.60	−34.24
$2C_8H_{16} \rightleftharpoons n\text{-}C_{16}H_{32}$	−18.59	−34.23
$2C_5H_{10} \rightleftharpoons n\text{-}C_{10}H_{20}$	−18.69	−34.29
$2C_4H_8 \rightleftharpoons n\text{-}C_8H_{16}$	−18.69	−33.49
$2C_3H_6 \rightleftharpoons n\text{-}C_6H_{12}$	−18.61	−33.61
$\rightleftharpoons$ 2-methylpentene-1	−20.84	−33.58
$\rightleftharpoons$ 3-methylpentene-1	−18.45	−33.48
$\rightleftharpoons$ 4-methylpentene-1	−19.71	−38.71
$\rightleftharpoons$ 2-methylpentene-2	−23.36	−35.99
$\rightleftharpoons$ 3-methylpentene-2 cis	−22.88	−35.99
$\rightleftharpoons$ 3-methylpentene-2 trans	−23.10	−35.18
$\rightleftharpoons$ 4-methylpentene-2 cis	−20.60	−36.15
$\rightleftharpoons$ 4-methylpentene-2 trans	−21.03	−36.16
$\rightleftharpoons$ 2,3-dimethylbutene-1	−21.05	−36.18
$\rightleftharpoons$ 2,3-dimethylbutene-2	−23.85	−40.54

The first four reactions refer to the dimerization of n-alkenes with double bonds in position 1 to dimers with the same structure. The rest of the reactions refer to the dimerization of propene to all the possible hexane isomers, except those that contain a quaternary carbon atom.

For reactions in the adsorbed layer, one takes into account the thermodynamic data for the liquid phase. The following reaction was selected as representative:

$$2C_8H_{16} \rightleftharpoons C_{16}H_{32}$$

for which the heat of reaction and the variation of the entropy are:

$$(\Delta H^0_{298})_{rl} = -18.30 \text{ kcal/mol}$$

$$(\Delta S^0_{298})_{rl} = -24.40 \text{ cal/mol·degree}$$

All these data were used for plotting the dimerization equilibrium in Figure 6.2.

It must be remarked that the equilibrium conversions of the propene dimerization reactions show important differences depending on the isomer produced. Similar differences or possibly even more pronounced ones should exist for alkenes having a larger number of carbon atoms. This statement cannot be verified directly due to the lack of corresponding thermodynamic constants.

The graph of Figure 6.2 proves that in catalytic cracking (temperatures of about 500°C and pressures lower than 3 bar), the polymerization reactions in vapor phase are practically absent (conversions of about 1%). Concomitantly it is shown that these reactions could achieve conversions of about 50% if they were

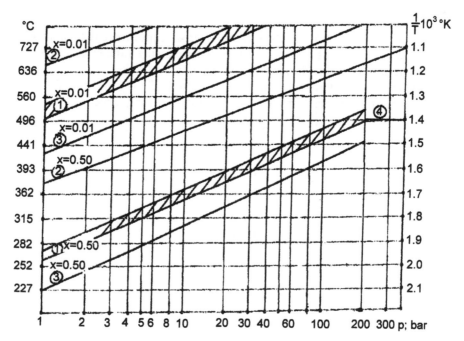

Figure 6.2 Equilibrium of alkenes dimerization. 1 – n-alkene $1C_4$-$1C_{10}$, 2 – $2C_3H_6 \rightarrow 3$-methylpentene 2-trans, 3 – $2C_3H_6 \rightarrow 4$-methylpentene 1. Between 2 and 3 $2C_3H_6$-other isomers, 4 – equilibrium in adsorbed layer or in liquid state, x – conversion in molar fraction.

carried out in liquid phase or at pressures of the order 100 bar. This means that they may take place in the adsorbed layer on the surface of the catalyst.

The polymerization of alkenes is also possible during the catalytic cracking of heavy feedstocks in liquid or partial liquid phase.

6.1.4 Cycloalkanes Decyclization–Alkenes Cyclization

This analysis is limited to the breaking of rings with 5- and 6-carbon atoms, the only present in significant amounts in crude oil fractions.

The reactions that are taken into account and the values of the respective thermodynamic constants are presented in the following table:

	$(\Delta H^0{}_{800})_r$ kcal/mol	$(\Delta S^0{}_{800})_r$ cal/mol·degree
$\square \rightleftharpoons C_5H_{10}(1)$	14.89	15.96
$\square^C \rightleftharpoons$ 2-methyl-pentene-1	14.43	13.38
4-methyl-pentene-1	15.56	14.88
$\bigcirc \rightleftharpoons C_6H_{12}(1)$	20.12	22.68
$\square^{C-C} \rightleftharpoons C_7H_{14}$	16.53	13.30
$\square^{C_3H_7} \rightleftharpoons C_8H_{16}$	16.80	13.30
$\square^{C_{15}H_{31}} \rightleftharpoons C_{20}H_{40}$	16.54	13.35
$\bigcirc^{C} \rightleftharpoons C_7H_{14}$	22.08	19.85
$\bigcirc^{C-C} \rightleftharpoons C_8H_{16}$	21.23	19.49
$\bigcirc^{C_{14}H_{29}} \rightleftharpoons C_{20}H_{40}$	20.91	19.52

Taking into account that on acid catalysts tertiary ions are preferably formed, only the breaking of the bond in the β-position to the tertiary carbon was taken into account when examining the cracking of methyl-cyclopentane.

The lack of thermodynamic data for the higher methyl- and dimethyl-alkenes prevented the use of the same reasoning for the hydrocarbons in the two final groups of the table. In their case the data for *n*-alkene-1 were used, which constitutes of course an approximation.

The following reaction was selected as being illustrative for reactions in liquid phase:

$$\bigcirc^{C_{10}H_{21}} \rightleftharpoons C_{16}H_{32}$$

$(\Delta H^0_{298})_{rl} = 22.69$ kcal/mol

$(\Delta S^0_{298})_{rl} = 18.80$ cal/mol·degree

Since the breaking of the ring takes place without a change in the number of moles, the pressure does not influence the equilibrium. For this reason, x-t coordinates were selected for these equilibria.

Figure 6.3 depicts this equilibrium for some vapor-phase reactions.

Calculations show that in liquid phase reactions, temperatures of 505 and 539°C are required in order to obtain a 1% equilibrium concentration of alkenes, in the conversion of alkyl-cyclopentanes and alkyl-cyclohexanes, respectively. These conversions correspond to those obtained in vapor phase at almost the same temperatures. The graph of Figure 6.3 is actually valid for reactions in both vapor and liquid phases.

From this graph it follows that the naphthenic rings are much more stable than the alkanes and the alkenes chains, the extent of ring breaking having thermodynamic limitations.

The lack of thermodynamic data hinders the extension of this analysis to bi- and polycyclic cyclanes. Cyclization of the alkenes constitutes the reverse reaction to

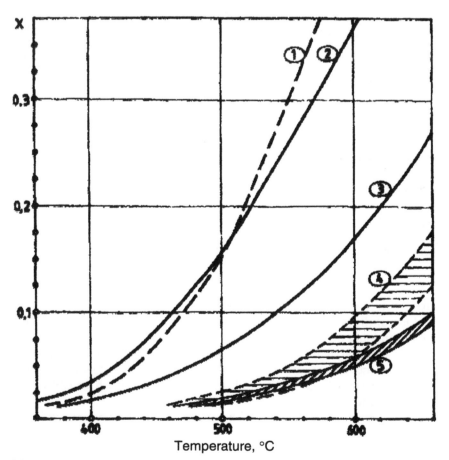

Figure 6.3 Cyclanes decyclization equilibrium. 1 – cyclohexane → hexene 1, 2 – cyclopentane → pentene 1, 3 – methylcyclopentane → 2 or 4 methylpentene 1, 4 – higher-alkylcyclohexanes → alkenes 1, 5 – higher-alkylcyclopentanes → alkenes 1.

the breaking of the rings. Examination of Figure 6.3 allows the formulation of the following conclusions: 1. Thermodynamic calculations indicate that at temperatures of catalytic cracking the cyclization of the alkenes can take place with high conversion, about 80% for cyclization in methylcyclopentane or cyclohexane and up to 94–98% for cyclization in their homologues; 2. The probability of formation for rings with five and six carbon atoms is basically the same.

6.1.5 Dealkylation of Cycloalkanes

Taking into account the reaction mechanism and the available thermodynamic data, heats and the entropies of reaction were calculated for the following reactions:

	$(\Delta H^0_{800})_r$ kcal/mol	$(\Delta S^0_{800})_r$ cal/mol·degree
⬠–C_3H_7 ⇌ C_2H_4 + ⬠–C	21.63	32.53
⬠–C_4H_9 ⇌ C_3H_6 + ⬠–C	18.48	33.86
⬠–$C_{15}H_{31}$ ⇌ $C_{14}H_{28}$ + ⬠–C	18.49	34.17
⬡–C_3H_7 ⇌ C_2H_4 + ⬡–C	20.66	32.25
⬡–C_4H_9 ⇌ C_3H_6 + ⬡–C	17.44	33.61
⬡–$C_{15}H_{31}$ ⇌ $C_{14}H_{28}$ + ⬡–C	17.43	33.88

For reactions in liquid phase, the heaviest hydrocarbon was taken into account for which thermodynamic data were available:

⬡–$C_{10}H_{21}$ ⇌ C_9H_{18} + ⬡–C

$(\Delta H^0_{298})_{rl} = 18.96$ kcal/mol

$(\Delta S^0_{298})_{rl} = 24.02$ cal/mol·degree

The graph of Figure 6.4 is based on these data.

In order to correctly compare the equilibrium conversions in the two phases, the equilibrium conversion in gaseous phase for the same hydrocarbon and reference temperature was also calculated:

⬡–$C_{10}H_{21}$ ⇌ C_9H_{18} + ⬡–C

$(\Delta H^0_{298})_r = 18.78$ kcal/mol

$(\Delta S^0_{298})_r = 36.49$ cal/mol·degree

The insignificant differences between these values and the thermodynamic constants for 800 K, which were calculated before, prove that no errors were introduced when, due to the lack of thermodynamic data for the liquid state at 800 K different reference temperatures for the equilibrium in the liquid phase (298 K) and in the vapor phase (800 K) were used.

Examination of Figure 6.4 allows the conclusion that according to thermodynamic calculations the dealkylation of cycloalkanes can be carried out to completion in vapor phase at the temperatures and pressures of catalytic cracking but it is limited to a conversion of about 70% for the hydrocarbons adsorbed on the catalyst or for reactions in liquid phase.

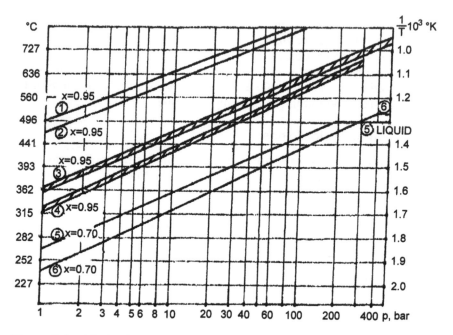

Figure 6.4 Alkylcyclanes dealkylation equilibrium.

6.1.6 Dehydrogenation of Cyclohexanes

The following reactions were taken into account:

	$(\Delta H^0{}_{800})_r$ kcal/mol	$(\Delta S^0{}_{800})_r$ cal/mol·degree
$\bigcirc \rightleftharpoons \bigcirc + 3H_2$	52.70	96.31
$\bigcirc^{-C} \rightleftharpoons \bigcirc^{-C} + 3H_2$	51.75	94.89
$\bigcirc^{-C-C} \rightleftharpoons \bigcirc^{-C-C} + 3H_2$	50.82	94.82
$\bigcirc^{-C_3} \rightleftharpoons \bigcirc^{-C_3} + 3H_2$	50.57	95.25

Reaction	(ΔH^0)	(ΔS^0)
cyclohexane–C_5 ⇌ benzene–C_5 + $3H_2$	50.14	95.32
cyclohexane–C_{14} ⇌ benzene–C_{14} + $3H_2$	50.15	95.34

There are only minimum differences between the thermodynamic constants of these reactions. Therefore the graph for the dehydrogenation equilibrium of methyl-cyclohexane given in Figure 2.7 can be considered as representating all these reactions also. Thus, in the conditions of catalytic cracking the equilibrium is completely shifted towards dehydrogenation.

For the dehydrogenation of butyl-cyclohexane in liquid phase, one obtains:

$$cyclohexane\text{–}C_4H_9 \rightleftharpoons benzene\text{–}C_4H_9 + 3H_2$$

$$(\Delta H^0_{298})_{rl} = 47.63 \text{ kcal/mol}$$
$$(\Delta S^0_{298})_{rl} = 87.95 \text{ cal/mol·degree}$$

These values are similar to those for the vapor phase reactions and, accordingly also here, the equilibrium is completely displaced towards dehydrogenation. In fact, the escaping into the vapor phase of the hydrogen formed in the liquid phase reaction displaces this equilibrium further to the right.

6.1.7 Dealkylation of Alkylaromatics

The following reactions were taken into consideration:

Reaction	$(\Delta H^0_{800})_r$ kcal/mol	$(\Delta S^0_{800})_r$ cal/mol·degree
benzene–C–C ⇌ benzene + C_2H_4	24.48	29.31
benzene–C_3 ⇌ benzene + C_3H_6	21.70	30.24
benzene–C_5 ⇌ benzene + C_5H_{10}	21.99	30.50
benzene–C_{14} ⇌ benzene + $C_{14}H_{28}$	21.95	30.47
benzene–C_3 ⇌ benzene–C + C_2H_4	21.84	31.89
benzene–C_5 ⇌ benzene–C + C_4H_8	19.07	33.08
benzene–C_{14} ⇌ benzene–C + $C_{13}H_{26}$	19.02	33.45
benzene–C_5 ⇌ benzene–$C{=}C$ + C_3H_8	17.97	31.35
benzene–C_{14} ⇌ benzene–$C{=}C$ + $C_{12}H_{26}$	17.65	31.95

The following reactions were selected as representative of the chemical transformations in liquid phase:

Reaction	$(\Delta H^0_{298})_{rl}$ kcal/mol	$(\Delta S^0_{298})_{rl}$ cal/mol·degree
benzene–C_{10} ⇌ benzene + $C_{10}H_{20}$	21.57	18.75
benzene–C_{10} ⇌ benzene–C + C_9H_{18}	18.82	22.43
benzene–C_{10} ⇌ benzene–$C{=}C$ + C_8H_{18}	16.67	18.77

For the liquid phase the values are less accurate than for the gas phase. The entropy value $S°_{298} = 124.24$ cal/mol·degree for decyl benzene was obtained by extrapolation. Since this value is the same in calculations for all the examined reactions, the comparative results are not affected.

All these values were used for plotting the equilibrium parameters for the considered reactions in Figure 6.5.

The examination of this graph leads to the conclusion that the dealkylation reactions with the formation of benzene and alkene are thermodynamically less probable that those that lead to the formation of toluene and alkene or styrene and alkane.

If the last two reactions were carried out in vapor phase, at the conditions of temperature and pressure for catalytic cracking, the equilibrium conversion would reach between 95% and 99%.

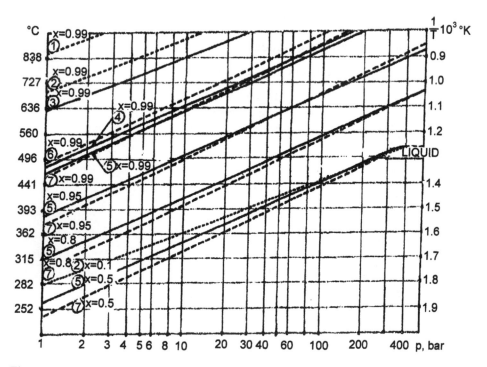

Figure 6.5 Alkyl-aromatics dealkylation equilibrium.

x – conversion in molar fraction.

In liquid phase, which is equivalent to pressures of the order of 200–300 bar when operating in vapor phase, the conversions at equilibrium are 50% for dealkylation with formation of toluene or styrene and only 10% if benzene is formed.

From here, an important conclusion is that dealkylation produces styrene and derivatives adsorbed on the catalyst, which may be important coke precursors.

6.1.8 Dealkylation of Polycyclic Hydrocarbons

The only available thermodynamic data are for alkyl-naphthalenes. On their basis, the following reactions may be analyzed:

	$(\Delta H^0_{800})_r$ cal/mol	$(\Delta S^0_{800})_r$ cal/mol·degree
(naphthalene-C_5H_{11}) $\rightleftharpoons$ (naphthalene-CH_3) $+ C_4H_8$	18.80	32.41
$\rightarrow$ (naphthalene) $+ C_5H_{10}$	21.61	31.28
(naphthalene-C_5H_{11}) $\rightleftharpoons$ (naphthalene-CH_3) $+ C_4H_8$	18.80	32.40
$\rightarrow$ (naphthalene) $+ C_5H_{10}$	21.97	31.60

The data show that the position of the alkyl group exercises little influence upon thermodynamic properties.

The calculations show that at 490°C and atmospheric pressure the conversions reach 99% if methyl-naphthalenes are formed and of 90% if dealkylation results in the formation of naphthalene.

As in the case of alkyl-benzenes, thermodynamics favor the formation of methyl-naphthalene derivatives. Also, as in the case of alkyl-benzenes, the formation of hydrocarbons with unsaturated side chains similar to styrene is expected. No data is available for confirming this.

6.1.9 Cracking of S-Containing Compounds

6.1.9.1 Decomposition of Sulfides

The sulfides are decomposed according to the reaction:

Sulfide $\rightleftharpoons$ Mercaptan + alkene

The following reactions were considered and the parameters needed for thermodynamic analysis were calculated:

	$(\Delta H^0_{800})_r$ cal/mol	$(\Delta S^0_{800})_r$ cal/mol·degree
$CH_3-S-C_{19}H_{39} \rightleftharpoons CH_3SH + C_{19}H_{38}$	17.54	33.59
$CH_3-S-C_5H_{11} \rightleftharpoons CH_3SH + C_5H_{10}$	17.57	33.63
$CH_3-S-C_4H_9 \rightleftharpoons CH_3SH + C_4H_8$	17.57	33.21
$CH_3-S-C_3H_7 \rightleftharpoons CH_3SH + C_3H_6$	17.80	33.91

$$CH_3-S-C_2H_5 \rightleftharpoons CH_3SH + C_2H_4 \qquad 20.53 \qquad 32.51$$
$$C_2H_5-S-C_2H_5 \rightleftharpoons C_2H_5SH + C_2H_4 \qquad 20.67 \qquad 33.99$$
$$C_3H_7-S-C_3H_7 \rightleftharpoons C_3H_7SH + C_3H_6 \qquad 17.44 \qquad 35.01$$
$$C_4H_9-S-C_4H_9 \rightleftharpoons C_4H_9SH + C_4H_8 \qquad 17.97 \qquad 35.22$$
$$C_5H_{11}-S-C_5H_{11} \rightleftharpoons C_5H_{11}SH + C_5H_{10} \qquad 18.06 \qquad 35.96$$
$$C_{10}H_{21}-S\cdot\cdot C_{10}H_{21} \rightleftharpoons C_{10}H_{21}SH + C_{10}H_{20} \qquad 18.03 \qquad 35.92$$

The plots of Figure 6.6 are based on these values. The thermodynamic results indicate that the conditions of catalytic cracking favor the complete decomposition of sulphides to mercaptans and alkenes.

6.1.9.2 Decomposition of Mercaptans

The following reactions were selected for analyzing the behavior of mercaptans:

	$(\Delta H^0_{800})_r$ cal/mol	$(\Delta S^0_{800})_r$ cal/mol·degree
$C_2H_5SH \rightleftharpoons H_2S + C_2H_4$	15.50	30.67
$C_3H_7SH \rightleftharpoons H_2S + C_3H_6$	12.66	31.63
$C_5H_{11}SH \rightleftharpoons H_2S + C_5H_{10}$	12.30	31.32
$C_{20}H_{41}SH \rightleftharpoons H_2S + C_{20}H_{40}$	12.25	31.29

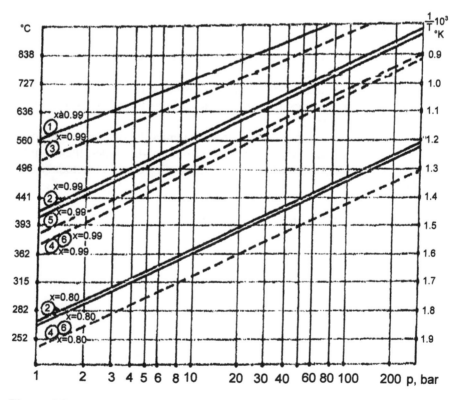

Figure 6.6 Equilibriums of sulfites decomposition to mercaptans and alkenes. The sulfites: 1. $CH_3-S-C_2H_5$; 2. $CH_3-S-C_3H_7$- - -$CH_3-S-C_{19}H_{39}$; 3. $(C_2H_5)_2S$; 4. $(C_3H_7)_2S$; 5. $(C_4H_9)_2S$; 6. $(C_5H_{11})_2S$- - -$(C_{10}H_{21})_2S$. x – conversion in molar fraction.

Following the same procedure as in the previous cases, the parameters for thermodynamic equilibrium were plotted in Figure 6.7.

From the plotted data it results that during the catalytic cracking, the thermodynamics favor the complete conversion of mercaptans to alkenes and hydrogen sulfide.

6.1.9.3 Decomposition of the Cyclic Compounds With Sulfur

The available thermodynamic data allow the examination of the thermodynamic equilibrium of the following reactions:

	$(\Delta H^0{}_{800})_r$ cal/mol	$(\Delta S^0{}_{800})_r$ cal/mol·degree
$\begin{smallmatrix}S\\\square\end{smallmatrix} \rightleftharpoons H_2S + 1.3C_4H_6$	30.48	44.31
$\begin{smallmatrix}S\\\hexagon\end{smallmatrix} \rightleftharpoons H_2S + 1.3C_5H_8$	28.24	48.93

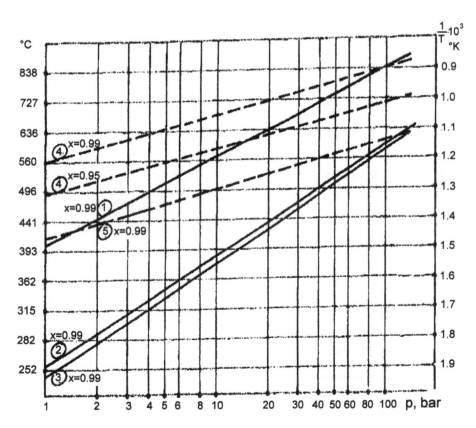

Figure 6.7 Mercaptans and cyclic sulfur compounds decomposition equilibrium. 1. $C_2H_5SH \rightleftharpoons C_2H_4 + H_2S$; 2. $C_3H_7SH \rightleftharpoons C_3H_6 + H_2S$; 3. C_nH_{2n+1} $SH \rightleftharpoons C_nH_{2n} + H_2S$ for $n = 5$–20; 4. $\begin{smallmatrix}S\\\square\end{smallmatrix} \rightleftharpoons 1,3\ C_4H_6 + H_2S$; 5. $\begin{smallmatrix}S\\\hexagon\end{smallmatrix} \rightleftharpoons 1,3\ C_5H_8 + H_2S$; x – conversion in molar fraction.

The results of the calculations made on the basis of these data are presented in the graph of Figure 6.7.

It results that the cyclic compounds with sulfur are much more stable than the aliphatic compounds. Thiophen and its derivatives are even more stable.

6.1.10 Cracking of Nitrogen-Containing Compounds

6.1.10.1 Decomposition of Primary Amines

The following reactions were considered:

	$(\Delta H^0{}_{800})_r$ cal/mol	$(\Delta S^0{}_{800})_r$ cal/mol·degree
$C_2H_5NH_2 \rightleftharpoons NH_3 + C_2H_4$	12.42	30.31
$C_3H_7NH_2 \rightleftharpoons NH_3 + C_3H_6$	10.76	31.61
$C_4H_9NH_2 \rightleftharpoons NH_3 + C_4H_8$ (1)	10.58	31.59
sec-$C_4H_9NH_2 \rightleftharpoons NH_3 + C_4H_8$ (1)	13.40	34.36
$\approx NH_3 + C_4H_8$ (2)	11.05	31.77
$tert$-$C_4H_9NH_2 \rightleftharpoons NH_3 + $ i-C_4H_8	12.71	33.88

These data and the plots of Figure 6.8, indicate that from a thermodynamic point of view, the primary amines are very reactive and may decompose completely at temperatures as low as 200–300°C.

6.1.10.2 The Decomposition of Diethyl- and Triethyl-amine

The following reactions were considered:

	$(\Delta H^0{}_{800})_r$ cal/mol	$(\Delta S^0{}_{800})_r$ cal/mol·degree
$\begin{array}{c}C_2H_5\\\end{array}NH \rightleftharpoons C_2H_5NH_2 + C_2H_4$	18.26	35.49
$\begin{array}{c}C_2H_5\\C_2H_5-N\\C_2H_5\end{array} \rightleftharpoons (C_2H_5)_2NH + C_2H_4$	18.25	38.40

These compounds are slightly more stable than the primary amines. However, their complete decomposition (99%) is thermodynamically possible at temperatures of 385°C to 322°C (Figure 6.8).

6.1.10.3 The Decomposition of Pyridine

Pyridine is decomposed according to the reaction:

	$(\Delta H^0{}_{800})_r$ cal/mol	$(\Delta S^0{}_{800})_r$ cal/mol·degree
$\text{N} \rightleftharpoons NH_3 + 1.3 C_4H_6$	18.03	42.85

According to these data, the decomposition of pyridine is thermodynamically possible with essentially complete conversion at temperatures of above 240°C (Figure 6.8).

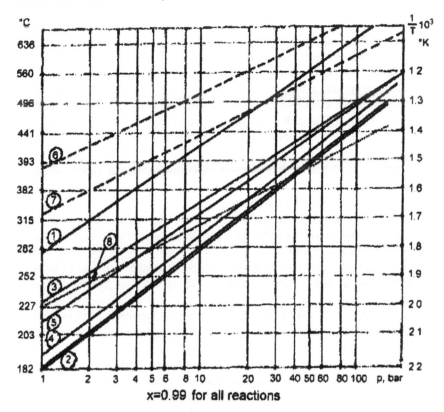

x=0.99 for all reactions

Figure 6.8 Nitrogen compounds decomposition equilibria. 1. $C_2H_5NH \rightleftharpoons C_2H_4 + NH_3$; 2. $C_3H_7NH_2 \rightleftharpoons C_3H_6 + NH_3$; or $C_4H_9NH_2 \rightleftharpoons C_4H_8(1) + NH_3$; 3. sec-$C_4H_9NH_2 \rightleftharpoons C_4H_8(1) + NH_3$; 4. sec-$C_4H_9NH_3 \rightleftharpoons C_4H_8(2) + NH_3$; 5. tert-$C_4H_9NH_2 \rightleftharpoons$ i-$C_4H_8 + NH_3$; 6. $(C_2H_5)_2NH \rightleftharpoons C_2H_5NH_2 + C_2H_4$; 7. $(C_2H_5)_3N \rightleftharpoons (C_2H_5)_2NH + C_2H_5$; 8. ⬡$\rightleftharpoons 1,3$ $C_4H_6 + NH_3$. x – conversion in molar fraction.

6.1.11 Conclusions

The above analysis of the reactions taking place in catalytic cracking allow the formulation of several conclusions on the directions of the thermodynamically possible transformations of substances in vapor phase, and of those which are adsorbed on the catalyst and lead to the gradual formation of coke.

Concerning the reaction in vapor phase, simple conclusions can be made on basis of the graphs of Figures 6.1–6.8, which take into account the process conditions in catalytic cracking reactors: temperatures ranging from 470–520°C, and pressure slightly above atmospheric.

In these conditions, thermodynamics do not limit the decomposition of alkanes and alkenes. The thermodynamic probability of decomposition is actually the same for both classes of hydrocarbons.

Also, neither the breaking off of alkylic chains attached to aromatic rings, nor the dehydrogenation of the cyclo-alkane rings of six carbon atoms to aromatic rings are limited from the thermodynamic point of view.

The cracking of the cyclo-alkane rings has a low thermodynamically probability and occurs with equilibrium conversions of only 3–15%. On the other hand, reaction conditions favor the dehydrogenation of cyclo-alkanes with 6 carbon atoms. The cyclo-alkane rings of 5 atoms will be either isomerized and dehydrogenated, or cracked. Therefore the cracking occurs with preference for the 5 carbon atoms rings.

The decomposition of alkane-thiols, dialkyl-sulphides, and alkyl-amines is not limited by thermodynamics. The sulfur-containing heterocyclic compounds and the dialkyldisulfides seem to be more resistant, but the available thermodynamic data are not sufficient for a satisfactory analysis.

The polymerization of the alkenes is essentially impossible in vapor phase, in catalytic cracking conditions. It can take place only in liquid phase or in the adsorbed layer on the surface of the catalyst.

In all the cases where the reactions are not limited by thermodynamics, the conversions of various compounds will be determined by the relative reaction rates, that is, by the kinetics of the process.

All these conclusions are in agreement with those expressed previously in other works [1,4,5] and have as their purpose only to complete and emphasize some quantitative aspects.

More complex and so far less explained is the thermodynamics of the reactions within the layer adsorbed on the catalyst. They lead to substances which are not desorbed but generate coke.

Thermodynamics show that the cracking of alkanes and alkenes, that are adsorbed on the catalyst can not exceed a conversion of 60% while the polymerization of alkenes can proceed with conversions of up to 50%. The breaking off of the side chains to the cyclo-alkanes with formation of alkenes can reach conversions of 70%. The de-alkylation of alkyl-aromatic hydrocarbons in the adsorbed layer can reach conversions of 50%. The breaking-off of side chains with formation of benzene has low thermodynamic probability. The dealkylation in the case of aromatics with unbranched side-chains leads, with thermodynamically equal chances, to the formation of toluene or of styrene. The dehydrogenation of cyclo-alkanes can exceed conversions of 99%.

These reactions that are fundamental different than those occurring in vapor phase explain fully the thermodynamically favored formation (by polymerization, dehydrogenation, and condensation reactions) of hydrocarbons with high molecular mass that do not desorb from the surface of the catalyst and lead eventually to the formation of coke.

The assumption made in the above deductions and calculations that the conditions within the adsorbed layer and those within the liquid phase are similar to those for vapors at pressures of the order of 100–200 bar seem realistic. In fact, the graphs show that the equilibrium conversions in the liquid phase are identical with those in vapor phase at pressures of the order 200–400 bar. The exact value depends on the reaction taken into consideration.

The thermodynamic analysis of the phenomena occurring in the adsorbed layer could be more detailed if thermodynamic data were available for bi- and polycyclic aromatic and naphthen-aromatic hydrocarbons and other more complex structures.

Thermodynamic analysis of the catalytic cracking of residues is much more difficult, first of all because no the thermodynamic constants are available for specific compounds contained in such fractions.

The fact that in this case the reactions take place in liquid phase allows one to extend to catalytic cracking of the residues the conclusions previously established concerning the direction of the reactions and the thermodynamic limitations that are specific to the reactions in liquid phase and in the adsorbed layer.

Thus, different from the processes in vapor phase, the decomposition reactions of the aliphatic hydrocarbons and the dealkylation of the cyclic ones will encounter thermodynamic limitations corresponding to conversions of 50–70%. The polymerization of alkenes can take place with conversions up to 50%. The dehydrogenation and possibly condensation reactions will be probably favored.

6.2 CRACKING CATALYSTS

The catalytic activity of aluminum chloride in the cracking of crude oil fractions was established in 1915–1918 by A.M. McAfee in the U.S. [16] and simultaneously by N.D. Zelinskyi in Russia [7]. Following the construction of a pilot plant at Kuzovsk [8] in the years 1919–1920, the process was abandoned owing to the excessive consumption of aluminum chloride [9].

The catalytic action of clays was discovered in 1911 by Ubbelhode and Voronin [10] and was followed by the implementation in 1928 by A.J. Houdry of their acid activation [6]. The difficulties related to the large deposits of coke on the catalyst delayed commercialization until 1936, by A.J. Houdry for Socony-Vacuum Oil. The coking problem was solved by the cyclic regeneration of the catalyst by means of burning the coke deposited thereon.

From that date on, cracking catalysts knew a rapid evolution, marked by the development and by the application at commercial scale of steadily improved types of catalysts:

1936	The use of activated natural clays
1940	The first catalyst of synthetic silica alumina (Houdry Process Corporation)
1946	First time use of microspherical catalysts
1950	Development and general use (1956) of catalysts with more than 25% Al_2O_3
1958	Development and commercialization (1960) of catalysts with 25–35% kaolin incorporated in the silica-alumina
1959	The synthesis and commercialization of Y zeolites
1962	First time use of zeolitic catalyst in catalytic cracking
1964	Inclusion of zeolite in a matrix
1964	Development of ultrastable catalysts (USY) and of those promoted with rare earths (REY)
1974	Additives of promoters for the combustion of CO to CO_2 in regenerator
1974	Additives for the fixation of SO_2 in the regenerator and its elimination as H_2S, in the reactor
1975	Catalyst passivation against nickel poisoning
1978	Catalyst passivation against vanadium poisoning
1983	Performance improvement by treating the catalyst with $(NH_3)_2SiF_6$
1986	Use of ZSM-5 – type additives for octane number enhancing
1988	Silicon enrichment by means of silicones (catalysts of the type LZ 210)[11]
1992	The experimentation in the U.S. and in Europe of the ALPHA and BETA catalysts.

Simultaneously, starting in 1980, in the U.S., residues were incorporated in catalytic cracking feed. Thus, in 1989–1990 the feed to FCC units could have 5% Conradson carbon and 10 ppm Ni + V [12]. These limits were almost doubled during the following years.

The catalytic cracking of residues poses special problems for the catalysts, mainly with reference to pore structure, passivation against poisonings, etc. The specific issues of these catalysts will be treated separately towards the final part of this chapter.

6.2.1 Activated Natural Clays

Despite the fact that activated natural clays have not been used as cracking catalysts for a long time, awareness of their characteristics is important since natural clays continue to be included in the composition of synthetic catalysts in order to reduce their cost. This technique, initially used in the production of silica alumina catalysts, is used today on a large scale in the production of zeolitic catalysts.

Montmorillonites, were the first natural clays used as cracking catalysts. They were activated by treating with diluted sulfuric acid or hydrochloric acid in order to increase the specific surface and the porosity. In the same time, the alkaline metals were eliminated together with a portion of the iron, which by reacting with sulfur compounds present in the feedstock, would decrease the activity of the catalyst [13].

Despite all the improvements brought to the activation including the use of special treatments [1], the stability of montmorillonite catalysts is insufficient and they were replaced by kaolin-based catalysts.

The kaolins, with a content of max. 2.5% Fe_2O_3, are activated by calcination in reducing medium, often in a fluidized bed in an ascendant current of flue gases, generator gas, or a mixture of methane and steam [1].

Catalysts obtained in this way have satisfactory stability and mechanical resistance.

6.2.2 Synthetic Silica-Aluminas

Silica is completely deprived of catalytic activity. The activity appears and increases after aluminum atoms are incorporated in its structure [11]. The Si–O–Al bonds that are formed confer acidity and therefore activity to the catalyst.

The first synthetic catalysts produced contained about 13% Al_2O_3.

The desire to increase the activity and the stability of the catalysts led to the production, at the end of the year 1950 of catalysts with about 25% Al_2O_3 content.

In the same time, catalysts of magnesia-silica were developed that produced higher gasoline yields with a lower octane number.

A comparison between the performances of natural catalysts and of classic synthetic ones containing aluminum or magnesium is given in Table 6.1 [14].

In order to reduce the price of the catalyst, kaolin was incorporated in its structure by dispersing 25–35% kaolin in the gel of silica-alumina. Such catalysts, called semisynthetic, were cheaper but were less active and had lesser mechanical resistance. Besides, a significant amount of unreacted products remained adsorbed on the catalyst after stripping and had to be burnt in the regenerator, which

Table 6.1 Pilot Plant Fluid Catalytic Cracking at Some
Conversion on Natural and Synthetic Catalysts

Yields and gasoline quality	Synthetic catalysts		Natural catalysts	
	13% Al_2O_3	Mg-Si	Filtrol 58	Filtrol SR
C_1C_3, wt %	6.4	5.0	7.0	6.8
C_4H_{10}, vol %	9.0	5.1	6.9	7.8
C_4H_8, vol %	7.0	4.9	7.1	6.3
Gasoline, vapor tension, 517 mm, vol %	46.9	57.2	49.3	49.0
Gas oil, vol %	40.0	40.0	40.0	40.0
Coke, wt %	3.4	3.4	3.6	3.4
Gasoline ON research	93.7	90.3	89.9	92.8
Gasoline ON motor	81.0	78.6	79.2	80.5

Feed: gas oil, $d = 0.882$.

decreased the process performance. A catalyst of such a type (Sm-3-S/S) was commercialized by the firm Davidson in 1958 [14].

Socony-Mobil Co. used additions of chrome oxide for activating coke burning, a process used on large scale in the following years for the activation of zeolitic catalysts.

Several methods for the production of the synthetic silica-alumina catalysts were used [15,17]. One of the methods performs the coprecipitation of sodium silicate by soluble aluminum salts, usually aluminum sulfate. The two solutions mix by converging into a Y-junction, then discharged into a heated oil bath, obtaining a catalyst shaped as balls, or the mixture is pulverized in an oil solution to produce a microspherical catalyst.

The following steps are washing, elimination of the Na^+ ion, drying, and calcination. Details concerning this method were described earlier by the author [1].

In a variation to this method, the silica-alumina gel formed from mixing the two solutions is aged, followed by filtration, washing, and ion exchange for removing the Na^+ ion. A predrying step follows, during which various promoters are added: metal promoter for regeneration and hydrofluoric acid, or fluorides for increasing the activity [4]. Finally, the microspheric shape is obtained by the atomization of the catalyst slurry in a rising flow of warm air.

Still, another method impregnates the silica hydrogel with a solution of aluminum sulfate followed by hydrolysis and the precipitation of aluminum with an aqueous ammonia solution. The resulted silica-alumina hydrogel is washed, dried, and calcined [18].

The catalysts obtained by all the above methods have a content of about 13% Al_2O_3. An increase of the Al_2O_3 up to 25–30% is obtained by impregnation with soluble aluminum salts, precipitation with ammonia, washing, drying, and calcination.

6.2.3 Nature of Acid Sites

From the facts presented in the previous paragraph one may deduce that the catalytic activity of synthetic silica-alumina is due to the simultaneous presence of aluminum oxide and silicon oxide in a structure obtained by concomitant precipitation or by the formation in other ways of a mixed gel that contains both oxides.

The silicon dioxide has a tetrahedral structure with the oxygen atoms occupying the vertices of the tetrahedron, while those of silicon, the centers. The inclusion in the network of aluminum atoms, by the substitution of some of the tetravalent silicon atoms by trivalent aluminum atoms results in negative charges at the aluminum atoms. These are compensated by sodium cations, which were contained in the salts used for preparing the hydrogel (Figure 6.9).

The sodium cations are exchanged by ammonium cations. During the calcination step, the latter are decomposed, releasing ammonia, while the protons remain in the oxide lattice and constitute the acid centers, which are catalytically active.

The acid properties of the silica-alumina catalysts were first noticed by Gayer in 1933, while the explanation of the inclusion of the aluminum atoms in the tetrahedral network of silicon dioxide was given by Ch. Thomas in 1949 [19]. The acid character of natural clays was demonstrated as early as in 1891 by V.I. Vernadski [20].

Despite the agreement that prevails concerning the acid character of the silica-alumina catalysts, the structures that confer this character are still debated.

Several of the proposed representations for the acid sites of silica-aluminas are depicted in Figure 6.10. The representation of Figure 6.10a corresponds to the structure imagined by Ch. Thomas, described previously but which was not confirmed by spectroscopic studies.

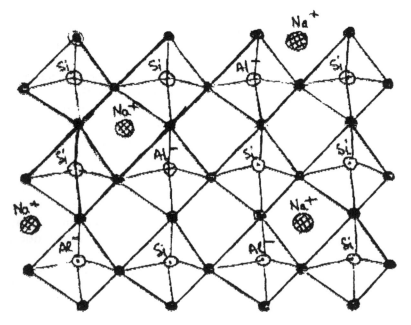

Figure 6.9 Alumo-silica gel structure. ● = Oxygen atoms.

Much more probable are considered the crypto-ionic forms, isomers of structure **a** of Figure 6.10, in which the aluminum and the silicon atoms or only the silicon atom are bound to acid hydroxyl groups (Figures 6–10b and c), susceptible to behave as hydrogen donors, thus as acids of the Brönsted type. The electronic deficiencies of the trivalent aluminum atoms contained in structures of the type of Figure 6.10c can generate acid centers of the Lewis type, Figure 6.10d. There are other representations, related to defects in the network of tetrahedrons, which explain the presence of the acid centers on the surface of silica-alumina [4], and still other representations concerning the possible structures of the acid centers [8].

The existence of both types of centers was experimentally proven by using the chemisorption of 2,6-dimethyl-pyridine and of the Hammett indicators, with the determination of the corresponding adsorption isotherms [21]. The fact that the 2,6-dimethyl-pyridine is adsorbed preferably on Brönsted centers and after heating to 380°C remains adsorbed exclusively on these centers, whereas the Hammett indi-

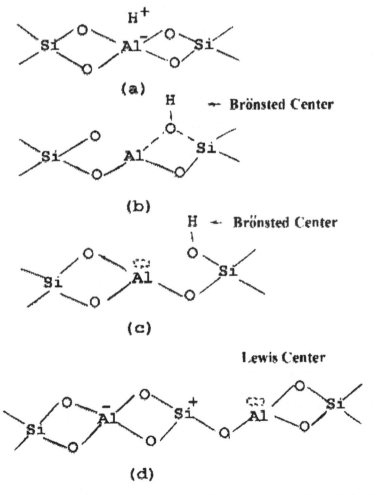

Figure 6.10 Opinions about alumo-silica acid centers. (From Ref. 4.)

cator is adsorbed on both (Lewis and Brönsted) types of acid centers, made possible their separate identification. From the adsorption isotherms the distribution of the acidity of the centers was determined. The conclusion was reached that both types of acid centers, are present: the Brönsted proton centers show a broader distribution of the acidity, while the aprotic centers of the type Lewis have a narrower one.

6.2.4 Zeolite Catalysts

Catalysts containing Y zeolites were first introduced in 1962 by Mobil Oil Co. and found widespread utilization. They displaced completely the natural and synthetic catalysts used previously*.

The properties and the performance of zeolitic catalysts has undergone continuous improvements with respect to the synthesis of zeolites, to the matrices used, to the additives, and to improvements by means of treatment of the produced catalyst. This activity goes on at a fast pace and is the object of a considerable number of publications and patents.

The characteristics of the zeolitic catalysts, the methods used for their improvement, and the additives used are closely tied to the nature of the process and the operation conditions of the fluid catalytic cracking. For a more ample documentation in the domain of the zeolites, the monographs published by J. Scherzer [6] and B.C. Gates [22] are recommended.

The catalysts for catalytic cracking on the basis of zeolites contain several components:

The zeolite Y, usually together with rare earths oxides

The matrix which may be inert or catalytically active

The promoters and additives that improve the performance of the catalyst and that may be introduced during the actual synthesis of the zeolite or of the matrix.

6.2.4.1 The Zeolite

The zeolite is mainly responsible for the activity, selectivity, and stability of the catalyst.

Among the synthetic zeolites the faujasites X and Y, the ophertites, the mordenites, and the erionites showed catalytic activity in the cracking process. Among these only the first two, but especially the faujasites Y which has a superior stability, have been used for the production of cracking catalysts. Thus, a catalyst of the classic type such as faujasite Y, which was submitted to ionic exchange with rare earths, preserves its crystalline structure after treatment during 12 hours with 20% steam at a temperature of 825°C, whereas a faujasite X prepared in the same way loses its crystalline structure after this treatment [11].

The basic structure of the zeolites X and Y is, as in classic synthetic catalysts, formed by groups of tetrahedrons with the aluminum and silicon atoms occupying their centers and the oxygen atoms the vertexes. In the case of zeolites, the groups of tetrahedrons form in fact regular structures of stumped octahedrons, known under

---

* In 1968, 85% of the catalytic cracking plants in the U.S. and Western Europe used catalysts on the basis of zeolites. Today, zeolites are used exclusively.

the name of *sodalite cage*. Each sodalite cage has 8 hexagonal sides, constructed of 6 tetrahedrons and 6 square sides constructed of 4 tetrahedrons, 24 vertexes and 36 edges (see Figure 6.11a).

The A-type molecular sieves have the basic element formed of four sodalite structures, the square faces of which are joined by prisms (see Figure 6.11b). In the type X- and Y-zeolites, the basic element is formed of six sodalite structures, the hexagonal faces of which are joined by prisms (see Figure 6.11c).

As a result of this difference in their structures, the cavities inside the X and Y zeolites have a mean diameter of 13 Å that communicate with the outside through 7.4 Å openings, whereas the type A molecular sieves have the inlet openings of 3 Å for the potassium form and 4 Å for the sodium form.

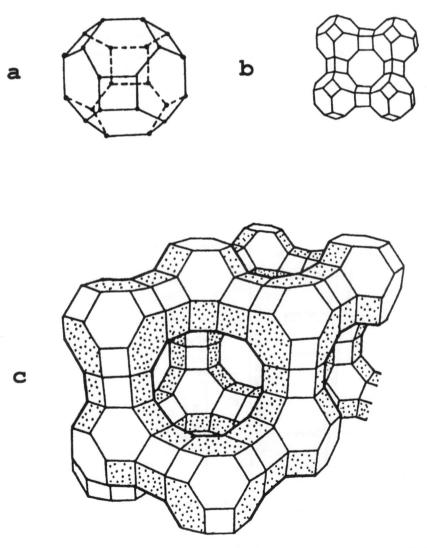

Figure 6.11   Structure of zeolites. a – sodalite, b – molecular sieve A, c – faujasite.

The inlet orifices in the X- and Y-zeolites allow the access of naphthalene molecules, whereas the type A molecular sieves allow exclusively the access of n-alkanes. Thus, basically the hydrocarbons contained in the atmospheric gas oil and the vacuum gas oil have access to the acid centers situated inside the cages of the X- and Y-zeolites.

Figure 6.12 [23] shows the sizes of the cavities and of the access openings of various types of zeolites, together with the necessary size for access through the pores of different types of hydrocarbons. This allows a more detailed examination of the size restrictions that can occur.

The raw formula of an elementary cell of a X- or Y-zeolite can be written as:

$$Na_n\left[(Al_2O_3)_n(SiO_2)_{192-n}\right] \cdot mH_2O$$

wherein m is approx. 250–260, and n has values ranging from 48–76 for Y-zeolites and between 77–96 for X-zeolites.

The study of zeolites by X-ray diffraction and by nuclear magnetic resonance, allowed the determination of the location of the cations that compensate the negative charges of the AlO_4-tetrahedrons. These locations are shown in Figure 6.13 [13]. Four types of acid sites exist, depending on the location they occupy within the crystalline structure of the faujasite: Sites I are situated inside the hexagonal prisms and I' are situated inside the sodalite cages; sites II are situated inside the internal cavity of the zeolite in the proximity of the free hexagonal opening, while II' is located farther from these orifices.

It is easy to ascertain that sites II and II' being easy accessible, are strongly involved in the catalytic reactions, whereas sites I and I', being accessible with difficulty, have a more reduced catalytic role.

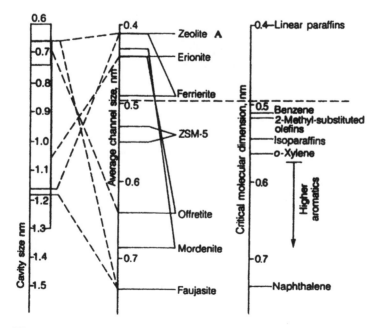

Figure 6.12 Cavities and orifices size, compared to molecules size (From Ref. 23.)

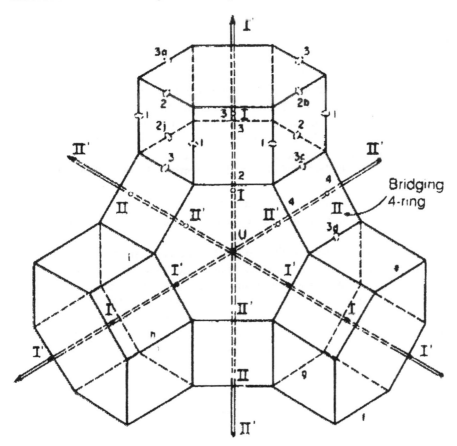

Figure 6.13   Positions of acid centers within elementary faujasite cell.

The synthesis of the X- and Y-zeolites is the object of a large number of publications and patents [24–35]; the examination of these methods of production exceeds the frame of the present book.

In principle, the zeolites of type Y are obtained by coprecipitation, starting from solutions of sodium silicate and aluminate in a strong basic medium (pH = 12–13) of the aluminosilicate gel. Maturation takes place during 5–10 hours at temperatures of 100–125°C, followed by filtration, washing, and drying at a temperature of about 150°C. The characteristics of the obtained zeolite depend to a large extent on the initial concentration of the solution of sodium silicate, on the amount of acid added to cause the gelation, on the manner and the conditions in which the aluminum salts are added, as well as on the conditions, the duration, and the maturation temperature.

Finally, the zeolite is obtained as sodium salts, which, contrary to the amorphous synthetic alumo-silica catalysts, possess an intrinsic catalytic activity. Ion exchange is necessary in order to increase activity.

Ion exchange for replacing sodium by ammonium, followed by its decomposition with elimination of ammonia, as practiced in the case of amorphous synthetic catalysts, cannot be applied for many zeolites. During such a treatment, the zeolites tend to decompose and to lose their crystalline structure.

As a result of this situation, the zeolite catalysts of high activity are obtained by substituting bivalent cations (Ca, Mg, Mn), but especially trivalent cations such as the rare earths for the sodium ions. The polyvalent ions contribute to the stabilization of the crystalline network, which becomes resistant to the prolonged exposure to temperatures of 850–900°C. Concomitantly, they contribute to the formation of strongly acidic sites that catalyze intense cracking reactions. The mechanism proposed for explaining these phenomena is given in Figure 6.14 [4]. As result of the ion exchange between the trivalent cation and the negative charges of the aluminum atoms important bipolar moments are formed, together with the generation of intense electric fields. The electrical induction, produced by these fields modifies the electronic distribution, emphasizing the acidic character of the active catalytic sites.

As result of the treatment with rare earths, the catalysts on the basis of Y-zeolites will acquire a much higher number of acid sites of stronger acidity than the amorphous synthetic alumo-silicate catalysts. Nevertheless, these differences cannot explain fully the much higher activity of the zeolite catalysts. But what is actually important is that by using the proper synthesis method and by making use of the ion exchange with rare earths, the resulting catalysts are up to 10,000 times more active than the amorphous, synthetic alumo-silicates.[*]

6.2.4.2 Matrices and Binders

There are two main reasons that determine the use of matrices in which zeolite is incorporated. First, the excessive catalytic activity of the zeolites makes them unsuitable for use in units designed for conventional catalysts. The enormous difference

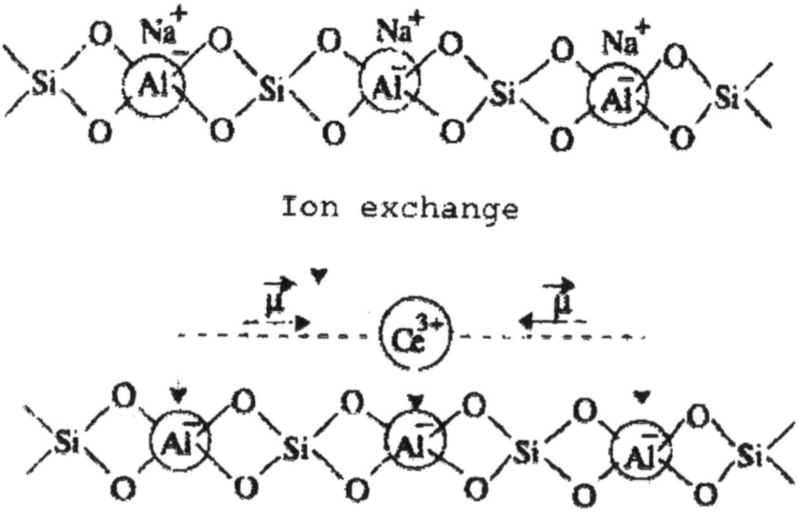

Figure 6.14 Catalyst active centers following sodium substitution by cerium. (From Ref. 4.)

---

[*] For comparison of activity the test of cracking of *n*-hexane was used.

between the activities of the two types of catalysts would require profound modifications to the working conditions and an expensive retrofitting of the units. The second reason is the relative high cost of the zeolite, which, owing to the losses by erosion would increase operation expenses. By incorporating the zeolite in a matrix, the cost of the catalyst is reduced by approximately a factor of 15 relative to the price of the zeolite while also obtaining a catalyst resistant to attrition.

The zeolite is incorporated into the matrix at a rate of 5–16%, as particles having a mean diameter between 2 and 20 μ. The matrix consists of solid particles, which may be natural components such as kaolin, or synthetic components. It represents 25–45% of the total weight of the catalyst. The balance is the binder, which ensures the uniform dispersion of the components within the catalyst particle, the final shape of the particles, and to a large extent, its resistance to attrition.

Recent studies [237] recommend to reduce the size of the zeolite crystals to 0.1 μ. A reduction in size from 1 μ to 0.1 μ increased the reaction rate and produced a gasoline with more alkenes and less aromatics [237].

The matrix may or may not have its own catalytic activity.

The molecules of gas oils with distillation end points below 480°C can pass through the 7.4 Å openings of the pores of the zeolite and therefore they will crack at the active sites within them. In this, the matrix does not have to possess its own catalytic activity and inert matrices are used.

Quite different is the situation of feeds with distillation end points in excess of 490°C, the heavy components of which have kinetic diameters of 10–100Å. Since these cannot enter the pores of the zeolite, and since the accessible active sites on its outer surface represents only about 3% of the total, it is necessary that the matrix is catalytically active on its own.

The catalytic activity of the matrix should not exceed the level required by the initial decomposition of the heavy molecules. The products generated in this first decomposition will diffuse into the pores of the zeolite where the reaction will continue on the catalytic sites. This will ensure improved selectivity (see Chapter 6.3).

The catalytic activity of the matrix may be due to the solid particles used for its production or to the binder.

The clays, which are usually the solid particles used, may be catalytically active or not, as discussed in Section 6.2.1, describing the natural catalysts. They may become catalytically active after the treatment to which the produced catalyst is submitted.

An example of a catalyst with a catalytically active matrix is that containing pseudo-boehmite [36]. This, at temperatures in excess of 320°C, is transformed into a form γ, very active, which possess both acid sites and hydroxy groups.

If the solid material used as a matrix is the only one having catalytic activity, the desired level of catalytic activity may be ensured by using adequate proportions of catalytic active material and inert material.

The binder may also be catalytically active. Thus, in the time period 1960–1972 the zeolite was incorporated in alumo-silica gels that had the composition of amorphous catalysts. Later on, the used mixtures of such gels with clay were similar to those used for the production of the amorphous semisynthetic catalysts.

At present, matrices containing a catalytically active binder are no longer in use [37]. Instead, catalytically inactive silica hydrosol is used, which ensures an improved resistance to attrition and a better selectivity for the produced catalysts. The better selectivity is the result of the fact that the cracking reactions will take place exclusively on the zeolite when the matrix is made of inert materials, or mostly on the zeolite when the presence of heavy fractions in the feed requires that the matrix also possess a catalytic activity.

The compositions and properties for several typical matrices are given in Table 6.2 [37].

Detailed information concerning the preparation and characteristics of matrices produced by various methods and using various binders are given in J. Scherzer's monograph [6].

In all cases, the matrices must possess sufficient porosity and pore size distributions to allow the access of the reactants to the zeolite particles and, if necessary, also the transport of the molecules with large kinetic diameters, including liquid components.

Moreover, the matrices often ensure completion of the ionic exchange within the catalyst by supplying the di- or trivalent ions that will substitute the sodium ions that are still contained in the catalyst. The matrix may also fulfill functions related to the presence in the feed of some elements or complexes that are poisons for the zeolite. Thus it may react with the nitrogen contained in the feed [41] or it may increase the resistance of the catalyst to metals, especially vanadium [42], by fixing it. In this way, the zeolite is protected from the damaging effects upon its structure caused by such elements present in the feed. Matrices that contain active alumina may reduce SO_2 and SO_3 emissions from the regeneration gases [43]. Also, the matrix may contribute to the increase of the cetane number of the gas oil. It does this by increasing the aliphatic character of the feed by the decomposition in an adequate manner of the heavy molecules contained in the feed, by performing their cracking on its the active centers.

Table 6.2 Properties of FCC Matrices

Description	Microactivity of steamed matrix*	Attrition Davison index (DJ)	Apparent bulk density (g/cm^3)	Year of matrix technology commercialization
Alumo-silica gel (14% Al$_2$O$_3$)	35	40	0.50	1963
Alumo-silica gel (25% Al$_2$O$_3$) and clay	35	35	0.52	1965
Silica hydrosol and clay	10	6	0.77	1972
Silica hydrosol with increased matrix activity	25	6	0.76	1978
Alumina sol and clay	25	5	0.80	1980
Clay based XP	58	4	0.72	1986
Silica hydrosol with novel matrix chemistry	25	6	0.76	1990

* Steamed 6 hours at 1400°F (760°C) 100% steam 5 psig (1.35 bar).
Source: Ref. 37.

6.2.4.3 Additives

The additives are used for promoting some of the reactions, such as the burning of the coke deposited on catalyst or for the passivation of the damaging action of some compounds contained in the feed.

Additives may be added to the system in a number of ways:

As solid particles that have the same characteristics of fluidization as the used catalyst.

A liquid additive may be added directly to the system or in mixture with the feed.

Additives may be introduced during the preparation of the catalyst.

All these methods are used in practice. The selection depends on the type of additive required and on the method developed for its use.

Improvement of the combustion within the regenerator. The improvement (promotion) targets a more complete burning of carbon monoxide to carbon dioxide. In this way, a larger amount of heat is produced in the regenerator and the amount of coke necessary for maintaining the thermal balance of the reactor is reduced. A detailed analysis of the effect of the promoter on the thermal balance of the system reactor/regenerator and on the performance of the process was published [44].

To promote the oxidation of carbon monoxide, small amounts of noble metals, especially platinum, have been added [45–47]. Chromium oxide was introduced for use as a promoter in an amount of 0.15% by weight in the zeolite catalyst [48], a process which was abandoned owing to the toxicity of the chromium salts.

Studies [48] on the amount of platinum and on the way it is introduced in a zeolite catalyst lead to the conclusion that 1 mg platinum per kg of catalyst is sufficient. The results of tests carried out in two industrial catalytic cracking units, nonpromoted and promoted catalyst, showed an increase of the CO_2/CO ratio from $0.9 - 1.0$ for the nonpromoted catalyst, to $5.0 - 7.0$ for the promoted one [48].

Similar results were obtained by the promotion of the catalyst spheres used in a moving bed catalytic cracking unit [49].

Reduction of the content of SO_2 and SO_3 in the regenerator flue gases. Such a reduction, very important for the protection of the environment is performed by means of reacting the SO_2 and SO_3 to produce sulfite, respectively sulfate, by interaction with a metallic oxide or hydroxide, within the regenerator. The sulfite and the sulfate are carried by the regenerated catalyst to the reactor and stripper, where they are reduced to H_2S. This is captured without difficulty from the reaction gases.

The additives used for this purpose are compounds of aluminum and magnesium [50] and they are generally protected by patents [51].

Considering aluminum hydroxide as the active component, the reaction of sulfur dioxide in the regenerator can be written as:

$$2Al(OH)_3 + 3SO_2 \rightarrow (SO_3)_3Al_2 + 3H_2O$$

and the reduction of the aluminum sulfite in the reactor and in the stripper as:

$$(SO_3)_3Al_2 + 9H_2 \rightarrow 3H_2S + 2Al(OH)_3 + 3H_2O$$

The reactions involving sulfur trioxide may be written in a similar manner.

The reduction of the content of nitrogen oxides from the flue gases. The elimination of nitrogen oxides from the combustion gases is more difficult than that of the sulfur oxides and until the present could not be solved by incorporating additives in catalyst formulation [6].

The adopted solution is to reduce the nitrogen oxides with ammonia, in the presence of a catalyst, to nitrogen and water [12] at 300–400°C. By using 0.6–0.9 moles NH_3 per mole of NO_x, 60–80% of the nitrogen oxides were eliminated; the resulting gases contained 1–5 ppm by volume of NH_3.

A very active catalyst (S-995) was developed recently by Shell; it allows the reduction of nitrogen oxides at a much lower temperature (150°C).

The passivation against nickel poisoning. The organo-nickel compounds contained in the feed lead to deposits of nickel, especially on the external surface of the catalyst particles. Owing to the dehydrogenating activity of the nickel, such deposits will increase the amount of coke and of gases on account of gasoline.

In the past, to fight against this effect the units with moving bed were operated in conditions in which the surface of the granules was eroded in the process; in this way a great proportion of nickel was eliminated [1]. Later on, zeolite catalysts with a high content of zeolite were used, which are less sensitive to poisoning [52].

The passivation of the dehydrogenating activity of nickel was shown to be more efficient. To this purpose, organometallic complexes of antimony and bismuth were added to the feed [52], beginning in 1977 and 1988 [50], respectively. Bismuth is generally preferred because it has a lesser toxicity. Also, the use of zirconium as a passivator was suggested, by adding to the catalyst of $ZrO(NO_3)_2$ [53].

The mechanism for the action of these promoters is not clear yet.

The passivation against poisoning with vanadium. The damaging action of the organo-vanadium compounds contained in the feed is manifested by the destabilization of the crystalline structure of the zeolite. It is supposed [54] that this destabilization is due to the formation of some compounds of zeolite-VO_4 that weaken the crystal structure, especially during the stripping of the catalyst.

Initially, calcium and magnesium were used [55]. Starting in 1980, tin [50] was used as a passivator. The passivation with tin must be made carefully, since if the amounts of tin necessary for passivation are exceeded, the opposite effect is produced.

The treatment of the catalyst with lanthanum chloride ($LaCl_3$) [53] was suggested as protection against poisoning by vanadium. This proposition does not seem to be applied commercially.

Compounds of bismuth and phosphor [6] were also suggested as passivation agents. It seems that iron also has a poisoning effect, which is however much less pronounced in comparison with nickel and vanadium.

6.2.5 Ultrastable Catalysts

Ultrastable zeolite catalysts are obtained by increasing the Si/Al ratio in the classic synthesis of zeolite catalysts Y. By increasing this ratio, the number of acidic centers is decreased while their acidity is increased. The decrease of the number of sites is the consequence of eliminating the trivalent aluminum atoms included in the network of tetrahedrons, which explains the increase of the stability of the crystalline network and therefore of the stability of the catalyst. The increase of the acidity of the remaining centers leads to the decrease of the role of the hydrogen transfer reactions and from here to the increase of the octane number of the gasoline [11].

It is to be remarked that the zeolites with a Si/Al ratio above 6 (the conventional Y-zeolites have a ratio of 5.0–5.5) obtained by direct synthesis are unstable products. The only way to obtain a Y-zeolite with a higher Si/Al ratio is by enriching in silicon a conventional catalyst [6].

A number of methods were developed for increasing the Si/Al ratio of a conventional Y-zeolite and obtaining ultrastable Y-zeolite, called USY-zeolites, which keep their structure intact up to temperatures of 1000°C.

The first method used [56] involves the stripping of the NH_4Y zeolite at temperatures of the order 760°C, which leads to the hydrolysis of the $Si-O-Al$ bonds with the corresponding decrease of the number of tetrahedrons wherein the aluminum occupies a central position. The alumina rests that are left behind act as nonselective catalytic agents.

A number of methods involving chemical treatment followed, that were used in combination or not with the hydrothermal treatment described above. These methods eliminate a portion of aluminum, with or without an introduction of silicon. To the first category belong treatments with F_2, $COCl_2$, BCl_3, or with a solution of NH_4BF_4, to the second category the treatment with a solution of $(NH_4)_2SiF_6$, or with vapors of $SiCl_4$ [6].

The treatments that involve the introduction of silicon achieve the substitution of aluminum atoms in the network with silicon atoms and are superior.

More extensively used is the treatment with a solution of ammonium fluorosilicate [57,58]; the aluminum is eliminated as soluble ammonium fluoro-aluminate:

However, not all the aluminum eliminated from the network is replaced by silicon. Thus, vacant places remain within the network. Eventually, zeolites are obtained with a ratio Si/Al = 12, marked usually in the literature as zeolites AFSY. The increase of the ratio Si/Al over the value of 12 leads to zeolites that are unstable, owing to the excessively high proportion of vacant sites remaining in the network.

The improvement of the treatment method with ammonium fluorosilicate leads to the production by Union Carbide Co. and Catalystics Co., of the zeolite LZ210 (perfected subsequently as LZ210K [59]) and of the series of HSZ (High Stability Zone) catalysts. The Si/Al ratio can reach values of 10, 20 or 30 creating a broad flexibility in the performances of catalyst. It seems that the refinements that were implemented consist in the treatment of the zeolite in which the ammonium ion exchange was only partially completed with a solution of ammonium fluorosilicate at a rigorously controlled pH-value. The main problem encountered is that aluminum is eliminated from the crystalline structure faster than the silicon can replace it. This leads to cavities and to the weakening of the network. It seems that this deficiency was avoided by acting on the following factors [6]: a) the pH-value was increased to the range 3–7 and the concentration of ammonium fluorosilicate, which reduces the rate of aluminium elimination, was reduced somewhat; b) the increase of the reaction temperature, which increases the inclusion rate of silicon in the network. The regulation accordingly of these factors allows therefore improvement of the stability of the produced zeolite and the increase of the Si/Al ratio. Following these refinements, several series of catalysts were produced that show high octane improving performance, especially for the series ALPHA and BETA, which are amply described [11].

6.2.6 Octane-Enhancing Catalysts and the C_4 Cut

The superior octane performances of the ultrastable catalysts are usually enhanced by the use of ZSM-5 additions.

In fact this is a zeolite having a completely different structure than the Y-zeolites and possessing its own catalytic activity. Thus, ZSM-5 is used as a catalyst in the processes of xylenes isomerization, disproportionation of toluene, in the production of ethyl-benzene [60], and in the processes for dewaxing of distillates [61].

In catalytic cracking ZSM-5 is used as H-ZSM-5, but also as Zn-ZSM-5 and Cd-ZSM-5. It could be used under the form of separate zeolite particles, in which case it is included in a proportion of 25% in an inert matrix or could be included in the catalyst particles.

Detailed results concerning the effects of the addition of ZSM-5 are available from several published papers [6, 62, 234, 235, 236]. Overall, the increase of the octane number as a result of using ZSM-5 is accompanied by a decrease of the gasoline yield. In alkylation, this decrease is compensated by the increased production of alkylate as a result of the larger amounts of iso-butane produced. It seems that the size of the ZSM-5 particle has an important influence on the distribution of products [236].

There has been great interest in recent years in increasing the amount of iso-butylene produced in catalytic cracking. New catalysts, such as IsoPlus 1000 of Engelhard were developed for the special purpose of producing increased amounts of iso-butene [35]. Initially, iso-butene served for the production of methyl-tert-butylether (MTBE), the addition of which to gasolines was considered indispensable, for satisfying the clean air regulations in many countries. (see Chapter 9).

The comparative performance of a typical USY catalyst, IsoPlus 1000 and IsoPlus 1000 containing 3% ZSM-5, was determined in a high performance UOP cat cracker of the riser type. Compared to the standard USY catalyst the increase in i-butylene was of 0.57% and of 0.84% for the IsoPlus 1000, respectively without and with ZSM-5. These effects correspond to an increase of about 50% of the MTBE potential [35].

Extensive research work currently underway will result in cracking catalysts of improved performance tuned to the selective synthesis of the products of high market demand. The interest in producing *iso*-butylene outlived the interest in MTBE and is kept alive by the increasing needs for alkylate gasoline.

6.2.7 Catalysts for Residue Cracking

The incorporation in the distillates fed to the cat crackers of up to 20% residual fractions does not require any change in the units; it was practiced in the U.S. since 1988 in 40% of catalytic cracking units. Catalytic cracking of the residues is covered in Chapter 7.

The inclusion of residues in the feed to cat crackers makes necessary the use of catalysts that are capable of handling the increased metal content and the larger sizes of the molecules within the feed. Thus, fractions with boiling temperatures of above 540°C contain molecules with more than 35 carbon atoms and sizes ranging from 10–25 Å. The vacuum residues contain molecules with molecular masses between 1,000–100,000, the sizes of which vary between 25–150 Å [63].

These dimensions require that the matrix used for the preparation of the catalysts for residue cracking has a certain structure and distribution of the pores, as well as an adequate ratio between the catalytic activity of the matrix and that of the zeolite. In order to correctly understand the issues at play here, it must be taken into account that molecules with kinetic diameters larger than those of the pore orifices (7.4Å) cannot penetrate the pores of the zeolite. The compounds contained in the vacuum gas oils or in residues, with the exception of *n*-alkanes, or of slightly branched *i*-alkanes, cannot be cracked on the active sites within the zeolite, but only on the active sites of the matrix. The products resulting from first-step cracking on the active centers of the matrix can penetrate in the pores of the zeolite, where the cracking is continued and gasoline is formed. Cracking on the active sites of the matrix should not exceed the conversion required for the formation of products having dimensions that allow them to penetrate inside the pores of the zeolite. Here, the cracking will continue in conditions of shape selectivity, ensured by the pores of the Y-zeolite.

The catalysts used in the cracking of residual fractions should show a certain optimal ratio between the catalytic activity of the zeolite and that of the matrix. In a first approximation, this can be expressed by the ratio of the respective specific surfaces. A decrease of this ratio, which means a too high activity of the matrix, leads to the increase of the conversion to dry gases (H_2, C_1, C_2) and to coke (see Figure 6.15) [64].

Another problem, of equal importance is the distribution of the pore sizes of the matrix. The matrix must possess [62]:

1. Large pores with diameters over 100Å. They allow traffic of components with large molecular mass, that remain liquid at the reaction temperatures. The acid sites on the walls of these pores should possess a low catalytic activity for the limitation of the conversion until gases and coke.

2. Pores of average size, with diameters ranging between 30–100Å. These pores have a more pronounced catalytic activity. Their role is to produce first-

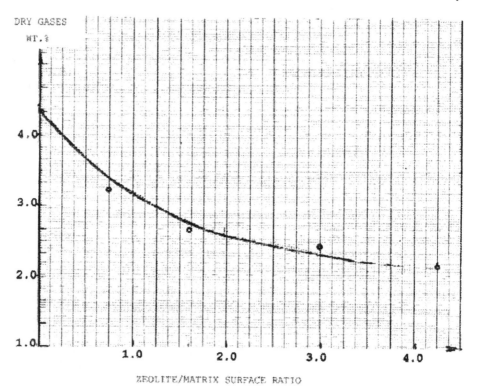

Figure 6.15 Conversion to dry gases function of zeolite/matrix surface ratio. Catalytic cracking at 500°C of a vacuum distillate $d = 0.9188$, $F_{UOP} = 11.5$. (From Ref. 64.)

pass cracking of the heavy components, especially of naphthenes and of aromatic rings.

3. Pores of small sizes below 20Å, have the most pronounced catalytic activity. Their role is to crack the light components possessing prevalent alkane structures but cannot penetrate through the orifices of the zeolite particles.

The ratios in which the three types of pores should be present depend on the ratio between the various cuts incorporated in the feed and chemical composition of the feedstock.

Of special concern is the high content of vanadium and nickel in feeds of residual origin. These metals are retained by the matrix in order to prevent their penetration inside the zeolite.

The vanadium is captured by tin compounds, by barium and strontium titanates, and by magnesium oxide, substances that in most cases are supported on inert particles.

The passivation of vanadium is done by means of antimony or bismuth compounds supported on the catalysts or introduced in the feed.

The reduction of sulfur oxides is performed by means of cerium or rare earths that may be incorporated in the catalyst, may form distinct particles, or may be supported on alumina by the spinel $(Mg \cdot Fe)O(Al_2Fe)_2O_3$ or by cerium supported on the spinel.

The catalysts used for the conversion of carbon monoxide to carbon dioxide during the regeneration use platinum or palladium supported on alumina or alumosilica [65–67], which may be added directly to the catalyst, in amounts of 5 ppm [45]. It must be mentioned that the use of antimony as a passivation agent for vanadium decreases the efficiency of the platinum and requires an increase of its concentration.

These protection measures are similar to those described previously for the cracking of distillates.

6.3 REACTION MECHANISMS

Different from the processes involving thermal cracking, where the active species that control the reactions are radicals, in catalytic cracking the active species are carbocations formed on the active sites of the catalyst. The difference between the results of the two processes is the consequence of the differences between the properties of the two types of active species.

It is to be mentioned that according to the IUPAC nomenclature, ions such as CH_3^+, which earlier were called carbonium ions, should be called *carbenium ions*, while those of the type CH_5^+ should be called *carbonium ions*. The term *carbocations* comprises both species and generally all the organic species that carry a positive charge. This nomenclature will be used in the present work.

6.3.1 Carbocation Formation

In general, there exist four possible ways to form carbocations:

1. The addition of a cation to an unsaturated molecule with the formation of a carbenium ion. Such a reaction occurs during the adsorption of an alkene at a Brönsted site of the catalyst; the formed ion remains adsorbed on the acid site.

$$R_1 - HC = CH - R_2 + HB \rightleftharpoons R_1 - H_2C - C^+H - R_2 + B^-$$

Such an interaction can take place also with an aromatic molecule, as was proved by use of calorimetric methods [68]:

In this case the electric charge is delocalized, i.e. it is distributed over the entire aromatic molecule.

The formation of carbenium ions by the addition of a cation to an alkene is unanimously accepted and, when compared with the other ways of formation of carbocations, appears to take place at the highest rate on catalytic cracking catalysts. If alkenes are not present in the feed, they are formed as a result of reactions of thermal decomposition that takes place at sufficiently high rates at the temperature of catalytic cracking (500°C).

2. The generation of carbenium ions by means of the extraction of a hydride ion by a Lewis site. The reaction can be illustrated by:

$$RH + L \rightarrow [R]^+ LH^-$$

In this case also, the carbenium ion formed remains adsorbed on the active site.

Since, as discussed in Section 6.2.3, both types of sites exist on catalytic cracking catalysts, it is accepted that the carbenium ions are formed both by the interaction of the alkenes with Brönsted-type sites and by the interaction of the alkenes with Lewis sites.

3. The intermediate formation of the carbonium ions. This mechanism was proposed in 1984 by Haag and Dessau [69]. The reaction is supposed to take place by the intermediate formation of carbonium ions, which split, to produce lower alkenes and carbenium ions. As an alternate path, the carbonium ions may convert into a carbenium ions by the loss of a hydrogen molecule.

The mechanism can be depicted by the reactions:

$$R_1- CH_2 - CH_2 - R_2 + HB \rightleftharpoons \left[R_1- CH_2 - CH_3^+ - R_2 \right] B^-$$

$$\searrow -H_2$$

$$[R_1^+] B^- + CH_3 - CH_2 - R_2 \qquad \left[R_1- CH_2 - CH^+ - R_2 \right] B^- + H_2$$

4. Heterolytic decomposition. The reaction is analogous to the initiation of thermal cracking reactions by the splitting of a molecule into two radicals. The difference is that neither fragment keeps two electrons; therefore, the two fragments carry electric charges of opposite signs:

$$R_1 R_2 \rightarrow R_1^+ + R_2^-$$

Heterolytic decomposition can take place in liquid phase processes; it does not seem to take place in catalytic cracking where the carbocations are formed as a result of the adsorption on the acid site of the catalyst.

6.3.2 Carbocation Reactions

The carbenium ions adsorbed on the active sites of the catalyst may initiate several types of reactions. A qualitative understanding of the direction of these reactions may be gained from the heats of the formation of the various ions involved.

It must be remarked that the these heats were not obtained from direct measurements, but by indirect calculations and evaluations performed in different ways by various authors. Thus, Evans and Polanyi [70] calculated the heat for the addition of a proton at an alkene. Pritchard [71] used the ionization energies and the affinities of the electrons at the carbon atom of the broken $C-H$ bond. Magaril [73] calculated the ionization potential of the radicals and the affinity for protons etc. The results are given in Table 6.3 and show that some of the values are quite different from one author to the next.

The second remark is that all the above values were calculated for gas phase and moderate temperatures, whereas the energies of the carbenium ions, which intervene in catalytic cracking, refer to ions adsorbed on the active sites of the catalyst and at temperatures of the order of 500°C. Also, the values of the heats of adsorption are significant and depend on the type of adsorbed ions and on their molecular mass [73,81].

Despite all these caveats, the values of heats of formation calculated for the carbenium ion supply useful and sometimes important information, concerning the directions taken by the chemical transformations. Thus, the comparison of the heats

Table 6.3 Heats of the Reactions RH → R⁺ + H⁻

Carbenium ion	Reaction heat (ΔH)$_5$, kJ/mol				
	EG Evans, M Polanyi [70]	HO Pritchard [71]	BS Greensfelder [72]	JR Franklin [74]	RZ Magaril [73]
CH_3^+	0	0	0	0	0
$C_2H_5^+$	132	130	146	138	142
$C-C-C^+$	214	126	209	167	180
$C-C^+-C$	243	226	276	268	264
$C-C-C-C^+$	–	126	199	196	214
$C-C-C^+-C$	184	–	310	284	285
C–Ċ–C (C)	301	293	351	351	360

Differences assuming CH_3^+ formation as null.

of formation for the primary, secondary and tertiary ions of the same hydrocarbon—the heats of adsorption being in this case identical—supply quantitative information concerning the relative stability and the direction of the isomerization reactions.

The heats of formation of the carbenium ions calculated by R.Z. Magaril [73] are given in Table 6.4. By providing information for more ions than other sources, this author makes it possible to use data having the same degree of accuracy (since it was obtained by application of the same method) for calculating the heats of reaction of a large number of transformations.

The formation of carbocations constitutes the initial step of a sequence of reactions, which gives to the process the character of a decomposition in an unbranched chain, similar to the thermal decomposition.

The reactions that follow the formation of the adsorbed carbocations are skeleton or charge isomerization, interactions with unadsorbed molecules, the transfer of a hydride ion, as well as breaking of carbon–carbon bonds. The combination of the last two reactions confers on the process the character of a chain decomposition. The carbocations generate also secondary reactions, such as polymerization, cyclization etc., and which lead eventually to the formation of coke deposited on the catalyst.

6.3.2.1 Charge Isomerization

The charge isomerization can be exemplified by the reaction:

$$\left[\begin{array}{c} \overset{+}{} \\ R-C-C-R' \\ | \ | \\ H \ H_2 \end{array}\right] \rightleftharpoons \left[\begin{array}{c} \overset{+}{} \\ R-C-C-R' \\ | \ | \\ H_2 \ H \end{array}\right]$$

The migration of the charge is accompanied by the migration in the opposite sense of a hydrogen atom.

Table 6.4 Heats of Formation of Carbenium Ions [73]

Ion	Heat of formation (kJ/mol)	Ion	Heat of formation (kJ/mol)
H^+	1537	$\begin{array}{c} C \\ \mid \\ C-C-C^+ \\ \mid \\ C \end{array}$	812
$CH_3{}^+$	1097		
$C_2H_3{}^+$	1185		
$C_2H_5{}^+$	955		
$C{=}C{-}C^+$	930		
$C{-}C{-}C^+$	917	$C{-}C{-}C{-}C^+{-}C$	754
$C{-}C^+{-}C$	833	$C{-}C{-}C^+{-}C{-}C$	762
$C{-}C{-}C{-}C^+$	883	$C{-}C{-}C^+{-}C$	682
$C{-}C{-}C^+{-}C$	812		
$\begin{array}{c} C \\ \mid \\ C{-}C{-}C^+ \end{array}$	846	$\begin{array}{c} C \\ \mid \\ C{-}C{=}C{-}C^+ \end{array}$	854
$\begin{array}{c} C \\ \mid \\ C{-}C^+{-}C \end{array}$	737	cyclopentyl$^+$	854
$C^+{-}C{-}C{-}C{-}C{-}C{-}C{-}C$	800	cyclohexyl$^+$	796
$C{-}C^+{-}C{-}C{-}C{-}C{-}C{-}C$	708	phenyl$^+$	1202
$C{-}C{-}C^+{-}C{-}C{-}C{-}C{-}C$	699	tolyl ($\mathrm{CH_3}$–phenyl)$^+$	925
$C{-}C{-}C{-}C^+{-}C{-}C{-}C{-}C$	695		

Source: Ref. 73.

For reasons of free energy, this reaction takes place in the direction towards formation of a secondary ion from a primary ion and/or the transfer of the electrical charge towards the middle of the molecule. The data of Table 6.5 for the propyl, *n*-butyl and octyl ions supply concrete energy data for justifying this direction of transformation.

Similar to the charge transfer reaction is that for the double bond transfer, which also involves a migration of a hydrogen atom [18, 75]. This reaction is explained by the formation of an intermediary compound absorbed on the catalyst [76]:

$$R_1{-}C{=}C{-}C{-}R_2 \;\rightleftharpoons\; \left[R_1{-}\overset{+}{C\text{=}C\text{=}C}{-}R_2 \right] H^- \;\rightleftharpoons\; R_1{-}C{-}C{=}C{-}R_2$$

This hypothesis is indirectly confirmed by the finding [77, 78] that in some cases, a migration takes place of the electrical charge between the carbons 1 and 3:

$$(CH_3)_2{-}CH_2{-}CH_2{-}\overset{+}{C}H{-}CH_3 \rightleftharpoons (CH_3)_2{-}\overset{+}{C}{-}CH_2{-}CH_2{-}CH_3$$

6.3.2.2 Skeletal Isomerization

Skeletal isomerization leads to the conversion of a secondary ion to a tertiary ion or, for hydrocarbons with at least 6 carbon atoms, to the migration of the methyl side-group along the chain.

Table 6.5 Isomerization and Cyclization Heats of Carbenium Ions

Reaction	Thermal Effect (kJ/mol)
Isomerization	
$\overset{+}{C}-C-C \rightarrow C-\overset{+}{C}-C$	84
$\overset{+}{C}-C-C-C \rightarrow C-\overset{+}{C}-C-C$	71
$C-\overset{+}{C}-C-C \rightarrow C-C\overset{C}{\overset{\mid}{-}}C^+-C$	75
$C-C-\overset{+}{C}-C-C \rightarrow C-C-C-\overset{+}{C}-C$	−8
$C-C-\overset{+}{C}-C-C \rightarrow C-C-C\overset{C}{\overset{\mid}{-}}C^+-C$	80
$C-C-\overset{+}{C}-C \rightarrow C-C-\overset{+}{C} \quad (C, C)$	−130
$\overset{+}{C}-C-C-C-C-C-C-C \rightarrow C-\overset{+}{C}-C-C-C-C-C-C$	92
$C-\overset{+}{C}-C-C-C-C-C-C \rightarrow C-C-\overset{+}{C}-C-C-C-C-C$	9
$C-C-\overset{+}{C}-C-C-C-C-C \rightarrow C-C-C-\overset{+}{C}-C-C-C-C$	4
Cyclization	
$C=C-C-C-C-C \rightarrow$ [cyclohexyl cation]	172

This isomerization type is of practical interest. It is justified by the favorable free energy effect for such a restructuring of the molecule (see Table 6.5).

A reaction mechanism that has found acceptance [18, 73] involves the intermediary formation [79] of cyclopropyl ions. The isomerization of the butyl ion may take place in this case according to the mechanism:

$$C-\overset{+}{C}-C-C \rightleftharpoons \overset{\overset{+}{C}}{\triangle}-C \rightleftharpoons C-\overset{C}{\overset{\mid}{C}}{}^+C$$

In fact, although any of the three carbon–carbon bonds of the cycle could break, only one of them leads to the formation of the isobutyl ion.

For the amyl ion the mechanism is:

$$C-\overset{+}{C}-C-C-C \rightleftharpoons C-\overset{+}{C}\overset{C}{\triangle}-C \rightleftharpoons C-\overset{C}{\overset{\mid}{C}}{}^+C-C$$

In this case two of the three possible ways of breaking the cycle lead to an iso-structure. This is a partial explanation for the easier isomerization of *n*-pentane compared to *n*-butane.

Since, as mentioned earlier, the heats of adsorption influence only to a small extent the energy balance of the isomerization the thermal effect for the isomerization of the carbenium ions may be calculated correctly on the basis of their heats of formation given in Table 6.5.

6.3.2.3 Hydride Ion Transfer

This transfer is illustrated by the reaction:

$$R_1-H + [R_2^+] \rightleftharpoons [R_1^+] + R_2H$$

in which, an ion initially adsorbed on the active site interacts with a molecule in the gas or liquid phase or with one that is only physically adsorbed, thereby extracting from it a hydride and escaping in the gas/liquid phase. The active site retains an ion from the molecule that lost the hydride.

This reaction is very important, because it is responsible for the character of chain reaction of the process.

The similar transfer of an alkyl group, which may be expressed by the reaction:

$$R_1-CH_3 + [R_2^+] \rightleftharpoons [R_1^+] + R_2-CH_3$$

was also observed [80] although the mechanism seems less sure; its existence should not bring any changes in the general mechanism of decomposition.

6.3.2.4 The Breaking of the Carbon–Carbon Bonds

As in thermal decomposition and for similar reasons, carbenium ions undergo breaking of the C−C bonds in β position to the carbon atom that carries the electric charge.

The differences between the decomposition mechanism on acid catalysts, and thermal decomposition are the consequence of the differences between the free energy values of the carbenium ions and those of the radicals.

Thus, the much higher heats of formation of the methyl and ethyl ions than of the higher ions, a difference which is much smaller when radicals are involved (Table 2.3), leads to much lower rates for the β-scissions and to a lower production of methyl and ethyl ions. There exists a much higher ratio $(C_3 + C_4)/(C_1 + C_2)$ in the gases of catalytic cracking than that from the thermal cracking.

As an illustration, the decomposition of the sec-octyl ion is considered:

$$C-C-C{+}C-C{+}C-\overset{+}{C}-C \longrightarrow C-C-C{+}C-\overset{+}{C} + C_3H_6 \longrightarrow$$
$$\longrightarrow C-C-\overset{+}{C} + C_2H_4 + C_3H_6 \longrightarrow \overset{+}{C} + 2C_2H_4 + C_3H_6$$

Since the heat of formation of the methyl ion is by 180 kJ/mole larger than for the propyl ion (Table 6.4), the last β-scission will not take place or will take place with a very low rate. In exchange, the propyl ion will suffer a charge isomerization, leading to the liberation of 84 kJ/mole (the same table).

The second difference from the behavior of the radicals is the consequence of very high differences between the heats of formation of the primary, secondary, and tertiary ions. These differences, taking into account the heats of adsorption, are estimated at: 42 kJ/mole between the tertiary ion and the secondary ion and at 105 kJ/mole between the secondary and the primary ion [18].

For radicals, such differences are minimal. As example, for the butyl radical there is no difference between the free energy tertiary and secondary radicals. The difference between the primary and secondary radicals is of only 5.4 kJ/mole (Table 2.5).

As consequence, the β-scissions that lead to the formation of primary ions will take place with low rates, much lower than those for the charge or skeleton isomerizations. In the case of molecules with a larger number of carbon atoms, repeated isomerizations may take place, so that the the scission to take place will generate a secondary ion. The above considerations may be exemplified by the following scheme for the isomerization and decomposition of the heptyl ion:

$$
\begin{array}{l}
\overset{+}{C}-\overset{+}{C}-\overset{}{C}-C-C-C-C \xrightarrow{\text{very slow}} C_3H_6 + \overset{+}{C}-C-C-C \\[2mm]
\overset{+}{C}-\overset{}{C}-C-C-C-C \xrightarrow{\text{very slow}} i\text{-}C_4H_8 + \overset{+}{C}-C-C \\
\quad | \\
\quad C \\
\qquad\qquad C-\overset{+}{C}-\overset{}{C}-C-C \xrightarrow{\text{very slow}} i\text{-}C_5H_{10} + \overset{+}{C}-C \\
\qquad\qquad\quad | \; | \\
\qquad\qquad\quad C \; C \\[2mm]
\overset{+}{C}-\overset{}{C}-\overset{}{C}-C \xrightarrow{\text{fast}} i\text{-}C_4H_8 + C-\overset{+}{C}-C \\
\quad | \quad | \\
\quad C \quad C
\end{array}
$$

The important practical consequence of the repeated isomerization of alkanes followed by their decomposition is the pronounced iso-alkanic character of the gasolines, which confers to them a high octane number and the prevalence of the iso structure in the C4 cut.

In conclusion, the decomposition of the adsorbed ions on the active sites of the catalyst and their interaction with nonadsorbed molecules leads to a chain of successive reactions, according to the following scheme, written for *n*-heptane.

Chain initiation on a Brönsted center:

$$C-C-C-C-C-C-C + BH \rightarrow \left[C-C^+-C-C-C-C-C\right]B^- + H_2$$

or on a Lewis site:

$$C-C-C-C-C-C-C + L \rightarrow \left[C-C^+-C-C-C-C-C\right]LH^-$$

Naturally, all other possible ion structures, especially the secondary ones will be formed.

Chain propagation by isomerization and β-scissions.

The 2-heptyl ion thus formed will undergo isomerization and decomposition reactions represented by the scheme given above for. The 3- and 4-heptyl ions formed by the initiation reaction will undergo similar reactions, but leading to the formation of other final species.

By generalization one may write:

$$\left[R_i^+\right] \rightarrow \sum M + \left[R_j^+\right] \tag{a}$$

where $\left[R_i^+\right]$ represents the various species of heptyl ions adsorbed on the catalyst; $\sum M$ = sum of the hydrocarbons produced by the decomposition, and $\left[R_j^+\right]$ = the ions of low molecular mass that are left adsorbed on the catalyst.

By the transfer of the hydride ion, heptyl ions will be reconstituted:

$$\left[R_j^+\right] + R_lH \rightarrow R_jH + \left[R_i^+\right] \tag{b}$$

where: R_lH is the molecule of n-heptane.

Reactions (a) and (b) will be repeated, giving a chain character to the decomposition process.

Chain interruption in the case of catalytic cracking takes place following progressive blocking of the active sites by the strong adsorption of ions of high molecular mass, generally produced by polymerization, cyclization, dehydrogenation, and condensation reactions. These ions are not capable of accepting the transfer of the hydride ion (reaction b), owing to very strong forces of attraction to the active site. As a consequence, they are not desorbed from the surface of the catalyst.

A more detailed analysis of the reaction mechanism taking place in the catalytic cracking of the alkanes was performed recently by S. Tiong Sie [81]. His analysis takes into account concomitantly the isomerization reactions by means of nonclassic carbenium ions of cyclo-propyl structure as well as by the cracking of the molecule. The formation of the cyclo-propyl ion is considered to precede the β-scission undergone by the molecule. In Figure 6.16 the classic mechanism and that proposed by S. Tiong Sie are compared.

According to this scheme, the classic mechanism leads to the formation of an alkene and an alkane molecule. The formation of iso-alkanes is explained by independent isomerization reactions that take place in parallel with those of decomposition. The mechanism proposed by S. Tiong Sie leads to the direct formation of iso-alkanes together with that of alkenes.

The comparison with the experimental data of n-alkanes cracking on acid catalysts supplies arguments in the support of the suggested mechanism [81]. However, one could argue that the two mechanisms do not exclude each other, and that the β-scission of the secondary ion could take place in parallel with the formation of the cyclo-propyl nonclassic ion, specific to the isomerization reactions. In this situation, the conversion in two directions will depend on their relative reaction rates.

Interesting attempts were made to predict the product distribution obtained in the catalytic cracking of a petroleum fraction [82]. The investigated cut was characterized in terms of the proportions of the various types of carbon atoms, by using mass spectroscopy and nuclear resonance techniques. The fraction was replaced by a number of pseudocomponents, which were assumed to undergo the catalytic reactions, leading to the distribution of products that would have resulted from the actual process. Although, as the authors recognize, the obtained results do not justify the use of this method for predicting the results from commercial operation, the adopted method allows comparative studies and presents an area for future study.

6.3.3   Catalytic Cracking of Various Compounds

The mechanisms of the catalytic cracking reactions presented above allow the examination of the conversions of various classes of hydrocarbon.

Alkanes.   The reactivity of n-alkanes increases and the activation energy decreases with chain length [83]. As a result, the conversion increases with the length of the chain. This is illustrated by the cracking in identical conditions of hydrocarbons on catalysts of Si-Al-Zr [84]:

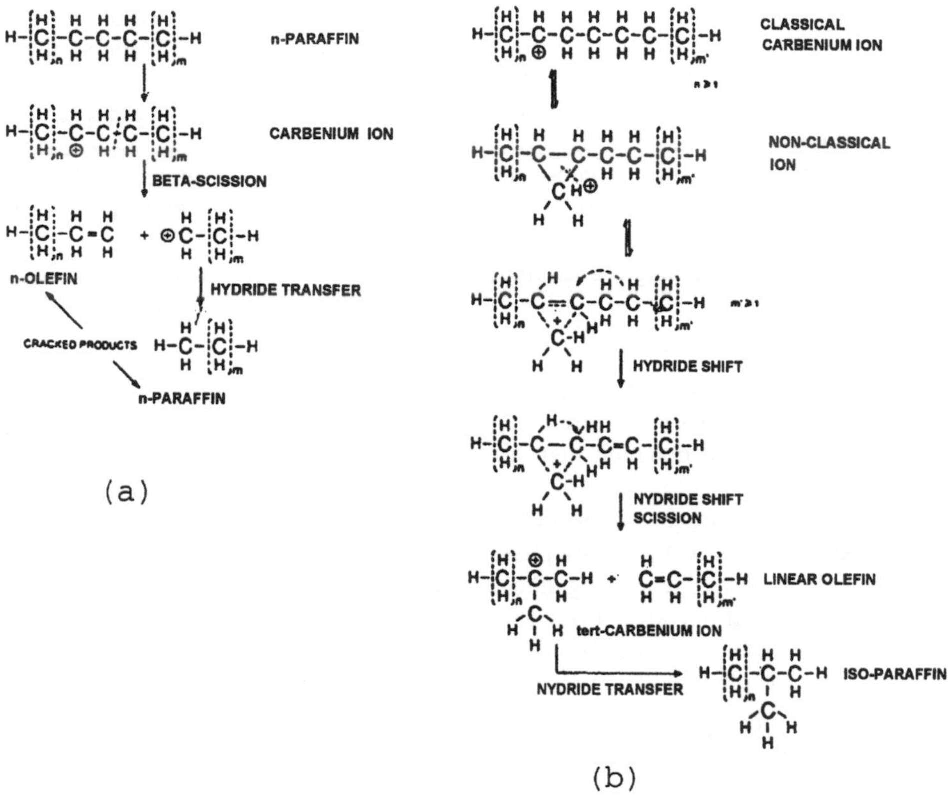

n-PARAFFIN

CARBENIUM ION

BETA-SCISSION

n-OLEFIN

HYDRIDE TRANSFER

CRACKED PRODUCTS

n-PARAFFIN

(a)

CLASSICAL CARBENIUM ION

NON-CLASSICAL ION

HYDRIDE SHIFT

NYDRIDE SHIFT SCISSION

LINEAR OLEFIN

tert-CARBENIUM ION

NYDRIDE TRANSFER

ISO-PARAFFIN

(b)

Figure 6.16 Mechanisms of *n*-alkanes catalytic cracking: classical (a) and proposed by S. Tiong Sie (b).

Hydrocarbon	Conversion, wt %
$n\text{-}C_5H_{12}$	1
$n\text{-}C_7H_{16}$	3
$n\text{-}C_{12}H_{26}$	18
$n\text{-}C_{16}H_{34}$	42

The reactivity increases also with the degree of branching. This is easily explained by the fact that tertiary ions are formed more easily than secondary ions. An exception is made by the hydrocarbons, which have side branches bound to a quaternary carbon atom. Thus, the conversions obtained for various hexanes at 550°C in identical conditions were [85]:

Hydrocarbon	Conversion, wt %
$n\text{-}C_6H_{14}$	13.8
2-methyl-pentane	24.9
3-methyl-pentane	25.4
2,3-dimethyl-butane	31.7
2,2-dimethyl-butane	9.9

The higher rate of skeletal isomerization than that of decomposition and the fact that the carbenium ions are formed preferably at the tertiary carbon atoms, leads to the *iso*-alkane character of the reaction products. This is true both for the C_4 fraction and naphtha fractions to which the *iso*-alkane character confers a high octane number.

Concerning the gases' composition, the higher heats of formation for the methyl and ethyl ions has as a result the prevalence of C_3 and C_4 hydrocarbons in the gases.

More data concerning the catalytic cracking of individual hydrocarbons and the distribution of the reaction products are contained in several studies [86] and are reviewed in the monograph of B.W. Wojciechowski and A. Corma [18].

Alkenes.   The alkenes form carbenium ions easier than alkanes by adsorption on a Brönsted site with the addition of a proton. The practical result is they show a higher rate of cracking than alkanes, the process following the same general rules.

The alkenes adsorb on the surface of the catalyst, as carbenium ions may interact with alkenes in the vapor phase or with those physically adsorbed on the surface of the catalyst, generating alkenes with a high molecular mass. Their participation by cyclization, aromatization and interactions with cyclic hydrocarbons from the vapor phase at the formation of coke on the surface of the catalyst is unanimously accepted.

Cyclo-alkanes.   By catalytic cracking, cyclanes produce a high proportion of *iso*-alkanes in addition to important amounts of aromatic hydrocarbons [86–88]. Such a distribution was determined earlier by V. Haensel [89] in the catalytic cracking of alkyl-cyclohexanes, dialkyl-cyclohexanes, dicyclohexyl-decaline, cyclohexane and methyl-cyclopentane. Other studies [90] have shown that higher proportions of hydrogen are produced in the catalytic cracking of cycloalkanes than of alkanes with the same number of carbon atoms.

These results show that the dehydrogenation of cycloalkanes to aromatic hydrocarbons is an important reaction, almost as important as that of ring breaking. In as much as such a dehydrogenation on acid catalysts passes through the intermediary phases of the formation of cyclohexenes and cyclohexadienes, their participation in the formation of coke by interaction with the alkenes (condensations, polymerizations, and dehydrogenations) may be important.

Besides the reactions that affect the ring, the alkyl-cyclanes undergo reactions of breaking the side chains, with the preferred formation of isoalkanes. Due to the high energy of formation of the corresponding ions, no methyl groups and very few ethyl groups are split from the rings.

Aromatic hydrocarbons.   The alkyl-aromatic hydrocarbons undergo breaking of the side chains in the same manner as the cyclanes.

The aromatic rings being very stable, no ring breaking occurs, but they may participate in condensation reactions with the formation of coke.

The methyl-aromatics may undergo isomerization reactions by the migration of the methyl groups and by disproportionation.

Detailed studies were performed on the catalytic cracking of cumene and are reviewed in the monograph of B.W. Wojciehowski and A. Corma [18].

Sulfur and nitrogen compounds. There are no published data concerning the decomposition mechanism of these compounds on catalytic cracking catalysts. The nitrogen compounds having a basic character are fixed (probably irreversibly) on the acid centers of the catalyst, in this way preventing their participation in cracking. The result is a deactivation of the catalyst, the degree of which depends on the content and the nature of nitrogen compounds present in the feedstock. Figure 6.17 shows the nitrogen compounds that are present in the crude oil with qualitative indication of their basicity [91].

NEUTRAL COMPOUNDS

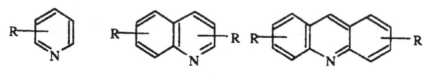

Indoles   Carbazoles

BASIC COMPOUNDS

Pyridines   Quinolines   acridines

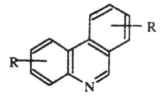

Pnenanthrenes

WEAK BASIC COMPOUNDS

Hydroxypyridines   Hydroxyquinolines

Figure 6.17 Basicity of nitrogen compounds found in crude oil. (From Ref. 91.)

In a recent study [92], a systematic study of the poisoning effect by 34 aromatic hydrocarbons and nitrogen compounds was undertaken and quantitative correlations were obtained. The research is based on a previous work [93], that determined experimentally the poisoning effect of these substances.

The authors define the poisoning effect by the equation:

$$y = 1 - \frac{\xi}{100 - \xi} \cdot \frac{100 - \xi'}{\xi'}$$

where y is the poisoning effect, which takes values between zero and one, ξ equals the percentage conversion in products with boiling temperature below 220°C, which is obtained in the absence of the contaminant, and ξ' is the conversion in presence of the contaminant.

Synthetic feeds were prepared with various contaminants. The amount of contaminant was adjusted to give overall 0.5 wt % nitrogen in the feed for contaminants containing one nitrogen atom, and 1.0 wt % in the feed for contaminants containing two nitrogen atoms. Aromatic hydrocarbons were added in the same proportions as the nitrogen compounds of similar structure in order to see the difference between the poisoning effect of the aromatic structure and that of the nitrogen atom.

A second series of experiments were performed by adding 0.3 wt % of contaminant substance.

The poisoning effect of the substances examined is recorded in Table 6.6 [92]. The comparison of the values obtained for the poisoning effect of similar structures, containing a nitrogen atom or not, such as: of quinoline ($y = 0.541$) with that of naphthalene ($y = 0.222$) and of acridine ($y = 0.621$) with that of anthracene ($y = 0.154$) proves the very strong influence of basicity on the poisoning effect. This effect is explainable by the neutralization of the acid centers of the catalyst by basic contaminants.

In the second study [93], in order to develop a quantitative correlation between the poisoning effect and the structure of the contaminant, 21 parameters were selected and their intervention in such a correlation was tested.

The processing of the whole experimental material and of the 24 parameters calculated for the examined substances lead to the correlation of the contaminant effect (y) with the proton affinity (PA) and the molecular mass (MW) of the contaminant by the equation:

$$y = 0.075 + 0.735(\text{MW})^2(\text{PA})^2 - 0.4067(\text{MW})^3$$

where the proton affinity (PA) is defined for the base (B) by the equation:

$$\text{PA}(\text{B}) = \Delta H_f(\text{H}^+) + \Delta H_f(\text{B}) - \Delta H_f(\text{HB}^+)$$

where ΔH_f are the heats of formation and (HB$^+$) is the acid corresponding to base B.

The values (PA) and (MW) are listed also in Table 6.6.

The authors [92] plotted the results (see Figure 6.18), correlating the values of y with PA and MW.

Table 6.6 Poisoning Effect of Some Nitrogen
Compounds and Aromatic Hydrocarbons

Compound	y	MW	PA
Aniline	0.206	93.13	211.1
Pirole	0.160	67.09	215.6
Pirolidine	0.279	71.12	216.9
Pirazine	0.210	80.09	204.1
Pyridine	0.247	79.10	215.1
Piperidine	0.302	85.15	219.9
Naphthaline	0.222	128.18	194.6
Indole	0.299	117.15	211.3
Chinoxaline	0.458	130.15	215.2
Chinoline	0.541	129.16	221.0
1,2,3,4-tetrahydrochinoline	0.552	133.20	215.8
5,6,7,8-tetrahydrochinoline	0.592	133.20	221.6
Antracene	0.154	178.24	196.3
Carbazole	0.238	167.21	210.8
1,2,3,4-tetrahydrocarbazole	0.288	171.24	209.9
Fenazine	0.484	180.21	220.2
Acridine	0.621	179.22	228.9
2-methylpiridine	0.347	93.13	219.0
2-ethylpiridine	0.363	107.16	219.9
2-methyl-5-vynilpiridine	0.368	119.17	220.5
2-vinylpiridine	0.381	105.14	220.1
2,4-dimethylpiridine	0.401	107.16	221.8
5-ethyl-2-methylpiridine	0.449	121.15	221.0
2,3-cyclopentapiridine	0.474	119.17	220.1
2p-tolylpiridine	0.491	169.23	223.3
2,6-di-tert-butylpiridine	0.565	191.32	228.6
3-methyl-2-phenilpiridine	0.592	169.23	223.5
1-ethylpiperidine	0.418	113.20	223.8
2-ethylpiperidine	0.446	113.20	220.7

Source: Ref. 92.

6.3.4 Mechanism of Coke Formation

The formation of coke belongs inherently to the reactions that take place during catalytic cracking. It results directly from maintaining the overall H/C ratio between the feed and the reaction products. The coke amount can vary in certain limits, depending mainly on the feed, used catalyst, and on the operating conditions. But coke formation can not be avoided completely.

The term coke comprises the total of the products that remain adsorbed irreversibly on the catalyst; in other words, those products not eliminated during the stripping that occurs when the catalyst is transferred from the reactor to the regenerator. Therefore, it contains a whole range of species of different chemical structure characterized by a relative low H/C ratio. Obviously, the species present depend to a large degree on the nature of the feedstock, the content of contaminants Ni, V, and

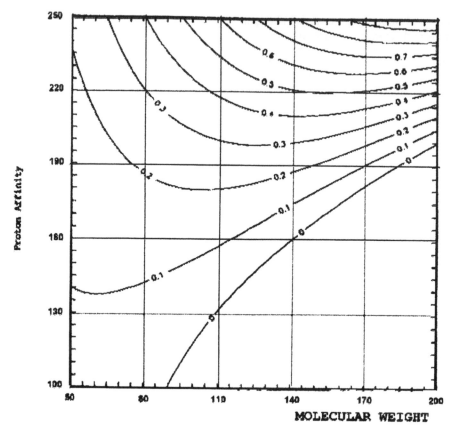

Figure 6.18 Correlation of the poisoning effect with the proton affinity (PA) and molecular mass (MW). (From Ref. 92.)

Fe that catalyze the dehydrogenation reactions, on the nature of the catalyst, and on operating conditions.

The atomic ratio H/C in coke varies in quite broad limits, between 1.0 and 0.3, and decreases during the process [94,95].

These ratios are obviously indicating the presence in the coke of polycyclic aromatic structures, the only ones which correspond to such ratios.

A large number of my own experimental studies [97] indicate that coke containing such aromatic structures is formed also during the catalytic cracking of gas oils that completely lack polycyclic aromatic hydrocarbons [96], as well as of white oil and of various individual hydrocarbons [93–100]. These studies prove that alkenes participate very actively in the formation of coke. Thus, propylene produces 2.65-times and 1-pentene 8.58-times more coke than cracking in the same conditions of n-hexadecane [98].

At present, there is no reaction scheme that can be formulated in a definitive manner. Nevertheless, it can be stated that the formation of coke is a consequence of oligomerization reactions of the alkenes, followed by cyclization, aromatization, alkylation, and condensation. Because these reactions are produced in the adsorbed layer on the catalyst, the thermodynamic analysis of the possible transformations must take into account concentrations in the adsorbed layer that are similar to those in the liquid state (see Section 6.1).

Oligomerization. Oligomerization commonly takes place as a result of the interaction of the adsorbed ions with unsaturated molecules with the alkenes produced by the decomposition reactions:

$$R^+ + \overset{|}{\underset{|}{C}} = \overset{|}{\underset{|}{C}} \longrightarrow R - \overset{|}{\underset{|}{C}} - \overset{|}{\underset{|}{C}}{}^+$$

$$R - \overset{|}{\underset{|}{C}} - \overset{|}{\underset{|}{C}}{}^+ + \overset{|}{\underset{|}{C}} = \overset{|}{\underset{|}{C}} \longrightarrow R - \overset{|}{\underset{|}{C}} - \overset{|}{\underset{|}{C}} - \overset{|}{\underset{|}{C}} - \overset{|}{\underset{|}{C}}{}^+, \text{ etc.}$$

Oligomerization produces ions of increasing molecular mass, which by desorption sets free the corresponding alkenes. The thermodynamic possibility for the occurrence of such reactions in the adsorbed layer is seen in the equilibrium graph of Figure 6.2a. From the same graph it follows that such reactions cannot take place in the conditions of catalytic cracking in gaseous phase.

Cyclization and formation of polycyclic hydrocarbons. The thermodynamic calculations presented in Section 6.1.4 prove that the cyclization of alkenes to cycles of 5 or 6 carbon atoms may take place with high conversions in the conditions of catalytic cracking, irrespective of the phase in which the reaction take place (vapor or liquid). Cyclization is the direction generally favored by thermodynamics and not the splitting of the cycle. The probability of forming cycles of 5 and 6 carbon atoms is actually the same.

The mechanism of this reaction may be represented in the following way:

$$C{=}C{-}C{-}C{-}C{-}C{-}C + L \longrightarrow \left[C{=}C{-}C{-}C{-}C{-}\overset{+}{C}{-}C \right] LH^- \rightleftharpoons$$

$$\rightleftharpoons \left[\overset{C}{\underset{+}{\bigcirc}} \right] LH^- \longrightarrow \overset{C}{\bigcirc} + L$$

Similarly, cycles of 5 carbon atoms may be formed:

$$C{=}C{-}C{-}C{-}C{-}C{-}C + L \longrightarrow \left[C{=}C{-}C{-}C{-}\overset{+}{C}{-}C{-}C \right] LH^- \rightleftharpoons$$

$$\rightleftharpoons \left[\overset{+}{\bigcirc}{-}C{-}C \right] LH^- \longrightarrow \bigcirc{-}C{-}C + L$$

The cyclopentyl and cyclohexyl ions may be formed not only by the interaction with a Lewis site, but also by the transfer of a hydride ion:

$$C{=}C{-}C{-}C{-}C{-}C{-}C + [R^+] \rightarrow RH + [C{=}C{-}C{-}C{-}C{-}C{-}\overset{+}{C}]$$

After charge isomerization, the adsorbed alkene may be cyclized to form a cyclohexyl or cyclopentyl ion.

The formed cyclic ions may desorb, producing the corresponding cycloalkenes [73,75]:

Cycloalkenes with rings of five or six atoms may easily undergo reciprocal isomerization. Those with six atoms in the cycle may be disproportioned with the formation of cycloalkanes and of aromatic hydrocarbons [73,75], or could be dehydrogenated directly and form aromatic hydrocarbons [73].

Both the cycloalkenes with five and those with six carbon atoms in the cycle may interact with an alkenyl ion adsorbed on the catalyst with the formation of a bicyclic hydrocarbon:

Such reactions may continue and lead to the formation of polycyclic condensed hydrocarbons. Aromatic or hydroaromatic hydrocarbons may also participate in such reactions.

Aromatization. According to the data presented in Section 6.1.6, the dehydrogenation of the 6 carbon atom rings in aromatic hydrocarbons may take place with high conversions in conditions of catalytic cracking. Combined with the previously given reactions, it could lead at the end to the formation of aromatic polycondensated hydrocarbons.

This aromatization process is favored by the catalytic effect of the Ni, V, and Fe deposits formed on the surface of the catalyst derived from the respective organometallic compounds contained in the feed. This phenomenon explains the increase of the conversion to coke when the catalyst is poisoned by metals.

If one accepts such a succession of reactions it obviously follows that the formed coke will contain not only polycyclic aromatic hydrocarbons, which are mostly insoluble in solvents, but also intermediary compounds participating in the successive reactions. Further studies on the various compounds that are soluble in organic solvents [101] will certainly contribute to the more complete elucidation of the paths for coke formation.

The detailed studies concerning the mechanism of coke formation must take into account the possibility of the formation in parallel, even on very active catalysts, of coke along thermal pathways as concluded by some recent studies [102].

6.4 KINETICS OF CATALYTIC CRACKING

The complete analysis of processes involving heterogeneous catalysis requires the examination of the mass transfer phenomena that precede the chemical steps. In processes involving solid catalysts these are:

> The diffusion of the reactants from the bulk fluid to the outer surface of the catalyst particles, as well as diffusion in the opposite sense of the reaction products—external diffusion
> The diffusion through the pores toward the active sites of the catalyst—internal diffusion

If the rates of the diffusion steps are lower or of the same order as the rate of the chemical reaction, the diffusion steps will influence the overall kinetics of the process but in different ways.

The external diffusion occurs before the reaction. For this reason, in the case when it is the slowest of the process steps it will determine the overall kinetics and will impose to it its rate equations. It is said that the process takes place "under external diffusion control."

The internal diffusion through the pores of the catalyst occurs in parallel with the chemical reactions. Its influence on the transformation rate of the feed molecules depends on the manner in which they travel through the pores before they are adsorbed on the active sites. Thus, internal diffusion decreases the reaction rate and influences the overall kinetics, without imposing a specific type of kinetic equation.

The heat transfer is similar to the diffusion phenomena: the external transfer takes place to and from the bulk fluid to the outer surface of the catalyst particles. The internal heat transfer is through the mass of the catalyst particle. Their influence on the overall process increases as the heat of reaction becomes more important.

Catalytic cracking presents some additional particularities. The reactions take place in the conditions of progressive deposition of coke on the active surface of the catalyst, a process so intense that in the formulation of kinetic expressions it is necessary to take into account the gradual decrease of the catalyst activity.

The regeneration by the repeated burning of the coke deposited on the catalyst is a process with the same importance as the reaction itself. The kinetics of coke combustion will be examined with the same degree of attention. Here, a specific trait is the gradual decrease of the rate of coke burning as the combustion penetrates inside the pores. Besides, the unburnt, residual coke left in the catalyst pores decreases catalyst activity.

Since the diffusion phenomena are similar for the reaction and for the regeneration, they will be examined before examining the reaction kinetics for cracking and for coke burning.

6.4.1 External Diffusion

The tangential fluid velocity at the surface of a particle is equal to zero and increases progressively with the distance, to reach at a distance δ the constant velocity in the free space between the particles. For the median section of a spherical particle this velocities profile is represented in the Figure 6.19.

For simplification, in the mathematical treatment of the external diffusion this velocity profile is often assimilated with a film of stationary fluid. From the hydrodynamic point of view, the thickness of such a film is considered to be equal to the *displacement thickness* (δ*) defined as the equality between the fluid flowrate that actually flows around the particle and that which would flow at the velocity of the free space if a hypothetical stationary film of thickness δ* would be present (see Figure 6.19).

The influence of the external diffusion depends obviously on the thickness of this film formed around the catalyst particle.

In the case of processes in a static or moving bed, the theoretical determination of film thickness meets major difficulties, making it necessary to resort to empirical or semi-empirical equations, or to use experimental methods.

The experimental determination of the possible influence of the external diffusion on the overall reaction rate is performed by using fixed bed reactors that allow the variation in sufficiently broad limits of the H/D (high/diameter) ratio for the catalyst layer, keeping constant the feed volume flowrate. In these conditions, the modification of H/D leads to the modification of the fluid linear rate and implicitly

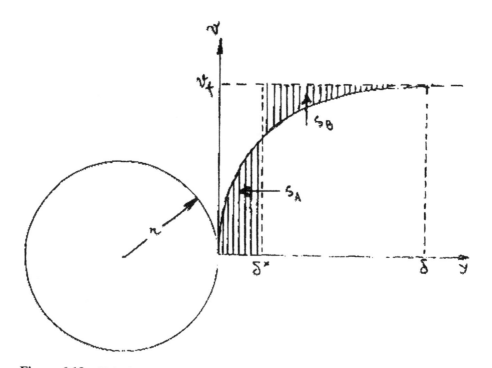

Figure 6.19 Velocity profile in the fluid outside the median section of a spherical particle. Graphical definition of the "displacement thickness" δ*: surfaces $S_A = S_B$.

ot the Reynolds number and of the thickness of the boundary layer. In conditions in which the external diffusion influences the overall process rate, higher rates of the fluid and therefore higher ratios H/D will lead to higher conversions (see Figure 6.20).

It must be mentioned that external diffusion modifies not only the conversion, but also the reaction order and the activation energy. If the external diffusion constitutes the rate determining step of the process, the reaction is of first order and the activation energy will take values that are typical for the physical processes of the order of several thousands calories/mol.

In the case of catalytic cracking, the real value of the activation energy for the catalytic process may be masked by the much larger activation energies of the thermal cracking process that occurs in parallel at the temperatures of catalytic cracking. Therefore, the determination of the activation energy is not to be used in this case as a criterion for establishing that the reaction proceeds under the control of external diffusion.

The facts presented above lead to very important conclusions of a general character concerning the manner of performing experiments in order to correctly model an industrial process. Since as a rule, such experiments use the same feed and the same catalyst (with the same particle size, in order to be situated in identical conditions concerning the diffusion through pores) as the commercial process, it results that in order to have identical conditions concerning the influence of the external diffusion, the linear velocity of the reactant in the pilot reactor must be identical or close to that practiced in the industrial plant. This means that the height of the catalyst layer in the pilot plant should be the same as in the industrial plant,

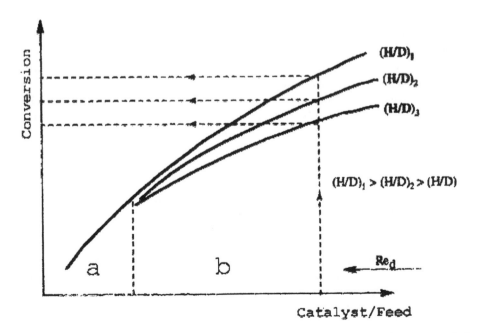

Figure 6.20 Experimental determination of effect of external diffusion. a – no influence; b – increasing influence.

and the cross section of the catalyst bed must be reduced in proportion to the decrease of the feedrate. Other, less expensive methods for measuring kinetic parameters (use of differential reactors and others), will also yield information concerning the effect of external diffusion on the kinetics process [256].

Also, one must avoid the error to consider that in a heterogeneous catalytic process it is sufficient to keep constant the ratio of the feedrate per unit weight of catalyst in order to obtain results that are reproducible at another scale.

In catalytic cracking, the most practiced is the fluidized bed processes, where the catalyst shaped as independent spherical particles is kept suspended in an ascending current of fluid. Accordingly, the problems of the external diffusion will be examined for the case of such particles.

The thickness of the boundary layer is dependent on the Reynolds number expressed by the relation:

$$Re = \frac{v \cdot d}{v} \tag{6.1}$$

For a spherical particle, the characteristic size that intervenes in Eq. (6.1) is the diameter of the particle.

At very low values of the Reynolds numbers, the Stokes domain, the boundary layer, formed around the spherical particle, has a practical uniform thickness. With increasing Reynolds numbers, the boundary layer is increasingly deformed until it detaches behind the particle when the Reynolds number reaches a value of about 30. The zone occupied by the detached boundary layer occupies a growing fraction of the external surface of the particle. The angle between the incidence point and the border of the detachment zone reaches a maximum value of $Q_D = 109.6°$, at $Re \geq 300$. At that point, the boundary layer has detached from 1/3 of the surface of the sphere. On the side of the particle from which the boundary layer has detached, vortices are produced that favor the direct access of the fluid to the external surface of the particle.

The change in the shape of the boundary layer with increasing Reynolds number is represented in Figure 6.21.

It is obvious that in the case of a porous catalyst, the areas of detached boundary layer will allow the direct access of the reactants to the external surface of the catalyst, facilitating their penetration into the pores within the catalyst particles. The barrier effect of the external diffusion in situations when the barrier is present will decrease, beginning with $Re = 30$ and will become practically zero when the Reynolds number approaches 300.

In view of the importance of knowing the variation with the Reynolds number of the detachment angle Q_D and of the surface fraction thus liberated—F_S—, they were plotted in the graph of Figure 6.22a on the basis of published experimental data [103–105].

Concerning boundary layer thickness, the published studies do not offer a general correlation that is valid for all the hydraulic domains. Most of the studies [244–251] refer to attempts to extend the Stokes equations by taking into consideration the inertia forces. The experimental verifications, by Maxworthy [252] proved that the obtained equations are exact only for $Re \leq 1$.

The study of Jenson [253] concerning the intermediary domain and those of Frössling [254] and Schokemeier [103] concerning the domain of the laminar bound-

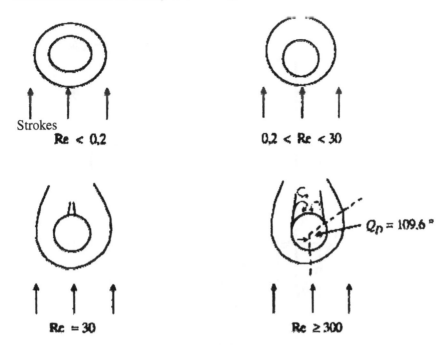

Strokes

$Re < 0.2$

$0.2 < Re < 30$

$Re = 30$

$Re \geq 300$

$Q_n = 109.6°$

Figure 6.21 Evolution of the boundary layer around a spherical particle in an ascendant flow.

ary layer do not provide analytical expressions for the variation of the fluid velocity, v, as a function of the distance from the wall, y (see Figure 6.19).

The equations deduced by Raseev [106] are based on the hypothesis that, in the middle section of the sphere perpendicular to fluid flow, the tangential tension decreases linearly with the distance, becoming equal to zero at a distance δ. Raseev estimated that this hypothesis is plausible, such a linear variation being valid for the laminar flow through tubes, for the gravitational flow along a plate and in many other cases.

This dependence was expressed by the differential equation:

$$-d\tau = \tau_{0_{90°}} \frac{dy}{\delta}$$

where τ is the tangential tension, τ_0 is the tangential tension at the surface of the sphere in the middle section ($\theta = 90°$).

The tangential tension can be expressed also as function of the dynamic viscosity by the relation:

$$d\tau = \mu d\left(\frac{dv}{dy}\right)$$

Equating the two expressions, it follows:

$$-\mu d\left(\frac{dv}{dy}\right) = \tau_{0_{90°}} \frac{dy}{\delta} \tag{6.2}$$

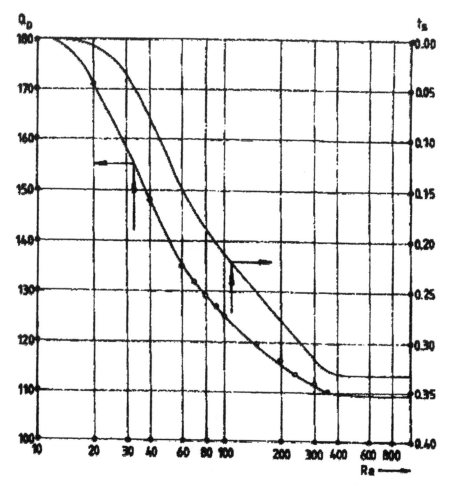

Figure 6.22a Correlation of detachment angle Q_D and of fraction of free surface F_S with Reynolds number.

Integrating this expression and using the boundary conditions $y = 0$ $v = 0$ and for $y = \delta$, $v = v_p$, where v_p is the terminal free-falling velocity, we get:

$$v = -\frac{\tau_{0_{90°}}}{\mu\delta}\frac{y^2}{2} + \frac{v_p}{\delta}y + \frac{\tau_{0_{90°}}}{2\mu}y \tag{6.3}$$

Concerning the second boundary condition, it must be remarked that for a free sphere suspended in an ascending fluid current, the velocity of the fluid beyond the boundary layer, therefore at a distance of at least δ relative to the sphere, will be always equal to the terminal free fall velocity.

By differentiating Eq. (6.3) in terms of y, it is obtained:

$$\frac{dv}{dy} = -\frac{\tau_{0_{90°}}}{\mu\delta}y + \frac{v_p}{\delta} + \frac{\tau_{0_{90°}}}{2\mu}$$

Since, for $y = \delta$, $\dfrac{dv}{dy} = 0$, it follows:

$$\frac{v_p}{\delta} = \frac{\tau_{0_{90°}}}{2\mu}$$

or:

$$\frac{\delta}{r} = \frac{2}{\dfrac{r \cdot \tau_{0_{90°}}}{v_p \mu}} \tag{6.4}$$

Substituting in Eq. (6.3) and performing simplifications gives:

$$\frac{v}{v_p} = -\frac{1}{4}\left(\frac{r \cdot \tau_{0_{90°}}}{v_p \mu}\right)^2 \frac{y^2}{r^2} + \left(\frac{r \cdot \tau_{0_{90°}}}{v_p \mu}\right)\frac{y}{r} \tag{6.5}$$

Introducing the dimensionless parameter:

$$\mathfrak{R} = \frac{r \tau_{0_{90°}}}{v_p \mu} \tag{6.6}$$

Eqs. (6.4) and (6.5) become:

$$\frac{\delta}{r} = \frac{2}{\mathfrak{R}} \tag{6.7}$$

$$\frac{v}{v_p} = -\frac{1}{4}\mathfrak{R}^2 \frac{y^2}{r^2} + \mathfrak{R}\frac{y}{r} \tag{6.8}$$

The dimensionless parameter $\mathfrak{R}$ is the *vorticity* identical to that calculated by Kawagurti [246] and of Jenson [253] and depends on the Reynolds number.

For the domain $0 < \text{Re} < 40$ the calculation of the parameter $\mathfrak{R}$ as a function of the Reynolds number was made on the basis of Stokes equations and of those suggested by different authors [244,246,248,250,253]. For the domain $\text{Re} > 350$ the calculation was made on the basis of the laminar boundary layer [103,254,255]. The obtained values were ploted in Figure 6.22b. For for the domain $40 < \text{Re} < 350$ the curve was drawn by interpolation [106].

For a certain angle θ, considering that the tangential tension decreases linearly with the distance from the sphere wall, analogously to the decrease in the middle section of the sphere, Eq. (6.2) has the form:

$$\mu d\left(\frac{dv}{dy}\right) = \tau_{0\theta}\frac{dy}{\delta} \tag{6.9}$$

Integrating this equation with the boundary condition:

$$v_\theta = 0 \text{ for } y = 0$$
$$v_\theta = v_p \sin\theta \text{ for } y = \delta_\theta$$

it is obtained analogously to Eq. (6.3):

$$v_\theta = -\frac{\tau_{0\theta}}{\mu\delta_\theta}\frac{y^2}{2} + \frac{v_p \sin\theta}{\delta_\theta}y + \frac{\tau_{0\theta}}{2\mu}y \tag{6.10}$$

Differentiating with respect to y gives:

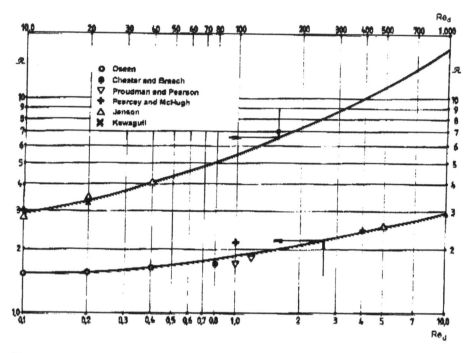

Figure 6.22b   Correlation of A with Reynolds number.

$$\frac{dv_\theta}{dy} = -\frac{\tau_{0\theta}}{\mu\delta_\theta}y + \frac{v_p \sin\theta}{\delta_\theta} + \frac{\tau_{0\theta}}{2\mu}$$

Since for $y = \delta_\theta$, $\dfrac{dv_\theta}{dy} = 0$, one obtains finally equations similar to those corresponding to the middle section:

$$\frac{\delta_\theta}{r} = \frac{2}{\left(\dfrac{r\tau_{0\theta}}{v_p\mu\sin\theta}\right)} \qquad (6.11)$$

and

$$\frac{v_\theta}{v_p \sin\theta} = -\frac{1}{4}\left(\frac{r\cdot\tau_{0\theta}}{v_p\mu\sin\theta}\right)^2 \frac{y^2}{r^2} + \left(\frac{r\cdot\tau_{0\theta}}{v_p\mu\sin\theta}\right)\frac{y}{r} \qquad (6.12)$$

The dimensionless criterion $\Re$ may be written here as:

$$\Re_\theta = \frac{r\cdot\tau_{0\theta}}{\mu\cdot v_p\sin\theta} \qquad (6.13)$$

which simplifies somewhat the writing of Eqs. (6.11) and (6.12) to the shape of Eqs. (6.7) and (6.8).

These equations are of course valid only up to the detaching point θ_D.

Once the velocity profile around the spherical particles is known, the problem of displacing thickness, which, as it was shown, intervenes in the calculation of the external diffusion phenomena, may be addressed.

In this calculation one must take into account that for a spherical particle the equating of the surfaces represented in Figure 6.19, which define the displacing thickness δ^x, must be done by taking into account the curvature of the sphere. Thus, for the middle section of the sphere, the ring-shaped surfaces must be set equal, corresponding to the distances δ^x and δ from the surface of sphere, using the relation:

$$v_p S^x = 2\pi \int_0^\delta (v_p - v)(r + y)dy \tag{6.14}$$

where S^x, the displacement section, can be correlated with the displacement thickness, δ^x, by using the equation:

$$S^x = \pi(2r\delta^x + \delta^{x2}) \tag{6.15}$$

For a certain angle θ these equations become:

$$v_p S_\theta^x \sin\theta = 2\pi \int_0^{\delta_\theta} (v_p \sin\theta - v_\theta)(r + y)\sin\theta \, dy \tag{6.16}$$

$$S_\theta^x = \pi \sin\theta (2r\delta_\theta^x + \delta_\theta^{x2}) \tag{6.17}$$

The distance that should be used in the diffusion calculations is the mean displacement thickness, δ^x. Since its calculation by using the Eq. (6.11) is difficulty in the neighborhood of the detaching point $\tau_{0\theta} \to 0$ and therefore $\delta_\theta^x \to \infty$, in the following the indirect calculation of the $\overline{\mathfrak{R}}_\theta$ was used, a much more exact calculation.

For the Stokes domain the velocity profile does not vary with the angle θ:

$$\overline{\mathfrak{R}}_\theta = \mathfrak{R}$$

For the intermediary domain using the values $\zeta_{0\theta}$ (vorticity) for different values of the angle θ and Reynolds, given by Jenson [253],

for: $\mathrm{Re}_d = 10$ $\overline{\mathfrak{R}}_\theta = 1.006\mathfrak{R}$
for: $\mathrm{Re}_d = 20$ $\overline{\mathfrak{R}}_\theta = 1.013\mathfrak{R}$
for: $\mathrm{Re}_d = 40$ $\overline{\mathfrak{R}}_\theta = 1.061\mathfrak{R}$
For the boundary layer domain, $\mathrm{Re}_d > 350$ the correlation:

$$\overline{\mathfrak{R}}_\theta = 1.536\mathfrak{R}$$

is obtained [106] allowing the graphical interpolation for the interval $40 < \mathrm{Re}_d < 350$.

In order to express δ^x as a function of $\overline{\mathfrak{R}}_\theta$ one replaces v_θ given by Eq. (6.12) in (6.16). Carrying out the integration of δ_θ using the expressions (6.11) and (6.13), it follows:

$$S^x = \pi \sin\theta \left(\frac{4}{3} r^2 \overline{\mathfrak{R}}_\theta^{-1} + \frac{2}{3} r^2 \overline{\mathfrak{R}}_\theta^{-2} \right)$$

By replacing S^x with the expression (6.17), simplifying, and isolating δ_θ^x we get:

$$\frac{\delta_\theta^x}{r} = -1 + \sqrt{1 + \frac{4}{3}\frac{1}{\overline{\mathfrak{R}}_\theta} + \frac{2}{3}\frac{1}{\overline{\mathfrak{R}}_\theta^2}} \tag{6.18}$$

a relation valid for all the values of θ, inclusively for that of the middle section for which $\overline{\mathfrak{R}}_\theta$ must be replaced by $\mathfrak{R}$.

Since, as shown above, for the domain $\mathrm{Re} \geq 40$, $\overline{\mathfrak{R}}_\theta \approx \mathfrak{R}$, it follows that for this domain Eq. (6.18) may be written as:

$$\frac{\bar{\delta}_\theta^x}{r} = \frac{\delta_{90°}^x}{r} = -1 + \sqrt{1 + \frac{4}{3}\frac{1}{\mathfrak{R}} + \frac{2}{3}\frac{1}{\mathfrak{R}^2}} \tag{6.19}$$

the average displacement thickness around the sphere $\bar{\delta}_\theta^x$ is therefore actually equal to the displacement thickness in the middle section.

For the domain Re > 350 the correlation between $\overline{\mathfrak{R}}_\theta$ and $\mathfrak{R}$ is:

$$\overline{\mathfrak{R}}_\theta = 1.536\mathfrak{R} \approx 1.5\mathfrak{R}$$

An approximative calculation for this domain can be made, by accepting for the values $\bar{\delta}_\theta^x$ and $\overline{\mathfrak{R}}_\theta$ an equation that is similar to (6.18). Such a calculation is of interest for the study of external diffusion. Since the boundary layer is detached over 1/3 of the external surface of the spherical catalyst particle, the reactants have access to the active sites without having to cross the boundary layer.

For the domain 40 < Re < 350, $\bar{\delta}_\theta^x$ can be determined by interpolation.

On this basis, the dependence between $\dfrac{\bar{\delta}^x}{r}$ and Re, which is necessary for the calculation of external diffusion, was plotted in Figure 6.23. This figure gives the mean boundary layer around a spherical particle as a function of the Reynolds number given by equation (6.1).

As shown above, to analyze the influence of external diffusion on reaction and regeneration in catalytic cracking, it is necessary to know the Reynolds domain in which the process takes place, or, according to Eq. (6.1), the diameter of the catalyst particles, the velocity, and the kinematic viscosity of the fluid.

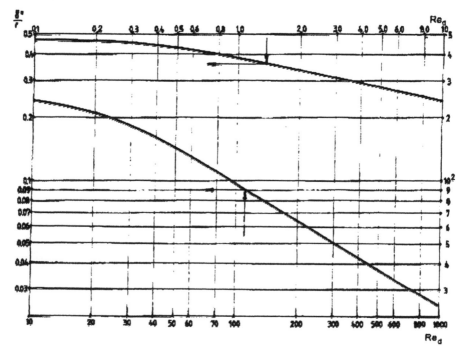

Figure 6.23 Mean boundary layer thickness (δ^*/r) ratio, function of Reynolds number. (From Ref. 106.)

Current fluid catalytic cracking processes use exclusively synthetic micro-spherical catalysts having the characteristics [107]:

diameter $d_p = 55 \times 10^{-6} - 70 \times 10^{-6}$ m

bulk density $\rho_a = 1,120 - 1,700$ kg/m^3

Since the Reynolds number depends on the properties of the fluid one has to examine the influence of external diffusion separately on the reaction process and on catalyst regeneration.

6.4.1.1 Effect of External Diffusion on Cracking

Besides the size of the catalyst particles, the influence of external diffusion depends on the relative velocity of the fluid particle and on the viscosity of the feed and of the products.

In the early versions of catalytic cracking systems that operated in dense phase, the velocity of the fluid above the catalyst bed was of the order 0.9–1.1 m/s. Since inside of the bed a portion of the flow cross section is occupied by the catalyst, the ascending movement of which is relatively low, the relative velocity between fluid and catalyst particles (the slip velocity) is larger and is estimated at 1.4 m/s.

In modern units such as the riser type, the linear fluid velocity ranges between 3.0 and 4.0 m/s, which is practically equal to the slip velocity.

Assuming as feedstock a heavy gas oils of molecular mass of 300–400, the fluid viscosity for vapors at 500°C and 1.6 bar may be obtained from Figure 6.24.

$$\rho = M\frac{p}{RT} = 7.6 \div 10.1 \text{ kg/m}^3$$

$$\mu_v = 0.0068 \div 0.005 \text{ cP} = 0.68 \cdot 10^{-5} \div 0.5 \cdot 10^{-5} \text{ kg/m} \cdot \text{s}$$

For the products: $M \approx 90$, $p = 1.4$ bar.

$$\rho = 1.98 \text{ kg/m}^3$$

$$\mu = 0.015 \text{ cP} = 1.5 \cdot 10^{-5} \text{ kg/m} \cdot \text{s}$$

Taking into account these data, one may calculate for a reactor operating in dense phase at the reactor inlet:

For distillates with $M = 300$:

$$\text{Re} = \frac{1.6 \cdot 60 \cdot 10^{-6} \cdot 7.6}{0.68 \cdot 10^{-5}} = 94$$

For heavy distillates with $M = 400$:

$$\text{Re} = \frac{1.6 \cdot 60 \cdot 10^{-6} \cdot 10.1}{0.5 \cdot 10^{-5}} = 170$$

For the outlet conditions it follows:

$$\text{Re} = \frac{1.4 \cdot 60 \cdot 10^{-6} \cdot 1.98}{1.5 \cdot 10^{-5}} = 11.2$$

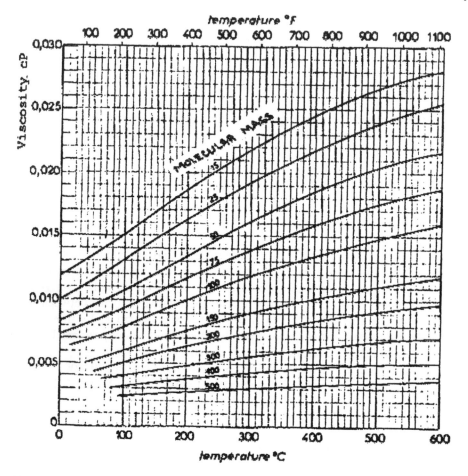

Figure 6.24 Viscosity of hydrocarbons in vapor phase at 1 bar. (From Ref. 108.)

As a result of intense internal mixing, the conditions in a reactor with a dense fluidized bed are close to perfect mixing, i.e., the conditions inside the bed correspond to those at the outlet: that means Re = 11.2. The boundary layer therefore will not be detached in the back of the particles (Figure 6.21) and its mean thickness will be, according to the graph of Figure 6.23: $\delta^* = 0.236 \times r = 0.236 \times 30 = 7.1 \mu m$ The external diffusion will thus influence the overall rate of the process.

The total thickness of the boundary layer, is given by Eq. (6.11) and the graph (Figure 6.22b).

For Re = 11.2, Figure 6.22b gives $\Re = 3$ and the relation (6.11):

$$\delta = \frac{2r}{\Re} = \frac{60}{3} = 20 \ \mu m$$

In order to see which method should be used for calculating external diffusion, it is necessary to check whether the boundary layers of neighboring particles are or are not interpenetrating. Only in the second case the theoretical calculation deduced for free spherical particles may be used.

According to Figure 8.14, the fraction of free volume inside the dense phase fluidized bed of catalytic cracking reactors varies between the limits:

$$1 - \eta_a = 0.160 \div 0.390$$

Because the fluidized bed is equivalent to a mean statistical position in the cubic network of the solid particles, the cubes have sides equal to 2a. One may write:

$$1 - \eta_a = \frac{\frac{4}{3}\pi r^3}{(2a)^3}$$

from which:

$$a = \frac{r}{2}\sqrt[3]{\frac{\frac{4}{3}\pi}{1 - \eta_a}} \tag{6.20}$$

Replacing $(1 - \eta_a)$ and r, it follows:

$$a = 44.5 \ \mu m \text{ and } a - r = 14.5 \ \mu m$$

Since $a - r < \delta$, the boundary layers are overlapping and the only calculation methods that may be used are those based on nondimensional criteria or those directly based on experimental data.

For the riser systems, the pressures at the inlet and outlet of the riser will be of the order of 2.2 respectively 1.6 bar and the Reynolds numbers will be:

For the distillates with $M = 300$:

$$Re = \frac{3.0 \cdot 60 \cdot 10^{-6} \cdot 10.4}{0.68 \cdot 10^{-5}} = 275$$

For the distillates with $M = 400$:

$$Re = \frac{3.0 \cdot 60 \cdot 10^{-6} \cdot 13.8}{0.5 \cdot 10^{-5}} = 490$$

For the outlet condition one obtains:

$$Re = \frac{4.0 \cdot 60 \cdot 10^{-6} \cdot 2.27}{1.5 \cdot 10^{-5}} = 36.3$$

It follows that in the conditions of the riser reactor, external diffusion will not influence the overall reaction rate since the boundary layer will be detached behind the catalyst particles.

6.4.1.2 Effect of External Diffusion on Catalyst Regeneration

The regeneration of the catalyst by the combustion of coke is performed in dense phase in a fluidized bed reactor. In modern units the regeneration is performed in a riser similar to that used for the cracking.

The linear velocities of the fluid are practically the same in the analogous equipment, while in the regeneration the pressure is somewhat lower than in the reactor.

The density and the viscosity of the air and of the flue gases are similar and have the following values*:

For the systems in dense phase, $t = 650°C$ and $p = 1.2$ bar:

density of air $= 0.457$ kg/m^3

$\mu = 0.01716 \times 2.309 = 0.0396$ cP $= 3.96 \cdot 10^{-5}$ kg/m·s

For the mean conditions in the riser: $t = 650°C$; $p = 2$ bar:

density $= 0.761$ kg/m^3

$\mu = 3.96 \times 10^{-5}$ kg/m·s

Based on these data and on the mean linear velocities of the fluid for the two systems, it follows:

For regenerators in dense phase:

$$Re = \frac{1.4 \cdot 60 \cdot 10^{-6} \cdot 0.457}{3.96 \cdot 10^{-5}} = 0.97$$

For regenerators of the riser type:

$$Re = \frac{4.0 \cdot 60 \cdot 10^{-6} \cdot 0.761}{3.96 \cdot 10^{-5}} = 4.61$$

As in the case of the reactors, according to the graph of Figure 6.23 and Eq. (6.11), it follows for:

$Re = 0.97 \qquad \delta^* = 11.6$ μ $\qquad \delta = 32.4$ μ

$Re = 4.61 \qquad \delta^* = 8.46$ μ $\qquad \delta = 24.0$ μ

For the regeneration in dense phase the fraction of free volume of the fluidized bed varies according to the graph of Figure 7.14, between the limits: $1 - \eta_a = 0.20 - 0.25$.

According to Eq. 6.20, the distance between the microspheres of the catalyst within the bed is:

$a - r = 11.3 \div 8.4$ μ

The comparison of these values with $\bar{\delta}$ * and δ proves that there is a significant overlap or interpenetration of the boundary layers that surround the catalyst particles. This situation precludes the use of the theoretical calculation method that was developed for independent microspheres. The only calculation methods that can be used are those based on nondimensional numbers developed by the processing of specific experimental data.

In regenerators of the riser type, the fraction of free volume in the ascending flow varies between the limits:

---

* The viscosities were calculated using Sutherland's law·

$$\mu = \mu_o \frac{273 + C}{T + C} \left(\frac{T}{273}\right)^{1.5}$$

in which for air: $\mu_o = 0.01716$ cP and $C = 111.0$ for temperatures between 289 and 1098 K [109].

$$1 - \eta_a = 0.035\text{--}0.080$$

and according to Eq. (6.20) it follows:

$$a \quad \prime = 43.9\text{--}26.1 \ \mu$$

In all cases $a - r > \delta$; therefore, for regenerators of the riser type the external diffusion rate may be calculated on basis of the theoretical classical methods of reactant diffusion through the boundary layer that surrounds the catalyst microspheres.

The equation that expresses the diffusion of a reactant A towards the external surface of the catalyst particle and that takes into account the diffusion in the opposite direction of the reaction products has the general form [110]:

$$\frac{dp_A}{dy}\frac{p}{RT} = \frac{\phi_{A_y}}{D_A}(p + \beta p_A) \tag{6.21}$$

where:

p_A = the partial pressure of the reactant A in ata
y = the distance from the surface in cm
p = the pressure in the system in ata
R = the gas constant in $cm^3 \cdot$ ata/mol·degree
ϕ_{A_y} = the reactant flow A at the distance y from the surface in moles/s·cm^2
D_A = the diffusion coefficient of the reactant A in cm^2/s
β = the number of moles of gases resulting from the reaction per mole of reactant A.

As an example, for the burning of a coke containing 10% H_2 and producing flue gases in which the ratio $CO_2/CO = 1.5$ (data typical for fluid catalytic cracking), $\beta = 0.3$. Since for air, $p_A = 0.21p$, the term βp_A in Eq. (6.3) can be neglected. It follows:

$$-\frac{dp_A}{dy} \cdot \frac{1}{RT} = \frac{\phi_{A_y}}{D_A} \tag{6.22}$$

Since for a microspherical catalyst, the displacement thickness $\bar{\delta}$ * is significant compared to the radius of the particles, ϕ_A can be expressed by the equation:

$$\phi_{A_y}(r + y)^2 = \phi_{A_0}r^2$$

Replacing in Eq. (6.21), after the separation of the variables, it follows:

$$\phi_{A_0}r^2\frac{dy}{(r + y)^2} = -\frac{D_A}{RT}dp_A$$

By the integration of this equation between the limits:

For $y = 0$ $\quad p_A = p_{A_S}$

For $y = \bar{\delta}*$ $\quad p_A = p_{A_f}$

where the subscript S refers to the surface of the catalyst microsphere and f to the flow, the end result is:

$$\Phi_{A_0} = \frac{-D_A}{RT\bar{\delta}^*} \cdot \frac{r+\bar{\delta}^*}{r}\left(p_{A_f} - p_{A_S}\right)$$

In order to express the diffusion rate r_{D_A} in moles/s·g cat as is customary in kinetic calculations, the right side of Eq. (6.22) must be multiplied by S—the external surface of all the microspheres contained in a gram of catalyst. It results:

$$r_{D_A} = \frac{-D_A \cdot S}{RT\bar{\delta}^*} \cdot \frac{r+\bar{\delta}^*}{r}\left(p_{A_f} - p_{A_S}\right) \tag{6.23}$$

wherein:

$$S = \frac{4\pi r^2}{\frac{4}{3}\pi r^3 \cdot \gamma_a} = \frac{3}{r\gamma_a} \tag{6.24}$$

r being the radius of the microspheres of catalyst.

The use of Eq. (6.23) requires the value of the diffusion coefficient. To this purpose, published values [111], theoretical calculations [112], or empirical methods such as that of Wilke and Lee [113] or of Fuller et al. [114–116] may be used, which are satisfactory for engineering calculations and also can be found in the monograph of Smith [112].

In the following calculations, the diffusion coefficient of oxygen through air was used, which has the value [117] 0.138 cm^2/s.

At 650°C the corrected value of the diffusion coefficient is:

$$D_T = D_0\left(\frac{T_T}{T_0}\right)^{1.5} = 0.138\left(\frac{923}{293}\right)^{1.5} = 0.772 \text{ cm}^2/\text{s}$$

Using Fuller's equation:

$$D_{AB} = \frac{0.00143\,T^{1.75}}{pM_{AB}^{0.5}\left[\left(\sum v\right)_A^{1/3} + \left(\sum v\right)_B^{1/3}\right]^2} \tag{6.25}$$

where:

D_{AB} = binary diffusion coefficient in cm^2/s
M_{AB} = $2[1/M_A + 1/M_B]^{-1}$, M_A and M_B = molecular masses of the substances
p = pressure in bar
$\sum v$ = sums of the atomic diffusion volumes.

For the diffusion of oxygen towards the external surface of the microspheres of the catalyst, considering that the diffusion takes place through the prevalent nitrogen layer at the temperature of 650°C and a pressure of 2 bar one obtains:

$$M_{AB} = 2\left[\frac{1}{32} + \frac{1}{28}\right]^{-1} = 29.87$$

$$D_{O_2N_2} = \frac{0.00143 \cdot 923^{1.75}}{2.0 \cdot 29.87^{0.5}\left[16.3^{1/3} + 18.5^{1/3}\right]^2} = 0.7535 \text{ cm}^2/\text{s},$$

a result which is very close to the one obtained before.

For the catalyst having $\overline{d} = 60 \cdot 10^{-4}$ cm and the apparent density $\gamma_a = 1.4$ g/cm^3, according to relation (6.6):

$$S = \frac{3}{r \cdot \gamma_a} = \frac{3}{30 \cdot 10^{-4} \cdot 1.4} = 714 \text{ cm}^2/\text{g}$$

the external diffusion rate for the regeneration in the riser system will be, according to the relation (6.5):

$$r_D = -\frac{0.7535 \cdot 714}{0.0821 \cdot 10^3 \cdot 923 \cdot 8.46 \cdot 10^{-4}} \cdot \frac{60 + 8.46}{60} \left(p_{A_f} - p_{A_S} \right)$$

$$= -9.57 \left(p_{A_f} - p_{A_S} \right) \text{ mols/s} \cdot \text{g cat}$$

The very high value of the mass transfer coefficient proves that in the conditions of the riser, the external diffusion does not exercise any influence on the overall coke burning rate.

A calculation for the regenerator in dense phase is provided in Chapter 8. It must be remarked that for such regenerators the estimate shows that the external diffusion strongly influences the burning of coke [118,119], while, as shown, it is weaker or missing for the riser systems.

In the case of a moving catalyst bed, the influence of external diffusion on reaction and regeneration may be evaluated by means of the methods used for processes with a stationary catalyst [112]. Since the cracking processes with moving a catalyst bed have no current interest, these problems aren't dealt with here.

6.4.2 Pore Diffusion

Internal diffusion through the pores of zeolitic catalyst is treated differently from the classical catalysts.

Classical methods are fully applicable to pore diffusion in classical catalysts. In zeolites, one takes into account specific effects due to the regular structure of the pores that leads to the phenomena called *shape selectivity*.

This selectivity is the consequence of the impossibility of hydrocarbon molecules, which exceed the pore diameter, penetrating the zeolite micropores. The result is that some components of the feed will be cracked selectively with much higher rates than others.

This situation is enhanced in the cracking of residual feedstocks, the very heavy components of which can penetrate only to the active sites of the matrix.

The most adequate method for expressing the effect of the diffusion influence through the pores of the classic catalyst is by means of the effectiveness factor η, which is defined as the ratio of the observed reaction rate to that which would occur in the absence of diffusion effects within the pores of the catalyst:

$$\eta = \frac{r'}{r} \tag{6.26}$$

where:

r' = the experimentally observed reaction rate
r = the rate that would occur in absence of diffusion effects, or corresponding to the use of the whole catalyst internal surface.

Eq. (6.26) allows the experimental determination of the effectiveness factor, using for the determination of the rate r very small catalyst particles for which the internal diffusion cannot decrease the rate of the process.

The effectiveness factor can be calculated using the Thiele modulus, h, using for plan particles the relation:

$$\eta = \frac{1}{h} th \cdot h \tag{6.27}$$

and for spherical particles the equation:

$$\eta = \frac{1}{h}\left(\frac{1}{th \cdot 3h} - \frac{1}{h}\right) \tag{6.28}$$

The graph of Figure 6.25a replaces such calculations and demonstrates among others that there are but small differences the between the value of the effectiveness factors of catalyst particles of different shapes. It is estimated [112] that these differences are smaller than the precision in the evaluation of the diffusion coefficient that intervenes in the calculation of module h.

The general expression of the module h is [120–123]:

$$h = L\frac{r_{AS}}{\sqrt{2}}\left[\int_{C_{Ae}}^{C_{AS}} D_e \cdot r_A dC_A\right]^{1/2} \tag{6.29}$$

where:

r_{AS} and r_A = reaction rates at the external surface, respectively in the pores, both expressed on basis of volume of catalyst particle
C_{AS} = concentration of the reactant at the external surface
C_{Ae} = concentration of the reactant at equilibrium; it is equal to zero for irreversible reactions.

For irreversible reactions of the nth order, Eq. (6.29) becomes:

$$h = L\left(\frac{n+1}{2}\right)^{1/2}\left[\frac{k\rho_S C_{AS}^{n-1}}{D_e}\right]^{1/2} \tag{6.30}$$

and for irreversible reactions of the order I:

$$h = L\left[\frac{k\rho_S}{D_e}\right]^{1/2} \tag{6.31}$$

In Eqs. (6.29)–(6.31), the terms are defined as

k = reaction rate constant
ρ_S = particle density

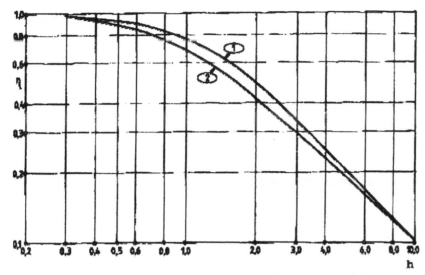

Figure 6.25a Effectiveness factor η as function of Thiele modulus. 1 – planar particle (Eq. 6.9), 2 – spherical particle (Eq. 6.10).

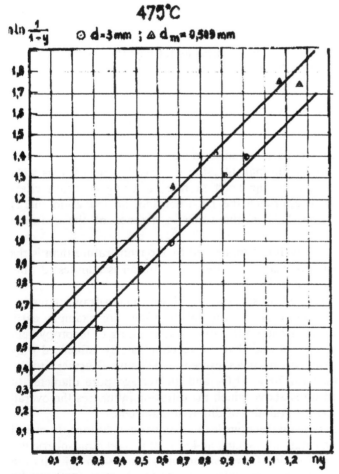

Figure 6.25b Effect of catalyst particle diameter on reaction rate [136]. $\Delta - d = 3.0$ mm, $o - d = 0.6$ mm.

D_e = effective diffusion coefficient

L = characteristic length, given by the ratio of the volume of the particle and its geometric surface [120]. For spheres $L = d/6$.

For zeolites with a 1 μm crystal size [237]:

$$L = \frac{10^{-6}}{6} \, \text{m}^{-1}$$

and for 0.1 μm crystal size (submicron zeolites):

$$L = \frac{0.1 \times 10^{-6}}{6} \, \text{m}^{-1}$$

It follows that the decrease of the size of the zeolite crystals increased the reaction rate. This was confirmed experimentally [237].

The effective diffusion coefficient is expressed by the equation:

$$D_e = \frac{1}{1/D_{AB} + 1/(D_K)_A} \tag{6.32}$$

where D_{AB} is the coefficient for the reactant A diffusing through the substance B, and $(D_K)_A$ is the Knudsen diffusion coefficient of reactant A:

$$(D_K)_A = 9.7 \cdot 10^3 \cdot a \sqrt{\frac{T}{M_A}} \tag{6.33}$$

In this last equation a is the particle radius in cm.

Concerning the influence of internal diffusion on overall kinetics, it is well-known that besides the decrease of the reaction rate, a strong diffusion effect will cause the reaction order to approach the 1st order and the activation energy to decrease, sometimes to only half of the value in absence of the diffusion barrier.

Since the diffusion rate is much less influenced by the temperature than is the reaction rate, there are situations where the effect of diffusion is manifested only after a certain temperature is exceeded.

Studies of catalytic cracking on classic catalysts allowed one to draw interesting conclusions concerning the influence of the internal diffusion. For example, early studies [124] on the influence of internal diffusion at the cracking of isopropylbenzene and of other alkylbenzenes showed that the reaction rate is influenced only when the diameter of the catalyst particles exceeds 0.080 mm. Further studies [125] using a classic catalyst with 10% Al_2O_3 and an average pore diameter of 30 Å, determined the temperatures above which the diffusion influences the overall rate of the process (Table 6.7) for granules of different diameters.

Studies on cracking of crude oil fractions lead to similar conclusions. In one of our studies [1] we investigated catalytic cracking at 450°C of several atmospheric gas oils with a distillation range of 250–360°C and of vacuum gas oils with a distillation range 300–450°C on a classic catalyst containing 12% Al_2O_3. The results showed a doubling of the reaction rate constant when switching from catalyst particles of 3

Table 6.7 Temperature Above Which Internal
Diffusion Influences Reaction Rate

Catalyst particles diameter (mm)	Temperature °C
0.056	690
0.290	380
0.630	345
1.750	307

Source: Ref. 125.

mm to particles having sizes in the range of 0.4–0.6 mm as depicted in Figure 6.25b
[136]. Other authors [126,127] obtained similar results.

These studies show that internal diffusion strongly influences the reaction rate
in moving bed processes that use catalysts as granules of 3–5 mm diameter and do
not influence the process rate in reactors with the catalyst in fluidized bed.

It would be wrong to believe that when using classic catalysts selectivity is not
related to the penetration of some components of the feed into the pores. This effect
on selectivity is related only to the difficulties of pore penetration by some components with a very high molecular mass such as the asphaltenes, resins, and organometallic compounds. It is much less specific than with zeolite catalysts.

This effect is illustrated by the data from the Table 6.8, which give the nickel
content of the catalyst powder obtained by the progressive erosion of the external
surface of used catalyst particles [128]. The table shows that the metal (nickel) content of the surface layer is four times the average concentration of metal in the whole
particle.

In addition to the problems seen in classic catalysts, zeolites have specific ones
related to steric hindrances for the penetration of some feedstock components in the
micropores. The dimensions of the access openings into the pores of various zeolites
are shown (Figure 6.12) compared to the sizes of the hydrocarbons of various
structures, illustrating steric limitations. The data of Figure 6.26 includes the values
of the corresponding diffusion coefficients [129]. The figure shows that important
variations of the diffusion coefficients may result for small variations of the molecular diameters.

Table 6.8 Nickel Content in the Dust Obtained by
External Surface of Erosion of Catalyst Granules

Dust obtained by erosion (wt %)	Ni content in dust
0.5	0.0403
1.5	0.0408
4.5	0.0318
100.0	0.0105

Source: Ref. 128.

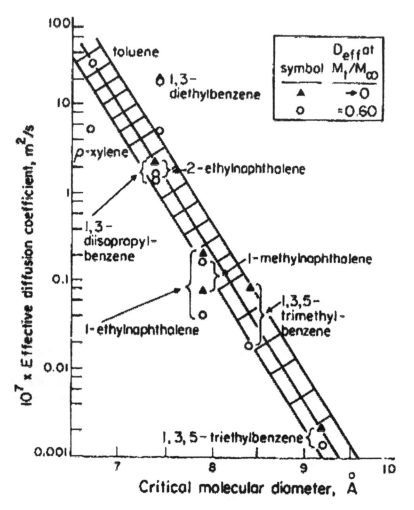

Figure 6.26 Diffusion coefficient as function of molecule diameter. ▲– far from equilibrium, ○ – 0.6 of equilibrium. (From Ref. 129.)

As a result, some components of the feed have access to all active sites of the catalyst, while others have access only to a fraction of them. Therefore, the relative reaction rates will be influenced by the shape of the components that will react, an example of *shape selectivity*.

As an illustration, whereas with classic catalysts the cracking rate of the cycloalkanes increases with the number of rings in the molecule, with zeolite catalysts it decreases because the molecules of larger sizes do not have access to the active sites. The data of Table 6.9 [130] illustrate this difference.

Similarly, with alkanes the rate of reaction depends on the degree of branching, as shown in the data of Table 6.10 [131]. Besides the limitations due to branching, the table illustrates also the much higher reactivity of alkenes compared to alkanes.

It is obvious that the selectivity obtained from various zeolites depends on their specific structure, on the size, and the shape of the pores. On the basis of this

Table 6.9 Relative Reaction Rate of Different Hydrocarbons on Classic Silica-Alumina Catalysts and on Zeolites

Hydrocarbon	Classic Si–Al catalyst	Zeolite REHx catalyst	Reaction rate ratio zeolite catalyst/classic catalyst
C_2H_5 — ring — C_2H_5 (with C_2H_5)	140	2370	17.0
CH_3, CH_3, CH_3 substituted bicyclic	190	2420	13.0
(anthracene-type structure)	205	953	4.7
(pyrene-type structure)	210	513	2.4

Source: Ref. 130.

knowledge, the zeolites to be incorporated in the matrix are selected according to the objectives of the process using the catalyst.

6.4.3 Reaction Kinetics

6.4.3.1 Overall Kinetic Equations

Among the first kinetic equations for the catalytic cracking of gas oil were those published in 1945 by A. Voorhies [132].

His equation correlates the average conversion $\bar{x}$ obtained on a stationary bed of catalyst with the volume flowrate w and the contact time τ, and has the form:

$$\bar{x} = \left(\frac{A'_c}{A'_f w}\right)^{1/N_f} \tau^{(N_c-1)/N_f} \tag{6.34}$$

where A'_c, A'_f, N_c, and N_f are the parameters that depend on the feedstock and on the catalyst, the subscripts indicating if they refer to the catalyst (c) or to the feed (f).

On basis of experimental measurements, Voorhies also established an empirical equation for coke formation:

$$C_c = A'_c \tau^{N_c} \tag{6.35}$$

where C_c is the coke amount as weight percent, on the catalyst.

The experiments allowed the calculation of the parameters A'_c and N_c, and then, by means of Eq. (6.14), of the parameters A'_f and N_f, lead to a second empirical equation:

Table 6.10 Reaction Rate Constants and Efficiency Factors for Some Alkanes and Alkenes at 538°C and 1 Bar

Hydrocarbon		Efficiency factor (η)	Reaction step rate, (s^{-1})
n-hexane	C–C–C–C–C–C	1.000	29
3 methyl-pentane	C–C–C–C–C–C (with C branch)	1.000	19
2,2-dimethyl-butane	C–C–C–C (with two C branches)	0.300	12
n-nonane	C–C–C–C–C–C–C–C–C	1.000	93
2,2-dimethyl-heptane	C–C–C–C–C–C–C (with two C branches)	0.130	63
Hexene-1	C=C–C–C–C–C	0.860	7530
3-methyl-pentene-2	C–C=C–C–C–C	0.500	7420
3,3-dimethylbutene-1	C=C–C–C (with two C branches)	0.028	4950

HZSM-5 catalyst with pores length 1.35 μm.
Source: Ref. 131.

$$v = A'_f \bar{x}^{N_f} \tag{6.36}$$

where v is the weight percent of coke in relation to the amount of feed.

From among these equations, the one relating coke formation to residence time (6.35) is still being recommended [11].

In 1946, by considering catalytic cracking as a first order process, A.V. Frost [133] suggested the use of the kinetic equation deduced by T.V. Antipina [134] for first order heterogeneous catalytic reactions in plug flow. This equation assumes that all participating substances follow the Langmuir adsorption law, and that the reaction step is rate controlling.

Catalytic cracking is written as:

$$A \rightarrow v_1 A_1 + v_2 A_2 + \cdots + v_n A_n \tag{a}$$

and the reaction rate r_A is expressed by:

$$r_A = k_S \theta_A \tag{6.37}$$

where k_S is the rate constant of the surface reaction and θ_A is the surface fraction occupied by the reactant A.

By using b_i for the adsorption coefficients and p_i for the partial pressure, and by accepting the Langmuir adsorption law, Eq. (6.17) becomes:

$$r_A = \frac{k_S b_A p_A}{1 + b_A p_A + b_{A_1} p_{A_1} + b_{A_2} p_{A_2} + \cdots + b_{A_n} p_{A_n}} \tag{6.38}$$

Using x for the conversion and δ for the increase of the number of moles resulted from the stoichiometry of the reaction:

$$\delta = \sum v_i - 1$$

one obtains for the partial pressures the expressions:

$$p_A = p \frac{1 - x}{1 + \delta x}$$

$$p_{A_i} = p \frac{v_i x}{1 + \delta x}$$

Substituting in (6.38) and performing the simplifications, it follows:

$$r_A = \frac{k_S b_A p (1 - x)}{1 + b_A p + (\delta - b_A p + p \sum b_i v_i) x}$$

Using w for the volume feed rate, for an integral plug flow reactor it follows:

$$\frac{1}{w} = \int_0^x \frac{dx}{r_A}$$

Substituting r_A and integrating, one obtains eventually:

$$\alpha = w \left[\ln \frac{1}{1 - x} - \beta x \right] \tag{6.39}$$

where α and β are given by the equations:

$$\alpha = \frac{k_S b_A p}{1 + \delta + p \sum b_i v_i} \qquad \beta = \frac{\delta - b_A p + p \sum b_i v_i}{1 + \delta + p \sum b_i v_i} \tag{6.40}$$

Taking into account that the adsorption of coke is very strong, G.M. Pancencov estimated [135] that the adsorption of the other components may be neglected from the denominator of Eq. (6.38) and this equation becomes:

$$r_A = \frac{k_S b_A p_A}{1 + b_c p_c} \tag{6.41}$$

which, after making the substitutions and integrating, leads to the final equation:

$$k = w \left(\ln \frac{1}{1 - x} - x \right) \tag{6.42}$$

where the reaction rate constant k is given by the expression:

$$k = \frac{k_S b_A}{b_c v_c}$$

For the case $\beta = 1$, Eq. (6.42) is identical with (6.39).

A large number of studies have confirmed this equation. In our experimental work [97,136,137] feedstocks were prepared by solvent refining and had aromatic carbon content determined by the method n-d-M comprised between 0 and 60% and distillation range between 200–450°C. In all cases, the deviation of the constant β from unity was below the limits of the experimental errors.

The simplification brought by G.M. Pancencov [135] can thus be considered correct and Eq. (6.42) is recommended.

An equation identical to that of Frost was suggested in 1958 by Voge [84] and is used by Decroocq et al. [4].

6.4.3.2 Correlation of Kinetic Constants with Feed Composition

As early as 1966, Raseev showed that compounds with aromatic structures that strongly adsorbed on the catalyst surface decrease the activity of the catalyst and the overall rate of the catalytic cracking reaction [136].

Four fractions with distillation ranges of: 200–250°C, 250–300°C, 300–350°C, and 350–450°C, were prepared using a Romanian crude oil. From each fraction, aromatic concentrates were prepared by means of solvent extraction. The content of aromatic carbon atoms (C_A) in the raffinate and extract were 50–57% C_A and 0.0–2.1% C_A respectively. By mixing them in different proportions, five feeds were obtained with various contents of aromatic carbons. The cyclo-analysis was performed by the n-d-M method [178] while, for the fractions with high C_A content, the Hazelwood method was used [179].

Each feed was cracked in a fixed bed micropilot unit, using a classical Si–Al catalyst [136,137,153].

The kinetic Eq. (6.39) was fitted to the results and the values of the constants α and β were obtained.

For all the feeds and cracking temperatures (450°C, 475°C, and 500°C) the value of the constant β was β = 1.

The rate constant α was be correlated with the content of aromatic carbons C_A determined by cyclo-analysis, and the general equation was obtained:

$$\alpha = a(1 - 1.38C_A) \qquad (6.43)$$

where C_A is expressed in weight fractions and α is a constant dependent on the distillation range of the feed and on the cracking temperature (see Table 6.11).

Table 6.11 Constant a Values, Eq. (6.43)

Feed distillation limits (°C)	Temperature (°C)		
	450	485	500
200–250	0.0909	–	0.2464
250–300	0.1775	–	0.3990
300–350	0.1985	–	–
350–400	0.3850	0.5122	–

Source: Ref. 1.

The feed characteristics were correlated with the percentage of coke v by using the equation:

$$v = dw^{-b} \tag{6.44}$$

By processing the experimental data the value of $b = 0.6$ for the atmospheric distillates and $b = 0.7$ for the vacuum distillates were obtained.

The constant d was correlated with the percentage of aromatic carbons C_A in the atmospheric distillate, as determined by cyclo-analysis, obtaining the equations [136,153]:

Fraction	200–250°C	$d = (1.635 + 2.93 \times C_A)10^2$
	250–300°C	$d = (3.182 + 2.83 \times C_A)10^2$
	350–400°C	$d = (4.074 + 3.68 \times C_A)10^2$

For the vacuum distillate, a correlation could be obtained with the content of hydrogen of the feed H_2 expressed by fractions by weight using the relations [97,137]:
For cracking at 450°C:

$$d = 0.1546 - 0.798 \; H_2$$

and for cracking at 485°C:

$$d = 0.1741 - 0.843 \; H_2$$

Similar studies using a white paraffin oil with additions of different proportions of α-methylnaphthalene [97] resulted in same type of correlations with the percentage of aromatic carbon atoms, C_A.

6.4.3.3 Catalyst Decay

The difficulties encountered in the analysis of kinetics of catalytic cracking derive mainly form the rapid decrease of catalyst activity during the reaction or catalyst decay, which affects the interpretation of the experimental results.

Two hypotheses were advanced concerning the cause of this decay:

1. The coke deposits forming on the catalyst
2. The duration of the catalyst use or the time on stream.

Various authors [18,138] take into account several variants concerning each of the two hypotheses of deactivation and formulate the corresponding kinetic expressions.

The decrease in catalyst activity as result of coke deposits is considered [137–141] to be a process of the first order, that may be expressed by the equation:

$$\phi_c = e^{-\alpha_c C_c} \tag{6.45}$$

where ϕ_c is the fraction of the catalyst activity remaining after a weight fraction C_c of coke was deposited on the catalyst, and α_c is the deactivation constant.

If the irreversible adsorption of coke is admitted as a cause of catalyst deactivation, one may write, similar to Eq. (6.41):

$$\phi' = \frac{1}{1 + \beta C_c} \tag{6.46}$$

As in (6.45), if deactivation is correlated to the duration of use of the catalyst τ_t, it will be expressed by the equation:

$$\phi_t = e^{-\alpha_t \tau_t} \tag{6.47}$$

Most of the studies [142–144] consider that in this case too, deactivation is a first order process. Nevertheless, some studies produced correlations considering the process to be of order zero [145], second order [146,147] or fractional [132,148]. In the latter case, the differential equation will have the form:

$$\frac{d\phi}{d\tau_t} = \alpha_m \phi^m \tag{6.48}$$

where the exponent m varies between 0 and 1.

The problem of deactivation was also approached by using expressions of the form (6.48) [18,149]. but they lead to additional mathematic difficulties and are not used much.

There are various opinions concerning the degree to which the deactivation of the catalyst affects the rate of the various reactions taken into consideration. In some studies, [11,150,151] it is assumed that since the reactions are catalyzed by the same active sites, the factor ϕ affects them identically and has thus the same value for all the reactions. In other studies [138], one considers it necessary to use various values of ϕ modifying the value of the constant α, depending on the reaction taken into consideration.

6.4.3.4 Three Components Modeling

The practical need to know not only the overall conversion, but also the coke amount and separately the yields of gasoline, $C_3 + C_4$ etc., imposed the use of increasingly complex kinetic models. The different behavior of the alkenes, cycloalkanes, and of the aromatic hydrocarbons imposed that some models also take into account the proportions in which these different classes of hydrocarbons are contained in the feedstock, which of course complicates the model.

Among these, the most simple is the model with three components, represented by the scheme:

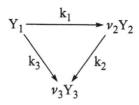

where Y_1 is the gas oil, Y_2 is the gasoline, Y_3 is the gases and the coke.

The first and the better known kinetic treatment based on this scheme was given by Weekman and Nace [152] and it is expressed by the following system of differential equations:

$$\frac{dy_1}{d(1/n)} = -k_0 y_1^2 \phi_t \tag{6.49}$$

$$\frac{dy_2}{d(1/n)} = \left(k_1 v_2 y_1^2 - k_2 y_2\right)\phi_t \tag{6.50}$$

$$\frac{d\phi_t}{d(1/n)} = -\alpha\phi_t \tag{6.51}$$

where:

y_1 and y_2 are the weight fractions of gas oil, and gasoline respectively

n is the mass reaction rate expressed as (mass of feed)/(mass of catalyst) $\cdot$ (hour), $k_0 = k_1 + k_3$, v_2 and v_3 stoichiometric coefficients

ϕ_t is the degree of deactivation of the catalyst expressed by Eq. (6.48), with m = 1

The results were fitted by using a 2nd order kinetics for the decomposition reaction of the gas oil and 1st order for the decomposition of gasoline (6.49).

The integration of Eqs. (6.39)–(6.51) for plug flow conditions through a fixed bed of catalyst leads to the following system of equations:

$$y_1 = \frac{1}{1 + \dfrac{k_0\phi_t}{n}} \tag{6.52}$$

written also as:

$$x = 1 - y_1 = \frac{\dfrac{k_0\phi_t}{n}}{1 + \dfrac{k_0\phi_t}{n}} \tag{6.53}$$

$$y_2 = r_1 r_2 e^{-r_2/y_1}\left[\frac{1}{r_2}e^{r_2} - \frac{y_1}{r_2}e^{r_2/y_1} - \text{Ein}(r_2) + \text{Ein}\left(\frac{r_2}{y_1}\right)\right] \tag{6.54}$$

where

$$r_1 = \frac{k_1 v_1}{k_0} \qquad r_2 = \frac{k_2}{k_0} \qquad \text{Ein}(x) = \int_{-\infty}^{x}\frac{e^x}{x}dx$$

Eqs. (6.52) and (6.54) give the instantaneous conversions that decrease with increasing deactivation of the catalyst. In processes with a stationary catalyst bed, such as those used in laboratory studies, one measures the final conversions of a complete cracking cycle. These conversions correspond thus to average conversions, which can be determined by the integration of Eqs. (6.52–6.54).

For the average gasoline conversion $\bar{y}_2$, the equation will be:

$$\bar{y}_2 = \int_0^1 y_2 d\left(\frac{\tau_t}{\tau_s}\right) \tag{6.55}$$

where τ_t is the time on stream of the catalyst, and τ_s is the total duration of the cycle.

The average conversion $\bar{y}_2$ of Eq. (6.54) can be calculated by using numerical methods.

The average total conversion $\bar{x}$ is given by:

$$\bar{x} = \frac{1}{\alpha_t} \ln \left[\frac{1 + \dfrac{k_0}{n}}{1 + \dfrac{k_0}{n} e^{-\alpha_t \tau_s}} \right] \qquad (6.56)$$

The system of equations (6.48–6.56) remains the same if ϕ_c—deactivation depending on the coke deposited on the catalyst (Eq. 6.45)—is used in place of ϕ_t.

Eqs. (6.52) and (6.54) allow the calculation of the total conversion and of the conversion to gasoline. The conversion to gases C_1–C_4 and coke is obtained by difference.

In order to obtain the conversion to coke by itself, it is necessary to use equations of the type (6.35) or (6.36). For systems in which the catalyst circulates between the reactor and the regenerator, the residence time of the catalyst in the reactor τ is easy to determine and the use of a type (6.35) equation is recommended. It is usually written in a simplified form, where C is the weight percent coke on the catalyst:

$$C = A\tau^N \qquad (6.57)$$

Note [11] that the value of the constant A depends on the catalyst, the feed, and the operating conditions, whereas the constant N varies very little with these factors and usually has a value of about 0.5.

The three components model, using first or second order kinetics for the cracking reactions of gas oil was utilized in many studies [151,155].

Thus, J.M. Kolesnicov, I.N. Frolova, and H.A. Lapshina [180] studied the cracking of a petroleum gas oil fraction with a density of 0.869 on a natural silimanite catalyst activated with sulfuric acid. The data were correlated by using kinetic equations of the 1st order of the form (6.39) and a three components model. The following values were obtained for the rate constants:

at 450°C $k_1 + k_3 = 0.095$ g/g·h $k_2 = 0.260$ g/g·h
at 480°C $k_1 + k_3 = 0.131$ g/g·h $k_2 = 0.326$ g/g·h
at 510°C $k_1 + k_3 = 0.208$ g/g·h $k_2 = 0.480$ g/g·h

D.W. Kraemer and H.J. Lasa [181] used a similar model but where the formation of coke was not included. They used 2nd order kinetics for the gas oil decomposition reaction. The results obtained in this study do not seem plausible. By processing the data, the value $k_2 = 0$ was obtained for the decomposition of gasoline to gases. Also, a strong influence of temperature was seen on the constant k_3 while its effect on the constant k_1 was almost insignificant.

N.N. Samoilova, V.N. Erkin, and P.J. Serikov [151] use the same three components kinetic model but took into account the chemical composition of the feed. They also took into account the time rate deactivation of the catalyst, using to this purpose Eq. (6.47). The study proved that the constant α_t in Eq. (6.47) does not depend on the nature of the feed.

The experimental data was processed using a 1st order kinetic equation for the decomposition of the vacuum distillates and also 1st order kinetics for the decomposition of gasoline.

The decomposition of the vacuum distillate (sum of the constants $k_1 + k_3$) was correlated with the ratio of the content of aromatic hydrocarbons to that of alkanic hydrocarbons in the feed—A/P by means of the exponential relation:

$$k_1 + k_3 = a_0(A/P)^{-b} \tag{6.58}$$

The values of the constants a and b for different temperatures are given in Table 6.12.

The apparent activation energies for the decomposition of vacuum distillates is situated between 24.7 and 28 kJ/mole.

M. Larocca, S. Ng, and H. de Lasa [159] make a distinction between the alkanic, cyclo-alkanic, and aromatic hydrocarbons contained in the feed in the framework of the kinetic model of Figure 6.29.

The work uses the exponential equation to express the deactivation of the catalyst as a function of the amount of deposited coke.

The gas oil feed was injected in the He stream that fluidizes the microspherical catalyst in the reactor. The rate constants were determined for three commercial catalysts, separately for alkanes, cycloalkanes, and aromatic hydrocarbons at various temperatures. On this basis, the behavior of these classes of hydrocarbons were compared.

The composition of the feed was also taken into account in other studies that dealt with the kinetics of catalytic cracking [18,156,237].

6.4.3.5 Three Components Modeling Applied to Moving Bed and Fluidized Bed Systems

The kinetic reactions deduced by Weekman and Nace were successfully applied to the design of systems involving moving beds, fluidized beds in dense phase [11], and in riser systems [154] presented in this section.

The accepting of a kinetic equation of 2nd order for the cracking of gas oil (E.6.49–6.51) was (Krishnaswamy and Kittrell [150]) compared with an identical model wherein the cracking reaction was of the 1st order, but another equation for the deactivation of the catalyst was considered. The system of differential equations used in their work was:

$$\frac{dy_1}{d(1/n)} = -k_0 y_1 \phi \tag{6.59}$$

Table 6.12 Eq. (6.58) Constants

Temperature (°C)	a_0	b
420	1.16	0.36
440	1.24	0.39
460	1.29	0.44
480	1.33	0.48
510	1.39	0.50
530	1.46	0.50

$$\frac{dy_2}{d(1/n)} = (k_1 v_1 y_1 - k_2 y_2)\phi \tag{6.60}$$

$$\frac{d\phi}{d(1/n)} = -\beta y_1 \phi \tag{6.61}$$

Different from Eq. (6.51) used by Weekman and Nice, Eq. (6.61) for the catalyst deactivation contains y_1, which is a weight fraction of feed.

Both systems of equations were integrated for systems of the riser type and compared with the experimental data [154]. For the usually used conversions, practically identical results were obtained. It is therefore possible to use in calculations two equation systems:

In the system described by Eqs. (6.49–6.51) the cracking of gas oil is a reaction of the 2nd order and the deactivation of the catalyst is independent of the feed concentration.

In the equation system (6.59–6.61), all the reactions are of the 1st order while the deactivation of the catalyst depends on the feed concentration.

In connection with this agreement it is to be remarked that Eq. (6.59) is in fact an equation of the 2nd order, since ϕ depends on y. This explains the similarity between the expressions for the integral equations obtained with the two models.

It must be emphasized that both equation systems are formulated for quasi-homogeneous reaction systems, which confers them a semi-empirical character. From this point of view, an interesting paper [156] reports that kinetic processing uses Langmuir adsorption isotherms and makes a comparison of the results with those obtained when using 2nd order, quasihomogeneous kinetics, expressed by the equations (6.49–6.51).

Another study [156] used a pulsed reactor, which made the assumption that the deactivation phenomena for the catalyst were negligible.

By accepting as in the previously cited works [135] that the adsorption of one of the reaction products is very strong, it leads to the expression:

$$-[\ln(1-x)] - x = k \frac{RT}{p_0} \left(\frac{b}{b_p}\right) \frac{G}{F} \tag{6.62}$$

where:

p_0 is the partial pressure of the feed, b and b_p are the adsorption constants for the feed and for the products
G is the amount of the catalyst,
F is the feedrate in gas phase

It is easy to understand that by including the adsorption constants in the rate constant and by using the volume feedrate, an expression identical with (6.42) is obtained.

Eq. (6.62) gives good agreement with the experimental data, including those that served to verify Eqs. (6.52–6.54). This agreement is illustrated in Figure 6.27, which shows an almost linear correlation between the corresponding expressions for the two types of equations for conversions ranging between $x = 0.3$ and $x = 0.8$. This finding led to the conclusion [156] that homogeneous 2nd order kinetics is only

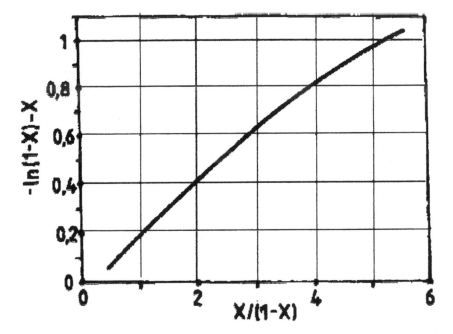

Figure 6.27 Conversion for second order reaction rate in homogenous system, versus for first order rate, in heterogeneous system, with strong adsorption of one of the reaction products. (From Ref. 156.)

a mathematical approximation of heterogeneous 1st kinetics, hindered by the coke which is formed in the process.

The mathematical form of Eq. (6.62) makes it difficult to treat models made of a number of simultaneous reactions. Therefore, in such cases the system of Eqs. (6.52–6.54) is used.

Eqs. (6.49–6.51) were used to derive the integral form not only for systems involving a stationary catalyst, but also for: (a) cracking in a fluidized bed in dense phase and (b) in reactors of the riser type, which were treated as systems with a cocurrent moving bed.

Dense phase fluidized bed reactors*. The most simple treatment is that of B. Goss, D.M. Nace, and S.E. Voltz [157], which applies the same effectiveness factor K_f to all the rate constants. Thus, Eqs. (6.53) and (6.56) become:

$$x = \frac{K_f \dfrac{k_0 \phi_t}{n}}{1 + K_f \dfrac{k_0 \phi_t}{n}} \tag{6.63}$$

$$\bar{x} = \frac{1}{\alpha_t} \ln \left[\frac{1 + K_f \dfrac{k_0}{n}}{1 + K_f \dfrac{k_0}{n} \phi_t} \right] \tag{6.64}$$

---

* The treatments that follow imply the existence of good fluidization.

Eq. (6.54) remains unchanged since it contains ratios of the rate constant and not absolute values.

The more exact equations deduced by Weekman and Nace [152] are also recommended in review articles [11]. The authors consider that the catalyst in the fluidized bed is perfectly mixed and that the ascending flow of reactants has a constant velocity across the entire cross section of the reactor, as it passes through the bed.

In this situation, in the equations system (6.49–6.50), ϕ_1 is constant, Eq. (6.51) disappears, and ϕ can be expressed by Eq. (6.46).

The Eq. (6.49) becomes:

$$\frac{dy_1}{d(1/n)} = -\frac{k_0}{1+\beta C_c} y_1^2 \tag{6.65}$$

Integrating between the limits $y = 1$ for $n = \infty$ and y for n, it results:

$$y_1 = \frac{1+\beta C_c}{1+\beta C_c + k_0/n} \tag{6.66}$$

and

$$\varepsilon = \frac{k_0/n}{1+\beta C_c + k_0/n} \tag{6.67}$$

where ε is the total conversion: $\varepsilon = 1 - y_1$.

Weekman and Nace [152] introduce an extended variable for the reaction time of the form:

$$\frac{x}{(1+\alpha_t \tau_t)w}$$

where:

x is the height, expressed as fraction of the height of the fluidized bed
$\alpha_t \tau_t$ is the exponent of Eq. (6.47) that expresses the deactivation of the catalyst function on the time on stream
w is the volumetric rate.

Introducing this variable, the Eq. (6.49) becomes:

$$\frac{dy_1}{dx} = \frac{k_0}{(1+\alpha_t \tau_t)w} y_1^2 \tag{6.68}$$

which, after integration gives:

$$y_1 = \frac{1+\alpha\tau}{1+\alpha\tau + k_0/w} \tag{6.69}$$

and

$$\varepsilon = \frac{k_0/w}{1+\alpha\tau + k_0/w} \tag{6.70}$$

These two equations differ from (6.66) and (6.67) only by the manner of expressing the deactivation of the catalyst, i.e., depending on the time on stream τ

and not on the amount of deposited coke C_c, and by expressing the feedrate in terms of volume instead of mass flowrate.

Concerning the conversion to gasoline y_2, since the integrated Eq. (6.54) involves only the ratios of the rate constants and assuming that all the constants are affected by the same deactivation factor ϕ, it results that this equation remains valid also for cracking in a fluidized bed.

The equation that expresses the maximum conversion to gasoline y_{2max} is obtained by equating to zero the differential equation (6.50), from which it results:

$$y_{2\,max} = \frac{k_1 v_2}{k_2}(1 - \varepsilon_{y\,max})^2 \tag{6.71}$$

If y_{2max} is expressed as a weight fraction, the stoichiometric coefficient v_2 must be omitted.

In order to express the volume rate corresponding to the maximum of gasoline, one writes w of Eq. (6.70) in explicit form. In the obtained equation ε is substituted by $\varepsilon_{y\,max}$ written from Eq. (6.71) in explicit form. Eventually, one obtains:

$$w_{y_{2\,max}} = \frac{k_0}{(1 + \alpha\tau)\left[\sqrt{\dfrac{k_1 v_1}{k_2 y_{2\,max}}} - 1\right]} \tag{6.72}$$

Similarly, using the expression (6.67) in the place of (6.70) it results:

$$n_{y_{2\,max}} = \frac{k_0}{(1 - \beta C_c)\left[\sqrt{\dfrac{k_1 v_1}{k_2 y_{2\,max}}} - 1\right]} \tag{6.73}$$

Riser and cocurrent moving bed reactors. These reactors are characterized by the cocurrent moving of the catalyst and the reactant, which leads to kinetic equations of the same form.

Generally, the simplifying assumption is accepted that the residence time of the catalyst in the reactor is identical to that of the vapors that circulate through the system [154]. In these conditions the residence time of the catalyst τ_1 contained in the exponent of Eq. (6.47) may be expressed by:

$$\tau_1 = \frac{1}{w} = \frac{\overline{\rho}}{n} \tag{6.74}$$

with the condition that w should be expressed in the same time units as used in Eq. (6.47); $\overline{\rho}$ represents the mean density of the vapor stream flowing through the reactor.

Substituting in (6.47) and then in (6.49), one obtains:

$$\frac{dy_1}{d(1/n)} = -k_0 y_1^2 e^{-\alpha\overline{\rho}/n} \tag{6.75}$$

By integrating this equation within the limits $y = 1$ for $n = \infty$ and y for n, it follows:

$$\frac{1}{y_1} - 1 = \frac{k_0}{\alpha\overline{\rho}}\left(1 - e^{-\alpha\overline{\rho}/n}\right) \tag{6.76}$$

which is identical with the equation obtained by Paraskos [154], and confirmed by his experimental determinations. Weekman and Nace [18] using a different path obtained at the end an equivalent equation.

Concerning the equations that express the conversion to gasoline (6.54) and the maximum of gasoline (6.71), they remain the same because in these equations only the ratios of the rate constants intervene, which are considered to be affected in identical manner by the catalyst deactivation phenomenon.

In order to obtain the feedrate corresponding to the maximum of gasoline, in Eq. (6.76) one substitutes $y_1 = 1 - \varepsilon$ and $\bar{p} = \tau_1 n$. One obtains:

$$1 - \varepsilon = \cfrac{1}{1 + \cfrac{k_0}{\alpha n \tau_t}(1 - e^{-\alpha \tau_t})}$$

For the conditions of maximum conversion to gasoline, one replaces $(1 - \varepsilon)$ by $(1 - \varepsilon_{y\max})$ from Eq. (6.71). After regrouping it follows:

$$n_{y_2 \max} = \frac{k_0(1 - e^{-\alpha \tau_t})}{\alpha \tau_t \left[\sqrt{\dfrac{k_1 v_2}{k_2 y_2 \max}} - 1 \right]} \tag{6.77}$$

Since gasoline represents the main product of catalytic cracking, the analysis of the influence of process factors and their selection to the purpose of maximizing gasoline yield is a problem of utmost importance. But, the obtained Eqs. (6.54) and (6.71) combined with (6.73) for the fluidization in dense phase, or with (6.77) for systems with riser reactor, make such an analysis very difficult.

By simplifying the system to one represented by:

$$A \xrightarrow{k_1} B \xrightarrow{k_2} C$$

Wojciechowski and Corma [18,158] deduced equations that allowed computer simulation of the variation of the conversion to gasoline, as a fraction of the total conversion. The results were plotted as curves corresponding to various reaction conditions. Despite the fact that the published representations [18] have qualitative character, they allow some conclusions with general character which are given in the Section 6.4.4.

The quantitative dependency of the conversion to gasoline versus the conversion of the gas oil was calculated by the above equations [152], and the results are compared to experimental data in Figure 6.28.

6.4.3.6 Systems With Four and More Components

A model with three components (lumps) was developed by Larocca et al. [159] in a study that focused especially on the modeling of catalyst deactivation. The study used a pulsed reactor. For the processing of the data the model with three components, described previously, and a model with five components reproduced in Figure 6.29 were applied. This latter model considers that in the cracking process, the alkanes (paraffins, P), cyclo-alkanes (naphthenes, N), and aromatic hydrocarbons (A) in the feed behave differently from each other.

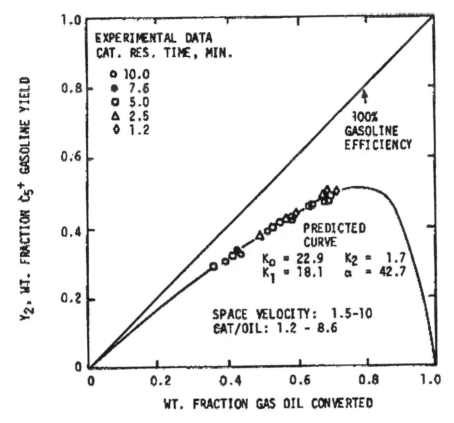

Figure 6.28 Comparison of the kinetic model with experimental data for a moving bed system. (From Ref. 152.)

The model is described by a system of differential equations corresponding to the pulsed flow reactor, considering that all the reactions are of 1st order, and that the deactivation of the catalyst is expressed by the equation:

$$k = k_0 \tau_t^{-m} \tag{6.78}$$

where:

k_0 and k are the rate constants at the beginning of the run and after a time on
 stream τ_t
m has the value 0.1–0.2.

The most important development of the three components model is a model with four components (Figure 6.30) [160] that separates the conversion to gases from the conversion to coke.

Expressing the concentrations Y in weight fractions, accepting 2nd order kinetics for the three cracking reactions of gas oil and 1st order kinetics for the other reactions, the authors formulate the following system of differential equations:

$$\frac{dY_1}{d\tau} = -(k_{12} + k_{13} + k_{14})\phi Y_1^2 \tag{6.79}$$

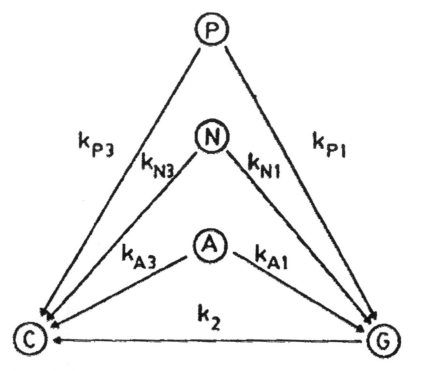

Figure 6.29 Five components (lumps) model. P – alkanes (paraffins), N – cyclanes (naphthenes), A – aromatics in the feed. G – gasoline, C – gases (C_1–C_4) and coke. (From Ref. 159.)

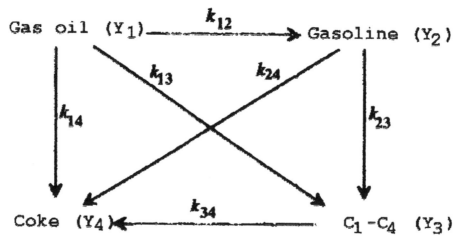

Figure 6.30 Four components (lumps) model. (From Ref. 160.)

$$\frac{dY_2}{d\tau} = k_{12}\phi Y_1^2 - (k_{23} + k_{24})\phi Y_2 \tag{6.80}$$

$$\frac{dY_3}{d\tau} = k_{13}\phi Y_1^2 + k_{23}\phi Y_2 - k_{34}\phi Y_3 \tag{6.81}$$

$$\frac{dY_4}{d\tau} = k_{14}\phi Y_1^2 + k_{24}\phi Y_2 + k_{34}\phi Y_3 \tag{6.82}$$

where ϕ is the catalyst deactivation function and the subscripts to the rate constants k are those indicated in Figure 6.29.

The system was solved in the same manner as the model with three components, obtaining the same type of integrated equations for gas oil and gasoline.

For the conversion to coke, the equation has the form:

$$Y_4 = 1 - a(1 - r_1)(1 - \varepsilon) - \frac{r_1(r_{34} - br_2)}{(r_{34} - r_2)}\exp\left(r_2 - \frac{r_2}{1 - \varepsilon}\right)$$

$$- \frac{r_1 r_2(r_{34} - br_2)}{(r_{34} - r_2)}\exp\left(-\frac{r_2}{1 - \varepsilon}\right)\left[\text{Ein}\left(\frac{r_2}{1 - \varepsilon}\right) - \text{Ein}(r_2)\right]$$

$$- \left[(1 - a)(1 - r_1) - \frac{r_1 r_2(1 - b)}{(r_{34} - r_2)}\right]\exp\left(r_{34} - \frac{r_{34}}{1 - \varepsilon}\right)$$

$$- \left[(1 - a)(1 - r_1)r_{34} - \frac{r_1 r_2 r_{34}(1 - b)}{(r_{34} - r_2)}\right]\exp\left(-\frac{r_{34}}{1 - \varepsilon}\right)\left[\text{Ein}\left(\frac{r_{34}}{1 - \varepsilon}\right)\right.$$

$$\left. - \text{Ein}(r_{34})\right] \tag{6.83}$$

where:

$$\varepsilon = 1 - Y_1, a = k_{14}/(k_{13} + k_{14}), b = k_{24}(k_{23} + k_{24}),$$
$$r_1 = k_{12}/(k_{12} + k_{13} + k_{14}), r_2 = (k_{23} + k_{24})/(k_{12} + k_{13} + k_{14}),$$
$$r_{34} = k_{34}/(k_{12} + k_{13} + k_{14}), \text{Ein}(x) = \int_{-\infty}^{x}\frac{e^x}{x}dx$$

The constants r_1 and r_2, which contain the expressions for the rate constants, result from the processing of the model with three components. The constants a, b, and r_{34}, which appear only in the model with four components, were obtained by processing data obtained in a Kellogg riser pilot plant.

The agreement between the conversion to coke calculated by Eq. (6.83) with that measured in the pilot plant or in a industrial plant is presented graphically in the original paper [160]. The maximum deviations are of the order ± 15–20%.

Another article [161] used the model with four component lumps but ignored, which seems logical, the direct formation of coke from gases.

Models with a much higher number of components were also suggested, either in order to detail the composition of the gases or to take into account the differences in the behavior of the feed components.

To the first category belongs the model proposed by John and Wojciechowski [162], reproduced in Figure 6.31 and that suggested by Corma et al. [163]. To the

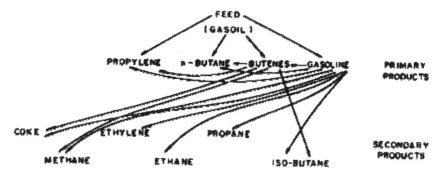

Figure 6.31 Model proposed by John and Wojciechowski. (From Ref. 162.)

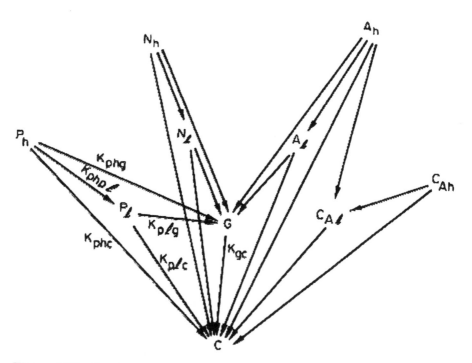

Figure 6.32 Ten-lump kinetic model scheme. P_l = wt % paraffinic molecules, 222–342°C; N_l = wt % naphthenic molecules 222–342°C; C_{Al} = wt % carbon atoms among aromatic rings, 222–342°C; A_l = wt % aromatic substituent groups (222–342°C); P_h = wt % paraffinic molecules, >342°C; N_h = wt % naphthenic molecules, >342°C; C_{Ah} = wt % carbon atoms among aromatic rings, >342°C; A_h = wt % aromatic substituent groups (>342°C); G = G lump (C_5-222°C); C = C lump (C_1–C_4 + coke); C_{Al} + P_l + N_l + A_l = LFO (222–342°C); C_{Ah} + P_h + N_h + A_h = HFO (>342°C).

second category belongs the model with 10 components suggested by Jacob et al. [164,165], depicted in Figure 6.32.

This latter model (Figure 6.32) takes into account in a detailed manner the chemical composition of the feed, as well as its structure in terms of distillation fractions, which allows its use for feeds containing components of a variety of natures. However, it has the disadvantage that it calculates only the overall yield of gases C_1–C_4 + coke resulted from the process. In order to compensate for this disadvantage, Olivera and Biscaia Jr. [166] completed the Jacob's model by separately modeling the formation of coke and the formation of primary and secondary gases.

The kinetic model of Jacob was further compared with the three above models. Finally four models resulted, as follows:

Model 1. From Figure 6.31, without completions, where the deactivation of the catalyst is described by a hyperbolic equation similar to Eq. (6.24):

$$\phi(t_c) = \frac{1}{1 + \beta t_c^\gamma} \tag{6.84}$$

where t_c is the duration of catalyst use, and β and γ are constants.

The other three models are based on the model of Figure 6.31 completed with the separate accounting for gas and coke formation, and which differ from each other by the catalyst deactivation equation considered.

Model 2. Uses the deactivation equation (6.84) and has the same value for the constants β and γ for gas oil and gasoline.

Model 3. Uses different values of the deactivation rate ϕ for gasoline and for gas oil, both being expressed by exponential equations of the form (6.47).

Model 4. Uses also different values for ϕ for gasoline and gas oil, but the deactivation is expressed, depending on the coke amount deposited on the catalyst, by exponential expressions of the form (6.45).

The experimental data for gasoline cracking obtained in an isothermal reactor with a fixed bed of catalyst were processed by means of each of the four models obtaining the following conclusions:

The reaction rates for gasoline cracking to secondary gases and to coke have rates equal to zero.

The catalyst activity decreases faster during the cracking of primary gas than for gasoline cracking.

Model 4, in which the deactivation of the catalyst depends on the deposited coke, gives the best agreement with the experimental data.

Thus, model 4, is recommended for modeling the catalytic cracking of gas oil.

A recent study [239] compared the 3-, 4-, and 5-lumps kinetic models with experimental data obtained at 480°C, 500°C, and 520°C in a micro-activity reactor (ASTM D 3907–92). The best results were obtained by using the 5-lumps kinetic model (Figure 6.33).

For catalyst deactivation, the exponential expression (6.45) was used, with $\alpha = 0.0875$.

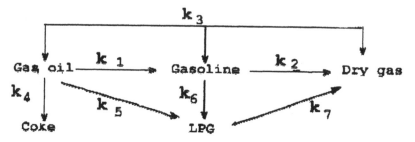

Figure 6.33   5 lumps kinetic model. (From Ref. 239.)

The obtained kinetic constants and activation energies (10^3 kcal/mol) were:

$k_1 = 0.1942$ $E_1 = 13.7$
$k_2 = 0.0032$ $E_2 = 10.8$
$k_3 = 0.0001$ $E_3 = 11.8$
$k_4 = 0.0140$ $E_4 = 7.6$
$k_5 = 0.0357$ $E_5 = 12.5$
$k_6 = 0.0061$ $E_6 = 17.5$
$k_7 = 0.0020$ $E_7 = 9.5$

In another study [161], a kinetic model with four components, similar to that of Figure 6.30 but without the reaction for coke formation from gases ($k_{34} = 0$), served for the determination of the kinetic constants of the 2nd order reaction for gas oil cracking and of 1st order gasoline cracking. The deactivation of the catalyst was considered the same for all the reactions and was expressed by an exponential equation where the variable was the time the catalyst was on stream (6.47).

The constants were derived by processing published data concerning the performance of a catalyst with high activity in a riser type reactor. The conversion was defined in terms of the weight percent of gasoline, gases, and coke.

The kinetic parameters deduced for a single feed and a catalyst are collected in Table 6.13.

A seven-lump model used by Al. Khattaf and de Lasa [237] gives the kinetic equation for the formation of alkanes, alkenes, naphthenes, aromatics, coke, and methane.

A new treatment of the kinetics of gas oil catalytic cracking [167], takes into account in detail the reaction mechanism involving carbenium ions. This approach is close to the "mechanistic modeling" used in the pyrolysis of the gas oil, discussed in Section 5.14.

Before the model was actually formulated, several necessary simplifications were brought to the reaction mechanism. For example, the rate of extraction of a hydride ion by a Lewis site was considered to be independent of the nature of the alkane hydrocarbon from which the extraction took place. Only the extractions from secondary or tertiary carbon atoms were considered.

The equations for the rates of formation of the carbenium ions were written on the basis of the classic interaction reactions of the alkanes with the active Lewis sites

Table 6.13 Kinetic Constants for 4-lump Model

Kinetic constants	Temperature (°C) 482.2	548.8	615.5	A (h^{-1})	E (kJ/mol)
α^a	1.944	10.140	31.102	$3.017 \cdot 10^8$	117.7050
k_{12}	15.644	39.364	79.408	$9.778 \cdot 10^5$	68.2495
k_{13}	3.323	9.749	28.020	$4.549 \cdot 10^6$	89.2164
k_{14}	1.297	3.302	6.102	$3.765 \cdot 10^4$	64.750
$k_1 = k_{12} + k_{13} + k_{14}$	20.264	52.415	113.440	$1.937 \cdot 10^6$	72.2526
k_{23}	0.711	1.370	2.470	$3.255 \cdot 10^3$	52.7184
k_{24}	0.411	0.753	1.384	$7.957 \cdot 10^3$	63.4580
$k_2 = k_{23} + k_{24}$	1.122	2.123	3.854	$4.308 \cdot 10^3$	51.6726

[a] α, constant from the equation $\phi = e^{-\alpha\tau}$
Source: Ref. 161.

and of the alkenes with the Brönsted centers. From here, the equations for the rates of formation by way of β scissions of the alkanes and alkenes were deduced.

By application of the steady states theorem to the carbonium ions produced on the two types of sites, the rate equations for the catalytic cracking of the considered hydrocarbons were derived.

This method of calculation was applied for the determination of the rate constants of the elementary reactions considered in the catalytic cracking of alkanes. This approach can be extended to the catalytic cracking of cyclo-alkenes and alkyl-aromatic hydrocarbons.

The determined reaction rates are initial rates that do not take into account the catalyst deactivation as a consequence of coke deposits.

It seems that this study opens new promising ways for the treatment of the kinetics of catalytic cracking of petroleum fractions.

The effect of adding residues derived from different crude oils in proportions of up to 20% to the vacuum distillates feed was studied. The study however, does not supply the necessary elements for the generalization of the conclusions and for their application to the catalytic cracking of other feeds.

6.5 EFFECT OF PROCESS CONDITIONS

The process conditions influencing the catalytic cracking may be grouped as follows:

Temperature
Pressure
Feed composition
Feed recycling
Catalyst behavior
Catalyst/feed ratio

The influence of the volume feedrate results from Section 6.4 (on the kinetics of catalytic cracking).

6.5.1 Temperature

In catalytic cracking as in other processes, the influence of temperature on the reaction rate is expressed by the Arrhenius equation:

$$k = Ae^{-E/RT} \tag{6.85}$$

It must be remarked that in heterogeneous catalytic process, the activation energy experimentally measured (in conditions where the diffusion phenomena do not influence the reaction rate) is an apparent energy E^X that includes the heats of adsorption of the reactants and of desorption of the products. The connection between the apparent activation energy and the activation energy of the reaction itself E is given by the relation:

$$E^X = E - \lambda_A + \lambda_B \tag{6.86}$$

where λ_A and λ_B are the heats of adsorption of the reactants, respectively of the products.

The second remark refers to the very large differences between the activation energies calculated by various researchers on basis of experimental data that vary between the limits 20–125 kJ/mole. In a previous work [1] examples were given and the causes of these differences were analyzed.

There are two main reasons that lead to such large differences:

The form of the kinetic equation used for calculating the rate constants at two temperatures

The influence of the diffusion phenomena on the conditions in which the determinations were carried out

The first cause of errors can be avoided by using the Arrhenius equation written as:

$$\ln \frac{k_{T_1}}{k_{T_2}} = \frac{E}{R}\left(\frac{1}{T_2} - \frac{1}{T_1}\right) \tag{6.87}$$

and replacing the ratio of the rate constants by the ratio of the reaction times required for obtaining the same degree of conversion at the two temperatures.

Indeed, since any kinetic equation may be written as:

$$k = w \cdot f(x),$$

it results that for two temperatures and for a specified conversion x constant, one may write:

$$\frac{k_{T_1}}{k_{T_2}} = \left(\frac{w_{T_1}}{w_{T_2}}\right)_{x=\text{const}} \tag{6.88}$$

which allows replacing in Eq. (6.67) the ratio of the rate constants with the ratio of the volume rates at constant conversion. There is no need to know the expression for the reaction rate equation.

Thus the activation energies for the formation of coke deposits on the catalyst can be calculated even if certain kinetic equations are missing.

Concerning the influence of diffusion phenomena, the only solution is to use experimental techniques that eliminate such influences. This approach is used in numerous studies [154,156,157,158]. Before accepting the obtained values for the activation energies, it is necessary to perform an analysis of the experimental technique and of the calculation methods.

The analysis of the published data allows one to conclude that the average values of the activation energies vary with the characteristics of the catalyst and of the feed.

For the lower alkanes, the activation energies [83] show a decrease of the activation energy with the molecular mass.

n-C_6H_{14}	153 kJ/mol
n-C_7H_{16}	123 kJ/mol
n-C_8H_{18}	104 kJ/mol

For atmospheric and vacuum petroleum fractions, the majority of the published studies [4] suggest values ranging between 40 and 60 kJ/mole. Our own studies of the catalytic cracking of a vacuum distillate using a microspherical, classic, equilibrium catalyst extracted from an industrial plant lead to values of about 60–70 kJ/mole [97]. It was also verified that the addition of α-methyl-naphthalene did not sensibly modify the value of the activation energy.

The variation of the apparent activation energy with the molecular mass shows a higher value of the activation energies corresponding to the decomposition of the gasoline to gases as compared to the formation of gasoline from the feed [1,127].

Since the formation and the decomposition of gasoline constitute two successive reactions, it results that, similarly to the thermal cracking, the increase of the temperature in catalytic cracking leads to the decrease of the maximum of gasoline yield. This conclusion was deduced theoretically [1] and was experimentally confirmed, among others, by determinations performed in a riser type pilot plant, depicted in Figure 6.34 [169].

The variation of the conversion to gasoline as a function of the overall conversion as shown in Figure 6.35, demonstrates that this conversion goes through a maximum that decreases as the riser outlet temperature increases.

With increasing temperature, the proportion of cycloalkanes dehydrogenation and thermal cracking reactions will also increase, since they have larger activation energies than those of catalytic cracking. The result is the increase of the unsaturated and aromatic character of the gasoline and as a consequence the increase of the research and motor octane numbers.

These conclusions are confirmed also by the experimental data obtained in the pilot plant (see Figure 6.33) [169]. Thus, Figures 6.36 and 6.37 show the variation with the temperature of the yield (as a percentage) of isobutene and isopentenes, Figure 6.38 shows the production of benzene and Figures 6.39 and 6.40 give the Research and Motor octane numbers for gasoline.

The data in these graphs were obtained for two feeds and two equilibrium catalysts, the main characteristics of which are reproduced in Tables 6.14 and 6.15.

There is no data concerning the activation energy for the formation of coke. This is due to the absence of kinetic reliable equations describing the formation of coke in the conditions of the process.

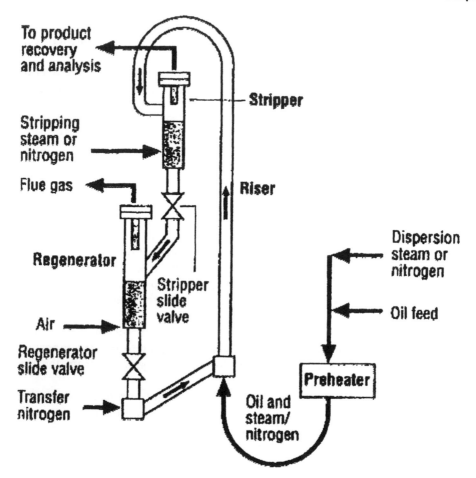

Figure 6.34   Pilot plant with riser reactor. (From Ref. 169.)

Eq. (6.87) combined with (6.88), which eliminates the need to know the form of the kinetic equations, leads to the value $E = 25$ kJ/mole [97], which appears to decrease as the aromatic character of the feed becomes stronger.

The lower value of the activation energy compared to that of the overall conversion of the feed explains the decrease of the percentage of coke at higher reaction temperature, in conditions of constant total conversion (see Table 6.16) [170].

Although the data in the table refers to a classic catalyst, it shows the same trend for the performance of a zeolite catalyst, discussed earlier. The increase in temperature produces the increase of the octane number, the unsaturated character of the gases, and the aromatic character of the gasoline, as illustrated by the density increase.

The variations of gasoline characteristics with the reaction temperature are depicted in detail in Figure 6.41 [171].

With increasing temperature the activation energy may decrease. This is the effect of the diffusion phenomena, which however are less influenced by temperature than the reaction kinetics (see Figure 6.42).

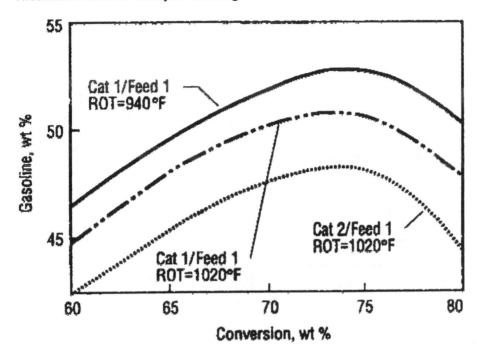

Figure 6.35 Conversion to gasoline function of the overall conversion at two temperatures and two catalysts. ROT riser outlet temperature. (From Ref. 169.)

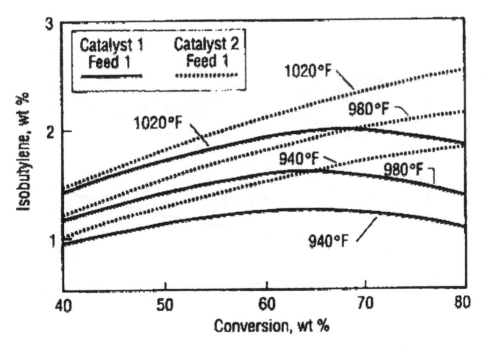

Figure 6.36 Conversion to isobutene for two catalysts, at various riser outlet temperatures. (From Ref. 169.)

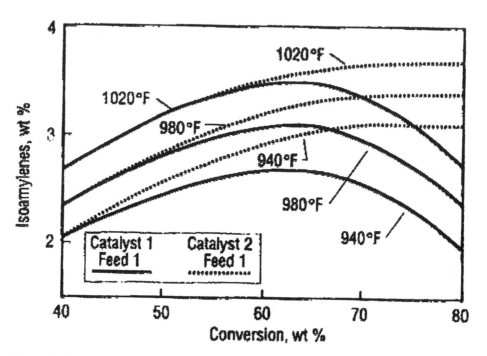

Figure 6.37 Conversion to isopentene for two catalysts and various ROT values. (From Ref. 169.)

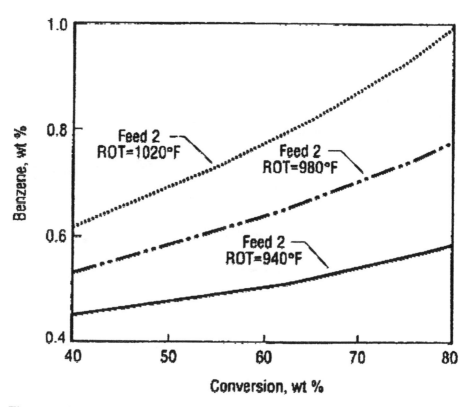

Figure 6.38 Conversion to benzene function of overall conversion for two outlet riser temperatures ROT. (From Ref. 169.)

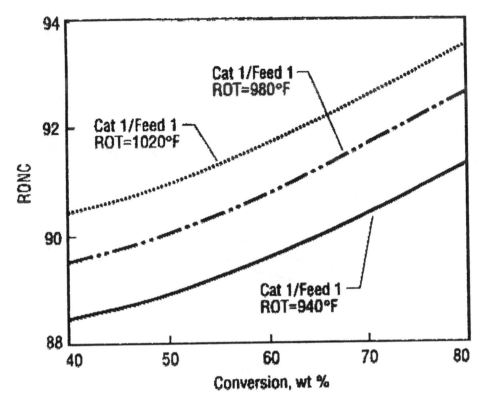

Figure 6.39 Research octane number as function of conversion for various riser outlet temperatures. (From Ref. 169.)

Taking into account the narrow range of temperatures, ranging from 470–545°C where the catalytic cracking is performed, it is difficult to verify if any change in the activation energy occurs within this tight interval. The low values of the previously cited activation energy [1] allow the assumption that the external diffusion influences in some cases the reaction rate. Thus, R.Z. Magaril [73] considers that for spherical catalysts with a diameter of 3–5 mm used in the moving bed processes, the transition to control by external diffusion takes place at temperatures of 480–510°C, while for microspherical catalysts used in the fluidized bed processes, takes place at 540–560°C. Since these estimates refer to classic synthetic catalysts with 12–25% Al_2O_3, one can assume that for zeolite catalysts, which are more active, the transition to control by the external diffusion will occur at still lower temperatures.

It must be remarked that the temperature for the transition from reaction control to diffusion control is different for the decomposition reaction of the feed to gasoline and that of gasoline to gas. The difference is a result of the apparent activation energy that is higher for the decomposition of gasoline than for its formation, and of a higher diffusion coefficient for the molecules of gasoline than for the molecules of feed. Accordingly, the diffusional barrier will intervene at higher temperatures for the gasoline formed in the process than for the feed submitted to catalytic cracking. The influence of the external diffusion can be a decrease of the maximum of gasoline yield.

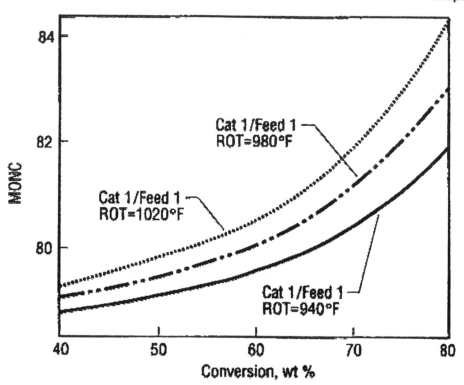

Figure 6.40 Variation of motor octane number function of conversion for various riser outlet temperatures. (From Ref. 169.)

The influence of the internal diffusion is similar to that of the external diffusion, with the difference that it is strongly influenced by the structure and size distribution, and by the size of the feed molecules. Besides, even the very strong influence of the internal diffusion cannot reduce the value of the apparent activation energy by more than half of the activation energy of the actual reaction.

The high differences among the structures of the catalysts and the variety of feeds used in catalytic cracking makes it difficult to formulate generally valid conclusions concerning the influence of the internal diffusion on the overall rate. It can however be stated that such influences are minimal when microspherical catalysts are used and may become important for granular catalysts used in the moving bed processes.

6.5.2 Pressure

In catalytic cracking the pressure varies within very narrow limits determined by the type of unit and by the catalyst circulation system. It doesn't constitute a process parameter to act on in the selection of the operating conditions of the unit.

However, it is very important to understand the effect of pressure upon catalytic cracking reactions. This effect explains the differences in the performances of various types of units and allows to make improvements to the design of units of a given type.

Table 6.14 Feed Used in Experiments. Fig. 6.35–6.40 and 6.46, 6.47

Characteristics	Feed 1	Feed 2
Density, 15°C	0.9188	0.9554
Distillation (°C)		
5%	278	286
50%	455	423
90%	565	504
S total, wt %	0.55	0.29
N total, wt %	0.18	0.36
N basic, wt %	0.056	0.12
C, wt %	86.72	87.72
H, wt %	12.27	11.61
Conradson coke, wt %	0.89	0.15
V, ppm	0.6	0.02
Ni, ppm	0.4	0.04
Fe, ppm	7.0	0.03
Na, ppm	1.8	0.50
Refraction index at 67°C	1.4950	1.5069
Aniline point (°C)	80.0	60.0
Molecular mass	371	322

Data refers to Figures 6.35–40, 6.46, 6.47.

Table 6.15 Equilibrium Catalysts used in Experiments

Characteristics	Catalyst 1	Catalyst 2
Specific surface m^2/g		
zeolite	41.7	109.3
matrix	109.1	46.4
Rare earths content		
cerium	0.72	0.04
lanthan	1.07	0.17
neodim	0.51	0.05
parazeodim	0.16	0.03
Metals content (ppm)		
iron	3,410	10,800
nickel	647	2,220
vanadium	447	2,045
natrium	2,387	6,100
bismuth	–	10
stibium	–	< 10
Microactivity MAT	74	66

Data refers to Figures 6.35–40, 6.46, 6.47.

Table 6.16 Temperature Influence on the Catalytic Cracking of a
Gas Oil

Characteristics	Mean reactor temperature °C		
	454	482	510
Volumetric rate, h^{-1}	0.8	1.3	2.0
Conversion, wt %	55.1	55.1	55.1
Yields, wt %			
H_2	0.04	0.05	0.06
CH_4	0.71	0.85	1.29
C_2H_4	0.40	0.55	0.75
C_2H_6	0.60	0.75	1.05
C_3H_6	2.40	3.35	4.40
C_3H_8	2.10	2.15	2.15
$i\text{-}C_4H_{10}$	5.10	4.20	3.35
$n\text{-}C_4H_8$	2.90	4.00	5.60
$n\text{-}C_4H_{10}$	1.40	1.30	1.25
Gasoline debutanised	34.6	33.5	32.2
Light gas oil	15.8	13.8	12.4
Heavy gas oil	29.1	31.1	32.5
Coke	4.85	4.20	3.70
Gasoline characteristics			
vapor pressure (torr)	374	379	384
Density	0.7511	0.7579	0.7649
Octane number F_1 with vapor			
Tension 517 torr	91.2	94.0	95.0
Tension 517 torr, with 0.8 ml/l TEP	97.6	98.6	99.0

$d = 0.882$.
Source: Ref. 170.

In the thermodynamic analysis of cracking processes (Section 6.1) it was shown that the pressure favors the polymerization reactions that take place in the layer adsorbed on the catalyst surface and lead to the formation of coke.

These theoretical conclusions are fully confirmed by the experimental data [172] on the influence of the pressure on the reactor performance of a catalytic cracking pilot plant presented in Table 6.17.

The data show the obvious increase of the coke percentage formed as the pressure in the reactor increases. Pressure has no significant effect on the yield and the quality of other obtained products, excepting the decrease of the unsaturated character of the products illustrated in the table by the relative yield of butylenes.

Since in the industrial units the pressures in the reactor and in the regenerator are interdependent, an increase of the pressure in the reactor leads to an increase of the pressure in the regenerator and thus to a higher burning rate of the coke. It results in a decrease of the residual coke and an increase of the mean activity of the catalyst. For this reason the increase of the pressure in industrial systems must be analyzed by taking into account the modifications of performance that will be produced in both vessels.

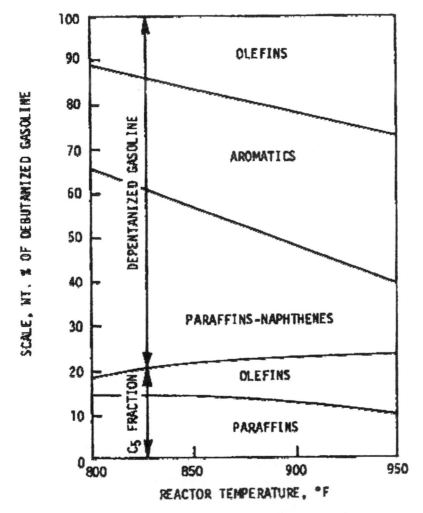

Figure 6.41 Effect of temperature on the debutanized gasoline and composition of C_5 fraction. (From Ref. 171.)

6.5.3 Feed Composition

The basis of the feed for catalytic cracking is the wide cut fraction obtained from the vacuum distillation of crude oil, and occasionally with the addition of some atmospheric gas oil.

As a rule, the product from the bottom of the fractionating column of the catalytic cracking unit is added to the feed with the purpose of recovering the entrained catalyst. This product has generally a distillation initial boiling point of about 420–450°C.

The feed can include also gas oils from visbreaking and coking and also the extracts from solvent refining of lube oils and the deasphalted oils, which sometimes may have been hydrotreated.

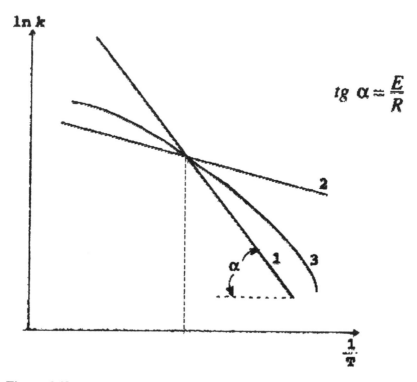

Figure 6.42 Effect of temperature on the external diffusion and reaction rates: 1 – reaction; 2 – external diffusion, 3 – resulting rate.

The improvements brought to the catalytic cracking make it possible to include in the feed variable proportions of residual fractions in general residue from the atmospheric distillation on the condition of not exceeding some limits for the Conradson carbon and the content of metals. Atmospheric residues also may be directly submitted to catalytic cracking.

The recycling of gas oils, especially of the heavy ones, strongly influences the results of the process due to their aromatic character. Since recycling constitutes an independent process parameter, it is examined separately in the next section.

Table 6.17 Pressure Influence on the Yields in a Catalytic Cracking Pilot Unit at Equivalent Technological Conditions [172]

	Hydrocarbons partial pressure (bar)		
	0.69	1.72	2.76
Global conversion (vol %)	69.30	70.40	75.70
Gasoline (vol %)	53.10	52.60	51.20
Coke (wt %)	7.40	9.60	12.40
Butylenes relative yield	1.00	0.86	0.72

Source: Ref. 172.

Currently, for estimating the performance of catalytic cracking, the characterization factor of the feed is used, as suggested by Watson and Nelson [173]. Expressed in metric units [174] it is given by the expression:

$$K = \frac{1.216\sqrt[3]{T}}{d} \tag{6.89}$$

where:

T = molar mean boiling temperature in degrees K
d = the density at 15.56°C.

The data of Table 6.18 show the correlation between the characterization factor and the yields to gasoline and to coke for fractions obtained from the crude oil distillation and for the recycled gas oils [175].

Since the units are generally obliged to work in conditions that should lead to a constant conversion to coke, the data of Table 6.19 supply correlations between the characterization factor, the conversion, and the yield to gasoline at a constant conversion of 5.3% to coke [175].

The influence of the distillation limits of the feed on the yield in products is illustrated in Figure 6.43 [176]. From this figure, the increase of the mean boiling temperature of the feed increased the conversion and the yields for all the products. However, the yield in gasoline increased with the alkane character of the feed and it may vary in the opposite direction if the feed is strongly aromatic (Figure 6.44) [177].

The data of tables 6.17 and from the Figures 6.43 and 6.44 correspond to a conventional catalyst and to the fluidization in dense phase.

The conversion of different classes of hydrocarbons depending on the severity of the process is given in the graphs from Figure 6.45a–c [177]. These results were obtained in a pilot plant using a conventional catalyst, reaction temperature of 482°C, and a feed with 50% vol distillate at 371°C.

In these graphs the severity was expressed by the ratio:

$S/100n$

Table 6.18 Correlation Between Characterization Factor UOP and Gasoline Yields at Constant 60% Conversion

	Fresh feed		Recycle	
K_{UOP}	vol % gasoline	wt % coke	vol % gasoline	wt % coke
11.0	–	–	(35.0)	–
11.2	(49.5)	(12.5)	37.0	(11.5)
11.4	47.0	9.1	39.0	9.0
11.6	45.0	7.1	40.0	7.2
11.8	43.0	5.3	41.09	6.0
12.0	41.5	4.0	(41.5)	(5.3)
12.2	(40.0)	(3.0)	–	–

Estimated values in parenthesis.
Source: Ref. 175.

Table 6.19 Correlation Between the UOP
Characterization Factor and Gasoline Yield at
Constant (5.1%) Conversion in Coke

	Fresh feed		Recycle	
K_{UOP}	Conversion %	vol % gasoline	Conversion %	vol % gasoline
11.0	–	–	30.0	20.0
11.2	50.0	39.0	39.0	26.5
11.4	52.0	40.0	45.0	31.5
11.6	56.0	41.5	51.0	35.5
11.8	60.0	43.5	57.0	39.0
12.0	(70.0)	(52.0)	60.0	41.0
12.2	–	(66.0)	–	–

Estimated values in parenthesis.
Source: Ref. 175.

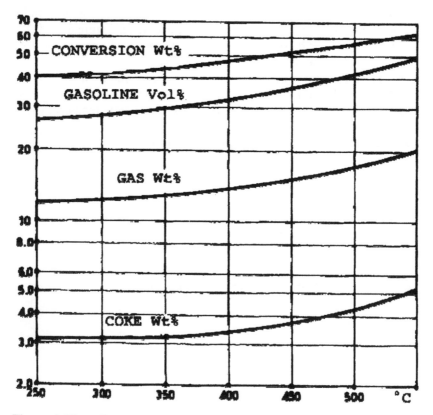

Figure 6.43 Effect of the average boiling point temperature of the feed on the conversion
and products yields. Cracking at 482°C. (From Ref. 176.)

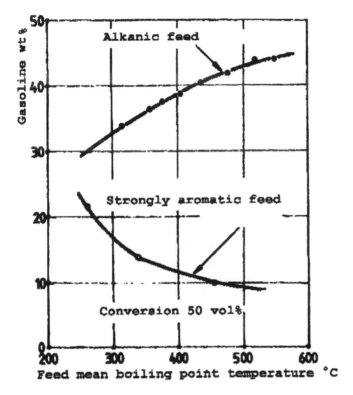

Figure 6.44 Influence of the feed volumetric mean boiling point temperature on gasoline yields. (From Ref. 177.)

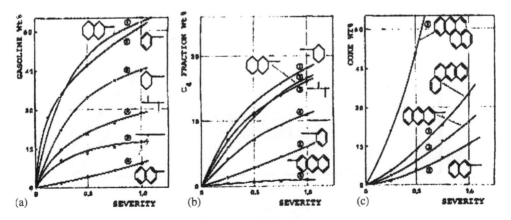

Figure 6.45 Conversion of different classes of hydrocarbons depending on process security. Weight percent of products: (a) gasoline; (b) C4 fraction; (c) coke.

where:

> S = catalyst surface in m^2/g
> n = feeding rate expressed by the kg liquid feed on the kg of catalyst per hour.

For riser systems, the zeolite equilibrium catalysts have the characteristics reproduced in Table 6.15 and two representative feeds from Table 6.14. The results of cracking are reproduced in Figures 6.35–6.40 and 6.46–6.47 [169].

6.5.3.1   Effect of Sulfur and Nitrogen Compounds

The sulfur compounds don't seem to exercise a direct action on the catalytic cracking process. Indirectly indications exist [1] that they should favor the deposit of the heavy metals and especially of iron under a dispersed form on the catalyst surface, emphasizing in this way its noxious effect.

The influence of sulfur and nitrogen on the reaction kinetics of catalytic cracking was recently analyzed [238], considering also the feed n-d-M composition and the kinetic equations proposed. It was found that sulfur content has a strong influence on kinetic constants while the effect of nitrogen is less important.

The distribution of the sulfur among the products is shown in Table 6.20 [183], from which it results that almost half ends up in the gases as hydrogen sulfide and a small part in the gasoline and gas oil. The portion that ends up in the residue and coke vary in large limits, possibly depending on the nature of the sulfur compounds in the feed.

For a unit of the riser type and a zeolite catalyst, the sulfur distribution among the products is given in the graphs of Figure 6.48. The cracking runs were carried out

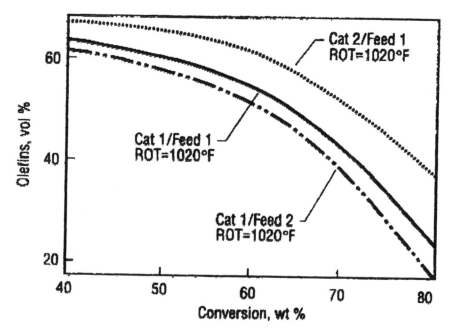

Figure 6.46   Alkenes content of gasoline for two feeds and catalysts. Riser outlet temperature: 549°C. (From Ref. 169.)

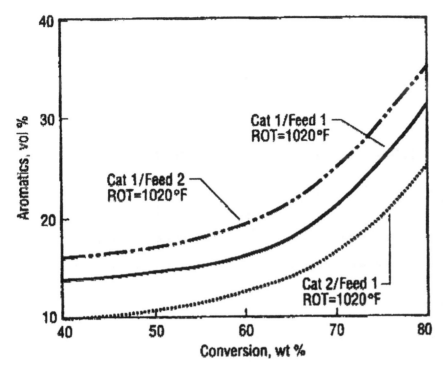

Figure 6.47 Aromatics content of gasoline for two feeds and catalysts. Riser outlet temperature: 549°C. (From Ref. 169.)

at temperatures of 555°C and 516°C, at conversions ranging from 76.3–84.1%. The feed was a Kuwait gas oil the sulfur content of which was decreased by hydrofining or deasphalting.

Despite the fact that the gasoline produced in catalytic cracking contains only 3–5% of the sulfur contained in the feed, this is almost exclusively as mercaptans requiring purification of the gasoline.

The influence of nitrogen compounds was discussed in Chapter 6.3.3.

Table 6.20 Sulfur Distribution in Catalytic Cracking Products

Feed		Sulfur distribution, wt %				
Origin	S wt %	Gases + H_2S	Gasoline	Gas oil	Fuel	Coke
Straight-run residue, Cabinda	0.21	53.6	6.8	10.9	9.4	19.3
Gas oil, South Louisiana	0.46	46.5	4.4	15.0	27.5	6.6
Gas oil, California	1.15	60.2	9.5	20.7	6.8	2.8
Gas oil, West Texas	1.75	42.9	3.5	28.0	20.5	5.1
Gas oil, Kuwait	2.66	46.5	3.8	21.1	17.3	11.3
Desasphalted straight run residue + Kuwait gas oil	3.14	50.0	6.9	17.3	15.3	10.5

Source: Ref. 183.

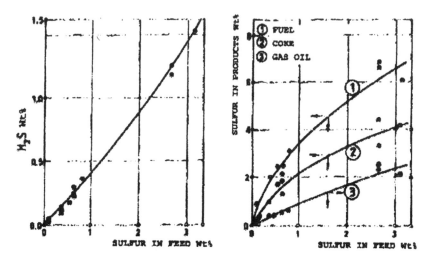

Figure 6.48 Effect of sulfur content in the feed on the sulfur content in the products. (From Ref. 176.)

6.5.3.2 Effect of Sodium and Heavy Metals

The residues fed to catalytic cracking may contain sodium from the NaOH introduced in the crude for fighting acidic corrosion in the atmospheric distillation tower.

Sodium is also introduced in the steam generating boilers and is also used in the treatments applied to the recycled gas oil [243].

The sodium will neutralize a portion of the active sites of the catalyst and thereby it will reduce its stability. In larger amounts, sodium can lead to the sinterization of the catalyst and the partial closing of the access to the pores. Vanadium has a similar effect on pore closing.

The volatile compounds of the heavy metals: Ni, V, Cu, and Fe are decomposed in contact with the catalyst and the metal is deposited on its surface as such or as sulfide.

The deposited nickel has a "parasite" catalytic activity, promoting dehydrogenation, aromatization, and coke formation. This parallel catalytic activity is very damaging because a portion of the hydrocarbons that are present will undergo reactions different from those that are typical for catalytic cracking. Thus, the alkyl-cycloalkanes, instead of forming iso-alkanes on the acid sites of the catalyst, will be dehydrogenated on the metallic sites, leading finally to formation of coke [207].

Finally, the metals poisoning leads to a decrease of the conversion. At a constant conversion, the result is a higher yield of light gases, including hydrogen, a decrease of the gasoline production, and an increase in coke [184, 185] (Table 6.21).

The poisoning effect is not the same for all four metals. (Fe seems to be less poisoning) and it depends to some extent also on the composition of the feed on the operating conditions and on the type of catalyst. Thus, the presence of sulfur in the feed emphasizes the noxious effect of iron (see also Section 6.5.3.2).

The poisoning effect of the metals deposited on the surface of the catalyst decreases in time possibly due to the progressive loss of their dehydrogenation

Table 6.21 Metals Poisoning Effect at Constant 70% Conversion

N, V, Fe on catalyst (ppm)	180	1130	3500
kg feed/kg catalyst per hour	16.5	10.2	5.8
Issues			
C_3 (wt %)	5.8	6.6	7.1
C_4 (wt %)	14.0	14.0	13.0
Gasoline (wt %)	61.0	59.0	54.0
Coke (wt %)	2.4	3.1	7.3

Arco pilot plant fluidized bed cracking; feed mid-Continent 262–570°C fraction; zeolite catalyst; Temperature 499°C, catalyst/feed circulation ratio = 8, 0.05% coke; on regenerated catalyst.
Source: Ref. 177.

catalytic effect and in numerous cases it becomes zero, after the unit is fed for 1–2 weeks with uncontaminated feed.

In the distillates, the amount of the heavy metals generally does not exceed 1 ppm if the distillation end point is not higher than 600°C. They are contained in higher amounts in the oils produced in the deasphalting of the residues, from contaminations with residues of the fractions produced by vacuum distillation, and from leaks in the heat exchangers. The contamination due to the metals resulted from the erosion of the equipment is negligible.

The most efficient solution for preventing the contamination with metals is the hydrofining of the feed, a problem explored in Section 10.1.4.

6.5.4 Feed Recycling

The recycling coefficient K is defined by the relation:

$$K = \frac{A}{A_0} = \frac{A_0 + F}{A_0} = \frac{B}{Z} \tag{6.90}$$

where:

A_0 is the fresh feed
F is the recycled feed
B is the yield in gasoline
Z is the conversion to gasoline at the outlet from the reactor.[*]

A small portion of the recycle, 3–5% of the feed, is made of "decant oil," i.e., the bottom fraction of the fractionating column containing the catalyst dust entrained from the reactor. The small amount of this "compulsory recycle" does not have a significant effect on the yield of the process.

A recent article [186] refers to results obtained in the laboratory and verified at the industrial scale that an addition of 1.5–18% of this product leads to an increase

---

[*] In some publications [4] the notion of recycle flow is used, which represents the ratio between the recycled amount and the fresh feed: F/A_0.

of about 2.2% in the conversion to gasoline and an increase of about 2.4 units for the motor octane number. This favorable effect is difficult to explain since it disappears when a larger amount of product is added. The recycled product has a density of 0.9959 and contains 52.5% polycyclic aromatic hydrocarbons.

The main portion of the recycle is made of the fraction that distills within the range 340–455°C.

In the case of the classic catalysts, the recycling displaces the maximum gasoline yield towards higher conversions with the net result of higher yields. Results obtained in this manner are illustrated in Figure 6.49 [187].

For zeolite catalysts of high activity, the maximum gasoline yield is situated at overall conversions that are higher than those for classic catalysts. In this condition, recycling becomes less interesting and is either completely eliminated or reduced to not more than 15% of the fresh feed.

6.5.5 Catalyst Behavior

Specific issues of the behavior of the catalyst during the operation of industrial units are discussed below.

6.5.5.1 Comparative Performances

The performance of amorphous catalysts are compared to that of zeolite catalysts, in Figures 6.50 and 6.51 [188].

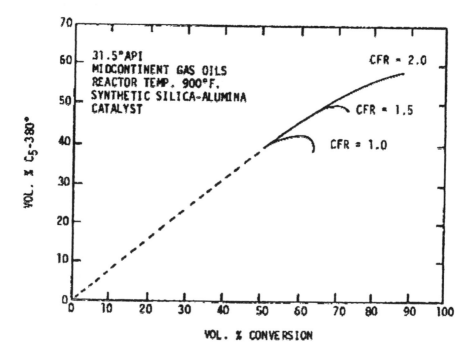

Figure 6.49 Effect of recycle on gasoline yield. Mid-continent gas oil $d = 0.868$, cracking on synthetic Si-Al catalyst at 482°C. (From Ref. 187.)

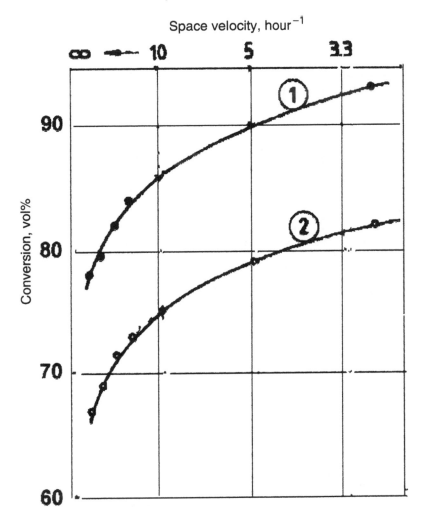

Figure 6.50 Conversion versus space velocity, for two catalysts. 1 – amorphous catalyst, 2 – zeolite catalyst. (From Ref. 188.)

These plotted results were obtained in a pilot plant with a fluidized bed reactor using: (a) an amorphous catalyst with 28.1% Al_2O_3, specific surface of 119 m^2/g, pore volume 0.35 cm^3/g and (b) the zeolite catalyst Davison XZ-25, having 28.2% Al_2O_3, specific surface 105 m^2/g, pore volume 0.28 cm^2/g. The feed was a gas oil $d = 0.857$, $K = 12.9$, S wt% = 0.19, distillation range 375–611°C. The unit was operated without recycle, at the temperature of 510°C using a ratio catalyst/feed = 7.

The effect of the zeolite content on the product yield and on the octane number at a constant conversion of 70% is shown in Figure 6.52 [189]. In these experiments, a gas oil from West Texas and a REY catalyst were used. The catalyst was pretreated by heating during 12 hours at 827°C in air, with ±20% steam. The test conditions were: $t = 493°C$ and catalyst/feed ratio = 5.

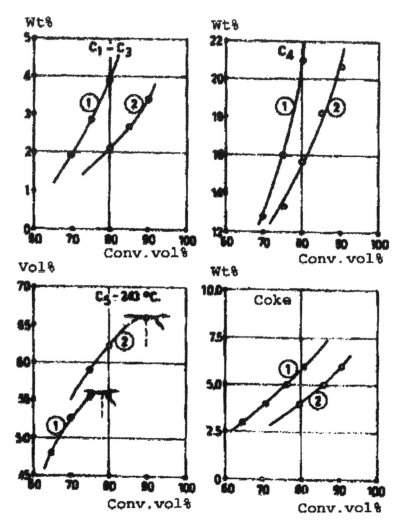

Figure 6.51 Products yields function of conversion for: 1 – amorphous catalyst, 2 – zeolite catalyst. (From Ref. 188.)

More detailed information concerning the performances of the catalysts produced with different types of zeolites and matrices are given in J. Scherzer's monograph [6] and in other publications [64,190,191,240]. Among others, these improvements resulted in a substantial increase of the resistance to poisoning by vanadium [64]. Another study [34] presents the improvements brought to the zeolite catalysts by work performed in the French Institute of Petroleum.

The continuing improvements of the catalysts led to process improvements and to a decrease of the percentage of coke produced, with a concomitant increase of conversion. Figure 6.53 presents this evolution [6].

The nature and the structure of the catalyst also influences the research and motor octane number of the gasoline [192,193]. The variations are between 2–3 octane units.

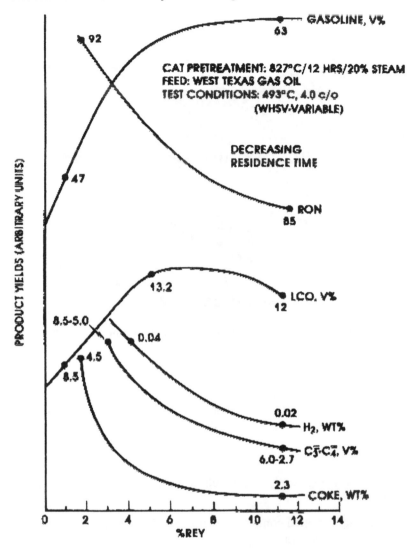

Figure 6.52 Effect of zeolite content on yields. (From Ref. 189.)

6.5.5.2 Residual Coke on Regenerated Catalyst

The percentage of residual coke remaining on the catalyst after regeneration influences sensibly the conversion. This influence is illustrated by the graph of Figure 6.54 [194]. The graph plots the results obtained in a pilot plant using Davison catalyst in a fluidized bed, fed with a West Texas gas oil having $d = 0.893$, distillation range 354–502°C, characterization factor 12.1, and sulfur content 0.30%. The catalyst, AGZ-50 of Davison, was pretreated during 12 hours with steam diluted by 20% air at 826°C. Following this treatment, the catalyst had a specific surface of 120 m²/g and a pore volume of 0.32 cm³/g.

The influence of the residual coke on the product yields at different values of the conversion is given in Figure 6.55 [195]. The graph shows that the negative influence of the residual coke increases with increasing conversion.

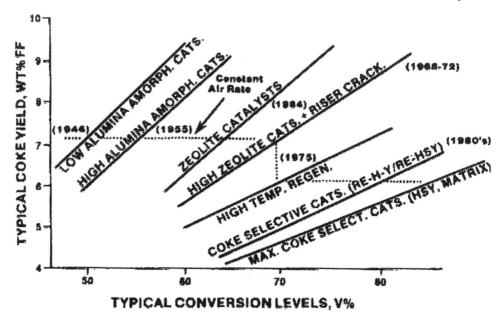

Figure 6.53 Evolution of the decrease in coke yield with conversion in catalytic cracking. (From Ref. 6.)

The larger amount of coke obtained when using a catalyst with a higher level of residual coke appears to be due to its catalytic action, which drives the reactions to higher degrees of decompositions to produce coke and gases [195,196].

6.5.5.3 Catalyst Aging

In operation, catalytic cracking catalysts are submitted to an aging phenomenon, which is due mainly to:

> Hydrothermal aging: repeated contact with steam at high temperatures, which takes place at each regeneration cycle
> Deposition of heavy metals, especially Ni and V, on the surface of the catalyst.

To these main effects, one can add the action of sodium, which may be present accidentally in the feed or in the water drops entrained by the steam.

Hydrothermal aging. A term that defines the combined action of steam and of high temperatures. It is of textural nature for the matrix and of structural nature for the zeolite [197]. It is manifested strongest during the initial contact cycles of the new catalyst and its intensity is highest at higher temperature levels.

The effect on the matrix is manifested by the decrease of the specific surface and of the catalytic activity of its active sites. At the same time the porosity may be affected, which decreases the accessibility of the reactants to the zeolite crystals situated within the matrix.

The effect on the zeolite is manifested especially by the replacement of the tetrahedral aluminum atoms located on the walls of the zeolite cells with silicon atoms, probably originating in the amorphous structures. The result is an increase

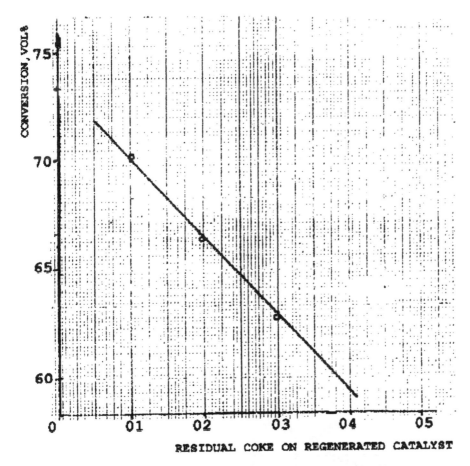

Figure 6.54 Effect of residual coke on conversion. (From Ref. 194.)

of the Si/Al ratio within the structures that are accessible to the reactants and as consequence, a decrease of the catalytic activity.

These modifications lead to the formation of structures that are more stable from the hydrothermal point of view. At temperatures above 850°C the zeolite is completely destroyed.

The zeolite and the matrix have different aging rates. For this reason, by expressing the aging of the catalyst by an Arrhenius equation, one obtains the following overall equation:

$$k = A_m e^{-E_M/RT} + A_z e^{-E_Z/RT} \tag{6.91}$$

where:

$k =$ the constant of the deactivation rate, expressed generally in units of days^{-1}

E_M and $E_Z =$ the deactivation energies for the matrix and for the zeolite respectively

A_m and $A_z =$ the respective pre-exponential factors.

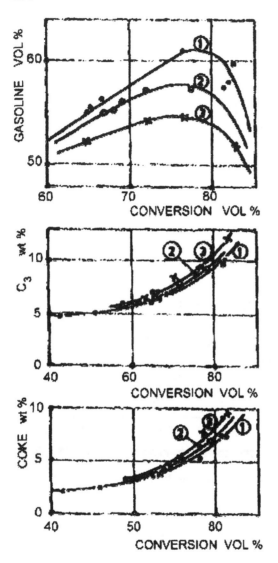

Figure 6.55 Effect of residual coke on product yields. 1 – 0.1 wt % coke, 2 – 0.2 wt % coke, 3 – 0.4 wt % coke. (From Ref. 195.)

According to the studies of Chester and Stover [198], $E_Z > E_M$, and therefore, the deactivation of the matrix is rate determining at moderate temperatures, while at higher temperatures, the deactivation of the zeolite becomes rate determining. The two rates are equal at about 780°C. The variation of the deactivation rate constant with the temperature is shown in Figure 6.56.

The presence of the rare earths hinders the extraction of the aluminum atoms from the walls of the zeolite cells, thus increasing the hydrothermal stability of the catalyst [199,201]. An opposite effect, the decrease of the stability, is produced by Na_2O. These influences are depicted in Figure 6.57. Catalyst REY contains zeolite Y

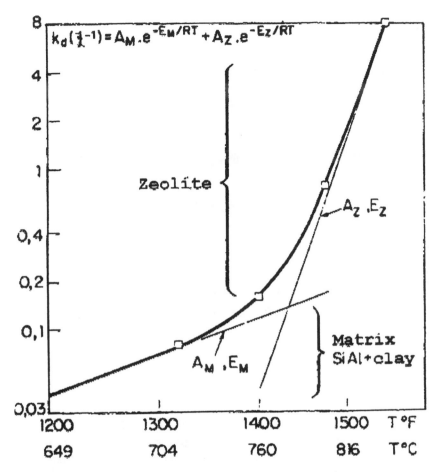

Figure 6.56 Rate constant for thermal deactivation. Catalyst was steam treated for 12 hours before test. (From Ref. 198.)

and rare earths, whereas HY does not. A better hydrothermal stability have the ultrastable catalysts promoted with rare earths, USY [197,201].

Aging due to metals.   Vanadium and nickel are the metals with the strongest poisoning action.

The effect of nickel is straightforward; it consists of the formation on the surface of the catalyst of metallic sites that are active in dehydrogenation reactions. As a result, the poisoning with nickel leads to an increase of the hydrogen content in the gases and of the coke deposited on the surface of the catalyst. The consequence is a decrease of catalyst activity. The data of Table 6.22 illustrate these effects of the poisoning by nickel and, in parallel, the quite different effects of catalyst degradation by the action of vanadium [202].

Other metals, such as the iron, have, especially in the presence of sulfur, effects similar to those of nickel. These effects are much weaker, and they are generally neglected in the evaluation of the aging phenomenon.

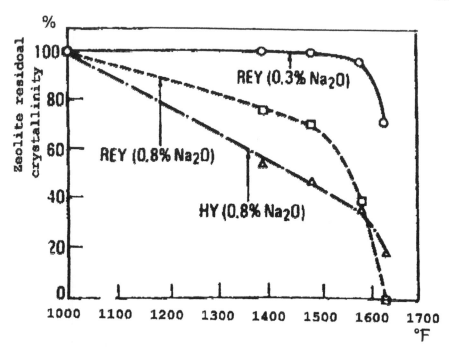

Figure 6.57 Hydrothermal stability of several catalyst (REY stands for Rare Earths Y zeolite). (From Ref. 202.)

While the aging due to nickel does not modify the crystalline structure of the zeolite and does not decrease the conversion, the aging caused by vanadium very strongly influences the crystal structure and strongly decreases the conversion (see Table 6.22).

This effect is due to the fact that the vanadium, which is deposited on the catalyst after its contact with the feed, migrates during regeneration to the zeolite crystals. In order that this migration takes place, the temperature must exceed 550–560°C, the medium must be weakly oxidizing, in order to maintain the vanadium as oxide, and steam must be present.

Table 6.22 Catalyst Aging – Ni and V Action

Contaminants	0 %	0.33% Ni	0.67% V	0.33% Ni + 0.67% V
% crystallinity dry air at 677°C		Reference basis		
% crystallinity after 8 hours stripping at 723°C and 2 bar with 100% H_2O	84	84	38	38
conversion, % vol[a]	80	82	61	61
H_2 (wt %)	0.014	0.274	0.109	0.244
coke (wt %)	3.3	5.8	1.9	2.5

[a] Test at 482°C, $n = 16$ h^{-1}, catalyst/feed ratio = 3.
Source: Ref. 202.

The destruction of the zeolite, after the vanadium has reached by migration the zeolite crystals is explained by the most authors [197,203–206] in terms of the formation of an eutectic with a low melting temperature between V_2O_5 and the zeolite. This explanation is in agreement with the known effects of the interactions between V_2O_5 and silica. The presence of sodium emphasizes the effect of the vanadium.

The V_2O_5 interacts also with the oxides of the rare earths that are present in the zeolites Y and REY, especially with the lanthanum. The ultrastable zeolites Y, without rare earths of the USY type seem to be less affected.

Electron microscographs showed the modifications of the catalyst surface structure following the action of vanadium [206].

Aging caused by the action of vanadium and nickel on some industrial catalysts were the object of experimental studies to determine the effect of aging on kinetic constants. The experimental results were fitted to both the model with 3 components (lumps) and that with 5 components of Figure 6.29. The kinetic constants obtained for the catalysts impregnated with Ni and V were compared with those corresponding to the fresh catalyst [207].

Another study analyzes aging caused by the action of nickel and of vanadium in parallel with the protective action of antimony. The kinetic treatment uses the model with 4 components in which the gases and the coke form separate components. Also, the effects of aging on the values of the kinetic constants are obtained in this study [208].

Both cited studies are of high interest for the detailed examination of the aging phenomenon.

It is to be remarked that caution should be exercised when extrapolating the results obtained in the laboratory (where the poisoning of the catalyst occurs usually with Ni and V naphthenates), to the conditions of industrial plants, where poisoning is due to organometallic compounds, such the porphyrins. In fact differences between the effects of the two types of poisonings have been pointed out [209].

6.5.5.4 Equilibrium Activity

The analysis of the means used for decreasing the aging effects must take into account that in industrial plants, the activity of the catalyst is maintained at the desired level by the addition of fresh catalyst, a level called "equilibrium activity."

The rate of catalyst addition varies generally in the range 0.4–0.57 kg/m^3 feed for the catalytic cracking of gas oils and of vacuum distillates and in the range 1.0–1.2 kg/m^3 feed for residue cracking.

The correlation between the metal content of the feed, the amount of catalyst added, and the content of metals of the equilibrium catalyst is given as a graph in Figure 6.58a [210].

From this graph it follows that the decision concerning the added amount of fresh catalyst towards a given feed depends on the metal content of the feed and on the tolerance towards the metals of the equilibrium catalyst.

The rate of hydrothermal aging, can be decreased by: reducing the sodium content of the zeolite, the incorporation of rare earths in the structure, and the use of high stability catalysts, of the type USY.

When fighting catalyst aging as a consequence of metal deposits, a method other than the use of additives, which was mentioned in Section 6.2.4.3, is recommended. The method is to decrease the zeolite content and to use high stability

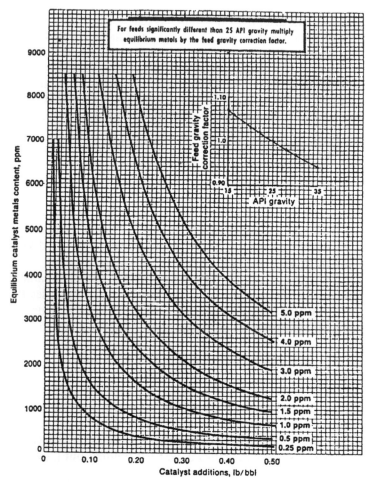

Figure 6.58a Metals content of equilibrium catalyst function of the fresh catalyst to feed ratio and the metals content in the feed. The plot is for sp. gr. = 0.904. For other sp. gravity, multiply result by correction factor. (From Ref. 210.)

catalysts. The latter resist at temperatures even above 760°C in the regenerator and tolerate, without the destruction of the crystalline network, a vanadium content of 6000 ppm and of nickel of 2000 ppm, figure 6.58b and 6.58c [211].

A favorable effect also follows from the increase of pore diameter, together with the decrease of the specific surface of the matrix.

The larger pores make easier the diffusion of the heavy components of the feed and limit the amount of liquid that remains in the catalyst after stripping, therefore reducing the thermal load of the regenerator. These favorable process effects are accompanied by the less important dispersion of the nickel and by a reduction of lower production of gases and of coke.

6.5.5.5 Test Methods

Testing methods for catalytic cracking catalysts refer to their physical-mechanical characteristics and to those directly related to selectivity and activity.

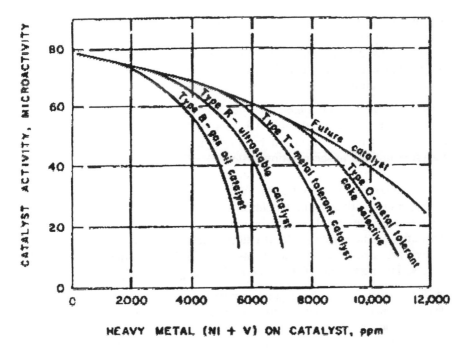

Figure 6.58b   Catalyst activity loss due to heavy metals. (From Ref. 211.)

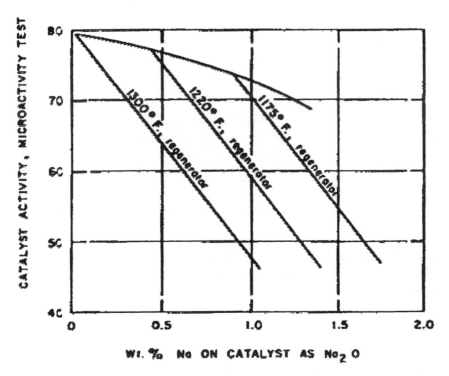

Figure 6.58c   Catalyst activity loss due to sodium contamination. (From Ref. 211.)

The physical-mechanical, characteristics, much of them common to other catalysts, have some aspects that are specific to the catalyzed reactions, but especially to the process in fluidized bed with continuous transport of the catalyst between the reactor and the regenerator. In the following, only the specific standardized methods of testing will be examined.

The attrition resistance. This determines to a large extent the amount of fresh catalyst that must be added (makeup). It is related to the catalyst dust that leaves the system and influences fluidization by modifying the dimensional distribution of the catalyst contained in the reaction system.

Generally, the resistance to attrition decreases with increasing zeolite content, limiting it to a maximum of about 35%. A decrease of the zeolite crystal size and their good dispersion within the matrix improves the resistance at attrition. It could be improved also by the use of specific binders and other similar measures [6].

Quantitatively, the resistance to attrition is determined by the method ASTM D-4058. There are also other frequently used methods recommended by catalyst manufacturers. The perfecting of these methods is ongoing [212].

Pores size distribution. This has a great importance for adequate performance of the catalyst and must be correlated with the nature of the feed [6].

The determination of the pore size distribution, is based on the usual methods for measuring adsorbtion isotherms: The adsorption of nitrogen is used for pores with diameters of 20–600 Å (ASTM D-4222 and D-4641) and porosimetry by mercury penetration for pores of 600–20,000 Å (ASTM D-4284).

The specific surface area is generally determined by the BET method, which measures the adsorption isotherm of nitrogen (ASTM D-3663).

Particle size distribution. Playing an important role in fluidized bed processes that use exclusively microspherical catalysts with sizes comprised between 60–80 µm, the particles with diameters below 40 µm and especially below 20 µm are not retained by most cyclone systems. Those above 140 µm can cause fluidization difficulties.

For size characterization the screening through standard sieves are used as well as correlations between the dispersion of laser beams and the size of the particles (ASTM D-4464) or an electronic counter (ASTM D-4438).

Fluidization characteristics. It was stated [213] and became generally accepted that only a direct characterization of the fluidization qualities of the catalyst is reliable for catalyst evaluation, since data supplied by particle size analysis are considered insufficient.

The information published [213] concerning the type of equipment used operation and calculation methodology, and the obtained results supply some interesting elements. But so far, no information is available on a standardized test that should be accepted and introduced in practice.

Thermal and hydrothermal stability. Catalyst stability is also the object of specific tests, consisting of thermal and hydrothermal treatment at different severities. The treatment is followed by the determination of specific physical characteristics (crystal structure by X-ray techniques, specific surface area and pore size distribution) that are compared with the respective properties before the treatment.

For fresh catalysts, the deactivation of the catalyst with steam is described in ASTM D-4463.

Activity and selectivity indexes. Since 1940, when the first testing method of the activity of catalytic cracking catalysts appeared, standard all other testing methods advanced together with the evolution of catalytic cracking technology [1,11].

Fluidized bed processes require specific methods. Popular current applications use low amounts of catalyst: Micro-Activity Test (MAT) and Micro Catalytic Cracking (MCC) in the U.S. In Russia and the new Caucasus republics favor the OCT 38.01161–78 method.

The methods differ by the amount of catalyst used for testing: 5 g in the MAT and OCT methods and between 4–20 g in the MCC method. The methods differ also by the duration, amount, and nature of the feed, and by the reaction temperature: 482°C in the MAT method, 510°C in the MCC method, and ranging between 494–518°C in the OCT method.

Results of studies comparing the MAT and OCT methods [214] favored using the MAT method, which is the standardized method ASTM 3907–92. This method refers to the testing of the equilibrium catalysts or of the fresh ones treated by steam stripping in the laboratory before performing the test. This procedure correlates the results of the test with identical measurements on fresh catalyst and, what is more important, correlates with the results obtained in the industrial units.

The analysis of these problems [215] lead to a series of recommendations for improving stripping conditions of the fresh catalysts, the increase of the temperature for the MAT test from 482 to 515°C, the increase of the feedrate, etc., aimed at bringing the test conditions closer to those encountered in commercial operation. The MAT method and its possible improvement is analyzed in other publications [216,217].

Since the activity index defines the total conversion, which is not sufficient for estimating the performance of the catalyst, one also makes use of the term "selectivity" which defines the proportion of a specified product (usually gasoline and the fraction C_4) obtained at a given conversion. Usually, to define the selectivity, the yields to all products are given (including gases and coke) allowing a complete evaluation of the performance of the catalyst. The term "selectivity index" which is used occasionally, represents the ratio of gasoline or gasoline + the C_4 fraction, to the total conversion.

Interesting information on the correlation of the MAT test data with the results obtained at semipilot and pilot scale, are given in the monograph of McKetta [11].

6.5.6 Effect of Catalyst/Feed Ratio

In catalytic cracking with moving or fluidized bed, the catalyst circulates continuously between the reactor and the regenerator. The intensity of this circulation, is expressed by the ratio between the weights catalyst G_{cat} and feed G_{feed} that circulate through the reactor in unit time:

$$a = \frac{G_{cat}}{G_{feed}} \tag{6.92}$$

The increase of the contacting ratio decreases the amount of coke on the catalyst measured at the outlet of the reactor. By decreasing the content of coke on the catalyst, the activity of the catalyst (and therefore the conversion) increases if the other operating parameters remain unchanged. The data of Table 6.23, obtained in a Davison pilot plant in fluidized bed for three catalysts and a feed with distillation range of 260–427°C illustrate this effect. In all the measurements, the mean reactor temperature was 482°C, the feed rate $r = 2h^{-1}$.

In Figure 6.59 the influence of the feedrate on the conversion is represented for different values of the contacting ratio. The data were obtained in the pilot plant in fluidized bed (Amoco), at the mean reaction temperature of 495°C and at a pressure of 1.38 bar. The used feed had: $d = 0.883$, distillation end point $= 393°C$ and the following composition (wt %): 27.1 alkanes, 43.9 cyclo-alkanes, 29.0 aromatics, and 0.59 sulfur. The residual coke on the catalyst in all cases was $= 0.10\%$.

At constant conversion, increasing the contacting ratio leads to a decrease of the yields of hydrogen and of the C_1–C_4 gases, leaves unchanged the yield in gasoline, and leads to an increase of coke.

This last result is explained by the amounts of liquid products remaining in the pores of the catalyst and are sent to the regenerator. These transported amounts increase with increasing contacting ratio. No generally valid correlation can be given since the amounts of liquid remaining in the pores depend on pore size distribution and stripping conditions. A large number of experimental data obtained in pilot plants and in the industrial units allowed the formulation of the following empirical correlation [4]:

$$\frac{x}{1-x} = C \cdot n^{-0.35} \cdot a^{0.65} \cdot e^{-7000/T} \tag{6.93}$$

where:

x is the conversion
$C =$ a constant depending on the nature of the feed and of the catalyst
$n =$ specific feedrate (kg feed on kg catalyst/hour within the reactor)
$a =$ catalyst/feed ratio.

The catalyst/feed ratio should not exceed values that correspond to the minimum residence time of the catalyst in regenerator, necessary to ensure the level of desired residual coke.

Table 6.23 Catalyst Feed Ratio Influence on the Conversion

Catalyst	$a = 3$	$a = 6$
XZ-40	84.8%	90.2%
DZ-7	83.1%	86.8%
XZ-25	72.4%	84.6%

Source: Ref. 218.

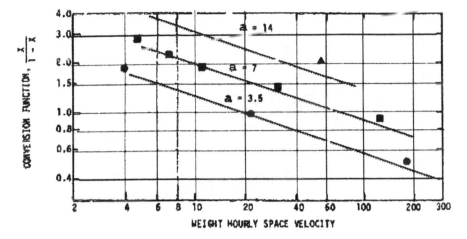

Figure 6.59 Effect of feedrate and catalyst/feed ratio on conversion. (From Ref. 219.)

6.6 CATALYST REGENERATION

Regeneration of catalysts is performed in the regenerator by continuously burning the coke deposited on the catalyst.

6.6.1 Coke Composition

In fluid catalytic cracking the content of hydrogen in the coke is in the range of 5–10% and can be obtained by using the graph of Figure 6.60 [220]. Analytical expressions are given in the same article [220].

In the moving bed units, the content of hydrogen in the coke ranges between 2.5–8%.

The amount of hydrogen in the coke may be determined directly by elemental analysis. But the analysis must differentiate between the water resulted from the combustion of hydrogen contained in coke and water content of the catalyst or water adsorbed by the catalyst.

Such a method was proposed and tested with good results [221]. It is based on the determination of the weight loss of the catalyst during elemental analysis.

Noting with h and c the amounts of hydrogen and carbon in the coke, $a =$ the water contained in the catalyst, $\Delta G =$ the weight loss of the catalyst submitted to elemental analysis, and H_2O and $CO_2 =$ the measured amounts of water and carbon dioxide, one may write:

$$H_2O = a + \frac{18.016}{2.016} \cdot h$$

$$CO_2 = \frac{44}{12} \cdot c$$

$$\Delta G = a + h + c$$

yielding the final equations:

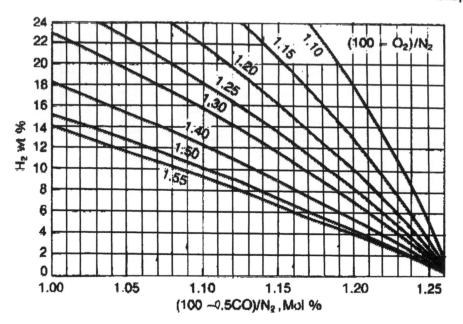

Figure 6.60 Correlation of hydrogen content of coke with the composition of flue gases. [220]. O_2, CO, and N_2 must be expressed as mole %.

$$c = 0.273CO_2$$
$$h = 0.126(H_2O + 0.273CO_2 - \Delta G) \tag{6.94}$$

6.6.2 Flue Gas Composition

It is of great interest to compare the composition of the flue gases corresponding to thermodynamic equilibrium with the one actually measured.

In Figure 6.61 [212] the curves for thermodynamic equilibrium at 660°C are plotted as a function of the contribution of oxygen calculated for the burning of a coke having the following composition (weight %):

Carbon   85.90
Hydrogen 10.00
Sulfur   1.15
Nitrogen   2.99

The scale on the abscissa of the graph, relative oxygen input (ROI), is defined as the amount of oxygen supplied to the regenerator, divided by the amount of oxygen theoretically required to burn all the coke hydrogen to H_2O, half of the coke carbon to CO, and the other half to CO_2.

For ROI = 0.8 all the hydrocarbons are burnt;
For ROI = 1.2 the CO is completely burnt and excess O_2 begins to appear in the flue gases.

For the thermodynamic calculation, the reactions of the nitrogen contained in the coke were simulated by the reaction between NH_3 and NO from which N_2 and

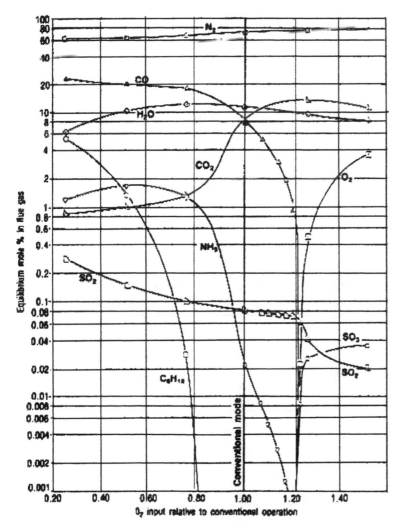

Figure 6.61 Equilibrium composition of flue gases at 660°C as a function of oxygen input. (From Ref. 222.)

H_2O are formed. The comparison of this equilibrium calculation with typical conditions for coke burning corresponding to ROI = 1.04 is given in Table 6.24. The table shows important differences between the calculated equilibrium composition and that measured experimentally. The differences are due to the differences between the burning rates but possibly also to the hindrance exerted by the catalyst on some of the reactions, such as the combustion of CO to CO_2. Such an effect could explain the combustion of CO→ CO_2 above the catalyst bed, which was observed occasionally.

Concerning the behavior of the sulfur contained in the feed, a graphic correlation was established between the SO_2 content in the gases and the sulfur content of the feed (see Figure 6.62 [183]).

Table 6.24 Comparison of Equilibrium and
Representative Coke Burning Flue Gas Composition ROI =
1.04

Component	Equilibrium composition	Real composition
O_2, % mol	0.0	0.10
CO, % mol	8.8	10.46
CO_2, % mol	7.5	8.65
H_2O, % mol	11.7	~ 10.0
NH_3, ppm	308	1285
SO_2, ppm	824	561
SO_3, ppm	0	5
NO, ppm	0	33

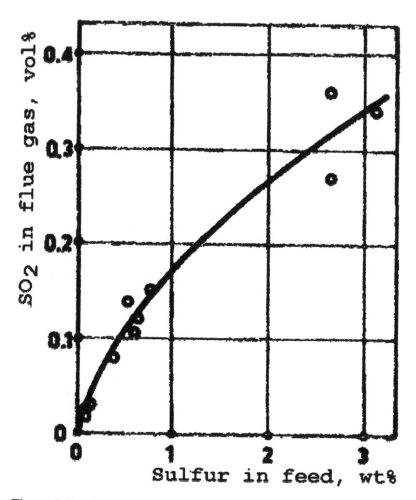

Figure 6.62 Correlation between the concentration of sulfur in the feed and of SO_2 in flue
gases. (From Ref. 183.)

In practical conditions used for burning coke in fluid catalytic cracking units, the CO_2/CO ratio in the flue gases is about 1.5. It increases if the catalyst is poisoned by heavy metals. The more recently promoted zeolite catalysts perform the complete combustion of CO to CO_2 (see Section 6.2.4.3).

6.6.3 Thermal Effect in Regeneration

The thermal effect of regeneration is dependent on the hydrogen content of the coke and on the CO_2/CO ratio in the flue gases. It can be calculated on the basis of the thermal effects of the following reactions:

$$H_2 + 1/2O_2 \rightarrow H_2O \qquad 121,000 \text{ kJ/kg}$$
$$C + O_2 \rightarrow CO_2 \qquad 32,741 \text{ kJ/kg}$$
$$C + 1/2O_2 \rightarrow CO \qquad 9,111 \text{ kJ/kg}$$
$$S + O_2 \rightarrow SO_2 \qquad 11,300 \text{ kJ/kg}$$

In order to avoid these calculations, interpolations can be made based on the data of Table 6.25.

6.6.4 Kinetics of Regeneration

6.6.4.1 Effect of the Diffusion Barriers

For processes in a moving bed, which use as catalysts granules with 3–6 mm diameters, the studies of Adelson and Zaitzeva [223] prove that the diffusion limitation is manifested at temperatures above 550–600°C. This influence is accounted for in equations for the overall burning rate developed by the same authors and discussed in the next section.

From the facts presented in Section 6.4.1.2, it followed that external diffusion strongly influences the regeneration of the catalyst in a fluidized bed operating in dense phase. This fact was also confirmed by other studies [118,119]. According to the represented calculations, in riser systems, the influence of external diffusion is negligible.

The influence of internal diffusion depends to a great extent on the structure and size distribution of the pores so that no valid generalizations can be given. It is possible that internal diffusion influences the overall rate for catalysts shaped as granules with diameters of 3–6 mm, especially when coke situated in the central

Table 6.25 Thermal Effect of Coke Burning, kJ/kg Burned Coke

CO_2/CO ratio in flue gas	Hydrogen content in coke wt %		
	4.0	8.0	12.0
0	14,590	19,010	23,400
1	24,075	28,070	32,090
2	27,215	31,085	34,960
4	29,745	33,535	37,300
10	31,820	35,505	39,210

part of the granule is combusted. In fluidized bed processes, that use microspherical catalysts with diameters of 40–70 μm, the influence of internal diffusion is less probable. This consideration is most probably valid for zeolite catalysts with channel macropores. For these processes, internal diffusion would play a part only when conditions are used for drastically reducing the residual coke.

6.6.4.2 Kinetics of Coke Burning

For regeneration of catalyst particles used in catalytic cracking processes with moving bed, Adelson and Zaitzeva suggested the equation:

$$\tau = \frac{c\gamma_v\delta}{\beta C_o V} + \frac{c\gamma_a R^2[3 - 2\delta - 3(1-\delta)^{2/3}]}{6\beta C_o n\bar{r}f\sqrt{T}} - \frac{\ln(1-\delta)}{\beta C_o k} \tag{6.95}$$

where:

τ = total burning time in s
c = weight fraction of coke on the catalyst
δ = weight fraction of burnt coke to total coke
β = amount of burnt coke in g/cm^3 of oxygen at the specified ratio CO_2/CO in the flue gases
C_o = fractions in volume oxygen in the regeneration air
V = cm^3 air/cm^3 catalyst
R = mean radius of the catalyst granules in cm
$\bar{r}$ = mean radius of the pores in cm
f = volume of the pores referred to the volume of the granule
T = regeneration temperature in Kelvin degrees
k = apparent rate constant for the burning of coke in $cm^3/g \cdot s$
n = constant depending on the percentage of coke on the catalyst, having the value of 162 for 2% coke

The apparent constant of the reaction rate k is given by the equation:

$$k = 7.33 \cdot 10^{10} e^{-146,850/RT} \tag{6.96}$$

Eq. (6.95) was verified by using a catalyst containing 2% coke from an industrial plant and by applying various regeneration conditions. A number of these results are collected in Table 6.26, where are written separately the three terms of Eq. (6.95) corresponding to the durations of the external diffusion, the internal diffusion, and of the reaction. The total is compared with the experimentally measured duration, the result being quite satisfactory.

The table shows that at temperatures below 500°C, the slowest process is the reaction. At higher temperatures, the diffusion phenomena increasingly influence the overall rate and become rate controlling at temperatures above 600°C. The external diffusion becomes rate determining at lower rates of the regeneration air, due to the increased thickness of the boundary layer at the lower Reynolds numbers.

Since Eq. (6.95) was verified only on one synthetic catalyst and only for an initial coke content of 2%, caution must be applied in its use.

For regeneration in fluidized bed the following equation was suggested by Johnson and Maryland [224]:

$$v = k_1 P_o(1 + k_2 P_s)C^n \tag{6.97}$$

Table 6.26 Regeneration Duration Calculated with Eq. (6.95) Compared with Experimental Data

Regeneration		τ calculated with Eq. (6.95)				τ
Temp. (°C)	Vol. air/vol catalyst · hour	Diffus. ext. τ (min)	Diffus. int. τ (min)	React. τ (min)	τ total (min)	experimental (min)
453	1500	4.2	5.7	98.1	109.0	109.0
471	1500	4.2	5.7	55.1	65.0	65.0
504	1500	4.1	5.5	18.4	28.0	28.0
556	200	31.5	5.4	4.8	41.7	39.5
600	200	31.2	5.2	1.6	38.0	36.2
615	1500	4.0	5.0	1.1	10.1	9.0
660	1500	4.0	4.8	0.4	9.2	8.2

Granules radius 1.65 mm. Initial coke on catalyst 2%. Burned coke/initial coke = 0.7.

where:

$$v = \text{burning rate of coke in kg coke per 1,000 kg of catalyst per hour}$$
$$P_o \text{ and } P_s = \text{partial pressures of oxygen and steam in kg/cm}^2$$
$$C = \text{average coke content on the catalyst within the regenerator}$$
$$n = \text{constant with value close to unity}$$
$$k_1 \text{ and } k_2 = \text{kinetic constants shown in the graph of the Figure 6.63 [1], based on the data of [224].}$$

From the values of the rate constants, the apparent activation energies are of the order of 171,600 J/mol.

Eq. (6.97) was deduced and verified for the burning of coke deposited on natural bentonite catalysts with an initial coke content of 0.13–0.60%, the regeneration being performed at temperatures of 510–565°C [224–225]. The equation has an empirical character and should not be used without additional plant verifications and on catalysts used in current operation.

A study by Tone, Niura, and Otaka [226] reports the results of experimental tests carried out in a differential reactor, which eliminates the influence of the diffusion steps, at temperatures of 500–560°C.

A synthetic catalyst was used, containing 25% Al_2O_3, having a porosity of 0.76, and a specific surface of 496 m²/g.

The reactions were represented by the scheme:

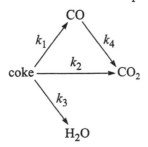

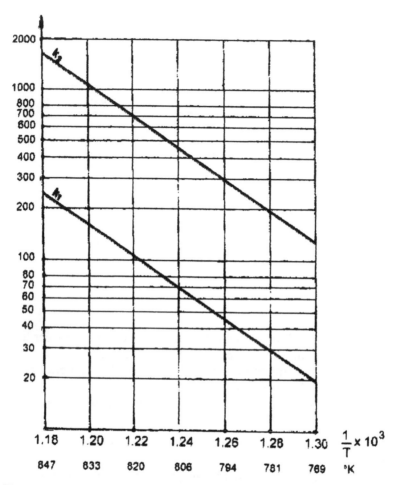

Figure 6.63 Temperature dependence of the constants k_1 and k_2 in Eq. (6.97).

and the following four rate equations were obtained:

$$
\begin{aligned}
r_1 &= k_1 n_C p_{O_2} \\
r_2 &= k_2 n_C p_{O_2} \\
r_3 &= k_3 n_H p_{O_2} \\
r_4 &= k_4 p_{CO}
\end{aligned}
\tag{6.98}
$$

where:

n_C and n_H represent the content of carbon and hydrogen in the coke, expressed in moles carbon or hydrogen per g of catalyst

p_i = partial pressures in atm

The formation of the final products is expressed by the equations:

for CO formation $\qquad F \cdot \dfrac{\Delta y_1}{\Delta w} = r_1 - r_4$

for CO_2 formation $\qquad F \cdot \dfrac{\Delta y_2}{\Delta w} = r_2 + r_4$ $\qquad\qquad$ (6.99)

for H_2O formation $\qquad F \cdot \dfrac{\Delta y_3}{\Delta w} = r_3$

where:

Subscripts 1, 2, and 3 correspond to CO, CO_2 and water; Δy_i = difference expressed in molar fractions between the inlet and outlet of the reactor
Δw = weight of the catalyst in g
F = feedrate in moles/minute.

It results that the reaction rates r_i are expressed in moles of carbon or hydrogen converted per g of catalyst per minute.

The linearization of the experimental data, including those of other researchers, led to the following equations of the reaction rate constants:

$$k_1 = 9.923 \cdot 10^9 \cdot \exp\left(-\frac{153,000}{RT}\right) \qquad 1/\min \cdot bar$$

$$k_2 = 2.578 \cdot 10^6 \cdot \exp\left(-\frac{107,350}{RT}\right) \qquad 1/\min \cdot bar$$

$$k_3 = 6.557 \cdot 10^4 \cdot \exp\left(-\frac{76,240}{RT}\right) \qquad 1/\min \cdot bar$$ $\qquad$ (6.100)

$$k_1 = 4.003 \cdot 10^9 \cdot \exp\left(-\frac{57,650}{RT}\right) \qquad mol/g \cdot cat \cdot \min \cdot bar$$

It is to be remarked that the burning of hydrogen contained in the coke takes place at a higher rate than that of carbon (Figure 6.64), which leads to a lower value for the hydrogen in the residual coke than in the initial one.

In a recent study, a system of differential equations was proposed for modeling coke burning and the composition of the flue gases [241]. This equations system is based on a large number of experiments realized on OCTADINE 1169 BR Engelhard catalyst with MAT activity of 71 wt %.

In another paper [242] one and two steps riser regeneration systems were studied to find the best way of introducing and distributing air, with the idea to improve the regenerator system. The result of such optimization was a carbon content of less than 0.1 wt % in the regenerated catalyst and operation at riser temperature lower than 730°C.

6.6.5 Effect of the Regeneration Conditions

6.6.5.1 Temperature

The lowest temperature level for regeneration is limited by the rate of coke burning. The highest level was limited in the past by the ability of the catalyst to resist high temperatures. Following refinements made to the catalysts, the temperature is now limited by metallurgical considerations, namely by the resistance at high temperatures of the metals of the regenerator and of the transport lines for the flue gases.

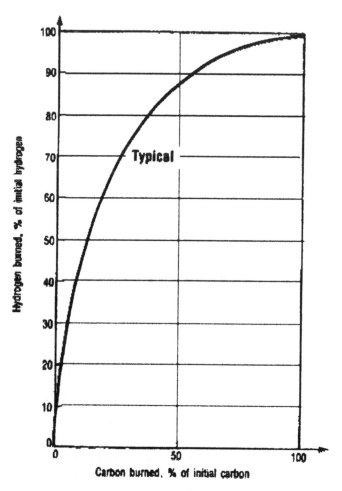

Figure 6.64 Relative combustion velocity of coke carbon and hydrogen at 700°C. (From Ref. 227.)

Increasing the temperature while maintaining constant the other working parameters increases the burning rate of coke and accordingly decreases the residual coke. Since the decrease of residual coke leads to the increase of the average activity of the catalyst, the regenerator is operated at the maximum possible temperature while taking into account the limitations mentioned above.

6.6.5.2 Pressure

As it results from Eqs. (6.97) and (6.98) the rates for burning carbon in CO and CO_2, as well as for burning hydrogen, are directly proportional to the partial pressure of oxygen.

The initial partial pressure may be increased by increasing the total pressure in the regenerator or by the addition of oxygen in the air fed to regeneration.

The pressure in the regenerator depends on the type of unit and does not constitute a controllable process parameter in the operation of the plant. The addition of oxygen in the air fed to the regenerator is an efficient measure suggested as early as 1980 [228] and was applied beginning with 1986 in the Gibraltar refinery [229]. The extension of this measure is limited by economic reasons.

The average oxygen partial pressure in the regenerator depends also on the air excess, a parameter on which the operator may act to the extent allowed by the performance of the turbo-blower and by the requirement of not exceeding reasonable linear velocities for the air and flue gases in the regenerator.

Finally, the increase of pressure and excess air, while keeping constant the other parameters, leads to the decrease of residual coke, which is illustrated in Figure 6.65 [230].

6.6.5.3 Catalyst Characteristics

The structure and the size distribution of the catalyst pores have a strong influence on the regeneration rate. This observation is in agreement with the fact that at the high regeneration temperatures used at present, even if the internal diffusion is not controlling the rate, it influences it to a large extent.

At otherwise constant parameters, the catalyst structure strongly influences the residual coke. The existence of the channeling macropores makes easier the access of

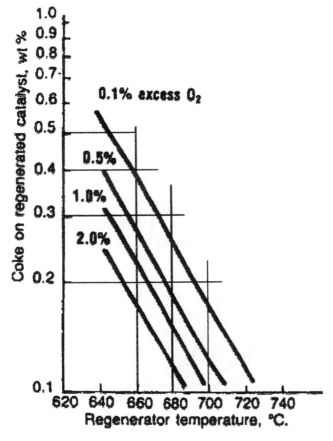

Figure 6.65 Residual coke as a function of temperature and excess air. (From Ref. 230.)

the oxygen to the coke particles located in the center of the granule and decreases the residual coke. For this reason the presence of the macropores is a desired feature of catalyst structure.

The characterization of the catalysts from the point of view of regenerability requires special methods [1].

Ni and V content is an important factor for the performance of the equilibrium catalyst present in the reaction system. In addition to the unfavorable influence on the cracking reactions discussed earlier, these metals can modify the thermal balance of the regenerator by the catalytic action of the combustion of CO to CO_2, due especially to nickel. In older units, the CO/CO_2 ratio was maintained at a certain level in order not to thermally overload the regenerator. The effect of the mentioned metals was to cause especially combustion in the diluted phase, above the catalyst bed. It seems that the alumosilica had an inhibiting effect and prevented $CO \rightarrow CO_2$ combustion inside the bed.

In modern units the situation is completely different. They are designed for a maximum heat production in the regenerator, in order to eliminate the supplementary equipment for the combustion of the CO contained in the flue gases. In this situation it is desired to achieve as completely as possible a combustion of the CO in the regenerator. Additives to promote burning contribute to this goal.

In this case the burning also takes place above the catalyst bed. This leads to a difference between the temperature of the dilute and the dense phases within the regenerator. This difference is illustrated in Figure 6.66 [230].

The control of the regeneration regime by increasing the level of the catalyst bed was tested at a refinery in Peru [231]. The disadvantage of the method is the necessity to increase the catalysts inventory, which has a negative impact on the economics of the operation.

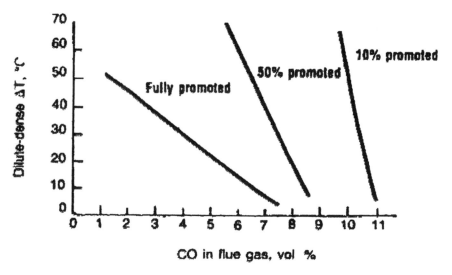

Figure 6.66 Effect of promoters on the CO content in flue gases. (From Ref. 230.)

REFERENCES

1. S Raseev. Procese distructive de prelucrare a titeiului, Editura Tehnica, Bucuresti, 1964.
2. S Raseev Stud Cercet Chim 5: 261,2S5, 1957.
3. DR Stull, EF Westrum Jr., GC Sinke. The Chemical Thermodynamics of Organic Compounds, Malabar, Florida: Robert E. Krieger Publ. Co., 1987.
4. D Decroocq, R Bulle, S Chatila, JP Franck, Y Jacquir, Le craquage catalytique des coupes lourdes, Technip, Paris, France 1978.
5. DG Tajbl, UOP Fluid Catalytic Cracking Process, Handbook of Petroleum Refining Processes, RA Meyers, ed., New York: McGraw-Hill, 1986.
6. J Scherzer, Octane-Enhancing Zeolitic FCC Catalysts, New York-Basel: Marcel Dekker, Inc., 1990.
7. ND Zelinski, Neftianoie i slanzevoie hoziaistvo, 2, No 9–12: 3, 1921.
8. SR Sergheenko Ocerk razvitia himii i pererabotki nefti, Akad URSS, Moscow, 1955.
9. AN Sachaven, Conversion of Petroleum, New York: Reinhold Publ. Co. 1948.
10. L Ubbelhode, N. Voronin, Petroleum, 7: 9, 1911.
11. Petroleum Processing Handbook; John J McKetta, ed. New York-Basel-Hong-Kong: Marcel Dekker, Inc., 1992.
12. JE Naber, M Akbar. "Shell's residue fluid catalytic cracking process", In: Hydrocarbon Technology International-1989/90, P Harrison, ed. Sterling Publ. Intern. London, 1990.
13. W Franz, P Gunther, CE Hofstandt, Fifth World Petroleum Congress, Section III,123, New York, 1959.
14. EJ Gohr. Development of the fluid catalytic cracking process (in Russian), Fourth World Petroleum Congress, Rome 1955. Gostoptehizdat Moskow, Vol. 4, 1956 p. 139.
15. P Courty, C Marcilly. Preparation of Catalysts III, 485, Elsevier, Amsterdam, 1983.
16. LLB Ryland, MW Tamele, JN Wilson. Catalysis, Editor. PH Emmett, ed. vol. VII, p. 1, New York: Reinhold, 1960.
17. HP Boehm, H Knozinger. Catalysis-Science and Technology, vol. 4, JR Anderson, M Boudart, Springer, ed. Berlin, 1983.
18. BW Wojciechowski, A Corma. Catalytic Cracking, New York: Marcel Dekker, Inc., 1986.
19. L Thomas Ch., Ind Eng Chem 41: 2564, 1949.
20. VI Vernadski. Ocerki gheochimii, Gorneftizdat 1934, p. 108.
21. K Hashimoto, T Masuda, H Sasaki. Ind Eng Chem 27: 1792, 1988.
22. BC Gates. Catalytic Chemistry, New York: John Wiley & Sons Inc., 1991.
23. EG Derouane. Intercalation Chemistry, MS Wittmgham, AJ Jacobson, eds. New York: Academic Press, 1982, p. 10.
24. PK Maher, CV McDaniel. U.S. Patent 3,402,996, 1968. DW Breck. U.S. Patent 3,130,007, 1964; JC Pitman, LJ Raid, U.S. Patent 3,473,589, 1969.
25. DW Breck. Zeolite Molecular Sieves, New York: John Wiley & Sons, Inc., 1974.
26. DW Breck. EM Fanigen, RM Milton. Abstr. 137th Meeting Am. Chem. Soc., Apr. 1960.
27. H Robson. Chem. Techn. 8: 176, 1978.
28. LD Rollman. Adv. Chem. Ser. (Inorg. Compd. Unusual Prop.-2) 173 387, 1979.
29. EM Flanigen. Pure Appl Chem 52: 2191, 1980.
30. LB Sand. Pure Appl Chem 52: 2105, 1980.
31. M Mengel. Chem. Tech. (Heidelberg) 10: 1135, 1981.
32. RM Barrer, Zeolites 1: 130, 1981.
33. LD Rollmann. NATO ASI, Ser, E., 1984, 80 (Zeolites Sci. Technol.), Portugal: 109, 1983.

34. C Marcilly. Raport nr. 14, Catalytic cracking-evolution of FCC catalysts. Published in OAPEC-IFP Workshop 5–7 July 1994, French Petroleum Institute, Rueil-Malmeson, France.
35. S Benton. Oil and Gas Journal 93: 98, 1 May 1995.
36. A Humphries, JR Wilcox. Oil and Gas Journal 87: 45, 6 Feb. 1989.
37. K Rajagopalan, ET Habib Jr. Hydrocarbon processing 61: 43, Sept. 1992.
38. AT Lengade. Hydrocarbon Conversion Catalysts, Can. Pat. 967, 136, 1975.
39. WA Welsh, MA Seese, AW Peters. Catalyst Manufacture, U.S. Patent 4,458,023 (1984).
40. RJ Nozemack, JA Rudesil, DA Denton, RD Feldwick. Catalyst Manufacture, U.S. Pat. 4,542,118, 1985.
41. J Scherzer, DP McArthur. Ind. Eng. Chem. Res. 27: 1571, 1988.
42. L Upson, S Jams. Meeting NPRA, Mars 1982.
43. WA Blanton, RL Flanders. U.S. Patent 4,115.249 (1978) and 4,071,436 (1978).
44. LL Upson, H Van der Zwan. Oil and Gas J 85: 23 Nov., 64, 1987.
45. L Rheaume, RE Ritter, JJ Blazek, JA Montgomery. Oil and Gas J 74(20): 103, 1976.
46. AW Chester, AB Schwartz, VA Stover, JP McWilliams. Prepr Div Pet Chem, Am Chem Soc. 24: 624, 1979.
47. FD Harzell, WA Chester. Hydrocarbon Processing 59 (7): 137, 1979.
48. VR Zinoviev, MJ Levinbuc, UK Sapieva. Himia i tehnologhia topliv i masel. (2): 9, 1993.
49. VR Zinoviev, MJ Levinbuc, VE Varsaver. Himia i tehnologhia topliv i masel. (4): 6, 1993.
50. AS Krishna, CR Hsieh, AR English, TA Pecoraro, CW Kuehler. Hydrocarbon Processing 70 (11): 59, 1991.
51. RJ Bertolacini, GM Lehmann, EG Wollaston. U.S. Patent 3,835,031, 1974; S Jin, JA Jaecker, U.S. Patent 4,472,267, 1984; JW Bryne, Nat. Petr. Ref. Assoc., AM-84-55, 1984.
52. U Alkemade, S Cartlidge, JM Thomson. Oil and Gas J. 88: 52, 1 Oct 1990.
53. LA Jandieva, SM Gairbekova. Himia i tehnologhia topliv I masel, (4): 5, 1993.
54. MW Anderson, ML Occelli, SL Suib. Journal of Catalysis, 122 (2): 374, 1990.
55. U.S. Patent 4, 791, 083.
56. EV McDaniel, PK Mahler. U.S. Patent 3,292,192, 1966; 3,449,070, 1968.
57. GW Skeels, DW Breck. Proc. 6th Interllat. Zeolites Conf. Reno, Nevada, 1983.
58. DW Breck, H Blass, GW Skeels. U.S. Patent 4,503,023, 1985.
59. LL Upson, RJ Lawson, WE Cormier, FJ Baars. Oil and Gas J. 88: 64, 1 Oct. 1990.
60. NY Chen, WE Garwood. J Catal 47: 249, 1977.
61. NY Chen, LR Gorring, HR Irland, TR Stein. Oil and Gas J 75 (23): 165, 1977.
62. AA Avidan, Oil and Gas J: 90, 59, May 18 1992.
63. P O'Connor, F van Houtert. Ketjen CataL Symposium, Scheveninsen Holland, 1986.
64. P O'Connor, LA Gerritsen, JR Pearce, PH Desai, A Humphries, S Janik. Catalyst Development in Resid FCC, AKZO Catalyst Symposium, Mai 1991, Scheveningen, Holland.
65. JE Penick. U.S. Patent 4,064,639, 1977.
66. AB Schwartz. U.S. Patent 4,072,600, 1978.
67. AW Chester. U.S. Patent 4,107,032, 1978.
68. EM Arnett, JW Larsen. J Am Chem Soc 91: 1438, 1969.
69. W Haag, MR Dessau, Proc. 8th Intern. Congr. Catal., Berlin 1984, vol 2, p. 305.
70. AG Evans, M Polanyi, J Chem Soc 69: 252, 1947.
71. HO Pritchard, Chem Rev 52: 529, 1953.
72. BS Greensfelder. The Chemistry of Petroleum Hydrocarbons, vol. II, BT Brooks, ed. New York: Reinhold Publ. Co., 1955, p. 137.

73. RZ Magaril, Teoretceskie osnovy himiceskih protesov pererabotki nefti, Himia, Moscow, 1976, p. 165.
74. JL Franklin, Thermodynamic Aspects, Carbonium, Ions, GA Olah, PR von Schleyer eds. New York: Interscience, 1968, vol. I, p. 85.
75. JE Germain, Catalytic Conversion of Hydrocarbons, London-New York: Academic Press, 1969.
76. J Turkevich, RK Smith, J Chem Phys 16: 466, 1948.
77. OA Reutov, TN Shatkina, Dokl Akad Nauk SSSR 133: 606, 1960.
78. PS Skell, RJ Maxwell, J Am Chem Soc 84: 3963, 1962.
79. DM Brouwer, JM Oelderik, Recl. Trav. Chem. Holland 87: 721, 1968.
80. H Hogeveen, AF Bickel, Reel Trav Chim Holland 88: 371, 1969.
81. TS Sie, Ind Eng Chem 31: 1881, 1992; 32: 397, 1993.
82. DK Liguras, DT Alien, Ind Eng Chem 28: 665, 1989; 28: 674, 1989.
83. GM Pancenkov, AS Kazanskaya, Zh. Fiz. Chim. 32: 1779, 1958.
84. HH Voge, Catalysis, PH Emmett, ed. vol. VI, 1958, p. 407, 466.
85. GM Good, HH Voge, BS Greensfelder. Ind Eng Chem 39: 1032, 1947.
86. E Angelescu, P Gurau, G Pogonaru, G Musca, Pop Gr, Pop Ec, Rev Roum de Chimie 35: 229, 1990.
87. A Corma, A Lopez Angudo, React Kinet Catal Lett 16: 253, 1981.
88. VV Kharlamov, TS Starostina, KhM Minachev, Izv. Acad.-Nauk USSR 10: 2291, 1982.
89. V Haensel, Adv Catal 3: 179, 1951.
90. BS Greensfelder, HH Voge, GM Good. Ind Eng Chem 41: 2573, 1949.
91. J Scherzer, DP McArthur, Oil and Gas J 84 76: 27 Oct. 1986.
92. TC Ho, AR Katritzky, SY Cato, Ind Eng Chem 31: 1589, 1992.
93. CM Fu, AM Schaffer, Ind Eng Chem 24: 68, 1985.
94. AC Oblad, T Milliken, GA Mills, The Chemistry of Petroleum Hydrocarbons, vol. II, BT Brooks, ed. New York: Reinhold Publ. Co., New York, 1955.
95. BV Klimenok, EA Andreev, VA Bordeeva, Izv. Akad. Nauk SSSR 5: 526, 1956.
96. TM John, RA Pachovsky, BW Wojciechowski. Adv Chem Ser 133: 422, 1974.
97. KR Al Latif, Ph.D. Thesis directed by S Raseev, Inst. Petrol Gaze, Ploiesti, 1975.
98. JW Hightower, PH Emmet, J Am Chem Soc 87: 939, 1965.
99. DA Best, BW Wojciechowski, J Catal 47: 11, 1977.
100. CC Lin, SW Park, W Hatcher, Ind Eng Chem Process Des Dev 22: 609, 1983.
101. M Guisnet, P Magnoux, Appl Catal 54: 1, 1989.
102. CT Ho, Ind Eng Chem Res 31: 2281, 1992.
103. FW Scholkemeier, Arch d Math 1: 270, 1949.
104. VG Jenson, Proc Roy Soc London A249: 346, 1959.
105. S Taneda, Repts Res Inst Appl Mech 4: 99, 1956.
106. S Raseev, Ph.D. Thesis, Bucarest, 1972.
107. B Jazayeri, Hydrocarbon Processing 7Q (5): 93, 1991.
108. P Wuithier, Le Petrole, Rafinage et Geme Chimique, I Tome, Technip., Paris, 1972.
109. Rascioty osnovnyh protsesov i aparatov neftenererabotki, Editor EN Sudacov, Himia, Moscova, 1979.
110. OA Hougen, KM Watson, Chemical Process Principles, vol. Ill, In Kinetics and Catalysis, New York: John Wiley & Sons, Inc., 1947.
111. RC Reid, JM Prausnitz, BE Poling, The Properties of Gases and Liquids, 4th ed. Chap. 11, New York: McGraw-Hill Book Co., 1987.
112. JM Smith, Chemical Engineering Kinetics, 3rd ed. New York: McGraw-Hill Book Co., 1981.
113. CR Wilke, CY Lee, Ind Eng Chem 47: 1253, 1955.
114. EN Fuller, JC Giddings, J Gas Chromatogr, 3: 222, 1965.

115. EN Fuller, K Ensley, JC Giddings, J Phys Chem 75: 3679, 1969.
116. EN Fuller, PD Schettler, JC Gilddings, J Ind Eng Chem 58 (5): 18, 1966.
117. Enciclopedia de Chimie, vol. 5, p. 132, Editura Stiintifica si Enciclopedica, Bucuresti, 1988.
118. J Pansing, J Am Inst Chem Eng 2: 71, 1956.
119. A Babicov, B Solar, L Glazov, J Liberzon, R Basov, T Melik-Ahnazarov, A Elsin, V Zarubin, Himia tehnologhia topliv i masel 9, No. 1, 1993.
120. R Aris, Chem Eng Sci, 6: 262, 1957.
121. R Aris, Ind Eng Chem Fund 4: 227, 1965.
122. EE Petersen, Chemical Reaction Analysis, Englewood Cliffs: Prentice-Hall, New Jersey, 1965.
123. GF Froment, KB Bischoff, Chemical Reactor Analysis and Design, New York: John Wiley & Sons, Inc., 1979.
124. TE Comgan, Chem Eng Propr 49: 603, 1953.
125. CD Prater, RM Lago, Advances in Catalysis and Related Subjects, WG Frankenburg, ed. vol. 8, 1956.
126. MFL Johnson, Ind Eng Chem 49: 283, 1957.
127. GM Pancencov, JM Jarov, Trudy Moskv Inst Neftianoy I Gazovoi Prom 37: 3, 1962.
128. EV Bergstrom, VO Bowles, LP Evans, JW Payne, Fourth World Petroleum Congress, sect. III, Rome, 1955.
129. RM Moore, JR Katzer, AlChEJ 18: 816, 1972.
130. DM Nace, Ind Eng Chem Prod Res Dev 9: 203, 1970.
131. WO Haag, RM Lago, PB Weisz. Faraday Disc Chem Soc 72: 317, 1981.
132. A Voorhis Jr., Ind Eng Chem 37: 318, 1945.
133. AV Frost, Vestnik Mosc Gos Univ., nr. 3, p. 11 and 4,117, 1946.
134. TV Antipina, Dokl Akad Nauk SSSR 53: 47, 1946.
135. GM Pancenkov, VP Lebedev, Himicescaia Kinetika i Kataliz, ed. Univ. Mosc., 1961.
136. V Tescan, S Raseev, Bul. Instr. Petr. si Gaze 14: 93, 1966.
137. S Raseev, HA Hassan, Bul Inst Petr si Gaze 15: 73, 1967.
138. LL Oliveira, EC Biscaia Jr., Ind Eng Chem 28: 264, 1989.
139. GF Froment, KB Bischoff, Chem Eng Sci 16: 189, 1961.
140. RP DePauw, GF Froment, Stud Surf Sci Catal 6: 1, 1980.
141. JW Beeckman, GF Froment, Ind Eng Chem Fundam 18: 245, 1979.
142. EFG Herington, EK Rideal. Proc Roy Soc (London), A184, 434, 1945.
143. VW Weekman Jr, Ind Eng Chem Process Des Dev 7: 90 (1968).
144. A Corma, A Lopez Agudo, J Nebot, F Tomas, J Catal 77: 159, 1982.
145. EB Maxted, Adv Catal 3: 129, 1951.
146. HV Maat, L Moscou, Proc 3th Int. Congr. Catal, Amsterdam, 2, 1277, 1964.
147. AL Pozzi, HF Rase, Ind Eng Chem 50: 1075, 1958.
148. FH Blanding, Ind Eng chem 45: 1186, 1953.
149. J Gustafson, Ind Eng Chem Process Des Dev 11: 4, 1972.
150. S Krishmaswamy, JR Kittrell, Ind Eng Chem Process Des Dev 17: 200, 1978.
151. NN Samoilova, VN Erkin, PJ Serikov, Himia i Tehnologhia Topliv i Masel, No 10: 11, 1990.
152. VW Weekman, DM Nace, AlChE J 16: 397, 1970.
153. V Tescan, S Raseev, Bul Inst Petrol Gaze 15: 81, 1967.
154. JA Paraskos, YT Shah, JD McKinney, NL Carr, Ind Eng Chem Process Des Dev 15: 165, 1976.
155. RA Pachowsky, BW Wojciechowski, Canad J Chem Eng 49: 365, 1971.
156. JF Coopmans, P Mars, RL de Groot, Ind Eng Chem Res 31: 2093, 1992.
157. B Gross, DM Nace, SE Voltz, Ind Eng Chem Process Des Develop 13: 199, 1974.
158. AN Ko, BW Wojciechowski, Progr React Kinet, 12:201 1984.

159. M Larocca, S Ng, H de Lasa, Ind Eng Chem Res 29: 171, 1990.

160. LC Yen, RE Wrench, AS Ong, Oil Gas J, Jan. 11, 67 (1988).

161. L Liang-Sun, C Yu-Men, H Tsung-Nieu, Canad J Chem Eng, 67: 615, 1989.

162. TM John, BW Wojciechowski, J Catal, 37: 240, 1975.

163. A Corma, J Juan, J Martos, JM Soriano, 8th International Congress on Catalysis, Berlin, vol. II, (1984), p. 293.

164. SM Jacob, B Gross, SE Voltz, AIChE J., 22(4): 701, 1976.

165. PB Weisz, "Kinetics and Catalysis in Petroleum Processing", 10th World Petroleum Congress vol. 4, 1980, p. 325.

166. LL Oliveira, EC Biscaia, Jr., Ind Eng Chem Res 28: 264, 1989.

167. W Feng, E Vynckier, GF Froment, Ind Eng Chem Res 32: 2997, 1993.

168. JW Dean, DB Dadyburjor, Ind Eng Chem Res 28: 271, 1989.

169. M Schlossman, WA Parker, LC Yen, Chemtech 41: Febr. 1994.

170. PH Emmett, Catalysis, New York: Reinhold Publ Co, 1956.

171. CR Olsen, MJ Sterba, Chem Eng Progr 45: 692, 1949.

172. CL Hemler, UOP Process Div, Sc Petr Council Yugoslav Academy, 13 June 1975.

173. KM Watson, EF Nelson, Ind Eng Chem 25: 880, 1933.

174. S Raseev, Proprietatile fizice si fizico-chimice ale hidrocarburilor, titeiului si fractiunilor de tite, Published in "Manualul chimistului, Vol. 2, ed. AGIR, Bucuresti, 1948.

175. WL Nelson, Oil and Gas J 60: 11 June 1962.

176. HE Reif, RF Kress, JS Smith, Petr Refiner 40: 237, 1961.

177. PJ White, Oil and Gas J 66: 112, 20 May 1968.

178. K Van Nes, HA van Westen, Aspects of Constitution of Mineral Oils, New York: Elsevier, 1951.

179. N Hazelwood, Analitical Chem 163, June, 1954.

180. LM Kolesnicov, IN Frolova, HA Lapsina, Nauka Kiev 6:111, 1981.

181. DW Kraemer, HJ Lasa, Ind Eng Chem 27: 2002, 1988.

182. LE Kruglova, SN Hodjiev, SL Andreev, VV Tcheremin, Himia tehnologhia topliv masel, No. 10, 1991, p. 11.

183. GP Huling, JD McKinney, TC Readal, Oil and Gas J 73: 13, 19 May 1975.

184. RN Cimbalo, RL Foster, SJ Wachtel, Oil and Gas J 70: 112, 15 May 1972.

185. VJ Pentcev, AJ Bontcev, Himia i tehn. topliv masel 26 (3): 10, 1991.

186. MS Useinova, AM Guseinov, VM Kapustin, VA Jaurov, Himia i tehnologhia topliv masel, 28, No 11: 13, 1993.

187. JB Pholenz, Oil and Gas J 61: 124, 1 Avr. 1963.

188. JJ Blazek, Oil and Gas J 71: 66, 8 Nov 1973.

189. JB Magee, JJ Blazek, Zeolite Chemistry and Catalysis, Editor JA Rabo, ed. American Chem Soc, Washington, 1976.

190. AK Rhodes, Oil and Gas J 92: 41, 10 Oct 1994.

191. CC Wear, RW Matt, Oil and Gas J 86: 71, 25 July 1988.

192. PH Desai, RP Haseltine, Oil and Gas J 87: 68, 23 Oct 1989.

193. WS Letzsch, JS Magee, LL Upson, P Valeri, Oil and Gas 7: 57, 86, Oct 31, 1988.

194. RE Ritter, WR Grace, Oil and Gas 7: 41, 73, 8 Sept 1975.

195. SJ Watchel, LA Bailie, RL Foster, HE Jacobs, Oil and Gas J 70: 104, Apr 10, 1972.

196. JL Giandjonts, PJ Serikov, VN Erkin, TH Melik-Ahnazarov, Himia Tehn Topl Masel (9): 9, 1992.

197. C Marciuy, Revue de l'Institut Francois du Petrole 42: 481, 1992.

198. AW Chester, WA Stever, IEC PRD 16 (4): 285, 1977.

199. LA Pine, RJ Maher, WA Wachter, Ketjen Catalysts Symp, Amsterdam, 1984, p. 19.

200. J Magnusson, R Pudas, Katalistic's 6th Annual FCC Symp, Munchen, Germany, 22–23 May 1985.

201. JB McLean, EL Morehead, Hydrocarbon Processing 70 (2): 41, 1991.

202. RE Ritter, L Rhéamne, WA Welsh, JS Magee, Oil and Gas J 79: 103, July 1981.
203. RJ Campana, AS Krishna, SJ Yanik, Oil and Gas J 81: 128, Oct 31, 1983.
204. L Upson, S Jaras, J Dalin, Oil and Gas J, 80: 135, Sept 20 1982.
205. CJ Groenenboom, FW Van Houtert, J Van Maare, H Elzeman, Ketjen Catalysis Symp, Amsterdam, (1984), p. 55.
206. ML Occelli, SAC Gould, Chemtech 24:24–27, May 1994.
207. M Larocca, H Farag, S Ng, H de Lasa, Ind Eng Chem Res 29: 2181, 1990.
208. H Farag, S Ng, H de Lasa, Ind Eng Chem Res 32: 1071, 1993.
209. BM Jitomirski, EM Soskin, VA Stankevitch, EA Klimtseva, Himia i tehnol topliv masel, (7): 11, (1989).
210. PG Thiel, Oil and Gas J 78: 132, 18 Aug, 1980.
211. DF Tolen, Oil and Gas J 79: 90, 30 March 1981); 83: 91, 1984.
212. D Geldart, AL Radke, Powder Technology 47: 157–165, 1986.
213. GW Brown, Oil and Gas J, 88: 46, 15 Jan, 1990.
214. L Maksimiuk, EJ Tselidi, Neftepererabotka i Neftehimia 1: 14, 1991.
215. RJ Campegna, JP Wick, MF Brady, DL Fort, In: Refining and Petrochemical Technology Yearbook, PennWell Books, Tulsa, Oklahoma, 1987 p. 70.
216. GDL Carter, G McElhiney, Hydrocarbon Processing 68: 63, 1989.
217. DA Keyworth, WY Turner, TA Reid, Oil and Gas J 86: 65, 14 March 1988.
218. JA Montgomery, WS Letzsh, Oil and Gas J 69: 60, 22 Nov 1971.
219. C Wollaston, WJ Haflin, WD Ford, CJ D'Souza, Hydrocarbon Processing 54 (9): 93, 1975; Oil & Gas J 73: 87, 22 Sept 1975.
220. R Sadeghbeigi, Hydrocarbon Processing 70: 39, Feb. 1991.
221. S Raseev, I Georgescu, Petrol si Gaze 22. (2): 108, 1971.
222. DP McArthur, HD Simpson, K Baron, Development in Catalytic Cracking, published by Oil and Gas J, 1983.
223. SV Adelson, AJ Zaiteva, Himia i tehnol. Topiv i masel 7(1): 25, 1962.
224. MFL Johnson, HC Maryland, Ind Eng Chem 47: 127, 1955.
225. WF Pansing, J Am Inst Chem Eng 2: 71, 1956.
226. S Tone, S Miura, T Otake, Bulletin of Japan Petroleum Institute 14(1): 76, May 1972.
227. JL Mauleon, SB Sigaud, Oil and Gas J 85: 52, 23 Febr 1987.
228. DP Bhasin, MS Liebenson, G Chapman, J Hydrocarbon Processing 62, Sept 1983.
229. JP de Haro, AJ Gonzales, N Schroder, O Stemberg, Oil and Gas J 90: 40, 11 May 1992.
230. LL Upson, H van der Zwan, Oil and Gas J 85: 65, 23 Nov 1987.
231. RE Wong, Hydrocarbon Processing 72: 59, Nov 1993.
232. A Pekediz, D Kraemer, A Blaseth, H De Lasa, Ind Eng Chem Res 36: 4516–4522, 1997.
233. JI Mauleon, JC Curcelle, Oil and Gas J 64–70, 21 Oct 1985.
234. X Zhao, TG Robene, Ind Eng Chem 38: 3847–3854, 1999.
235. X Zhao, TH Harding, Ind Eng Chem 38: 3854–3859, 1999.
236. L Nalbandian, IA Vasalos, Ind Eng Chem 38: 916–927, 1999.
237. S Al-Khataff, H De Lasa, Ind Eng Chem Res 38: 1350–1356, 1999.
238. J Ancheyta-Juarez, F Lopez-Isunza, E Angular-Rodriguez, Ind Eng Chem Res 37: 4637–4640, 1998.
239. J Ancheyta-Juarez, F Lopez-Isznza, E Angular-Rodriguez, JC Moreno-Mayorga, Ind Eng Res, 36: 5170–5174, 1997.
240. K-H Lee, B-H He, Ind Eng Chem 37 (5): 1761–1768, 1998.
241. JM Arandes, I Abajo, I Fernandez, D Lopez, J Bilbao, Ind Eng Chem (9) 3255–3260, 1999.
242. D Bai, J-X Zhu, Y Jin, Z Yu, Ind Eng Chem 36, (11) 4543–4548, 1997.
243. JR Harris, Hydrocarbon Processing 75: 63, 1996.

244   CW Osen, Arch Math Asu Phys 6: 29, 1910; 7: 1, 1911.

245.   S Goldstein, Proc Roy Soc London A123: 225, 1929.

246.   M Kawaguti, Repts Inst Sci Techn Un Tokyo 2: 4, 1948.

247.   S Timotika, T Aoi, Quart Journ Mech and Appl Math 3 (Pt. 2): 140, 1950.

248.   T Pearcey, B McHugh, Phil Mag 46(7): 783, 1955.

249.   J Proudman, JRA Pearson, J Fluid Mech 2: 237, 1957.

250.   W Chester, DR Breach, J Fluid Mech 37: 751, 1969.

251.   M Van Dyke, J Fluid Mech 44: 365, 1970.

252.   T Maxworthy, J Fluid Mech 23: 360, 1965.

253.   VG Jenson, AIChE Meeting, Chicago, III, Dec. 1957.

254.   N Frossling, Lunds Universitèts Arsskrift, NF Avd 2, 35: 4, 1940.

255.   T Oroveanu, Mecanica fluidelor viscoase, Ed. Tehnica, Bucharest, 1967.

256.   JM Smith, Chemical Engineering Kinetics, International Student Edition, Singapore: McGraw-Hill, 6th Printing 1987, p. 535–537.

7

Industrial Catalytic Cracking

7.1 FEED SELECTION AND PRETREATMENT

Since the 1970s, the basic feedstock for catalytic cracking has been crude oil vacuum distillate. Depending on the conjuncture and market demands for fuel oils, visbreaking or coking distillate as well as deasphalted oil were also used as feed. In very rare cases a portion of the straight-run gas oil was included in the catalytic cracker feed.

The increase of the price of crude oil starting in the 1970s and the increase of gasoline consumption led to the trend of converting crude as completely as possible to motor fuels. Thus, residual fuel was replaced by natural gases, hydroelectric, and nuclear energy, and sometimes even with coal. This situation led to the extension and even the general use of catalytic cracking and to a significant increase in the amounts of vacuum residue processed by visbreaking and coking. Concomitantly important investments were made in hydrocracking units as tools for the complete conversion of vacuum residue into light products.

The situation changed fundamentally in the middle of the 1980s. The difference between the price of residual fuel and of gasoline, which was 125 $/t at the beginning of the 1980s dropped within a few years to 20 $/t. Concomitantly, the difference between the cost of the light and heavy crude oils decreased from 50–60 $/t to 12 $/t.

In these conditions the investments for new hydrocracking units became non-profitable and it became necessary to find other less expensive solutions for the complete conversion of residues to light products.

The solution adopted was to use as feed to the catalytic cracking units, straight run residue, initially as supplement to the traditional feeds and subsequently by itself.

The direct cracking of the straight run residue required the development of catalysts with adequate characteristics, and of units capable of burning the much larger amounts of coke which form now on the catalyst.

The use in catalytic cracking of the mentioned feeds accentuates the importance of ensuring the lowest possible concentrations of heavy metals, especially of Ni and V, and of Conradson carbon, in order to make the process economically feasible.

The following sections will discuss the requirements set for the selection of the feed and the pretreatment processes used for obtaining raw materials of the specified quality.

7.1.1   Vacuum Distillation of Straight Run Residue

The two current trends, i.e., to produce vacuum cuts with 565°C end point, and to increase the production of heavy crude oils with high Conradson carbon values and high Ni and V concentrations, made it necessary to improve the vacuum distillation columns.

The procedures considered depend to a large extent on the characteristics of the processed crude oil. Thus, the process and constructive recommendations mostly refer to the processing of crude oils of a specified quality [1–3]. However, the reported information allows one to draw conclusions concerning recent developments and future trends.

The evaluation of the quality of a vacuum distillate intended as feed for catalytic cracking is made by using, besides the standard analyses and the true boiling points (TBP) curve, the distribution curves for Conradson carbon and the concentrations of nickel and vanadium. The curves showing the distribution of nickel for three typical crude oils, are shown in Figure 7.1 [2].

A detailed analysis of the effect of various process solutions on the quality of the vacuum distillate used as feed for catalytic cracking was made by S. W. Golden and G. R. Martin [1].

The main characteristics of the two vacuum columns used are represented in Figures 7.2a and 7.2b, in which the numbers of theoretical plates for each specific section of each column are indicated.

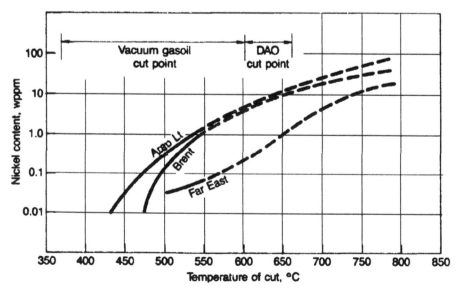

Figure 7.1   Nickel distribution for three representative crude oils. (From Ref. 2.)

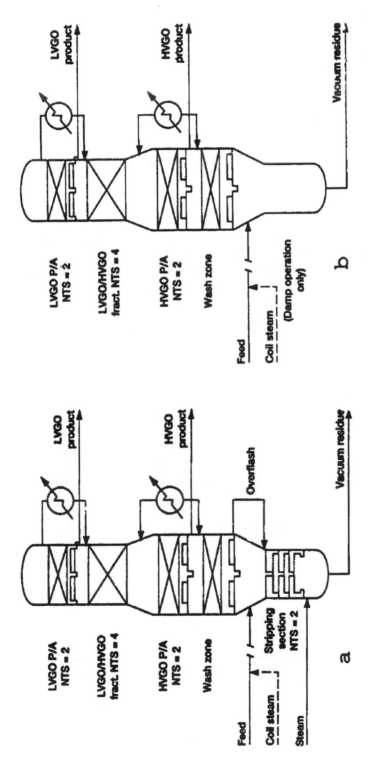

Figure 7.2 Typical vacuum columns configuration for catalytic cracking feed preparation. (a) Wet and damp (with stripping) operation mode. (b) Dry and damp (no stripping) operating mode. (From Ref. 1.)

The light distillate, obtained as the top product in the two columns, had in all cases an end point (TBP method) of 390°C and was used as component for the Diesel fuel. Only the heavy distillate obtained was used as feed for catalytic cracking.

For a feed having the characteristics:

Initial TBP	360°C
Density	0.9965
Molecular mass	505
Conradson carbon	8.3 wt %
Nickel	28 ppm

The feed was processed in four operating conditions reported in Table 7.1.

In all cases the feedrate used was $6.65 \, m^3/s$ and the internal reflux in the separation section between the distillates, provided with 4 theoretical trays, was $0.98 \, m^3/s$. The recovered light distillate was $0.82 \, m^3/s$ in all cases.

The concentrations of vanadium, nickel, and Conradson carbon of the heavy distillate destined to the catalytic cracking, depending on its final value for the four operating conditions, are plotted in the Figures 7.3a, b, and c.

The best results correspond to operating condition 4 and to an end point TBP of 569°C. The product obtained in these conditions was used in the studies that followed.

The variables were the number of theoretical trays in the stripping zone of the residue and the overflash in volume % of the feed. For three sets of operating conditions, the distillate with the end point of 569°C was redistilled and the content of metals and coke were plotted against the % distilled (see Figure 7.4a, b, c) [1].

Attempts to reduce from 3 to 2 or to 1 the number of washing stages, keeping the overflash at 3%, led to a substantial increase in metal content, especially of vanadium, in the distillate. This proved that 3 wash stages were necessary.

The contacting efficiency achieved by the materials used for the trays of the washing zone is also important. The height equivalent to a theoretical plate (HETP) was found to be for:

High efficiency metallic mesh	65–200 cm
Random packing	125–150 cm
Ordered packing	100–120 cm

Table 7.1 Conditions for Four Modes of Column Operation

Conditions	1	2	3	4
Operating conditions	Dry	Wet	Damp no stripping	Damp w/stripping
Column type Fig. 7.2 a or b	b	a	b	a
Steam:				
coil (t/h)	—	1.34	2.86	1.34
stripping	—	1.82	—	1.82
Top pressure (mm Hg)	8	50	20	20
Flash zone pressure (mm Hg)	18	66	35	35
Overflash (%)	3	3	3	3

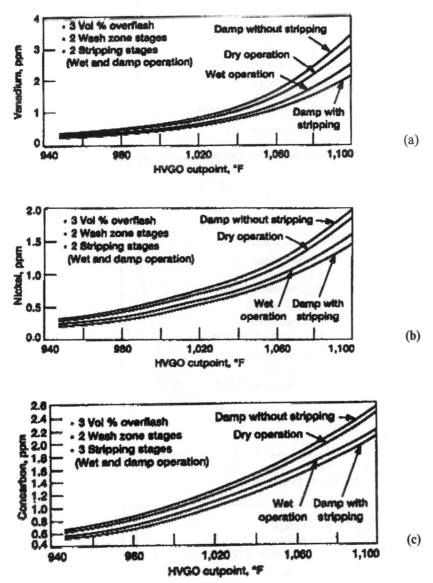

Figure 7.3 Effect of operation on HVGO. (a) vanadium, (b) nickel, (c) Conradson carbon. Content—variable mode of operation. (From Ref. 1.)

The efficiency of a fractionating plate was found to be equivalent to 0.25 theoretical plates.

The operating conditions that were finally selected were:

Operating conditions	No. 4 in Table 7.1
The end TBP of the heavy distillate	575°C
Overflash	3.0 vol %
Theoretical plates in the washing zone	2

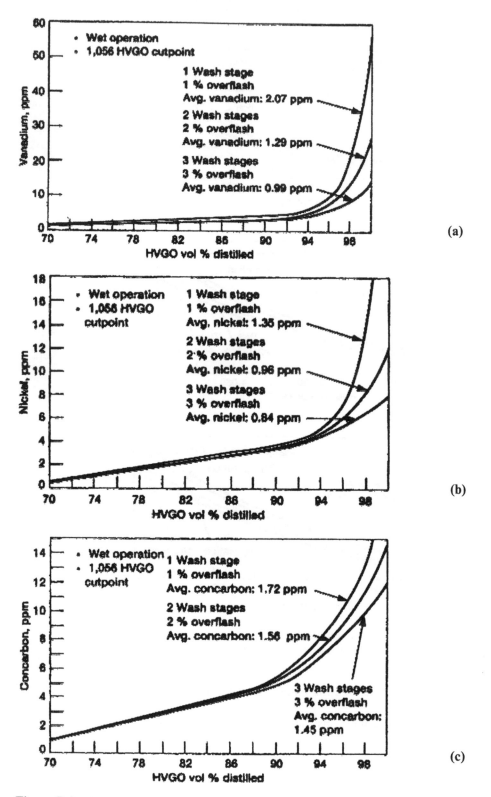

Figure 7.4 Contaminants distribution in HVGO for operating conditions in Fig. 7.2. (a) Vanadium, (b) nickel, (c) Conradson carbon.

The results obtained in these conditions:

20.9% vol. of the feed as heavy distillate, containing:
   0.85 ppm Ni
   1.15 ppm V
   1.5 ppm coke

Besides the problems of the number of theoretical trays in each section of the column, of the side reflux, and of the residue stripping, the measures taken for decreasing the formation of a fog of liquid drops in the vaporization zone of the column are also important. Fog formation is favored by the high inlet velocity (about 90 m/s) of the feed. From this point of view, the tangential inlet of the feed stream in the column is recommended, since it favors the separation of the liquid drops under the effect of the centrifugal force, and of other constructive means implemented in the design of the vaporization zone [4].

7.1.2 Processing of the Vacuum Residue

At the beginning of this chapter, several processes were indicated for obtaining catalytic cracking feedstocks from vacuum residues: visbreaking, coking, deasphalting with lower alkanes, and also hydrocracking. Several synergies between fluid catalytic cracking and hydroprocessing were also mentioned [77].

The main issues concerning the processes of visbreaking and coking were discussed in Chapter 4 and those of hydrocracking in Chapter 11.

The deasphalting process applied here is different from the "classical" propane deasphalting used for producing lubricating oils, since it uses higher alkanes and results in significantly higher yields of deasphalted product. As shown in Table 7.2, the content of Ni, V, and coke in the deasphalted product increases with the molecular weight of the alkane solvent [5]. The data of this table are orientative only, since the yields and the metal content in the deasphalted product is to a large extent

Table 7.2 Deasphalting of Arabian Light Vacuum Residue

Characteristics and yields	Feed	Deasphalted with		
		C_3H_8	C_4H_{10}	C_5H_{12}
Density (d_4^{15})	1.003	0.935	0.959	0.974
Viscosity at 100°C, cSt	345	34.9	63	105
Conradson carbon (wt %)	16.4	1.65	5.30	7.90
Asphaltenes (insoluble C_7) (wt %)	4.20	<0.05	<0.05	<0.05
Nickel (ppm)	19	1.0	2.0	7.0
Vanadium (ppm)	61	1.4	2.6	15.5
Sulfur (wt %)	4.05	2.55	3.30	3.65
Nitrogen (ppm)	2,875	1,200	1,950	2,170
Yields, wt %	100	45.15	70.10	85.50

Source: Ref. 5.

depending on the properties of the crude oil from which the vacuum distillate was obtained and on the particularities of the deasphalting process.

Overall, the product from the classical propane deasphalting mixed with the respective vacuum distillate may be submitted without other treatments to the catalytic cracking in classical plants. The deasphalted product obtained from butane deasphalting may be submitted directly to the catalytic cracking in mixture with vacuum distillate only in units designed for processing a heavy feed, whereas the product from pentane deasphalting needs to be hydrofined prior to catalytic cracking.

It is to be remarked that mixtures of light alkane hydrocarbons are often used as solvents. Besides the C_3–C_5 alkanes, light gasolines containing no aromatic hydrocarbons may be used as deasphalting solvents.

The selection of the type of deasphalting unit depends on the hydrocarbons used. It is recommended that the plant should use supercritical conditions for the recovery of the solvent (as in the ROSE process) that lead to important energy savings. The detailed examination of the performance of various deasphalting units is beyond the framework of this book.

A comparison between the results obtained by the catalytical cracking of the vacuum distillate by itself and in mixture with the products from visbreaking, coking, or deasphalting is given in Table 7.3. In the same table also the yields obtained by direct cracking of straight run residue with or without previous hydrofining are also given. These processes will be discussed in the next chapter.

7.1.3 Direct Use of the Straight Run Residue

The use of this feed became possible after catalytic crackers with two-step regenerators were developed. Such units can burn much larger amounts of coke than the earlier ones and have means for recovering the excess heat produced in the regenerator. Also, catalysts with better tolerance towards contaminating metals were developed.

The use of the straight run residue as feed for the catalytic cracking and the necessity for its pretreatment depend upon the level of coke and especially of metal content. The limits accepted as guiding values are given in Table 7.4 [7–9].

Results obtained in the direct cracking of the straight run residue are compared with those of a distillate in Table 7.5.

The large amounts of metals present on the equilibrium catalyst make compulsory the use of passivators, especially of antimony. The passivation leads to an increase by about 4% of the gasoline/conversion ratio and to a decrease by about 16% of the deposited coke. The reduced extent of the secondary reaction catalyzed by the metals (dehydrogenation, demethylation etc.) leads to a decrease by about 45% of the hydrogen and by about 25% of the amount of C_1–C_2 hydrocarbons produced.

Quite different technologies were obtained by the combination of the catalytic cracking of residues with the contacting in a "riser" system of the feed with an inert solid material, heated to high temperatures by the burning of the deposited coke—the ART [9,10] and 3D [11] processes.

These processes achieve in fact a coking on a heat carrier, in a riser system, at very short contact time and high temperatures. The presence of butadiene in the

Table 7.3 Yields by Different Catalytic Cracking Feeds for Arabian Light Crude

Products (vol %)	Vacuum distillate	Vacuum + visbreak. distillates	Vacuum + cracking distillates	Vacuum distillate + deasphalt. oil	Atmospheric residue	
					without hydrofining	with hydrofining
Liquified gases	2.2	2.2	2.3	2.6	2.6	2.8
heavy gasoline	4.6	4.8	4.8	4.1	3.5	3.6
gasoline	48.9	49.6	52.7	56.7	58.9	65.1
jet fuel	6.3	6.2	6.2	6.2	6.3	6.3
diesel gas oil	20.4	21.6	27.4	20.6	24.1	25.1
Total motor fuels	82.4	84.4	93.4	90.2	95.4	102.9
Products (wt %)						
residual fuels	23.2	20.9	5.5	15.2	8.7	3.9
coke production	—	—	5.3	—	—	—
coke burnt in the regenerator	1.8	1.8	1.8	3.0	4.5	2.9
Total heavy products	25.0	22.7	12.6	18.2	13.2	6.8

Source: Ref. 6.

Table 7.4 Feed Pretreatment as Function of Metal and Coke Content

Ni + V ppm	Conradson carbon wt %	Recommended pretreatment
< 30	5–10	Without pretreatment
30–150	10–20	Hydrofining
> 150	> 20	Coking

Source: Refs. 7–9.

reaction products in the case of the ART process [9] proves that the reaction temperature exceeds 600–650°C.

In the ART process, the entire amount of asphaltenes present in the feed, 95% of the organometallic compounds and 30–50% of those with sulfur and nitrogen are destroyed, without any effect on the hydrogen contained in the hydrocarbons. The amount of coke deposited on the support called ARTCAT and burnt in order to reheat it, represents 80–90% from the Ramsbottom carbon, compared to 130–170% for the traditional coking. The sulfur is eliminated mainly as SO_2 and SO_3 together with the flue gases.

After the contacting in the riser and the separation of ARTCAT the products are cooled by injection of a cold liquid and separated by fractionation. The recovered gasoline contains 50% alkenes, has an octane number of 70 F_2 and 80 F_1. The fraction distilling above 343°C, which goes into the feed of the catalytic cracking

Table 7.5 Results of Direct Cracking of Primary Residue and a Distillate

	Distillate	Residue
Feed		
density, d_{15}^{15}	0.8927	0.9267
Ramsbottom coke, wt %	0.11	5.1
Ni, ppm	—	4.6
V, ppm	—	10.5
Yields: conversion, vol %	75.7	74.4
hydrogen, Nm^3/m^3 liquid feed	0.135	0.455
$C_1 + C_2$, Nm^3/m^3 liquid feed	0.765	1.26
alkenes, $C_3 + C_4$, vol %	11.8	14.0
alkanes, $C_3 + C_4$, vol %	9.0	4.4
gasoline, vol %	60.1	56.5
light GO, vol %	13.6	16.0
column bottom, vol %	10.7	10.6
coke, wt %	4.6	11.7
Equilibrium catalyst (ppm)		
nickel	177	4290
vanadium	426	5490

Source: Ref. 8.

or of the hydrocracking, is similar to a vacuum distillate. The process makes possible the conversion of very heavy vacuum residues and even of the natural bitumens [9]. A portion of the ARTCAT is consumed in the process.

The solid inert contact material used in the 3D process [11] is not consumed in the process. It allows the processing of very heavy feeds having densities of the order 0.975, and containing about 43% components with boiling temperatures above 540°C and 10.7% Conradson carbon. Since it contains 2.6% S and 440 ppm metals, the product resulting from the contacting requires a preliminary hydrofining before being submitted to catalytic cracking.

A 3D industrial plant with a capacity of 500,000 t/year was started in 1989 [11].

The catalytical cracking of the liquid product obtained by the recovery of crude oil by means of underground combustion is to some extent similar to the above processes. Raseev et al. studied the cracking of such a feed, with the following characteristics [11]:

Density	$d_4^{20} = 0.9274$
Characterization factor	$K = 11.4$
Molecular mass	$M = 347$
Metals content	8.9 ppm

The comparative tests performed in a catalytic cracking unit with fixed bed of catalyst showed that this feed was similar to a coking distillate.

7.1.4 Feed Hydrofining

The hydrofining of the feed can significantly improve the performance of the catalytic cracking [12,13,77,78].

A systematic study was published [13] on the efficiency of the hydrofining of heavy distillates in mixture with gas oils from coking and visbreaking. Table 7.6 shows the improvements of the feed quality and yields, and the decrease in the SO_2 emissions that were obtained in different working conditions. These data were obtained for heavy distillates obtained from U.S. and Canadian crudes, which are similar to distillates obtained from Saudi Arabian and Russian crudes. To these distillates 20% by weight of coking gas oil was added.

Straight run residues may also be hydrofined prior to catalytic cracking. Such a combination of the processes was performed for the first time in 1981 in the Sweeny, Texas and Borger, Texas, refineries with good results. The efficiency of such treatment, especially concerning the decrease of the yield of hearth fuel and of coke is shown in Table 7.3.

The economic efficiency of these treatments depend upon the difference between the cost of gasoline and that of the straight run residue, and on the source and the cost of hydrogen. All factors must be analyzed case by case.

As general guideline, the hydrofining of the residue used as feed for catalytic cracking is recommended if the nickel and vanadium content is comprised between 30–150 ppm, and the Conradson carbon is between 10–20%.

For feeds exceeding these limits, special coking processes are recommended with very short contact times—ART and 3D—as described in the final part of Section 7.1.3.

Table 7.6   FCC Results from Untreated and Treated Feeds Using
Hydrotreater/FCC

HT severity operating conditions	Untreated max. naphtha	Low severity max. naphtha	max. diesel	Moderate severity max. naphtha	max. diesel
Feedstock					
dens., g/cm^3	0.9123	0.8866	0.8927	0.8745	0.8805
sulfur, w ppm	17,500	100	178	33	53
nitrogen, w ppm	1,050	140	185	8	10
UOP K index	11.75	11.96	12.10	12.12	12.27
VAPB, °C	416	376	433	376	437
Products (wt %)					
dry gas	4.0	2.9	1.8	3.1	2.0
LPG	14.0	16.6	9.8	19.4	11.4
gasoline	46.2	55.9	36.6	58.6	38.5
LCO	20.0	13.6	43.3	11.2	41.7
decant oil	9.8	4.7	4.4	3.3	3.3
coke	6.0	4.6	2.6	3.4	2.0
Sulfur (wppm)					
gasoline	2,700	6	—	<3	—
gas oil	27,400	146	—	70	—
decant oil	33,700	443	—	156	—
SO$_2$ emission, g/kg of feed	4.235	0.109	—	0.037	—

Source: Ref. 13.

The hydrofining of the distillates is applied in a large number of catalytic cracking units. An important result is the almost complete reducing of SO_2 emissions to the atmosphere, a very important factor for the protection of the environment [78].

7.2   PROCESS HISTORY, TYPES OF UNITS

7.2.1   Fixed Bed and Moving Bed Units

The catalytic cracking process on alumosilica catalysts was implemented commercially for the first time by F. J. Houdry and Socony Vacuum Oil Co. with a fixed bed of catalyst in cyclic operation in a unit having a processing capacity of 320 m^3/day.

The unit, which started operation on April 6, 1936 in the Paulsboro, New Jersey refinery, was provided with 3 reactors. This ensured, by means of an automatic valve system, a cyclic operation: 10 min reaction and 10 min regeneration, separated by 5 min of stripping.

Besides the technical difficulties in operation, including the removal of the regeneration heat, the process had the major disadvantage of continuous change of the effluent composition along the cycle. The conversion decreased strongly as the amount of coke deposited on the catalyst increased.

In order to eliminate this disadvantage, Socony Vacuum Co. developed a unit with a moving bed of catalyst, the Thermofor Catalytic Cracker (TCC) in Figure 7.5, and simultaneously the Houdry process with moving bed shown in Figure 7.6. As shown in the two figures, the difference between the two processes consists in separate vessels or in a unique body for the reactor and the regenerator.

Initially, the catalyst was shaped as 3 mm pellets while subsequent particles were of spherical shape with the same diameter. The elevator system used at the beginning was replaced in 1949 by pneumatic transport for the used catalyst, which permitted high catalyst/feed ratios and later on, the processing of heavier feeds. In Figure 7.7, the device located at the basis of the transport is sketched, which makes possible changing the contact ratio by acting on the flowrate of injected primary air.

Operating data for a Thermofor unit was given in Table 7.7 [14]. They refer to the processing of a gas oil from mid-Continent crude, having $d_4^{20} = 0.8984$, 50% distillation at 400°C, characterization factor $K = 11.9$. Two catalysts were used:

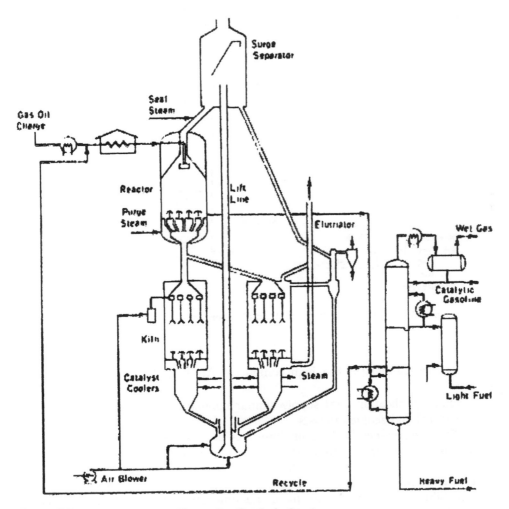

Figure 7.5 Sacony Vacuum Thermofor Catalytic Cracker.

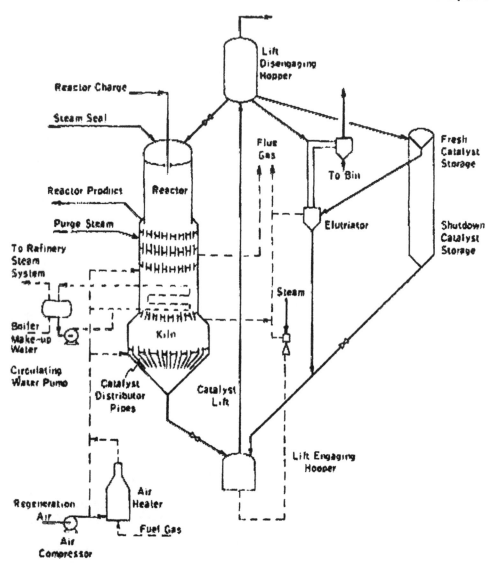

Figure 7.6   Houndry Moving Bed Catalytic Cracker.

3A—classical catalyst with a high content of Al_2O_3 and Durahead 5—a zeolite catalyst.

Other publications [15] supply similar comparisons between the operations with classical and zeolite catalysts.

The performance of the units improves at higher catalyst/feed ratios, as the mean coke content on the catalyst decreases and the mean activity increases. Thus, at a contact ratio of 5.5 the coke content on the catalyst at the inlet of the regenerator is of 1.1 wt % and the residual coke on the catalyst going to the reactor is of 0.05 wt %. The inability to further increase the contacting ratio was one of the reasons why units with fluidized bed became preferred. Indeed, depending on the unit capacity,

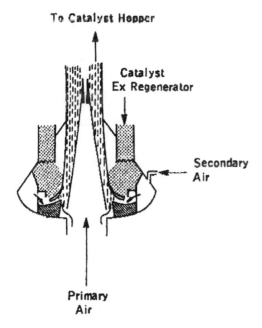

Figure 7.7   The bottom of a pneumatic transport system.

the amount of catalyst circulating through the system reached amounts of 200–1000 t/hour, which lead to high catalyst losses by erosion as well as by wear of the equipment and of the pneumatic transport system.

In addition, the burning of the coke deposited on the granules being a relatively slow process led to a residence time of about 1 hour for the catalyst in the regenerator.

Table 7.7   Operating Data of a Typical Thermofor Plant

Catalyst	A3	D5
Catalyst consumption, kg/day	750	350
fresh feed, m³/day	922	918
reactor temperature, °C	495	510
contacting ratio catalyst/feed	1.43	1.33
conversion, vol %	65.1	72.6
Yields		
combustion gas, wt %	6.0	6.4
oligomerization feed, vol %	16.5	17.5
gasoline, vol %	46.3	57.3
gas oil, vol %	21.9	18.0
decant oil, vol %	7.6	7.0
coke, wt %	7.6	7.0
octane F1 unblended	—	93.6

Source: Ref. 14.

The sizes of the reactor also became important. The reactor for a unit of 1 million t/year has a diameter of 5 m and a height of the catalyst bed of 4–5 m.

For all these reasons and other technical difficulties, the capacity of the plants with moving bed did not exceed 1.5 million t/year and they were or are gradually being replaced by fluidized bed units.

More details on this process, its various variations, examples of performance data and its estimation, and design methods for the reactor–regenerator system are presented in an earlier work of the author [4] and in other studies [16–18].

7.2.2 "Classical" Fluid Bed Units

The first catalytic cracking unit in fluidized bed was started in May 1942 at the Baton Rouge refinery of Standard Oil Co. as a result of a conjugated effort of a great number of American petroleum companies, determined by the state of war. For the same reason, in a relatively short time, the number of the units in the U.S. increased considerably. The processing capacities in the occidental countries reached the following values:

Year	m^3/day
1945	160,000
1950	270,000
1960	830,000
1970	1320,000
1978	1575,000

In 1995 in the U.S. alone, the number of the units exceeds 350 and the processing capacity is over 1,600,000 m^3/day.

The first catalytic cracking unit, called Model I, is characterized by the ascending circulation of the catalyst through the reactor and the regenerator, together with the reaction products, and the flue gases respectively. The separation of the catalyst is performed in external cyclone systems (see Figure 7.8). The plant had a processing capacity of 2400 m^3/day and required 6000 tons steel, 3200 m pipes, 209 control instruments, and 63 electromotors. A complete description of these beginnings was made by A. D. Reichle [17].

Only 3 Model I units were ever built, being followed by the Model II, characterized by the internal placement of the cyclones, fluidization in dense phase, and descendent pipes for catalyst transport at its exit from the reactor and regenerator (see Figure 7.9). These solutions decreased the diameter of the reactor and the regenerator and the necessary amount of metal. The first Model II unit was started at the end of 1942, also in the Baton Rouge refinery, where several months before the operation of the Model I unit was started.

The Model II units were reproduced in the former Soviet Union under the name of A-1 and then B-1. A type B-1 unit was installed in the Onesti Refinery in Romania.

The Model III unit differs from Model II by the location at the same level of the reactor and of the regenerator, which required the use of a higher pressure for the

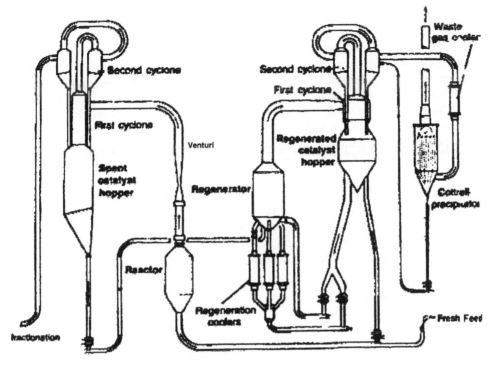

Figure 7.8 Catalytic cracking Model I.

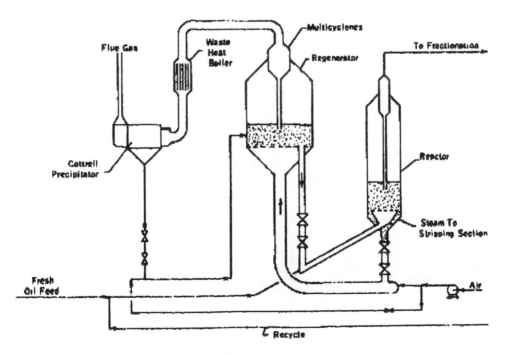

Figure 7.9 Catalytic cracking Model II.

regeneration air. Higher pressure became possible due to improved compression equipment.

The Model IV developed in 1952 made important progress over the previously described units. A number of units of this type are still in operation, including the Esso refinery in Port-Jérome (France). A detailed description of these units and of the exploitation experience is given in the monograph published by Decroocq et al. [20].

The unit is characterized by the location at the same level of the reactor and of the regenerator and the transportation of the catalyst between the two vessels in dense phase through semicircular pipes. The dense phase in these pipes is maintained by injections of air and steam respectively.

The circulation sense of the solids is determined by steam and air injections respectively (see Figure 7.10). They produce a diluted phase of a lower volume density in the ascending portions of the transport pipes above the valves.

A similar transport system is used also in some fluid coking units and was presented in the final part of Section 5.3.5. Figure 5.8 and Eqs. (5.69–5.72) give the conditions that ensure correct circulation.

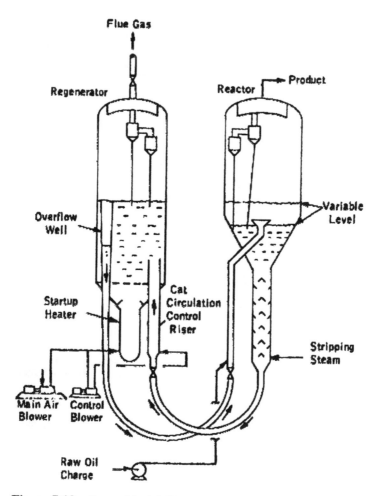

Figure 7.10   Exxon Model IV.

Another characteristic of this unit is the increase of the diameters of the reactor and regenerator in their upper part, which ensures a dense phase of fluidized catalyst between the distributor and this section with a bulk density of about 0.3 g/cm³. The increase of the diameter in the upper part decreases the amount of catalyst which is in this zone and accordingly decreases the rate of the overcracking reactions that lead to the decomposition of the gasoline to gases. Figure 7.11 gives the variation of the bulk density along the reactor height for different operating condition [21].

Another design adopted in the same period was to locate the reactor and the regenerator, coaxially, one above the other. This resulted in metal economy and the decrease of investment costs.

Thus units were built with the reactor overlapping the regenerator as in the UOP "stacked unit" (1947) and the Kellogg Orthoflow A (1951), and wherein the regenerator is overlapped to the reactor: Orthoflow B. The first option corresponds to a higher pressure in the regenerator, which increases the burning rate of the coke and leads to the decrease of the equipment diameter, but in exchange, needs a larger consumption of energy for air compression. The schematic flow sheet of the Orthoflow system A is given in Figure 7.12.

A more detailed description of the catalytic cracking units of the classical type including the sizes and the characteristic operating parameters, the yields, and the quality of the products was presented earlier by Raseev [4] and by other authors, such as Wuithier [22].

7.2.3 Units with Riser Reactors

The reactors of units of the classical type, which are described in the previous section, are characterized by intense backmixing of the catalyst and of the hydrocarbons, the behavior of the reactor being similar to that of a perfectly mixed reactor. On the other hand, the reactor of the "riser" type is in fact a tubular plug-flow reactor.

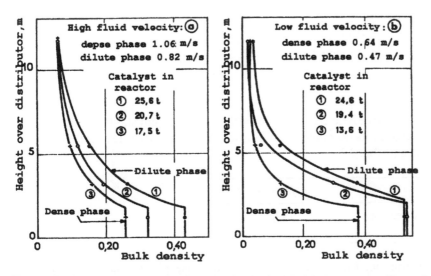

Figure 7.11 Variation of catalyst bulk density in Model IV reactors. (From Ref. 21.)

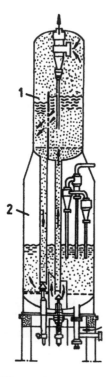

Figure 7.12. Kellogg Orthoflow A catalytic cracking. 1 – reactor, 2 – regenerator.

Since gasoline constitutes the intermediary product of a process made of successive steps, the maximum conversion of the gasoline will be higher in a riser reactor than in one containing a mixed fluidized bed.

In Sections 3.1 and 2.3.5 the equations of the maximum yield of an intermediary product were deduced for the two types of reactors. They are:

$$z_{max} = \frac{1}{(1 + \sqrt{k_2/k_1})^2} \tag{7.1}$$

for the perfectly mixed reactor and

$$z_{max} = (k_1/k_2)^{\frac{1}{1-k_1/k_2}} \tag{7.2}$$

for the plug-flow reactor.

Since for the catalytic cracking the ratio of the rate constants for the gasoline formation k_1 and for its decomposition k_2 is of the order 3–10, it results that z_{max} will have the values (Table 3.1):

		$k_1/k_2 = 3$	$k_1/k_2 = 10$
Perfectly mixed reactor	$z_{max} =$	0.402	0.577
Plug-flow reactor	$z_{max} =$	0.577	0.774

Despite the fact that in the reactors with dense phase fluidized beds of the classical units a perfect mixing is not achieved and thus, the maximum of the yield will have somewhat larger values and the kinetic of the catalytic cracking is more complex than that for two successive reactions, the advantages of the reactor of the "riser" reactors are obvious.

Therefore, after 1960, when the first riser type unit was developed by Kellogg, all new units are provided with a reactor of this type. Moreover, the classical plants in operation were revamped and the reaction system was modified to riser.

7.2.3.1 Revamping of Classical Units

The revamping of classical catalytic cracking units by the incorporation of the "riser" reactor was applied to all plants in operation. The manner in which the revamping was implemented depends on the type of the classical unit.

The following examples refer to plants IA/IM, which are similar to the classical Model II, of the former Soviet Union and to the Model IV, which was revamped at the ESSO Refinery of Port-Jérome (France) and the revamping of a riser plant in Caltex Refinery in Kurnell (NSW, Australia).

Revamping of the plant of the type IA/IM. A complex program for the modernization of the IA/IM units, with the purpose of bringing them to the performances of the modern plants, was decided in 1985 and comprised of 8 tasks, of which 5 were implemented by the year 1992 [23].

The process scheme of the system reactor–regenerator, as it will appear after all the modifications have been implemented is shown in Fig. 7.13.

The performed modifications refer to: a) use of high efficiency dispersers for achieving a good mixing of the feed with the regenerated catalyst; b) increase of temperature in the catalyst-feed mixing point, which leads to the increase of the temperature in the transport line to the reactor, which fulfils thus the part of a riser; c) improvement of the valves on the transport lines; d) use of promoters for the $CO \rightarrow CO_2$, conversion, which increases the temperature in the catalyst-feed mixing point; e) additives for decreasing the emissions of SO_2 and SO_3. Concomitantly, the catalysts were replaced with others that are more active and more resistant to attrition.

Until the end of 1992, the following modifications were implemented: f) the modification of the devices through which the catalyst-feed mixture enters the reactor; g) the two-step regeneration of the catalyst, the first taking place in a central fluidization chamber, marked by 10 in the figure; h) the completion of the system for the retaining of the catalyst fines by installing a group of external cyclones and the use of existing Cottrell filters, without reintroducing into the system the dust retained therein. The last measures limited the concentration of the catalyst in the flue gases to $0.15 \ mg/m^3$, compared to $0.6 \ g/m^3$ reported previously.

The results following the revamping are compared in Table 7.8, with the previous situation and the final results obtained after the completion of the revamping.

Revamping of the Model IV unit. The Model IV unit was presented in Figure 7.10 and described, in its classical version, in Section 7.2.2.

The modifications focused on the reactor and they were performed in succession as indicated by a, b, and c in Figure 7.14 [20]. The modifications are justified by the fact that for the zeolitic catalyst used, the maximum gasoline yield corresponds to the composition at the inlet in the vessel marked "reactor", the transfer pipe fulfilling the part of riser (see Figure 7.15) [24].

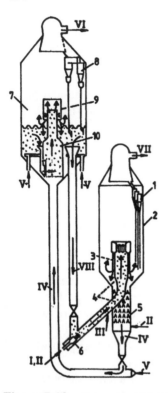

Figure 7.13 Revamped 1A/1M plant. 1,8 – cyclones, 2 – separator, 3,9 – inertial separators, 4 – riser, 5 – stripper, 6 – feed injection, 7 – regenerator, 10 – central fluidization chamber. I – feed, II – steam, III – recycle, IV – catalyst to regeneration, V – air, VI – flue gases to external cyclones, VII – products, VIII – regenerated catalyst.

In these conditions it was correct to reduce to a minimum and then to eliminate completely the cracking in the dense phase of the vessel that was initially the reactor and at the end, will act as catalyst separator and stripper.

The modifications indicated in Figure 7.14 were accompanied by the revamping of the feed dispersion system when coming in contact with the catalyst, in order to ensure a uniform cracking.

Revamping of a riser unit. The revamped unit of type UOP "straight run" (Figure 16a) at the Caltex Refinery in Kurnell, Australia came on-stream in 1961. The revamping realized in 1998 included a new air distributor, a new spent-catalyst distributor and new stripper internals. In addition, an inertial separator installed during the riser revamp in 1989 was replaced with riser cyclones.

The revamping resulted in a significant improvement in regenerator and stripper performance as well as improved yields and less catalyst deactivation [7].

7.2.3.2 Further Units of the "Riser" Type

The "riser" reactor was invented by Shell in 1956 [19] after extensive studies at pilot and semi-industrial scale. The unit was conceived so that the cracking reactions that take place in the riser are continued inside the reactor in dense phase.

Table 7.8   Results of the Plant 1A/1B Revamping in Romania

Parameters	Before revamping	After partial revamping (1991)	Second revamping (estimation)
Operating conditions			
capacity, % from the project	110	120	133
temperature after cat./feed mixing, °C	490–500	510–520	560
final reaction temperature, °C	460–470	470–480	510
catalyst/feed ratio	4–5	5–6	6–7
feed, h^{-1}	10	12	14
catalyst residence time in reactor, s	80	60	15
regeneration temperature, °C	535–550	600–615	645–665
residual coke on catalyst, wt %	0.4	0.25	0.1
flue gases: vol %			
CO_2	7.5	13.5	14.5
CO	5.5	0.05	0.05
O_2	6.5	4	3
SO_2	0.08	0.03	0.03
dust in flue gases, g/m^3	0.95	0.6	0.15
catalyst consumption, kg/t feed	1.5	0.9	0.5
Yields (wt %)			
H_2S	0.3	0.3	0.4
$C_1 + C_2$	3.0	2.8	3.2
$C_3 + C_4$	3.0	2.9	5.6
$C_3'' + C_4''$	2.4	2.7	9.0
gasoline, FBP 195°C	31	38	48
LGO, 195–350°C	26	30	18
HGO, >350°C	32.2	20.1	11.1
coke, wt %	3.1	3.2	4.5
conversion	41.8	49.9	70.7
gasoline F2 octane	79.6	80.1	81

The advantages of the system were evident and other construction firms adopted this principle and designed units that differed by the relative positions of the reactor and of the regenerator, the position of the lines for the catalyst circulation, etc.

The units from the first period were characterized by the fact that the cracking reactions took place partially in the riser and were continued in a reactor, in dense phase, in conditions of backmixing.

Two units of this type are presented as illustration, in Figures 7.16a and b. The unit of Figure 7.16c is provided with two independent risers, one for the cracking of the feed and the other for the cracking of the recycle*. Such a measure is justified by the large difference in the cracking rates of the two flows. This requires quite different reaction times in order to achieve for each of the flows the maximum yield of the

---

*The system with two risers is used also in the units by Kellogg-Orthoflow and Texaco.

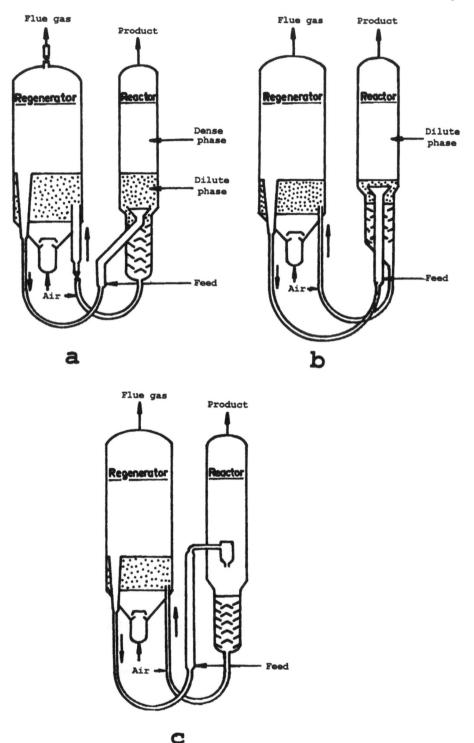

Figure 7.14 Revamping of the Model IV unit at Esso Port-Jérome refinery (France). (From Ref. 20.)

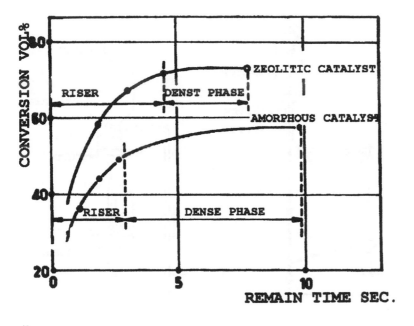

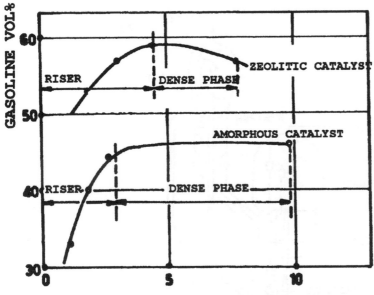

Figure 7.15 Conversion and gasoline yield evolution using amorphous and zeolitic catalysts. (From Ref. 24.)

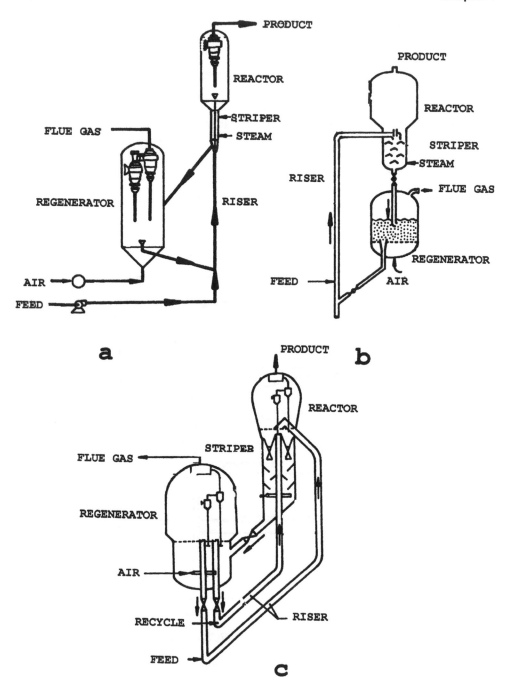

Figure 7.16 First generation catalytic cracking riser units. (a) UOP "Straight riser," (b) Esso Flexicracking, (c) Texaco, two risers system.

intermediary product—gasoline. In subsequent units, this objective was achieved by feeding the two flows into the riser at two different heights.

The fact that the reactions continued in the reactor in dense phase, where the prevailing backmixing led to longer average residence times, led to excessive cracking and to the decrease of the gasoline yield.

In order to remedy this disadvantage, several solutions were suggested: the decrease of the height of the catalyst bed in the reactor, devices that decreased the internal mixing etc., arriving finally at the solution adopted in the modern units: directly connecting the exit from the riser to a cyclone dedicated to this purpose (Figure 7.17) or to the cyclone system of the reactor. In this latter case, the separated reaction products pass directly to the fractionation tower without entering the reac-

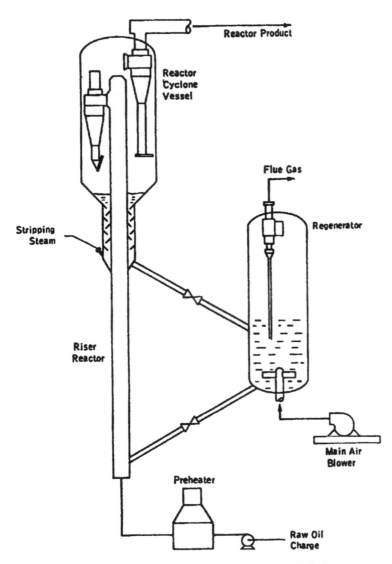

Figure 7.17   UOP unit without dense phase cracking in reactor.

tor, the role of which remains only to ensure the operation of catalyst stripping. These solutions require, of course, substantial improvements of the performance of the cyclone system.

The complete elimination of the dense phase leads to a substantial improvement of process performances. The data of Table 7.9 compares the performance of the UOP unit of Figure 7.17 with that of Figure 7.16a.

Improvements were brought also to the regeneration system by introducing two regeneration zones. The result was a reduction of the residual coke and thus, an increase of the process performances. The Orthoflow, model F units of Kellogg (Figure 7.18), is an example of the application of this concept.

A more advanced solution is the system where the first step of coke calcination is performed in a system type riser and only the second step takes place in dense phase in a classical fluidized bed. Together with these constructive improvements, the process conditions were also optimized: the pressure in the system reactor/regenerator was increased in order to increase the partial pressure of oxygen and accordingly, the burning rate of the coke. Thus, in the UOP unit of Figure 7.19, which applies these concepts, the pressure in the regenerator is of 1.75 bar compared to a maximum of 1.4 bar in the previous units, with the regenerator being located below the level of the reactor. The pressure in the reactor reaches 1.4 bar compared to 0.4 bar in earlier units. Overall, since the regenerators of this type were capable of increasing the amount of burned coke by means of controlling the CO_2/CO ratios in the two zones, they are currently used for the cracking of noncontaminated residues or of distillates with residue addition.

Table 7.9 Effect of Complete Elimination of the Dense Phase Cracking in Reactor

	UOP Fig. 7.16a*	UOP Fig. 7.17 Middle severity	UOP Fig. 7.17 High severity
Feeding with Mid-Continent GO			
density	0.876	0.886	0.888
UOP characterization factor	12.14	12.01	12.13
Conradson carbon, wt %	0.20	0.20	0.20
Conversion, vol %	80.2	77.5	89.6
Yields			
C_3'', wt %	7.7	5.7	10.6
C_4, wt %	10.9	10.0	13.3
gasoline (90% at 193°C), vol %	63.0	67.1	70.2
GO (90% at 310°C), vol %	15.0	14.5	6.9
heavy GO and residue, vol %	4.9	8.0	3.5
coke, wt %	4.9	4.5	5.7
Gasoline + alkylate			
yield, vol %	84.4	89.3	103.4
F1 octane number	90.1	89.2	92.5
F2 octane number	82.1	81.9	84.4

*Reactor temperature 505°C with diminished catalyst level.
Source: Ref. 24.

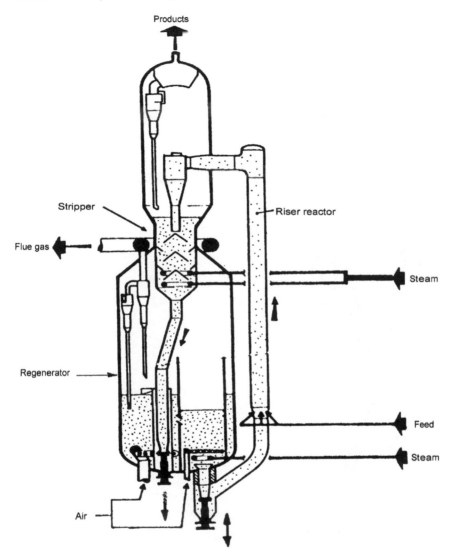

Figure 7.18   Kellogg Orthoflow F unit.

Since the constructive details and the operation conditions are common for all units of this type, they will be covered in Section 7.3.

7.2.4   Units for Residue Cracking

As indicated earlier, the direct catalytic cracking of residues is possible in units specially designed for this purpose when the content of Ni + V does not exceed 30 ppm and the Conradson carbon is below 5–10% by weight. Feedstocks with 150 ppm metals and 20% coke have to be submitted to a preliminary hydrofining. It must be mentioned that the above limits are not rigid, since they depend on the

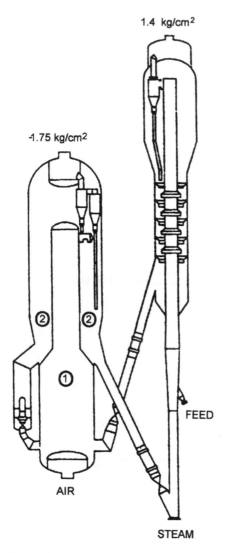

1.4 kg/cm²

-1.75 kg/cm²

② ②

①

FEED

AIR

STEAM

Figure 7.19 UOP unit with two steps regeneration.

type of unit, on the catalyst and on the passivators used for controlling the effect of the metals deposited on the catalyst.

The catalytic cracking of residues raises interest in estimating the content in metals (Ni + V) and Conradson coke in the straight run residues (> 370°C +) of the known crude oil reserves. These estimations are depicted in Figure 7.20 [25, 91].

Based on the data from this figure it is estimated [26] that 25% of the world reserves of crude oil allow the direct processing of the straight run residues by catalytic cracking. With a preliminary hydrofining and with the improvements brought to the process, this percentage could reach 50%.

The first residue catalytic cracking unit, with a capacity of 4,700 m³/day, was built by Kellogg in the Borger, Texas refinery and came on-stream in 1961. 20 years later saw the beginning of the intense building of such units, so that at the end of

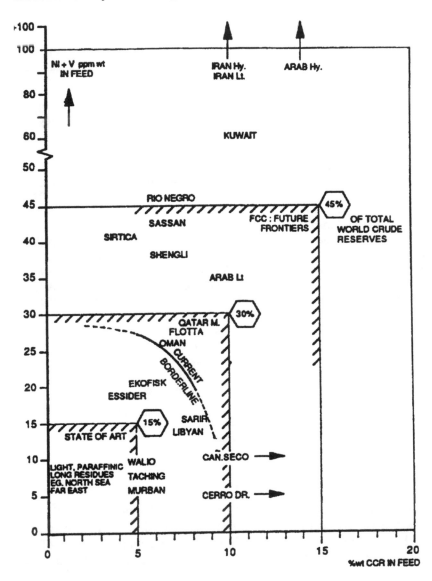

Figure 7.20 Ni+V and Conradson carbon contents in the >370°C fractions of several crude oils. (From Ref. 25.)

1983 processing capacity reached 38,800 m³/day, at the end of 1989, 81,500 m³/day, and at the end of 1994, it was 132,400 m³/day.

Table 7.10 shows the units put on-stream up to 1994 [26].

The successful catalytic cracking of the residues requires the solution of two problems: (1) coping with the noxious effect of the heavy metals, especially Ni and V, contained in larger amounts in residues than in the distilled fractions and (2) the much larger amounts of coke produced in the process, the burning of which generates a much larger amount of heat required for maintaining the thermal balance of the process.

Table 7.10 Residue Catalytic Cracking Plants Put on Stream up to 1994

Refinery, country	Licenser	Capacity (m^3/day)	Go in stream year
Philips 66 Co., Borger, Texas	Kellogg	4,700	1961
Philips 66 Co., Sweeny, Texas	Kellogg	8,000	1981
Total Petroleum Inc., Arkansas City, Arkansas	Stone & Webster	3,000	1981
Total Petroleum Inc., Ardmore, Okla	Stone & Webster	6,400	1982
Valero Refining Co., Corpus Christi, Texas	Kellogg	10,300	1983
Ashland Petroleum Co., Catlettsburg, Kentucky	UOP-RCC	6,400	1983
Shell Canada Ltd., Montreal, Canada	Stone & Webster	4,000	1987
Petro-Canada Products Inc., Montreal, Canada	Stone & Webster	3,000	1988
Shell U.K. Ltd., Stanlow, Great Britain	Shell	10,200	1988
Idemitsu Kosan CL, Tokyo, Japan	IFP	5,600	1988
BP Australia, Kwinana, Australia	Stone & Webster	4,000	1988
4 plants in China	Stone & Webster	15,900	1989
Chinese Petroleum Corp, Taiwan	Kellogg	4,000	1990
Statoil A/S, Norway	UOP-RCC	6,400	1990
Schell Eastern Petroleum Ltd., Singapore	Schell	4,700	1990
Mitsubishi Oil Cl, Japan	Kellogg	4,000	1992
Schell Refining PL, Geelong, Australia	Schell	4,700	1992
Nippon Petroleum Refining CL, Japan	Stone & Webster	4,700	1992
Suncor Inc., Sarina, Canada	Kellogg	3,200	1994
Pertamina, Indonesia	UOP-RCC	13,200	1994
Caltex Petroleum Corp., Map Ta Phut, Thailand	Stone & Webster	6,000	—
	Total	132,400	

Note: The reconstructed plants are not indicated in this table.
Source: Ref. 26.

The first problem was solved by the development of Zeolite catalysts that are resistant at much larger concentrations of metals in the feed [27]. The use and the improvements brought to the passivators decreased the damaging effect of the metals. Hydrofining of the feed [28] is practiced for reducing the metals and coke concentrations to those tolerated by the catalysts.

The second problem was more difficult to solve. The decrease of the inlet temperature of the feed to the riser could reduce the heat excess only to a small extent. More efficient was to decrease the CO_2/CO ratio in the flue gases leaving the regenerator, decreasing in this way the heat of combustion (see Figure 7.21). This measure was however limited by the increase of residual coke, which strongly decreased the performance of the unit.

The burning of coke in the conditions of reduced CO_2/CO ratio in the flue gases, while holding the residual coke at the same or lower values, was achieved without difficulties in units with two regeneration zones, such as are the Orthoflow F, units of Figure 7.18. In this case, in the first zone the largest fraction of the coke is burned while using a reduced amount of air, at a minimum CO/CO_2 ratio. In the second zone, the remaining coke is burned in excess air until the desired residual coke level is reached. In units of the type shown in Figure 7.18, the flue gases from

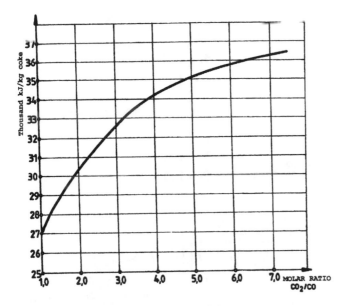

Figure 7.21 Coke (with normal hydrogen content) heat of combustion function of CO_2/CO ratio in flue gases.

the two zones are mixed and pass through a common cyclone system. In such systems, special measures must be taken in order to prevent the burning (in the upper part of the generator, in the cyclones system, or in the transport pipe) of the CO coming from the first zone, with the O_2 left over from the air excess used in the second zone.

The application of these two measures in the existing catalytic cracking units made it possible to include in the feed 10–20% straight run residue having a reduced content of Conradson carbon and metals.

Since the use of two regeneration steps was shown to be an efficient solution, the revamping of older units was accomplished by modifying the existing regenerator, and when this was not possible, by adding a supplementary regenerator. Such a solution applied by UOP is shown in Figure 7.22 [29].

The radical solution, applied to plants especially designed for the processing of residue, is to remove the excess heat directly from the regenerator by an adequate heat exchanger.

To this purpose two main systems are used:

1. Cooling coils are located in the lower part of the regenerator. They remove the heat from the dense phase of the fluidized bed. The boiler feed water which is fed to the coils generates steam as it expands into an external vessel. Such coils are sometimes placed in the regenerator also when the revamping of the units is made by the addition of a supplementary regenerator [9].

2. Tubular heat exchangers (vertical tubes) are located in the regenerator. A portion of the regeneration air is used for ensuring circulation of the catalyst through the tubes.

Details concerning these systems are given in Section 7.3.2.2.

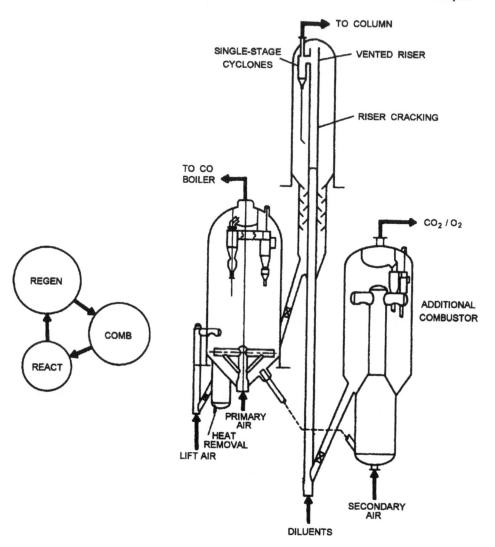

Figure 7.22 UOP unit with supplementary regenerator for residues processing. (From Ref. 29.)

It must be mentioned that both systems for heat removal were used in the first catalytic cracking units Model II and type B-1 respectively [4]. This was being justified by the larger amounts of coke that resulted when the natural or synthetic catalysts containing 12% Al_2O_3 were used.

The system was abandoned as a result of the improvement achieved in the feed pretreatment of the use of more active catalysts.

Various constructors use one or the other of the two systems and sometimes even both of them in order to remove a larger amount of heat or for a higher degree of safety in operation.

Figure 7.23 depicts the Kellogg Heavy Oil Cracker unit, which uses a coil as heat recovery system [30].

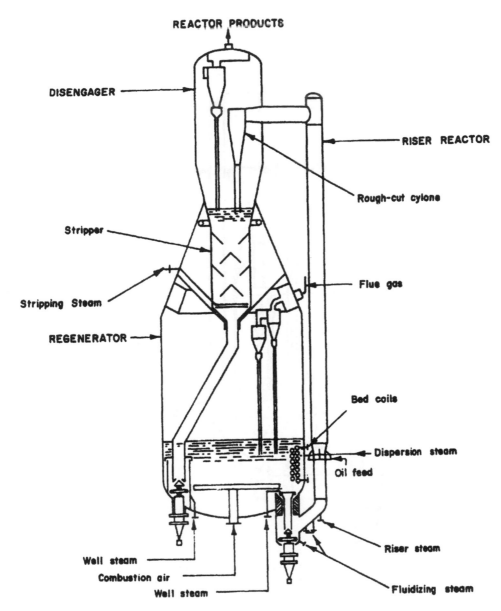

Figure 7.23   Kellogg Heavy Oil Cracker. (From Ref. 30.)

In Figure 7.24 a residue cracking plant is presented, which applies to both systems of heat recovery [30].

A special design is represented by the Total R2R process built by the French Petroleum Institute, where, following the second regeneration step the regenerated catalyst enters a vessel where it is submitted to a last contact with air in dense phase, the vessel serving also as catalyst buffer (see Figure 7.25) [31–34]. A unit of this type is in operation in Japan at the Aichi refinery.

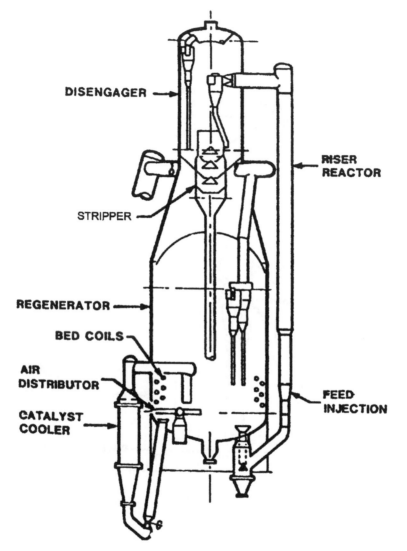

Figure 7.24   Catalytic cracking unit with dual coil and external heat-exchange heat recovery systems from regenerator. (From Ref. 30.)

7.3   CHARACTERISTIC EQUIPMENT

This section examines exclusively the equipment issues that refer to modern units of the riser type, including those for the catalytic cracking of residues.

7.3.1   Reaction and Stripping Equipment

The riser reactors used in modern plants require finding adequate solutions for a number of operating problems, that determine the efficiency of the process. These are: the feed dispersion system; the separation of the product vapors from the catalyst at the top of the riser; the location of the feed flows that enter the riser

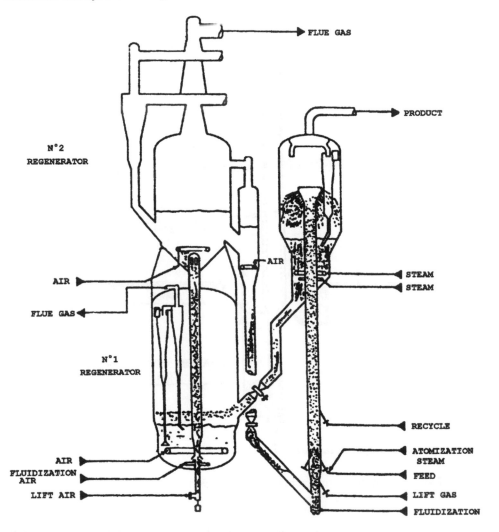

Figure 7.25   Total R2R Residues Catalytic Cracking plant.

and the dynamics of the temperatures and of the velocities along the riser. Issues related to the stripping must be added. An overall picture of the characteristic zones of a riser reactor is given in Figure 7.26.

7.3.1.1   Feed Dispersion

The system that disperses the feed in the ascending flow of catalyst has a direct influence on the final yields. A good dispersion leads to an increase of up to $+0.9$ wt % in the gasoline yield and to a decrease of the conversion to dry gases and coke [23,35]. A decrease of the latter is more important for feeds with a greater Conradson carbon, such as the residual feeds.

These differences are explained by the fact that a deficient dispersion creates regions with unequal (high and low) catalyst/feed ratios that cause returning

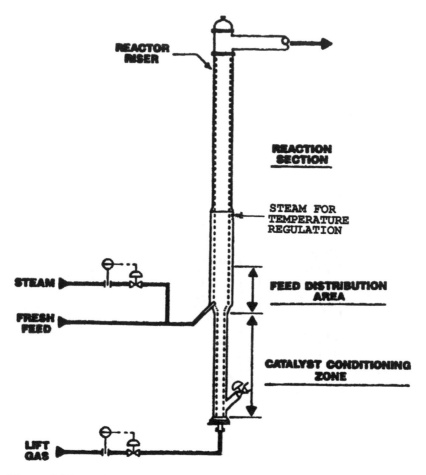

Figure 7.26   Characteristic riser zones. (From Ref. 90.)

streams. The returning streams lead to damaging effects similar to those due to backmixing.

A good injection system must contribute to the creation of as uniform as possible catalyst to feed ratios. Therefore, the distribution of the catalyst over the cross section of the riser before it comes in contact with the feed must be as uniform as possible, and the bulk density must be as high as possible in order to achieve the complete adsorption in the shortest time of the injected feed.

The proper injection system must achieve:

The atomization of the feed in drops as small as possible, having the narrowest possible size distribution

Uniform feed distribution over the cross section of the riser

Drops having sufficient velocity in order to penetrate through the flow of catalyst, without however exceeding the limits that would lead to erosion of the walls of the riser and the catalyst particles

An intimate mixing of the feed with the injection steam

Operation at the lower possible pressure drop

In order to satisfy as much as possible these requirements, several injection devices were developed. They are uniformly distributed on the circumference of the riser (see Figure 7.27) so that the injectors ensure the formation of fine drops of narrow size distribution.

The shape of the injectors has known a remarkable evolution. From the open tube (Figure 7.28a) it was changed to tubes ending with a slit (Figure 7.28b), used by Kellogg in 1980, but both producing a broad size distribution of the drops. The impact system (Figure 7.28c) achieves a good pulverization but requires a high pressure drop. The last system, a Venturi tube (Figure 7.28e), produces drops having diameters comprised between 30–50 μ that vaporize completely at a distance of only 0.5 m from the injection point [16].

Following joint studies carried out in 1990 by Mobil Research & Development Corp. and M. W. Kellogg Co., the system Atomax (U.S. patent 5,305,416) was developed. It was called a "third generation injector" (Figure 7.29) [35].

Comparison of the performances of this type of injector are compared in Table 7.11 to those of the slit injector and that with impact.

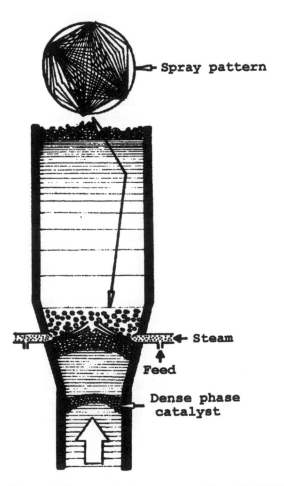

Figure 7.27   Injectors distribution. (From Ref. 36.)

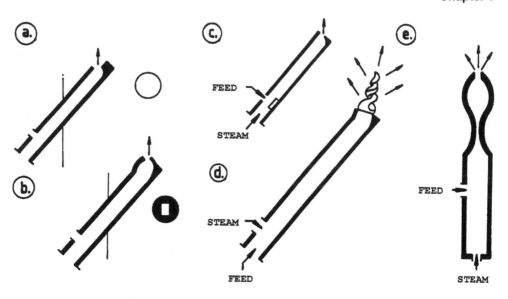

Figure 7.28 Injectors evolution.

The introduction of the Atomax system in a catalytic cracking unit on the Gulf Coast, which was previously provided with the injectors represented in Figure 7.27e, led to the following changes expressed in % by weight:

Dry gases	−0.3
C_3–C_4	−3.3
Gasoline	+5.0
Gas oil	−1.0
Residue and coke	0.0

Details concerning other modern nozzles: Optimax (UOP), Micro-Jet (Lummus), Stone and Webster are given and compared with Atomax nozzle in the excellent monograph of J. W. Wilson: Fluid Catalytic Cracking Technology and Operations [80].

In order to prevent the formation of backflows in the zone of feed introduction the diameter of the riser is narrowed just below this zone (Figure 7.27) and supplementary steam is injected below this zone (CCS-Catalyst Centering Steam device). Such a measure is provided at the R2R unit of the French Institute of Petroleum [33].

Usually the injectors are orientated upward, so that the feed jet is orientated in the direction of the movement of the catalyst particles. However, the idea of the reversed orientation of the jet was also tested. The results obtained at industrial scale seem to indicate that such a solution is also of interest.

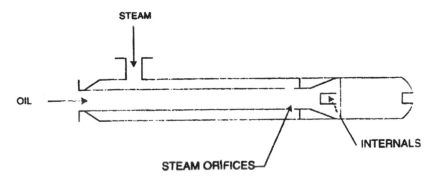

Figure 7.29   M. W. Kellogg Atomax nozzle.

7.3.1.2 Riser Process Conditions

The general temperature regime in the riser depends on the targeted objectives. From this point of view there are three typical regimes, depending on the product whose yield is maximized: medium distillate, gasoline, or lower alkenes [87].

The conditions for the three regimes, the material balance, and the quality of the products are listed in Table 7.12 [29]. The feed in the three cases was the vacuum distillate obtained from a Middle-Eastern crude oil.

Several zones may be identified along the height of the riser (Figure 7.26).

The first zone situated below the feed introduction point has the role of ensuring uniform repartition of the catalyst over the riser cross section and to provide the particles desired ascending velocity that determines the bulk density in the riser.

Two approaches have been followed for achieving these requirements: the first consists in conferring the catalyst particles a large ascending rate, using steam injectors oriented upwards, located below the section of the riser where the feed is atomized. Such a solution is adopted in the IFP-R2R unit, depicted in Figure 7.25. Thus, the formation of backflow is avoided, the feed rapidly leaves the high temperature zone, and a good penetration and distribution of the feed over the catalyst is easier to obtain. The disadvantage is that since the minimum fluidization velocity is significantly exceeded the formation of bubbles of transport fluid within the catalyst mass will occur and the feed contained in these bubbles be thermally cracked.

Table 7.11   Comparison Between Atomax and Other Injection
Systems

	System b	System c	Atomax
Relative mean drops diameter (SMD)	2.35	1.08	1.00
Vol % > 1.6 SMD	81	44	31
Vol % > 8.0 SMD	24	4	0
Relative pressure drop	0.2	7.4	1.0

Systems b and c refer to Figure 28.

Table 7.12 The Three Riser Operation Modes

	Middle distillate mode	Gasoline mode	Light-olefin mode
Operating conditions			
reactor temperature, °C	449–510	527–538	538–560
residence time	< Base	Base	> Base
catalyst/oil ratio	< Base	Base	> Base
recycle, CFR	1.4	optional	optional
	(HCO)	(HDT LCO)	(heavy naphtha)
Catalyst formulation			
zeolite type	ReY	CSDY*	USY
zeolite level, wt %	15	30	40
rare earth, wt %	1–2	0.5–1.5	0
Product yields			
H_2S, wt %	0.7	1.0	1.0
C_2-, wt %	2.6	3.2	4.7
C_3, LV %	6.9	10.7	16.1
C_4, LV %	9.8	15.4	20.5
C_5+, Gasoline, LV %	43.4	60.0	55.2
LCO, LV %	37.5	13.9	10.1
CO, LV %	7.6	9.2	7.0
Coke, wt %	4.9	5.0	6.4
Product properties, vol/vol			
C_3 olefin/saturate	3.4	3.2	3.6
C_4 olefin/saturate	1.6	1.8	2.1
Gasoline			
ASTM 90% Pt., °C	193	193	193
RON clear	90.5	93.2	94.8
MON clear	78.8	80.4	82.1
Light cycle oil			
ASTM 90% Pt., °C	350	316	316
viscosity, cSt (50°C)	3.7	3.1	3.2
sulfur, wt %	2.9	3.4	3.7
cetane index	34.3	24.3	20.6
Clarified oil			
viscosity, cSt (100°C)	10.9	9.0	10.1
sulfur, wt %	5.1	6.0	6.8

*Chemically Stabilized and Dealuminated Y Zeolite.
Source: Ref. 29.

The second system achieves the catalyst flow to the feed injection points in dense phase, eliminating the danger of occurrence of thermal cracking, but making more difficult a good repartition of the feed over the mass of flowing catalyst. Since cracking reactions take place with increased volume, the linear flow velocity increases and the density of the bed decreases as it flows through the feed introduction section. This second system is sketched in Figure 7.27.

Following the feed dispersion and vaporization section is the reaction section (Figure 7.26). To avoid erosions, the riser must be perfectly vertical, and the flow velocity should not exceed 18 m/s if the inside walls of the riser are not provided with

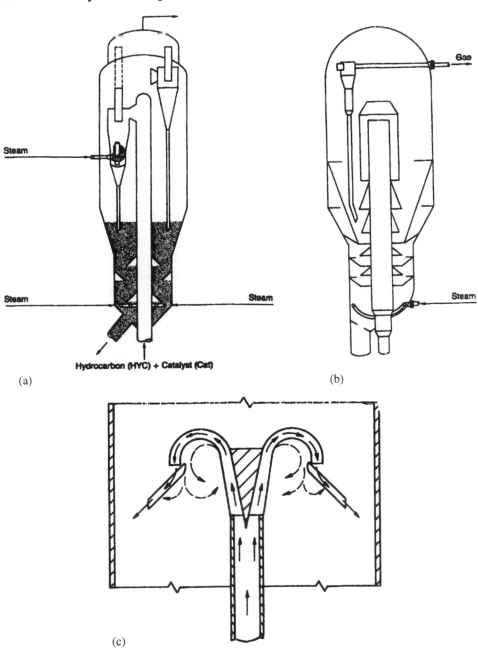

Figure 7.30 Riser top catalyst/products separator systems. (a) Direct connection of the riser with the cyclones system (U.S. Patent 4,043,899), (b) inertial separation (Chevron U.S. Patent 4,721,603–1988), (c) (U.S. Patent 4,664,888).

refractory protection. If the walls are lined with refractory material (usually about 12 cm thick), the linear velocity may be higher.

In order to control the temperature regime in the riser, recycle material is injected at a location above the injection zone of the feed at a distance where the feed is completely vaporized as a result of the contact with the hot catalyst. In this way, without changing the contact ratio or the inlet temperature of the feed, one can control the operation conditions and severity respectively [81].

7.3.1.3 Catalyst-Product Separation

The rapid separation of the catalyst from the reaction products is necessary in order to avoid over cracking and thermal cracking due to the high temperatures used in the modern risers. Thermal cracking would cause also a cooling of the effluent after it separated from the largest portion of the catalyst.

In older systems, the upper part of the riser was of the open type, the mixture of vapors and catalyst being released into an expansion zone, which was connected to the cyclone system. Later on, a system was developed where the products left the riser through a side tube, which made the separation of the catalyst easier. Eventually several improved solutions were developed.

The first improved solution consists in the direct connection of the riser with the cyclone system (Figure 7.30a). The disadvantage is that in this case the whole amount of catalyst passes through the first step cyclones, the size of which has to be significantly increased.

The second solution is that of the inertial separation of the catalyst by imposing a sudden change of direction to the product and catalyst flow (Figure 7.30b,c). Such systems proved to be very efficient, the separation reaching 98% [82] and backmixing currents being completely eliminated.

After the separation, but especially in the case of operation at high temperatures, a cooling liquid is injected in the stream of products immediately after the separation of the catalyst. This measure comes mandatory if the temperature exceeds 535°C [82].

As seen, special attention during the design and operation of the risers is given to the elimination of backmixing, which leads to decrease of the yields.

It was found that backmixing is absent in the descending portion of the inertial separators. This led to the idea to use risers with descending circulation, which will be discussed in Section 7.3.3.

7.3.1.4 Stripping

A rule of thumb is that good stripping must produce a coke containing not more than 6–9 wt % hydrogen in coke. The stripping is very poor when this percentage reaches or exceeds 10 wt % [37].

In the old plants that used less active catalysts and a high catalyst to feed ratio, the residence time of the catalyst in the stripper was short and the usual steam consumption was 6 kg per 100 kg circulating catalyst [36]. As a result of successive improvements, current steam consumption is $\frac{1}{4}$ of that value, whole hydrogen content in coke is only 5–6 wt %.

This spectacular increase of stripping efficiency is the result of several constructive measures:

1. The size of the orifices in the steam distributors was significantly decreased on the basis of accumulated experience with air distributors in the regenerators. In this way, the size of the steam bubbles became much smaller, which favored stripping efficiency.

2. The use of two steam distributors was adopted: one, located near the inlet of the catalyst in the stripper is used to eliminate the product vapors from the space between the catalyst particles; the second, near to the outlet from the stripper, in order to eliminate the vapors from within the catalyst pores or adsorbed on the particles surface.

3. The size of the baffles was decreased to prevent the accumulation of steam, and their location was reduced to the role of preventing the flow of catalyst along the walls.

4. A perfect symmetry of the locations of the catalyst inlet to, and outlet from the stripper was ensured. This measure is needed for efficient stripping by avoiding formation of stagnant spaces, since the dispersion of the catalyst normal to the direction of flow is very limited. Otherwise, a higher stripping zone and a significant increase of the number of baffles would be required.

7.3.2 Equipment for Catalyst Regeneration

The adequate design of the catalyst regeneration equipment provides answers to the following main issues: the design of the injection system for the air and for catalyst distribution; measures for avoiding local superheating within and above the catalyst bed; means for removing the excessive heat developed when residues are incorporated in the feed; systems with two regeneration steps used in the catalytic cracking of straight run residues.

7.3.2.1 Air injection and Catalyst Distribution

In older units the distribution of air and catalyst over the cross section of the regenerator was ensured by gratings with gauged orifices [4]. In large units, the gratings became very heavy and had to be made by assembling several elements together. The main difficulty was ensuring tight seals between the elements and with the wall of the regenerator. The thermal expansion that occurred in operation lead to formation of non–tight spots mostly along the walls, which resulted in preferential flow for the air, bypassing, and flow maldistribution.

For these reasons in the new designs, the injection of air used distribution systems that were formed of tubes provided with injection nozzles (Figure 7.31). The uniform repartition of the catalyst over the cross section of the regenerator is achieved by separate devices. The main problem is that the dispersion of the catalyst particles is very intense along the direction of the gas flow, while it is very small in directions normal to it.

For systems where the spent catalyst is free flowing from the reactor, which is situated at a high level, devices are provided such as that of Figure 7.31. The catalyst is distributed uniformly between several (generally 6) distribution troughs located above the air injection level.

For systems with the reactor and the regenerator placed side by side, the spent catalyst is lifted by using a portion (10–15%) of the regeneration air. In this case, the

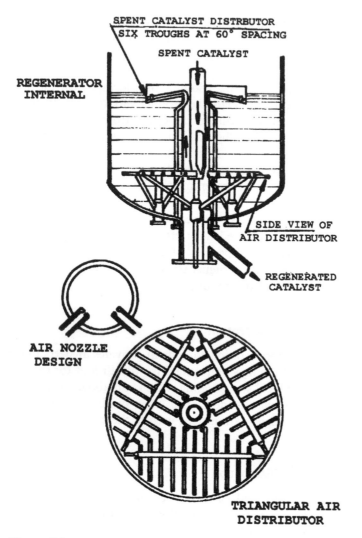

SPENT CATALYST DISTRBUTOR
SIX TROUGHS AT 60° SPACING

SPENT CATALYST

REGENERATOR
INTERNAL

SIDE VIEW OF
AIR DISTRIBUTOR

REGENERATED
CATALYST

AIR NOZZLE
DESIGN

TRIANGULAR AIR
DISTRIBUTOR

Figure 7.31   Air distributor with nozzles. (From Ref. 36.)

conical shape of the base of the regenerator ensures a good horizontal repartition of the catalyst.

7.3.2.2 Prevention of Overheating

In older units, the excess of oxygen needed for obtaining acceptable values for the residual coke led in many cases to auto-ignitions of CO in the freeboard of the regenerator. In such cases temperatures above those acceptable for the catalyst were produced and catalyst deactivation followed. Such auto-ignitions were catalyzed among others by the nickel deposited on the catalyst.

Several methods were proposed for controlling this phenomenon [4], of which the most efficient proved to be the installation of a bypass valve for the regeneration air, which was activated by the temperature above the bed.

Following the commercial introduction of the zeolite catalysts, which made it possible to increase the regeneration temperatures up to 730°C and even above this temperature, it became possible to completely burn CO to CO_2 by using promoters. This eliminated the danger of uncontrolled overheating above the bed. Despite the fact that the operation is now carried out without any excess of oxygen beyond the amount necessary for converting the CO to CO_2, the residual coke is situated generally below 0.2%.

A special problem is due to local burning that may occur when the total flow of fresh air comes in contact with the spent catalyst, strongly loaded with coke, such as in the processing of residual feeds. In such cases, local combustion may take place that could lead for some catalyst particles to temperature increases of 150–200°C above the one in the bed, leading to their deactivation.

In units with the reactor and regenerator placed side by side, the catalyst is pneumatically lifted to the regenerator by means of air. The air used for lifting represents 10–15% of the total amount necessary for regeneration and the mild combustions that take place in the transport line, removing the danger of local overheating.

In units where the catalyst discharges from the reactor under gravity, designs such as in the Orthoflow units, model F (Figure 7.18), or of an even simpler construction, may be used for achieving a first, mild combustion of the coke with only a portion of the total air.

7.3.2.3   Heat Removal

The possibility of using residue as a component of the feed is dependent on the ability to burn a larger amount of coke in the regenerator. Otherwise, the only possibility would be to decrease the feedrate of the residue containing feed, so as to maintain the amount of coke burned equal to that prior to addition of residue.

The amount of the burnt coke may be limited by the capacity of the blower or by the excess heat produced in the regenerator. In the second case, debottlenecking involves the addition of cooling devices.

Two systems are practiced for the removal of heat from the regenerator: cooling coils which are located inside the bed and tubular heat exchangers, located outside the regenerator.

The actual construction of the cooling coils placed inside the bed of catalyst is shown in Figure 7.32 [39]. It is a simple design that avoids the presence of joints, branchings, or changes in the diameter etc., which are the locations where intense erosions appear first. Their design is based on the lengthy experience of using cooling coils, starting in the years 1942–1948, when the use of the natural catalyst led to deposits of up to 12 wt % coke on the catalyst.

At current practiced densities of the dense phase, of 400–500 kg/m^3 and at the low air velocities in the regenerator of 0.6–0.9 m/s, the erosions do not affect the operation life of the coils.

The second solution is the installation of external coolers (Figures 7.33, 7.24).

It is to be observed that the circulation in current coolers is different from that used in the period 1942–1948, when the ascendant mixture air/catalyst circulated through tubes provoking intense erosions. In current coolers the circulation is descendent through the shell side. The catalyst is maintained fluidized by air injection at the lower end of the cooler. This injection may be also used, within some limits, for

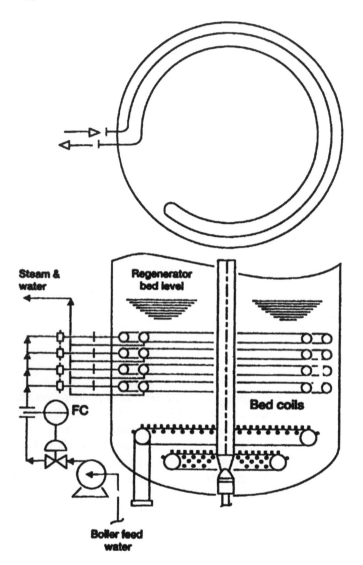

Figure 7.32   Cooling coil inside regenerator. (From Ref. 39.)

the control of the heat transfer. An increase of the air flow decreases the catalyst density but increases the heat transfer.

7.3.2.4   Two-step Regeneration

Catalytic cracking units with two regeneration steps were depicted in Figures 7.19, 7.22, and 7.25.

These systems make it possible to reach residual coke concentrations below 0.05 wt %, even during the cracking of heavy residues that generate a large percentage of coke [33].

The hydrogen present in the coke burns in the first regeneration step. Therefore, in the second step there is no steam present and the temperature can

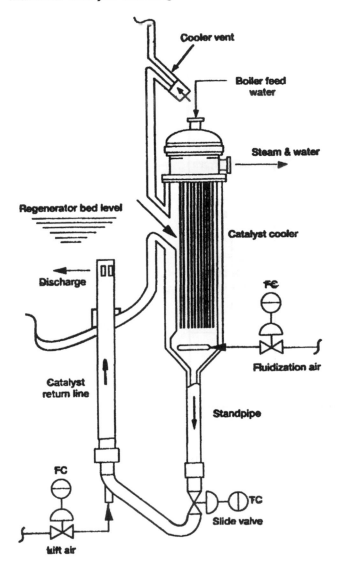

Figure 7.33 Regenerator external cooler. (From Ref. 39.)

reach 850–900°C without danger of destroying the catalyst since, in absence of steam, the main culprit for catalyst deactivation, vanadic acid, is not formed. At these temperatures the reaction rate is increased greatly so that coke burning is complete at short contact times in regenerators of the riser type, where the back-mixing is virtually absent (Figure 7.19).

The pattern of catalyst circulation in the regenerator is of importance. In the first regeneration step, the preferred pattern is one in which the air flows countercurrent to descending catalyst. In this manner, the strongly coked catalyst does not come into contact directly with the fresh air and thus, local uncontrolled combustion and over-heating of the catalyst particles are prevented. The influence of the flow pattern on the increase of the temperature of catalyst particles is plotted in Figure 7.34 [40].

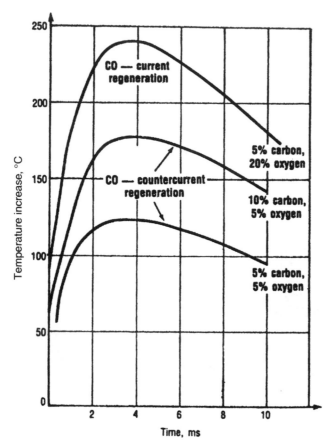

Figure 7.34 Effect of catalyst circulation pattern on the temperature increase of catalyst particles. (From Ref. 40.)

7.3.3 Future Outlook

The information on the reaction/regeneration systems presented above is the basis for ideas on possible directions for the future development of this technology. These ideas are generally accepted by the people involved in the development, design, and operation of catalytic cracking units [36].

Reactors having the overall structure of a riser, but in which the catalyst and feed flow cocurrently downwards, will completely eliminate the back-mixing and stagnation zones. In these conditions, a shorter reaction time can be achieved, since the vapor flowrate can be varied independent of the catalyst flowrate. Since in these conditions the diameter/height ratios become impractically high, it will become beneficial to use a series of parallel "risers" with descending flows, which will be fed with a variety of streams submitted to catalytic cracking.

The concept of using two steam injection points in the stripper will be maintained. The first steam injection immediately downstream of the catalyst entry line, will displace most of the product vapors between the catalyst particles. The second steam, at the exit of the catalyst from the stripper will remove product adsorbed on the catalyst particles as completely as possible.

The design of the regenerator must ensure the initial combustion of the coke in mild conditions by using only 10–15% of the incoming air. This may occur in the catalyst transport line. It will be followed by as complete combustion of the coke as possible. A regenerator containing a diluted catalyst phase over its whole length is preferred for this duty.

These ideas were concretized by J. R. Murphy [36] in the scheme of Figure 7.35.

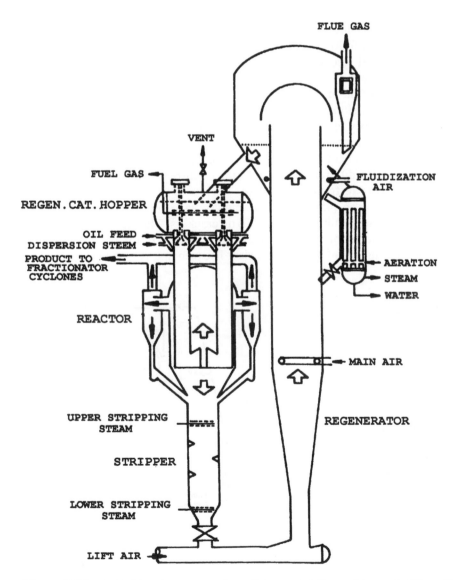

Figure 7.35 D.R. Murphy concept for a future catalytic cracking unit. (From Ref. 36.)

7.3.4 Energy Recovery from Flue Gases and Emissions Control

The energy recovery from flue gases by the means of expansion turbines requires an additional stage for the recovery of traces of catalyst situated after the regenerator (Figure 7.36), called the third separation stage.

Today's popular separator (Figure 7.37) was designed and patented by Shell Oil Co., UOP and M. W. Kellogg Co. The main feature of the separator is the tubes, which cause a sudden change of the direction of flow by 180°, thus producing the inertial separation of the gas from the catalyst particles. The number of tubes varies between 50 and 150 depending on the size of the unit [42].

Another option is the use of classical cyclones with large diameters. The high temperature of the flue gases leads to protecting with refractory material the walls of the cyclones.

The expansion turbine, which serves as a compressor for the air used in the regeneration, is located following the last separation step. Downstream of the turbine are located the heat recovery and the purification of the flue gases prior to their release in the atmosphere.

The operation of the turbine–compressor system are the object of specialized papers [43]. Their discussion exceeds the framework of this book.

The degree of purity of the flue gases is determined by the regulations imposed in each country for the protection of the environment. Some of the limit concentrations are given for several countries in Table 7.13. The table shows that the acceptable concentrations are different in the various countries, but in all the cases the standards become more restrictive with time.

Chevron [44], Exxon [45] and other refiners have published designs for the purification of flue gases resulting from catalytic crackers. The latter [45] gives interesting details concerning contacting systems for the neutralization of acid gases with NaOH and for the elimination of solid particles. The principle is to use venturi contac-

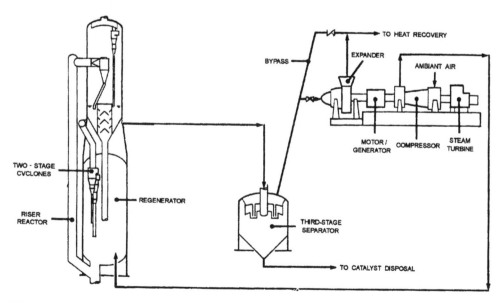

Figure 7.36 Ultra-Orthoflow. Energy recovery from flue gases. (From Ref. 41.)

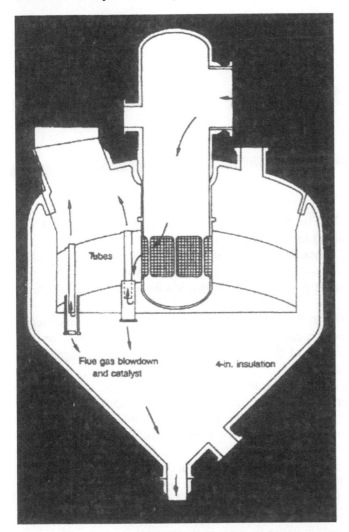

Figure 7.37 External third separation stage. (From Ref. 40.)

tors where the pulverized liquid retains the solid particles and neutralizes the acid gases.

Such a system removes 93–95% of the SO_2 and 85% of the catalyst dust. Excellent efficiency is also achieved in the elimination of nitrogen oxides [83,84] and hydrocarbons.

7.3.5 Products Recovery; Process Control

Product recovery equipment is designed with specific features as a consequence of the high temperature and low pressure (slightly above the atmospheric) at which the effluent leaves the reactor. The temperature in the vaporization zone is lowered by recycling the cooled product recovered at the bottom of the column and means of several interval refluxes, which discharge the thermal load in the upper part of the

Table 7.13   Regenerator Flue Gas
Pollution Limits (g/Nm^3)

Country	SO$_x$	NO$_x$	Dust
U.S. 1970	4.50	0.00	0.150
U.S. 1990	0.90	0.00	0.100
California 1995	0.06	0.06	0.006
Germany 1991	1.70	0.70	0.050
Germany 1995	0.50	0.50	0.050
Japan	0.90	0.50	0.040

fractionating column [46,47]. The heat of the reactor effluent is best utilized if it enters the fractionation column in its lowest section.

The configuration of a typical fractionating column and that which resulted following its thermal optimization are sketched in Figure 7.38 [47].

The separation of gases is performed by adsorption-fractionating using classical methods.

Much attention is given to the control of the fractionating system in order to ensure in all cases the proper processing of all products that leave the reactor [48]. This issue, as well as the control of the compression of the air for the regenerator [43], exceed the scope of this book.

In several studies, the operating parameters of the fluid catalytic cracking units are correlated with the help of computers, in order to achieve optimization of the process [49,50,88]. The problem seems to be extremely complex. Even before a complete solution was found, several ways were identified for improving the operation of existing units [50].

Of special interest is the processing of the product at the bottom of the fractionating column. Normally, it was separated in two layers by settling: the upper layer of free catalyst particles (the decant oil) was used as a component for hearth fuels or as raw material for the production of carbon black. The lower layer was recycled to the reaction system in order to recover the catalyst it contained.

The bottom product has a strong aromatic character and, its recycling decreases gasoline yield. To avoid this disadvantage, starting in 1979 the electrostatic separation of catalyst particles was studied and improved, which eliminated the requirement for recycling [51].

The system is very efficient; two oils with ash contents of 1.22% and 1.89% respectively, after electrostatic separation had ash contents of 46 ppm and 85 ppm respectively.

The aromatic character of the recovered oil and the significant amounts recovered by electrostatic separation suggest that recovered oil could be an excellent raw material for needle coke, either by itself or in a mixture with heavy coking gas oil.

7.4   OPERATION ASPECTS

Three specific aspects related to fluidized bed catalytic cracking will be discussed: the maintenance of catalyst properties, equipment erosion, and the quality of fluidization.

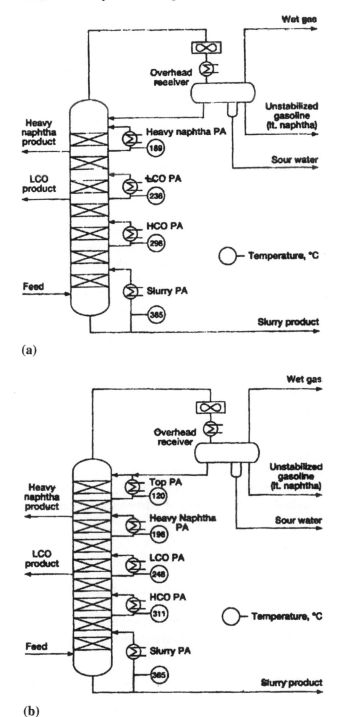

Figure 7.38   Catalytic cracking fractionation column. (a) before, (b) after revamping.

7.4.1 Maintenance of the Catalyst Properties

The maintenance of adequate activity and physical characteristics of the catalyst inventory is not limited to the control of the general phenomena of aging and poisoning by metals, which are common to all catalytic systems and which were discussed in Section 6.5.5.3. In the fluidized bed systems, maintenance of the size characteristics and of the particle size distribution are necessary for good fluidization and for not exceeding the admissible entraining losses through the cyclones. Besides, in view of the high cost of the catalysts it is important to have catalysts with high resistance to attrition so as to decrease mechanical losses.

The particle size distribution of the microspherical catalysts used presently in fluid catalytic cracking units is characterized by a Gauss curve having a maximum corresponding to diameters between 60 and 80 µm. It is important that the form of this curve should correspond to a minimum content of particles below 40 µm and especially below 20 µm, while the particles with diameters above 140 µm should be completely absent, since they have a negative effect on the quality of the fluidization [52].

A typical distribution of the sizes of the catalysts particles [53,54] corresponds to a content of below 15 wt % for particles below wt 40 µm, of 30–35% below 60 µm, about 60 wt % below 80 µm and of about 80 wt % below 100 µm.

Over time, the size repartition of the catalyst will change as result of the catalyst losses, which affect especially the particles below 40 µm, and of the shrinkage and attrition suffered by the particles. These losses are compensated by the addition of fresh catalyst, the size of which is selected in order to maintain the size distribution, indicated above as being necessary for a good quality of fluidization.

The shrinkage phenomenon and the attrition were studied [55] by taking into consideration the following size fractions expressed in microns: 0–10, 10–20, 20–40, 40–60, 60–80, > 80.

For any one of these fractions, the equations describing the shrinkage were written assuming that the kinetics of the process is of the 1st order, and the variation of its rate constant with the temperature may be expressed by the Arrhenius equation. The shrinkage was characterized by the fraction of particles which remained unchanged after the shrinkage:

$$F = \frac{W}{W_0} \tag{7.3}$$

where W is the amount of the size fraction remaining after shrinkage, and W_0 the added amount.

The time was expressed by the age of the catalyst, θ, given by the ratio between the inventory of catalyst within the system and the amount of catalyst that is added per day.

Using these notations, the shrinkage equations become:

$$-\frac{dF}{d\theta} = k_T F \tag{7.4}$$

$$F = e^{-k_T \theta} \tag{7.5}$$

$$k_T/k_{T_0} = e^{E(1/T_0 - 1/T)} \tag{7.6}$$

where k_{T_0} has the value of 0.00693 at $T_0 = 992 \, \text{K}$, k_T is measured at temperature T and $E = 5,944$. The temperature is expressed in K.

For erosion the calculation may be carried out only for the fractions above 40 μm.

The attrition is described by an expression identical to (7.3) but where W_0 expresses the weight of the 40 μm fraction of the added catalyst W is the weight of the 40 μm fraction remaining after attrition.

The erosion process being of the 1.5 order it will be expressed by the equations:

$$-\frac{dF}{d\theta} = kF^{1.5} \tag{7.7}$$

$$F = \left(\frac{2}{2+k\theta}\right)^2 \tag{7.8}$$

The rate constant k is independent of the temperature and has the value $k = 0.00828$.

On basis of equations (7.3–7.8) the graph of Figure 7.39 was plotted.

Equations (7.3)–(7.8) are expected to be valid also for other catalysts than those studied in 1981 [55]. The two rate constants k_{To} and k have to be determined for each individual catalyst.

7.4.2 Equipment Erosion

The equipment erosion caused by the catalyst particles depends on their velocity, diameter, and density. Depending on these parameters, the erosion may be correlated by the empirical equation:

$$\text{Erosion} = C \cdot \gamma_a^2 \cdot d^3 \cdot v^3 \tag{7.9}$$

where:

$C = \text{constant}$
$\gamma_a = \text{density of the particles in g/cm}^3$
$d = \text{their mean diameter in μm}$
$v = \text{velocity in m/s}.$

As a result of the correlation of the experimental data, Eq. (7.9) was written as [54]:

$$\text{Erosion} = C \cdot D^3 \cdot \gamma_v^2 \cdot v^3 (E/D^2)^{0.36} \tag{7.10}$$

where:

$D = \text{mean weighed diameter of the fractions with diameters above 40 μm,}$
$\quad \text{or corresponding to 50\% of these fractions, in μm}$
$\gamma_v = \text{bulk density of the catalyst}$
$E = \text{erosion of a standard catalyst of a refractory standard material in}$
$\quad \text{wt \%, determined according to the Engelhard method}$
$v = \text{velocity in m/s}.$

Without knowing the values of the constant that intervene in Eqs. (7.9) and (7.10), or the values that are not available such as E, they may be used with success for determining the effect modifying some parameters on the erosion of the equipment. Thus, if the bulk density of the equilibrium catalyst increases from 0.82 to 0.92, and the mean diameter of the particles from 0.70 to 0.76, the rest of the parameters remaining the same, the erosion will increase by 52%.

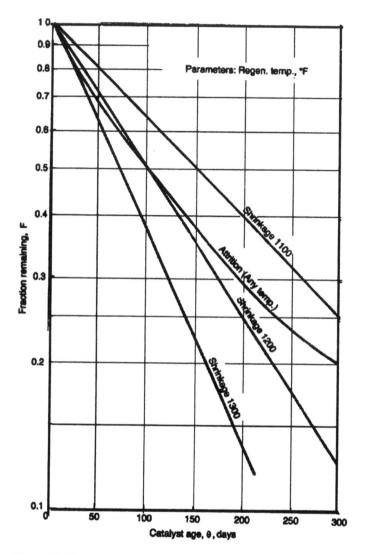

Figure 7.39 Shrinkage and erosion of catalyst particles above 40 μm. (From Ref. 89.)

The shape of the particles has a large influence on the erosion. Particles with a wholly irregular form, as those which result from the crushing operation cause in identical conditions, erosions about 10 times larger than those of spherical shape. Even the degree to which the surface of the spheres is or not smooth influences the erosion. Thus, the equilibrium catalyst which has a rougher surface on the balls, provokes stronger erosions than the fresh catalyst.

These influences are manifested in the value of the constant C in Eqs. (7.9) and (7.10), but do not affect the form of the equations.

The regions strongest affected by erosion are the outlet from the riser and the inlets to the first cyclone of the reactor and of the regenerator [85]. In the regenerator itself, the erosions are very reduced, because the linear velocities are below 1.5 m/s.

In order to avoid erosion, it is recommended [37] that the velocities in the two steps of the reactor cyclones should not exceed 18 m/s, 18.5 m/s in the first cyclone step of the regenerator while in the second step, 23 m/s. Similar values are recommended in other publications [54]. But also recommendations exist [56] to use higher velocities, of up to 28 m/s in the inlet of the cyclones if modern refractory materials of high resistance are used. While such velocities lead to a more efficient operation of the cyclones, the erosions become stronger.

Also, other parts of the equipment of the unit are submitted to erosion, which imposes the prescription of maximum pressure drops, in those locations where exceeding them may lead to abnormal erosions. Such prescriptions are:

The control valves for the catalyst flow	41 kPa
The valves for pressure control	103 kPa
The nozzles of the regenerator distributor grate through which also the catalyst flows	7 kPa
The nozzles of the regenerator distributor grate, when through them only the air passes	21 kPa
The nozzles of the reactor grill	14 kPa

The pressure drops shown for the grill nozzles do not eliminate the erosion of the catalyst particles.

In the case of feeds containing above 0.5% sulfur, the erosion effects are combined with those of the corrosion, which requires the use of alloyed steels with 12% Cr at reduced sulfur content and 18/8 Cr-Ni at higher contents.

The use of the alloyed steels is recommended in the areas where the temperature exceeds 480°C in order to decrease the wear. The austenitic stainless steels, strongly alloyed, become compulsory in the areas where the temperature (in the regenerator) exceeds 650°C.

Details concerning the design of the equipment, the maintenance, and the periodic revisions are given in the paper of Luckenbach et al. [37].

7.4.3 Quality of Fluidization

A good fluidization is very important for obtaining optimal performances. Four states may be distinguished as air or vapors pass upwards through a layer of small catalyst particles: moving dense bed, fluidized bed, bubbling fluidized bed, transported diluted bed.

All these four states are used in the catalytic cracking. It is important to avoid operating in the intermediary domain between two states, because such a situation could lead to unstable operation of the unit. Thus, in the descendent pipes an excessive flow rate of the aeration fluid may produce bubbles which, in their ascending motion, could prevent the descent circulation with the necessary flowrate of the catalyst and eventually could block up the normal operation of the unit. Such blocking could be provoked also by an insufficient flowrate of the fluid in the ascending pipes.

It is desirable that in the ascending and descending pipes the catalyst particles should be in a fluidized state. However, if the aeration is not correct, the pressure drop in the descending pipes could lead to states close to settling in the lower portion and to the formation of bubbles in the upper portion of the pipe.

The key values that determine the selection of the working parameters are given by the following ratios and values, expressed by empirical equations [54]:

$$\frac{v_b}{v_i} = \frac{6.9}{(V_{80} - V_{40})^{0.2} \cdot [D + 0.5 \cdot V_{80}/100]^{3.6}} \qquad (7.11)$$

$$\tau_B = \frac{38}{[(D + 0.5 \cdot V_{80})/100]^{6.8}} \qquad (7.12)$$

where:

V_i and V_b are the fluid velocities in m/s at which the fluidization and the bubbling begins respectively,

V_{80} and V_{40} = volume % of catalyst particles with sizes below 80 µm and 40 µm respectively

D = mean weighed diameter of the fractions with diameters above 40 µm, or corresponding to 50% of these fractions in µm

τ_B = blocking time by settling from a rate of the fluid through the bed of 10 cm/s in s/m of bed (the reverse phenomenon to aeration)

The values given by the Eqs. (7.11) and (7.12) may be directly measured in the laboratory using samples of equilibrium catalyst from the unit.

Equation (7.11) shows that the ratio v_b/v_i will increase and thus the fluidization will be better as the fraction of catalyst particles with diameters below 80 µm will be larger. This confirms the observation that the larger particles worsen the quality of fluidization. By increasing the v_b/v_i ratio, the situation is reached that in the conditions of the dense phase process, no large fluid bubbles are formed or only small bubbles are formed, which ensure good conditions for the process of chemical transformation.

According to Eq. (7.12), the increase of the fraction of particles with diameters below 80 µ decreases the duration for blocking by settling of the particles, which is a disadvantage for the transport pipes.

One should note that the density of the particles is not present in Eqs. (7.11) and (7.12), which means that it does not influence the quality of fluidization.

7.5 CATALYST DEMETALLATION

The demetallation of the spent catalyst may be performed continuously in a separate unit, attached to the catalytic cracking unit. The result is the decrease of the metal content of the equilibrium catalyst, the decrease of the amounts of fresh catalyst addition, and the improvement of the unit yields. Its performance become similar with those obtained with hydrofined feed.

The demetallation processes, which address especially the nickel and the vanadium, are based on a large number of studies that proposed a number of possible routes: extraction of the previously oxidized metals by means of organic acids; various acid–basic treatments; chlorination, which allowed the partial extraction of nickel; treatment with CO etc. In the same time studies were carried out on the removal of sodium from the catalyst [57].

As a result of these studies the processes MET-X [58–60] and DEMET [51–63] were developed.

The emergence of passivators, which are a solution much easier to implement and did not require any supplementary investments, caused a substantial reduction in the interest in demetallation units and even opinions were expressed that demetallation technologies had been surpassed [64].

However, they reappear under an improved form in 1989–1990 [65–69], a period where the construction of two industrial units was announced [69], of which the second one, for 20 t/day (Wichita, Kansa, U.S.) was directly coupled to a fluid catalytic cracking unit. The investment cost for this unit was $7.4 million and the exploitation costs of $130 per ton of treated catalyst.

The process flowsheet is shown in Figure 7.40 [69].

The efficiency of the demetallation is about 75% for nickel and of 46% for vanadium. At the same metal content in the catalyst, no differences are observed in the yields in products between the approaches of maintaining of a constant metal content by demetallation, versus the addition of fresh catalyst.

However, the order and use of the demetallation processes are of concern especially for the following reasons:

There is no data on the possible influence of demetallation on the metals used as promoters.

Technology for catalyst production and its characteristics advance very quickly and there is no insurance that the new catalysts will resist the treatments of demetallation or that they will be efficient.

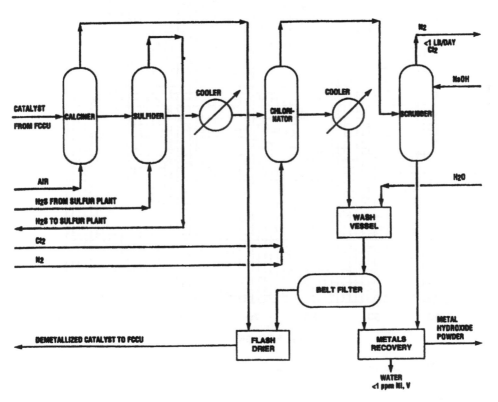

Figure 7.40   Catalyst demetallization plant. (From Ref. 69.)

The experience of industrial use of the process is too short for the formulation of reliable conclusions concerning the duration in exploitation and efficiency of the catalyst.

7.6 YIELD ESTIMATION

Methods for the estimation of yields from processes with moving bed of catalyst and for classical processes of fluid catalytic cracking on alumo-silica catalysts are given in a study by Raseev [4]. Also, other graphs for quick calculations of a more specialized interest were published [70].

For modern units using zeolite catalysts and distilled feeds, the graphs for the estimation of the yields and of the product qualities are given in Figures 7.40–46 [71]*.

On the basis of feed density and K_{UOP}, the UOP characterization factor indicated on the curves of the graphs and of the selected conversion, the estimation of the yields is effected in the following way:

1. From Figures 7.41 and 7.42 the wt % coke and fuel gases are obtained.
2. From Figures 7.43 and 7.44 the C_3 and C_4 components in vol % are obtained. The total LPG must be adjusted, as shown by Figure 7.45 using the feed K_{UOP} characterization factor.

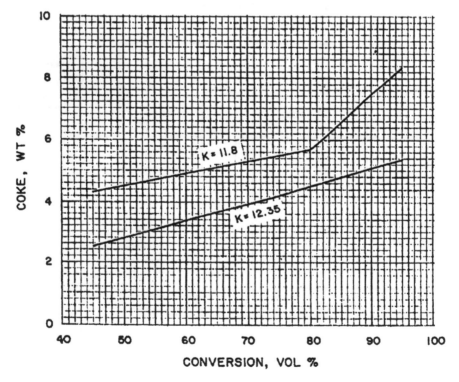

Figure 7.41 Coke yields on zeolite catalyst. (From Ref. 71.)

---

*The graphs for amorphous catalysts given in the same work [71] are not reproduced here.

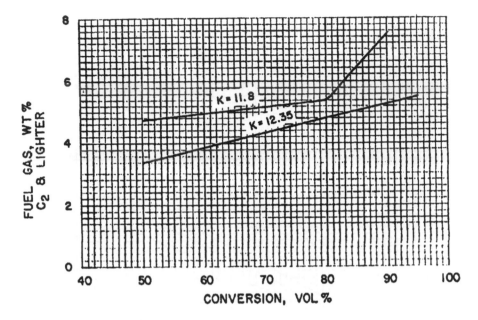

Figure 7.42   Fuel gas yield on zeolite catalyst. (From Ref. 71.)

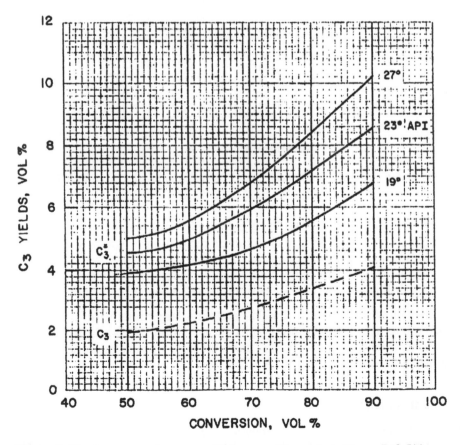

Figure 7.43   Propane and propene yields on zeolite catalyst. (From Ref. 71.)

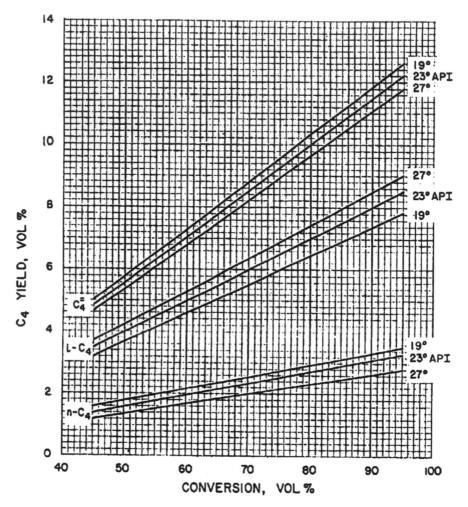

Figure 7.44   Butane, *i*-butane and butenes yields on zeolite catalyst. (From Ref. 71.)

3.   From Figure 7.46, C_5^+ gasoline yield in vol % is obtained.
4.   Figures 7.47 and 7.48 give all products repartition in vol % and give the yields of light and heavy gas oil.
5.   Figure 7.49 gives the sulfur distribution in the products. Figure 7.50 gives the gasoline and the gas oil densities for vol %, wt % calculations.

Software allowed estimation of yields, so that a larger number of parameters may be taken into account than in a graphical calculation [72–74]. But it must be mentioned that works [37] that developed these methods of calculation recommend using graphic methods from Figures 7.41–7.50 for a first approximation.

7.7   ECONOMIC DATA

The investments in a fluidized bed catalytic cracking unit of $1,150,000 \, \text{m}^3/\text{year}$, processing a feed that gives 5.5% coke, was estimated at year-end 1983 to be [75]:

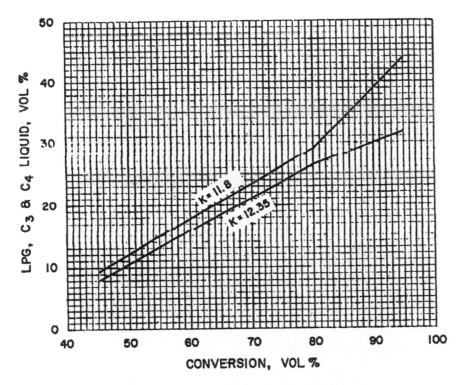

Figure 7.45 Total LPG yield on zeolite catalyst. (From Ref. 71.)

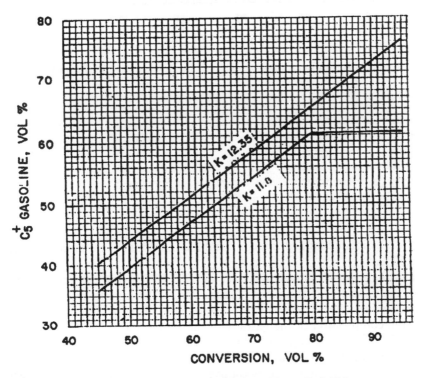

Figure 7.46 Gasoline yield on zeolite catalyst. (From Ref. 71.)

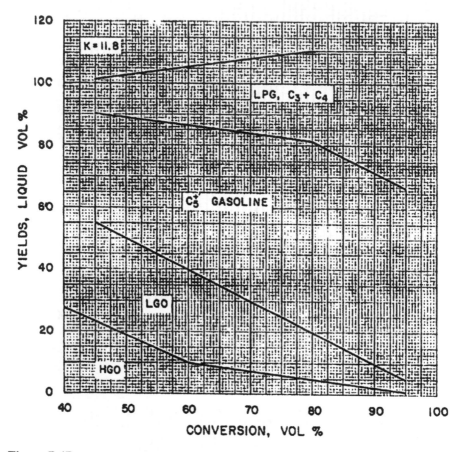

Figure 7.47   Products distribution on zeolite catalyst. Feed K_{UOP} = 11.8. (From Ref. 71.)

Catalytic cracking + gas concentration	31.0 million $
Catalytic cracking + concentration of gases and gases recovery	37.5 million $
Catalytic cracking, gas concentration, energy recovery, and electrostatic precipitation	38.5 million $

The authors estimate that depending on stable conditions the investment could vary within ±30%.

Other authors supplied concurring data in 1992 [71] giving the investments depending on the capacity of the plant (Figure 7.51). The investments from the graph contain, besides the system reactor–regenerator, also the fractionating of the products: the compression of the gases with the recovery of 95% C_4 and 80% C_3, the heat exchange and the cooling of the products to the surrounding temperature, and the central control system.

The utilities consumption in two alternatives with and without energy recovery is given in the Table 7.14 [75].

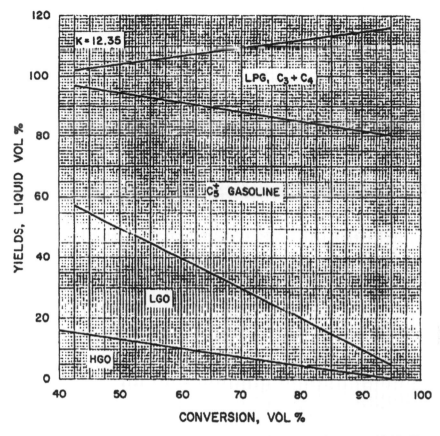

Figure 7.48 Products distribution on zeolite catalyst. Feed $K_{UOP} = 12.35$. (From Ref. 71.)

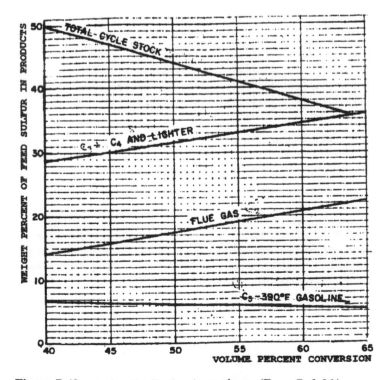

Figure 7.49 Sulfur distribution in products. (From Ref. 86.)

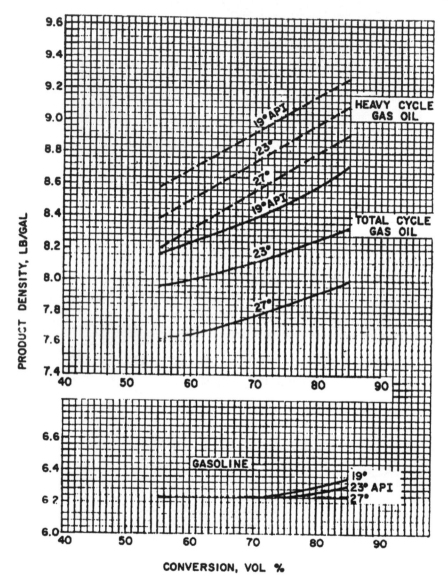

Figure 7.50   Densities of catalytic cracking gas oil and gasoline. (From Ref. 71.)

Taking into account more severe measures taken during recent history for the protection of the environment, investment and operating costs corresponding to the various measures that could be considered were studied [76]. Since these costs depend on regulations from the respective country, they are not presented here.

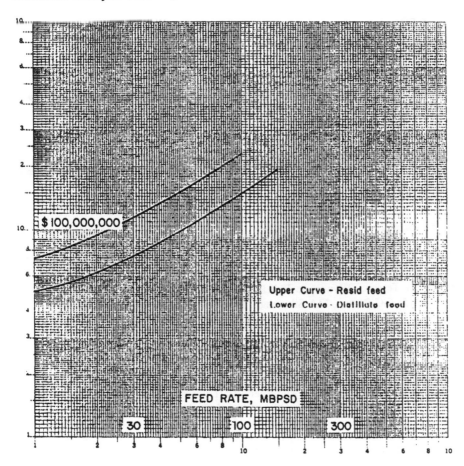

Figure 7.51 Investment cost for fluid catalytic cracking unit, 1992 U.S. Gulf Coast.

Table 7.14 Utilities Consumption

Utilities	m³ feed	With energy recovery*	Without energy recovery*
electricity	kWh/1000	−1900	7550
steam 42 bar	t/1000	34.30	91.40
steam 10.5 bar	t/1000	−40.00	−40.00
steam 3.5 bar	t/1000	6.60	6.60
demineralized water	t/1000	97.10	97.10
cooling water	t/1000	4.80	4.80
Catalyst consumption	kg	0.46	0.46

*Minus for energy production.
Source: Ref. 75.

REFERENCES

1. SV Golden, GR Martin. Hydrocarbon Processing 70: 69–74, Nov. 1991.
2. KV Krikorian, JC Brice. Hydrocarbon Processing 66: 63, Sept. 1987.
3. EC Herthel, EH Reynolds, JH Roe. Oil and Gas J 58: 86, 1960.
4. S Raseev. Procese distructive de prelucrare a titeiului, Editura Tehnica Bucuresti, 1964.
5. JF Le Page, SG Chatila, M Davidson. Raffinage et conversion des produits lourds du petrole, Technip Paris, 1990.
6. MJ Hunbach, DM Cepla, BW Hedrick, CL Hemler, HJ Niclaes. UOP-Residue Conversion. 1990.
7. JR Murphy. Katalistiks, Third Annual Fluid Catalytic Cracking Sym., Amsterdam, 1982.
8. GH Dale. Heavy Oil Cracking. In: Proceedings of the 11th World Petroleum Congress, vol. 4, 83. New York: John Wiley and Sons.
9. GE Weismantel. Heavy Oil Cracking. In: JJ McKetta, ed. Petroleum Processing Handbook. New York: Marcel Dekker, Inc. 1992.
10. DB Bartholic, RP Haseltine. Development in Catalytic Cracking and Hydrocracking. Oil and Gas J: 80, 10–13, 1982.
11. S Raseev, V Tescan, D Ivanescu. Bull Inst Petrol, Gaze si Geologie 18, 25, 1971.
12. BE Reynolds, EC Brown, MA Silverman. Hydrocarbon Processing 71: 43, 1992.
13. T Nguyen, M Skripek. Hydrocarbon Technology International. P Harrison, ed. London: Sterling Publ. Ltd 1993 p. 33–43.
14. GD Hobson. Modern Petroleum Technology. 4th ed. New York: John Wiley & Sons 1973, pp. 288–309.
15. SC Eastwood, CJ Plank, P Weisz. Proceedings of the eighth World Petroleum Congress, vol. 4. London: Applied Science Publishers, 1971.
16. AA Avidan, E Michael, H Owen. Oil and Gas J 88: 33, 8 Jan. 1990.
17. AD Reichle. Oil and Gas J 90: 41, 18 Mai 1992.
18. AA Murcia. Oil and Gas J 90: 68, 18 Mai 1992.
19. AA Avidan, R Shinnar. Ind Eng Chem 29: 931, 1990.
20. D Decroocq, R Bulle, S Chatila, JP Franck, Y Jacquir. Le craquage catalytique des coupes lourdes. Technip, Paris 1978.
21. AL Saxton, AC Worley. Oil and Gas J 68: 82, 18 Mai 1970.
22. P Wuithier. Le Pétrole-Raffinage et Genie chimique. Technip, Paris 1965.
23. AF Babicov, BZ Soliar, LS Glazov, JM Liberson, RV Basov, TH Melik-Ahnazarov, AV Elsin, VM Zarubin. Himia Tehnol Topliv Masel 38: 9, 1993.
24. CW Strother, WL Vermilion, AJ Connes. Oil and Gas J. 70: 102, 15 May 1972.
25. FHH Khow, GV Tonks, KW Szetch, ACC Van Els, A Van Hatten. The Shell Residue Fluid Catalytic Cracking Process. Akzo Catalysis Symposium, Scheveningen, Netherlands, May 1991.
26. AA Avidan. Oil and Gas J 90: 59, 18 May 1992.
27. CR Santner. Hydrocarbon Processing 69: 75, Dec. 1990.
28. FM Hibbs, O Genis, DA Kauff. Hydrocarbon Technology International 1994, P Harrison, ed. London: Sterling Publ. Ltd., 1994.
29. DA Lomas, CA Cabrera, DM Cepla, CL Hemler, LL Upson. Controlled Catalytic Cracking, UOP, 1990.
30. JR Murphy, EL Whittingen. Development of Reduced Crude Catalytic Cracking Technology. In: Proceedings of the 11th World Petroleum Congress, vol. 4. New York: John Wiley & Sons, 1990.
31. L Upson. Hydrocarbon Technology International 1987. P Harrison, ed. London: Gibbons Sterling Publ. Ltd., p. 53.
32. L Upson. Hydrocarbon Processing 66: 67, Sept. 1987.

33. JL Mauléeon, G Heinrich. Revue de l'Institut Français du Pétrole 49: 509, 1994.
34. K Inai. Revue de l'Institut Français du Pétrole 49: 521, 1994.
35. DL Johnson, AA Avidan, PH Schepper, RB Miller, TE Johnson. Oil and Gas J 92: 80, 24 Oct. 1994.
36. JR Murphy. Oil and Gas J 90: 49, 18 May 1992.
37. EC Luckenbach, AC Worley, AD Reichle, EM Gladrow. Cracking Catalytic. In: Petroleum Processing Handbook, JJ McKetta, ed. New York: Marcel Dekker, Inc., 1992.
38. KE Londer, L Kulapaditharom, EJ Juno. Hydrocarbon Processing 64: 80, Sept. 1985.
39. TE Johnson. Hydrocarbon Processing 70: 55, Nov. 1991.
40. JL Mauleon, JB Sigaud. Oil and Gas J 85: 52, 23 Feb. 1987.
41. AD Scheiman. Development in Catalytic Cracking and Hydrocracking. Oil and Gas 53, 60–64, 1982.
42. HL Franzel. Hydrocarbon Processing 64: 51, Jan. 1985.
43. MCMM Campos, PSB Rodrigues. Oil and Gas J 91: 29, Jan. 11, 1985.
44. WA Blanton. Development in Catalytic Cracking and Hydrocracking. Oil and Gas J 53, 1982.
45. JD Cunic, R Diener, GE Ellis. Hydrocarbon Technology International. P Harrison, ed. London: Sterling Publ. Co., 1994, p. 49.
46. SW Golden, AW Stoley, PB Fleming, S Costanzo. Hydrocarbon Technology International 1994. P Harrison, ed. 1994, pp. 73–79.
47. SW Golden, AW Stoley, PB Fleming. Hydrocarbon Processing 72: 43, 1993.
48. G Rowlands, A Knouk, F Kleinschrod. Oil and Gas Journal 89: 64, Nov. 25, 1991.
49. H Dhulesia. Refining and Petrochemical Technology Yearbook 1987. Tulsa, Oklahoma: Penn Wellbooks, 1987, p. 241.
50. SS Elshishini, SSEH Elnashaie. Chemical Engineering Science 45: 553, 1990.
51. GR Fritsche, AF Stegelman. Development of Catalytic Cracking and Hydrocracking. Oil and Gas Journal 53, 1982.
52. J Scherzer. Octane Enhancing Zeolitic FCC Catalysts. New York: Marcel Dekker, Inc., 1990.
53. LL Upson. Hydrocarbon Processing 60: 253, 1981.
54. MA Murphy. Oil and Gas Journal 92: 54, Feb. 21, 1994.
55. RB Ewell, G Gadmer, WJ Turk. Hydrocarbon Processing 60: 103, 1981.
56. JL Mauleon. Revamping of FCC Units. Seminar on Fluid Catalytic Cracking, organized by Engelhard on Sept. 1989 at Brugge.
57. JE Boevink, CE Foster, SR Kumar. Hydrocarbon Processing 60: 123, Sept. 1981.
58. RJ Dilliplane, GP Middlebrooks, RC Hicks, EP Bradley. Oil and Gas Journal 61: 119, Aug. 5, 1963.
59. PT Atterding, J Hunible. Hydrocarbon Processing and Petroleum Refining 42: 167, Apr. 1963.
60. DH Stormont. Oil and Gas Journal 61: 89, Jan. 14, 1963.
61. RA Sanford, H Erickson, EH Burk, EC Gossett, SC Van Petter. Hydrocarbon Processing 41: 103, July 1962.
62. NR Adams, MJ Sterba. Hydrocarbon Processing 42: 175, May 1963.
63. NR Edison, JO Somessen, GP Masoligites. Hydrocarbon Processing 55: 133, May 1976.
64. C Marcilly. Revue de l'Institut Français du Pétrole 42(4): 481, 1987.
65. FJ Elvin. Oil and Gas Journal 85: 42, March 2, 1987.
66. FJ Elvin. Hydrocarbon Processing 68: 71, Oct. 1989.
67. FJ Elvin, SK Pavel. Oil and Gas Journal 89: 94, July 22, 1991.
68. FJ Elvin, O Pallotta. Hydrocarbon Technology International 1992, vol 51. P Harrison, ed. London: Sterling Publ. Int., 1992.

69. FJ Elvin, SK Pavel. 1993 NPRA Annual Meeting, Sept. 21–23, 1993. Convention Center, San Antonio, Texas.
70. BP Castiglioni. Hydrocarbon Processing 62: 35, Feb. 1983.
71. JH Gary, GE Handwerk. Petroleum Refining Technology and Economics, 3rd ed. New York: Marcel Dekker, Inc., 1994.
72. WL Pierce, et al. Hydrocarbon Processing 51: 92, May 1972.
73. EG Wollaston, et al. Oil and Gas Journal 73: 87, Sept. 22, 1975.
74. GWG McDonald. Oil and Gas Journal 87: 80, July 31, 1989.
75. DG Tajb. UOP Fluid Catalytic Cracking Process, published in Handbook of Petroleum Refining Process, RA Meyers, ed. New York: McGraw-Hill, 1986.
76. MJ Humbarch, DM Cepla, BW Hedrick, CL Hemler, HJ Niclaes. Residue Conversion, edited by UOP, 1990.
77. CW Stanger, R Fletcher, Jr., C Johnson, TA Reid. Hydroprocessing/FCC Synergy. Hydrocarbon Processing 75: 89–96, Aug. 1996.
78. SW Shorey, DA Lomas, WH Keesom. Use FCC Feed Pretreating Methods to Remove Sulfur. Hydrocarbon Processing 78: 43–51, Nov. 1999.
79. JW Wilson. FCC Revamp Improves Operations at Australian Refinery. Oil and Gas Journal 97: 63–67, Oct. 1999.
80. JW Wilson. Fluid Catalytic Cracking Technology and Operations. Penn Well Books, 1997, pp. 97–100.
81. G de la Puente, G Chiovetta, U Sedran. FCC Operation with Split Feed Injection. Ind Eng Chem 38: 2, 1999, pp. 368–372.
82. B Jazayeri. Optimize FCC Riser Design. Hydrocarbon Processing 70, May 1991, pp. 93–94.
83. X Zhao, AW Peters, GW Weatherber. Nitrogen Chemistry and NO_2 Control in Fluid Catalytic Cracking Regenerator. Ind Eng Chem 36(11): 4535–4542, 1997.
84. KL Dishman, PK Doolin, LD Tullock. NO_2 Emissions in Fluid Catalytic Cracking Catalyst Regenerator. Ind Eng Chem 37(12): 4631–4636, 1998.
85. JB McLean. Fluid Catalyst Properties can affect cyclone erosion. Oil and Gas Journal 98: 33–36, Jan 3, 2000.
86. MJ Fowle, RD Bent. Petroleum Refiner 26(11): 719–727, 1947.
87. MJ Fowle, RD Bent. Hydrocarbon Processing 78: 124, March 1999.
88. G Xu, B Chan, X He. Hydrocarbon Processing 80: 81, April 2001.
89. MB Ewell, G Gardmer, WJ Turk. Hydrocarbon Processing 60: 103–112, Sept. 1981.
90. CA Cabrera, D Knepper. Hydrocarbon Technology International 1990/91. P Harrison, ed. London: Sterling Publ. Int.
91. JE Naber, M Akabar. Shell's residue fluid catalytical cracking process. Hydrocarbon Technology International 1989/90. P Harrison, ed. London: Sterling Publ. International, pp. 37–43.

8

Design Elements for the Reactor–Regenerator System

This chapter presents some specific design elements for fluidized bed catalytic cracking. Catalytic cracking with moving bed is currently of low interest and was discussed in an earlier work of the author [1]. Actually, the issues that are specific to moving bed systems in regard to thermal processes were discussed in Section 5.3 and can be used for computations pertaining to catalytic cracking in such systems.

8.1 SOME FLUIDIZATION PROBLEMS

The fluidization phenomenon is well-known. A fluid introduced at the lower end of a bed of solid particles will pass upwards, through the voids between the particles. With increasing fluid flowrate, the friction between fluid particles will increase as measured by the higher pressure drop encountered by the fluid. When the pressure drop equals the weight of the bed, the contact between the particles is no longer continuous and fluidization occurs. The particles acquire irregular movements around some equilibrium position. The corresponding linear velocity, calculated as through the empty vessel, is the minimum fluidization velocity. With further increase of the fluid velocity, the pressure drop remains basically constant over a broad range of velocities. In fluidization with gases, the gas flow in excess of that corresponding to the minimum fluidization velocity will pass through the bed as bubbles. With further increase of the velocity, the space between the particles increases until they may be entrained out of the vessel. Fluidization may be performed with liquids or gases and the solid particles may be of the same size or of a range of sizes.

Fluidization presents obvious advantages in processes where an intimate contact is desired between the solid particles and the fluid. Moreover it can be easily combined with the transport of the solids between vessels having various functions. For these reasons, fluidization is much used in numerous industrial processes and is in virtually general use in catalytic cracking [16,29,34,37].

A first issue of importance for the processes carried out in fluidized bed is the homogeneity of the fluidization. This means the achievement of identical contacting conditions between solids and fluid within the whole space occupied by the fluidized bed. Homogeneous fluidization implies the absence of bubbles passing through the bed.

The demarcation between homogeneous fluidization and heterogeneous fluidization is shown in Figure 8.1 [37].

This figure explains why fluidization with liquids is homogeneous over the whole domain of bed expansions while for the fluidization with gases, where the ratio of the densities is of the order of 10^3, inhomogeneous fluidization occupies a large domain. Only small ranges in the domains of the dense phase and dilute phase are quasihomogeneous.

With decreasing sizes of the particles, the heterogeneous domain becomes smaller. By adequate selection of particle size distribution the quasihomogeneous domain can be somewhat expanded. These possibilities are used when the fluidized solid is made of catalyst particles.

The fact that at small ratios of phase densities, fluidization is homogeneous in the whole domain of expansion, makes it possible to study homogeneous fluidization by using liquids as the fluidization agent.

Our studies in this domain [11–15], proved the determinant role of the boundary layer formed around the fluidized particles in the correlation between the fluid velocity and the bed expansion.

Since the parameters of the boundary layer are dependent on the particles Reynolds number, with the hydraulic equivalent diameter as length parameter (see Section 6.4.1), the correlation between the fluid velocity and the bed expansion depends on the Reynolds number. The analytical deduction of this dependence follows.

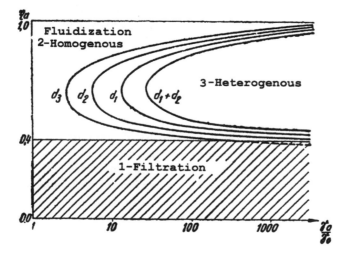

Figure 8.1 Borders of homogeneous fluidization. 1 – Filtration, 2 – Homogeneous fluidization, 3 – Heterogeneous fluidization: d – particles diameter; $d_1 < d_2 < d_3$. (From Ref. 37.)

8.1.1 Homogeneous Bed Expansion Equations

The early studies of Raseev on homogeneous fluidization [11–14] and the equations deduced therein, were based on the following assumptions:

> In a fluidized bed the mean statistic position of the particles corresponds to a cubic lattice.
>
> In order that the particles have a stationary statistically average position, the maximum velocity of the fluid in the bed $v_{s\ max}$ (the velocity in the smallest cross section) has to be equal to the final free fall velocity, v_p:

$$v_{s\ max} = v_p \tag{8.1}$$

On this basis, for a cubic bed structure the following equation was obtained:

$$v_p = \frac{v_1}{1 - b(1 - \eta_{af})^{2/3}} \tag{8.2}$$

where:

> v_1 = superficial velocity in the total cross section of the apparatus
> η_{af} = the apparent void fraction of the fluidized bed
> b = a constant given by the equation:

$$b = \frac{S}{V^{2/3}} \tag{8.3}$$

where:

> S = particle cross section, normal to the moving direction of the fluid
> V = the particle volume.

For spherical particles $b = 1.2091$.

Similar equations were suggested by Martens and Rhodes [42], Kadymova [43] and Razumov [44] and their analysis was made by Ghosal and Mekherhjea [45].

The experimental verification of Eq. (8.2) was done subsequently in a large range of conditions [15] and proved that the equation is correct only for large Reynolds numbers[*]. The deviations increase with decreasing Re number.

As shown by Raseev [15] these deviations result when the boundary layer thickness is not included. In order to obtain correct equations for all the Re number ranges the displacement thickness δ^x should be taken into consideration (Figure 6.19).

Since δ^x decreases with the Reynolds number (Figure 6.23) it is expected that Eq. (8.2) is verified for high Reynolds numbers, for which the thickness δ^x is negligible. The difference between the experimental data and theory increases with decreasing Re number, which leads to the increase of δ^x.

This conclusion leads to the requirement to replace Eq. (8.1) by the expression:

$$v_{s\ max} = v^x \tag{8.4}$$

where v^x represents the average fluid velocity in the smallest cross section in the bed needed for maintaining a statistically stationary positions of the fluidized particles.

---

[*]In all this account the Reynolds number is defined by the expression:

$$Re = v_p \cdot d/v$$

where d = diameter of the particle and v = the kinematic viscosity.

Taking into account this definition, v^x can be expressed by the equation:

$$v^x \cdot f_{1\,min} = \int\int(r+y)v\,dy\,d\varphi \tag{8.5}$$

which gives the fluid flow drained through the free minimum space corresponding to a single particle. The integration limits depend on the distance between particles, influenced by the degree of expansion η_{af} and on the geometric form of the minimum free section corresponding to a particle.

Since owing to the law of the continuity:

$$v^x \cdot f_{1\,min} = v_1 \cdot f_t \tag{8.6}$$

Eq. (8.5) may be written as:

$$v_1 \cdot f_t = \int\int(r+y)v\,dy\,d\varphi \tag{8.7}$$

or, dividing by $v_p \cdot f_t$:

$$\frac{v_1}{v_p} = \frac{1}{f_t}\int\int(r+y)\frac{v}{v_p}\,dy\,d\varphi \tag{8.8}$$

In these equations

$f_{1\,min}$ = minimum free cross section through the fluidized bed
r = the radius of the particle
y = the distance from the particle wall in the median section of the particle, normal to the direction of the fluid movement
v = the velocity of the fluid in the median plane at various distances y
φ = the angle in the median plane
f_t = the free board above the fluidized bed
v_1 = the velocity in the free board section.

Replacing $\dfrac{v}{v_p}$ by the expression (6.8) it results:

$$\frac{v_1}{v_p} = \frac{1}{f_t}\int\int(r+y)\left[-\frac{1}{4}\mathscr{R}^2\frac{y^2}{r^2}+\mathscr{R}\frac{y}{r}\right]dy\,d\varphi \tag{8.9}$$

where the dimensionless criterion $\mathscr{R}$ is given by the equation:

$$\mathscr{R} = \frac{r \cdot \tau_{o90}}{v_p \cdot \mu} \tag{6.6}$$

and is correlated with the Reynolds number by the graph shown in Fig. 6.22.2.

Equation (8.9) was integrated [15] for various assumed structure forms and degress of expansion of the fluidized bed made of spherical particles having the same diameter. The following structures were taken into account: cubic, orthorhombic, and rhombohedral and assumptions of bi- and tridimensional expansion.

The only hypothesis that was experimentally verified is the cubic structure and the tridimensional expansion. For these alone the deduction of the equations is presented below.

8.1.1.1 The domain $\eta a \geq 0.476$ ($\eta = 0.476$) corresponds to the densely packed cubic structure

For the integration of Eq. (8.9) three cases are considered, as represented in Figure 8.2. These correspond to the three situations that occur during the expansion of the cubic lattice. The following parameters are used:

$$h = \frac{a}{\cos \varphi} - r \tag{8.10}$$

and

$$\psi = \arccos\left(\frac{a}{\delta + r}\right) \tag{8.11}$$

Besides, for the correlation of δ with $\mathscr{R}$, Eq. (6.7) must be taken into account:

$$\delta = 2r/\mathscr{R} \tag{6.7}$$

The form taken by Eq. (8.9) for the three cases (Figure 8.2), their integration, and their domain of validity are given below.

Case 1. The boundary layer δ extends beyond the surface element afferent to the fluidized microsphere:

$$r + \delta \geq \sqrt{2}a \geq \sqrt{2}r$$

or, using expression (6.7) and dividing by r, one obtains the validity domain:

$$(2/\mathscr{R} + 1) \geq \sqrt{2}\,\frac{a}{r} \geq \sqrt{2} \tag{8.12}$$

Equation (8.9) becomes:

$$\frac{v_1}{v_p} = \frac{2}{a^2} \int_0^{\frac{\pi}{4}} \int_0^h (r + y)\left(-\frac{1}{4}\,\mathscr{R}^2\,\frac{y^2}{r^2} + \mathscr{R}\,\frac{y}{r}\right) dy\, d\varphi \tag{8.13}$$

which, by integration, gives:

$$\frac{v_1}{v_p} = \frac{2}{a^2} \int_0^{\frac{\pi}{4}}\left[-\mathscr{R}^2 \cdot \frac{1}{4r}\int_0^h y^2\,dy - \mathscr{R}^2 \cdot \frac{1}{4r^2}\int_0^h y^3\,dy + \mathscr{R}\int_0^h y\,dy + \mathscr{R}\frac{1}{r}\int_0^h y^2\,dy\right] d\varphi$$

$$= \frac{2}{a^2} \int_0^{\frac{\pi}{4}}\left[-\mathscr{R}^2 \cdot \frac{h^3}{12r} - \mathscr{R}^2 \cdot \frac{h^4}{16r^2} + \mathscr{R} \cdot \frac{h^2}{2} + \mathscr{R} \cdot \frac{h^3}{3r}\right] d\varphi$$

Substituting h with the expression (8.10), integrating and simplifying, leads to:

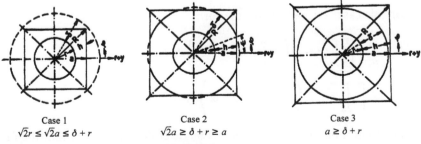

Case 1   Case 2   Case 3
$\sqrt{2}r \leq \sqrt{2}a \leq \delta + r$ $\sqrt{2}a \geq \delta + r \geq a$ $a \geq \delta + r$

Figure 8.2 The expansion of the cubic lattice.

$$\frac{v_1}{v_p} = -\mathcal{R}^2 \cdot \frac{a^2}{6r^2} + [\sqrt{2} - \ln(\sqrt{2} - 1)]\left(\frac{1}{6}\mathcal{R}^2 + \frac{1}{3}\mathcal{R}\right)\frac{a}{r} - \left(\frac{1}{4}\mathcal{R}^2 + \mathcal{R}\right)$$
$$+ \pi\left(\frac{1}{96}\mathcal{R}^2 + \frac{1}{12}\mathcal{R}\right)\frac{r^2}{a^2} \tag{8.14}$$

Case 3. (Case 2 will follow.) The surface element afferent to the fluidized micro-sphere extends beyond the boundary layer δ:

$$a \geq (\delta + r)$$

or, by using the expression (6.7) and dividing by r, the validity domain results:

$$a/r = (2/\mathcal{R} + 1) \tag{8.15}$$

Equation (8.9) becomes:

$$\frac{v_1}{v_p} = \frac{2}{a^2}\left[\int_0^{\frac{\pi}{4}}\int_0^{\delta}(r+y)\left(-\frac{1}{4}\cdot\frac{y^2}{r^2}\mathcal{R}^2 + \frac{y}{r}\mathcal{R}\right)dyd\varphi + \int_0^{\frac{\pi}{4}}\int_{\delta}^{h}(r+y)dyd\varphi\right] \tag{8.16}$$

The integration of this equation gives:

$$\frac{v_1}{v_p} = \frac{1}{a^2}\int_0^{\frac{\pi}{4}}\left[-\frac{\delta^3}{12r}\mathcal{R}^2 - \frac{\delta^4}{16r^2}\mathcal{R}^2 + \frac{\delta^2}{2}\mathcal{R} + \frac{\delta^3}{3r}\mathcal{R}\right]d\varphi$$
$$+ \frac{1}{a^2}\int_0^{\frac{\pi}{4}}\left(rh - r\delta + \frac{1}{2}h^2 - \frac{1}{2}\delta^2\right)d\varphi$$

Substituting h and δ by the expressions (8.10) and (8.11), integrating and simplifying leads to:

$$\frac{v_1}{v_p} = 1 - \pi\left(\frac{1}{6}\mathcal{R}^{-2} + \frac{1}{3}\mathcal{R}^{-1} + \frac{1}{4}\right)\frac{r^2}{a^2} \tag{8.17}$$

Case 2. This case represents the intermediary situation between case 1 and case 3, defined according to the Figure 8.2, by the conditions:

$$\sqrt{2}a \geq r + \delta \geq a$$

which, taking into account Eq. (6.7) defines its validity domain:

$$\sqrt{2}a/r \geq (2/\mathcal{R} + 1) \geq a/r \tag{8.18}$$

For this case, Eq. (8.9) takes the form:

$$\frac{v_1}{v_p} = \frac{2}{a^2}\left[\int_0^{\psi}\int_0^{h}(r+y)\left(-\frac{y^2}{4r^2}\mathcal{R}^2 + \frac{y}{r}\mathcal{R}\right)dyd\varphi\right.$$
$$\left. + \int_{\psi}^{\frac{\pi}{4}}\int_0^{\delta}(r+y)\left(-\frac{y^2}{4r^2}\mathcal{R}^2 + \frac{y}{r}\mathcal{R}\right)dyd\varphi + \int_{\psi}^{\frac{\pi}{4}}\int_{\delta}^{h}(r+y)dyd\varphi\right] \tag{8.19}$$

The integration of this expression is made easier by the fact that the first term differs from Eq. (8.13) only by the upper limit of the first integral, while the second and the third terms are different from Eq. (8.16) only by the lower limit of the first integral.

Finally, it follows:

$$
\frac{v_1}{v_p} = -\left(\frac{1}{2} + \frac{\mathscr{R}^2}{8} + \frac{\mathscr{R}}{2}\right)\sqrt{\left[\frac{r(2+\mathscr{R})}{a\mathscr{R}}\right]^2 - 1} - \frac{\mathscr{R}^2 a^2}{12r^2}\sqrt{\left[\frac{r(2+\mathscr{R})}{a\mathscr{R}}\right]^2 - 1}
$$

$$
-\left(\frac{\mathscr{R}^2}{6} + \frac{\mathscr{R}}{3}\right)\frac{a}{r}\ln\left[\frac{r(2+\mathscr{R})}{a\mathscr{R}} - \sqrt{\left[\frac{r(2+\mathscr{R})}{a\mathscr{R}}\right]^2 - 1}\right] + 1
$$

$$
-\frac{\pi}{4}\left(\frac{2}{3\mathscr{R}^2} + \frac{4}{3\mathscr{R}} + 1\right)\frac{r^2}{a^2}
$$

$$
+\left(\frac{\mathscr{R}^2}{24} + \frac{\mathscr{R}}{3} + \frac{2}{3\mathscr{R}^2} + \frac{4}{3\mathscr{R}} + 1\right)\frac{r^2}{a^2}\arccos\left[\frac{a\mathscr{R}}{r(2+\mathscr{R})}\right]
$$

$$(8.20)$$

In order to use the obtained final equations (8.14), (8.17), and (8.20), the term a/r, which intervenes in these equations, must be replaced by the expression of the cubic expanded tridimensional net considered here, having the form:

$$
\frac{a}{r} = \left[\frac{\pi}{6(1 - \eta_a)}\right]^{1/3}
$$

$$(8.21)$$

8.1.1.2 The domain $0.476 \geq \eta_a >$ fluidization initiation

Since the most compact cubic structure corresponds to an apparent free volume $\eta_a = 0.476$, the assumption of the mean statistic cubic structure was completed by specifying how this structure is deformed in order to reach values of the order $\eta_a = 0.385$ that correspond to the initiation of fluidization (or minimum fluidization conditions).

By comparing the experimental data with various assumptions [15], the conclusion was that the most plausible deformation is the transformation of the cubic lattice to a rhombohedral one.

In order to analyze the effect of the boundary layer, the conditions in a plane normal on the direction of displacement of the fluid is taken into consideration, as shown in Figure 8.3.

In a similar manner to the deduction of the equations for the expansion of the cubic structure, the following equations result [15]:

CASE 1 ($\delta \geq h_M$)

$$
\frac{v_1}{v_p} = \frac{4}{1/2 d_1 d_2}\int_{\psi=0}^{\psi=\psi_M}\int_{y=0}^{y=h_1}(r+y)\left(-\frac{y^2}{4r^2}\mathscr{R}^2 + \frac{y}{r}\mathscr{R}\right)dy\,d\psi
$$

$$
+ \frac{8}{1/2 d_1 d_2}\int_{\varphi=0}^{\varphi=\varphi_M}\int_{y=0}^{y=h_2}(r+y)\left(-\frac{y^2}{4r^2}\mathscr{R}^2 + \frac{y}{r}\mathscr{R}\right)dy\,d\varphi
$$

$$(8.22)$$

CASE 2 $\left(h_M \geq \delta \geq \frac{d_1}{2} - r\right)$

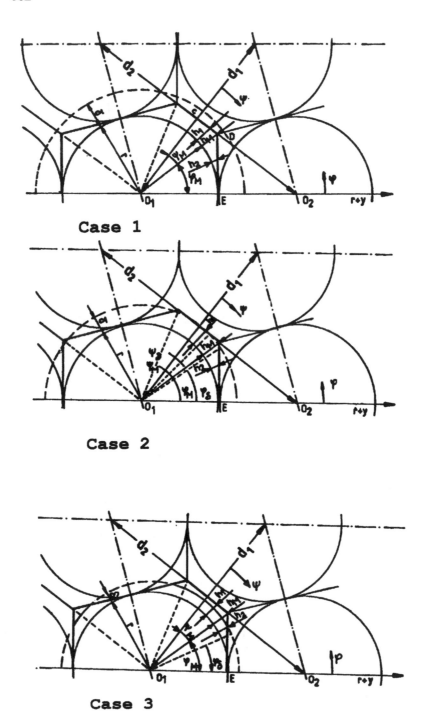

Figure 8.3 The compression of the cubic lattice.

$$\frac{v_1}{v_p} = \frac{4}{1/2 d_1 d_2}\left[\int_{\psi=0}^{\psi=\psi_\delta}\int_{y=0}^{y=h_1}(r+y)\left(-\frac{y^2}{4r^2}\mathscr{R}^2 + \frac{y}{r}\mathscr{R}\right)dy\,d\psi\right.$$

$$+ \int_{\psi=\psi_\delta}^{\psi=\psi_M}\int_{y=0}^{y=\delta}(r+y)\left(-\frac{y^2}{4r^2}\mathscr{R}^2 + \frac{y}{r}\mathscr{R}\right)dy\,d\psi + \left.\int_{\psi=\psi_\delta}^{\psi=\psi_M}\int_{(y=0)}^{y=h_1}(r+y)dy\,d\psi\right]$$

$$+ \frac{8}{1/2 d_1 d_2}\left[\int_{\varphi=0}^{\varphi=\varphi_\delta}\int_{y=0}^{y=h_2}(r+y)\left(-\frac{y^2}{4r^2}\mathscr{R}^2 + \frac{y}{r}\mathscr{R}\right)dy\,d\varphi\right.$$

$$+ \left.\int_{\varphi=\varphi_\delta}^{\varphi=\varphi_M}\int_{y=0}^{y=\delta}(r+y)\left(-\frac{y^2}{4r^2}\mathscr{R}^2 + \frac{y}{r}\mathscr{R}\right)dy\,d\varphi + \int_{\varphi=\varphi_\delta}^{\varphi=\varphi_M}\int_{y=0}^{y=h_2}(r+y)dy\,d\varphi\right]$$

$$(8.23)$$

CASE 3 $\left(\dfrac{d_1}{2} - r \geq \delta\right)$

$$\frac{v_1}{v_p} = \frac{4}{1/2 d_1 d_2}\left[\int_{\psi=0}^{\psi=\psi_M}\int_{y=0}^{y=\delta}(r+y)\left(-\frac{y^2}{4r^2}\mathscr{R}^2 + \frac{y}{r}\mathscr{R}\right)dy\,d\psi\right.$$

$$+ \left.\int_{\psi=\psi_\delta}^{\psi=\psi_M}\int_{y=0}^{y=h_1}(r+y)dy\,d\psi\right]$$

$$+ \frac{8}{1/2 d_1 d_2}\left[\int_{\varphi=0}^{\varphi=\varphi_\delta}\int_{y=0}^{y=h_2}(r+y)\left(-\frac{y^2}{4r^2}\mathscr{R}r^2 + \frac{y}{r}\mathscr{R}\right)dy\,d\varphi\right.$$

$$+ \left.\int_{\varphi=\varphi_\delta}^{\varphi=\varphi_M}\int_{y=0}^{y=\delta}(r+y)\left(-\frac{y^2}{4r^2}\mathscr{R}^2 + \frac{y}{r}\mathscr{R}\right)dy\,d\varphi + \int_{\varphi=\varphi_\delta}^{\varphi=\varphi_M}\int_{y=\delta}^{y=h_2}(r+y)dy\,d\varphi\right]$$

$$(8.24)$$

The integration of equations (8.22–8.24), which express the deformation in the horizontal plane by the equation:

$$u = \frac{d_1}{2r} \tag{8.25}$$

leads to the following final equations and to their validity limits:
For the domain:

$$\frac{2+\mathscr{R}}{\mathscr{R}} \geq \frac{2}{\sqrt{4-u^2}} \tag{8.26}$$

$$\frac{v_1}{v_p} = -\frac{\mathscr{R}^2}{12} - \mathscr{R} + \mathscr{R}^2\frac{(u^4-12)}{12(4-u^2)} + \frac{1}{3\sqrt{4-u^2}}(\mathscr{R}^2 + 2\mathscr{R})$$

$$+ \left(\frac{\mathscr{R}^2}{6}+\frac{\mathscr{R}}{3}\right)\left[\frac{u^2}{\sqrt{4-u^2}}\ln\frac{\sqrt{4-u^2}}{u} + \frac{2}{u\sqrt{4-u^2}}\ln\frac{\sqrt{4-u^2}}{2-u}\right] \tag{8.27}$$

$$+ \left(\frac{\mathscr{R}^2}{48}+\frac{\mathscr{R}}{6}\right)\frac{\pi}{u\sqrt{4-u^2}}$$

For the domain:

$$\frac{2}{\sqrt{4-u^2}} \geq \frac{2+\mathscr{R}}{\mathscr{R}} \geq u \tag{8.28}$$

$$\frac{v_1}{v_p} = -\left[\left(\frac{1}{2}+\frac{\mathscr{R}}{2}+\frac{\mathscr{R}^2}{8}\right)\frac{u}{\sqrt{4-u^2}}+\frac{1}{12}\,\mathscr{R}^2\,\frac{u^3}{\sqrt{4-u^2}}\right]\sqrt{\left(\frac{2+\mathscr{R}}{u\mathscr{R}}\right)^2-1}$$

$$+1-\left(\frac{\mathscr{R}^2}{6}+\frac{\mathscr{R}}{3}\right)\frac{u^2}{\sqrt{4-u^2}}\ln\left[\frac{2+\mathscr{R}}{u\mathscr{R}}-\sqrt{\left(\frac{2+\mathscr{R}}{u\mathscr{R}}\right)^2-1}\right]$$

$$+\frac{1}{u\sqrt{4-u^2}}\left\{\left(\frac{\mathscr{R}^2}{12}+\frac{2\mathscr{R}}{3}+\frac{4}{3\mathscr{R}^2}+\frac{8}{3\mathscr{R}}+2\right)\left[\frac{1}{2}\,\text{arctg}\sqrt{\left(\frac{2+\mathscr{R}}{u\mathscr{R}}\right)^2-1}\right.\right.$$

$$\left.+\text{arctg}\sqrt{\left(\frac{2+\mathscr{R}}{\mathscr{R}}\right)^2-1}\right]-\left(\frac{1}{3\mathscr{R}^2}+\frac{2}{3\mathscr{R}}+\frac{1}{2}\right)\pi-\left(1+\mathscr{R}+\frac{5}{12}\,\mathscr{R}^2\right)$$

$$\left.\times\sqrt{\left(\frac{2+\mathscr{R}}{\mathscr{R}}\right)^2-1}-\left(\frac{\mathscr{R}^2}{3}+\frac{2\mathscr{R}}{3}\right)\ln\left[\frac{2+\mathscr{R}}{\mathscr{R}}-\sqrt{\left(\frac{2+\mathscr{R}}{\mathscr{R}}\right)^2-1}\right]\right\}$$

$$\tag{8.29}$$

For the domain:

$$\frac{2+\mathscr{R}}{\mathscr{R}} \leq u \tag{8.30}$$

$$\frac{v_1}{v_p}=1-\frac{1}{u\sqrt{4-u^2}}\left\{\left(\frac{1}{3\mathscr{R}^2}+\frac{2}{3\mathscr{R}}+\frac{1}{2}\right)\pi+\left(1+\mathscr{R}+\frac{5\mathscr{R}^2}{12}\right)\sqrt{\left(\frac{2+\mathscr{R}}{\mathscr{R}}\right)^2-1}\right.$$

$$+\left(\frac{1}{3}\,\mathscr{R}^2+\frac{2}{3}\,\mathscr{R}\right)\ln\left[\frac{2+\mathscr{R}}{\mathscr{R}}\sqrt{\left(\frac{2+\mathscr{R}}{\mathscr{R}}\right)^2-1}\right]-\left(\frac{1}{12}\,\mathscr{R}^2+\frac{2}{3}\,\mathscr{R}+\frac{4}{3\mathscr{R}^2}\right.$$

$$\left.\left.+\frac{8}{3\mathscr{R}}+2\right)\text{arctg}\sqrt{\left(\frac{2+\mathscr{R}}{\mathscr{R}}\right)^2-1}\right\}$$

$$\tag{8.31}$$

To use these equations more easily in Figure 8.4 their validity domains are shown as function of η_a and $\mathscr{R}$.

8.1.2 Experimental Verification of the Equations

The final equations obtained above were submitted to extensive experimental verification [15] and comparison with other 9 assumptions concerning the structure of the bed and equations for the boundary layer. They were fully confirmed by these verifications.

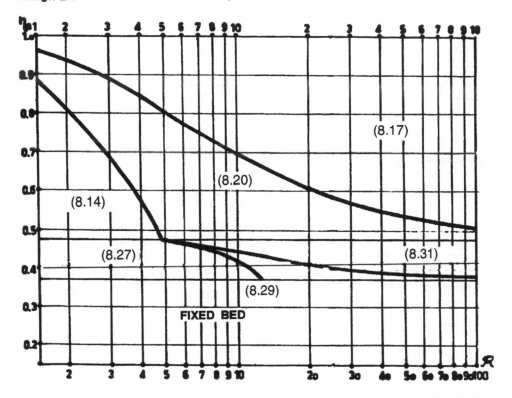

Figure 8.4 The validity domains of equations (8.14), (8.17), (8.20), (8.27), (8.29), (8.31).

The verifications were carried out for Reynolds numbers ranging between 0.035 and 1040 and for degrees of expansion of the fluidized beds running between the minimum fluidization and $\eta_a = 0.95 \div 0.98$.

The experimental work performed the fluidization with liquids in order to ensure a homogeneous fluidization over the entire domain of expansion. The fluidized solids were of spherical shape, for which the terminal free fall velocity was computed and used in the deduced equations.

In order to cover the whole domain of Reynolds numbers, the viscosity of the fluid, the density (metal or glass), and the diameter of the particles were varied.

Several of the obtained results are plotted in Figures 8.5–8.7. They show very good agreement between the deduced equations and the experimental data.

In view of the good agreement, the deduced equations were used for calculating the void fraction in homogeneous fluidization, function of the particle Re number, and the v_1/v_p ratio. The results are plotted in Figure 8.8.

The use of these equations, and of the plots of Figure 8.8 for the homogeneous fluidization of particles of unequal diameters and irregular shape was also investigated.

The experiments proved that the above equations and the plots are valid also for particles of irregular shapes, if the used diameter is that of a sphere having the

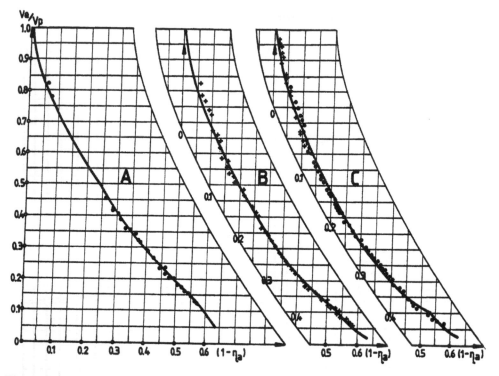

Figure 8.5 Experimental verifications of the deduced equations (spherical particles). $A - \mathrm{Re} = 1040;\ B - \mathrm{Re} = 156;\ C - \mathrm{Re} = 95.$

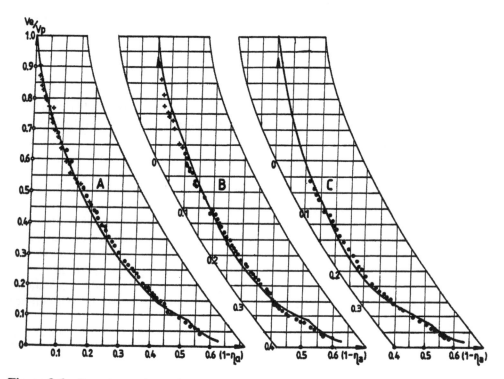

Figure 8.6 Experimental verifications of the deduced equations (spherical particles). $A - \mathrm{Re} = 40.2;\ B - \mathrm{Re} = 28.6;\ C - \mathrm{Re} = 8.25.$

506

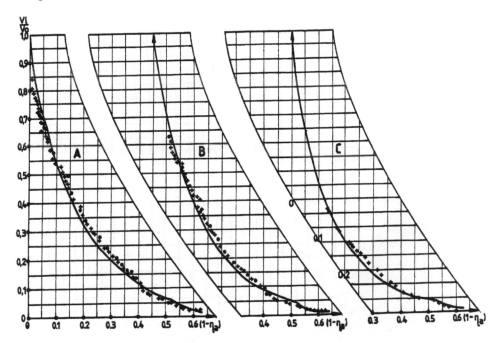

Figure 8.7 Experimental verifications of the deduced equations (spherical particles). $A - \text{Re} = 2.8$; $B - \text{Re} = 0.98$; $C - \text{Re} = 0.0345$.

volume equal to that of the particle and the terminal free fall velocity is the fluid velocity corresponding to the entrainment of 50 wt % of the particles.

For illustration, the graph of Figure 8.9 shows two of the experimental series carried out with quartz particles.

It is also of interest to compare our results to the semi-empirical or empirical correlations suggested by other authors. The curves of Figure 8.10 give the results for our equations for $\text{Re} \to 0$, compared to those obtained by using the equations of Happel [47], Jottrand [48], Johnson [49], Lewa [50], Suciu [46]. The curve representing our results is seen to have a median location among the results of the quoted authors.

An interesting comparison is that also that of Figure 8.11 where the curves computed from the equations deduced by us and traced for various Reynolds numbers are compared to those for exponential equations of the form:

$$\frac{v_1}{v_p} = \eta_a^n \tag{8.32}$$

where the exponent n takes values comprised between 3.5 and 6. Note that increasing the Re number is equivalent to decreasing values of the exponent n and that important differences exist between the two correlations.

The terminal free falling velocity v_p, which appears in our equation and in the ordinate of Figure 8.8, may be calculated with the approximate equations:

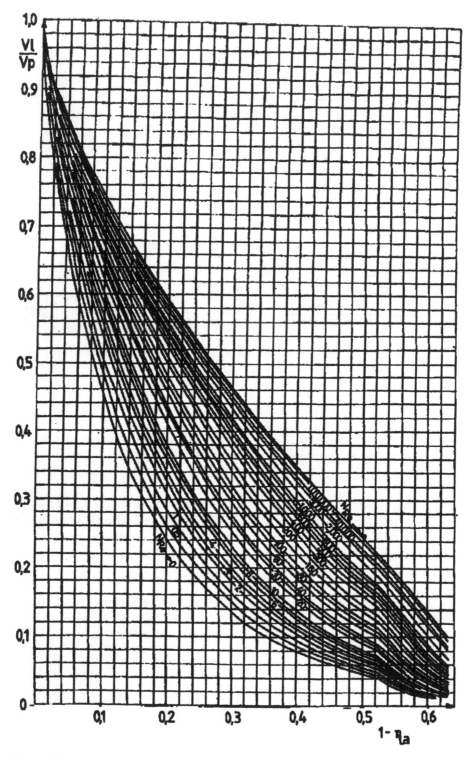

Figure 8.8   Determination of void fraction for the homogeneous fluidization, function of the Reynolds number and of the ratio v_1/v_p. (From Ref. 15.)

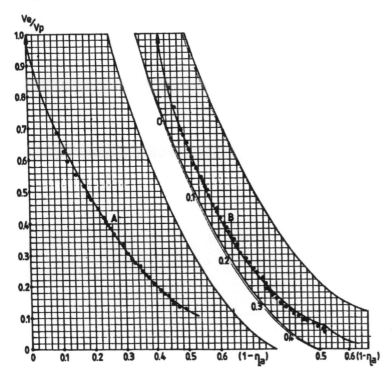

Figure 8.9 The fluidization of quartz particles. A – density $= 2.6093\,\text{g/cm}^3$; size range $= 0.64 - 0.80$, $\text{Re} = 82.1$. B – density $= 2.6384\,\text{g/cm}^3$, size range $= 0.25\text{–}0.315\,\text{mm}$, $\text{Re} = 10$.

for $\text{Re} < 2$ $\qquad\qquad$ $v_p = 0.545\, \dfrac{d_e^2}{\nu} \cdot \dfrac{\gamma_a - \gamma_0}{\gamma_0}$

for $2 < \text{Re} < 1000$ $\qquad$ $v_p = 0.782\, \dfrac{d_e^{1.14}}{\nu^{0.428}} \left(\dfrac{\gamma_a - \gamma_0}{\gamma_0} \right)^{0.714}$ $\qquad\qquad$ (8.33)

for $\text{Re} > 1000$ $\qquad\quad$ $v_p = \sqrt{29.7 d\, \dfrac{\gamma_a - \gamma_0}{\gamma_0}}$

where: ν is the kinetmatic viscosity in m^2/s and d is the particle diameter in m. For more exact results, the following equations are recommended:

$$v_p = \sqrt[3]{\frac{13.08 \cdot v \cdot \text{Re}(\gamma_a - \gamma_0)}{C\gamma_0}} \qquad\qquad (8.34)$$

and

$$d_e = 0.0765 \cdot C \cdot v_p^2\, \frac{\gamma_0}{\gamma_a - \gamma_0} \qquad\qquad (8.35)$$

which are used together with the graph of Figure 8.12. The figure correlates the resistance coefficient C with the Reynolds number. The calculation starts by taking from the graph of Figure 8.12 pairs of values C, Re for which, with the aid of Eqs. (8.34) and (8.35), the corresponding values v_p and d_e are obtained. These calculations make it possible to trace the curve $v_p - f(d_e)$ from which the exact value of the terminal velocity is obtained.

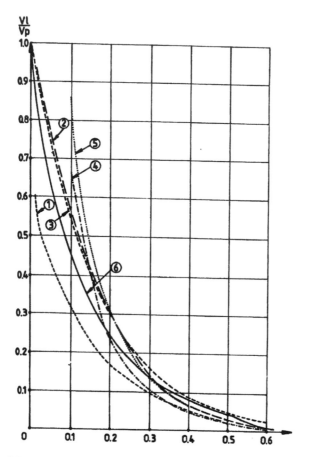

Figure 8.10 Comparison of the curves for Re → 0 obtained with the equations proposed by: 1 – Happel; 2 – Jottrand; 3 – Johnson; 4– Lewa; 5 – Suciu; 6 – Raseev.

8.1.3 Bed Homogeneity

In the case of the nonhomogeneous fluidization, the phenomena are much more complex. The increase of the fluid velocity leads at the beginning, identical to the case of homogeneous fluidization, to a progressive expansion of the bed. But as a certain velocity is exceeded, the bed ceases to expand and its bulk density remains almost constant, while the fluid in excess to that corresponding to the homogeneous fluidization passes through the bed as bubbles. The aspect of the bed becomes similar to that of a liquid through which gas bubbles pass. The increase of the fluid velocity increases the number and the size of the bubbles without modifying the density of the fluidized bed.

The generation of the gas bubbles marks the beginning of the nonhomogeneous fluidization. The nonhomogeneity increases to the same extent as the amount of fluid passing through the bed as bubbles, the latter being in excess to the gas flowrate that generates the fluidized state of the solid particles. The degree of the nonhomogeneity may thus be expressed by the ratio between the velocities of the two fluid flows by the equation:

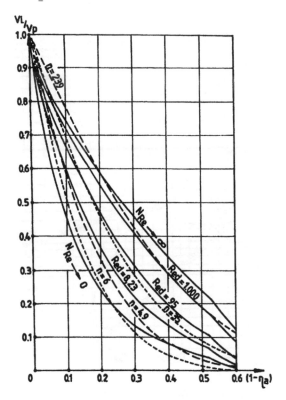

Figure 8.11 Comparison of the equations deduced by Raseev [15] with the exponential Eqs. (8.32).

$$IH = \frac{v_t - v_1}{v_1} \tag{8.35a}$$

where:

v_t is the fluid velocity in the free board of the bed (superficial velocity), v_1 is the fluid superficial velocity corresponding to the homogeneous fluidization (see section 8.1.1.)

IH is the degree of the bed nonhomogeneity, which may vary from 0 for the homogeneous fluidization to a maximum value near 1.

For fluidization in equipment of small diameter, the bubbles may produce pulsations and even slugging phenomena.

It is obvious that the efficiency of the solid/fluid contacting and thus the efficiency of the process carried out in fluidized bed decreases with the increase of nonhomogeneity. Ongoing efforts are searching for ways to increase the homogeneity of the fluidized beds.

Since the densities of the two phases are physical properties of the participants in the process, their ratio is not a parameter that can be acted on for improving the quality of the fluidization.

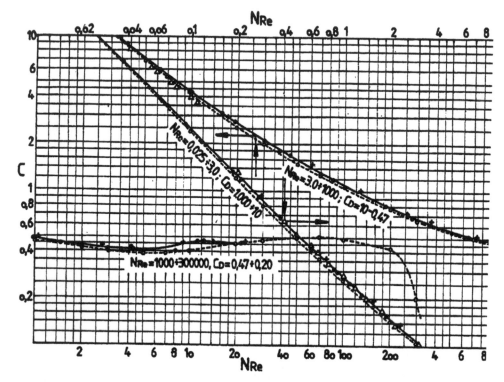

Figure 8.12 Correlation between the flow resistance coefficient C and Reynolds number. (From Ref. 33.)

Reducing the sizes of the solid particles is an efficient means of diminishing heterogeneity of the fluidization (see Figure 8.1). By reducing the size of the particles, their terminal velocity decreases sensibly and the increase of the diameter of the reactor becomes necessary in order to accommodate the design production capacity. This measure is limited however by economic considerations. The use of mixtures of particles of various sizes (see Figure 8.1) makes it possible to a certain extent to reduce the heterogeneity of the fluidization.

Accordingly, the size and the size distribution of the fluidized solid particles is selected depending on the type of process without however using particles with diameters in the range of 40–60 μ.

Another important factor in the operation of fluidized beds is the entrainment from the bed of small particles. These are usually recovered by one or more cyclone stages and recycled to the bed.

Of great importance is also the uniform distribution of the fluid producing the fluidization over the cross section of the reactor. One of the solutions to this problem is to use a distributor made of a perforated plate what also supports the catalyst bed in slumped state. The cross section of all the openings in the perforated plate does not exceed a few percent of the cross section of the reactor. This measure ensures a pressure drop of 0.1–0.2 bar through the distributor. The velocity of the fluid through the various orifices is no more influenced by local unevenness or local fluctuations of the density of the fluidized bed.

Practice teaches that one has to use a certain minimum ratio between the height of the fluidized bed and the diameter of the apparatus. Since the actual values of these sizes depend upon the nature of the process, there is no general manner in which they can be reported here.

The up and down moving solid particles also carry with them significant amounts of fluid, especially that which constitutes the boundary layer formed around the solid particles. Owing to this effect the dense phase fluidized bed reactors approach perfect mixing conditions. The design of such reactors follows the methods developed for the reactors with ideal mixing.

Such condition has a marked negative impact on the maximum yields of the intermediary product obtained in successive-parallel reactions (see Section 6.4.3).

A specific result of the backmixing is the maintaining of a uniform temperature inside the fluidized bed. The differences between different points of the bed don't exceed 1–2°C. Even in process with important thermal effects, the reactions take place in practically isothermal conditions.

As mentioned, the intense mixing within the fluidized bed has a negative effect on the yield obtained in the process. In some applications, in order to reduce the scale of the backmixing, the height of the fluidized bed is divided by means of several horizontal baffles that act as redistributors of the fluidizing fluid. The linear velocity in the orifices of these plates prevent the return of the solid particles to the bed below the plate, from which they were entrained. Such reactors are no longer isothermal. In each zone located between two perforated plates the prevailing temperature is determined by the thermal effect of the reactions taking place in the respective zone. Such a reactor is similar in fact to a series of perfect mixed reactors and can be modeled as such.

Much more efficient is the riser system—a vertical tube wherein the reaction takes place, the solid particles being transported upwards by the fluid. In such a system backmixing is almost completely eliminated. The thermal conditions in a riser are those in an adiabatic reactor.

At the lower part of the riser, the solid is introduced directly into the ascendant flow of the fluid, and in the upper end the riser discharges into a vessel of a larger diameter, usually provided with cyclones, for achieving the separation solid/fluid.

8.2 FLUIDIZATION WITH SOLIDS CIRCULATION

As it is illustrated in Figure 6.19, the velocity profile in the median section of the fluidized particle is a function of the relative velocity between fluid and particle and reaches the maximum value v_p, at the distance δ from the particle wall.

In order that the velocity profile and the bed expansion remain the same when the particle possesses the velocity v_g, the relative velocity between fluid and particle must remain unchanged.

Relative to a fixed reference system, the value of the fluid velocity is:

$$v_{1c} = v_1 \pm v_g \tag{8.36}$$

where the subscript c indicates that the velocity refers to a solids circulation system and is measured relative to a fixed reference system (for example the wall of the apparatus in which the fluidization takes place). The plus sign corresponds to the ascending circulation of the particles; it is minus for their descent circulation countercurrent to the fluid.

Noting that $\bar{v}_g$ is the average velocity of the fluidized particles, Eq. (8.36) may be written:

$$\frac{v_{1c}}{v_p} \mp \frac{v_g}{v_p} = \frac{v_1}{v_p} = (\mathscr{R}, \eta_a) \tag{8.37}$$

where the function on the right-hand side of the equality represents the equations that express v_1/v_p, for various batch fluidization conditions.

In order to apply Eq. (8.37) it is necessary to express $\bar{v}_g$ as a function of measurable parameters.

The amount of solid that passes upwards through an arbitrary cross section f_t in unit time can be expressed as:

$$G_s = f_t \cdot \bar{v}_g (1 - \eta_a) \gamma_a \tag{8.38}$$

Expressing the weight circulation ratio solid/fluid by the equation:

$$\frac{G_s}{G_f} = a \tag{8.39}$$

and the fluid weight flow by the equation:

$$G_f = f_t \cdot v_{1c} \cdot \gamma_0 \tag{8.40}$$

Equation (8.39) becomes:

$$\bar{v}_g (1 - \eta_a) \gamma_a = a \cdot v_{1c} \cdot \gamma_0$$

or

$$\bar{v}_g = \frac{a \cdot v_{1c} \cdot \gamma_0}{(1 - \eta_a) \gamma_a} \tag{8.41}$$

Introducing this result in (8.37), after simplifications it follows:

$$\frac{v_{1c}}{v_p} \left(1 - \frac{a \cdot \gamma_0}{(1 - \eta_a) \gamma_a} \right) = \frac{v_1}{v_p} \tag{8.42}$$

If instead of expressing the circulation ratio by Eq. (8.39) one uses the volume circulation ratio:

$$\frac{D_s}{D_f} = a_v \tag{8.43}$$

Equation (8.42) becomes:

$$\frac{v_{1c}}{v_p} \left(1 - \frac{a_v}{1 - \eta_a} \right) = \frac{v_1}{v_p} \tag{8.44}$$

For the case of the downward circulation of the solids in countercurrent with the fluid, the sign inside the parenthesis on the left-hand side of Eqs. (8.42) and (8.44) will be +.

The graph of Figure 8.13 was developed on the basis of Eq. (8.44) and allows the direct determination of the necessary fluid velocity v_{1c}/v_p on the basis of the values a_v and $(1 - \eta_a)$.

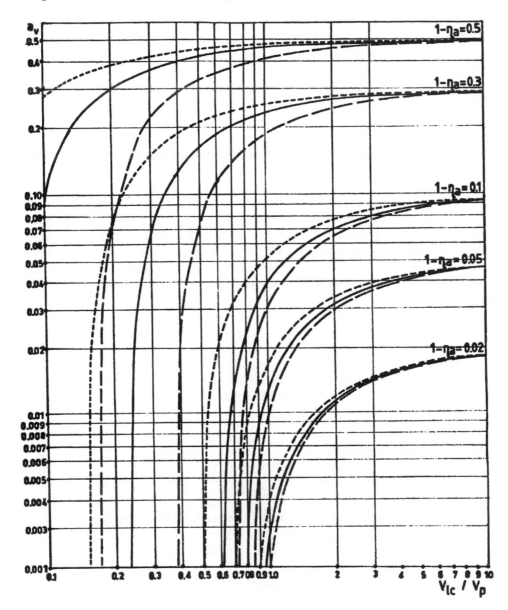

Figure 8.13 Plot of Eq. (8.44), fluidization with particles in ascendent circulation. - - - Re = 0.05, ——— Re = 10, – – – Re = 300.

One must remark that when selecting the value $(1 - \eta_a)$ one should take into account not only the efficiency of the circulation, but also the total pressure drop in the transport line and the riser respectively. This pressure drop may be considered equal to the hydrostatic pressure difference, while the pressure drop due to the friction of the walls may be neglected [52].

The hydrostatic pressure drop is a function of the height of the riser and the value of $(1 - \eta_a)$, the latter being the only parameter that one can vary. In order to reduce the pressure drop to values acceptable for the process technology, the parameter $(1 - \eta_a)$ may be decreased. For this reason the riser and the ascending transport lines of the catalytic cracking units operate at low solid concentration and small values of $(1 - \eta_a)$ respectively.

The curves of Figure 8.13 show that, at low solids concentration, the fluid velocity relative to the vessel walls, v_{1c} may be several times larger than the terminal velocity (it is the very domain of operation of the risers and pneumatic transport).

Related to this, one should note that the use of Eq. (8.44) for computing the term v_1/v_p for obtaining $f(\mathscr{R}, \eta_a)$ in Figure 8.8 does not present sufficient accuracy for the domain of low solids concentration. Therefore, the use of Eq. (8.17) combined with (8.21) is recommended.

The phenomena that take place in the riser are very complex: pressure and temperature decrease, the chemical reactions are accompanied by volume increases, and coke deposits are formed on the catalyst particles. Equation (8.44) should be used exclusively for orientation calculations or for a qualitative understanding and analysis of the phenomena. For design purposes, it is recommended to use data obtained from existing operating units without neglecting to incorporate improvements. (Section 7.3).

It must be mentioned that in addition to the pneumatic transport presented above in the catalytic cracking units, the catalyst may be transported under the effect of the difference in the hydrostatic pressure on the two branches of a U-shaped fluidized bed (the Model IV unit is a typical example for this method). The injection of fluid along the transport line has the exclusive role of maintaining the fluidized state. In these conditions, the fluidized bed behaves as a liquid that flows as result of a difference in hydrostatic pressure.

8.3 REACTION SYSTEMS

As indicated in the previous chapter, two reaction systems are used in fluid catalytic cracking units:

> Reactors with catalyst in dense phase, characterized by intensive backmixing of
>   the catalyst and the reactants
> Reactors of the riser type, with negligible backmixing

8.3.1 Reactors With Catalyst in Dense Phase

The typical example for this system is the reactor of the Model IV unit, which is the most advanced implementation of the dense phase reactor.

Characteristic of these reactors is the very intense backmixing of the two phases (catalyst and vapors).

The mixing intensity for the catalyst phase is proportional to the fluid velocity and the diameter squared of the vessel, so that it is practically complete even in tubes of 25 mm diameter [2]. The achievement of perfect mixing was confirmed by studies in industrial catalytic cracking reactors by following the flow patterns of catalyst particles tagged with radioactive isotopes [3].

The backmixing of the vapor phase in tubes of small diameter (25–75 mm) and bed heights of 1–2 m is insignificant [4]. The intensity of the mixing increases with the square or even with the third power of the diameter [2]. Gas samples were taken in several points in industrial catalytic cracking reactors having diameters of the order of 12 m and bed heights of 4–5 m. The compositions were identical, which proved the perfect mixing within the vapor phase [5–7].

Recent studies [8] attempted to limit the extent of backmixing by use of various devices introduced in the reactor. A decrease by 75–80% of backmixing was obtained.

The mixing of the two phases has as result a uniform temperature in the fluidized bed of the industrial reactors.

Based on the above facts, it would seem that the design of a catalytic cracking reactor in dense phase would present no difficulties, considering it an isothermal reactor with perfect mixing in which the activity of the catalyst and the composition of the reactant phase in the fluidized bed are those corresponding to the outlet from the bed. Such calculation does not lead to correct results since it neglects the non-homogeneity of the fluidized bed, the formation of gas bubbles containing very little catalyst, etc. The effect of the presence of the bubbles is equivalent to a portion of the reactants bypassing the reaction zone, or at least to a shorter residence time for a large portion of the feed.

Gross et al. [9] attempted to perform kinetic calculations and to design fluidized bed reactors by applying coefficients to the rate constants used in the kinetic equations deduced for fixed bed reactors. However, since the kinetic equations describing the reactions in a fixed bed (quasi–plug flow) reactor are quite different from those for a fluid bed the result obtained were not correct.

More adequate equations were deduced by Weekmann and Nace [35] and Wojciechowski and Corma [36]. These kinetic problems and the proposed equations were presented in Section 6.4.3.5 for dense phase and for riser reactors.

This type of calculation can be used for order of magnitude evaluations. For design applications, it is safe to base the sizing on practical data collected from existing commercial reactors, taking into account the following:

> The velocity of the vapors in the reactor free board is set at a value that will not exceed the admissable flowrate of catalyst entrained into the cyclones. Thus the diameter of the reactor, the investment, and the catalysts inventory are minimized.
>
> The height of the catalyst bed and the reaction temperature are set for reaching the desired conversion, while allowing for some reserve in the catalyst level.

The height of the catalyst bed is generally 4–6 m.

The graph of Figure 8.14 gives the density of the fluidized bed as a function of the superficial velocity of the vapors for several densities of the settled bed (or of the bed at the minimum fluidization velocity). The graph is valid for microspherical catalyst and for normal fluidization in industrial reactors. In such reactors, the fluidized bed contains vapor bubbles of moderate size, while large bubblers are absent. It is mentioned [10] that deviations from this graph denote a maldistribution of the vapors within the bed and the presence of large bubbles.

In order to distinguish between the fluid that passes through the bed as bubbles and that which actually causes the fluidization, it is necessary to compare the data of Figure 8.14 for industrial reactors, with that for homogeneous fluidization using Figure 8.8 and Eq. (8.42).

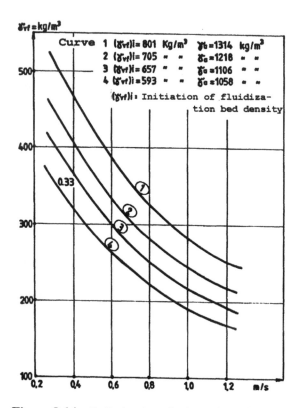

Figure 8.14 Bulk density of microspherical catalyst as function of the vapors superficial velocity. $1 - \gamma_a = 1314\,\text{kg/m}^3$; $2 - \gamma_a = 1218\,\text{kg/m}^3$; $3 - \gamma_a = 1106\,\text{kg/m}^3$; $4 - \gamma_a = 1058\,\text{kg/m}^3$. (From Ref. 10.)

For homogeneous fluidization, the use of the graph from the Figure 8.8 requires knowledge of the Reynolds number. This was already calculated in Section 6.4.1.1 and has the following values: for the outlet from the reactor Re = 11.2 and for the inlet Re = 94–170. Taking into account the almost perfect mixing prevailing in the dense phase reactors, the value Re = 20 is accepted.

For the two cases: of normal operation in commercial unit (Figure 8.14) and perfectly homogeneous fluidization (Figure 8.8), the ratio between the variation of superficial vapors velocity and bed density can be obtained by graphic or numerical differentiation.

For the calculation vapors velocity over the bed of 0.3; 0.4; 0.8 and 1.2 m/s was used, which cover the dense phase reactors operation.

For normal operation in a commercial unit, the calculations for Figure 8.14 are collected in Table 8.1.

For the γ_{vf}/γ_a values of Table 8.1, the graph of Figure 8.8 for Re = 20, allows one to determine by graphic differentiation, the values, corresponding to the homogeneous batch fluidization (Table 8.2, column 4). By using Eq. (8.42), the values for homogeneous continuous fluidization (Table 8.2, column 4) were computed.

The comparison of these values with those of the last column of Table 8.1, allows one to establish (last column of Table 8.2) the fraction of the fluid flowrate

Table 8.1 Calculation by Using Curve 2 from the Graph of Figure 8.14

$$\left(\gamma_{vf} - \text{col. 2 and } \frac{dv_1}{d\gamma_{vf}} - \text{col. 4 are determined by using the graph}\right)$$

Fluid velocity in the free section v_{1c}, m/s	Bed density γ_{vf}, kg/m^3	$\dfrac{\gamma_{vf}}{\gamma_a}$	$\dfrac{dv_1}{d\gamma_{vf}}, \dfrac{\text{m/s}}{\text{kg/m}^3}$ *	$\dfrac{dv_1}{d\left(\dfrac{\gamma_{vf}}{\gamma_a}\right)}, m/s$
0.3	448	0.368	$-2.03 \cdot 10^{-3}$	-2.47
0.4	405	0.333	$-2.52 \cdot 10^{-3}$	-3.07
0.8	283	0.232	$-4.49 \cdot 10^{-3}$	-5.47
1.2	220	0.180	$-8.28 \cdot 10^{-3}$	-10.08

*Minus sign resulted from graphical differentiation. The bed density decreases as fluid velocity increases.

that performs the fluidization. The balance to 100% represents the vapors flowrate that passes through the bed as small and partially large bubbles.

Despite the fact that this is an estimate, one can observe that in the domain of the fluid flowrates of 0.3–0.8 m/s the contact between the vapors and the catalyst is maintained in basically constant conditions. The participation of the vapors in the actual fluidization decreases as the flow velocities reach values of the order of 1.0–1.2 m/s, which are close to the conditions for entraining the catalyst out of the reactor.

The hydrocarbons contained in the vapors that pass through the bed as bubbles are not completely protected from catalytic transformations. The vapor bubbles contain catalyst particles, and between the bubbles and the dense phase there is a continuous mass transfer [16] so that the vapors inside the bubbles are also undergoing catalytic reactions. The smaller the bubbles, the more extensive the degree of the reactions of the vapors within. This stresses the importance of the design of the vapors distributor in the reactor.

8.3.2 Riser Reactors

As it was described previously, the reactors of this type are close to the ideal plug flow reactors, with ascendant cocurrent flow of the vapor feed and catalyst.

Measurements by means of radioactive tracers were used [17] to determine the radial and axial dispersion coefficients in the riser of an industrial unit. The conclu-

Table 8.2 Percentage of Fluid Flow Performing the Fluidization

$\dfrac{\gamma_{vf}}{\gamma_a}$	$\dfrac{dv_1}{d\left(\dfrac{\gamma_{vf}}{\gamma_a}\right)}, m/s$	$1 - \dfrac{a\gamma_0}{\gamma_{vf}}$	$\dfrac{dv_{1c}}{d\left(\dfrac{\gamma_{vf}}{\gamma_a}\right)}, m/s$	Fluid performing fluidization (%)
0.368	-0.900	0.900	-1.000	40.5
0.333	-1.071	0.889	-1.205	39.2
0.232	-1.704	0.841	-2.026	37.0
0.180	-1.947	0.795	-2.447	24.3

sion of the study is that the radial dispersion is very small; the axial flow contains returning currents causing backmixing, the importance of which must be taken into account in exact calculations.

The effects in the riser of the lower catalyst velocity relative to that of the vapors in the riser (slipping factor) and the backmixing justifies the suggestion [18] to design reactors with descending circulation of both phases. This will completely eliminate the returning currents and will result in the catalyst operating at an average activity of the catalyst, which is higher than for the upflow riser, thus improving the performances of the unit.

The design of the riser requires the simultaneous solution of several problems connected to the complexity of the phenomena that occur.

In modern units, the contact of the heavy feeds with the hot catalyst leads to their vaporization, which is often completed after cracking reactions started taking place. The produced vapors accelerate the ascending motion of the catalyst particles so that the complete vaporization and the stabilization of the movement of the two phases is achieved only after a certain length of the riser, estimated generally at 6–12 m [10].

The cracking reactions lead to an important increase of the number of moles, which represents a continuous increase of the volume flowrate of the vapor phase in the riser, with the concomitant acceleration of the catalyst particles and with the decrease of the solid phase concentration. Along the riser, a simultaneous decrease of the pressure takes place, mainly as a result of the hydrostatic pressure decrease and frictional losses. In the same time, the temperature decreases due to the endothermal character of the reactions. Besides, the cracking reactions lead to an increase of the dynamic viscosity of the vapors, as a result of the decrease of the molecular mass and of the Reynolds number.

The finding of an exact calculation requires taking into consideration all these complex phenomena, correlated with the kinetic equations of the catalytic cracking of the feed and expressing them by a system of differential equations. Their integration for the boundary conditions inside the riser will provide the solution to the problem.

For approximate evaluations it is useful to know some average values of the parameters involved in commercial operation of riser reactions [19]:

Fluid density	1.12–1.76 kg/m^3
Fluid viscosity	0.018–0.027 cP
Apparent density of the catalyst	1120–1760 kg/m^3
Catalyst particle size	55–70 μ

For a heavy feed, such as the residual feed, with $M = 350$–400, the viscosity value is smaller, namely 0.005 cP at 550°C.

One of the main problems with these calculations is the need to know the concentration and the instantaneous catalyst inventory in the riser. The suggested equation for estimating this parameter is [10]:

$$\Delta H = \frac{D_{\text{cat}} \cdot (\text{SF}) \cdot \Delta l}{v_1}$$

where:

ΔH is the catalyst inventory in kg present in the length Δl of the riser
Δl being expressed in meters
D_{cat} is the flowrate of catalyst circulating through the system in kg/s
v_1 is the superficial fluid velocity in the riser, in m/s
(SF) is the slip factor expressed by the equation:

$$(SF) = v_1/v_g \tag{8.46}$$

v_g being the velocity of the catalyst particles.

A more exact determination of the solids concentration in the riser may be made by means of the graph of Figure 8.15, which was drawn mostly on basis of experimental data [19].

The average value of the slip factor over the entire length of the riser is esti-mated at 1.5 [10], while at the outlet from the riser, the value is 1.1. These values correspond to the situation when the equilibrium state was reached between the fluid velocity and the velocity of the solid particles, i.e., after their acceleration ended. Thus, it becomes necessary to know the point in the riser or the moment when the equilibrium was reached. Beyond this point, the slip velocity may be determined by the indicated methods.

The initial acceleration of the regenerated catalyst that is descending from the regenerator is obtained by means of steam injection at the basis of the riser (Figure 7.26).

Considering that the pressure at the injection point is 1.6 bar and the tempera-ture of the superheated steam at the contact point with the regenerated catalyst reaches 950 K, the characteristics of the steam will be:

$\gamma_0 = 0.328 \, \text{kg/m}^3$
$\mu = 0.275 \cdot 10^{-5} \, \text{kg} \cdot \text{s/m}^2$ calculated using the Sutherland and Sudakov method

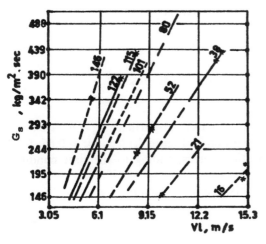

Figure 8.15 Particles concentration in the riser in kg/m^3. Dotted lines—interpolated data. (From Ref. 19.)

Accepting the characteristics $d = 70\,\mu$ and $\gamma_s = 1700\,kg/m^3$ for the microspheres of catalyst and a steam velocity $v_1 = 5\,m/s$ (Figure 8.15), by accepting the first of the Eqs. (8.33) it follows:

$$v_p = 0.545 \frac{(70 \times 10^{-6})^2}{0.275 \cdot 10^{-5} \cdot 9.81} \cdot 1700 = 0.170\,m/s \tag{8.47}$$

The Reynolds number is:

$$Re = \frac{5.0 \times 70 \times 10^{-6} \times 0.328}{0.275 \times 10^{-5} \times 9.81} = 4.25 \tag{8.48}$$

And according to Figure 8.12, $C = 6.2$.

The ascending force exercised by the fluid on a microsphere is given by the relation:

$$F = \pi \cdot r^2 \cdot C \frac{(v_1 - v_g)^2}{2g} \gamma_0 \tag{8.49}$$

where the term within the parentheses represents the difference between the velocity of the fluid and that of the microsphere.

This force will produce an acceleration, correlated with the force by the equation:

$$F = -m \frac{d(v_{1c} - v_g)}{d\tau} \tag{8.50}$$

where the mass of the microsphere is expressed by the equation:

$$m = \frac{4}{3} \pi r^3 \frac{\gamma_s}{g} \tag{8.51}$$

Substituting in (8.50):

$$F = \frac{4}{3} \pi r^3 \frac{\gamma_s}{g} \frac{d(v_{1c} - v_g)}{d\tau} \tag{8.52}$$

Equating with (8.49) and separating the variables yields:

$$0.375 \frac{C\gamma_0}{r\gamma_s} d\tau = \frac{d(v_{1c} - v_g)}{(v_{1c} - v_g)^2} \tag{8.53}$$

Integrating this equation with the boundary conditions:

for $\tau = 0$ $v_g = 0$ $v_{1c} - v_g = v_1$

and for τ corresponding to a stabilized movement fluid/particle: $v_{1c} - v_g = 0.65\,v_p$.

The coefficient 0.65 results from the graph of Figure 8.13 $v_{1c}/v_p = 0.65$, for Re = 4.25 and $(1 - \eta_a) = 0.122$, concentration corresponding to the riser condition (Figure 8.15).

The integration of Eq. (8.53) with this boundary condition gives:

$$0.375 \frac{C\gamma_0}{r\gamma_s} \tau = \frac{1}{v_{1c} - v_g} - \frac{1}{v_1} = \frac{1}{0.65\,v_p} - \frac{1}{v_1} \tag{8.54}$$

By substituting the numerical values, and performing the calculations it follows:

$$\tau = 1.38\,\text{s}$$

This means that the acceleration of the catalyst particles is ended and movement of steam/catalyst particles is stabilized after 1.38 s or at a point $1.38 \times 5 = 6.8\,\text{m}$ higher than the mixing point between the steam and the regenerated catalyst.

If the first contact of the regenerated catalyst at the entrance in the riser is with feed and not with steam, accepting the data of Jeszyeri [19], only γ_0 changes and becomes:

$$\gamma_0 = 1.5\,\text{kg/m}^3$$

Equation (8.47) gives: Re = 19.4 and according to Figure 8.12, $C = 1.2$. The v_p value (Eq. 8.47) remains the same; $v_p = 0.170\,\text{m/s}$. Equation (8.53) becomes:

$$5.67\,d\,\tau = d(v_1 - v_g)/(v_1 - v_g)^2 \tag{8.55}$$

After integration within the same limits:

$$5.67\,\tau = 1/0.65 v_p - 1/v_1$$

Substituting v_p with (8.47) and $v_1 = 5\,\text{m/s}$, one obtains:

$$\tau = 1.56\,\text{s}$$

which corresponds to a height in the riser of $1.56 \times 5 = 7.8\,\text{m}$.

In the choice of the superficial velocity, v_1, it is necessary to assume that it will be well above the choke velocity. The choke velocity is the minimum superficial velocity below which the riser will not remain in dilute phase transport. Below this velocity, the catalyst/vapor mixture collapses abruptly into a dense phase and the riser pressure drop shows a sudden increase.

The choke velocity can be estimated using the equation [40]:

$$\frac{V_{ch}}{\sqrt{g \cdot d}} = 9.07 \left(\frac{\gamma_a}{\gamma_0}\right)^{0.347} \left(\frac{Wdg}{\mu}\right)^{0.214} \left(\frac{d}{D_R}\right)^{0.246} \tag{8.56}$$

where:

V_{ch} = choke velocity
g = gravitational constant
d = particle diameter
D_R = riser diameter
W = solid mass velocity, $\text{lb/ft}^2 \cdot \text{s}$ or $\text{kg/m}^2\text{s}$, the rest of the notations, as before, in consistent units.

Since all the expressions in parenthesis are dimensionless, the values can be expressed in American units as in the original work, or in metric units.

It must be remarked that the phenomena that take place in the riser are much more complex than a simple entrainment of catalyst microspheres by the fluid.

In a first portion the contact of the steam with the regenerated catalyst takes place, accompanied by the temperature equalization and by the increase of the velocity of the catalyst particles. The injection of the feed follows (or two injections at different heights of the feed and of the recycled material) accompanied by their vaporization in a portion of the riser. The cracking reactions that take place are accompanied by the increase of volume and by the decrease of the viscosity.

For these reasons, Eq. (8.54) serves only for estimates and not for performing exact calculations.

Such estimates can be easily performed by using the velocity and the conversion variations for a typical riser given in Figure 8.16.

The specific feature of the kinetic design of the riser is that the catalyst and the reactants are flowing in cocurrent. The cocurrent flow is influenced to a certain extent by the slippage phenomenon.

An exact calculation of the riser is to a large extent similar to that of a coil of a pryolysis furnace, i.e., it involves the simultaneous solution of a system of differential equations that expresses the thermal, momentum, and chemical transformation phenomena taking place. As a catalytic process the following must be taken into account when determining the reaction rates: (1) the decrease of the catalyst activity as a result of coke deposits, and (2) the decrease of the catalyst concentration as a result of the increase of the number of moles and consequently of the velocity of the fluid. Backmixing must also be taken into account in calculations.

For simplification, the riser is divided in elements, short enough so that the conditions within each element can be represented by the average between those at the two ends of the element. Within each element, the calculations use the average between the inlet and outlet conditions. This is a trial and error method easily carried out on a computer.

A more simple calculation method uses as an independent variable the amount of catalyst within the riser, instead of the residence time or of the length of the riser

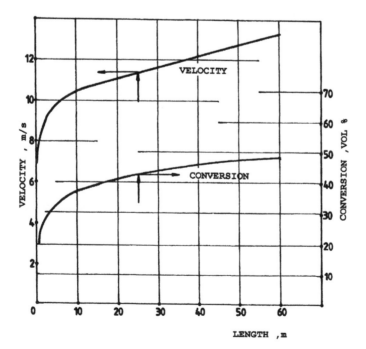

Figure 8.16 Variation of conversion and velocity in a typical riser. (From Ref. 10.)

corresponding to it. Such a treatment, proposed by Fogler [21] was applied to catalytic cracking by Gray [22].

In this treatment the decomposition rate of the feed is expressed by the equation:

$$F_{a0} \frac{dx}{dW} = a(-r_a) \tag{8.57}$$

where:

F_{a0} = feed in kg/s
x = conversion
W = amount of catalyst contained in the riser in kg
a = activity of the catalyst
r_a = the reaction rate in kg/kg catalyst × s.

For the reaction rate, the authors accept the suggestion of Weekman [23] to use an equation of the second order, written for a homogeneous system, as a quantitative expression of the decomposition rate of a complex feed. The feed is made of chemical components that have various cracking rates.

It follows:

$$-r_a = k \cdot C_a^2 \tag{8.58}$$

where the instantaneous concentration of the reactant C_a, may be expressed as a function of the initial concentration, C_{a0}, by the equation:

$$C_a = C_{a0}(1 - x) \tag{8.59}$$

both being expressed in kg/m^3.

Expressing the decrease of catalyst activity by an equation of type (6.47), where the decrease of the activity is a function of time and expressing the residence time of the catalyst in the riser, τ, by the ratio W/m_{cat}, where m_{cat} is the catalyst flow rate in kg/s, it results:

$$a = e^{-\alpha_t W/m_{cat}} \tag{8.60}$$

a being the rate constant for the catalyst deactivation.

Substitution of Eqs. (8.58–8.60) into (8.57) and separating the variables gives:

$$\frac{dx}{(1-x)^2} = \frac{kC_{a0}^2}{F_{a0}} e^{-\alpha_t W/m_{cat}} dW \tag{8.61}$$

Integrating this expression between the limits $W = 0$ and $x = 0$ at the inlet to the riser and W and x at outlet, it results:

$$\frac{x}{1-x} = \frac{kC_{a0}^2}{F_{a0}} \cdot \frac{m_{cat}}{\alpha_t} (1 - e^{-\alpha_t W/m_{cat}}) \tag{8.62}$$

Since the process in the riser is adiabatic, the rate constants in this equation are for the temperature corresponding to the equivalent adiabatic mean rate (Eqs. 2.163 and 2.164).

The riser outlet temperature necessary for this calculation is determined by means of the heat balance for the riser, which accounts for the enthalpies of the incoming and outgoing streams and for the heat of reaction.

The calculation method using Eq. (8.58) may serve for the kinetic modeling of the riser in which more complex kinetic equations are used.

An additional element that intervenes in the calculation is the pressure drop in the riser. The main components of pressure drop in a riser are: the solids static head, wall friction, exit loss, and solids acceleration. Of these, the solids static head is by far the most important component. The solids holdup can be obtained directly from Figure 8.15 [19].

As compared to the term for the static head, wall friction in the risers of commercial catalytic crackers is often negligible. The actual acceleration and exit losses are dependent on the system design.

Accurate riser pressure drop calculation based on the studies of Knowlton on pneumatic transport are given in the monograph of Wilson [40].

Other papers [38,39] have analyzed the hydrodynamics of riser units and its impact on cracker operation. The conclusion is that the riser reactors may be described by bidimensional models. The vertical flow of the gases and catalyst involves no backmixing, while there are radial profiles of velocity and concentration. A single radial dispersion coefficient accounts for the gas behavior over the whole cross section of the riser.

The authors estimate that in this process the mixing behavior of the catalyst is not important since the catalyst reaches very quickly its final coke content and activity.

Analyses of samples collected from commercial units confirmed the predictions of the model. The model allows one to estimate a 3 wt % loss of conversion and of gasoline yield when the riser is compared to a plug-flow reactor. The quality of the products is not significantly affected.

8.4 CATALYST REGENERATION

For the regenerators in dense phase, a good fluidization quality corresponds to the correlation between the fluid velocity and the apparent density of the fluidized bed depicted in the graph of Figure 8.17 [10]. Data points that do not correspond to this graph denote poor fluidization quality, which corresponds to an increase of the residual coke and excessive carryover of the catalyst from the bed, having as consequence a shortening of the operating life of the cyclones.

By comparing the data obtained from this graph with those for the same bulk densities given in the graph of Figure 8.8, which is valid for conditions of homogeneous fluidization, it is possible to evaluate the flowrate of air, and flue gases, that pass through the bed as bubbles or microbubbles. A similar evaluation was used in Section 8.3.1 for the reactor in dense phase.

Here too, average conditions may be selected, as represented by curve 2 in the graph of Figure 8.17 ($\gamma_a = 1218\,kg/m^3$).

For gas superficial velocities of 0.4, 0.8, and 1.2 m/s, by applying the same methodology as it was used in the case of the reactors and using the graph of Figure 8.17, one obtains the data presented in Table 8.3.

The fraction of the total fluid flow that participates in the homogeneous fluidization may be determined by using, as in the case of the reactor, the graph of Figure 8.8 and the Eq. (8.37).

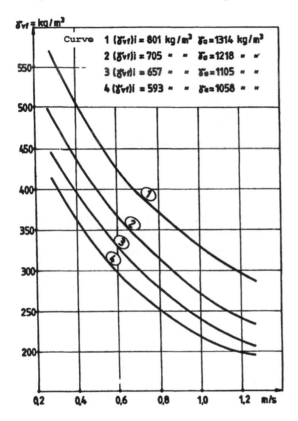

Figure 8.17 Bulk density of microspherical catalyst in regenerator as function of superficial velocity of the flue gases. $1 - \gamma_a = 1314\,\text{kg/m}^3$; $2 - \gamma_a = 1218\,\text{kg/m}^3$; $3 - \gamma_a = 1106\,\text{kg/m}^3$; $4 - \gamma_a = 1058\,\text{kg/m}^3$. (From Ref. 10.)

Taking into account that for regenerators Re = 2 (Section 6.4.1.2) and using the same computing methodology as for the reactors, the data in Table 8.4 are obtained.

In the third column of this table, γ_0 represents the density of the air for each of the flue gases, which may be estimated at: $\gamma = 0.457\,\text{kg/m}^3$ (see the Section 6.4.1.2).

The catalysts/air contacting ratio a results from the amount of coke that must be burnt and the amount of air required for the combustion. For a catalyst/feed ratio of 5, accepted earlier, and for 3.5 wt % formed coke, 7 g coke/kg of catalyst should be burnt in the regenerator. This will require approximately 90 g air, from which a contact ratio $a = 11$ results. This value used in the calculations given in Table 8.4.

The values shown in the last column of the table allow one to estimate that in the case of the regenerators in dense phase, a larger fraction of the fluid passes through the bed as microbubbles or bubbles than in the case of the reactors.

The modeling of coke combustion in the regenerator and the kinetic modeling of the regeneration are similar to those for the reaction systems using the elements given in Section 6.6.

The successive steps of this calculation are: (1) the mass balance of the coke combustion is written on basis of the hydrogen content of the coke and the composi-

Table 8.3 Calculation Using the Graph of Figure 8.17

Air velocity in the free section v_1, m/s	Bed density Fig. 8.8 γ_{vf}, kg/m^3	$\dfrac{\gamma_{vf}}{\gamma_a}$	$\dfrac{dv_1}{d\gamma_{vf}}$, $\dfrac{\text{m/s}}{\text{kg/m}^3}$*	$\dfrac{dv_1}{d\left(\dfrac{\gamma_{vf}}{\gamma_a}\right)}$, m/s
0.4	0.437	0.359	$-2.68 \cdot 10^{-3}$	-3.27
0.8	0.315	0.259	$-4.30 \cdot 10^{-3}$	-5.24
1.2	0.242	0.199	$-6.55 \cdot 10^{-3}$	-10.40

*Minus sign resulted from graphic differentiation. The bed density decreases as fluid velocity increases.

tion of the flue gas; (2) the thermal effect of the combustion is computed and used for writing the heat balance of the regeneration from which results the amount of heat that must be possibly removed by heat exchange, (3) the kinetic modeling of the regeneration and the design of the regenerator.

With reference to the removal of heat from the regenerator by means of cooling coils located within the fluidized bed, a typical value for the heat transfer coefficient is $1226 \, \text{kJ/m}^2 \cdot \text{hour} \cdot °\text{C}$. This coefficient is frequently exceeded.

Calculations for the basic design of a regenerator are presented by Walas [25]. The treament of the kinetics of coke combustion is of interest.

On the basis of previous publications [26], which showed that external diffusion has a determinant role in the regeneration process of the catalytic cracking catalyst, Walas equates the reaction rate with the rate of external diffusion, from which it results:

$$r = k_r C_r p_i = k_D(p - p_i) \tag{8.63}$$

where:

k_r and k_D are the reaction rate constant, and the rate constant respectively for external diffusion
C_r is the weight fraction of the residual coke
p_i is the partial pressure of oxygen at the surface of the catalyst particles
p is the oxygen partial in the air stream.

The two rate constants are expressed by the equations:

Table 8.4 Percentage of Fluid Flow Performing the Fluidization

$\dfrac{\gamma_{vf}}{\gamma_a}$	$\dfrac{dv_1}{d\left(\dfrac{\gamma_{vf}}{\gamma_a}\right)}$, m/s^a	$1 - \dfrac{a\gamma_0}{\gamma_{vf}}$	$\dfrac{dv_{1c}}{d\left(\dfrac{\gamma_{vf}}{\gamma_a}\right)}$, m/s	Fluid performing the fluidization (%)
0.359	-0.715	0.987	-0.724	22.1
0.259	-1.117	0.983	-1.136	21.6
0.199	-1.747	0.977	01.788	17.2

^{a}Minus sign resulted from graphical differentiation. The bed density decreases as fluid velocity increases.

$$\log k_r = 10.3 - \frac{7930}{T} \tag{8.64}$$

$$k_D = \frac{G^2}{d_p^{1.5}} \cdot 2.87 \cdot 10^{-4} \tag{8.65}$$

where:

T is the absolute temperature in K
G is the regeneration air flow in $kg/m^2 \cdot hour$
d_p is the diameter of the catalyst particles in microns.

Using the data from the example of Walas: $T = 839\,K$, $G = 71.7\,kg/m^2 \cdot hour$, and $d_p = 44.5\,\mu$ and the Eqs. (8.64) and (8.65), one obtains:

$k_r = 7.04, k_D = 0.00495$

from which, using the Eq. (8.63), it results:

$r = 0.00073\,kmol/hour \cdot kg\,cat$

$p_i = 0.05\,bar$

The value of p_i proves that indeed the external diffusion is the slowest phenomenon.

An equation that correlates the operating data of industrial regenerators and is recommended by several authors [10,22–25] is:

$$C_r = \frac{A \cdot D \cdot T \left[-\ln\left(\frac{O_{2fin}}{21}\right) \right]}{k \cdot p \cdot G_c} \tag{8.66}$$

where:

C_r = residual coke in wt %
A = constant obtained from the operating data of the unit
D = the air flowrate in m^3/min
T = the regeneration temperature in K
O_{2fin} = the molar percentage of oxygen in the flue gases
k = the rate constant in s^{-1} obtained from the graph of Figure 8.18
p = the absolute pressure in kg/cm^2
G_c = the catalyst weight in the regenerators in tons

Introducing to this equation the operating data of an existent unit, the constant A is determined, which makes possible to compute how modifications to some of the operating variables will influence the residual coke. In a similar manner, the effects may be determined that would be obtained in a different unit.

8.5　CATALYST ENTRAINMENT

8.5.1　Disengagement Height

The catalyst entrainment from the fluidized bed is a phenomenon common to the reactor as well as to the regenerator. It is treated here for both.

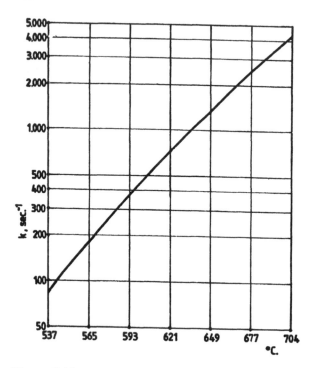

Figure 8.18   Rate constant k – Eq. (8.66).

The gases or the vapors that leave the dense phase entrain a considerable amount of catalyst above the fluidized bed, largely due to the gas bubbles that break at the surface of the bed. A large portion of the catalyst entrained in this way falls back, but some remains entrained. Above a certain height above the bed, called total *disengagement height* (TDH) the concentration of the catalyst contained in the fluid remains constant. TDH can be determined approximately using the graph of Figure 8.19 [10].

The amount of catalyst entrained above the TDH is needed for the design of the cyclones. For the conditions in the regenerator it may be determined by using the nomogram of Figure 8.20.

The nomogram may be used also for determining the amount of catalyst entrained above the "disengagement height" of the reactors. In this case, the value obtained for the entrained amount must be divided by 2, since the viscosity of the fluid is approximately half that in the regenerator [10].

When using the graphs of Figures 8.19 and 8.20, the effective superficial velocities of the gases must be used, derived by dividing the volume flowrates to the free sectional area at the level of TDH. Therefore, from the total cross section of the vessel, one must subtract the cross section occupied by the legs of the cyclones.

There exist calculation methods of the particles content at the inlet in the first stage cyclones, as follows:

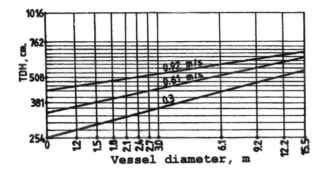

Figure 8.19 Disengagement height (TDH) as function of vessel diameter. (From Ref. 10.)

1. For a given gas and particle density, the entrained amount of each of the catalyst particle diameters is determined by using the nomogram.
2. The entrained amounts are multiplied by the weight fractions of the catalyst of the respective size within the dense phase, for obtaining the amount of catalyst that will be entrained in cyclones.
3. The sum of these amounts, equals the total entrained amount.

As indicated previously, the amount of entrained catalyst can be much higher if the distribution of the fluid velocities through the fluidized bed is not uniform over the cross section of the vessel. The preferential channels formed within the bed by the fluid lead to entrained amounts lead are much larger than those estimated. The consequence is a rapid wear of the cyclones.

8.5.2 Cyclones System

The design of cyclones is a classical chemical engineering problem covered in specialized textbooks [10,31–33]. In the following, only aspects that are specific to the cyclones systems used in the reactors and the regenerators of the fluidized bed catalyst cracking units will be discussed.

For the recovery of the catalyst, entrained in the upper part of the reactor or the regenerator, two and sometimes three cyclone stages are used. A system with two stages is shown in Figure 8.21 and the characteristic parameters in Figure 8.22.

The pressure difference $p_1 - p_2$ (Figure 8.22) corresponds to the difference between the hydrostatic pressures within the bed (dense phase). The difference $p_3 - p_2$ corresponds to the dilute phase above the fluidized bed.

The flow through the cyclone is produced by a pressure drop. Thus, $p_5 < p_3$ and $p_7 < p_5$. The pressures p_5 and p_7 are lower than the pressure p_1 in the fluidized bed. In order to provide a hydrostatic balance to these pressure differences, the catalyst levels L_{D1} and L_{D2} (Figure 8.22) are maintained in the legs of the respective cyclones.

A sufficient distance should exist between the upper level of the dense phase in the cyclone legs and their lower ends, in order to buffer fluctuations that might appear during normal operation. Upsets in operating conditions may lead to increased pressure drops through the cyclones over the design values. As result,

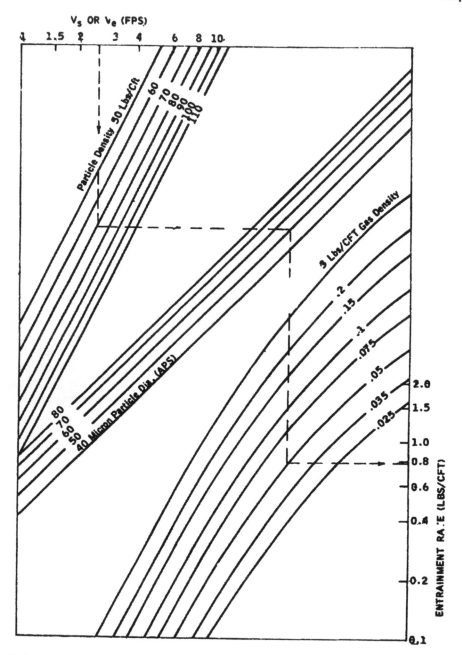

Figure 8.20 Nomogram for catalyst entrained function of the disengagement height. (Courtesy Emtral Corp). (From Ref. 10.)

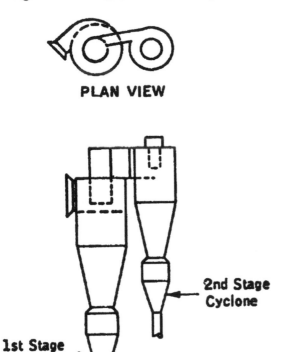

PLAN VIEW

2nd Stage
Cyclone

1st Stage
Cyclone

Figure 8.21 Two-stage cyclone.

the pressure differences $p_1 - p_5$ and $p_1 - p_7$ increase and may lead to "flooding" (complete filling) of the cyclone legs by a dense bed of catalyst. Since this will produce excessive entrainment and losses of catalyst, the unit throughput must be reduced.

Besides issues concerning the pressure balances discussed above, another important factor is the efficiency of the cyclone system, especially in view the high cost of the catalyst. Without presenting here the details of the efficiency of the calculations, given elsewhere [31], it suffices to say that it is preferred to use larger cyclones than smaller ones operated in parallel. The latter lead to catalyst losses of 20–50% larger than the former ones and to higher maintenance costs.

8.6 CATALYST CIRCULATION, TRANSPORT LINES

The main issues of catalyst circulation between the reactor and the regenerator were discussed in previous chapters. Thus, thermal balance problems that determine the contact ratio and the correlation of the catalyst circulation with the pressure regimes within fluidized bed systems were discussed in Section 5.3.5. Although it referred to thermal cracking, the referred to discussion can be adapted easily to the specifics of catalytic cracking.

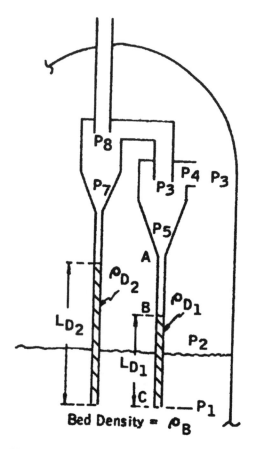

Figure 8.22   Characteristic parameters in two-stage cyclones.

Some specific remarks are needed with reference to the dense phase transport in descending pipes, namely the calculation of the amount of "aeration" fluid necessary for assuring correct transport conditions, Figure 8.23.

The amount of the aeration fluid may be determined by means of the empirical Eq. [10]:

$$W_A = \frac{20.42 \cdot e \cdot G_{cat} \cdot \Delta p}{T} \qquad (8.67)$$

where:

W_A = amount of steam injected in the descending pipe, in kg/h per meter downcomer length

G_{cat} = flowrate of circulating catalyst in t/h

Δp = pressure drop in the pipe in kg/m^2 per meter of downcomer length and is equal to the bulk density of the catalyst in the pipe expressed in kg/m^3

T = the temperature in K

e = free volume in the line expressed in m^3/kg catalyst, calculated by means of equation:

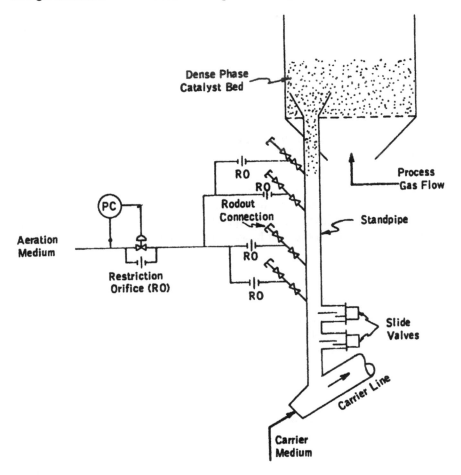

Figure 8.23   Catalyst transport in descending pipes.

$$e = \frac{1}{\rho_f} - \frac{1}{\rho_s}$$

where ρ_f and ρ_s are the density of the fluid and the catalyst respectively.

The correct control of the aeration steam is very important, since on one side, it must maintain within the line a dense phase of suitable fluidity that ensures the correct flow of the catalyst, while on the other side it should not exceed the strict necessary amount. Excessive steam could lead to the formation of steam bubbles, which, by their ascending tendency, might decrease or even block the downwards flow of catalyst [10,34].

The circulation between the reactor and the regenerator is regulated by means of special valves situated in the descending pipes, called control valves. In order to ensure a fine control of the catalyst flowrate generally for each pipe, two valves in series are provided.

The sizing of these valves may be done by using the equation [10]:

$$A = 5.07 \cdot G_c \cdot C \sqrt{\rho \cdot \Delta P} \tag{8.68}$$

where:

A = free section of the valve in cm^2

G_c = circulation rate of the catalyst in t/h

C = flow coefficient, between 0.85 and 0.95

ρ = the density of the flowing stream in kg/m^3

ΔP = pressure drop in kg/cm^2.

A recent study [41] deduces the equation for the nonfluidized flow of solids through pipes. This flow is broken in three areas (see Figure 8.24).

The first is for solids flowing into a pipe in nonhindered flow (Figure 8.24A). As an example is the flow from the dust bowl of a nonflooded cyclone, described [41] by equation:

$$W_{NH} = 1.75 \rho_B D^{1/2} \tag{8.69}$$

The second type is the hindered (friction) flow of solids in a pipe (Figure 8.24B). An example of this is the nonfluidized flow of solids from hoppers or from a flooded cyclone bowl. For this flow, the deduced equation is:

$$W_{HIN} = 0.768 \rho_B D^{1/2} \tag{8.70}$$

The third type is the transport of solids in a pipe with no restrictions (Figure 8.24C). The equation is:

$$W_{core} = 6.22 \rho_B D^{1/2} \tag{8.71}$$

In Eqs. (8.69–8.71): ρ_B is the solids bulk density in lb/ft^3 and D is the pipe diameter in inches.

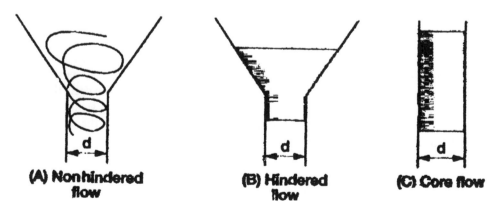

(A) Nonhindered flow   (B) Hindered flow   (C) Core flow

Figure 8.24 The three types of nonfluidized dense flow of solids.

REFERENCES

1. S Raseev. Procese distmctive de prelucrare a titeiulur, Ed. Tehnica, Bucarest, 1964.
2. S Stemerding. Koninklijke Shell Lab. Amsterdam. In: SM Walas Reaction Kinetics for Chemical Engineers. New York: McGraw-Hill, 1959.
3. E Singer, DB Todd, VP Guinn. Ind Eng Chem 49: 11, 1957.
4. ER Gilliland, EA Mason. Ind Eng Chem 41: 1191, 1949.
5. JW Askins, GP Hinds, F Kunrenther. Chem Eng Prog 47: 401, 1951.
6. PV Danckwerts, JW Jenkins, G Place. Chem Eng Sci 3: 26, 1954.
7. AE Handles, RW Kunstman, DO Schissler. Ind Eng Chem 49: 25, 1957.
8. G Papa, FA Zenz. Hydrocarbon Processing 74: 81, Jan. 1995.
9. B Gross, DM Nace, SE Voltz. Ind Eng Chem Process Des Develop 13(3): 199, 1974.
10. EG Luckenbach, AC Worley, AD Reichle, EM Gladrow. Cracking Catalytic. In: JJ McKetta, ed. Petroleum Processing Handbook. New York: Marcel Dekker Inc., 1992.
11. S Raseev. Studii si cercetari de chimie ale Academiei Romane 5: 569, 1957; 5: 581, 1957; 6: 295, 1958.
12. S Raseev. Buletinul Institutului de patrol si Gaze 3: 117, 1957; 4: 121, 1958; 4: 153, 1958.
13. S Raseev. Revue de Chimie de l'Academie Roumaine 7: 439, 1962.
14. S Raseev, K Clemens. Buletinul Institutului de Petrol, Gaze si Geologie 8: 137, 1962.
15. S Raseev. PhD disertation, Influence of the boundary layer in fluidization. University of Bucarest, 1972.
16. JF Davidson, D Harrison. Fluidization. New York: Academic Press, 1971.
17. PJ Viitanen. Ind Eng Chem 32: 577, 1993.
18. JR Murphy. Oil and Gas Journal 90: 49, 18 May 1992.
19. B Jazayeri. Hydrocarbon Processing 70: 93, May 1991.
20. EN Sudakov. Spravotshnik osnownyh protsesow i aparatow neftepererabotky, Himia, Moscow, 1979.
21. HS Fogler. Elements of Chemical Reactors Design. Englewood Cliffs, NJ: Prentice-Hall, 1986.
22. MR Gray. Upgrading Petroleum Residues and Heavy Oils. New York: Marcel Dekker Inc., 1994.
23. VW Weekman. Lumps, Models and Kinetics in Practice. AIChE Monograph Series 11: 75, 1979.
24. TE Johnson. Hydrocarbon Processing 70: 55, Nov. 1991.
25. SM Walas. Reaction Kinetics for Chemical Engineers. New York: McGraw-Hill, 1959.
26. J Pansing. J Am Inst Chem Eng 2: 71, 1956.
27. PB Weisz, RB Goodwin. J Catal 6: 227, 1966.
28. S Tone, S Miura, T Otake. Bulletin of Japan Petr Instit 14(1): 76, May 1972.
29. MF Johnson, HC Maryland. Ind Eng Chem 47(1): 127, 1955.
30. A Tesoreiro. Hydrocarbon Processing 63: 739, Nov. 1984.
31. RB Ewell, G Gadmer, WJ Turk. Hydrocarbon Processing 60: 103, Sep. 1981.
32. T Oroveanu. Transferul de impuls si aplicatii. In: G Suciu, ed. Ingineria prelucrarii hidrocarburilor. Vol 2. Tehnica, Bucarest, 1985.
33. J Ciborowski. Inzyneria Chemiczna. Translation, edited in Russian by PG Romancov, Gosteh Himizdat, Leningrad, 1958.
34. JM Matsen. Powder Technol 7: 93, 1973.
35. VW Weekmann, DM Nace. AIChE Journal 16: 397, 1970.
36. BW Wojciechowski, A Corma. Catalytic Cracking. New York: Marcel Dekker Inc., 1986.
37. FA Zenz. Fluidization. FF Othmer, ed. London: Reinhold Publishing Co., 1956.
38. C Derouin, D Nevicato, M Foressier, G Wild, JR Bernard. Hydrodynamics of Riser Units and their Impact on FCC Operation. Ind Eng Chem Res 36: 4504–4515, 1997.

39. L Godfrey, GS Patience, J Chaouki. Radial Hydrodynamics in Riser. Ind Eng Chem 38: 81–89, 1999.
40. JW Wilson. Fluid Catalytic Cracking Technology and Operations. Tulsa, Oklahoma: PennWell Publishing Co., 1995, 122–124.
41. PG Talavera. Calculate nonfluidizated flow in cyclone diplegs and transition pipes. Hydrocarbon Processing 74: 89–92, Dec. 1995.
42. TS Martens, HB Rhodes. Chem Eng Progr 51: 429, 517, 1995.
43. Kadymova. Neft Hoz 3: 24, 1953.
44. IM Razumov. Psevdoojyjenie I pnevmotransport syputshih materialov. Himia, Moscow, 1964.
45. SK Ghosal, RN Mukhejea. Chem Eng Tech 42: 81(2), 1970.
46. G Suciu, V Schor. Revista de Chimie 9: 482, 1958.
47. J Happel. AJCh Journal 4: 197, 1958.
48. RJ Jottrand. Appl Chem 2, Suppl I, (17): London, 1952.
49. E Johnson. Inst Gas Engrs, Publ 378/179, Rept. 1949–1950.
50. M Leva, M Weintraub, M Gummer, M Pollichik. Ind Eng Chem 41: 1206, 1949.

9

Other Processes on Acid Catalysts

This chapter examines two processes: the oligomerization of alkenes C_3-C_5 and the alkylation of iso-alkanes with alkenes. Both processes are used for producing high-octane components used in the production of superior gasoline.

Both processes were developed during the Second World War to fill the need for high-octane aviation gasoline.

The alkylation involves higher investments and operation costs but produces saturated high-octane components in amounts that are twice those obtained per unit olefin feed by means of oligomerization. Besides, the use of oligomerization product is limited by the maximum content of alkenes accepted in the gasoline, unless the oligomers are hydrogenated. After hydrogenation, the product is almost identical with the alkylate. For this reason the oligomerization + hydrogenation process is sometimes called *indirect alkylation*.

The feedstocks for this process derive mainly from catalytic cracking, which generates important amounts of C_3-C_5 hydrocarbons, with the prevalence of the iso-structures. Additional amounts of iso-alkanes derive from hydrocracking while alkenes derive from thermal cracking processes.

The development of alkylation and oligomerization occurred simultaneously with the expansion of catalytic cracking. The economics of this process imposed the use of large amounts of the C_3-C_4 fraction, that were obtained with the addition of gasoline (the fraction C_4 by itself represents about 30% of the gasoline).

The development of catalytic reforming increased the number of methods for obtaining high octane gasoline and for many countries, e.g. in Western Europe, provided a more economical way for its production. For many years, the excess of high-octane gasoline required by the growth of automotive transport was ensured by catalytic reforming of heavy gasoline.

The situation changed as result of the increasing attention paid to the protection of the environment, which has to be taken into account when formulating gasoline blends.

Adding tetra-ethyl lead was first limited and then eliminated in most countries. This lead to the increase of the octane rating of clear gasoline, which increased interest in catalytic processes. The rate at which the usage of ethyl lead was eliminated depended directly on the rate of construction of new catalytic cracking and alkylation units.

The toxic effects of the nitrogen oxides formed during combustion are enhanced by their reaction in sun light with some of the hydrocarbons (for example the isoamylenes) to form ozone, which even in minimal amounts is very noxious. The toxic effect is increased also by benzene, which is carcinogenic, and in general by the excess of aromatic hydrocarbons.

Beginning of 1990, when the Clean Air Act Amendment (CAAA) became law in the U.S., a number of measures became compulsory. They include: the inclusion of oxygenated compounds in the gasoline composition, so that the oxygen represents at least 2% by weight; the decrease of the benzene content to a maximum of 1%, and that of the content of aromatic hydrocarbon to a maximum of 25%, of alkenes to 10%; a decrease of vapor pressure. Other industrialized countries have similar regulations.

The oxygenated compounds incorporated in the gasoline to assure 2% oxygen content were especially the methyl-tert-butyl-ether (MTBE), to a lesser extent the higher ethers ethyl-tert-butyl ether (ETBE) and tert-amyl-methyl-ether (TAME), and only in Brazil, ethyl alcohol. The MTBE has a blending octane number $F_1 = 118$, $F_2 = 102$ and a content of 18.2 wt % O_2.

After 1990, the addition of MTBE to the high octane number gasoline was standard in the U.S. and in the other developed countries.

Beginning in late 1998, the use of MTBE started to be criticized due to the large amounts of CO_2 emitted during the production of the raw materials for MTBE from methane and butane [78]. Also, several cases of MTBE leaking from underground storage tanks into the drinking water systems were reported.

Several states are seeking to ban the use of MTBE and in March 2000 the Clinton Administration proposed to replace the 2% oxygenates requirement by a renewable fuel standard for the whole gasoline pool [75].

Without addressing here issues of blending in the production of finished gasoline, a topic covered elsewhere [1–4,75], important issues related to the content of this book are:

1. The increasing limitations on the content of benzene and of aromatic hydrocarbons in gasoline and progressive elimination of MTBE cause alkylation and oligomerization to have a high priority in the improvement of the gasoline.

2. The limitation of alkenes content makes it necessary to hydrogenate the oligomerization products.

3. The high vapor tension of isoamylenes (0.97–1.86 bar) and their contribution to the formation of ozone are reasons to use them as precursors for alkylation and TAME production.

9.1 OLIGOMERIZATION

Oligomerization converts the C_3–C_4 alkenes produced in thermal or catalytic cracking to C_6–C_8 iso-alkenes used as octane enhancing components in the production of high octane gasoline.

9.1.1 Thermodynamics of the Process

The dimerization of the C_3–C_4 alkenes if favored by low temperatures and high pressures. In Figure 9.1 the conditions are plotted for achieving an equilibrium conversion of 99% of propene is *iso*-butene. The graph takes into account the dimers, which are obtained in the process in high proportions. Before being dimerized, the *n*-butenes are isomerized to a large extent. For this reason they were not included in this graph.

From Figure 9.1 shows that at temperatures below 210–230°C and pressures of the order of 30–40 bar, the equilibrium conversion of the C_3–C_4 alkenes is essentially complete (99%). Since in industrial conditions the C_3–C_4 fractions submitted to the process contain an important proportion of alkanes, the equilibrium is less favorable than shown by the graph.

Thermodynamic calculations for the dimerization of *iso*-butene [5] indicate that at temperatures ranging between 25–190°C the preferred isomers produced are trimethyl-pentenes. Their content in the dimerized product can reach 90%. Another conclusion of this study, which is confirmed also by the representation of the propene dimerization equilibrium given in Figure 1.3, is that the high pressures used in industrial processes are determined mainly by the requirements to reduce the size of the reactor, rather than by thermodynamic considerations.

9.1.2 Oligomerization Catalysts

The classical catalyst used in the oligomerization of the alkenes is phosphoric acid.

The range of concentrations used in the industrial process is comprised between 72.4% P_2O_5, which corresponds to ortho-phosphoric acid H_3PO_4, and 79.7% P_2O_5, which corresponds to pyro-phosphoric acid $H_4P_2O_7$.

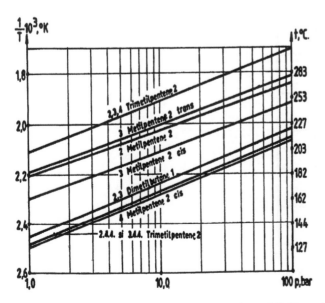

Figure 9.1 The 99% equilibrium conversion of C_3H_6 and *i*-C_4H_8 to dimers.

Actually, in both cases one uses an equilibrium mixture of acids, as is generally the case in the family of phosphoric acids. Thus, the pyro-phosphoric acid represents in fact, a mixture of 14% ortho-phosphoric acid, 38% pyro-phosphoric acid, 23% tri-, 13% tetra-, 7% penta-, 2% hexa-, 1% hepta and traces of octa-phosphoric acids.

Sometimes the concentration of the used acid is expressed in percentage of ortho-phosphoric acid. This manner is not recommended, since it leads to percentages that exceed 100%. Thus the pyro-phosphoric acid will be expressed as an ortho-phosphoric acid having a concentration of 110%.

The ortho-phosphoric acid has a density of 1870 kg/m^3, a fusion temperature of 42.3°C, and a boiling temperature of 255.3°C while the pyro-phosphoric acid has a fusion temperature of 61°C and a boiling temperature of 427°C. The Hammett acidity [6,7] is −4.66 for the ortho-phosphoric acid and −5.72 for the pyro-phosphoric acid.

The catalytic activity of the phosphoric acid increases with concentration. The variation of the reaction rate constant with the concentration expressed in % H_3PO_4 is reproduced in the Figure 9.2.

The optimal concentration for performing the oligomerization ranges between 108 and 110% H_3PO_4; at lower concentration the catalytic activity decreases strongly while at higher concentration high polymers are formed. They have a resinous structure and block the active sites of the catalyst.

The desired concentration of the phosphoric acid is obtained by maintaining an adequate partial pressure of steam within the reactor. This partial pressure is deter-

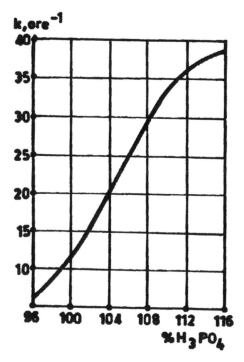

Figure 9.2   Variation of the reaction rate constant with H_3PO_4 concentration.

mined as a function of the temperature inside the reactor by using the graph of Figure 9.3.

The phosphoric acid catalyst is used as tablets obtained from a Kiesselguhr powder, impregnated with H_3PO_4, calcined at 300–400°C (the UOP process), or as a film of phosphoric acid deposited on quartz sand.

In the first case the phosphoric acid is partially combined chemically with silica, partly physically adsorbed. The composition of the catalyst corresponds approximately to the chemical formula $P_2O_3 \cdot SiO_2 \cdot 2H_2O$.

In the second case, quartz sand with particle sizes of 0.7–0.9 mm is used; smaller sizes would increase too much the pressure drop through the catalyst bed. The advantage of this system is the easy regeneration of the catalyst by extraction of the phosphoric acid with water, burning of the resin deposits on the support, and re-impregnation with acid. The disadvantage is the large losses of phosphoric acid. In both cases, the specific surface of the catalyst is of 2–4 m^2/g.

New oligomerization processes were developed that use as catalyst high acidity zeolites such as ZSM-5 [8–10]. Also, catalysis based on natural zeolite catalysts was tested, such as ion exchanged mordenites [11]. The zeolite catalysts are less active than those based on phosphoric acid. Higher reaction temperatures are needed, which leads to more rapid formation of resin deposits and leads to the need for more frequent regeneration of the catalyst.

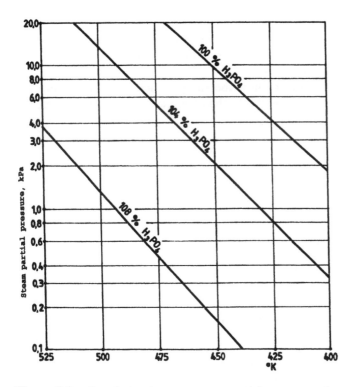

Figure 9.3 Correlation between steam partial pressure and temperature for various H_3PO_4 concentrations.

A quite different approach is the use of organometallic catalysts in processes with either homogeneous or heterogeneous catalysts. In the latter case, the organo-metallic compound is deposited on supports of alumina or of alumosilica.

Beginning in 1979, Angelescu et al. studied the dimerization of ethene on a complex organometallic $Ni(C_5H_7O_2)_2-AlCl(C_2H_5)_2-P(C_6H_5)_3$ catalyst in homogeneous and heterogeneous catalytic systems [12–14]. In the preferred system the catalyst was deposited on a metallic oxide support.

The French Petroleum Institute used as catalysts nickel organometallic compounds in a homogeneous liquid phase system [15–19]. These studies led to the development of the Dimersol process.

Oligomerization of *n*-butenes on the ferrite surface of a solid catalyst was also tested [93].

9.1.3 Reaction Mechanisms

For phosphoric acid catalysts, the reaction mechanism involves the intermediate formation of carbenium ions on the protonic Brönsted sites.

This mechanism is presented for the dimerization of isobutene in Structure 1. The dimerization reaction of propene takes place in a similar way (see Structure 2).

The reactions of the (b) type proceed by similar interactions of the formed ions, producing trimers and higher polymers. These are successive reactions, in which the yield of the polymer depends on the duration of the contact with the catalyst.

The heats of the ionization reactions and the heats of formation of the carbenium ions were given in Tables 6.4 and 6.5. The data indicate that on the acid sites of the catalyst, the formation of the tertiary ions is preferred. In the absence of tertiary carbon, secondary ions result. Thus, by polymerization of alkenes on acid catalysts, the product has exclusively iso structure.

The data of the cited tables also show the energy efficiency of the chain isomerization of the *sec*-butyl ion to *iso*-butyl. In the oligomerization of *n*-butenes, this

Structure 1

a) $CH_3-CH=CH_2 + HA \longrightarrow \left[CH_3-\overset{+}{C}H-CH_3\right] A^-$

b) $\left[CH_3-\overset{+}{C}H-CH_3\right] A^- + CH_2=CH-CH_3 \longrightarrow \left[\begin{array}{c} CH_3 \\ | \\ CH_3-CH-CH_2-\overset{+}{C}H-CH_3 \end{array}\right] A^-$

c) $\left[\begin{array}{c} CH_3 \\ | \\ CH_3-CH-CH_2-\overset{+}{C}H-CH_3 \end{array}\right] A^-$

$\xrightarrow{\sim 80\%} \begin{array}{c} CH_3 \\ | \\ CH_3-CH-CH_2-CH=CH_2 \end{array} + HA$

$\xrightarrow{\sim 20\%} \begin{array}{c} CH_3-C-CH=CH-CH_3 \\ | \\ CH_3 \end{array} + HA$

Structure 2

reaction may occur before an interaction of the (b) type. This contributes also to the final formation of the highly branched iso structures [93].

The resinous deposits produced as secondary products on phosphoric acid catalysts are favored by higher concentrations of P_2O_5 and by higher temperatures.

Owing to the lower activity of the zeolite catalysts, the process requires higher temperatures, which increase the secondary reactions, including those of hydrogen transfer that lead to the formation of aromatic and saturated hydrocarbons. These reactions also cause intense formation of deposits on the surface of the catalysts. The development of zeolites with strong acidity improved their overall performance.

The active species that occur in oligomerization by homogeneous catalysis on nickel complexes are considered [18] to have also ionic structure, such as:

$$[L_n - NiH]^+A^-$$

where A is an anion derived from the second component of the catalytic system, and L_n may be constituted of various molecules present (monomer, dimer, or solvent).

For such a structure of the active species, the mechanism of the oligomerization reactions is analogous to that on phosphoric acids. The difference between the two catalysts is that while the nickel complexes can catalyze the dimerization of ethene in conditions, which allowed the development of an industrial process, this was not possible for phosphoric acid catalysts [12–14].

9.1.4 The Kinetics of the Process

The rate of the oligomerization process is determined mainly by the external diffusion [20], which makes the reaction close to first order. The reaction rate is proportional to the partial pressure of the alkenes. This finding is confirmed by numerous papers [21] that recommend for process calculations the use of a first order kinetic equation written for a homogeneous process.

The determining role of diffusion is due to the fact that in reaction conditions the polymers are in liquid phase, whereas the propene and the butenes are in the gas phase, since the reaction temperature exceeds their critical temperature. In these conditions, in order to reach the active sites of the catalyst, the reactants have to diffuse through a film of reaction products. Further detailed studies of the kinetics of oligomerization on alumosilica catalysts [22] and on transition metals [23] have considered the process as heterogeneous, the access of the feed to the catalyst surface being controlled by Langmuir adsorption. The results were compared to experimental data obtained in conditions where the reaction rate is not influenced by diffusion.

The first of these papers studied the dimerization of *iso*-amylenes, when both the reactants and the reaction products were in liquid phase. By making some approximations based on the assumption that the adsorption of the dimmer is strong while that of the monomer is weak, the following expression was obtained:

$$\frac{k}{w} = \frac{x}{1-x} + x + 2\ln(1-x) \tag{9.1}$$

where:

k = rate constant
w = space velocity
x = conversion

Another study [24] suggests the following empirical equation for the dimerization of propene on a catalyst of phosphoric acid supported as a film, on quartz particles:

$$\frac{k}{w} = \int_0^x \frac{(1-Bx)^2}{(1-x)^2 + 0.3x(1-x)} dx \tag{9.2}$$

where

$$B = y \cdot \frac{M_M}{M_\rho} \tag{9.3}$$

and

y = molar fraction of monomer in the feed
x = monomer conversion in molar fraction
M_M and M_P = the molecular mass of the monomer and of the polymer.

In Eq. (9.2) w must be expressed in volumes of gaseous feed per hour at the pressure and the temperature of the reactor reported to the void volume of the catalyst bed, which is 42% of the total volume.

The variation with temperature of the rate constant k given by Eq. (9.2) is plotted in the graph of Figure 9.4, while the effect of the size of the quartz particles on the rate constant is given in the graph of Figure 9.5. The concentration of the phosphoric acid may be determined by means of the graphs of Figure 9.2, knowing that the value of the rate constant as determined by means of Figure 9.4 corresponds to an acid concentration of 105% H_2PO_4.

In view of the reaction mechanism, oligomerization should be treated as a successive process for which the conditions that lead to maximum yields of dimer, trimer etc. are deduced. In order that such a treatment be of practical use, the values for the kinetic parameters for each of the reaction stages should be known. Presently, this is not the case.

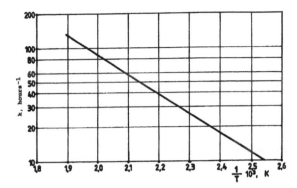

Figure 9.4 Variation with temperature of the rate constant k (Eq. 9.2).

9.1.5 Effects of Process Conditions

The effect of process conditions will be examined in the following for the classic, most used phosphoric acid process. For other catalysts, the data needed for such an analysis is presently not available.

9.1.5.1 Temperature

The industrial oligomerization process on phosphoric acid use the temperature range 175–245°C. Since diffusion is a controlling factor for the overall reaction rate, the activation energy is relatively small, i.e. between 21 and 31 kJ/mol. For this reason the reaction rate increases only 3–5 times when the temperature increases from 175–245°C.

At temperatures below 175°C, the reaction rate becomes too slow and below 130°C the polymerization of propene does not take place any more while phosphoric ethers are formed.

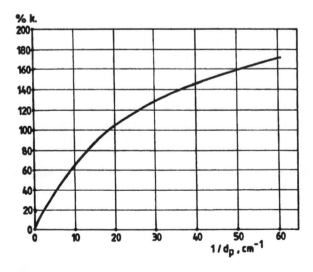

Figure 9.5 Influence of the size of quartz particles on the rate constant (Eq. 9.2).

At higher temperatures the decomposition of the carbenium ions is accelerated and the polymerizate becomes lighter. In the same time, secondary reactions of hydride ion transfer occur, which lead to dienes, aromatic hydrocarbons, and finally to catalyst coking. Simultaneously, the losses of phosphoric acid increase. For all these reasons it is recommended that the temperature in the commercial process should not exceed 205–220°C.

The composition of the deposits formed on the catalyst surface was studied by dissolving them in carbon tetrachloride at high temperature and by the use of infrared spectroscopy [25]. Besides the insoluble carbon deposits left on the catalyst, the presence of aromatic hydrocarbons including polynuclear aromatics, resins, and asphaltenes also were confirmed. The proportion of the latter reached 10–12%.

9.1.5.2 Pressure

The reaction temperatures exceed the critical temperatures for the C_3–C_4 hydrocarbons contained in the feed. Therefore, they will be in vapor phase, regardless of the system pressure.

Since the feed usually contains about 50% propane and butanes, the hexenes will be found partly in liquid phase only if the pressure is higher than 40 bar. If the pressure is lower than 40 bar, the amount of liquid phase in the reactor is small and the liquid polymer will be colorless, since the resinous substances remain on the surface of the catalyst.

At pressures of the order of 60 bar, most of the polymerization products will be in liquid phase and will dissolve the resinous substances. The catalyst life will be 2–3 times longer than at lower pressure, but the polymer will be discolored and will have to be redistilled.

At a constant conversion, the degree of polymerization is not influenced by the pressure increase.

9.1.5.3 Feedstock

The data of Tables 6.4 and 6.5 indicate that the formation of the ethyl ion takes place at much lower rates than of the C_3 and C_4 ions. For this reason the fraction C_2 is not included in the feed to the oligomerization units. The participation of ethene to these reactions would require temperatures above 250°C, where the intense formation of resinous compounds would make the process not economical.

The relative reaction rate for the C_3–C_4 alkenes increases in the order: C_3H_6, n-C_4H_8, i-C_4H_8 and has the value: 1,2,10. However, in mixture, the i-butene participates in copolymerization reactions and increases the rate of conversion rate of propene and n-butenes.

The much larger dimerization rate of i-butene makes it possible to achieve a selective reaction by operating at low temperatures:

$$
\underset{\underset{C}{|}}{\overset{+}{C-C}}-C + C=\underset{\underset{C}{|}}{C}-C \longrightarrow C-\underset{\underset{C}{|}}{C}-C-\underset{\underset{C}{|}}{\overset{+}{C}}-C
$$

even in the presence of propene and of *n*-butenes. This is in fact the reaction by which, following hydrogenation, one produces 2,2,4-trimethylpentane, which is used as the standard for determining the octane number.

In order to maintain the concentration of phosphoric acid at the required level, the feed must contain $(3.5–4.0) \cdot 10^{-2}\%$ water. This moisture content corresponds to the solubility of water in liquid $C_3–C_4$ alkenes at 20–25°C and can be easily obtained by contacting the feed stream with water in these conditions.

The presence of hydrogen sulfide in the feed leads to an undesired presence of mercaptans in the polymerization gasoline. The removal of H_2S from the feed must be made in conditions that ensure the complete absence of traces of basic compounds in the $C_3–C_4$ fraction, which would deactivate the catalyst by the neutralization of its acidity. Also, oxygen must be completely absent from the feed since it intensifies the formation of resinous substances.

9.1.5.4 Space Velocity

The operation in conditions of the complete conversion of the alkenes leads to the increase of the extent of the secondary reactions and to a less efficient process. For this reason, the space velocity is controlled so as not to exceed a conversion of about 90%. A function of the operating conditions, the practiced space velocity is in the range 1.7–4 hours^{-1}.

9.1.6 Oligomerization on Phosphoric Acid Catalyst

The activity of the catalyst made by calcining phosphoric acid supported on kieselguhr in the oligomerization reaction was discovered in the 1920s by V. N. Ipatieff. The commercial process was developed with his participation by UOP and was given the name of *Catalytic Condensation Process* [26].

Units for the oligomerization of gases produced in thermal cracking were commercialized in 1935. The oligomerization of the alkenes on phosphoric acid supported on kieselguhr was thus the first catalytic process used in the processing of crude oil. In fact, it is the most common oligomerization process; over 200 units were in operation in 1985, with processing capacities ranging between 21.6 and 2010 $m^3/24$ hours.

The UOP process was developed in two versions, differing by the way in which one removes the heat of the strongly exothermic process.

1. Reactors of shell and tube type (usually two reactors in series, with cooling in between), the catalyst being placed in tubes. Boiling water is circulating in the shell.
2. Reactors with the catalyst divided in layers. The cooling is achieved by introduction of cooling liquid between the catalyst layers. The coolants are usually propane or *n*-butane recycled from the separation section of the unit.

The multiple-bed reactors require lower investment and catalyst replacement is easier than for multitubular reactors. However, the consumption of catalyst and utilities is higher.

An advantage of the multiple-bed reactors, which at one time could determine their selection, was the possibility to perform in the same unit and on the same catalyst, the alkylation of benzene with propene or with dodecene [27]. In the first case isopropyl-benzene (cumene) was obtained, which was used in the production of phenol and acetone. In the second case, dodecyl-benzene was produced, used for the fabrication of detergents.

The performances of the process depend to a large extent on the pressure inside the reactor, the modern processes using pressures of the order 40–60 bar (see Table 9.1).

The scheme of a polymerization plant with multiple-bed reactor, able to perform either the oligomerization of gasoline or the synthesis of cumene or dodecyl-benzene is shown in Figure 9.6.

The polymerization gasoline produced by the processing of the C_3-C_4 fraction obtained from the catalytic cracking has the following mean characteristics [26]:

Density at 15°C		0.738
Distillation ASTM	10% vol	85°C
Distillation ASTM	50% vol	132°C
Distillation ASTM	10% vol	193°C
Research octane number (RON) clear		95
Motor octane number (MON) clear		83

The blending research octane number is situated between the limits 106–126, the measuring method depending on the nature of the components in the mixture.

Of the C_3-C_4 fraction from the catalytic cracking used as feed, the polymerized gasoline and C_5 represent 58.7% by weight, the rest being liquefied gases.

The limitations on the alkenes content in gasoline, a maximum of 9.2% in the U.S. since 1990 and 6.0% since 1996 in California, make necessary the hydrogenation of the *iso*-octene obtained by polymerization [75,76]. This two-step process is sometimes named *indirect alkylation* [75].

There are two olefin hydrogenation processes. One uses a supported noble metal (Pt) catalyst and requires feed having a low level of contaminants especially sulfur. The second uses a non–noble metal oxide catalyst, similar to that used in hydrofining processes, which is rather insensitive to contaminants. This last process requires a higher hydrogen partial pressure. Hydrogen recycling and the capital cost are also slightly higher.

Table 9.1 Effect of Pressure on Process Performance

	Reactor pressure, bar		
	17.5	35.0	63.0
Temperature, °C			
reactor inlet	204	190	204
reactor outlet	246	232	213
Alkenes conversion, %	80	85.90	92
Liters of polymer obtained/kg catalyst	417	625–1050	1250–1670

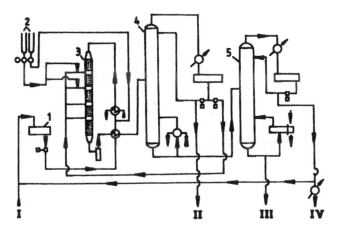

Figure 9.6 Flowsheet of a phosphoric acid oligomerization unit with multiple bed reactor. 1 – feed, 2 – water, 3 – reactor, 4 – depropenizer column, 5 – debutanizer column. (From Ref. 27.)

The consumptions and the costs are exemplified by the following values, reported for 1 m^3 of polymer-grade C_5+ gasoline [26]:

Electric energy, kWh	12.5–18.5
Steam, kg	570–850
Cost of catalyst, $	3.7–4.4

Another use of the process is the production of jet fuel by means of a more extensive oligomerization, followed by hydrogenation. The produced fuel has a low smoke point and a freezing temperature of −70°C. Different from the classic process the production of jet fuel requires the recycling of the light polymerizate to the reactor.

The use of oligomerization for producing components for Diesel fuel is of no interest, since the corresponding fraction has a cetane number of only 28 [26].

9.1.7 Other Oligomerization Processes

Despite the fact that oligomerization on phosphoric acid catalyst is widely applied, other processes are emerging.

9.1.7.1 Oligomerization on Zeolites

The studies referred to previously [8–11] on the catalytic activity of alumosilicas led Badger Co. to the development of the oligomerization process on zeolite catalyst ZSM-5 [9,77].

FCC off-gas, light olefin cuts, and light FCC gasoline are passed through a fluidized bed of ZSM-5 zeolite catalyst operating at 7–14 bar. Over 90% of C_2 to C_7 olefins are converted to form gasoline and LPG. At least 40% of any incoming benzene is alkylated while 75% or more of the C_5^+ olefins are converted.

A slipstream of catalyst is withdrawn and coke is burned off in a small dense fluid bed regenerator. Reactor products are condensed and separated [28].

The process flow diagram is given in Figure 9.7.

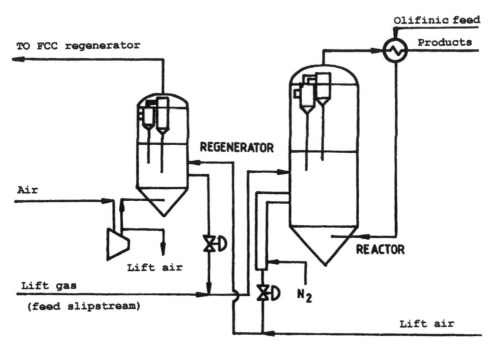

Figure 9.7   Oligomerization on ZSM-5 zeolite catalyst. (From Ref. 10.)

The yields and the quality of the gasoline are illustrated by the following values:

Yields % by weight	
fuel gas	2.0
propene-butenes	25.5
gasoline	72.0
coke	0.5
Gasoline properties	
density	0.751
RON, nonethylated	95–98
MON, nonethylated	82–84
vapor tension, mm Hg	370
PONA, % vol (typical)	34/30/8/28
Process economics	
Investment (basis: 20 mill. scfd FCC off-gas plus 16,000 bpsd light FCC gasoline, 1992 U.S. Gulf Coast)	\$10 million
Utilities (typical per bbl C_5^+ gasoline)	
electricity, kWh	12
nitrogen, scf	15
water, cooling, m^3	20–21
steam (MP or HP), kg	11–14

9.1.7.2 The Octol Process

The process was especially developed for processing the C_4 fraction obtained after the isobutene was removed for the fabrication of MTBE.

The unit has two shell and tube reactors with intermediary cooling, similar to those used for oligomerization on phosphoric acid. The operating temperature is below 100°C and the pressure is that required for maintaining a liquid phase in the reactor. This is possible since, different from the oligomerization on phosphoric acid, one operates below the critical temperature of the butenes present in the feed.

Two types of catalyst A and B were used, which were not further disclosed. The catalyst A leads to branched alkenes, which are of interest for the fabrication of the gasoline, whereas catalyst B favors the formation of the linear chains, of interest for plasticizers [29,30].

The oligomer obtained on the type A catalyst has an average content of 66% octenes and 17% dodecenes and has RON = 97 and MON = 85.

Utilities [29] per ton of oligomer:

Steam, 20 bar	0.85 t/t
Cooling water	57 t/t
Power	21 kWh/t

9.1.7.3 The Dimersol Process

The Dimersol process, developed by the French Institute of Petroleum, seems to be one of the most promising among the new oligomerization processes. There were more than 25 units in operation with a total capacity of 3.4 mill. t/year [18,19], of a total of 34 units with a total capacity of 4.3 mill. t/year, sold until Oct. 1994 [18].

Dimersol is a process using a homogeneous catalyst in liquid phases. The catalyst system is of the Ziegler type, based on a nickel derivative, and activated by means of an organometallic compound [18].

The process is realized in three options:

Dimersol E, by which the ethene and propene from the C_2-C_3 fraction are converted to high octane gasoline

Dimersol G, specific for the essentially complete dimerization of propene to a gasoline with a very high blending octane number

Dimersol X, used for the oligomerization of C_4 fractions, from which the isobutenes were converted to MTBE.

The reaction mechanism is of the "degenerated polymerization" type, which for the case of ethene is:

$$NiH + CH_2{=}CH_2 \longrightarrow NiCH_2CH_3 \xrightarrow{\;+C_2H_4\;} NiCH_2{-}CH_2{-}CH_2{-}CH_3$$

$$NiH + CH_2{=}CH{-}CH_2{-}CH_3$$

$$\updownarrow$$

$$CH_3{-}CH{=}CH{-}CH_3$$

$$NiCH_2{-}CH_2{-}CH_2{-}CH_3 \xrightarrow{\;+C_2H_4\;} Ni(CH_2)_5CH_3 \longrightarrow$$

As it results from this scheme, the reaction may stop at the dimer stage or may continue as a chain reaction.

The dienes and alkynes which with nickel will form coordination species, which are more stable than those formed by the alkenes and thus inhibit the reaction, are poisons for the catalyst. Also, compounds containing oxygen, sulfur, and nitrogen, which form even more stable coordination compounds, are also poisons for the catalyst.

The maximum tolerated level of impurities expressed in ppm is given in Table 9.2. In order to achieve these levels of purity, the feed undergoes a special treatment.

The flowsheet, common for all three process options, is given in Figure 9.8.

The catalyst precursors made of a nickel and an organometallic compound are injected in the recycling loop where they form the active catalyst. The reaction system may consist of one or two reactors in series.

Ammonia is injected at the exit of the reaction system for preventing the formation of chlorinated derivatives. After neutralization, the product is washed with water, whereby the traces of catalyst are destroyed. Since the consumption of catalyst is very low no regeneration is attempted.

The process Dimersol G produces a gasoline that is used only in proportion of 5–15% mixed in other gasoline, and has the following typical characteristics [18]:

Density	0.7
Distillation ASTM	
initial temperature	60°C
50%	68°C
70%	80°C
95%	185°C
Reid vapor pressure	0.5 bar
RON	96
Blending RON	105
MON	81
Blending MON	85

Dimersol X is used for converting the C_4 fraction that results after the isobutene was converted to MTBE. It must be purified by selective hydrogenation of the dienes and the acetylenes, removal of traces of dimethylether, and drying.

Table 9.2 Concentrations of Feed Impurities Tolerated by the Dimersol Process (ppm)

	C_3 fraction	C_4 fraction
Sulfur	2	2
Nitrogen	2	2
Water	5	5
Dienes and alkynes	30	30
Oxygenated compounds	–	30
Chlorine	–	1

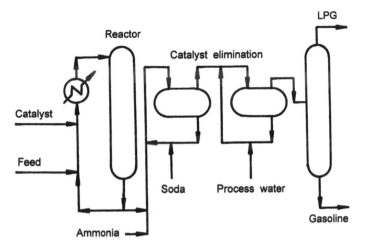

Figure 9.8   Flowsheet of a Dimersol unit.

Owing to the lower activity of the butenes the reaction system comprises 3–4 reactors in series.

The proportion of the more or less branched octenes in the final product depends on the concentration of isobutene in the feed (see Table 9.3).

Data on the process economics for Dimersol G and Dimersol X are given in Table 9.4.

In 1994, a number of 27 Dimersol units were built or were under construction [19].

9.1.1.4   Oligomerization on Resin Catalysts

The resin catalyst used in nearly all MTBE units is a synthetic, organic crosslinked, sulfonated, cation-exchange resin. The most common resin is polystyrene crosslinked with divinylbenzene. The same catalyst and reaction system may be used for the isomerization of i-C_4'' to trimethylpentene, with about 10% C_{12} olefins produced as byproduct.

Thus, it is possible to revamp MTBE units for i-C_4'' oligomerization [75].

Table 9.3   Distribution of Octenes in the Dimersol Product

Isobutene content in the feed, wt %	15.0	47.0
Octenes distribution in the product, wt %		
n-octene	5.5	4.0
methyl-heptenes	53.5	40.0
dimethyl-hexenes	39.0	50.0
other octenes	2.0	6.0

Table 9.4 Economic Data of the Dimersol G Process

	Dimersol G	Dimersol X
Feed	$C_3 (75\%\ C_3H_6)$	$C_4 (75\%\ n\text{-}C_4H_8)$
Capacity, t/year	100,000	50,000
Products		
liquefied gases	29,000	18,200
gasoline	71,000	–
octenes	–	27,000
dodecene	–	3,200
fuel	–	1,600
Investments, US $ (1994)	5.9 mill.	4.6 mill.
Consumptions	Per hour	Per t. octenes
steam, t	1.6	0.5
cooling water, t	330	50
process water, t	0.4	–
power, kWh	125	36
catalyst and chemicals, US $	7.7–9.4	54

9.2 ISOPARAFFIN-OLEFIN ALKYLATION

The main advantage of alkylation over oligomerization is that it produces a higher quantity of high-octane gasoline of a saturated character. In this situation oligomerization is practiced owing exclusively to its lower investment and operation costs. But if the produced octenes are hydrogenated, such in indirect alkylation technology, the economic data are practically the same as for alkylation [75].

The lower vapor pressure of the alkylate, allows for more *n*-butane to be added, without exceeding the prescribed vapor pressure of the gasoline.

New regulations implemented in the U.S. [31], the principles of which are being extended to other countries, have increased the importance of the alkylation process. Thus, these regulations drastically limit the accepted concentrations of iso-amylenes in the final gasoline. The iso-amylenes are increasingly incorporated in the alkylation feed. It is easy to estimate the increase in alkylate capacity produced by this measure if one takes into account that the catalytic cracking gasoline contains almost 10% C_5 alkenes [32].

Table 9.5 Equilibrium Constants of Selected Alkylation Reactions

Reaction	$K_{300\ K}$	$K_{500\ K}$
$i\text{-}C_4H_{10} + C_2H_4$	1.64×10^9	2.99×10^2
$i\text{-}C_4H_{10} + C_3H_6$	3.90×10^8	34.5
$i\text{-}C_4H_{10} + i\text{-}C_4H_8$	1.43×10^5	0.617
$i\text{-}C_4H_{10} + 2\text{-methyl-butene}$	4.30×10^3	7.45×10^{-2}

Source: Ref. 20.

The high octane rating of the alkylate allows regulations for the drastic decrease of benzene concentration in the finished gasoline without loss in the octane rating.

The above facts led to the situation that from 1985–1990 the alkylate represented 11% of the gasoline produced in the U.S. [33].

The total 1987 capacity of alkylation plants in the U.S. was of 38.6 million tons. It reached 44.7 million tons in 1992, which represented nearly 70% of world capacity. In order to satisfy obligations concerning the quality of the gasoline required by the Clean Air Act [31], alkylation capacity will continue to increase. In 2003, U.S. alkylation capacity is estimated to be 160,000–180,000 m^3/day [34].

9.2.1 Thermodynamic Aspects

The alkylation of isobutane with alkenes is strongly exothermal. The heat of reaction is 75–96 kJ/mole function on the alkene used in the reaction.

The equilibrium constants depend on the temperature and type of the alkene. Some representative values are given in Table 9.5 [20].

As shown in the table, the equilibrium constant decreases strongly with temperature and the molecular mass of the alkene. This explains in part, the need to operate the process at relatively low temperatures in liquid phase. A second reason favoring the use of low temperature is to avoid competitive polymerization reactions.

A favorable equilibrium conversion in the alkylation with ethene was obtained when neohexane (2,3- and 2,3-dimethylbutanes) were produced by a thermal reaction at high pressures (300 bar), in a process no longer used.

9.2.2 Alkylation Catalysts

For a long time, the catalysts used exclusively in industrial alkylation units were sulfuric acid and hydrofluoric acid. The characteristics of these catalysts are given in Table 9.6.

There are serious difficulties when these catalysts are used for liquid-liquid heterogeneous processes: corrosion, catalyst consumption, regeneration difficulties etc. Beginning in the 1960s studies were initiated on the use of solid catalysts. The use of solid superacidic catalysts if of interest also for other processes of organic synthesis [35].

The alkylation on zeolite catalysts was studied [36] on a CeY catalyst, where in a Y zeolite, the sodium was replaced with cerium. A mixture of 10/1 isobutane/butenes at 80°C and 31 bar reacted within 30 minutes to a conversion of 100%. Then, it decreased quickly to 30% due to the carbon deposits on the surface of the catalyst.

Compared with this zeolite with macropores, the zeolites with mezopores such as HZSM-5 and HZSM-11 require higher temperatures of the order of 200°C. The formation of coke on these catalysts is less intense. However, no highly branched iso-alkanes are formed due of the relatively narrow pores of these catalysts.

Other catalysts were also tested without achieving success: ion exchange resins, chlorinated alumina, sulfated zirconia, BF_3, associated with zeolites, or on porous supports.

Table 9.6 Properties of HF and H_2SO_4 Catalysts and Hydrocarbons Solubilities

	HF	H_2SO_4
Molecular mass	20.01	98.08
Boiling temperature, °C	19.4	332.4(98%)
		296.2(100%)
Crystallization temperature, °C	−82.8	0.1(98%)
		5.7(99%)
		10.4(100%)
Density	0.99	1.84
Viscosity, cP	0.256(0°C)	24.5(25°C)
		33.0(15°C)
Surface tension, dyne/cm	10.1(°C)	55.0(20°C)
	8.6(19°C)	
Acidity Hammett (H_0)	−10.2(100%)	−12.2(100%)
solubilities, wt %	−8.9(98%)	−9.85(95%)
i-C_4H_{10} at 26.6°C in acid	−	0.100(99.5%)
		0.070(98.7%)
		0.040(96.5%)
i-C_4H_{10} at 13°C in acid	2.7(100%)	−
acid in i-C_4H_{10} at 27°C	0.44	−
acid in C_3H_8 at 27°C	0.90	−

A list of processes for isobutane alkylation at research and development stages on basis of solid catalysts is given in Table 9.7.

Promising perspectives seem to be offered by catalysts of oxides, resins, and zeolites promoted with strong acids.

Thus, in a pilot plant of 1.1 m^3/day on experiments began in 1993 a solid catalyst of alumina/zirconia promoted with halogens [34]. The process was developed by Catalytica and Conoco together with the Finnish company Neste Oy.

Another process was developed by Exxon and Topsøe in cooperation with Kellogg. The process was tested since 1992 in a pilot plant of 80 t/day. The process uses perfluoro-methanesulfonic acid adsorbed on a porous support, forming a fixed bed of catalyst. The requirement for stirring is thus eliminated and no volatilization losses of the acid take place.

Chevron did studies in a pilot plant of 1.6 m^3/day [34] on a catalyst of antimony pentafluoride on acid-washed silica support.

Kerr McGee developed a homogeneous catalysis process (HAT), based on the use of aluminum chloride solubilized in hydrocarbons.

The catalyst (1%) is introduced in the isobutane feed line, downstream of which the alkenes are introduced. At the end of the reaction the spent catalyst is separated as a distinct phase and is regenerated [34].

The authors of these studies show that by replacing the reactors of the existing units that use H_2SO_4 or HF with the new studied types of reactors, no modifications of the fractionating systems will be required, which should significantly reduce the investment costs. The cited studies include economic calculations that compare the

Table 9.7 Isobutane Alkylation Processes on Solid Catalysts at Research and Development Stage

Catalyst	Process developers	Year	Reference
Lewis acid in combination with non-zeolitic solid inorganic oxide, large-pore crystalline molecular sieve, and/or ion exchange resin	Mobil Oil Corp.	1990	[80]
Fluorinated ion exchange resin	Allied Signal	1993	[81]
Sulfated zirconia	Texas A&M University	1993	[82]
SO_4^{2-}/ZrO_2 and SO_4^{2-}/TiO_2 H_3PO_4–BF_3–H_2SO_4 supported on SiO_2 and ZrO_2	Jilin University (China)	1994	[83,84]
Ultrastable H-Y zeolite with Si/Al ratio of 6.9	Princeton University	1994	[85]
Sulfated zirconia and Beta zeolite	Universidad Politecnica de Valencia (Spain)	1994	[86]
H_2SO_4 or Al or B halide and quaternary ammonium halide or amine hydrohalide, impregnated on an organic or a mineral support	Institute Français du Pétrole	1995	[87]
Sulfated zirconia	Université Laval	1995	[88]
Brönsted acid treated transition metal oxide	Hydrocarbon Technologies Inc. (HTI)	1996	[89]
H_2SO_4 impregnated on silica	Institute Français du Pétrole	1996	[90]
H_2SO_4 and $HB(HSO_4)_4$ impregnated on silica	Institute Français du Pétrole	1996	[91,92]

investment costs and operating costs for the new processes with those of the classic processes [75].

According to the literature [34], it seems that some of the new systems could replace the hydrofluoric acid and possibly the sulfuric acid. Solid catalysts will probably be used in the future, mostly in the new units.

9.2.3 Reaction Mechanisms

The mechanism of the alkylation of iso-alkanes with alkenes is complex and is the object of many studies [38–41,94].

The following mechanism is generally admitted [20,42,43]:

9.2.3.1 Initiation

The initiation step is the formation of a carbenium ion by the interaction of the alkene with the acid:

$$C-C=C + HA \longrightarrow \left[C-\overset{+}{C}-C\right]A^-$$

$$C-C=C-C + HA \longrightarrow \left[C-C-\overset{+}{C}-C\right]A^-$$

$$C-C-C=C + HA \longrightarrow \left[C-C-\overset{+}{C}-C\right]A^-$$ (a)

$$C-\underset{\underset{C}{|}}{C}=C + HA \longrightarrow \left[C-\underset{\underset{C}{|}}{\overset{+}{C}}-C\right]A^-$$

The secondary ions formed from *n*-butenes or higher *n*-alkenes may isomerize:

$$\left[C-C-\overset{+}{C}-C\right]A^- \longrightarrow \left[C-\underset{\underset{C}{|}}{\overset{+}{C}}-C\right]A^-$$ (b)

The formation of the tertiary butyl cation is also possible by way of the reactions:

$$\left[C-\overset{+}{C}-C\right]A^- + C-\underset{\underset{C}{|}}{C}-C \longrightarrow C-C-C + \left[C-\underset{\underset{C}{|}}{\overset{+}{C}}-C\right]A^-$$

$$\left[C-\overset{+}{C}-C-C\right]A^- + C-\underset{\underset{C}{|}}{C}-C \longrightarrow C-C-C-C + \left[C-\underset{\underset{C}{|}}{\overset{+}{C}}-C\right]A^-$$ (c)

The excess of *i*-butane favors these reactions.

9.2.3.2 Propagation

The tertiary butyl cations formed by the reactions (a)–(c) interact with the alkenes to form a larger carbenium ion, for example:

$$\left[C-\underset{\underset{C}{|}}{\overset{+}{C}}-C\right]A^- + C_4H_8 \longrightarrow \left[C-\underset{\underset{C}{|}}{\overset{\overset{C}{|}}{C}}-C-\overset{+}{C}-C\right]A^-$$ (d)

The resulting product is in fact a mixture of (the predominant) trimethylpentyl ions and dimethylpentyl ions.

The new ion formed in the reaction:

$$\left[C-\underset{\underset{C}{|}}{\overset{\overset{C}{|}}{C}}-C-\underset{\underset{C}{|}}{\overset{+}{C}}-C\right]A^- + i\text{-}C_4H_{10} \longrightarrow C-\underset{\underset{C}{|}}{\overset{\overset{C}{|}}{C}}-C-\underset{\underset{C}{|}}{C}-C + \left[C-\underset{\underset{C}{|}}{\overset{+}{C}}-C\right]A^-$$ (e)

continues the reaction chain.

Besides this main chain, secondary reactions take place that lead finally to a mixture of iso-alkenes, containing components with larger and smaller number of carbon atoms than that derived from the main reactions. Thus:

- The ion formed by way of reaction (d) can interact with another butene molecule giving a dodecyl ion, which as a result of a reaction of type (e), will give an isododecane;
- Since the described reactions are reversible, the formed hydrocarbons may be dealkylated following various mechanisms, producing isoalkenes and alkenes with a different number of carbon atoms than those from which they derived and which will lead to other alkylation reactions. For example:

$$i\text{-}C_8H_{18} \rightarrow i\text{-}C_5H_{12} + C_3H_6$$
$$i\text{-}C_5H_{12} + C_4H_8 \rightarrow i\text{-}C_9H_{20}$$
$$C_3H_6 + i\text{-}C_4H_{10} \rightarrow i\text{-}C_7H_{16}$$
$$i\text{-}C_9H_{20} \rightarrow i\text{-}C_6H_{14} + C_3H_6$$
$$i\text{-}C_6H_{14} + C_4H_8 \rightarrow i\text{-}C_{10}H_{22}$$

The intensity of the first type of reactions could be limited by regulation in a corresponding way of the residence time. The second reaction type that takes place with enough important rates leads to the situation that finally, a large fraction of iso-alkanes, called alkylate, is obtained from the process.

Besides, this group of alkylation reactions, polymerization reactions of alkenes may take place. The large excess of iso-alkanes and low temperatures serve to decrease their extent.

As result of experimental studies using sulfuric acid as catalyst, Albright [38,40], assigns a special importance to initiation reactions of type (a), which leads to the formation of sulfate by way of the reaction of the sulfuric acid with the alkene. He recommends that the process be carried out in two steps. During the first one, the formation of an ester takes place at somewhat higher temperatures, of the order of 20°C, and at a short reaction time. The second step is performed at the temperature of 10°C [38,40]. This way of performing the reaction is justified, according to the author, by the large exothermal effect of the first step.

The two-step operation produces an alkylate of better quality and having a higher octane number than the one-step process.

9.2.4 Process Kinetics

The reactions take place in a system that has the hydrocarbon drops dispersed in the continuous phase made of acid. Since the formation of the carbenium ions cannot take place in the hydrocarbon phase, the reaction is considered to take place [20,44] in the acid phase. The alkenes are much more soluble in the acid than the iso-alkanes. Consequently, in these conditions the reaction one would expect a high conversion to polymers, which is not in agreement with the practical data. This fact suggests [20] that the reaction takes place in the acid film near the wall of droplets of the hydrocarbon phase. This explanation cannot be considered as completely satisfactory.

Our opinion is that the species formed by the interaction of the acid with the alkene, such as:

$$\left[C - \overset{+}{C} - C \right] A^-$$

and which initiates the reaction chain, are found to a larger extent at the interface between the two phases, with the carbenium ions oriented towards the hydrocarbon phase. This explains the easy interaction between the carbenium ion with the iso-alkane present in the hydrocarbon phase in high concentration. Such a mechanism eliminates the contradiction between the large solubility of the alkenes in the acid phase relative to that of the iso-alkane and the fact that the oligomerization reaction may be almost completely eliminated by using a high excess of iso-alkanes.

This mechanism is confirmed by the fact that the intensity of mixing, which determines the size of the hydrocarbon drops, influences the reaction rate.

The situation may be somehow different in the case of the hydrofluoric acid in which the solubility of the isobutane is much higher than in sulfuric acid.

Many of the authors who studied the kinetics of the alkylation with sulfuric acid [20,44,45] consider the diffusion of the isobutane from the hydrocarbon drops in the acid phase to be the rate determining step of the reaction.

However, this opinion is not unanimous. Thus, Lee and Harriot [46] consider that the reactions take place in the acid phase in the immediate proximity of the drops and that both the reaction and the diffusion take place with comparable rates so that none of these two stages is rate determining. On this basis, the authors deduce a system of differential kinetic equations for the alkylation of iso-butane with 1-butene, catalyzed by sulfuric acid, in which the diffusion rates are equated with the reaction rates for both the alkylation and the oligomerization.

The reaction rates for the oligomerization and for the alkylation are written by the author, as:

$$r_A = k_A \cdot C_A^2 \tag{9.4}$$

$$r_B = k_B \cdot C_A C_B \tag{9.5}$$

where:

> subscript A refers to alkenes in oligomerization, and B to isobutane in alkylation.

The system of kinetic equations was solved [46] by a numerical method using a computer program. The values of the obtained reaction rate constants are given in Table 9.8.

Table 9.8 Reaction Rate Constants for Oligomerization and Alkylation

H_2SO_4, wt % concentration	Temperature, °C	k_B, cm^3/g·mol·s	k_A, cm^3/g·mol·s	k_A/k_B
98	15	–	3.9×10^8	–
98	25	2.25×10^8	9.5×10^8	4.2
94	15	–	1.8×10^8	–
94	25	1.32×10^8	5.3×10^8	4.0
90	25	6.64×10^7	2.3×10^8	3.5

Source: Ref. 46.

Taking into account the simplifying assumptions incorporated in the development of the model, one must accept that these values are only approximate.

In order to correctly estimate the diffusion rates in the experimental part of the research, measures were taken so that the drops of hydrocarbons in contact with the acid have the same size, the contact time is the same, and no coalescence takes place.

In an industrial reactor there will always exist a distribution of the particle sizes, a dispersion of the residence times, and the possibility of coalescence (and redispersion). In these conditions, the diffusion is not the rate determining step and therefore a design of the reactor on the basis of rigorous kinetic calculations is not justified. This problem solving approach encounters also other major difficulties, such as the determination of the diffusion coefficients of the alkenes in the acid phase.

If one accepts the suggestion formulated by us above, namely that the reactions take place on the hydrocarbon side of the acid–hydrocarbons interface, it is possible to treat the alkylation reaction as a homogeneous process, but where the reaction rate is proportional to the interfacial area. Therefore, the rate will increase with increasing degree of dispersion, that is with the decrease of the diameter of the hydrocarbon drops as confirmed experimentally [44].

Since the alkenes are soluble in sulfuric acid, the oligomerization reactions can take place also in the acid phase. Therefore, the reaction rate will be less influenced by the size of the interfacial area. This fact explains the improvement of the alkylate quality and the increase of its octane number produced experimentally by increasing the degree of dispersion as a result of intensifying the mixing within the reactor [44,46].

Since the alkylation and polymerization reactions are simultaneous, the number of the moles present in the reactor, initially and at a later time τ, may be expressed as follows:

$$2A \xrightarrow{k_A} P \qquad A \quad + \quad B \xrightarrow{k_B} I$$

τ_0	n_{0A}	0	n_{0A}	n_{0B}	0
τ	$(n_{0A} - 2n_p - n_I)$	n_p	$(n_{0A} - 2n_p - n_I)$	$(n_{0B} - n_I)$	n_I

Since for both above reactions at the process temperature the thermodynamic equilibrium is displaced to right, the formation rates of the polymers respectively of the alkylate will be:

$$\frac{1}{V}\frac{dn_p}{d\tau} = k_A \frac{(n_{0A} - 2n_p - n_I)^2}{V^2} \tag{9.6}$$

$$\frac{1}{V}\frac{dn_1}{d\tau} = k_B \frac{(n_{0A} - 2n_p - n_I)(n_{0B} - n_I)}{V^2} \tag{9.7}$$

It is to be remarked that since the process takes place in liquid phase, the volume remains constant.

By dividing the second equation by the first one, it results:

$$\frac{dn_I}{dn_P} = \frac{k_B}{k_A} \cdot \frac{n_{0B} - n_I}{n_{0A} - 2n_p - n_I} \tag{9.8}$$

an equation that expressed the relative rate of the two transformations.

The Eq. (9.8) illustrates in an obvious manner that in order to decrease the formation of the polymers the largest possible excess of isobutane must be present in the reactor. In the same time, by ensuring an intensive mixing within the reactor, the ratio k_B/k_A of the rate constants will increase.

Despite numerous publications in the recent years, only moderate progress was achieved in the alkylation on solid catalysts [95]. The product is a saturated alkylate produced by a carbonium ion chain mechanism. Catalyst deactivation is rapid and is caused by nondesorbed products, identified as polyalkyl-bicyclic olefinic products with 12–28 carbon atoms.

9.2.5 Effect of Process Conditions

The properties of the two catalysts shown in Table 9.6 make it easier to understand the differences between the two processes, the operating conditions, and the effect of the process factors.

The much larger density, viscosity, and surface tension of the sulfuric acid than of HF make it much more difficult to obtain a fine dispersion of the hydrocarbons in the former than in the latter. That leads to important differences in the performance of the reactors for the two processes.

The second important difference between the two catalysts is the solubility of isobutane in the acid phase. This higher solubility leads, in the case of hydrofluoric acid, to a much higher ratio of isobutane/alkenes in the acid side of the interface of acid and hydrocarbons, where the reaction takes place. This, reduces the extent of the secondary reactions, improves the quality of the alkylate, and allows the HF process to operate at higher temperatures.

The influence of the operating parameters on the process is analyzed below, by taking into account these differences.

9.2.5.1 Temperature

The increase of the operating temperature favors the decomposition of carbenium ions. This increases the content of light alkanes and decreases the octane rating of the alkylate. It is estimated that the octane number decreases by 0.1 points per 1°C increase of temperature [20].

The decrease of the viscosity as a result of the temperature increase improves the transfer between the phases, which increases the reaction rate.

In the alkylation with sulfuric acid, if the temperature of 15°C is exceeded, the oxidation and sulfonation reactions are intensified, which increases to a significant extent the acid consumption.

The decrease of the temperature below 5°C increases the viscosity significantly and makes it difficult to obtain a fine dispersion of hydrocarbons in acid. The optimal reaction temperatures for the sulfuric acid are situated between 5 and 10°C.

The laboratory studies performed at the temperature of 0°C [40] showed an important improvement of the alkylate quality. But the industrial units do not operate below +5°C because of the excessive viscosity of the sulfuric acid.

For hydrofluoric acid, the most favorable technical-economical conditions correspond to the use of water for the cooling of the reactor, which corresponds to a temperature ranging between 30 and 45°C.

9.2.5.2 Pressure

Since the process is taking place in liquid phase, the reactions are not influenced by the pressure. The pressure level is determined by the need to maintain the reactants in liquid phase.

9.2.5.3 Feeds and Catalysts

The alkylation of isobutane with ethene is endothermal and does not take place at the low temperatures at which the alkylation with sulfuric acid or with hydrofluoric acid are practiced.

The alkylation with propene takes place with a lower rate than with butenes and gives reaction products that have octane ratings that are 4–5 points lower. Overall, the alkylation takes place with more difficulty in sulfuric acid than in hydrofluoric acid.

Results of the alkylation of isobutane with the butenes are shown in Table 9.9. It may be remarked that the results of the alkylation with 1- and 2-butene are almost identical. This is explained by the high rate of isomerization of 1-butene to 2-butene. The alkylation with isobutene leads to products of lesser quality than with n-butenes. The use of i-butene in alkylation is today of low interest, since i-butene serves mainly for the production of MTBE.

In the refineries with their own units for the production of MTBE, the C_4 fraction is fed first to that unit where i-C_4H_8 is completely consumed, and then to alkylation.

In order to compensate for the consequential decrease of the available alkenes potential for the alkylation, pentenes were incorporated in the feed to the process [47].

Table 9.9 Alkylation with 98% Sulfuric Acid at 7°C

	Isobutene 158	Butene-2 180–185	Butene-1 180–185
Alkylate, wt %			
Alkylate composition, wt %			
isopentane	6.42	2.67	3.00
methylpentane	4.01	3.52	3.42
2,4-dimethylpentane	3.41	2.11	2.33
2,3-dimethylpentane	2.29	1.23	1.29
Total C_5–C_7	16.13	9.53	10.04
2,2,4-trimethylpentane	29.14	31.14	31.05
dimethylhexanes	9.68	9.16	11.23
2,2,3-2,3,4-2,3,3-trimethylpentanes	23.49	42.24	39.80
Total C_8	62.31	82.54	82.08
nonanes	7.50	2.74	2.81
decanes	14.03	5.19	5.07
Total > C_8	21.53	7.93	7.38

Source: Ref. 20.
Alkenes liquid hourly space velocity 0.22 h^{-1}.

The octane rating of the alkylate resulting when using pentenes is lower than for butanes, but close to that obtained with propene (see Table 9.10) [48].

The undesired effect of isopentene on the environment, in relation to the formation of nitrogen oxides and of ozone makes its incorporation as such in gasoline impractical and increased the interest for its processing. Similar to the procedure with the butenes, the C_5 fraction is first fed to a unit that converts the isopentenes (iso-amylenes to tert-amyl-methyl-ether (TAME). The unconverted normal pentenes are then fed to the alkylation unit.

A comparison between the products obtained with the two catalysts in the alkylation of isobutane with different alkenes is provided by the data of Table 9.11 [49]. The octane rating of the alkylate obtained with various pentenes is given in Table 9.12 [49].

Among the alkanes, actually only isobutane is used in alkylation. The n-butane does not participate to the reaction and isopentane is by itself, without further processing, a valuable component of the gasoline.

For good reaction efficiency, it is important to have in the reactor a high concentration of isobutane, i.e. a high molar ratio of isobutane to alkenes, and a low content of n-butene in the feed of mixed butanes.

As shown by Eq. (9.8), a high concentration of isobutane decreases the conversion of the polymerization reactions as well as of other secondary reactions, such as the formation of dienes in the case of sulfuric acid and of fluorides in the case of hydrofluoric acid. In both cases, the secondary reactions lead to increased acid consumption.

Isobutane excess is maintained also in the equipment for the separation of the acid by decantation, in order to preserve the quality of the alkylate. Thus, the concentration of the isobutane at the reactor outlet is about 60%.

Owing to the difference in solubility in the two acids, the concentration of the isobutane in the reactor must be higher in the case of sulfuric acid, than in the case of hydrofluoric acid. For these reasons, in the case of the sulfuric acid constructive measures are taken that lead to local isobutane/alkenes ratios, much higher than the overall ratio.

The presence in the feed of dienes has a very negative effect on the process and produces a strong increase of acid consumption. The presence of water, oxygenated compound, and sulfur produces a similar negative effect. Depending on these impu-

Table 9.10 Octane Ratings for i-Butane Alkylation with Different Alkenes

Alkene	H$_2$SO$_4$		HF	
	Research	Motor	Research	Motor
Butene-1	98–99.6	94–95	95–94	91–92
Butene-2	98–99	94–95	97–98	93–94
Isobutene	90–91	88–89	94–95	90–91
Propene	89–92	88–90	91–93	89–91
Pentenes	92–93	91	91–92	90

Source: Ref. 48.

Table 9.11   Effect of Catalyst on Alkylate Composition

	C$_3$ =		1-C$_4$ =		2-C$_4$ =		i-C$_4$ =	
	H$_2$SO$_4$	HF	H$_2$SO$_4$	HF	H$_2$SO$_4$	HF	H$_2$SO$_4$	HF
i-C$_5$	3.79	1	4.66	1	4.16	0.3	10.01	0.5
i-C$_6$	4.24	0.3	4.44	1.1	4.58	0.9	5.21	1
i-C$_7$	71.11	43.8	4.06	1.2	3.75	1.5	6.42	2
TMP	8.13	47.8	68.57	68.2	72.19	85.6	51.78	86.1
DMH	1.70	3.2	11.03	22.1	9.02	6.9	9.53	3.4
C$_9$+	11.03	3.7	7.24	5.7	6.30	4.1	17.05	5.3
COM	89	90	94.50	91.5	94.50	93.50	88.50	90.5

Source: Ref. 49.

rities, the consumption of catalyst is situated between 0.6–0.8 kg HF/t of alkylate and 68.5–103 kg H$_2$SO$_4$/t alkylate [50].

9.2.5.4   Interphase Contact and Reaction Time

In addition to the requirement of intense mixing for achieving a large interfacial area, it is important that the acid should constitute the continuous phase and the hydrocarbons the dispersed phase. The opposite situation leads to alkylate that has a research octane number by 1–2 units lower [49]. In order that the acid constitutes the continuous phase, the acid/hydrocarbons volume ratio must be comprised between 1 and 1.5 in the case of H$_2$SO$_4$ and between 1 and 4, in the case of HF. However, some later patents [51] indicate that in the case of HF, a continuous hydrocarbon phase leads to a decrease in acid consumption related to the reduced HF inventory.

The reaction time depends on the degree to which the phases are dispersed. For the sulfuric acid it is 20–30 minutes; for the hydrofluoric acid, the duration is 1.5–3 times lower.

9.2.6   Feed Preparation

The C$_4$ or of C$_3$ + C$_4$ fractions used in alkylation are pretreated for the elimination of various components or impurities that harm the process. Thus, the removal of

Table 9.12   (Research + Motor)/
2 Octane Rating Obtained with
Various Pentenes

Pentene	HF	H$_2$SO$_4$
1-Pentene	87.9	91.8
2-Pentene	89.0	90.9
2-Methyl-2-butene	90.3	91.5
2-Methyl-1-butene	90.0	90.9
3-Methyl-1-butene	90.7	91.4
Cyclopentene	–	86.6

Source: Ref. 49.

butadiene is performed by selective hydrogenation (on palladium catalyst), which follows the removal of mercaptanes [52], of traces of arsenic and mercury, which poison this catalyst [53], and by the elimination of oxygenated compounds left over after the conversion of the isobutene to MTBE by reaction with methyl alcohol [54].

The C_4 fractions produced in catalytic cracking contain 0.5 to 1% butadiene. The proportions are higher in the C_4 cuts from the high-severity cracking of vacuum distillates practiced in Western Europe. Butadiene is also contained in the C_4 fraction produced in pyrolysis.

If the feed to the alkylation with sulfuric acid is not previously submitted to the selective hydrogenation of the butadiene, the acid consumption increases and reaches values of 150–200 kg acid per ton of alkylate, which puts at risk the economics of the process. In alkylation with hydrofluoric acid, despite the fact that here also an increase of acid consumption is observed, the effect is less damaging.

The isomerization of butene-1 to butene-2 [48], which leads to the increase of the octane rating of the alkylate, can be performed simultaneously with the selective hydrogenation of butadiene.

The presence of arsenic in the olefins fractions used as feed to alkylation is due to the decomposition in the catalytic cracking process of arsenic-organic compounds. Since 95% of the arsenic contained in crude oil is in fractions having a boiling temperature above 330°C [53], the problem of the presence of arsenic in the C_3-C_4 cut is especially important in the catalytic cracking of vacuum distillate.

The distribution of mercury in the various crude oil fractions is different from that of arsenic. The largest proportion (about 60%) is found in the condensate, namely in the gasoline and petroleum fractions [53]. Mercury is present also in natural gases, but exclusively as metal.

The simultaneous elimination of arsenic and mercury is very easily achieved by contacting the alkene fraction with a contact mass (MEP 191 produced by Procatalyse or other, similar catalysts) before being fed to the reactors for the selective hydrogenation, on palladium.

Such a preliminary purification is necessary both for the fractions produced in catalytic cracking and for those resulting from pyrolysis.

As a result of the MTBE fabrication process, the principle scheme of which is presented in the Figure 9.9, the butenes fraction that leaves the MTBE unit contains traces of methyl alcohol and acid esters and neutral esters [54]. They are removed by passing the effluent over bauxite, which adsorbs not only the oxygenated compounds but also the water, eliminating in this way their damaging effects and the increases of acid consumption.

The alkylation process on solid catalysts requires also that contaminants such as diolefins, sulfur, oxygen, and nitrogen compounds are removed by the feed pretreatment [75].

9.2.7 Sulfuric Acid Alkylation

There are two main types of sulfuric acid alkylation units that differ mainly by the type of reactor used: (1) the units of Stratco with indirectly cooled, single reactors and (2) the units designed by M. W. Kellogg under license from Exxon Research & Engineering Co. with reactors in cascade and self-cooling. The first of these two

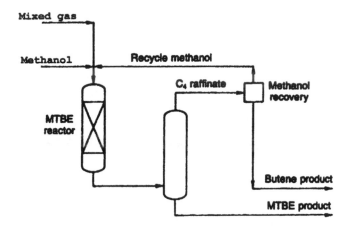

Figure 9.9 Flowsheet of an MTBE unit.

systems seems to account for 60% of the world production of alkylate obtained by using sulfuric acid as a catalyst [33,35].

A third type of unit, which continues to be exploited although none was built in the last 25 years, is that with "temporization vessel" (Time Tank Process). In this system, hydrocarbons and sulfuric acid are mixed in a pipe, from which a centrifugal pump sends them through tubular coolers to a vessel where the reaction is completed and the two phases are separated.

The process scheme of the Stratco system, with single reactors and indirect cooling, in given in Figure 9.10, and a cross section through the reactor in Figure 9.11.

The reactor has the cooling agent circulating through the tubes while the emulsion of the mixture of hydrocarbons and sulfuric acid is forced to circulate in the shell, along the cooling tubes with a high velocity, to ensure an efficient heat exchange.

The intensity of the stirring has a large influence on the performance of the process. The case is cited [33] where in the alkylation with butene-2, by increasing the rotation speed of the stirrer from 1000 to 3000 rotations/min led, the RON increased by 7.5 points. Also, it is important that within the reactor the sulfuric acid should be the continuous phase and that the alkenes should be fed premixed with isobutane.

The reactor effluent contains, besides isobutane and alkylate, also low amounts of sulphates that are soluble in the hydrocarbons, such as the di-sec-butyl sulfate and di-isopropyl sulfate.

There are three methods for their separation.

The first, indicated by the scheme of Figure 9.10, is an acid washing, which transforms the disulfates to acid sulfates, soluble in sulfuric acid, followed by a washing with caustic soda for removing the traces of acid.

The second method is the washing with caustic soda for decomposing the sulfates, with formation of Na_2SO_4, followed by washing with water for removing the traces of soda. This treatment is less frequently used since often it does not achieve the complete decomposition of the sulfates.

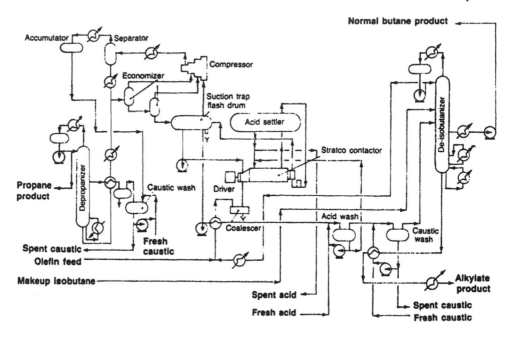

Figure 9.10 H_2SO_4 alkylation in reactor cooled by heat exchange (Stratco Inc.). Alkylation with effluent refrigeration.

The third and most preferred method, consists of an acid wash followed by contact with bauxite powder, which eliminates the corrosion hazard and dries the product at the same time.

The first and third methods have low acid consumption and allow the recovery of high quality alkylate.

Two methods are used for maintaining the activity of the acid catalyst. The first consists in the continuous addition of fresh acid and the discarding of spent acid. In the second method, the acid is recycled until its concentration decreases below a certain level, after which most of the acid is replaced. This second method leads to a better average quality of the alkylate.

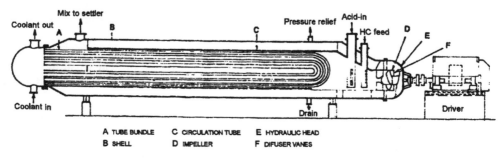

| A TUBE BUNDLE | C CIRCULATION TUBE | E HYDRAULIC HEAD |
| B SHELL | D IMPELLER | F DIFFUSER VANES |

Figure 9.11 Cross section through a reactor cooled by heat exchange.

The process flow diagram of the system with a cascade-type reactor and cooling by evaporation of a part of the isobutane present in the reactor is shown in Figure 9.12. Details of the reactor internals are shown in Figure 9.13.

The cascade-type reactor contains three to seven reaction compartments. The recycled isobutane and the acid flow successively through this system, whereas the fraction rich in alkenes is introduced in equal portions in each compartment in the center of each stirrer system. In this way, the isobutane/butenes ratio is in each step sensibly higher than the overall ratio, despite the fact that a part of isobutane is evaporated for cooling. Thus, in a system with 5 reaction compartments and an overall ratio of 8, it is of 35 in the first compartment, decreasing to 12 in the last one.

The decrease of the i-butane/butenes ratio along the system leads to the decrease of the octane rating of the alkylate from one compartment to the following, the total difference reaching 2 units [33,35].

It is considered that in this system too, the acid represents the continuous phase, despite the fact that a separation of the hydrocarbons in the upper part of the compartments was noted.

The composition of the alkylate, the octane rating, and the acid consumption depend mainly on the type of alkenes used in the process. In Table 9.13, typical data valid for both types of units are collected [47].

The butenes and pentenes fractions represent mixtures with compositions typical for the streams from the catalytic cracking. Further studies on the alkylation with mixtures of butenes and pentenes [56] demonstrate that the octane rating of the

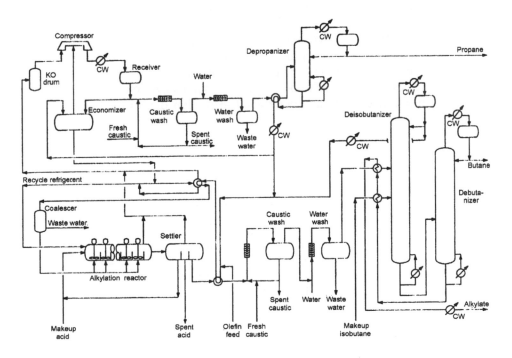

Figure 9.12 H_2SO_4 alkylation unit with cascade reactor. 1 – pretreatment, 2 – heat exchanger, 3 – reactor, 4 – alkylate treatment, 6 – compressor, 5,7 – columns.

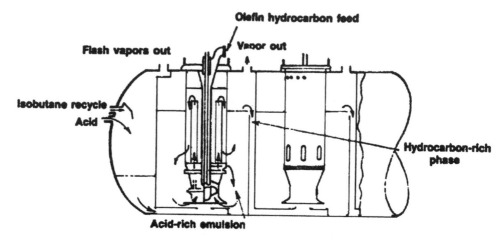

Figure 9.13 Exxon cascade reactor.

alkylate and the consumption of acid may be calculated approximately by addition on the basis of the values of Table 9.13. For the alkylation with mixtures of butenes and propene [57] the published data do not confirm that the octane rating is additive, but allow the formulation of conclusions concerning the sulfuric acid consumption.

Sulfuric acid consumption increases as a function of the impurities present in the hydrocarbon feed (see Table 9.14) [50].

Commercial-scale tests with precisely known compositions of the feed and alkylate [58] confirm the previously known facts that in the feed, n-butenes should be preferred to i-butene and that butadiene has a negative role (see also Table 9.9).

The overall i-butane/alkenes ratio may vary between the limits 5 and 8.

In either of the two systems, the reaction temperature should be maintained at +5°C. An increase above this value favors the polymerization reactions while a decrease creates difficulties in achieving good dispersion due to the increase of the acid viscosity.

Table 9.13 Alkylation with C_3H_6, C_4H_8, and C_5H_{10}

	C_3H_6	C_4H_8	C_5H_{10}
Alkylate composition			
i-C_5	3.1	4.0	8.2
n-C_5	0.0	0.0	0.0
C_6	3.8	4.0	4.2
C_7	71.3	3.9	2.2
C_8	10.8	77.5	33.2
C_9	3.1	2.6	35.1
$C_{10}+$	7.9	8.0	17.1
RON	89.6	95.3	90.2
MON	87.6	92.0	87.8
H_2SO_4 consumption, 98.5%→90% kg/m^{3+} alkylate	153	46	140

Table 9.14 Typical H_2SO_4 Consumption Values for Various Impurities in Hydrocarbon Feed

Impurity	kg H_2SO_4 per kg Impurity
Water	10.6
Butadiene	13.4
Ethene	30.6
Mercaptans (wt S)	17.0
Disulfides (wt S)	12.8
Methanol	26.8
Dimethylether	11.1
MTBE	17.3

Source: Ref. 50.

The spent acid contains polymers, which are recovered and are called *red oil*. This product finds applications due to its surface-active properties. The amounts produced reach several percent units and depend to a large extent on the acid concentration [33].

Several studies have examined the potential and claimed successes of use of additives for improving alkylation performance and reducing the acid consumption [59–63].

The consumption of utilities for the two alkylation processes using sulfuric acid is reported in Table 9.15 [64].

9.2.8 Hydrofluoric Acid Alkylation

Two types of units are in operation, each accounting for about half of the number of existent units: the Phillips Petroleum Co. unit and the UOP unit. More recent and less used is the Stratco unit.

Table 9.15 Utilities Consumption for the Two Sulfuric Acid Alkylation Processes

	Individual cooling reactors	Cascade reactor
Investment per m³ alkylate	$22,000	
Consumption per m³ alkylate		
electricity, kWh	85	65
steam 4.2 bar, kg	–	570
steam 10.5 bar, kg	510	–
cooling water ($\Delta t = 11°C$), m³	44	50
sulfuric acid, kg	43	54
caustic soda, kg	0.3	0.3

Source: Ref. 64.

The process scheme of the Phillips alkylation unit is given in Figure 9.14 and a sketch of the reactor in Figure 9.15 [65]. In 1994 there were 107 operating Phillips units worldwide.

The hydrocarbons fed to the unit are previously dried. This is very important for alkylation with hydrofluoric acid. Indeed, the presence of water induces strong equipment corrosion [43a] and also leads to increased acid consumption and to deterioration of alkylate quality [43]. Molecular sieves are more efficient drying agents than alumina or other adsorbents.

The hydrocarbon mixture (the feed and the isobutene recycle) is introduced with a pressure of 4–6 bar in the reactor recycle lines using for this purpose small diameter nozzles in order to generate high velocity and ensure a good dispersion of the hydrocarbons in the acid (see Figure 9.15).

The dispersion stream is introduced in the reactor. The temperature is maintained at the desired value by recycling the reaction mixture through external coolers. Details on the internal construction of the reactor were not published. It probably contains devices such as perforated plates for keeping the liquid phases well dispersed [37,65].

The advantage of the Phillips system is that recycling through the coolers takes place without the use of pumps. It is produced by the difference in density between the upflowing and downflowing streams, amplified by the level difference between the reactor and the coolers and by the entrainment effect generated by the high velocity stream of hydrocarbons fed through the injection nozzles.

The reactor diameter is about 1 m and the height of the whole system consisting of HF storage tank, reactor, and the hydrocarbon–acid separator is approximately 15 m.

The hydrocarbon phase that leaves the reactor contains isobutane, alkylate, small amounts of *n*-butane and propane, dissolved HF, and alkyl fluorides.

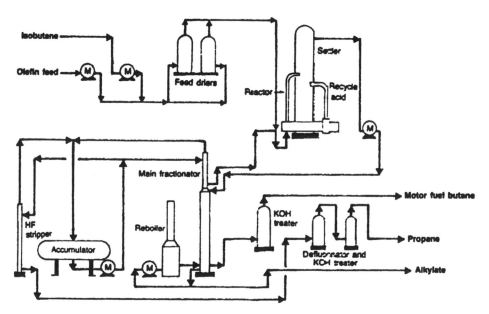

Figure 9.14 Phillips HF alkylation process.

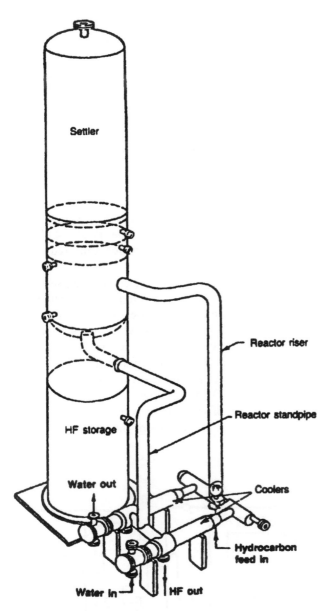

Figure 9.15   The Phillips alkylation reactor.

This mixture is separated by fractionation, after which the fluorides are neutralized.

In recent systems [37] a vessel is located below the reactor in which, if necessary,
all the hydrofluoric acid inventory may be collected. Another measure that increases
the safety of the unit is the small number of valves and pumps in the system. Actually,
a single pump is used for pumping hydrofluoric acid. The Phillips system reduces also
to a minimum the operations from which diluted HF results, in order to decrease the
size of the equipment needed for its concentration [66].

The process scheme of a UOP unit for the alkylation of butenes is given in the Figure 9.16. In this case also, the hydrocarbon feed is dried before entering the reactor. Isobutane mixed with alkenes is introduced in the reactor through a nozzle system located at several heights in order to ensure a uniform temperature distribution. The cooling is achieved by water circulating through coils situated at the upper end of the reactor.

The hydrofluoric acid enters through the lower end of the reactor and the products leave through the upper end and enter a separator. In large capacity units, such as that of Figure 9.16, two reactors are used. Reactor 1 is fed with the decanted acid from the separator of reactor 2. The feed is made of a mixture of alkenes and recycled isobutane. This mixture enters reactor 1 and the hydrocarbons from the separator of this reactor (the isobutane) enter reactor 2, together with the alkenes. In each reactor of a two reactors system, the isobutane/alkenes ratio is higher than if a single reactor were used, which improves the quality of the alkylate, allowing it to operate at a lower isobutane recycle ratio, which thus reduces the utilities consumption.

The circulation of the hydrofluoric acid from the lower ends of the separators to the reactors is ensured by pumps, which are not represented in the drawing.

An advantage of the UOP system over the Phillips process is the lower amount of hydrofluoric acid inventory, which is of 14 kg per m^3 alkylate, for systems with one reactor and 16–17 kg/m^3 alkylate for systems with two reactors. This difference is due mainly to the internal cooling system and to the fact that in the UOP process, a more intense circulation, produced by the pumps, increases the heat transfer rate.

In 1994 there were 95 units using the UOP technology.

No details concerning the construction of the reactor were published.

The regeneration of the acid is carried out in a section of the unit activated only when operating conditions require it.

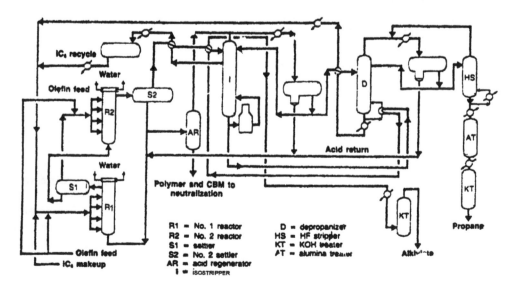

Figure 9.16   UOP process of HF alkylation of C_3–C_4 olefins.

A third process was proposed [67] by Stratco and uses the same type of reactor as that used for the alkylation with sulfuric acid (Figure 9.11) but using water cooling. No information exists concerning the number of operating units using this technology. The schema of the process is given in Figure 9.17.

Within the usual range of process variables [74] the properties of the typical alkylate obtained in a Phillips unit are given in Table 9.16 [66].

Since the quality of the products vary significantly with the composition of the feed, programs were developed [74] for the optimization of the process.

The production of alkylate is about 1.77–1.78 volumes per unit volume of alkenes in the feed.

The investment (end 1982) for a UOP unit with a capacity of 1400 m^3/day alkylate cost \$22 million [39]. The investment (end 1983) for a Phillips unit with a capacity of 550 m^3/day alkylate was \$9.2 million. More detailed data are given in the monograph of Gary and Handwerk (78).

Published data [43,66] on utilities consumption were collected and reported in Table 9.17 on the same reference basis—m^3 alkylate.

Hydrofluoric acid consumption is between 0.4–0.5 kg/m^3 [33]; Phillips indicates for its units [66] the HF consumption is only 0.3 kg/m^3 alkylate.

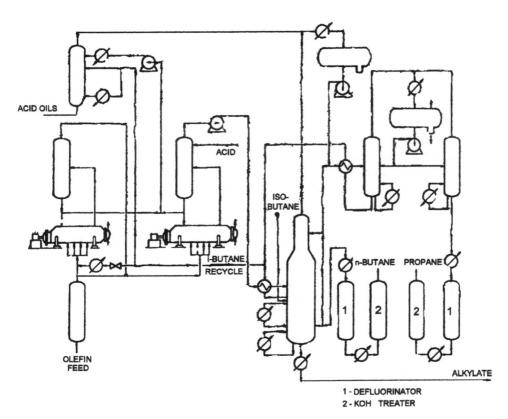

Figure 9.17 Stratco HF alkylation process.

Table 9.16 Alkylation Products from the Phillips Process

	Propene + butenes	Butenes
Density	0.693	0.697
Distillation, °C		
IBP	41	41
10%	71	76
30%	93	100
50%	99	104
70%	104	107
90%	122	125
end point	192	196
Clear F1 ON	93.3	95.5
F1 ON + 0.15 g Pb/liter	99.6	101.2
F1 ON + 0.84 g Pb/liter	104.5	107.1
Clear F2 ON	91.7	93.5
F2 ON + 0.15 g Pb/liter	99.6	100.6
F2 ON + 0.84 g Pb/liter	105.9	107.1

The major disadvantage of hydrofluoric acid is its toxicity. Concentrations of 2–10 ppm cause eye irritations, as well as of the skin and of the nasal passages. Concentrations above 20 ppm are a health and life hazard.

In 1986 tests were performed in the Nevada desert [68–70] to determine the mechanisms by which HF may disperse in the atmosphere. Despite the fact that the acid boils at 19.4°C, at the testing temperatures of 40°C only 20% of the acid was vaporized, while the rest dispersed at ground level as aerosols, which was transported with velocities of 3.8–5.6 m/s and reached distances over 8 km.

Besides the toxicity of the vapors, skin contact with hydrofluoric acid produces wounds that are very difficult to cure.

For these reasons, special attention is paid to safety measures and protection in the HF alkylation units. See Ref. 71 for details and Ref. 72 for special rules and regulations issued for the use of hydrofluoric acid.

The tendency of hydrofluoric acid to form aerosols may be efficiently controlled by additives that do not reduce its efficiency in the alkylation process [73].

Table 9.17 Utilities Consumption in Alkylation

	UOP	Phillips
Electricity, kWh	0.57	0.30
Cooling water, m^3	51.5	55($\Delta t = 11$°C)
Steam, medium pressure, kg	–	101
Steam, low pressure, kg	580	–
Fuel, mill. kJ	0.63	2.16

Source: Refs. 39, 66.

The hazards related to the use of HF, compared to those for sulfuric acid and the development of protective measures made the ratio of the units based on the respective technologies vary widely along the years.

In 1970 three-quarters of the alkylate was produced using sulfuric acid as catalyst, but later on many units using HF were built, so that in 1990 these units accounted for 47% of the alkylate produced in the U.S. [33]. After 1990, the priority was given again to sulfuric acid processes, which are generally preferred outside the borders of the U.S.

9.2.9 Alkylation on Solid Superacid Catalysts

Safety and convenience in operation are the major advantages of solid acid catalysts over liquid HF and sulfuric acid catalysts.

Several case studies investigate revamped scenarios of existing alkylation units [75]. The revamping of an existing HF alkylation unit for the use of solid catalyst will increase operation safety.

In the late 1990s at least four solid acids catalyzed processes for isobutane-olefin alkylation were being developed in pilot plant units (Table 9.18). In the same period, several other processes were in the research and development stages (Table 9.19) [79].

Other studies addressed the 1-butene/isobutene alkylation in supercritical conditions on a fixed bed of solid acid catalysts using CO_2 as diluent [96].

9.2.9.1 Topsøe/Kellogg Process [37]

The process has been tested in a 0.5 b/d pilot plant for two years, including a 6-months continuous run. It uses a simple plug-flow reactor, requires no agitation, has a low catalyst consumption, produces no volatile acid byproducts, and yields high-quality alkylate (98 RON and 94 MON). The process is operated at temperatures in the 0°C–20°C range.

Table 9.18 Catalyzed Processes for Alkylation

Process developers	Alkylate capacity, b/d	Catalyst	Process	References
Catalytica Inc. Neste Oy, Conoco Inc.	7	Proprietary catalyst	Recycle reactor with catalyst regeneration	[37,79]
Chevron Corp., Chemical Research and Licensing (CR&L)	10	SbF_5 in acid-washed silica	Slurry reactor	[79]
Haldor Topsøe Inc.' M. W. Kellogg Co.	0.5	Triflic acid on a porous support[a]	Fixed-bed reactor with recycle and catalyst regeneration	[37,79]
UOP	Not known	Proprietary, regenerable catalyst		[75]

Includes only processes at pilot plant stage.

[a] Trifluoro-methanesulfonic acid.

Table 9.19 Alkylation Processes on Solid Catalysts in Research and Development Stage 1996

Process developers	Catalyst
Allied Signal	Fluorinated ion exchange resin
Hydrocarbon Technologies Inc. (HTI)	Brönsted-acid-treated transition metal oxide
Institut Français du Pétrole (IFP)	H_2SO_4 or Al or B halide and quaternary ammonium halide or amine hydrohalide, impregnated on an organic or mineral support H_2SO_4 impregnated on silica H_2SO_4 and $HB(HSO_4)_4$ impregnated on silica
Jilin University (China)	SO_4^{2-}/ZrO_2 and SO_4^{2-}/TiO_2 H_3PO_4–BF_3–H_2SO_4 supported on SiO_2 and ZrO_2
Mobil Oil Corp.	Lewis acid in combination with a non-zeolitic solid inorganic oxide, large-pore crystalline molecular sieve, and/or ion exchange resin
Princeton University	Ultrastable H–Y zeolite with Si/Al ratio of 6.9
Texas A&M University	Sulfated zirconia
Universidad Politecnica de Valencia (Spain)	Sulfated zirconia and zeolite beta
Université Laval	Sulfated zirconia

Source: Ref. 79.

The process uses trifluormethanesulfonic acid on a porous support, in a fixed-bed reactor with recycle.

Economic estimations were published [37]. The advantage of the process as compared to sulfuric acid alkylation was estimated to be $1.07/bbl.

9.2.9.2 UOP Process

The UOP process is realized in a fluidized bed riser system, shown in Figure 9.18, with continuous catalyst reactivation. Recycling of excess isobutane from the

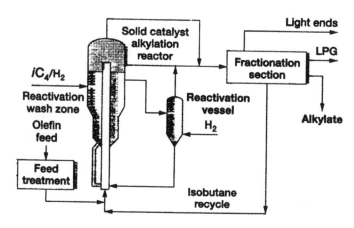

Figure 9.18 Flowsheet of UOP alkylation process with solid catalyst. (From Ref. 75.)

fractionation section controls the isobutane/olefins ratio. An optimal isoparaffin/olefin (I/O) ratio is used for maximizing the selectivity to desirable products.

The riser reactor achieves optimal contact between reactants and catalyst, without backmixing. High olefin conversion is obtained while minimizing secondary sections such as isomerization, which lowers the octane rating of the alkylate.

All the equipment used in the process is made of carbon steel to minimize equipment cost. The single vertical reactor minimizes the plot area requirements.

The reactor effluent is fractionated in distillation equipment similar to that used in conventional liquid-acid alkylation processes.

The alkylate quality from a typical FCC derived mixed-C_4 olefin feedstock is shown in Figure 9.19.

The capital and operating costs for the sulfuric acid unit and those estimated for a new solid catalyst alkylation unit are shown in Table 9.20 [75].

9.2.10 Consideration on the Design of Alkylation Reactors

The alkylation reactors for both the sulfuric acid and hydrofluoric acid catalysts can be considered as perfectly mixed reactors, which simplifies the solving of the system of kinetic equations. Thus, since the composition inside the reactor is identical to that at the outlet, no integration of the system of kinetic equations is necessary.

The main design difficulty is evaluation of the mass transfer rate, which requires among other factors, the size of the contact surface between the two liquid phases and the mean diameter of the hydrocarbon drops respectively, dispersed within the acid. The size of the droplets depends not only on the rotation rate of

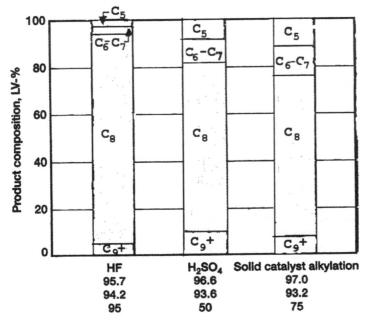

Figure 9.19 Alkylate composition.

Table 9.20 Comparison Between Grassroot Alkylation
Units

	H_2SO_4	Solid catalyst
Alkylate product, bpd	10,272	10,660
Alkylate C_5 octane, $(R+M)/2$	94.7	94.7
Capital costs, $MM		
ISBL estimated erected cost	49.1	36.4
OSBL estimated erected cost	18.4	16.8
Total capital investment, $MM	67.5	53.2
Production costs, $/bbl		
variable costs[a]	18.64	17.71
fixed costs[b]	3.86	2.22
capital charges[c]	2.81	2.75
Total production cost	25.31	22.68

[a] Feedstock, utilities, catalyst and chemicals.
[b] Labor, maintenance, overhead and interest.
[c] Depreciation and amortization.

the stirrer but also on the shape of the turbine blades, of the diffuser, and on the shape of the other internals of the reactor.

For these reasons, the safe design of an alkylation reactor is best made by scaling up results and models developed from tests performed at a conveniently small scale with feeds and in conditions anticipated for the commercial unit.

The design of alkylation reactors for solid catalyst is similar to other fixed bed or fluidized bed catalytic reactors.

REFERENCES

1. JE Johnson, FM Peterson Chemtech 296, May 1991.
2. JA Weiszmann, JH D'Auria, FG McWiliams, FM Hibbs, Octane Options. In: JJ McKetta, ed. Petroleum Processing Handbook, New York: Marcel Dekker, Inc., 1992.
3. RJ Schmidt, PL Bogdan, NL Gilsdorf. Chemtech 41, Feb. 1993.
4. JL Nocca, A Forestière, J Cosyns, Revue de l'Institut Français du Pétrole 49: 461, 1994.
5. OJ Kuznetsov, AMD Guseinov. Katalititsheskyie reactii prevrastshenia uglevodorodov, no. 6, 40, Nauka Kiev, 1981.
6. LP Hammett, AJ Deyrup, J Am Chem Soc 54: 2721, 1932.
7. LP Hammett, MA Paul, J Am Chem Soc 56: 327, 1934.
8. RJ Quann, LA Green, SA Tabac, FJ Krambeck, Ind Eng Chem 27: 565, 1988.
9. Anon. Hydrocarbon Processing 69: 134, Nov. 1990.
10. Anon. Hydrocarbon Processing 73: 142, Nov. 1994.
11. M Ojima, MV Rautenbach, CT O'Connor, Ind Ing Chem 27: 248, 1988.
12. E Angelescu, A Angelescu, IV Niculescu, Revista de chimie 30: 523, 1979; 32: 559, 1981; 32: 633, 1981.
13. E Angelescu, A Angelescu, M Udrea, G Poenaeu, Proc 6th Int Symp Heterogeneous Catalysis, Part 2, 55, Sofia, 1987.
14. IV Niculescu, L Botez, E Angelescu, Revue Roumaine de Chimie 35 (2): 193, 1990.

15. Y Chauvin, J Gaillard, J Leonard, P Bounifay, KW Andrews, Hydrocarbon Processing 61: 110, May 1982.
16. Hydrocarbon Processing 61: 175, Sept. 1982.
17. D Commerenc, Y Chauvin, J Gaillard, J Leonard, J Andrews, Hydrocarbon Processing 63: 118, Nov. 1984.
18. A Convers, D Commerenc, B Torec, Revue de l'Institut Français du Pétrole 49: 437, Oct. 1994.
19. Hydrocarbon Processing 73: 144, Nov. 1994.
20. RZ Magaril, Teoretisheskie osnovy himitsheskih protsesov pererabotci nefti, Himia Moscow, 1976.
21. SM Walas, Reaction Kinetics for Chemical Engineers, New York: McGraw-Hill, 1959.
22. GM Pancencov, AS Kazanscaia, LL Kozlov, Nauka Kiew no. 6, 1165, 1981.
23. L Forni, R Invernizzi, Ind Eng Chem Process Res Dev 12 (4): 455, 1973.
24. M Langlois, Petroleum Refiner 31: 79, Aug. 1952.
25. JF Galimov, MN Rahimov, Himia i Tehnolgia Topliv i Masel 18 (11), 1989.
26. DG Tajbl, UOP Catalytic Condensation Process for Transportation Fuels, In: RA Mayers, ed. Handbook of Petroleum Refining Process, New York, McGraw-Hill, 1986.
27. G Egloff, EK Jones, 4th World Petroleum Congress, Vol. V, 1956.
28. Anon. Hydrocarbon Processing 73: 142, Nov. 1994.
29. F Nierlich, Hydrocarbon Processing 71: 45, Feb. 1992.
30. Anon. Hydrocarbon Processing 73: 92, Nov. 1994.
31. The Clear Air Act Amendements of 1990, Title II, Sect. 219.
32. J Cosyns, JL Nocca, P Keefer, K Masters, Hydrocarbon Technology International 1993, P Harrison, ed. London: Sterling Publishing Co. Ltd., 1993.
33. LF Albright, Oil and Gas Journal 88: 79, 12 Nov. 1990.
34. LF Albright, Oil and Gas Journal Special 92: 49, 22 Aug. 1994.
35. M Misono, T Okuhara, Chemtech 23, Nov. 1993.
36. J Weitkamp, S Ernst, Proceedings of the Thirteenth World Petroleum Congress, vol. 3, 315, New York: John Wiley & Sons, 1992.
37. AK Rhodes, Oil and Gas Journal 92: 52, 22 Aug. 1994.
38. LF Albright, MA Spalding, JA Nowinski, RM Ybarra, Ind Eng Chem Res 27: 381, 1988.
39. LF Albright, MA Spalding, CG Kopser, RE Eckert, Ind Eng Chem Res 27: 368, 1988.
40. LF Albright, MA Spalding, J Faunce, RE Eckert, Ind Eng Chem Res 27: 391, 1988.
41. LF Albright, KE Kranz, Ind Eng Chem Res 31: 475, 1992.
42. JE Germain, Catalytic Conversion of Hydrocarbons, New York: Academic, 1969.
43. BR Shah, UOP Alkylation Process, In RA Meyers, ed. Handbook of Petroleum Refining Process, New York: McGraw-Hill, 1986.
43a. Anon. Refiners discuss HF alkylation process and issues, Oil & Gas Jour. 90: (Apr. 6) 67–72, 1992.
44. KW Li, RE Eckert, LF Albricht, Ind Eng Chem Process Devs Dev 9: 434, 1970.
45. FB Spow, Ind Eng Chem Process Des Dev 8: 254, 1969.
46. L Lee, P Harriott, Ind Eng Chem Process Des Dev 16(3): 282, 1977.
47. K Kranz, JR Peterson, DC Geaves, Hydrocarbon Technology International 1994, P Harrison, ed. London: Sterling Publishers 1994, p. 65.
48. G Chaput, J Laurent, JP Boitiaux, J Cosyns, P Sarrazin, Hydrocarbon Processing 71: 51, 1992.
49. JF Joly, E Benazzi, OAPEC-IFP Joint Workshop, 5–7 July 1994, Report 18, French Petroleum Institute, Rueil-Malmaison, France. 1994.
50. LF Albright. Oil and Gas Journal 88: 79, Nov. 12, 1990.
51. U.S. Patent 5,258,568, Mobil Oil Corp.
52. S Novalny, RG McClung. Hydrocarbon Processing 68: 66, Sept. 1989.

53. P Sarrazin, CJ Cameron, Y Barthel. Oil and Gas Journal 91: 86–90, Jan. 25, 1993.
54. KR Masters, EA Prahaska. Hydrocarbon Processing 67: 48, Aug. 1988.
55. H Lerner, VA Citarella. Hydrocarbon Processing 70: 89, Nov. 1991.
56. Oil and Gas J 90: 72, Feb. 17, 1992.
57. JT Tagarov, VT Sumanov, SN Hajdev. Himia Tehnol Topliv Masel 2: 14, 1988.
58. EA Nikiforov. Himia Tehnol Topliv Masel 6: p. 13, 1987.
59. GM Kramer. U.S. Patent 4,357,482, Exxon Res. Eng. Co., Nov. 2, 1982.
60. GM Kramer. U.S. Patent 4,560,825, Exxon Res. Eng. Co., Dec. 24, 1985.
61. MR Nicholson, RG Miller, RL Knickerbocker, C Go. Oil and Gas J 85: 47, Sept. 29, 1986.
62. VA Krylov, VG Riabov, AL Sviridenco, NP Uglov, JM Jorov. Himia Tehnol Topliv Masel 1: 19, 1990.
63. VG Riabov, AL Sviridenco, AV Kudinov, SA Stepanov. Himia Tehnol Topliv Masel 10: 2, 1992.
64. Hydrocarbon Processing 73: 89, Nov. 1994.
65. LF Albright. Oil and Gas J 88: 70, Nov. 26, 1990.
66. T Hutson, Jr., WC McCarthy. Phillips Alkylation Process. In RA Meyers, ed. Handbook of Petroleum Refining Processes, New York: McGraw-Hill, 1986, pp. 1–24.
67. LE Chapin, GC Liollos, TM Robertson. Hydrocarbon Processing 64: 67, Sept. 1985.
68. DN Biewett, JF Yohn, RP Koopman, TC Brown. Conduct of Anhydrous Hydrofluoric Acid Spill Experiments, International Conference on Vapor Cloud Modeling, Cambridge, MA, Nov. 1977.
69. DN Biewett, JF Yohn, DL Erwak. Evaluation of SLAB and Degadis Against the Hydrofluoric Acid Experimental Base, International Conference on Vapor Cloud Modeling, Cambridge MA, Nov. 1977.
70. Studies Cover HF Spills and Mitigation, Oil and Gas J 86: 58, Oct. 17, 1988.
71. B Scott. Identify alkylation hazards. Hydrocarbon Processing 71: 77, Oct. 1992.
72. South Coast Air Quality Management District, Guideline to Comply with Proposed Rule 1410 Hydrogen Fluoride Storage and Use, El Monte, CA June 6, 1991.
73. JC Sheckler, HU Hammershaimb, LJ Ross, KR Comey. Oil and Gas J 92: 60, Aug. 22, 1991.
74. GL Funk, JA Feldman. Hydrocarbon Processing 62: 92, Sept. 1983.
75. JM Meister, SM Black, BS Muldoon, DH Wei, CM Roeseler. Optimize Alkylate Production for Clean Fuels, Hydrocarbon Processing 79: 11: 63–75, May 2000.
76. L Kane, S Romanow, New Technology for High-Octane Gasoline, Hydrocarbon Processing: 79, 30–33, March 2000.
77. NPRA, Paper AM-92-55, San Antonio, Texas, March, 1992.
78. JH Gary, E Glenn. Handwork Petroleum Refining Technology and Economics, 3rd Edition, Marcel Dekker, Inc., 1994, p. 336.
79. P Rao, R Sorab, S Vatcha. Solid-acid alkylation process development is at crucial stage, Oil and Gas J 94: 56–61, Sept. 9, 1996.
80. A Huss, CR Kennedy. U.S. Patent 4,935,577, Mobil Oil Corp., June 19, 1990.
81. MB Berenbaum, TPJ Izod, DR Taylor, JD Hewes. U.S. Patent 5,220,087, Allied Signal Inc., June 15, 1993.
82. CH Liang, RG Anthony. Preprints, Div. Petroleum Chemistry, ACS 38(4): 892–894, 1993.
83. C Guo, S Yao, J Cao, Z Qian. Applied Catalysis A: General, 107(2): 229–238, 1994.
84. C Guo, S Liao, Z Qian, K Tanabe. Applied Catalysis A: General, 107(2): pp. 239–248, 1994.
85. M Simpson, J Wei, S Sundaresan. The Alkylation of Isobutane with 2-butene over ultrastable HY-Type Zeolite, Ind. Eng. Chem. Res. 35(11) 3861–3873 (1996).

86. A Corma, MF Juan-Rajadell, JM Lopez-Nieto, A Martinez, C Martinez. Applied Catalysis A: General, 111(2): 1994, pp. 175–189.
87. JF Joly, E Benazzi, C Marcilly, R Pontier, JF Le Page. U.S. Patent 5,444,175, Institut Français du Pétrole, Aug. 22, 1995.
88. X Song, Z Sayary. Chemtech, 25(8): 27–35, 1995.
89. A 'Clean' Catalyst for Alkylation, Chemical Engineering, 103(1): 23, January 1996.
90. JF Joly, C Marcilly, E Benazzi, F Chaigne, JY Bernhard. U.S. Patent 5,489,730, Institut Français du Pétrole, Feb. 6, 1996.
91. E Benazzi, JF Joly, C Marcilly, F Chaigne, JY Bernhard. U.S. Patent 5,489,728, Institut Français du Pétrole, Feb. 6, 1996.
92. E Benazzi, JF Joly, F Chaigne, JY Bernhard, C Marcilly. U.S. Patent 5,489,729, Institut Français du Pétrole, Feb. 6, 1996.
93. ML Petkovic, G Larszn, Oligomerization of n-Butene on a surface of a Ferite Catalyst: Real-Time Monitoring in a Packed Bed Reactor during Isomerisation to Isobutene, Ind Eng Chem 38: 1822–1829, 1999.
94. LF Albright, KV Wood. Alkylation of Isobutane with C_3–C_4 Olefins: Identification and Chemistry of Heavy-End Products, Ind Eng Chem 36: 2110–2112, 1997.
95. J Pater, F Cardona, C Canaff, NS Gnop, G Szabo and M Guisnet, Alkylation of Isobutane with 2-Butane over a HFAU Zeolite. Composition of coke and Deactivation Effect, Ind Eng Chem 38: 3822–3829, 1999.
96. MS Clark, B Subramanian. Extended Alkylation Production Activity during Fixed-Bed Supercritical 1-Butene/Isobutane Alkylation on Solid Acid Catalysts Using Carbon Dioxide as diluent, Ind Eng Chem 37: 1243–1250, 1998.

10

Hydrofining and Hydrotreating

Hydrofining and hydrotreating are processes in which selected petroleum fractions react with hydrogen in the presence of monofunctional catalysts: metallic sulfides or oxides. The purpose of hydrofining is the elimination of the heteroatoms by means of hydrogenolysis reactions. In hydrotreating, which uses more severe operating conditions and more active catalysts, the polycylic aromatic hydrocarbons also undergo a partial hydrogenation.

Hydrofining was developed in order to meet the low levels of sulfur content required by the quality norms for the final distillate products or, in order to make naphtha acceptable as feed to catalytic reforming on platinum catalysts. Attention was also paid to the removal of nitrogen, which decreases the oxidation stability of the products.

The hydrogenolysis of the sulfur- and nitrogen-containing compounds led inherently to the elimination of the more reactive compounds containing oxygen and to the hydrogenation of the unsaturated hydrocarbons, that are possibly contained in the fraction submitted to the process.

The terms of desulfuration and denitration currently used indicate the nature of the main heteroatom, the removal of which is sought. But in all cases all the heteroatoms are affected. The nature of the catalyst and the operating conditions influence only the degree to which they are removed.

The increase of quality demands concerning sulfur and nitrogen content in heavy and vacuum distillates led to the situation in which the hydrogenolysis of the easily accessible C–S and C–N bonds by classical hydrofining developed in the 1970s was no longer sufficient. Moreover, the need to remove the nickel and vanadium (contained within more complex chemical structures) appeared, especially in order to be able to submit the heavy vacuum distillates to catalytic cracking.

These new demands led to modifications of the catalytic system and of the operating conditions of the hydrofining to achieve a preliminary hydrogenolysis of some of the C–C bonds. In this manner, the S, N, V, and Ni heteroatoms contained in the feed became accessible and could be eliminated.

More difficult problems appeared as the norms concerning the sulphur, nitrogen, and metals content in the residual fuel became more strict. The difficulties are caused mainly by the fact that the process occurs in vapors/liquid mixed phase, wherein the hydrogen has to diffuse through a liquid film in order to reach the active sites of the catalyst.

As indicated above, hydrotreating belongs to a different category of processes, where besides the hydrogenolysis reactions, the bi- and polycyclic aromatics are hydrogenated to hydroaromatic hydrocarbons.

In the case of gas oils, a significant increase of the cetane number results, which makes it possible to convert highly aromatic gas oils to Diesel fuel.

The hydrotreating of lube oils leads to the increase of the viscosity index in conditions which are often preferred to selective solvent refining.

The partial hydrogenation of the polycyclic aromatic hydrocarbons is achieved by operating the hydrotreating in more severe conditions than of the hydrofining and by using more active catalysts.

Although as indicated above, there are large differences between the hydrofining of the distillate, of the residues, and the hydrotreating, they have a large number of similar features that allows the joint treatment of their theoretical aspects.

10.1 PROCESS THERMODYNAMICS

In the following, the thermodynamic equilibrium of the main reaction that takes place in the hydrofining and hydrotreating will be analyzed, within the limits set by the available thermodynamic constants.

In this analysis, the method for representing the equilibrium given in Section 1.2 is used whenever necessary.

It is to be remarked that contrary to the cracking processes, where the thermodynamic analysis shows in most cases only the direction of the transformations, in the case of hydrofining and hydrotreating it is capable of supplying important quantitative data.

10.1.1 Hydrogenolysis of Sulfur Compounds

The logarithms of the equilibrium constants of the hydrogenolysis reactions of some characteristic sulfur compounds for the temperature domain used in hydrofining and hydrotreating are given in Table 10.1 [1].

The Table shows that the thiols, alkyl sulfides, and disulfides have the equilibrium completely displaced towards the desulfurization, so that the reactions are not limited by thermodynamics.

For thiocyclopentane and thiocyclohexane, the equilibrium conversion at 482°C is 93% and 96% respectively and increases with the pressure.

For thiophen, 2-methyl-thiophen, and 3-methyl-thiophen the thermodynamic values suggested by various authors are contradictory [1–4]. In the calculations given below, the most recent values [5] were used.

Table 10.1 Equilibrium Constants for Selected
Hydrogenolysis Reactions

	$\log K_p$	
Reaction	371°C	482°C
$CH_3SH + H_2 \rightleftharpoons H_2S + CH_4$	+6.10	+4.69
$C_2H_5SH + H_2 \rightleftharpoons H_2S + C_2H_6$	+5.01	+3.84
$CH_3-CH-CH_3 + H_2 \rightleftharpoons H_2S + C_2H_8$ $\quad\quad\vert$ $\quad\quad SH$	+4.45	+3.52
$\quad\quad CH_3$ $\quad\quad\vert$ $CH_3-C-CH_3 + H_2 \rightleftharpoons H_2S + i\text{-}C_4H_{10}$ $\quad\quad\vert$ $\quad\quad SH$	+4.68	+3.81
$CH_3-S-CH_3 + 2H_2 \rightleftharpoons H_2S + 2CH_4$	+11.41	+8.96
$CH_3-S-C_2H_5 + 2H_2 \rightleftharpoons H_2S + CH_4 + C_2H_6$	+9.97	+7.74
$+2H_2 \rightleftharpoons H_2S + C_4H_{10}$	+5.26	+3.24
$+2H_2 \rightleftharpoons H_2S + C_5H_{12}$	+5.92	+3.97
$CH_3-S-S-CH_3 + 3H_2 \rightleftharpoons 2H_2S + 2CH_4$	+19.03	+14.97
$C_2H_5-S-S-C_2H_5 + 3H_2 \rightleftharpoons 2H_2S + 2C_2H_6$	+16.79	+13.23

For the three reactions it results:

$+4H_2 \rightleftharpoons H_2S + C_4H_{10}$

$(\Delta H^0_{700})_r = 67{,}060 \text{ cal/mol}$

$(\Delta S^0_{700})_r = -77.67 \text{ cal/mol} \cdot \text{grd}$

$+4H_2 \rightleftharpoons H_2S + C_5H_{12}$

$(\Delta H^0_{700})_r = -64020 \text{ cal/mol}$

$(\Delta S^0_{700})_r = -78.20 \text{ cal/mol} \cdot \text{grd}$

$-C +4H_2 \rightleftharpoons H_2S + C_5H_{12}$

$(\Delta H^0_{700})_r = -65620 \text{ cal/mol}$

$(\Delta S^0_{700})_r = -79.49 \text{ cal/mol} \cdot \text{grd}$

Based on these constants, in Figure 10.1 the equilibrium of the three reactions are plotted. These plots show that obviously the process must be performed at pressures of several tens of atmospheres in order to obtain conversions of an order of 98%–99% for the thiophen and its homologs. One may assume that the higher heterocyclic compounds have a similar behavior.

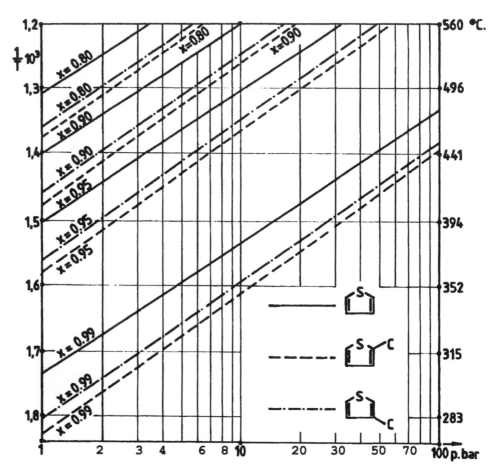

Figure 10.1 Hydrogenolysis equilibrium for thiophen, 2-methyl-thiophen and 3-methyl-thiophen.

10.1.2 Hydrogenolysis of Nitrogen Compounds

Table 10.2 contains the heats of reaction, entropies [5], and equilibrium constant values for several representative nitrogen compounds.

   The data of the table show that at the hydrofining temperature, the equilibrium for the secondary, primary, and tertiary alkylamines, for pyrolidine, aniline, and diphenylamine is displaced to the right, thus the potential exists for their complete conversion.

   A different situation exists for pyridine and picolines, for which the plots of the respective equilibrium is given in the graph of Figure 10.2.

   The graph indicates that at a pressure of several tens of atmospheres the equilibrium conversion will reach values of over 95%. One may assume that an analogous behavior will be valid for the polycyclic compounds of similar structures.

Table 10.2   Heats of Reaction Heats, Entropies, and Equilibrium Constants for the Hydrogenolysis of Selected Nitrogen Compounds

Reaction	$(\Delta H^0_{700})_r$ cal/mol	$(\Delta S^0_{700})_r$ cal/mol·grd	log K_p 700 K	800 K
$CH_3-NH_2 + H_2 \rightleftharpoons NH_3 + CH_4$	−24,280	−0.657	+7.45	+6.50
$C_2H_5NH_2 + H_2 \rightleftharpoons NH_3 + C_2H_6$	−21,600	−1.437	+6.43	+5.59
$C_3H_7NH_2 + H_2 \rightleftharpoons NH_3 + C_3H_8$	−19,840	−1.147	+5.95	+5.18
$C_4H_9NH_2 + H_2 \rightleftharpoons NH_3 + C_4H_{10}$	−20,420	−0.707	+6.23	+5.43
$C_2H_5-CHNH_2 + H_2 \rightleftharpoons NH_3 + C_4H_{10}$ (with CH$_3$)	−17,550	+2.113	+5.95	+5.26
$CH_3-CNH_2 + H_2 \rightleftharpoons NH_3 + i\text{-}C_4H_{10}$ (with two CH$_3$)	−16,190	+0.763	+5.23	+4.59
$CH_3-NH-CH_3 + 2H_2 \rightleftharpoons NH_3 + 2CH_4$	−44,700	+2.186	+14.45	+12.71
$C_2H_5-NH-C_2H_5 + 2H_2 \rightleftharpoons NH_3 + 2C_2H_5$	−37,280	+2.366	+12.17	+10.72
$CH_3-N-CH_3 + 3H_2 \rightleftharpoons NH_3 + 3CH_4$ (with CH$_3$)	−63,140	+8.019	+21.49	+19.08
$C_2H_5-N-C_2H_5 + 3H_2 \rightleftharpoons NH_3 + 3C_2H_5$ (with C$_2$H$_5$)	−52,960	+9.119	+18.55	+16.48
(pyrrolidine) $+ 2H_2 \rightleftharpoons NH_3 + C_4H_{10}$	−41,310	−18.474	+8.87	+7.26
(piperidine) $+ 5H_2 \rightleftharpoons NH_3 + C_5H_{12}$	−84,390	−105.245	+3.35	+0.053
(2-methylpyridine) $+ 5H_2 \rightleftharpoons NH_3 + C_6H_{14}$	−79,320	−105.515	+1.707	−1.393
(3-methylpyridine) $+ 5H_2 \rightleftharpoons NH_3 + i\text{-}C_6H_{14}$	−82,470	−106.835	+2.40	−0.82
(aniline, NH_2) $+ H_2 \rightleftharpoons NH_3 + C_6H_6$	−13,790	−1.257	+4.04	+3.50
(aminobiphenyl, NH_2) $+ H_2 \rightleftharpoons$ (biphenyl) $+ NH_3$	−15,600[a]	11.169[a]	+7.32	+6.71

[a] At 298 K.

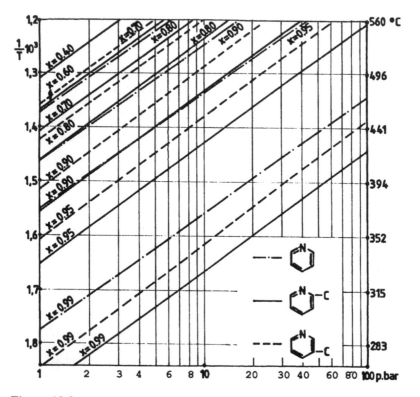

Figure 10.2 Hydrogenolysis equilibrium for pyridine and picolines.

10.1.3 Hydrogenolysis of Oxygenated Compounds

Table 10.3 gives the heats and entropies of reaction, as well as the logarithms of the equilibrium constants for several oxygenated compounds contained in the crude oil. From these data it results that, at least for the studied compounds, the hydrogenolysis reactions do not occur thermodynamic limitations.

10.1.4 Hydrogenation of Alkenes and Dienes

In the hydrofining of products from cracking or pyrolysis processes, the feed contains alkenes, dienes and possibly, unsaturated cyclic compounds.

In the presence of hydrogen and of the specific hydrofining catalyst, these hydrocarbons will be hydrogenated to the corresponding saturated compounds. The amount of hydrogen consumption and the heat developed depend on the content of alkenes and dienes in the feed and they may reach important values.

The thermodynamic equilibrium for the hydrogenation of alkenes for the lines of constant conversion at 90% and 99% in the coordinate log p vs. $1/T$ are represented in Figure 10.3.

For alkenes with double bond in the 1 position, the equilibrium is identical for the whole series of hydrocarbons $C_5H_{10}-C_{20}H_{40}$.

For isomers, the dispersion of the conversions was calculated using the thermodynamic constants for the 6 pentenes and for 14 of the 17 isomers of the hexene.

Table 10.3 Heats of Reaction, Entropies and Equilibrium Constants for the Hydrogenolysis of Selected Oxygen Compounds

Reaction	$(\Delta H^0_{700})_r$ cal/mol	$(\Delta S^0_{700})_r$ cal/mol · grd K	$\log K_p$ 700 K	800 K
(phenol) $C_6H_5OH + H_2 \rightleftharpoons H_2O + C_6H_6$	−16,660	−0.86	+5.02	+4.37
(o-cresol) $+ H_2 \rightleftharpoons H_2O + C_7H_8$	−16,890	+1.163	+5.53	+4.87
(m-cresol) $+ H_2 \rightleftharpoons H_2O + C_7H_8$	−15,690	+2.083	+5.36	+4.75
(p-cresol) $+ H_2 \rightleftharpoons H_2O + C_7H_8$	−17,310	+4.353	+6.36	+5.69
(benzoic acid) $C_6H_5COOH + 3H_2 \rightleftharpoons 2H_2O + C_7H_8$	−35,830	−18.311	+7.19	+5.79

The graph shows that at the temperatures and the pressures used in hydrofining, the hydrogenation of the alkenes is actually complete.

The equilibrium for the hydrogenation of dienes to alkanes is represented also in Figure 10.3. The calculations were made for the dienes C_4 and C_5 for which thermodynamic constants were available [5].

The graph shows that the equilibrium is much more favorable for dienes than for alkenes. For the former, the equilibrium conversions reach 99% in conditions in which for alkenes they do not reach but 90%.

10.1.5 Hydrogenation of Aromatics

It is known that the benzene ring is very difficult to hydrogenate. Accordingly the hydrogenation of alkylbenzenes does not occur in the hydrofining and hydrotreating processes.

The partial or complete hydrogenation of the polycyclic aromatic hydrocarbons is of interest, especially for the condensed polycyclic aromatic hydrocarbons, the derivatives of which are present in the crude oil fractions. The problem may be conveniently examined for naphthalene, mono- and dialkyl naphthalenes, for which thermodynamic constants are available [5,6].

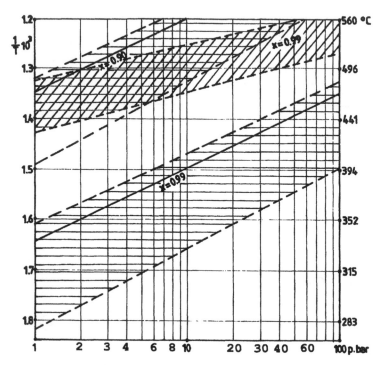

Figure 10.3 Hydrogenation equilibrium for alkenes and dienes.
— n-alkenes, 1, C_5-C_{20}, ▤ Limits for iso-alkenes, ▨ Dienes C_4-C_{10}.

For the hydrogenation of naphthalene to 1,2,3,4-tetrahydronaphthalene, the variation of the heat of formation and of the entropy during the reaction have the values:

$$(\Delta H^0_{298}) = -29480 \text{ cal/mole} = -123400 \text{ J/mol}$$

$$(\Delta S^0_{298}) = -53442 \text{ cal/mole·degree} = -223.75 \text{ J/mol·degree}$$

On the basis the graph of Figure 10.4 shows the lines corresponding to constant conversion for the hydrogenation of napthalene to tetrahydronaphthalene.

For mono- and dialkylnaphthalenes, the available data [6,7] do not allow an exact representation of the equilibrium. One may estimate that the equilibrium constants are less favorable than that for naphthalene.

In order to obtain a given conversion at the same pressure, the required temperatures should be about 25°C lower for mono-alkyl-naphthalenes and about 40°C lower for dialkyl-naphthalenes.

The graph of Figure 10.4, shows that the hydrogenation of naphthalene to tetra-, hydro-napthalene takes place with satisfactory conversion, at a moderate temperature when operating at several tens of atmospheres, preferably with an excess of hydrogen in the presence of an active catalyst.

The phenantrene and the higher aromatic hydrocarbons with isolated or condensed rings have less favorable equilibrium constants than naphthalene and therefore need more severe reaction conditions.

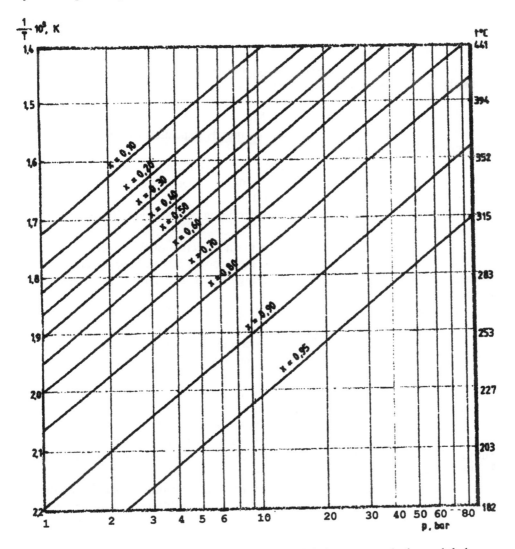

Figure 10.4 The hydrogenation equilibrium for naphthalene to tetra-hydronaphthalene.

10.2 CATALYSTS

The catalysts used in the hydrofining and hydrotreating processes are monofunctional catalysts containing metallic sulfides or oxides. As an exception, the second step of the hydrotreating processes may use metals, sometimes precious metals, supported on γ-Al$_2$O$_3$.

The used catalyst depends on the nature of the process: hydrofining or hydrotreating, on the nature of the feed (residue or distillate), and on the element or elements that have to be removed (sulphur or nitrogen in hydrofining of the distillates, sulphur and nickel-vanadium, in the case of residues).

In hydrofining, where the main objective is desulfurization, catalysts of Co-Mo supported on γ-alumina are mostly used. The metals are deposited as oxides, using to this purpose soluble salts of Co and Mo. During the process, the oxides are converted to sulfides or oxi-sulfides (with oxygen to sulfur ratios which are some-times nonstoichiometric). The molybdenium-sulphur bonds are believed to be the catalytic active sites [8]. In order to speed up the sulfiding process and to shorten the time for recovering the catalyst activity following regeneration, several sulfiding methods are suggested and used [9,35]. In general those are recommended by the catalyst manufacturer.

The atomic ratio between Co and Mo is about 0.3, the Co having in effect the role of a promoter. The total amount of metals deposited on the support varies between the limits 8–13%.

A favorable effect [7] is produced by a second promoter element, usually Ni or Fe. Thus, double-promoted catalysts are obtained: Co:Ni:Mo or Co:Fe:Mo, having atomic ratios of approximately 0.3:0.2:1.0.

The specific surface areas of the hydrofining catalysts are 200–300 m^2/g, and the bulk densities 480–800 kg/m^3 [94].

Ni-Mo catalysts are used when the main objective of the hydrofining is deni-tration. This is the case especially when the product has to have a better stability toward oxidation (for example in lubricating oils). The activity of a Co-Mo catalyst is not sufficient.

The metals content of these catalysts is 10–14% Mo and 2–4% Ni [8].

In addition, the Ni-Mo catalysts have sufficient hydrogenating activity to satu-rate the alkenes contained in the thermal cracking products. The selection of these catalysts is therefore determined also by the inclusion in the feed of cracking pro-ducts.

In the hydrotreating processes, the partial hydrogenation of the aromatic hydrocarbons requires a stronger hydrogenation activity that the Ni-Mo catalysts can occasionally provide. In such cases Ni-W catalysts are used.

Different from the Ni-Mo catalysts, those containing Ni-W are sensitive to sulfur. Their hydrogenating activity decreases strongly when the sulfur concentration in the feed exceeds some rather low values.

This situation may be illustrated by the relative catalytic activity of the two catalysts for the hydrotreating of two feeds with quite different sulfur contents (the activity given in the table represents the relative rate constant, considering the hydrogenation reaction to be of first order) [10].

Sulfur in feed (% wt)	1.7	0.14
Ni — Mo catalyst	100	100
Ni — W catalyst	90	225

From these data it results that at a low sulfur content of 0.14% the Ni-W catalyst is two times more active in the hydrogenation of the aromatic hydrocarbons, than the Ni-Mo catalyst. It becomes less active than Ni-Mo when the sulfur content in the feed is higher (1.7%).

For these reasons, at the hydrotreating of the gas oils one proceeds many times with two reaction steps: the first performs the desulfurization on a Ni-Mo or Co-Mo catalyst, while the second performs a partial hydrogenation of the aromatic hydro-carbons on Ni-W or Pt/alumina catalysts, with the elimination of the produced H_2S between the steps.

The Ni-Mo and Co-Mo catalysts eliminate also the oxygen, which results in a significant improvement of the product color.

The hydrofining and especially the hydrotreating of the residue feeds have major problems related to the diffusion of their heavy components and especially of the asphaltenes towards the central portion of the catalyst particles.

For such feeds, the preferred catalysts are Co-Mo and Ni-Mo on a porous alumina support, having a surface area of 200–300 m^2/g. Both types of catalysts promote both the desulfurization and the dematallation, the extent of the two activities being dependent on the dimensions of the pores. The larger pores favor the removal of vanadium and of nickel, while the small pores the desulfurization (see Figures 10.5 and 10.6 [11].

The distribution of the pore sizes for two typical catalysts, developed by the Petroleum French Institute [12], HDT for desulfurization and HDM for demetallation, are given in Figure 10.7.

In order to make easier the diffusion of the components with high molecular mass, such as those that contain metals, the catalyst IFP-HDM is shaped as needles, similar to the shape of chestnuts shell or of hedgehogs. Such a form ensures good

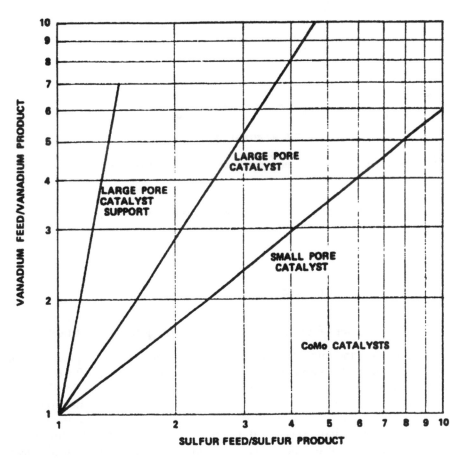

Figure 10.5 Removal of sulfur and vanadium on various Co-Mo catalysts.

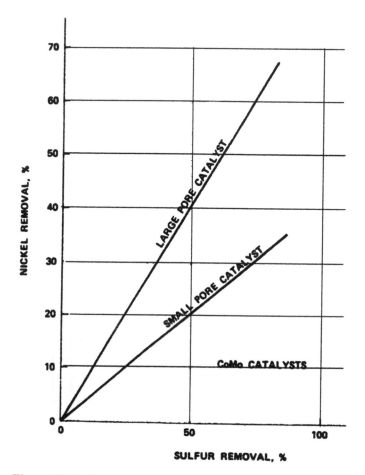

Figure 10.6 The removal of sulfur and nickel on various Co-Mo catalysts.

access to the internal surface for demetallation [13], while also eliminating to a large extent the radial distributions specific to the spherical granules of catalyst (see Figure 10.8).

10.3 REACTION MECHANISMS

10.3.1 Hydrogenolysis of Sulfur, Nitrogen and Oxygen Compounds

The range of the sulfur and nitrogen compounds with various structures present in the straight run gasoline, middle- and vacuum distillates is given in Table 10.4 on p. 602.

Several studies [95–97] have investigated the hydrogenolysis reaction in terms of the mechanism of the interaction of the catalyst with the heteroatomic molecules, the manner in which the hydrogen intervenes in these transformations, and the structure of the intermediate species. It is considered that molybdenum sulfide forms structures that comprise the active sites [14], separated by known distances [15]. This finding led to the hypothesis that hydrogenolysis reactions are the result of

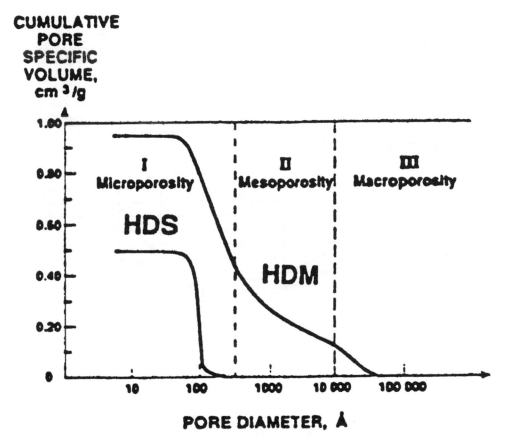

Figure 10.7   Pores size distribution for two typical catalysts.

the adsorption of the molecule on two neighboring active sites according to a doublet mechanism, similar to that which was described for the hydrogenation of ethene [16].

Mechanisms involving the participation of a single active site have been also formulated [8,17].

Without entering into the details of this hypothetical mechanism, overall reactions schemes may be formulated, leading to intermediary substances and products that were detected in the hydrogenolysis reactions:

Thiols

$$R{-}CH_2{-}SH \xrightarrow{+H_2} RCH_3 + H_2S$$

Sulfides

$$R{-}CH_2{-}S{-}CH_2{-}R' \xrightarrow{+H_2} RCH_3 + R'CH_2SH \xrightarrow{+H_2} RCH_3 + R'CH_3 + H_2$$
$$\xrightarrow{+H_2} RCH_2SH + R'CH_3 \xrightarrow{+H_2}$$

Disulfides

$$R{-}CH_2{-}S{-}S{-}CH_2{-}R' \xrightarrow{+H_2} RCH_2SH + R'CH_2SH \xrightarrow{+2H_2}$$
$$\rightarrow RCH_3 + R'CH_3 + 2H_2S$$

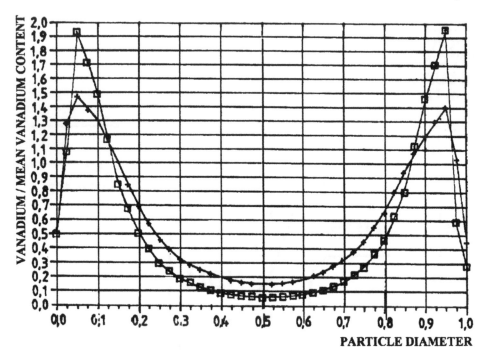

Figure 10.8 Vanadium distribution on a spherical catalyst particle. □ – 0.1% asphaltenes in feed, + 1.8% asphaltenes in feed.

Cyclic sulfides
$$H_2C{-}CH_2 \quad H_2C \quad CH_2 \quad \overset{+H_2}{\longrightarrow} C_4H_9SH \overset{+H_2S}{\longrightarrow} C_4H_{10} + H_2S$$

Thiophens
$$\overset{+3H_2}{\longrightarrow} C_4H_9SH \overset{+H_2}{\longrightarrow} C_4H_{10} + H_2S$$

Benzothiophens
$$\overset{+3H_2}{\longrightarrow} \quad {-}C_2H_5 + H_2S$$

Dibenzothiophens
$$\overset{+2H_2}{\longrightarrow} \quad + H_2S$$

Phenols
$$OH \overset{+H_2}{\longrightarrow} + H_2O$$

$$OH \overset{+H_2}{\longrightarrow} + H_2O$$

Naphthenic acids

$$R\text{-}C_6H_{10}\text{-COOH} \xrightarrow{+3H_2} R\text{-}C_6H_{10}\text{-CH}_3 + {}_2H_2O$$

Pyridines

$$\text{pyridine} \xrightarrow{+3H_2} \text{piperidine} \xrightarrow{+2H_2} C_5H_{11}NH_2 \xrightarrow{+H_2} C_5H_{12} + NH_3$$

Quinoline

$$\text{quinoline} \xrightarrow{+2H_2} \xrightarrow{+2H_2} \underset{NH_2}{C_3H_7} \xrightarrow{+H_2} C_3H_7 + NH_3$$

Indole

$$\text{indole} \xrightarrow{+2H_2} \underset{NH_2}{C_2H_3} \xrightarrow{+H_2} C_2H_3 + NH_3$$
$$\xrightarrow{+2H_2} C_2H_5 + NH_3$$

Carbazole

$$\text{carbazole} \xrightarrow{+H_2} \underset{NH_2}{} \xrightarrow{+H_2} + NH_3$$

The hydrogenation of the rings of the polycyclic aromatic derivatives can occur also before the hydrogenolysis, these reactions being competitive.

10.3.2 Hydrogenation Reactions

Compared to classical hydrogenation reactions, those occurring in hydrofining of polycyclic aromatic hydrocarbons and especially of the aromatic compounds with sulfur or nitrogen involve a competition between hydrogenation and hydrogenolysis.

The hydrogenation of several aromatic hydrocarbons on a sulfided catalyst of Co-Mo/γ-Al$_2$O$_3$, at a temperature of 325°C and 75 bar, is presented in Figure 10.9 [18]. For all reactions, the rate constants are indicated. These were calculated assuming pseudo–first order and are expressed in 1/g cat·s.

The hydrogenation of the polycyclic aromatic hydrocarbons: naphthalene, anthracene, pyrene, and chrysene was studied recently [19] using also sulfated catalysts of Co-Mo on γ-Al$_2$O$_3$ (1.7% Co, 7.0% Mo). The operating conditions were of 350°C and 191.6 bar. The obtained results, including the relative rate constants, were calculated and are given in Figure 10.10. In this work, the rate constants were calculated using the Langmuir-Hinshelwood-Hougen-Watson kinetic model, considering that the reactions are of the first order in hydrocarbon. The bold numbers in the figure represent the numerator of the Langmuir-Hinshelwood expressions and are given in 1/kg cat·sec. They are therefore larger by three orders of magnitude than those in Figure 10.9. Since in both studies first order kinetics were used, the numbers in the two figures are comparable if one takes into account the difference in the units

Table 10.4 Distribution of Sulfur Compounds in Straight-run Fractions

Alkyl compounds	Fractions		
	82–177°C	177–343°C	343–525°C
thiols	$C_1 - C_7$	$C_8 - C_{16}$	$C_{17}-$
sulfides	$C_2 - C_7$	$C_8 - C_{15}$	$C_{16}-$
disulfides	$C_2 - C_5$	$C_6 - C_{16}$	$C_{18}-$
	Number of carbons of the aliphatic substituent		
Thiophene	$C_0 - C_4$	$C_4 - C_{11}$	$C_{11} - C_{19}$
Benzothiophene	–	$C_0 - C_5$	$C_5 - C_{13}$
Dibenzothiophene	–	C_0	$C_1 - C_9$
Pyrole	$C_0 - C_2$	$C_2 - C_{10}$	$C_{10} - C_{20}$
Indole	–	$C_0 - C_4$	$C_4 - C_{11}$
Carbazole	–	–	$C_0 - C_7$
Pyridine	$C_0 - C_3$	$C_3 - C_{10}$	$C_{11} - C_{19}$
Quinoline	–	$C_0 - C_3$	$C_4 - C_{11}$
Acrydine	–	–	$C_0 - C_7$

used and the differences in the operating conditions. In order to make such a comparison possible, one may resort to the reaction rate constant for the hydrogenation of the naphthalene to tetraline, which has the value of 0.0578×10^{-3} l/g·cat·s, in Figure 10.9 and 0.394 l/kg·cat·s reported in the second study [19].

The rate constants given in Figure 10.9 must therefore be multiplied by 6,800 to be comparable with those in Figure 10.10 and for the analysis of the relative hydrogenation rate of various aromatic structures.

To illustrate hydrodesulphurization and the competitive hydrogenation of sulfur compounds in Figure 10.11 [20], the reactions of dibenzothiophene are shown, the kinetic constants expressed in l/g·cat·sec. The reaction conditions in this case

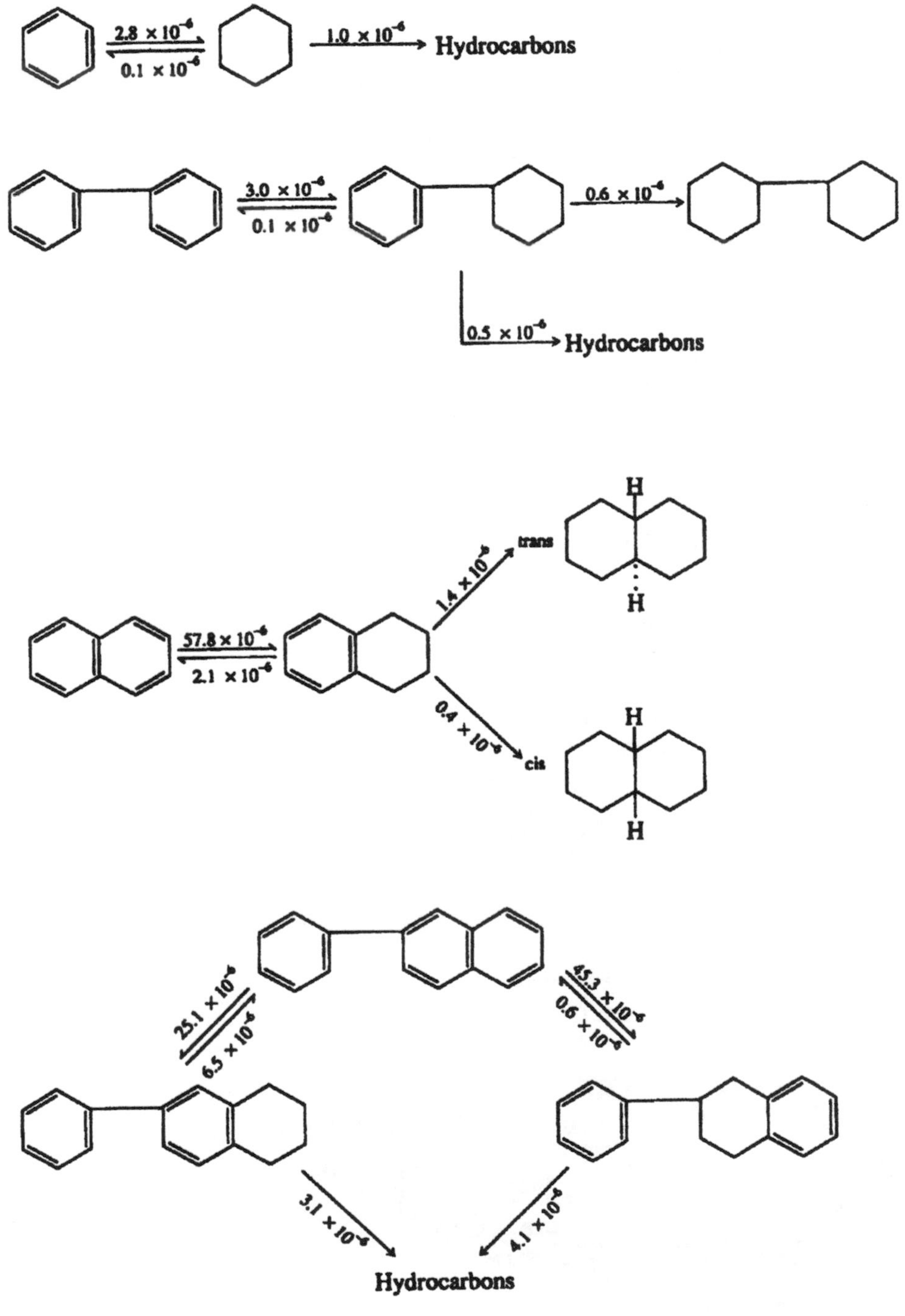

Figure 10.9 Hydrogenation of selected aromatic hydrocarbons on a Co-Mo/γ-Al$_2$O$_3$ catalyst at 325°C and 75 bar. (From Ref. 18.)

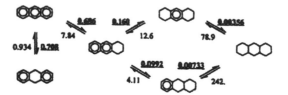

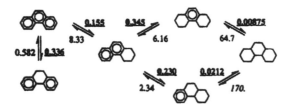

ANTHRACENE

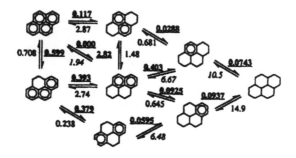

PHENANTHRENE

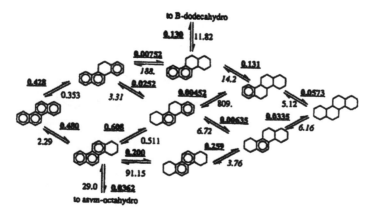

PYRENE

to B-dodecahydro

to asym-octahydro

CHRYSENE

Figure 10.10 Hydrogenation of several polynuclear aromatic hydrocarbons. (From Ref. 19.)

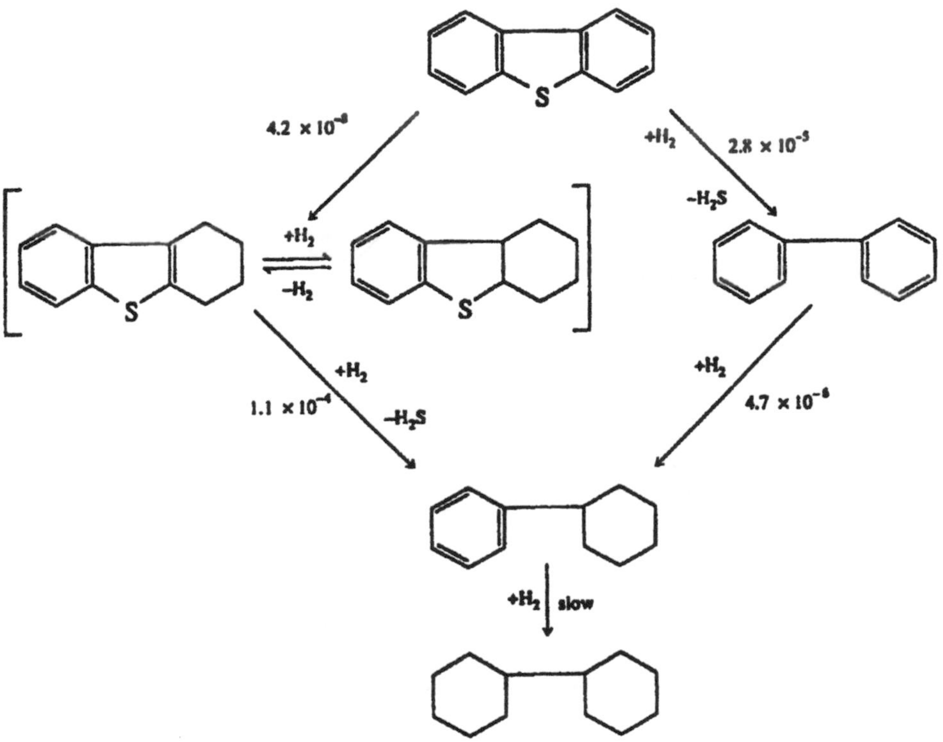

Figure 10.11 Desulfurization and dehydrogenation of dibenzothiophen, on a Co-Mo/γ-Al$_2$O$_3$ catalyst at 300°C and 102 bar. (From Ref. 20.)

were 300°C and 102 bar and a very low concentration of H$_2$S. A higher concentration of sulfur in the feed would significantly decrease the desulfurization rate.

The behavior of the nitrogen compounds is illustrated in Figure 10.12, by the hydrogenation and hydrogenolysis of quinoline on a catalyst of Ni-Mo/γ-Al$_2$O$_3$ at 350°C and 35 bar [14]. The expression of the rate constants is the same as in the Figures 10.9 and 10.11.

The analysis of the behavior of some representative compounds shown in the four figures makes possible some generalizations and estimates of the behavior of similar compounds.

10.3.3 Demetallation Reactions

Demetallation is a more difficult reaction. The organometallic compounds having molecular masses of the order of 450 and boiling temperatures higher than 565°C are contained exclusively in the residue. The micelles, which contain metal atoms, form planar structures with distances between them of about 12 Å. Despite the fact that they are not bound to each other by valence bonds, these structures can be separated only with difficulty.

In these structures the nickel atoms occupy central positions, whereas those of vanadium have mostly more peripheral positions (Figure 10.13).

Demetallation requires a hydrogenolysis and a preliminary hydrogenation in order to expose the metal atoms. Since the vanadium atoms are situated

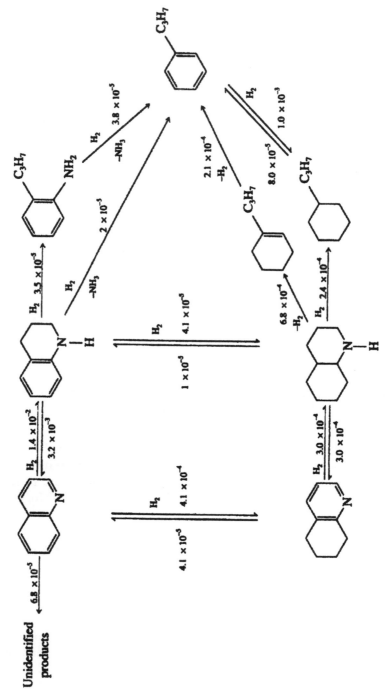

Figure 10.12 Hydrogenation and denitration of quinoline on a Ni-Mo/γ-Al$_2$O$_3$ catalyst at 350°C and 35 bar. (From Ref. 14.)

BEFORE DESULFURIZATION

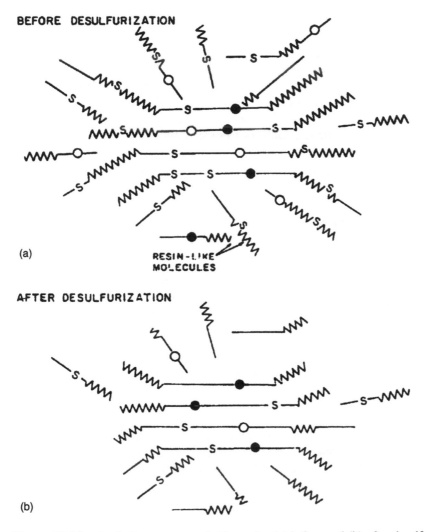

(a)

RESIN-LIKE
MOLECULES

AFTER DESULFURIZATION

(b)

Figure 10.13 Asphaltenes surrounded by resins (a) before and (b) after desulfurization. S – sulfur atoms, ○ – vanadium atoms, ● – nickel atoms, – – aromatic cycles, ᴡᴡᴡ – naphthenic rings.

closer to the periphery of the molecules, they are easier to remove than the nickel atoms.

Also, the place where the metal is deposited within the catalyst particle is different: generally the nickel is closer to the center, which the vanadium is located closer to the periphery, where it contributes often to block the inlet to the pores (see Section 10.5.5).

10.3.4 Coke Formation

Coke is formed on the surface of the catalyst as a result of the interaction of the unsaturated hydrocarbons with the aromatic structures, reactions which take place between the adsorbed molecules on the catalyst surface.

The rate of this process is very low. Often, these deposits become significant after 3–5 months and make it necessary to regenerate the catalyst.

The reasons for the low coking rate is on one hand the absence of acid sites and on the other hand, the high partial pressure of hydrogen. This, in the presence of the metallic sites, is efficient in saturating the alkenes and other unsaturated hydrocarbons before these may interact with each other and with the aromatic hydrocarbons to produce less active structures of high molecular mass.

Concerning the asphaltenes, contrary to older opinions, their concentration is not determining the amount of coke formed, despite the fact that they contribute to its formation. The greatest part of the coke is the result of the interaction between alkenes and aromatics indicated above.

10.4   PROCESS KINETICS

10.4.1   Effect of Diffusion

The diffusional barriers, determined by the diffusion through pores of the large molecules, intervene in the hydrofining and hydrotreating processes to a significant extent. If for the distilled fractions it is sufficient to resort to the effectiveness factor, using the methods presented in Section 6.4.2 for the heavy fractions and especially for the residual ones, additional sterical hindrances may intervene.

In their absence, the relationship between the bulk diffusion coefficient D_b and the effective diffusion coefficient through pores D_c, is given by the equation:

$$D_e = \frac{D_b \eta_p}{\zeta}$$
(10.1)

where η_p is the void fraction of the pores and ζ is the tortuosity factor of the pores.

When the hydrodynamic radius of the molecule a is close to the radius of the pores r_p, the additional restrictive factor $F(\lambda)$ intervenes, which depends on the ratio λ between these two radiuses:

$$\lambda = \frac{a}{r_p}$$
(10.2)

The theoretical value of the restrictive factor is:

$$F(\lambda) = (1 - \lambda)^4$$

As a result of the experimental studies, values other than 4 were proposed for the exponent of this equation, or other expressions for the restrictive factor $F(\lambda)$ [8,21].

Thus, for the hydrofining and hydrotreating of crude oil fractions, some of the proposed equations are:

On the basis of kinetic studies [22] for a Ni-Mo/γ-al$_2$O$_3$ catalyst at 52.7 bar and 350°C and $\lambda < 0.5$:

$$F(\lambda) = (1 - \lambda)^{4.9}$$
(10.3)

On the basis of adsorption studies in ambient conditions and $\lambda < 0.4$ [23]:

$$F(\lambda) = (1 - \lambda)^{3.7}$$
(10.4)

On the basis of kinetic studies [34] effected on a Co-Mo/γ-Al$_2$O$_3$ + P$_2$O$_5$ catalyst at 76 bar and 390°C.

$$F(\lambda) = (1 - \lambda)^{3.5} \quad \text{for hydrodesulfuration}$$
$$F(\lambda) = (1 - \lambda)^{3.8} \quad \text{for hydrodemetallation}$$

(10.5)

The expression suggested for asphaltenes [8]:

$$F(\lambda) = e^{-3.89\lambda}$$

(10.6)

Introducing the restrictive factor $F(\lambda)$, the Eq. (10.1) becomes:

$$D_e = \frac{D_b \eta_p}{\zeta} F(\lambda)$$

(10.7)

For γ-Al$_2$O$_3$, the values of the parameters in this equation are: $\eta_p = 0.64$–0.78 and $\zeta = 1.3$. Therefore, in the conditions wherein the restrictive factor does not intervene, $D_e = (0.49 \times 0.60)D_b$. The diffusion coefficient D_b generally has the values of 10^{-5} cm^2/s for liquids and 10^{-1} cm^2/s for gases at normal temperature and increases with temperature (see Section 6.4.2).

The diffusion of asphaltenes was studied [24] by using membranes with cylindrical pores of controlled diameter and comparing the results with the values given by Eq. (10.6).

Five types of asphaltenes were used, characterized by the values shown in Table 10.5.

The experimental results are shown in Figure 10.14, where the effective diffusion coefficients are plotted as a function of the pore radius. By comparing the molecular radius of asphaltenes, shown above, with the curves of the graph, one may estimate the extent to which the restrictive factor has a stronger effect as the pore radius decreases.

The comparison of the experimental results with Eq. (10.6) is given in Figure 10.15 and shows that this equation gives satisfactory results.

The diffusivity of nitrogen compounds measured on Ni-Mo/γ-Al$_2$O$_3$ bi-modal catalysts at 350°C and 52.7 bar having macropores of 10^{-3} mm and micropores of 6-$10 \cdot 10^{-6}$ mm led to the data in [25] reported in Table 10.6.

A study of the diffusion in bimodal catalysts with the development of a mathematical model was performed by Pereira and Beckman [26]. A comparison of the model with experimental data, using a vacuum residue having $d = 1.0336$, Conradson carbon 18.8%, and asphaltenes 25.6% gave satisfactory agreement.

Table 10.5 Specific Values for 5 Asphaltenes

Asphaltene type	Molecular mass	$D_b \cdot 10^7$cm^2/s	Molecular radius, Å
A	3,000	16.10	26
B	6,000	13.10	32
C	12,000	8.85	47
D	24,000	5.24	79
E	48,000	2.71	153

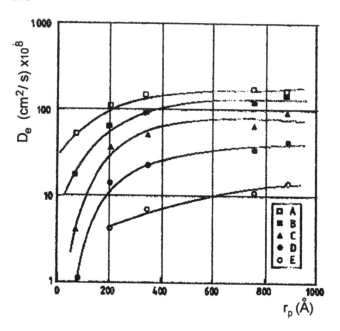

Figure 10.14 Effective diffusion coefficients as a function of pores radius for the asphaltenes A–E (Table 10.5).

Besides the presented calculation methods, which make it possible to estimate the influence of the pore diffusion, also of high interest are practical data obtained for representative catalysts and feeds.

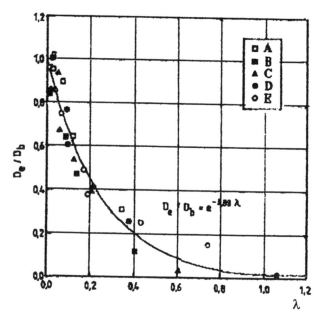

Figure 10.15 Agreement between Eq. (10.6) and experimental data.

Table 10.6 Diffusivity of Selected Nitrogen Compounds

	Pore diameters, nm			D_b
Reactant	Macro	Micro	D_e/D_b	$cm^2/s \cdot 10^5$
Indole	1217	17	0.25	25.6
	632	8.3	0.18	
	911	6.2	0.15	
9-Phenylacrydine	1217	17	0.18	16.7
	632	8.3	0.12	
	911	6.2	0.10	

Source: Ref. 25.

Thus, for a coking gas oil, which was hydrofined on a Ni-Mo/γ-Al_2O_3 catalyst, at 400°C, at a hydrogen partial pressure of 139 bar, and at a feedrate of 12.5 cm^3/g cat·h, the effectiveness factor was 0.701 for a degree of desulfurization of 76.4% and 0.848 for a degree of desulfurization of 53.4% [27]. The catalyst was shaped as pellets having Ø = 0.913 mm. The effectiveness factors were calculated according to the method of Thiele, for first order reactions.

For a residue feed, using a catalyst of Co-Mo/γ-Al_2O_3 effectiveness factors of the order 0.21–0.26 [28] were obtained.

10.4.2 Kinetics of Hydrogenolysis

In order to identify the elementary kinetic steps, numerous studies examined the conversion of some pure model compounds, usually assuming a Langmuir-Hinshelwood model.

Such a treatment of the hydrodesulfurization of dibenzothiophene was given by Sundaram, Katzer, and Bischoff [29]. In their work the kinetic scheme of Figure 10.16 was adopted, accepting the following simplifying assumptions:

The catalyst has two types of sites: some adsorb the sulfur compounds, including H_2S, while other adsorb only the hydrogen.
Since the hydrogen is in large excess, its partial pressure is considered constant.
All the adsorption-desorption phenomena are at equilibrium; the adsorption coefficients for the sulfur compounds are equal to each other ($b_{\mathrm{DBT}} = b_{THDBT} = b_{HHDBT}$);
The surface reaction between dibenzothiophene and the hydrogen adsorbed on the neighboring specific sites, namely the first stage of this reaction is rate determining.

Noting by θ the sites that adsorb the sulfur compounds and θ' those adsorbing the hydrogen, using the index t for the total of the sites and v for the vacant sites, one may write:

$$\theta_t = \theta_v + \theta_{\mathrm{DBT}} + \theta_{\mathrm{THDBT}} + \theta_{\mathrm{HHDBT}} + \theta_{\mathrm{H_2S}} \qquad (10.8)$$

Since the adsorptions is at equilibrium, its results:

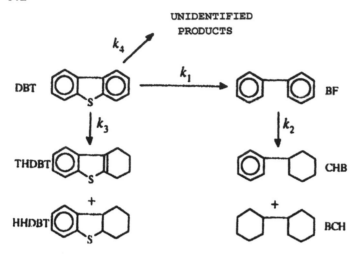

Figure 10.16 Desulfurization of dibenzothiophene. DBT – Dibenzothiophene, THDBT – Tetrahydro-benzothiophene, HHDBT – Hexahydro-benzothiophene, BT – biphenyl, CHB – cyclohexyl-benzene, BCH – bicyclohexane.

$$\theta_{DBT} = b_{DBT}[DBT] \cdot \theta_v$$
$$\theta_{THDBT} = b_{DBT}[THDBT] \cdot \theta_v$$
$$\theta_{HHDBT} = b_{DBT}[HHDBT] \cdot \theta_v \tag{10.9}$$
$$\theta_{H_2S} = b_{H_2S}[H_2S] \cdot \theta_v$$

Replacing in (10.8), it results:

$$\theta_v = \frac{\theta_t}{1 + b_{DBT}([DBT] + [THDBT] + [HHDBT]) + b_{H_2S}[H_2S]} \tag{10.10}$$

Similarly, the adsorption of hydrogen may be expressed by the equation:

$$\theta'_v = \frac{\theta'_t}{1 + b_{H_2}[H_2]} \tag{10.11}$$

According to the scheme of Figure 10.16 and to the last simplifying assumption, the rate of dibenzothiophene decomposition will be given by:

$$-r = (k'_1 + k'_3 + k'_4)\theta_{DBT} \cdot \theta'_{H_2} \tag{10.12}$$

Replacing θ_{DBT} and θ'_{H_2} with the corresponding expressions, one obtains:

$$-r = \frac{\{(k'_1 + k'_3 + k'_4)b_{DBT}b_{H_2}\theta_t\theta'_t\}[DBT][H_2]}{\{1 + b_{DBT})[DBT] + [THDBT] + [HHDBT]) + b_{H_2S}[H_2S]\}\{1 + b_{H_2}[H_2]\}} \tag{10.13}$$

where the straight brackets express the concentrations in moles/g catalyst.

The experimental data obtained at a constant hydrogen pressure of 35.5 bar and a temperature of 350°C permitted the determination of the apparent rate constants and of the adsorption coefficients:

$$\left.\begin{array}{l} k_1 = k_1' b_{DBT} b_{H_2} \theta_t \theta_t' [H_2] = 3.24 \text{ g/g cat·min} \\ k_2 = k_2' b_{DBT} b_{H_2} \theta_t \theta_t' [H_2] = 1.26 \text{ g/g cat·min} \\ k_3 = k_3' b_{DBT} b_{H_2} \theta_t \theta_t' [H_2] = 0.69 \text{ g/g cat·min} \end{array}\right\} \qquad (10.14)$$

$$\left.\begin{array}{l} b_{DBT} = 0.13 \cdot 10^4 \text{ g/mol} \\ b_{H_2S} = 0.83 \cdot 10^4 \text{ g/mol} \end{array}\right\} \qquad (10.15)$$

At a constant partial pressure of hydrogen, by taking into account the values of the adsorption coefficients and the low concentrations of the sulfur compounds as compared to that of H_2S, Eq. (10.13) may be simplified to the form:

$$-r = \frac{k[DBT]}{1 + b_{H_2S}[H_2S]} \qquad (10.16)$$

For the hydrogen partial pressure and temperature indicated above, the reaction rate constant has the value:

$$k = 5.19 \text{ g/g cat·min}$$

Other kinetic expressions based on the Langmuir-Hinshelwood mechanism may also be formulated by using different simplifying assumptions. Thus, Broderick [30] tested 14 kinetic expressions and reached the conclusion that the best agreement with the experimental data is obtained by assuming that the dibenzothiophene is adsorbed on two sites. Admitting this mechanism, the following expression is obtained:

$$-r = \frac{k[DBT][H_2]}{\{1 + b_{DBT}[DBT] + b_{H_2S}[H_2S]\}^2 \cdot \{1 + b_{H_2}[H_2]\}} \qquad (10.17)$$

The rate constants and the adsorption coefficients were determined and used for comparing and characterizing various types of catalysts.

Thus, catalysts of Co-Mo, Ni-Mo and Ni-W supported on Al_2O_3 were compared, by determining the kinetic constants for the thiophene hydrodesulfurization [31]. The adsorption of thiophene and of H_2S on Ni-W is weaker than on Co-Mo and Ni-Mo catalysts, from which a stronger hydrogenating activity results for this catalyst. The adsorption of hydrogen, is nondissociating and is weaker than of other reactants.

In another work [32], the kinetic study determined the reactivity of 61 sulfur compounds belonging to the benzothiophenes and dibenzothiophenes, on Co-Mo and Ni-Mo catalysts. The values of the rate constants obtained using a kinetic equation of the first order for a homogeneous system were grouped according to the structure of the compounds (see Table 10.7). In the case of dibenzothiophenes, the substituents in the 4 and 6 positions strongly decrease the reaction rates. No significant differences were observed between Co-Mo and Ni-Mo catalysts.

A further study [98] confirmed the conclusions concerning the reactivity of the benzothiophenes and dibenzothiophenes, in a light cycle oil, on Co-Mo/Al_2O_3 catalysts.

The desulfurization and denitrification of real-life petroleum fractions was studied by Raseev and Ionescu [33]. The purpose was to obtain overall kinetic equations for conditions close to those in the commercial operation.

Table 10.7 Reaction Rate Constants for Bezothiophenes and Dibenzothiophenes, Calculated as First Order Pseudo-homogeneous Kinetics

Component	k, min^{-1}
	> 0.10
	0.034–0.10
	0.013–0.034
	0.005–0.013

Source: Ref. 32.

The experiments used 4 thermal cracking gas oils having a sulfur content comprised between 0.21 and 2.28 wt %, nitrogen between 0.40 and 0.66%, and bromine number between 30 and 36 g/100 g. Also, the same gas oils with additional amounts of dibenzothiophene were used.

The measurements used an industrial Co-Mo/γ-Al$_2$O$_3$ catalyst in a continuously operating pilot unit with an isothermal reactor (Figure 10.17). The runs were carried out at a pressure of 105 bar, a molar ratio hydrogen/feed of 10, and temperatures of 325, 350, 375, and 400°C.

The reaction rate was written assuming that the process was pseudohomogeneous, using the general equation:

$$\frac{dp_i}{d(1/w)} = k_i \cdot p_{H_2}^m \cdot p_i^n \tag{10.18}$$

where the subscript i stood for the sulfur, the nitrogen, or the unsaturated hydrocarbon.

Since in the experimental conditions the partial pressure of hydrogen could be considered constant, Eq. (10.18) becomes:

$$\frac{dp_i}{d(1/w)} = k_i' \cdot p_i^n \tag{10.19}$$

The experimental data showed that in all cases the exponent n has values very close to one, so that the process may be considered of the first order.

By integrating Eq. (10.19) and by giving successively to the subscript i the meaning of sulfur or nitrogen content, the following equations were obtained:

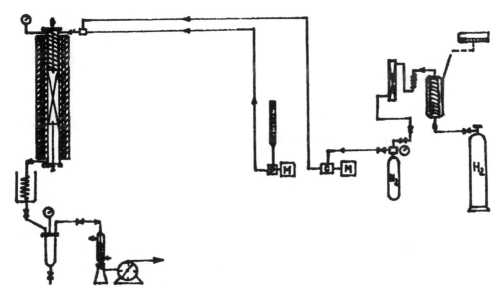

Figure 10.17 Continuous pilot plant used in hydrofining studies. (From Ref. 55.)

$$k_S \cdot \left(\frac{1}{w}\right) = \ln \frac{S_i}{S_f} \qquad (10.20)$$

$$k_N \cdot \left(\frac{1}{w}\right) = \ln \frac{N_i}{N_f} \qquad (10.21)$$

where initial and final contents are S_i and S_f for the sulfur and N_i and N_f for nitrogen.

These equations represented very well the experimental data up to a degree of desulfurization of 90–95% and of denitration of 65–70%, which is usually sufficient for the precision imposed to the operation of commercial units.

At higher conversions, the deviations correspond to a decrease of the reaction rate constants, which is easily explained by the presence in the gas oil of some sulfur and nitrogen compounds, which have a low reactivity. This fact was known from much earlier reports [34].

Equations (10.20) and (10.21) make it easy to obtain the overall rate constants for the desulfurization and denitration of a specific fraction by using the results from a small number or even of a single experimental determination. The determined constants are valid for the catalyst and for the operating conditions used in the experimental measurement.

The rate constants could be determined also on the basis of the measured sulphur and nitrogen content in the feed and in the effluent of a commercial unit where these analyses are currently performed. The rate constants thus obtained may be used to determine the degree of sulfurization and denitration that will be obtained when using a different feed.

Kinetic equations of the form (10.19) were used also in other studies [35–37].

10.4.3 Kinetics of Hydrogenation

A large number of studies [19,38–45] have examined the kinetics of the hydrogenation of aromatic hydrocarbons on metallic sulfides. In one of the studies [44], after the examining earlier models the authors adopted the Langmuir-Hinshelwood kinetics, with the surface reaction as the rate determining step. For the hydrogenation of 1-methyl naphthalene to methyl-tetraline, taking into account that in the conditions of hydrotreating the reaction must be considered as being reversible (see Figure 10.4), the rate equation has the form:

$$r_1 = \frac{k_1 b_{MN} \left[p_{MN} - \dfrac{1}{K_1} \cdot \dfrac{p_{MT}}{p_{H_2}^2} \right]}{1 + b_{MN} p_{MN} + b_{MT} p_{MT} + b_{MD} p_{MD} + b_z p_z} \tag{10.22}$$

where the subscripts MN, MT, and MD refer to 1-methyl naphthalene, methyl-tetrahydronaphthalene and methyl-decahydronaphthalene, and b_z and p_z to strongly adsorbed substances present in the reaction zone, such as NH_2 and H_2S.

In this equation, the reaction rate r is expressed in moles/h·m³ cat.

Taking into account that H_2S and especially NH_3 are very strongly adsorbed while MN, MT, and MD, only weakly:

$$1 + b_z p_z \gg b_{MN} p_{MN} + b_{MT} p_{MT} + b_{MD} p_{MD}$$

and the Eq. (10.22) may be written as:

$$r_1 \approx \frac{k_1 b_{MN} \left[p_{MN} - \dfrac{1}{K_1} \cdot \dfrac{p_{MT}}{p_{H_2}^2} \right]}{1 + b_z p_z} \tag{10.23}$$

The partial pressures of the various components may be expressed as function of the partial pressure p_{MN}^0 of the 1-methylnaphthalene at the inlet of the reactor, and of the mole fraction of the respective component in the reaction mixture (excluding the hydrogen), by using the equation:

$$p_i = p_{MN}^0 - \frac{N_i}{\sum N_i} = p_{MN}^0 \cdot x_i \tag{10.24}$$

where N is the number of moles of the various components i.

In Eqs. (10.22) and (10.23) r is expressed in moles MN/m³ catalyst $x \cdot h$. If the feed is expressed in m^3 liquid feed/m³ cat $x \cdot h = w$, Eq. (10.23) may be transformed using α_{MN} which gives moles MN/m³ feed.

Effecting these substitutions, Eq. (10.23) becomes:

$$\frac{d(x_{MN})}{d(1/w)} = -\frac{k_1 \cdot b_{MN} \cdot p_{MN}^0 \left[x_{MN} - \dfrac{1}{K_1 p_{H_2}} x_{MT} \right]}{\alpha_{MN}(1 + b_z p_{MN}^0 \cdot x_z)} \tag{10.25}$$

Introducing the notation:

$$k_1' = \frac{k_1 b_{MN} p_{MN}^0}{\alpha_{MN}(1 + b_z p_{MN}^0 \cdot x_z)} \tag{10.26}$$

which in the given experimental conditions has the meaning of the apparent reaction rate constant, Eq. (10.25) becomes:

$$\frac{d(x_{MN})}{d(1/w)} = -k_1' \left[X_{MN} - \frac{1}{K_1 p_{H_2}} x_{MT} \right] \tag{10.27}$$

an equation for a pseudo first-order.

The Eqs. (10.26) and (10.27) may be similarly written for the formation rate of MD:

$$k_2' = \frac{k_2 \cdot b_{MT} \cdot p_{MT}^0}{\alpha_{MN}(1 + b_z p_{MN}^0 \cdot x_z)} \tag{10.28}$$

$$\frac{d(x_{MD})}{d(1/w)} = -k_2' \left[x_{MT} - \frac{1}{K_2 p_{H_2}} x_{MD} \right] \tag{10.29}$$

The formation of MT will be given by the expression:

$$\frac{d(x_{MT})}{d(1/w)} = -\frac{d(x_{MN})}{d(1/w)} - \frac{d(x_{MD})}{d(1/w)} \tag{10.30}$$

The derived equations were verified [44] for the temperatures of 300, 320, 350, and 400°C and for a total pressure of 11.75 bar, p_{MN}^0 having the value of 9 bar and $\alpha_{MN} = 2.27 \cdot 10^3$, mol/m^3, obtaining very good agreement.

For the temperature of 400°C, the rate constants and the activation energies had the values:

$$k_1' = 100 \text{ h}^{-1} \quad E_1' = 96 \text{ kJ/mol}$$
$$k_2' = 0.56 \text{ h}^{-1} \quad E_2' = 63 \text{ kJ/mol}$$

In these verifications the used catalyst contained 2.7% NiO, 16.5% MoO$_3$, and 6.5% P$_2$O$_5$ supported on Al$_2$O$_3$. The specific surface of the catalyst was 170 m^3/g, and the bulk density, 3.4 g/cm^3.

These data confirmed what is well-known, that the hydrogenation to MT takes place much more faster than the hydrogenation of MT to MD. This fact allows one to neglect the second step of the hydrogenation and to express the conversion of MN and MT using the equation:

$$\frac{x_{MN} - x_{MN}^{ec}}{1 - x_{MN}^{ec}} = \exp[-k_1'(1 - x_{MN}^{ec})(1/w)] \tag{10.31}$$

where x_{MN}^{ec} is the mole fraction at the thermodynamic equilibrium:

$$x_{MN}^{ec} = 1/[1 + K_1 p_{H_2}^2]$$

10.4.4 Residues Conversion

The kinetics of the conversion of residues has important particularities and was the object of many studies with sometimes contradictory results, preventing definitive conclusions.

One of the causes that led to such contradictions is the fact that in the case of residues, desulfurization took place partly as a result of thermal reactions. Thus, in the absence of the catalyst, in the usual reaction conditions, hydrodesulfurization reaches 30–40% of the conversion that might be obtained in the presence of the catalyst [8].

Lack of agreement exists also concerning the reaction order, which seems to be higher than first. Thus, for an atmospheric residue with an initial boiling point of 343°C the reaction order ranged between 1.9 and 2.3 and the apparent energy of activation was 132–145 kJ/mole [28].

In another study [46], the hydrodesulfurization of asphaltenes and of a residual fraction was performed while paying special attention to the diffusion processes. The hydrodesulfuration process was accompanied by some cracking, necessary to make accessible the sulfur atoms included in complex molecular structures. The catalyst used was $Co\text{-}Mo/Al_2O_3$.

The use of the kinetic equation of the form (10.19) led to values of the exponent n of 2 and in some cases of 3. The latter value is difficult to understand and to accept.

Another research [47] used as feed an Athabasca bitumen having $d = 0.9809$; the atomic ratio $C/H = 0.7$, and the distillation range: 5% at 322°C, 50% at 429°C, and 80% at 504°C.

The processing of the experimental results at a desulfurization degree of 99% and of denitration of 86%, led to a reaction order (in Eq. (10.19)) of 1.5 for hydrodesulfurization and 2.0 for denitration.

Still another study [48] used a fraction with an initial boiling temperature of 424°C. The hydrodesulfurization was performed at conversions of 80–90%. The obtained reaction order was 1.5 and the apparent activation energy was 140 kJ/mol.

In other papers [8,49–51], it is specified that the use of a first-order equation written for homogeneous systems gives correct results for conversions not exceeding 60–70% [28]. At higher conversions, the reaction rate constant decreases. This is probably the reason for obtaining for conversions of 80–90% and higher, reaction orders above one.

Some recent works [99,100] study the kinetics of converting the Conradson Carbon Residue (CCR) in catalytic hydroprocessing. A half-order kinetics was established with the activation energy being 277.58 kJ/mol. No dependence of the rate constant on hydrogen pressure was observed. It was found that the CCR is linearly related to the content of asphaltenes, hydrogen content, H/C atomic ratio, and residue content (350°C). Gas yield was also linearly related to the CCR content.

For the kinetics of denitration and demetallation of residual fractions, conclusions are more difficult to formulate.

10.5 EFFECT OF PROCESS PARAMETERS

Hydrofining and hydrotreating comprise the treatment with hydrogen of a large range of feedstocks, from naphthas to the vacuum residues, with sometimes quite different objectives. There are few common features concerning the effect of the process parameters on performance that could be presented here. For this reason, many specifications will be given in the framework of the commercial implementation of the processes.

10.5.1 Temperature

The upper temperature limit is generally determined by the need to avoid parallel thermal decomposition reactions. Since their activation energy is higher than that of the catalytic reactions, the importance of the thermal decomposition increases with temperature and may have a negative effect on the quality of the final products.

The degree to which the thermal reactions may be tolerated depends on the targeted product. Lubricating oils are the most sensitive to these negative effects.

There are also situations when a limited degree of thermal decomposition is welcome. Thus, when processing vacuum residues, it makes the vanadium and especially the nickel atoms, which are included in complex structures, accessible to hydrogenolysis reactions.

As the rate of the thermal decomposition reactions increases with the molecular mass of the hydrocarbon, the highest accepted temperature depends also on the boiling range of the feedstock. In hydrofining processes the temperature of 400–420°C should not be exceeded.

In hydrotreating processes, thermodynamic limitations may occur. For this reason, a certain temperature level should not be exceeded. In each case, this temperature level depends on the partial pressure of hydrogen.

The lowest operating temperatures are determined by the need to obtain reasonable reaction rates and thus obtain economically justifiable sizes for the reaction equipment.

A factor that determines the temperature level is the relative importance of the denitration or desulfurization. Since the activation energy of the desulfurization reactions is higher than that of the denitration, higher temperatures favor the desulfurization, while lower ones, the denitration. In order that the removal of both sulfur and nitrogen be efficient, the temperature should not be lower than 350–360°C. An exception is the hydrogen treating of oils, where it is very important to preserve intact the structure of the hydrocarbons present in the feed.

Special cases are the selective hydrogenations, for example that of di-alkenes, in presence and without converting the alkenes, to ensure the stability of the cracked gasoline without hurting its octane number. Such processes are performed at much lower temperatures, about 250°C or in some processes even lower.

10.5.2 Pressure

The system pressure determines to a large extent the partial pressure of the hydrogen. Higher pressures favor both the hydrogenolysis and the hydrogenation reactions. Moreover, in hydrotreating, the thermodynamic limitations intervening in the hydrogenation of the aromatic cycles explain the higher pressures used in this process compared to that used in hydrofining, where the pressure level is determined by economic considerations.

With heavy feeds in liquid state, the pressure in the reaction zone favors the diffusion of the hydrogen through the liquid film to the surface with the catalyst, where the concentration of hydrogen must be sufficient to minimize the formation of coke.

The operating pressure is generally situated between 20 and 80 bar, being higher for heavier feeds and in hydrotreating processes.

10.5.3 H$_2$/Feed Ratio

Overall this ratio ranges between the limits 20 and 800 Nm^3/m^3, the actual value depending on the nature of the feed, on the catalyst, and on the hydrogen purity.

In hydrofining, which requires lower amounts of hydrogen than hydrotreating, one uses mainly hydrogen produced in the catalytic reforming units, if it contains at least 65–70% vol. hydrogen.

The limits imposed on the maximum content of H$_2$S, tolerated in the hydrogen fed to the reactor require the purification of the recycle hydrogen. These limits are of 4–5% H$_2$S when Co-Mo/γ-Al$_2$O$_3$ catalysts are used and more severe for the catalysts that contain Ni, W, or other metals.

The benefits of increasing the H$_2$/feed ratio show a maximum, beyond which the negative effect derived from the decrease of the catalyst contact time exceed the advantage of the large hydrogen excess, and the degrees of desulfurization and denitration decrease. Unless contrary economic considerations prevail, the above considerations determine the H$_2$/feed ratio at which each unit is operated.

10.5.4 Feedstock Composition

The behavior of various feedstocks in the hydrofining and hydrotreating processes depends first on their distillation ranges. In general, broader ranges decrease the efficiency of the processes which, in the case of the hydrogenolysis, is explained by the nature of the heteroatomic compounds contained in the feed.

However, some exceptions were remarked. Thus, the comparative study of the desulfurization of products with various boiling temperatures [52] indicates that the fraction boiling in the range 330–380°C is the most difficult to desulfurize. This is explained by the presence of the alkyl-benzothiophenes, which have a low reactivity and illustrated by the data of Table 10.8, obtained in a pilot plant on a Co-Mo/γ-Al$_2$O$_3$ catalyst in identical operating conditions. This conclusion is not necessarily valid for all the crude oils.

The products produced in the pyrolysis and cracking processes, are more difficult to be hydrofined and hydrotreated than the primary products [7,53,101].

A systematic study of the hydrofining on an industrial Co-Mo catalyst, of gas oils of thermal cracking, as such or with additions of dibenzothiophene, dodecyl-mercaptane, butyldodecylsulphide, quinoline, and hexadecene was reported by Raseev and Feyer-Ionescu [54–56].

The experiments were performed in the pilot unit presented in Figure 10.17, at 100 bar, a molar ratio H$_2$/gas oil = 10, volume hourly space velocity = 3 hours^{-1} $D_e = \frac{D_h \eta_p}{\zeta}$ and temperatures comprised between 300 and 400°C.

Table 10.8 Conversions of Desulfurization as Function of Distillation Ranges

Distillation ranges, °C	Desulfurization, wt %
< 220	91
220–330	77
330–380	45
> 380	87

It was verified that the addition of dodecyl-mercaptane, butyl-dodecyl-sulfide, and hexadecene do not influence the desulfurization, denitration, and the hydrogenation of the gas oil. On the other hand, the additions of dibenzothiophene or of quinoline strongly decrease the conversion for the desulfurization, denitration, and hydrogenation of the unsaturated hydrocarbons. This action may be explained by the strong adsorption of these compounds on the active sites of the catalyst, which does not occur for compounds with a linear structure, such as the alkenes, mercaptans, or alkyl-sulphides.

The study also indicated that a similar decrease of the desulfuration, denitration, and saturation of the alkenes is caused by the polycyclic aromatic hydrocarbons that are strongly adsorbed on the active sites.

It is much more difficult to establish quantitative relationships for the reciprocal effect of the heteroatomic substances or of the content in aromatic hydrocarbons on the reactivity. It is to be mentioned in this sense the work of Gray et al. [49] which, based on studies on Alberta bitumen, correlated the concentration of the alpha carbon atoms in the aromatic structures with the reaction rate constant.

A different result was obtained by Guttian et al. [50], which studied the vacuum residues of the crude oils from Venezuela, and ascertained that the reactivity increases with the sulfur content of the residue. The generalization of the above conclusions is not advisable since they refer to specific feedstocks.

10.5.5 Catalyst Deactivation

The performance of the catalysts used in hydrofining and hydrotreating was compared in Section 10.2. It was the object of a large number of studies aiming at optimizing their performance [13,57–63].

Of significant importance is the deactivation of these catalysts.

The deactivation has two causes: (1) accumulation of nickel and vanadium sulfides on the surface of the catalyst; (2) formation of coke deposits.

The time evolution of the two effects and their influence on the activity of the Ni-Mo or Co-Mo catalysts is presented in Figure 10.18 [64].

The coke deposits are the results of adsorption of coke precursors, followed by the combination of oligomerization and aromatization reactions [65]. In the first step, the adsorption is favored by the high molecular mass and the polarity of the compounds present in the reaction mixture. However, although the asphaltenes have the larger molecular mass and polarity, the formation of coke is not controlled by the amount of asphaltenes. Thus, their elimination by deasphalting causes [65] a reduction of only 20–30% in the amount of coke deposited on the catalyst.

The rate of coke formation depends largely on the hydrogen partial pressure, which stops the oligomerization reactions, and on the temperature, which favors them.

When the amount of formed coke reaches 15–35 wt % expressed in carbon, the catalyst is submitted to regeneration.

The accumulation of metals on the catalyst is a result of the demetallation reactions, which may be considered to be autocatalytic [8]. The metals are deposited on the surface of the catalyst as sulfides, which have some catalytic activity, until they plug the pores of the catalyst. The amount of vanadium so deposited may reach 30–50 wt % on the catalyst [60,66].

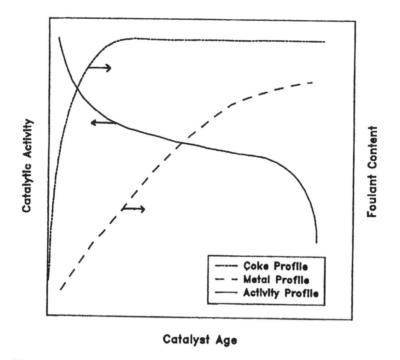

Figure 10.18 Activity of Co-Mo and Ni-Mo catalysts as function of the deposited coke and metals (Ni,V) amounts. (From Ref. 64.)

The kinetics of the deactivation is treated many times as the influence of the metal deposits on the effectiveness factor [8,67]. The deactivation may be expressed also as function on the time on stream (duration of use) of the catalyst using expressions similar to those used in catalytic cracking.

However, such a treatment is more difficult in the case of the hydrofining and hydrotreating processes due to the two parallel deactivation phenomena, leading to coke and metals. For a Kuwait residue, 60% of the deactivation is attributed to the metals and 40% to the coke deposits [68]. These proportions can vary depending on the specific feed used.

An expression suggested by Myers [66] that models the deactivation by taking into account both phenomena has the form:

$$\alpha_c = c_1 \left[\frac{PV_0 - c_2[C] - c_3[M]}{PV_0} \right] + c_4 \tag{10.32}$$

where:

α_c = ratio between the activity at the time t and the initial activity
V_0 = initial volume of the catalyst pores
[C] = concentration in carbon
[M] = concentration in metals of the feedstock
$c_1 - c_4$ = empirical constants, to be determined on the basis of the operating data of the unit.

The main objection is that the form of Eq. (10.32) does not describe the difference between the dynamics of the deposition of the coke of the metals illustrated in Figure 10.18. Whereas the metals are deposited gradually, the coke is deposited with a high rate at the beginning of the run (fresh catalyst), remains constant for a long time, and is deposited intensively again at the end of the cycle.

The regeneration of the catalysts is performed by the combustion of coke in controlled temperature conditions. The combustion does not effect the vanadium and nickel oxides. The passing of a solution of oxalic acid over the catalyst could restore most of its activity, but would require close supervision in order not to remove the Mo, together with Ni and V. This method, suggested in 1993 [69] was not used on a large scale yet.

A more intense washing with acid leads to the complete elimination of the vanadium, nickel, and molybdenum, leaving behind the inactive alumina [70,71]. The extracted metals are then separated and may be redeposited on the support.

10.6 INDUSTRIAL HYDROFINING

10.6.1 Hydrofining of Gasoline

In most cases, the hydrofining of gasoline is considered to be pretreatment in view of their use as feed for catalytic reforming.

To this purpose, besides the required distillation limits, the gasolines must satisfy the following purity conditions:

Sulfur, ppm	< 1
Nitrogen, ppm	< 0.5
Alkenes, % vol	< 0.5
Oxygen, ppm	< 5

A similar purity must be satisfied by the pure hydrocarbons or by the fractions destined to isomerization*.

In the case of straight run gasoline, these objectives may be reached by using a single hydrofining step, without the need to use a hydrogen stream for the cooling of the reactor.

The flow diagram of such a unit is given in Figure 10.19.

If gasolines produced in cracking processes are to be used as feed for catalytic reforming, the above purity specifications are more difficult to achieve due to both a high content of alkenes and to a much higher stability of the sulfur and nitrogen compounds contained therein. In order not to exceed the purity prescriptions of the feed, the hydrofining of these gasolines is carried out in mixture with primary gasolines. If the feed to catalytic reforming contains not more than 30–40% cracked gasolines, no changes are needed to the hydrofining unit, but only more severe operating conditions (see Table 10.9) [72].

If the cracked gasolines are to be fed to the catalytic reformer without being previously mixed with straight run gasoline, it is necessary to have a reactor provided

---

* Reformation and isomerization processes using catalysts that contain precious metals, especially platinum, are in general use today.

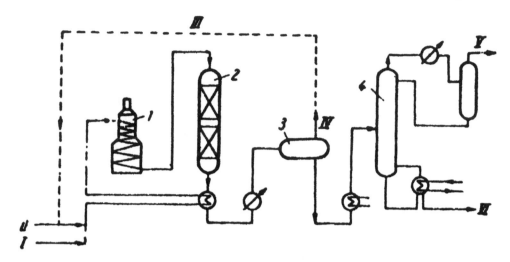

Figure 10.19 Flow diagram of a hydrofining unit for straight run gasoline. 1 – furnace, 2 – reactor, 3 – high pressure separator, 4 – stabilizing column; I – feed, II – fresh hydrogen, III – recycled hydrogen, IV, V – gases, VI – hydrofined gasoline.

with hydrogen injections for cooling. In order to obtain the required final purity, in most cases it is necessary to provide a second reactor, in series, for completing the desulfurization and denitration.

Straight run gasoline, not intended for catalytic reforming feed, is hydrofined in one reaction stage. No hydrogen injection is needed for cooling purposes. If the cracked naphtha is to be used as a component of automotive gasoline, only an incomplete hydrogenation is performed; otherwise, a strong reduction of the octane

Table 10.9 Operation Parameters for Typical Hydrofining

	Straight run gasoline	Straight run mixed with thermal cracking gasoline	Straight run kerosene	Straight run gas oil
Catalyst	CoMo/γ-Al$_2$O$_3$	CoMo/γ-Al$_2$O$_3$	NiMo/γ-Al$_2$O$_3$	NiMo/γ-Al$_2$O$_3$
Temperature, °C	250–315	290–360	340–370	350–380
Pressure, bar	20–30	25–45	30–50	30–60
Space velocity, h^{-1}	7–15	3–6	4–6	3–6
H$_2$/feed ratio, Nm^3/m^3	35–70	70–180	100–200	200–300
Hydrogen purity, % mol	70–90	70–90		
Cycle duration, months	6–24	6–18		

Source: Ref. 11,72.

rating would occur. Thus, only a selective hydrogenation of the dienes (to alkenes) and of the more active alkenes is carried out.

Such a selective hydrogenation is performed on a catalyst of supported palladium or, in the Shell process on nickel, in a single reaction step. In the case of the nickel catalyst the temperature is 80–130°C, the pressure is 60 bar, the space velocity in t/m^3 cat $\cdot h$ is 0.8–1.5, the amount of H_2 is 50–100 Nm^3/t feed, and that of recycled gases 200–500 Nm^3/t feed. For the palladium catalyst the smaller pressure is used.

The palladium catalyst achieves a reduction of the diene value from 20 to 1, of the resins from 35 to 4 $mg/100$ cm^3, of the bromine number from 75 to 56. The octane number and the sulfur content remain practically unchanged.

The utilities consumption for the two types of units were recalculated on the basis of the published data [11,73] for a plant with a capacity of 2000 m^3/day and are given in Table 10.10.

10.6.2 Hydrofining of Middle Fractions

The hydrofining of the kerosene and gas oil has two main purposes: desulfurization and color improvement. In the same time, a lowering of the nitrogen content occurs, which increases the stability of the products.

The hydrogenating activity of the hydrofining catalyst is low, which leaves the molar inventory of the hydrocarbons, different from hydrotreating, actually unchanged. In fact the operation conditions are not severe enough to determine the hydrogenation of the aromatic rings.

Straight run kerosene and gas oils are also submitted to hydrofining, especially those obtained from sulfurous crude oils. In some cases, middle fractions from delayed coking and from visbreaking are also added to the feed. Their proportion in the mixtures with the straight run fractions is limited by the exothermal heat of reaction for the hydrogenation of alkenes. For high concentrations of middle cuts in the feed, the reactor is provided with means for injecting cooling hydrogen between the catalyst beds. Besides, because the cracked products generally contain relatively stable heteroatomic compounds, their introduction in the feed requires a higher severity of operating conditions.

Such a severity means an increase by about 10°C of temperature, a slight increase of pressure, and of the $H_2/feed$ ratio, with a decrease of the feedrate by about 20–30%.

A high content of aromatic hydrocarbons in the catalytic cracking gas oils is undesirable. However, their inclusion in the hydrofining feed for improving their quality is inefficient. The reduction of the aromatics content is achieved in the more severe conditions prevailing in hydrotreating.

Table 10.10 Hydrogenation of Cracked Gasoline

	Catalytic reforming feed	Selective gasoline hydrogenation
Fuel, kJ/h $\cdot 10^6$	16.1	6.8
Steam, t/h	3.1	–
Medium pressure steam, t/h	–	30
Water, m^3/h	15.0	6.5
Power, kW	630	1030

When no cracked products are added to the feed, the process flow diagram of a hydrofining unit for kerosene or gas oil is similar to that of Figure 10.19, and the operating conditions are those shown in Table 10.10.

Most units use Co-Mo catalysts, while the Ni-Mo catalysts are used when a more complete denitrification is required.

10.6.3 Hydrofining of Lubricating Oils

The hydrofining of lubricating oils was developed initially for finishing the refining of motor oils, following solvent extraction and dewaxing, and replacing the sulfuric acid and clay finishing. Owing to yields of above 99%, to the process flexibility, and to the better susceptibility to additives, especially to antioxidants, the hydrofining has actually displaced the older finishing methods.

The hydrofining is used also with predilection for the production of transformer and of turbine oils, to which it confers a better stability to oxidation, compared to the oils obtained by traditional methods. More recently, the hydrofining proved itself in the production of industrial oils, replacing the traditional methods of acid refining.

As in other hydrofining processes, the reactions that take place in hydrofinishing of oils are: hydrodesulfurization, hydrodenitration, hydrodeoxygenation, and hydrodemetallation.

The French Petroleum Institute published the results of a study [74] on the effect of the operating parameters on the properties of the hydrofinished product. The temperature, pressure and (volume) space velocity were varied in the ranges 240–300°C, 32–38 bar, and 0.38–0.78 h^{-1}, respectively. The ratio hydrogen/hydrocarbons was maintained constant (265 Nm^3/m^3). Only a single parameter was varied in each experiment. The catalyst was Co-Mo/Al$_2$O$_3$, which is typical for hydrofining processes.

The optimal conditions established by this study were: $t = 300°C$, $p = 38$, and bar, $w = 0.38$ h^{-1}, for a H$_2$/hydrocarbons ratio of 265 Nm^3/m^3. The results obtained by hydrofinishing in these conditions of two heavy oils obtained from Egyptian crude oils are reported in Table 10.11.

The flow diagram of a unit for the hydrofinishing of oils is given in Figure 10.20.

In hydrofinishing, cold hydrogen is injected into the reactor to increase process flexibility. This measure is not determined by any significant reaction exotherm, but by the requirement for the preservation of the structure of the hydrocarbons by maintaining a moderate temperature in the reactor.

The working temperatures depend to a large extent on the catalyst. Catalysts operating at moderate temperatures are preferred, such as those having dual promotion, sometimes with Fe. These catalysts have a bimodal pore distribution, i.e., contain both micro- and macro-pores.

The working pressures are situated in the range of 30–60 bar.

The stability to oxidation of the produced oils increases with the degrees of desulfurization and especially of the denitration. Nevertheless, if some optimal value is exceeded, corresponding to a desulfurization of 90–95%, a decrease of the stability results [72,75].

Typical utilities consumptions (per ton of product) for the hydrofining of lubricating oils, wax, and of white oils are given in Table 10.12 [73].

Table 10.11 Results for the Hydrofining of Two Egyptian Oils

	Marine Belayim		Western Desert	
	feed	product	feed	product
Density, 15°/15°C	0.9105	0.906	0.9057	0.903
Color ASTM D-1500	4.5	3.5	3.5	2.5
Viscosity, cSt at 38°C	532	467	480	434
Viscosity, cSt at 99°C	29	28	28	27
Viscosity index	89.6	91.8	89.3	92.7
Acidity, mg KOH/g	0.016	0.009	0.012	0.005
Sulfur, wt %	1.55	1.19	1.23	0.80
Nitrogen, wt %	0.06	0.05	0.05	0.04
Vanadium, ppm	40	22	28	18
Nickel, ppm	30	16	24	14
Conradson carbon, wt %	0.78	0.56	0.66	0.41
Ash, wt %	0.016	0.006	0.009	0.004
Pour point, °C	-4	-4	-4	-4
Aromatic carbon	10.41	9.76	9.55	7.92
Naphthenic carbon	18.71	18.65	18.97	19.09
Aliphatic carbon	70.88	71.59	71.48	72.99
Aromatic rings	0.83	0.80	0.76	0.64
Naphthenic rings	1.96	1.98	2.04	2.37

Source: Ref. 75.

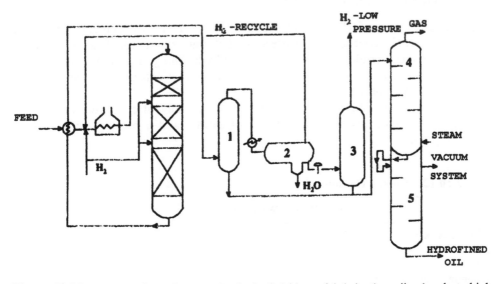

Figure 10.20 Process flow diagram for hydrofinishing of lubricating oils. 1 – hot, high pressure separator, 2 – cold, high presure separator, 3 – low pressure separator, 4 – stripper, 5 – dryer.

10.6.4 Hydrofining of Waxes and White Oils

As in the case of lubricating oils, the objective of hydrofining of waxes and of white oils is hydrofinishing.

Generally more active catalysts of the $NiO-MoO_3/Al_2O_3$ type are used, at pressures of 50–60 bar for wax and 120–170 bar for the white oils.

In order to comply with the requirement of complete absence of aromatic hydrocarbons, the white oils for medicinal use are submitted either to a preliminary selective, deep solvent refining, or to hydrotreating, instead of only a hydrofinishing.

A systematic study of the hydrofinishing of microcrystalline wax [76] identified the following optimal conditions: temperature = 300°C, p = 73 bar, w = 0.52 h^{-1}, H_2/H_c = 206.6 Nm^3/m^3. The obtained results are presented in Table 10.13.

The basic process flow diagram of a hydrofinishing unit for waxes and white oils is similar to that for lubricating oils shown in Figure 10.20.

The utilities consumptions are compared to those for oils in Table 10.12.

10.6.5 Hydrofining of Residues

The hydrofining of the very heavy distillates and of residues is inefficient. Severe hydrogenation conditions are required in order to make accessible the heteroatoms contained within the molecule. For this reason, the hydrodesulfurization of such fractions is carried out in more severe hydrogenation conditions similar to those of hydrotreating, as will be described in Section 10.7.4.

10.7 INDUSTRIAL HYDROTREATING

10.7.1 Hydrotreating of Jet Fuels

Hydrotreating of jet fuels was developed when the increased demand for jet fuel could no longer be covered by the selection of straight run jet fuel fractions. While a low sulfur content could be ensured by simple hydrofining, a partial hydrogenation by hydrotreating is required in order to satisfy the low concentration specified for aromatic hydrocarbons.

The removal of the aromatic hydrocarbons by solvent extraction has the disadvantage of producing important losses of extract and a product with a higher freezing point.

Table 10.12 Utilities Consumption

	Oil	Wax	White Oil
Electricity, kWh	10	15	35
Heat, kWh	40	60	65
Steam, kg	60	200	80
Cooling water 15°C, m^3	6	–	7
H_2/Nm^3 consumption	25	15	80

Source: Ref. 73.

Table 10.13 Hydrofinishing of Petroleum Wax

	Feed	Product
Density	0.8529	0.8409
Drop point, °C	81.5	80.0
Viscosity, cSt at 99°C	13.5	13.1
Sulfur, wt %	0.4	0
Acidity, mg KOH/g	1.854	1.334
Vanadium, ppm	1.1	0
Nickel, ppm	0	0
Lead, ppm	2.3	0.06
Nitrogen, ppm	155	43
Oil, wt %	0.5	0.84
Penetration at 25°C	11	13
Color	2	<0.5
Flammability, °C	250	288

Source: Ref. 76.

The more stringent quality requirements for the kerosene (smoke point) made it necessary to decrease its content of aromatic hydrocarbons, which extended the range of feeds processed by hydrotreating.

It is anticipated that the quality requirements for both types of kerosene will be more severe and the consumption of jet fuel will increase significantly, which will further extend the hydrotreating capacities.

The hydrotreating of jet fuels is usually performed on Ni-W catalysts, which have a good hydrogenation activity, at temperatures of 320–350°C, pressures of 30–70 bar and space velocities of 1–3 h^{-1}.

There exist processes, such as that of Shell [73] where, a first desulfurization step to a level of 1–5 ppm S is followed by a hydrogenation on a noble metal catalyst, to obtain a final product containing only 1.5 vol % aromatic hydrocarbons.

Since in all cases the process is strongly exothermic, the reactor contains several catalyst beds in series. The cooling is achieved by hydrogen injection between the beds – Figure 10.21.

The data of Table 10.14 are typical for the effect of hydrotreating on the changes undergone by the properties of a feedstock for jet fuel.

10.7.2 Hydrotreating of Gas Oils

The spectacular expansion of the hydrotreating of gas oils is the result of their growing market, within the frame of increasingly stringent quality requirements, imposed by environmental and conservation regulations.

The rapid increase of the consumption of Diesel fuels for cars in conditions of a stagnant or small increase of the crude oil production requires the broadening of the feedstock basis by the inclusion of gas oils of lower quality. That makes it more difficult to comply with the stricter quality specifications for the finished product.

This evolution is more obvious in Western Europe, where Diesel motors in small cars are in wider use than in the U.S.

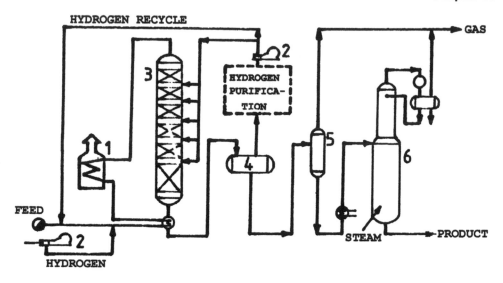

Figure 10.21   Flow diagram of a one-step gas oil jet fuel hydrotreating unit with hydrogen injection: 1 – furnace, 2 – compressor, 3 – reactor, 4 – high pressure separator, 5 – low pressure separator, 6 – column.

The change over recent years of the consumption of various types of fuels in Western Europe is shown in Figure 10.22. One may note that in the period 1980–1995, while the production of crude oil was basically constant, a doubling of gas oil consumption took place.

The selection of new feedstocks for Diesel fuel was largely determined by the difficulties in improving their properties by hydrotreating.

Table 10.15 shows the typical characteristics of several gas oils produced in some representative processes [77].

The quality requirements for Diesel fuels, specified in the European Norm EN 590 have been enforced since the year 1993 in 18 European countries.

The initially accepted sulfur content of 0.2% was reduced to 0.05% in Oct. 1996. In the U.S. the average sulfur content was 0.25% on basis of legislation from

Table 10.14   Hydrotreating of Jet Fuels and Kerosenes

	Feed	Product
Density	0.8373	0.8203
Distillation range, °C	159–275	158–277
Sulfur, ppm	10	< 1
Alkanes content	23	23
Cyclanes content	35	63
Aromatics content	42	14
Smoke point, ASTM, mm	15	23
Luminosity	30	48
Freezing point	−45	−46

Million
tons/year

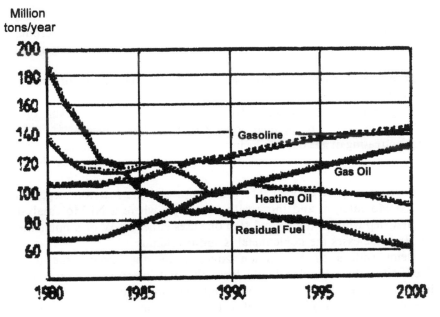

Figure 10.22   Evolution of fuel consumption in Western Europe.

1980 concerning the emissions of motor exhaust gases and became 0.05% in Oct. 1993.

Other provisions of the norm EN 590 present differences depending on the climate (arctic or temperate). The cetane number has a minimum range of 45 to 49, and range of the cetane index is 43 to 46. The percentage distilled is not more than 10% at 180°C and not less than 95% at 340°C for arctic climate, and not more than 65% at 250°C, not less than 85% at 350°C, and 95% at 370°C for the countries with temperate climate.

Table 10.15   Typical Characteristics of Gas Oils Produced in Various Processes

	DDA	LCO	HCO	VR	COK. R.	CLF
Density, kg/m^3	845	942	820	860	857	936
Sulfur, wt %	1.2	2.76	0.002	2.7	0.45	3.8
Nitrogen, wt %	0.03	0.063	0.0005	0.06	0.04	0.1
Cetane number	55	21	60	40	40	27
Bromine number, g/100g	0.5	15	0	20	17	28
Alkanes content, vol %	34	14	40	22	30	4
Naphthenes content, vol %	41	9	55	38	35	31
Aromatics content, vol %	25	77	5	40	35	65
Distillation 50% ASTM, °C	293	276	302	280	256	308
Distillation 95% ASTM, °C	353	347	360	345	338	360

DDA – straight run, LCO – catalytic cracking light gas oil, HCO – hydrocracked oil, VR – visbreaking, COK. R. – delayed coking, CLF - fluid coking.
Source: Ref. 77.

Sweden applies more severe norms: the minimum cetane index is 50 for the gas oil of class 1 and 47 for the two lower gas oils. The specified content of aromatic hydrocarbons is of maximum 5 vol% for the gas oil of class 1, and 20 vol % for class 2.

The accepted sulfur concentrations are 10, 50, and 1000 ppm for the three classes of gas oils.

Beginning in 1996, the European Norms became tighter as a result of the recommendations made by professional organizations.

The hydrotreating of gas oils can be implemented in a number of ways depending on the nature of the feedstock and on the quality specifications for the final product.

The processes may be divided into those with a single reaction step, which often use a Ni-Mo catalyst, and those with two steps, where a Ni-Mo catalyst is used in the first step while in the second step one uses either Ni-W or a noble metal Pt/γ-Al_2O_3. In some cases the noble metal is sulfur tolerant.

In systems with a single reaction zone, means for cooling the reactor are provided, either by removing liquid from the high pressure separator or by cold hydrogen injection (Fig. 10.21). Some units are provided with both systems, so that recycling of the gas oil submitted to the hydrotreating may be practiced if necessary.

Since the operation is in the presence of excess hydrogen, the recycled hydrogen is washed with amines for removing hydrogen sulfide, and thus increases the efficiency of the process.

In systems with two reaction steps that use a noble metal catalyst in the second step, and H_2S must be eliminated from the product resulting from the first step. The process scheme of such a plant is given in Figure 10.23 [78].

The two-step system is more complex, and the investment is higher, than in the one-step system. For this reason, the one-step system is often preferred. Its efficiency may be increased by using a more active catalyst and more severe operating conditions, especially concerning the reactor pressure. In the case of an existing plant, which doesn't allow more severe operating conditions, the addition of a second reaction step is often a preferred revamping solution [77].

The conditions for the hydrotreating of a straight run gas oil in one step and in two steps on various catalysts are shown in Table 10.16 [10]. The results of the hydrotreating of a light gas oil from catalytic cracking are given in Table 10.17 [77]. Further information is available in numerous data and graphs reported in several publications [77,78,80], which allow one to evaluate the efficiency of the process in various working conditions.

The economics for the hydrotreating of a mixture of 80% straight run gas oil and 20% catalytic cracking gas oil, containing 39.2% aromatic hydrocarbons, 1.16% S, and 220 ppm N, in order to achieve a content of 10% aromatic hydrocarbons are summarized in Table 10.18. As expected, utilities consumptions are higher if the feed is more aromatic or contains more sulfur. Also, the operating conditions are more severe if a lower final content of aromatic hydrocarbons is required.

The data of Table 10.18 shows that the system with one reaction step is more economic both from the point of view of the investments and of the cost of operation. The system with two steps is justified only for feeds with a higher content of

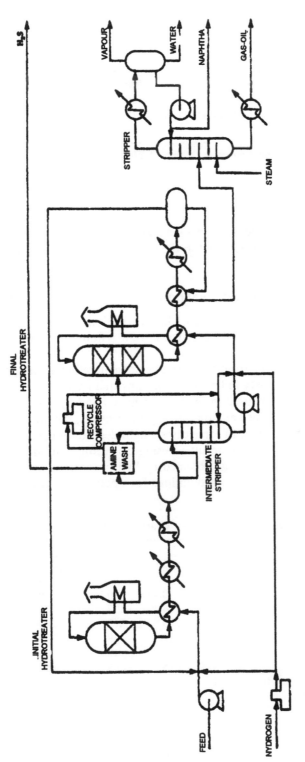

Figure 10.23 Two-step gas oil hydrotreating unit.

Table 10.16   Various Gas Oil Hydrotreating Systems

Catalyst for 1 step operation	NiMo	NiMo	NiMo
Catalysts for 2 step operation	–	NiW	Noble metal
H$_2$ pressure, bar	100	100	65
Mean temperature, °C	330	330	330
Aromatic hydrocarbons in the product, wt %	18	18	10
Relative reactors volumes	1.0	0.7	1.0

Feeding: $d = 0.865$; $S = 17,000$ ppm, $N = 225$ ppm, Aromatics 36.0%; Dist:
10% = 278°C, 50% = 329°C, 90% = 377°C.
Source: Ref. 10.

aromatics, such as the catalytic cracking gas oils, for which the required quality specifications may not be obtained in the system with one reaction step.

10.7.3   Hydrotreating of Lubricating Oils

The objective of the hydrotreating of lubricating oils is to increase the viscosity index, replacing in this way the selective solvent refining in which a portion of the feed is lost in the extract.

Moreover, different from the selective solvent refining, the process produces oils having a high viscosity index even from low-quality feeds.

The Ni-W catalysts used in the process have good hydrogenation activity.

Depending on the feed and on the target objectives, the process is performed in one or in two reaction steps. In some cases a light degradation, generating some light products is accepted. It is the result of using operating conditions that ensure the opening of the naphthenic rings obtained by hydrogenation.

An example of the process conditions for operation in two steps is:

Table 10.17   Hydrotreating of a Catalytic Cracking Gas Oil

		Feed	1st step NiMo	2nd step Noble metal
Density		0.945	0.896	0.850
Distillation ASTM	IBP	184	177	175
Distillation ASTM	50%	283	263	240
Distillation ASTM	FBP	372	359	350
Sulfur, ppm		21,800	18	<2
Nitrogen, ppm		625	5	<2
Aromatic hydrocarbons, total, wt %		80	69	10
monocyclic, wt %		17	48	9
bicyclic, wt %		39	19	1
tricyclic, wt %		11	2	0
others, wt %		13	0	0
Pressure, bar		60	60	60

Source: Ref. 77.

Table 10.18 Economic Data for Hydrotreating of a Gas Oil

	One step	Two steps First step	Two steps Second step	Total
Catalyst	NiMo	NiMo	Noble metal	
Pressure, bar	76	62	62	
Hydrogen consumption, wt %	1.45	0.75	0.70	1.45
Product:				
density	0.828	0.838	0.828	0.828
sulfur, ppm	< 50	< 50	< 5	< 5
aromatics, wt %	10	28	10	10
cetane number	57	53	57	57
Investment (1993)[a]	45	33	29	62
Operation cost, $/t (1993)[b]	23.6	14.6	13.5	28.2

80% straight run + 20% CC, with a content of 39.2 aromatic hydrocarbons, 1.16 wt % S and 220 ppm N
[a]For a plant with feed capacity of 4,750 m^3/day.
[b]Considering hydrogen cost 800 $/t.
Source: Ref. 77.

	Step I	Step II
Temperature, °C	370–425	260–315
Pressure, bar	210	210
Space velocity, h^{-1}	0.5–1.0	0.5
H$_2$ consumption, Nm^3/m^3	890	890

In this example (Engelhard Corp. license) both steps are performed on a Ni-W catalyst. The higher temperature used in the first step determines slight decomposition reactions. A vacuum distillation follows for the elimination of the light products formed. Following dewaxing, the oil is submitted to hydrotreating in a second step, which also uses Ni-W catalyst, at a lower temperature.

The viscosity index obtained depends on the severity of the process. It is of the order 105 for an average severity, and can reach 130 for a high severity [73, pp. 124].

10.7.4 Hydrotreating of Residues

The hydrotreating of residues for desulfurization, denitration, and especially demetallation is a process that recently underwent significant improvement. In combination with the catalytic cracking of the hydrotreated residues, it represents today one of the most attractive methods for converting heavy residues to motor fuels with high yields and relatively low hydrogen consumption.

Compared to the direct catalytic cracking of the residual fractions, the preliminary hydrotreating has two major advantages: it eliminates the poisoning of the catalytic cracking catalyst by Ni and V and makes it possible to produce without difficulties gasoline that meets the increasingly more severe specifications for sulfur [82,83].

Three types of processes exist for the commercial implementation of the residues hydrotreating:

- Systems with fixed catalyst bed
- Systems with moving bed
- Systems with ebullated bed

In all three systems the pressure is of the order of 180 bar and an average temperature of 450°C.

Systems with fixed catalyst bed. The general scheme of such a unit is given in Figure 10.24. In this scheme, several small reactors connected in series are used, in the place of one reactor with cooling between the beds as provided for the processing of distilled feeds. The reason for this design is the fact that the deposition of the metals on the catalyst takes place rapidly, so that they are concentrated on the first portion of the contact zone, therefore in the first reactor and especially at its upper portion. Since the catalyst poisoned with metals can not be regenerated in place, the use of several reactors in series allows taking the first reactor out of the circuit and to replace the catalyst, without having to stop the operation of the unit.

The flow of the feeds in the reactors is downwards, with the gaseous phase being the continuous phase.

An essential problem for a good operation of the unit is the uniform distribution and good dispersion of the liquid at the reactor inlet [84,85]. An example of such a distribution is depicted in Figure 10.25 [86].

In operation, high attention is paid to the temperature profile and to the injection of adequate amounts of cold recycle gases between the reactors, and possibly between the catalyst beds. Following each injection, the liquid must be redistributed by using a system similar to that of Figure 10.25. As the coke deposits

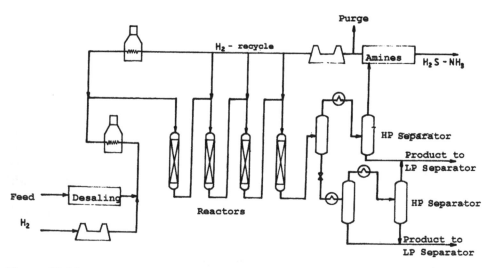

Figure 10.24 Fixed bed residue hydrotreating unit.

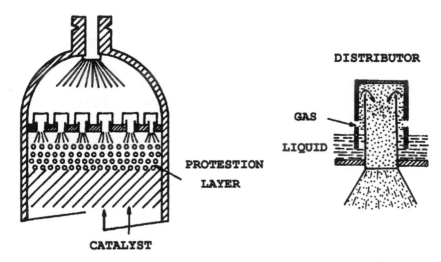

Figure 10.25 Liquid dispersion at the top of a residue hydrotreating reactor.

formed on the catalyst increase, the temperature is increased, without however being allowed to exceed 425°C at the end of the cycle.

The length of a cycle for a feed containing 150–200 ppm metals is of 11 months. For feeds with a metal content up to 400 ppm, two reactors that can be switched one with other on the top of the system are used. The length of the cycle is also 11 months. For a larger metals content, the employed reaction systems are the moving bed, or the ebullated bed.

In general, two types of catalyst are used: the first is for breaking open the asphaltene molecules and demetallation, while the second is for hydrofining. The proportion between the two catalysts depends on the properties of the feed and on the quality requirements of the finished product [87].

For a good operation of the fixed bed catalyst, a good two-step desalting of the crude oil is needed, followed by a filtration of the feed to the unit, in order to prevent salt plugging the pores of the catalyst.

In the HYVAHL process of the French Institute of Petroleum [88], feed with up to 500 ppm V + Ni can be used. The system uses two interchangeable guard reactors located ahead of the main reactors. The feed passes first through one of the guard reactors containing demetallation catalyst. The other reactors contain both demetallation and desulfurization catalysts.

The sizes of catalysts extrudates must be sufficiently large for maintaining a reasonable pressure drop through the bed and for avoiding coke deposits that prevent circulation through the bed.

The advantages of the process in fixed bed are the lack of catalyst attrition and low risk of plugging of the bed, which can be of concern in system with moving bed or ebullating bed.

These advantages led to the situation that worldwide, in 1993, there were 30 operating units using fixed beds representing 80% of the operating units.

The catalytic system used in the fixed bed units is rather complex. Two classes catalysts are used:

Hydrodemetallation catalysts (HDM),
Hydrodesulfurization catalysts (HDS), of hydrodenitration (HDN), of hydro-
  decarbonization (HDCC), etc.

Also, in the upper portion of the first reactor a filter is placed, which no or low
catalytic activity but containing macropores with diameters of about 100μ. The role
is to retain the impurities (especially iron) that may enter the reaction system.

Typical characteristics of the material used in the guard bed and of the HDM
and HDS catalysts are given in Table 10.19 [88].

The pore size distribution for the HDM and HDS catalysts is given in
Figure 10.7.

The first reactors in the system contain the HDM catalyst for the demetallation
and hydrogenolysis of asphaltenes and resins, to make "accessible" the contained
heteroatoms. The temperature in these reactors is higher than in the following ones
to favor such decomposition reactions. These type of catalysts have a broad pore size
distribution and contain mesopores and even macropores to allow the adsorption of
the molecules of large molecular mass (Figure 10.26). They are capable of retaining a
large amount of metals $(Ni + V)$ that at the end may represent a weight equal to that
of the catalyst.

Following the HDM catalyst, the reaction mixture contacts the catalysts for
desulfurization (HDS), denitration (HDN), and for reduction of the Conradson coke
(HDCC). The volume allocated to the HDM and to the other types of catalysts is
dependent on the concentrations of the heteroatoms contained in the feed and on the
quality requirements of the end product.

The temperature in the second reactor group is milder than in the first, which
favors both the hydrogenolysis reactions and those of partial hydrogenation of the
polycyclic aromatic structures.

Performance data for a process of this type is presented in Table 10.20 [88].

Systems with moving bed.   The system is similar to moving bed catalytic cracking,
from which it borrows many traits.

The circulation of the catalyst is downwards. The circulation of the feed and
hydrogen is cocurrent with the catalyst in the Shell system [89], and countercurrent in
the ASVAHL system of the French Institute of Petroleum [84]. In the latter case the
velocities of the gas and liquid phases should not exceed those that would disturb the
downwards plug flow of the catalyst particles of the bed (see Figure 10.26).

Table 10.19   Properties of Catalyst for Residue
Hydrotreating

Characteristics	Guard material	HDM	HDS
Surface, m^2/g	<0.1	120–170	180–220
Pore volume, cm^3/g	<0.25	0.7–1.0	0.4–0.6
Active component	–	NiMo	CoMo
wt % of total catalysts	<5	30–70	30–70

Source: Ref. 88.

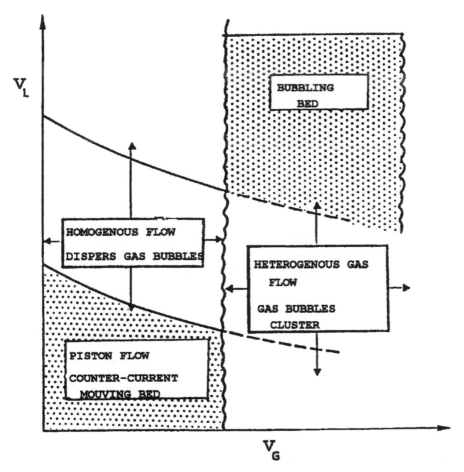

Figure 10.26 Moving bed system with descending catalyst, gas and liquid in countercurrent flow [84].

In both cases, the metals are gradually deposited on the catalyst particles. In the ASVAHL system, the highly contaminated catalyst, containing also the impurities retained from the feed, is immediately eliminated. Thus, the feed and the hydrogen are further contacted with uncontaminated catalyst. In the Shell process, the catalyst, which has retained the metals, passes through the entire reaction volume before leaving the system. Despite the lack of comparative data, one may suppose that in identical conditions the performance of the ASVAHL process is superior.

In many cases, the moving bed reactor is used as a first reaction step, followed by a fixed bed reactor. Thus, in the moving bed reactor takes place the decomposition of the molecules with complex structures and the separation of the metals (Ni + V). Both are inherently accompanied by formation of coke deposits. Therefore, the moving bed reactor protects the fixed bed reactor system and significantly extends the length of the cycle.

Table 10.20 Performance of HYVAHL Process for Residues Hydrotreating

| Characteristics and yields | Residue | | | |
| | Straight run | | Vacuum | |
	Arabian	Cold Lake	Kirkuk	Safaniya
Feed				
density	0.988	1.024	1.021	1.035
viscosity, cSt at 100°C	95	1095	880	3900
sulfur, wt %	3.95	5.05	5.14	5.28
conradson carbon, wt %	13.8	18.3	18.2	25.0
asphaltenes, wt %	5.7	10.5	7.5	13.2
Ni + V, ppm	104	325	189	203
Hydrotreated residue				
density	0.934	0.958	0.950	0.970
sulfur, wt %	0.50	0.80	0.47	0.70
Conradson carbon, wt %	4.0	8.9	6.4	10.0
asphaltenes, wt %	0.3	2.5	0.8	5.3
Ni + V, ppm	1.5	24	1.5	20
Yields, wt %				
$H_2S + NH_3$	3.88	5.08	5.28	4.92
$C_1 - C_4$	2.14	3.20	2.27	2.60
gasoline	3.47	5.20	4.02	4.60
gas oil	21.55	27.50	22.20	32.80
residue	70.50	61.02	68.01	70.26

Source: Ref. 88.

The difficulties in the operation of such a reactor are related to the operations of introducing and removing of the catalyst into and out of the reactor, in which a high pressure and an atmosphere rich in hydrogen prevail.

In general, catalysts of spherical shape are used. Despite the fact that it has a lower catalytic activity, this shape has the advantage of easier circulation and of a lesser attrition than other shapes. It must be mentioned that if attrition occurs in this reactor, the fines produced may be carried into the fixed bed reactor that follows and impact negatively the flow pattern through the fixed bed.

The moving bed are used for feeds that are very rich in metals—above 400 or even 500 ppm of Ni + V.

Only a limited number of units of this type were operating worldwide as of 2002.

Systems with ebullated bed. The process with ebullated bed is carried out in reactors of the type depicted in Figure 10.27 [8].

The concomitant presence in the reactor of the moving gas, solid, and liquid phases, in intimate contact, leads to difficult hydrodynamic conditions. Good liquid-solid mixing is performed by means of the recycling pump, while achieving adequate contact with the gas phase is much more complicated. The length of the contact time between the gas and liquid phases depends on the size of the gas bubbles, on the interfacial tension, and on the density of the gas

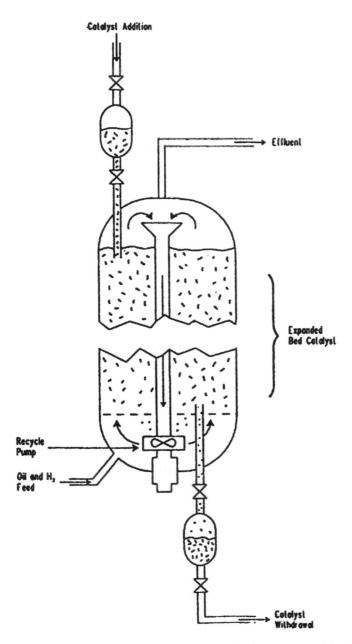

Figure 10.27 Ebullated bed reactor for residue hydrotreating (From Ref [8]).

phase. No systematic studies were published on the interaction between these parameters [8].

Pressure has an important effect on the operation in the three-phase system [90]. At pressures of 10 bar, the diameter of the gas bubbles was of 8 mm, and the fraction of the cross section occupied by gas is constant along the height of the apparatus. For solids with diameters of 0.46 mm, the pressure leads to the increase of the cross section

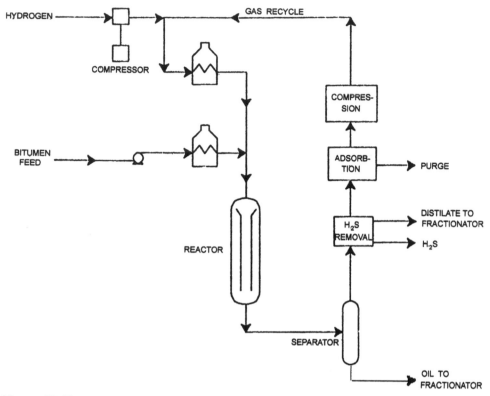

Figure 10.28   Process flow diagram for residue hydrotreating with one-step ebullated bed reactor.

occupied by the gas until ratio of the volume of the gas in the bed divided by the total volume becomes approx. 0.25. This effect due to the pressure increase becomes insignificant if the linear gas velocity is below 2 cm/s. These data are difficult to extrapolate to the pressures of 70–100 bar at which the commercial units operate.

There are two licenses concerning this process: H-Oil and L.C. Fining, with minimum differences in the design of the equipment. As per the end of 2002, some of the units built earlier were either shut off or converted to hydrocracking operation (Section 12.7). One unit for processing vacuum gas oil is being designed.

Two versions of the process exist. One has a single reaction step and performs a moderate conversion of the residue. The second operates in two steps and leads to a high conversion.

The process flow diagram of unit of the first type is represented in Figure 10.28.

The typical operating conditions for such a unit are: temperatures of 420–450°C, pressures of 70–150 bar, space velocity 0.1–0.5 h^{-1}, the recycling coefficient inside the reactor ranges between 5 and 10. The recycled hydrogen represents 3–4-times the amount consumed in the reaction.

Daily, about 1% of the catalyst inventory is replaced and the catalyst consumption is 0.3–1.5 kg/m^3 of hydrotreated residue.

Owing to the intense circulation within the reactor, the thermal gradient in the reactor does not exceed 2°C.

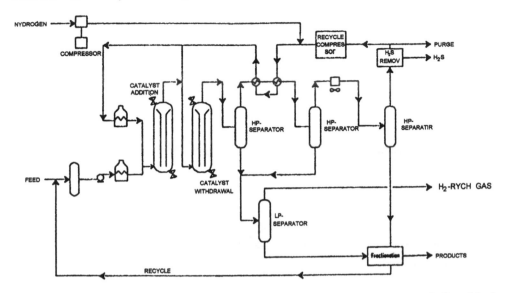

Figure 10.29 Process flow diagram for residue hydrotreating with two-step ebullated bed reactor.

Published data [9] indicate the following results obtained by treating of an Athabasca bitumen in the indicated conditions:

Conversion 524°C+, % vol	66.8
Decrease of Conradson carbon, wt %	42
Desulfurization, wt %	65
Vanadium removal, wt %	71
Nickel removal, wt %	60
Hydrogen consumption Nm^3/m^3	167

The combined liquid products represent 104 vol % to the feed volume, as a result of the density decrease by hydrogenation.

The second version, which produces a high conversion, involves a second reaction step (Figure 10.29). The process may take place with conversions of the order 60–80% to fractions with boiling temperatures below 524°C. At higher conversions, solid particles are formed, which must be eliminated by filtration of the recycle stream through a coke bed. The coke retains the most polar particles [92]. Another method for preventing the formation of solid particles is to mix in the feed strongly aromatic fractions [93], which will act as mediators in the hydrogenation process.

10.8   DESIGN ELEMENTS FOR THE REACTOR SYSTEM

10.8.1   Thermal Effects

Despite the fact that the hydrogenolysis of the heteroatomic compounds is a strongly exothermal reaction, their relative low concentration in the feed makes for a weak

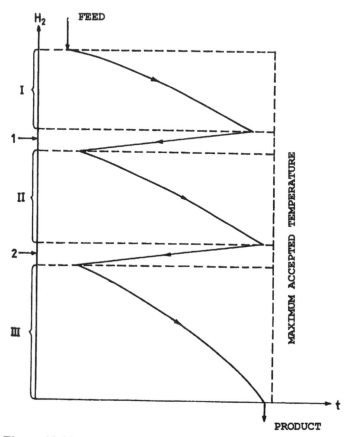

Figure 10.30 Temperature profile in a hydrotreating reactor. I, II, III – catalyst beds, 1, 2 – coolant injections.

thermal effect of the hydrofining process and does not require special measures for the removal of the heat of reaction.

The situation is different in the hydrotreating processes, which attain high conversion of the strongly exothermic hydrogenation of aromatic hydrocarbons, leading to the development of large amounts of heat.

In the commercial fixed bed units, the heat is removed by injection of cold streams of hydrogen or of recycled product.

To ensure a uniform removal rate of heat over the whole cross section of the reactor, the injections of cold product are made in the free space between the catalyst beds.

Accordingly, each catalyst bed behaves as an adiabatic reactor. The temperature profile along the reactor may be represented by the diagram of Figure 10.30.

The precise calculation of this temperature profile requires the kinetic modeling of each reaction zone and the thermal balances for the spaces between the beds. From this, one may calculate the amounts of coolant that must be injected for controlling the temperature.

The total amount of coolant may be estimated by using published data on the hydrotreating of similar feeds. The analytic characteristics of the feed and of the product allow evaluating the total thermal effect, using for this purpose the methods presented in Section 1.1. The value of the thermal effect and a thermal balance for the whole reactor will allow the computation of the coolant flowrate for the selected temperatures of inlet and outlet from the reactor, and the nature and temperature of the coolant flows.

10.8.2 Kinetics and Optimization

The kinetic calculation may be carried out by using the methods and the equations of Section 10.4, or recommended methods from the specialized literature.

It must be taken into account that the temperature in the reactor is variable and therefore the kinetic calculations for estimative purposes must be performed by using values of the kinetic constants corresponding to average conditions. For approximate calculations the arithmetic mean temperature may be used. For more exact calculations, the temperature that is equivalent to the adiabatic mean rate, given by Eqs. (2.163) or (2.164) should be used.

Frequently asked questions refer to improving the performance of existing units, evaluation of the impact of modified operating conditions, or of product specifications.

In such cases it is useful to use the overall kinetic constants, deduced from the operating data of the unit, and using in this purpose Eqs. (10.20) and (10.21) for the desulfurization and denitration, or similar equations for reducing the acidity and the Conradson carbon value.

The correct evaluation of the hydrogenation of aromatic hydrocarbons is more difficult. However, in this case also, an equation of pseudo–first order may be used as a first approximation, but preferably written similarly to Eq. (10.27) for a reversible process.

Since the kinetic constants used in the computations correspond to the actual catalyst and operating conditions of the particular unit examined, the errors will be lower than if more exact equations would be applied, but using kinetic constants from the literature as a first approximation.

REFERENCES

1. JB McKinley. Catalysis, Chap. 6, New York: Reinhold Pub. Cor., 1957.
2. RZ Magaril. Teoreticeskie osnovy himiceskih protsesov pererabotki nefti Himia, Moscow (1976).
3. Th C Douves. Erdöl Erdgas Zeitschrift 1: 25, 1966.
4. CWZ Langhout, FJG Stijnjes, JH Watermann. J. Inst. Petroleum 41: 263, 1955.
5. DR Stull, EF Westrum Jr, GC Sinke. The Chemical Thermodynamics of Organic Compounds, Malabar, Florida: RE Krieger Publishing Co., 1987.
6. AJ Gully, MP Ballard. Advances in Petroleum Chemistry, vol VII, pp. 2140, New York: Interscience Publ, 1963.
7. E Welther, S Feyer-Ionescu, C Ionescu, E Turjanschi-Ghionea. Hidrofinare. Hidrocracare, S Raseev, ed., Ed Tehnică, Bucharest, 1967.
8. MR Gray. Upgrading Petroleum Residues and Heavy Oils, New York: Marcel Dekker Inc., New York, 1994.
9. JG Welch, P Poyner, EF Skelly. Oil and Gas J 92: 56, Oct. 10, 1994.
10. BH Cooper, A Stanislaus, PH Hannerup. Hydrocarbon Processing 72: 83, June, 1993.

11. JJ McKetta. Petroleum Processing Handbook. New York: Marcel Dekker Inc., 1992.
12. A Billon, F Morel, ME Morrison, JP Péries. Revue de l'Institut Français du Pétrole 49: 495, 1994.
13. J Tanoubi, P Martinerie, P Bourseau, G Muratel, H Taulhoat. Revue de l'Institut Français du Pétrole 46: 389, 1991.
14. BC Gates. Catalytic Chemistry, John Wiley & Sons Inc., 1991.
15. Y Iwasawa, BC Gates. Chemtech, March 1989, pp. 173.
16. IG Murgulescu, E Segal, T Oncescu. "Introducere în Chimia Fizică", vol. II, "Cinetică chimică si cataliză", Editura Academiei, Bucharest, 1981.
17. R Prius, VHJ De Beer, GA Somorjai. Catal. Rev. Sci. Eng 31: 1, 1989.
18. AV Sapre, BC Gates. Ind. Eng. Chem. Proc. Des. Dev 20: 68, 1981.
19. SC Korre, MT Klein, RJ Quann. Ind Eng Chem Res 34: 101, 1995.
20. M Houalla, NK Nag, AV Sapre, DH Broderick, BC Gates. AIChE Journal 24: 1015, 1978.
21. MC Tsai, YW Chem Ind Eng Chem Res 32: 1603, 1993.
22. SY Lee, JD Seader, CH Tsai, FE Massoth. Ind Eng Chem Res 30: 29, 1991.
23. CH Tsai, FE Massoth, SY Lee, ID Seader. Ind Eng Chem Res 30: 22, 1991.
24. RE Baltus, JL Anderson. Chem Eng Sci 38; 1959, 1983.
25. SY Lee, JD Seader, CH Tsai, FE Massoth. Ind Eng Chem Res 30: 607, 1991.
26. CJ Pereira, JW Beeckman. Ind Eng Chem Res 28; 422, 1989.
27. LC Trytten. Thesis (M. Sc.), University of Alberta, 1989.
28. JM Ammus, GP Androutsopoulos. Ind Eng Chem Res 26: 494, 1987.
29. KM Sundaram, JR Katzer, KB Bischoff. Chem Eng comm 71: 53, 1988.
30. DH Broderick. Thesis (Ph. D.), University of Delaware, 1980.
31. S Ihun, SJ Moon, HJ Chai. Ind Eng Chem 29: 1147, 1990.
32. X Ma, K Sakanishi, J Mochida. Ind Eng Chem 33: 218, 1994.
33. S Raseev, S Ionescu. Buletinul Institutului de Petrol si Gaze, 6(XXVI), 79, 1976.
34. RM Masagutov, GA Berg, GM Kulinich. Proceedings 7-World Petroleum Congress PD, no. 20(7), Mexico, 1967.
35. JF Mosby, GB Hoekstra, TA Kleinhenz, JM Stroks. Hydrocarbon Processing 52: 93, 1973.
36. J Sudoh, Y Shiroto, Y Fukul, C Takeuchi. Ind Eng Chem Process Des Dev 23: 641, 1984.
37. N Papayannakos. Appl Catal 24: 99, 1986.
38. JF Le Page et al. Applied Heterogeneous Catalysis, Technip, Paris, 1987.
39. P Zeuthen, P Stoltze, UB Petersen. Bul Soc Chem Belg 96: 985, 1987.
40. LD Rollmann. J Catal 46: 243, 1977.
41. J Van Parys, GF Froment, D Delmon. Bul Soc Chim Belg 93: 823, 1984.
42. R Galiasso, W Garcia, MM Ramirez de Agueldo, P Andrew. Hydrotreatment of Cracked Gas Oil, In H Heinmann, GA Somorjai, eds Chemical Industries Catalysis and Surface Science, vol 20, Marcel Dekker, New York, 1984.
43. LM Magnabosco. In DL Trimm, S Akashah, M AbsiHalabi, A Bishara, eds. Studies in Surface Science and Catalysis: Catalysis in Petroleum Refining, vol. 53, 481, Amsterdam: Elsevier, 1990.
44. M Bouchy, P Dufresne, S Kasztelan. Ind Eng Chem Res 31: 2661, 1992.
45. MJ Girgis, BC Gates. Ind Eng Chem Res 30: 2021, 1991.
46. C Philippopoulos, N Papayannakos. Ind Eng Chem 27; 415, 1988.
47. RS Mann, IS Sambi, KC Khulle. Ind Eng Chem 27: 1788, 1988.
48. M Oballa, C Wong, A Krzywicki, S Chase, G Deania, W Vandenhengel, R Jeffries. In: C Han and C Hai eds. Proc. Int. Symp. on Heavy Oil and Residue Upgrading and Utilization, International Academic, Beijing, p. 67, 1992.

49. MR Gray, P Jokuty, H Yeniova, L Nazarewycz, SE Wanke, U Achia, A Krzywicki, EC Sanford, OKY Sy. Can J Chem Eng 69: 833, 1991.
50. J Guitian, A Sonto, R Ramirez, R Marzin, B Solari. In: C Han and C Hai, eds. Proc. Int. Symp. on Heavy Oil and Residue Upgrading and Utilization, Beijing: International Academic, 237, 1992.
51. WJ Beaton, RJ Bertolacini. Catal Rev Sci Eng 33: 281, 1991.
52. A Amorelli, YD Amos, CP Halsig, JJ Kosman, RJ Jonker, M de Wind, J Vrieling. Hydrocarbon Processing 71: 93, June 1992.
53. A Billon, G Heinrich, A Quinard. Hydrotreating of Middle and Vacuum Distillates from Coking and FCC, in "Hydrocarbon Technology International 1987", P Harrison ed., Sterling Publ., London (1987).
54. S Ionescu, S Raseev. Bul Inst Petrol Gaze Geol 17: 89, 1971.
55. S Raseev, S Ionescu. Bul Inst Petrol Gaze 21: 84, 1974; 22: 33, 1975.
56. S Ionescu. Ph. D. Thesis, Inst. Petrol, Gaze, Geol., 1970.
57. GT Adams, AA Del Peggio, H Schaper, WHJ Stork, WK Shiflet. Hydrocarbon Processing 68: 57, Sept. 1989.
58. AHAK Mohammed, AK Aboul-Gheit. Hydrocarbon Processing 60: 145, Sept. 1981.
59. MV Landau, LN Alekseenco, LJ Niculina, OV Vinogradova, BK Nefedov, LD Konovalcicov, VJ Stein, TP Podobaieva. Himia Tehnol Topliv Masel, no. 1, 1991, p. 8.
60. CF LeRoy, MJ Hanshaw, SM Fischer, T Malik, PR Koolman. Oil and Gas J 89: 49, May 27, 1991; 89: 90, June 3, 1991.
61. VN Podlesny, JJ Zadko, SL Mund, RK Nasirov. Himia Tehnol Topliv Masel no. 1, 1991, p. 2.
62. G Gualda, H Toulhoat. Revue de l'Institut Français du Pétrole 43: 567, 1988.
63. BK Nefedov. Himia Tehnol. Topliv Masel, no. 2, 1991, p. 13.
64. DS Thakur, MG Thomas. Appl. Catal 15: 197, 1985.
65. E Furinsky. Erdöl und Kohle 32: 383, 1979.
66. TE Myers, FS Lee, BL Myers, TH Fleisch, GW Zajac. AIChE Symp Ser. 273, New York: AIChE, 1989, p. 21.
67. CJ Pereira. Ind Eng Chem Res 29:512, 1990.
68. J Bartholdy, BH Coope. ACS Div Petrol Chem Prepr 38(2): 386, 1993.
69. A Stanislaus, M Marafi, M Absi-Halabi. ACS Div Petrol Chem Prepr 38(1): 62, 1993.
70. JSM Jocker. ACS Div. Petrol Chem. Prepr 38(1): 74, 1993.
71. K Inone, P Zhang, H Tsyuyama. ACS Div. Petrol. Chem. Prepr 38(1): 77, 1993.
72. A Nastasi. "Hidrofinarea", In G Suciu, ed. Ingineria Prelucrării Hidrocarburilor, vol. II, pp. 292–319, Editura Tehnică, Bucharest, 1974.
73. Hydrocarbon Processing 61: 123, 124, 126, 131, Sept. 1982.
74. J Hassabou, S Kandil, S El Khdrachi. "Hydrofinishing of Heavy Lube Oils", OAPEC-IFP Conference, July 5–7, 1994, Report no. 8, Rueil-Malmaison, France.
75. LR Menzl, LW Webb. Petroleum Refiner 44(5): 202, 1965.
76. H Marwan, M Hamad, S El Kharachi. Hydrotreating of Microcrystalline Wax, OAPEC-IFP Conference, July 5–7, 1995, IFP, Rueil-Malmaison, France.
77. G Heinrich, S Kasztelan, L Kerdradon. Revue de l'Institut Français du Pétrole 49: 475, 1994.
78. BH Cooper, P Sogaard-Andersen, PN Hannerup. Hydrocarbon Technology International 1994, P Harrison, ed., London: Sterling Publ. Ltd., London, p. 95.
79. AJ Suchanek. Oil and Gas J 88: 107, May 7, 1990.
80. N Marchal. Sulphur Resistant Catalysis for Deep Aromatics Reduction in Gas-Oil, OAPEC-IFP Conference, July 5–7, 1994, Report no. 9, IFP, Rueil-Malmaison, France.
81. Hydrocarbon Processing 71: 183, 1992.
82. FM Hibbs, O Gems, DA Kauff. In P. Harrison, ed. Hydrocarbon Technology International 1994, London: Sterling Publ. Ltd, p. 17.

83. JF Hohnholt, CY Fausto. Refining and Petrochemical Technology Yearbook 1987, Tulsa: Reunwell Books, p. 245.
84. JF Le Page, SG Chatila, M Davidson. Reffinage et convrsion des produits lourds du pétrole, Technip, Paris, 1990.
85. FL Plantenga, YJN Torihara. Oil and Gas J, 89: 74, Oct. 21, 1991.
86. RL Richardson et al. Oil and Gas J 77: 80, May 28, 1979.
87. A Hennico, G Cohen, T des Courières, M Espeillac. Hydrocarbon Technology International 1993, P. Harrison, ed. pp. 33, London: Sterling Publ. Ltd.
88. F Morel, JP Péries, S Kasztelan, S Kressmann. Hydroconversion Processes and Catalysts for Residues Upgrading, OAPEC-IFP Conference, July 5–7, 1994, Report no. 11, IFP, Rueil-Malmaison, France.
89. RH Van Dongen, KHW Groeneveld. Hydrocarbon Technology International 1987, P. Harrison ed., London: Sterling Publ. Ltd.
90. W Jiang, D Arters, LS Fan. Ind Eng Chem Res 31: 2322, 1992.
91. W Bishop LC. Finer Operating Experience at Syncrude, Proc. Symp. Heavy Oil: Upgrading to Revining, Can. Soc. Chem. Eng., Calgary, AB, 1990, p. 14.
92. RP Van Dreisen, J Caspers, AR Campbell, G Lumin. Hydrocarbon Processing 58(5): 107, 1987.
93. WJ Beaton, RJ Bertolacini. Catal. Rev. Sci. Eng 33: 281, 1991.
94. F Hamdy, DD Whitchurst, M Isac. Ind. Eng. Chem 37, (9): 3533–3539, Sept. 1998.
95. V Vanrysselberghe, R Le Gall, GF Froment. Hydrodesulfurisation of 4-Methyl-dibenzothiophene and 4,6-Dimethyl-dibenzothiophene on a CoMo/Al$_2$O$_3$ Catalyst. Reaction Network and Kinetics, Ind Eng Chem 37(4) 1235–1242, 1998.
96. S Hatanaka, M Yamada, O Sadkane. Hydrodesulfurization of Catalytic Cracked Gasoline 3. Selective Catalytic Cracked Gasoline Hydrodesulfurization on the CoMo/γ-Al$_2$O$_3$ Catalyst Modified by Coking Pretreatment, Ind Eng Chem Res, 37: 1748–1754, 1998.
97. H Toulhoat, R Ressas. L'hydrodéazotation de distilats issus de la conversion des hydrocarbures lourdes, Revue de l'Institut Français de Pétrole 41 (94), July–August, 1986.
98. V Vanrysselberghe, G Froment. Kinetic Modeling of Hydrodesulfurization of Oil Fractions: Light Cycle Oil Ind Eng Chem Res 37: 4231–4240, 1998.
99. S Trassobares, AM Callejas, Maria, MA Benito, TM Martinez, D Severin, L Brouwer. Kinetics of Conradson Carbon Residue Conversion in Catalytic Hydroprocessing of Maya Residue, Ind. Eng. Chem. Res 37: 11–17, 1998.
100. S Trassobares, AM Callejas, MA Benito, TM Martinez. Upgrading of a Petroleum Residue. Kinetics of Conradson Carbon Residue Conversion, Ind Eng Chem Res 38: 938–943 (3): March, 1999.
101. Y Sok, Oil and Gas Journal 97: 64, Sept. 6 1999.

11

Hydroisomerization of Alkanes

Two types of hydroisomerization processes of alkanes were developed, having different objectives and technologies:

1. The isomerization of lower *n*-alkanes: of C_5-C_7 for the production of high-octane components and of *n*-C_4 to *i*-C_4 as feed for the production of alkylate.
2. The isomerization of the *n*-alkanes contained in paraffinic oils in order to produce a significant decrease of the freezing temperature and thus eliminate the need for dewaxing, which is accompanied by yield losses.

The worldwide evolution of the plant capacities for the isomerization of lower alkanes is given in Table 11.1.

Despite the fact that these processes have quite different technologies, their theoretical bases are similar and are discussed in the first part of this chapter. References are made to which of the processes the discussion applies to.

11.1 THERMODYNAMICS OF HYDROISOMERIZATION

The thermodynamic equilibrium between the isomers for the C_4, C_5, C_6, and C_7 alkanes are represented in Figures 11.1–11.4.

These graphs show that a lowering of the reaction temperature favors the formation of highly branched hydrocarbons. As a result, the octane number of the equilibrium products increases by decreasing the reaction temperature.

For the C_5-C_6 hydrocarbons, this variation of the octane number is represented in Figure 11.5. It explains why it is important to convert the C_5-C_6 fractions at the lowest possible temperatures, which requires very active catalysts.

When comparing the equilibrium graphs for hexanes and heptanes (Figures 11.3 and 11.4) with the experimental data, one must observe that on the acid sites of the catalysts the reaction mechanism does not lead to the formation of isomers

Table 11.1   Worldwide Isomerization Capacities in Thousands BPSD

	1990	1995	1998	2000	2010[a]
North America	504	604	664	700	800
Western Europe	169	437	501	550	580
Pacific and South Asia	24	67	92	210	700
Eastern Europe and CIS	—	39	57	100	200
Latin America	23	41	116	120	135
Middle East	3	49	66	150	200
Africa	15	15	30	50	80
Total	738	1252	1526	1880	2695

[a] Estimate
Source: Ref. 41.

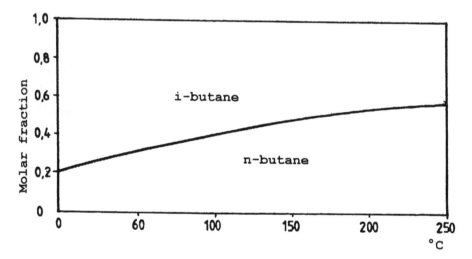

Figure 11.1   Thermodynamic equilibrium for nC_4H_{10}-iC_4H_{10} isomerization in vapor phase.

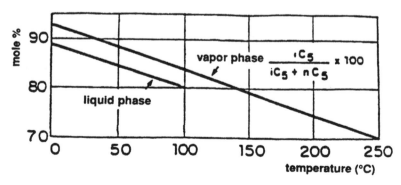

Figure 11.2   Thermodynamic equilibrium for isomerization of nC_5H_{12}-iC_5H_{12} [1].

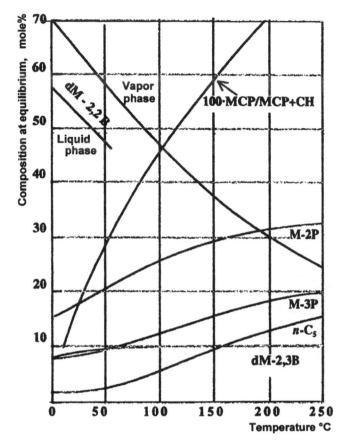

Figure 11.3 Thermodynamic equilibrium for hexane isomerization. (From Ref. 11.)

containing quaternary carbon atoms. Such isomers are formed in small quantities, as a result of secondary reactions and do not reach equilibrium.

Insufficient thermodynamic data are available for studying equilibrium compositions for the isomerization of the higher alkanes contained in the petroleum fractions. The approximate calculations made on the basis of extrapolated data as well as the experimental data indicate that in this case also, lower operating temperatures favor the formation of branched structures.

11.2 HYDROISOMERIZATION CATALYSTS

The first catalyst used for the isomerization of alkanes was aluminum chloride, as it was demonstrated for the first time by CD Nenitescu and A Drăgan [3] in 1933.

The aluminum chloride being very active, the process takes place at temperatures of 80–100°C, which ensures a high conversion to branched isomers. Its disadvantages are high corrosivity, as well as high consumption of catalyst and high processing costs. The isomerization of alkanes was developed on a large scale only after the development of bifunctional catalysts containing platinum on an acid support of Al_2O_3 or on alumosilica. These catalysts are less active and require reaction

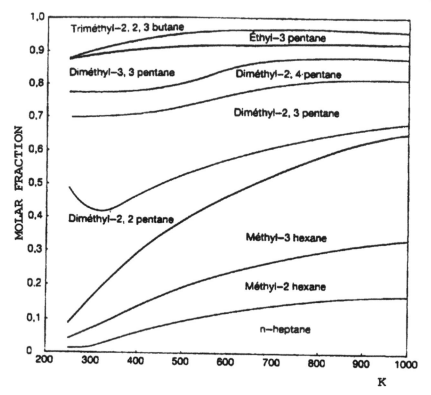

Figure 11.4 Thermodynamic equilibrium for the isomerization of heptane in vapor phase. (From Ref. 12.)

temperatures of 320–450°C. Such high temperatures do not favor the formation of branched isomers and require the separation and the recycling of the normal, not converted hydrocarbons.

This situation led to a new generation of platinum catalysts on alumina supports with continuous introduction of small amounts (ppm levels) of a chlorination agent. This technique, maintains a high acidity and allows operating at temperatures of about 150°C. However, the prepurification of the feed is needed since the catalyst is sensitive to poisons, especially water. Moreover, the continuous injection of chlorination agents (organic chlorides) leads to corrosion.

Despite the fact that this catalyst type is still being used, the trend is to gradually replace it by catalysts of platinum on zeolite, which do not require a continuous chlorination, do not have corrosion problems, and are less sensitive to poisons. Also, they do not need a preliminary purification of the feed.

The disadvantage of the catalyst of platinum on zeolite is the higher working temperatures (250–270°C), which leads to lower equilibrium concentrations of branched isomers.

Not much information is available on the catalysts for the hydroisomerization of lube oils, since the process is relatively recent. The bifunctional catalysts used here are of platinum or platinum-palladium supported on zeolite. It seems that the most important issue is a correct balancing of the two catalytic functions and the adequate

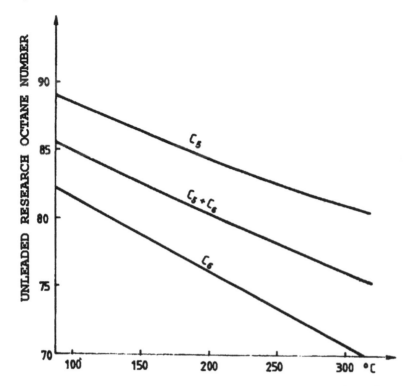

Figure 11.5 Research octane number for the C$_5$-C$_6$ equilibrium mixtures.

size distribution of the zeolite pores. The acidity must be moderate so as not to favor undesired hydrocracking reactions. Also, it is important to maintain a sufficient partial pressure of the hydrogen in order that the metallic function should tone down the decomposition reactions.

British Petroleum uses a platinum catalyst on mordenite [4]. The French Institute of Petroleum uses a "dual" system in two steps, wherein the catalyst of the first step ensures the hydrotreating of the feed, especially the denitration, to a level of 10 ppm and, in the second step, the hydrogenation of the polycyclic aromatic hydrocarbons such as coronene and ovalene, which may block the active sites of the catalyst [5]. The catalyst for the second step is a precious metal on a zeolite with a specially developed pore structure. Other companies that also use platinum/zeolite catalysts are CDTECH and Lyondell for C$_4$ isomerization, Texaco for lube oil dewaxing [45], and others [46].

A important characteristics of the platinum/zeolite catalysts are the very important tolerance for ring-containing hydrocarbons and a long catalyst life.

11.3 REACTION MECHANISM

The mechanism of the carbocation formation on the Lewis or Brönsted acid sites and of their isomerization was largely examined in the analysis of the reaction mechanism of catalytic cracking (Sections 6.3.1 and 6.3.2).

Some authors are of the opinion [6] that on the catalysts of platinum on chlorinated alumina, the mechanism is the extraction of a hydride ion by a Lewis site with the formation of a carbenium ion, which promotes isomerization reactions by an unbranched chain mechanism:

Initiation:

$$H_3C-(CH_2)_2-CH_2-CH_3 + L \longrightarrow LH + H_3C-(CH_2)_2-\overset{+}{C}H-CH_3$$

Chain propagation:

$$H_3C-(CH_2)_2-\overset{+}{C}H-CH_3 \longrightarrow H_3C-CH_2-\underset{\underset{CH_3}{|}}{\overset{+}{C}}-CH_3$$

$$H_3C-CH_2-\underset{\underset{CH_3}{|}}{\overset{+}{C}}-CH_3 + H_3C-(CH_2)_2-CH_2-CH_3 \longrightarrow$$

$$\longrightarrow H_3C-CH_2-\underset{\underset{CH_3}{|}}{CH}-CH_3 + H_3C-(CH_2)_2-\overset{+}{C}H-CH_3$$

In the isomerization of normal alkanes, a dialkylcyclopropane carbonium ion was suggested to be the intermediate, instead of a classical carbonium ion [47].

According to these researchers, the reaction mechanism on this type of catalyst is monofunctional, in which the role of platinum and of a significant partial pressure of hydrogen is only to block some of the secondary reactions that lead to the formation of coke.

This monofunctional mechanism however, does not seem plausible. The use of platinum and hydrogen appears unjustified and the significant increase of the reaction rate as a result of the deposition of platinum on the chlorinated alumina is not explained.

Our opinion and that of a number of researchers [7,8] is that on catalysts of this type the reaction mechanism is bifunctional, the same with that unanimously accepted for catalysts of platinum on zeolites.

This latter mechanism requires the dehydrogenation of alkanes on the metallic sites, the diffusion and the absorption of alkenes on the acid sites accompanied by the isomerization with propagation reactions (as above, in the unbranched chain), followed by the diffusion of the formed iso-alkenes, with their adsorption and hydrogenation on the metallic sites.

This mechanism is summarily shown in the scheme below:

In this complex mechanism, which comprises steps consisting of reactions on the two types of sites and diffusion stages, it is very important to decrease to the least possible the secondary reactions and to ensure concomitantly a satisfactory overall reaction rate. This requirement makes it necessary to have a ratio between the number of metal sites and acidic sites above a certain minimum value.

If this ratio is too low, the ions formed on the acidic sites may also undergo a series of reactions with the alkenes in the gaseous phase, which leads to the formation of polymers and finally to coke. The same thing happens when, even at a sufficient content of precious metal (Pt), the partial pressure of the hydrogen is not sufficiently high.

For a bifunctional catalyst of Pt on H-zeolite, Giannetto et al. [9,10] established that the optimal ratio between the number of metallic centers and the acidic ones should be greater than 0.15, for the optimal isomerization of n-heptane [5].

For the hydroisomerization of higher alkanes contained in the oil fractions, the mechanism is similar, with the difference that since the rate of the cracking reaction is higher, the hydrogenating function must be correspondingly stronger. This means that for a support of moderate acidity, the Pt content and the hydrogen partial pressure should be larger than for lighter feedstocks.

Studies of the hydroisomerization of C_8-C_{12} n-alkanes [48] show that the physical adsorption process determines the concentration of the reacting molecules on the active sites.

Other studies [23] confirm the occurrence of physical adsorption of reactants and products inside the zeolite cages and their chemisorption at the reaction sites. This explains the influence of the zeolite structure on the results of the hydroisomerization process and the attention given by the process developers to the use of zeolites having specific structures for catalysts preparation.

11.4 KINETICS OF HYDROISOMERIZATION

The kinetics of hydroisomerization was the object of a large number of studies concerning the controlling reaction step [9,24,48,49], containing many contradictory conclusions.

The assumptions that the physical adsorption process determines the concentration of the reacting molecules on the active sites and that the pores of the zeolite are practically completely filled [48] leads to a pseudo–zero-order of reaction, which was not confirmed experimentally.

Other studies of the kinetics of the hydroisomerization and hydrocracking of n-octane were performed in a CSTR reactor (of Berty type) at $t = 180$–240 °C and $p = 5$–100 bar, using 0.5 wt % Pt on USY-zeolite [49]. The developed kinetic model assumes that the physical adsorption of both the feed octane and cracked products determines the concentration at the active sites. The best results are obtained with the equation:

$$ r = \frac{k\left(\dfrac{p_A - p_B}{K}\right)}{p_{H_2}[1 + K_1(p_A + p_{MB} + p_{MTB} + ap_{CR})]} $$

where:

k = reaction rate constant

K = the equilibrium constant

K_1 = Langmuir constant for physical adsorption

p = partial pressures

The subscripts A, BMB, MTB, H_2 and CR, refer respectively, to the adsorption of n-C_8, methyl cyclopentane, monobranched alkanes, multibranched alkanes, hydrogen and cracked products.

The most important stage was found to be the adsorption of cracked products. This conclusion is applicable also to n-alkanes with higher boiling points [49].

Other studies [23] reached the conclusion that rate expressions for pseudo–first-order kinetics, which considers the physical adsorption of reactants and products inside the zeolite cages and chemisorption at the reaction sites, give good agreement with the experiments.

In our studies concerning the isomerization of n-pentane on bifunctional platinum catalysts (Octafining of Engelhard) a differential reactor with magnetic stirring was used, the catalyst being introduced in a basket that accomplished ascending–descending movements (Figure 11.6) [11–15].

Using an impulse (Dirac) tracer input and analyzing the response functions for the residence time distribution, it was determined that at a stirring frequency of over 50 oscillations per minute and a feedrate of at least 20 cm^3/min, the reactor behaved as with perfect mixing [11,12]. In these conditions, the calculated difference between the reactant concentration in the catalyst bed and in the reactor did not exceed 0.4%.

Preliminary tests suggested setting the maximum operating temperature at 400°C, where the conversion of the secondary reactions in hydrocracking does not exceed 0.5%. Also, the conditions were selected in which the external diffusion and the internal diffusion through the catalyst pores had no influence on the kinetics [13].

The experimental unit is presented in Figure 11.7. The feed was n-pentane, with a minimum purity of 98.5%. The effluent composition was determined by gas chromatography [14].

More than 250 experiments were performed at temperatures of 400 and 380°C, pressures between 4 and 25 bar, hydrogen/n-pentane molar ratios between 0.5 and 2.5 and space velocities between 4 and 20 g/g cat.hour [15].

For the kinetic processing of the experimental data, the following reaction scheme was used:

$$n\text{-}C_5 \rightleftharpoons [n\text{-}C_5]_M \overset{-H_2}{\rightleftharpoons} [n\text{-}C_5']_M \rightleftharpoons n\text{-}C_5' \overset{\cdots}{\underset{\text{Dif.}}{}} n\text{-}C_5' \rightleftharpoons [n\text{-}C_5']_A \rightleftharpoons$$

$$\rightleftharpoons [i\text{-}C_5']_A \rightleftharpoons i\text{-}C_5' \overset{\cdots}{\underset{\text{Dif.}}{}} i\text{-}C_5' \rightleftharpoons [i\text{-}C_5']_n \overset{+H_2}{\rightleftharpoons} [i\text{-}C_5]_n \rightleftharpoons i\text{-}C_5$$

A total of 21 reaction mechanisms were formulated. In some of these, one step was considered to rate controlling, while in others more steps were considered. The Langmuir law was used for describing the adsorption.

For the models where the reaction rate is expressed by a single equation, a nonlinear simplex, Bayesian regression was used for the estimation of the parameters [16].

For the models in which the reaction rate was expressed by several equations, the experimental data was processed by an original nonlinear regression method [15].

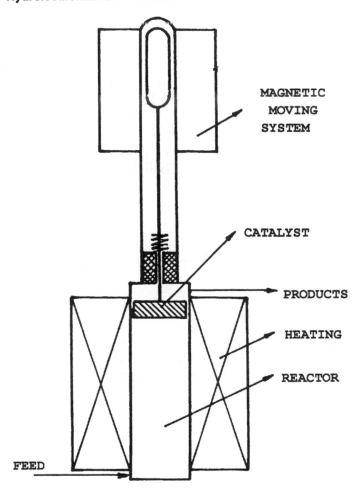

Figure 11.6 Experimental differential reactor.

Computations and analysis of the results led to the conclusion that the reaction rate is determined by the steps of n-pentane dehydrogenation and of the i-pentane hydrogenation on the metallic sites. On basis of this conclusion a Hougen-Watson-type equation for the reaction rate may be written:

$$r = \frac{\dfrac{k_2 k_9 K_{NP} K_I K_{IO} K_2}{k_2 K_{NO} + k_9 K_I K_{IO} K_2}\left(p_{NP} - \dfrac{p_{IP}}{K}\right)}{\left[\begin{array}{l} 1 + K_{NP} p_{NP} + K_{IP} p_{IP} + K_{H_2} p_{H_2} + \dfrac{k_1 K_2 K_{NP}(K_{NP} + K_I K_{IO})}{K_{H_2}(k_2 K_{NO} + k_9 K_I K_{IO} K_2)} \\[2ex] \dfrac{p_{NP}}{p_{H_2}} + \dfrac{k_1 K_2 K_{IP}(K_{NP} + K_I K_{IO})}{K_{H_2} K_9 (k_2 K_{NO} + k_9 K_I K_{IO} K_2)} \cdot \dfrac{p_{IP}}{p_{H_2}} \end{array}\right]^2} \qquad (11.1)$$

where: k are the rate constants, $K =$ the reaction equilibrium constants or the adsorption equilibrium constants, $p =$ the partial pressures, the indexes NP, IP,

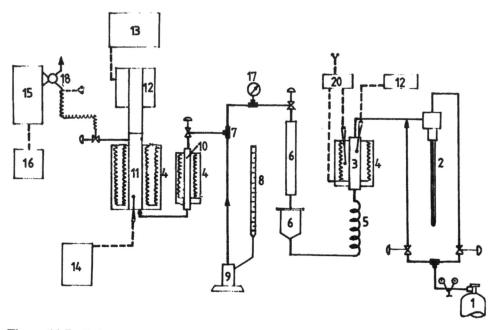

Figure 11.7 Laboratory differential reactor system for *n*-pentane isomerization studies. 1 - hydrogen, 2 - flowmeter, 3 - Reactor for removal of oxygen traces from hydrogen, 4 - electrical heating, 5 - cooling, 6 - silicagel drying, 7,8,9 - *n*-pentane/hydrogen mixing, 10 - preheater, 11 - differential reactor, 12,13 - mixing system, 14 - temperature recording, 15,16,18 - effluent chromatographic analysis.

NO, IO refer to the adsorption of *n*-C$_5$, *i*-C$_5$, *n*-C$_5^=$ and *i*-C$_5^=$, the subscript 2 = to the *n*-C$_5$ dehydrogenation reaction, the subscript 9 to the dehydrogenation of *i*-C$_4$, K_I = the overall equilibrium constant, $K_I = p_{IO}/p_{NO}$.

In order to simplify this equation, the following assumptions were made:

The adsorption coefficients on the metallic sites for *n*-C$_5$H$_{12}$ and *i*-C$_5$-H$_{12}$: ($K_{NP} = K_{IP} = K_P$) are equal.
The similar equality for *n*-C$_5$H$_{10}$ and *i*-C$_5$H$_{10}$: ($K_{NO} = K_{IO} = K_O$).
The equality of the hydrogenation rate constants of *n*-C$_5$H$_{10}$ and *i*-C$_5$H$_{10}$: ($k_{-2} = k_9$).
The equality of the constants for the dehydrogenation of *i*-C$_5$H$_{12}$ and *n*-C$_5$H$_{12}$ ($k_2 = k_{-9}$).
From the latter two, it results that $K_2 = 1/K_9$, which allows one to note:

$$K_D = \frac{p_{NO} \cdot p_H}{p_{NP}} = \frac{p_{IO} \cdot p}{p_{NO}} \tag{11.2}$$

Following the substitutions, Eq. (11.1) becomes:

$$r = \frac{\dfrac{k_2 K_P K}{1+K}\left(p_{NP} - \dfrac{p_{IP}}{K}\right)}{\left[1 + K_P(p_{NP} + p_{IP}) + K_H p_H + \dfrac{K_D K_O}{K_H} \cdot \dfrac{p_{NP} + p_{IP}}{p_H}\right]^2} \tag{11.3}$$

The partial pressures of the intermediary alkenes may be expressed by the equations:

$$p_{NO} = \frac{K_D}{K_H(1+K)} \cdot \frac{p_{NP} + p_{IP}}{p_H} \tag{11.4}$$

$$p_{IO} = \frac{K_D \cdot K}{K_H(1+K)} \cdot \frac{p_{NP} + p_{IP}}{p_H} \tag{11.5}$$

The processing of the data from more than 200 runs at a temperature of 400°C led to the following values of the kinetic and equilibrium constants:

$k_2 = 1.48 \, mol/g \, cat. \cdot hour$

$K_P = 0.0745 \, at^{-1}$

$K_O = 1.28 \, at^{-1}$

$K_{H_2} = 0.057 \, at^{-1}$

Eqs. (11.4) and (11.5) were used for calculating the values of the partial pressures of the alkenes:

$p_{NO} = 0.0027 \, at$

$p_{IO} = 0.0044 \, at$

which are in good agreement with the experimental measurements.

The fractions of the active surface occupied by the various species were obtained:

$$\theta_{H_2} : (\theta_{NP} + \theta_{IP}) : (\theta_{NO} + \theta_{IO}) = 1:0.13:0.016 \tag{11.6}$$

which confirm the results of other studies on the very strong adsorption of the hydrogen on the platinum.

By also performing measurements at a temperature of 380°C, the apparent activation energy for the direct reaction was obtained:

$E = 114.3 \pm 7.1 \, kJ/mol$

In another study [17], we investigated the competition between the isomerization and hydrocracking reactions of n-pentane, using a commercial catalyst (0.55% Pt on alumosilica), Atlantic 16 of Kali-Chemie Engelhard.

The experimental unit had a microreactor followed by an on-line gas chromatograph, Figure 11.8. The operating conditions were: temperature of 400, 425, and 450°C, pressures of 10 and 16 bar and with a constant molar ratio $H_2/C_5H_{12} = 10$.

Since the hydrogen is in great excess and is adsorbed on platinum much stronger than the hydrocarbons—Eq. (11.6)—one may treat the process as pseudo-homogeneous.

The reaction scheme used, which involves the competitive reactions of isomerization and hydrocracking of alkane hydrocarbons, has the form:

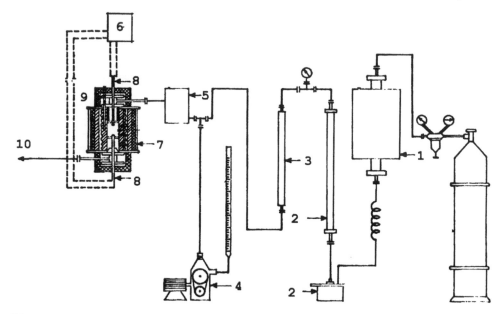

Figure 11.8 Laboratory integral reactor system. 1 – hydrogen purification, 2 – hydrogen drying, 3 – flowmeter, 4 – pump, 5 – preheater, 6 – temperature regulations, 7 – heating, 8 – thermocouples, 9 – reactor, 10 – products to chromatograph.

(a)

In the particular case of the pentane, the rate of i-pentane hydrocracking is very low in comparison with that of n-pentane. On the acid sites, the i-pentane forms preferably a tertiary ion, which when undergoing a β-cleavage, would produce a methyl ion. Owing to the high formation energy of the methyl ion, this cleavage takes place very slowly, or in any case, with a much lower rate than the β cleavage of the n-amyl ion (Section 6.3). Accordingly, in the case of n-pentane, $k_3 \ll k_2$ and scheme (a) is reduced to:

$$\text{hydrocracking products} \xleftarrow{k_2} n\text{-}C_5H_{12} \underset{k_{-1}}{\overset{k_1}{\rightleftharpoons}} i\text{-}C_5H_{12}$$

(b)

In our studies [17] the experimental data were processed using both reaction schemes (a) and (b) and fully confirmed the validity of scheme (b).

The processing of the experimental data obtained at a pressure of 16 bar by means of this scheme led to the values of the kinetic constants (expressed in s^{-1}) given in Table 11.2.

When expressed in mol/g cat. $\cdot$ h, the constant k_1 at 400°C, has the value

Table 11.2 Kinetic Constants for Reaction Model (b), n Pentane Isomerization and Hydrocracking (s^{-1}).

t, °C	k_1	k_{-1}	k_2
400	3.33×10^{-2}	2.03×10^{-2}	5.35×10^{-3}
425	6.44×10^{-2}	4.10×10^{-2}	4.32×10^{-2}
450	9.12×10^{-2}	6.0×10^{-2}	1.0×10^{-1}

$$\frac{3.33 \cdot 10^{-2} \cdot 3600}{72} = 1.665$$

which is very close to the value 1.48 obtained in a differential reactor in the previous study.

The following values were obtained for the apparent activation energies:

$$E_{isom} = 108.9 \pm 4.0 \, kJ/mol$$

$$E_{hcr} = 142.4 \pm 4.0 \, kJ/mol$$

The values of the apparent activation energy for isomerization, obtained in our studies [15,17], are also very close, 114.3 and 108.9 kJ/mol, as compared to other published data (Table 11.3).

The kinetics of n-hexane isomerization is somewhat more complex, owing to the larger number of isomers that may be formed: 2-methyl-pentane, 3-methyl-pentane, and 2,3-dimethyl-butane.

On bifunctional catalysts Pt on H-mordenite, the ratio between the two methyl-pentanes reaches quickly the thermodynamic equilibrium. The ratio 2-methyl-pentane/2,3-dimethyl-butane reaches the equilibrium a little later, which justifies the following formulation of the reactions scheme:

$$nC_6 \rightleftharpoons 2MP \rightleftharpoons 2,3 \, DMB$$
$$\searrow \quad \updownarrow$$
$$3MP$$

(c)

Table 11.3 Activation Energies for the Isomerization of n-Pentane

	Operation data	E_a, kJ/mol	Reference
Pt/Al$_2$O$_3$ chlorinated	150°C, 40 bar H$_2$/n-C$_5$ = 2	63	[18]
Pt Re/HY	300°C, 30 bar H$_2$/n-C$_5$ = 4	142	[19]
Pd/H-mordenite	280°C, 30 bar H$_2$/n-C$_5$ = 8	153	[20]

The 2,2-dimethyl-butane is formed much more slowly, apparently as a result of some secondary reactions.

Depending on the reaction conditions, hydrocracking reactions may or may not be added to this scheme.

The activation energies for the isomerization of n-hexane at temperatures of 250–300°C, pressures of 30 bars, and molar ratios H_2/n-C_6 of 10 range between 121 and 155 kJ/mol [18,21].

Considering that the rate-determining step is the isomerization on acid sites, the conversion of n-hexane on a catalyst Pt/H-mordenite fits the following overall reaction scheme [18]:

$$
\begin{array}{ccc}
nC_6H_{14} & & iC_6H_{14} \\
(1)\text{ -}H_2 \Big\| +H_2 & & \text{-}H_2 \Big\| +H_2\ (5) \\
nC_6^+H_{12} \underset{\text{-}H^+}{\overset{+H^+}{\rightleftharpoons}} nC_6^+H_{13} \rightleftharpoons iC_6^+H_{13} \underset{+H^+}{\overset{\text{-}H^+}{\rightleftharpoons}} iC_6^+H_{12} & & \\
(2) \qquad\qquad (3) \qquad\qquad (4) & &
\end{array}
\qquad\qquad (d)
$$

This scheme leads to the following expression of the reaction rate:

$$
r = k_3 C_m K_1 K_2 \frac{p_n C_6}{p_{H_2} + K_1 K_2 p_n C_6}
\tag{11.7}
$$

where:

C_m = concentration of the Brönsted sites on the surface of the catalyst

k = the rate constants

K = the equilibrium constants.

The subscripts correspond to the steps indicated in scheme (d). A similar kinetic equation was proposed for the isomerization of n-heptane [18].

Overall, our studies and other published data [7,22], indicate that the rate determining step depends on the ratio between the metallic and the acid sites of the catalyst. Therefore, prudence and good knowledge of the catalyst characteristics are recommended when using kinetic data obtained by various authors.

The hydro-isomerization of n-octane and its competition with that of hydrocracking were studied [23] using a precious metal on faujasite, leading to the successive formation of mono-, bi- or multibranched derivatives.

Similar studies were made for n-heptane on a catalyst of Pt on H-offertite [24].

11.5 INFLUENCE OF OPERATING PARAMETERS

In the analysis of the operating parameters of hydro-isomerization processes, one must take into account the influence on the equilibrium of the isomerization reaction of the relative rates of various chemical transformations that take place: isomerization, hydrocracking, and coking.

In many cases it is necessary to take into account the influence of the operating parameters on the individual reaction steps that take place on the acid or metallic sites of the catalyst.

11.5.1 Temperature

The influence of the temperature on the equilibrium of the isomerization reactions of n-alkanes produces the decrease of the iso structures as the temperature increases.

For processes that take place on the two types of sites, the increase of temperature displaces the equilibrium of the reactions taking place on the metallic sites toward dehydrogenation. The equilibrium on the acid sites is actually not influenced by the temperature, the increase of which only increases the reaction rate.

These influences lead to the following results:

1. The increase of temperature—while favoring dehydrogenation, which initiates the isomerization reactions—strongly increases their rate. At a certain level the thermodynamic limitations intervene, so that a temperature level exists that is optimal for the conversion to iso structures.

2. Hydrocracking is strongly favored by increasing temperature, which emphasizes the initiation reactions—the formation of alkenes that generate carbenium ions on the acid sites—and by the increase of the rates of the decomposition (cracking) reactions taking place on these sites.

3. Since the temperature increase favors the formation of alkenes and the acceleration of the reactions on the acid sites, while decreasing the hydrogenation of the alkenes, it leads to the increase of alkene concentration on the catalyst surface, and favors coking.

It results that hydroisomerization is favored by lower operating temperatures. The lowest operating temperature depends on the activity of the used catalysts and especially on the activity of the metallic sites.

11.5.2 Pressure

Since the overall isomerization reaction takes place without a change of the number of moles, the thermodynamic equilibrium should not be influenced by pressure.

However, by increasing the pressure, at constant values of the other parameters the rate of the conversion to branched isomers is reduced, probably due to the inhibition of the first stage of the process, that of dehydrogenation of the alkane on the metallic sites.

The increase of pressure reduces the rate of the hydrocracking and coking reactions, because it leads to an increase of the partial pressure of hydrogen and thus favors the hydrogenation of the alkenes.

It results that increased pressure slows down both the isomerization and the hydrocracking reactions, the influence on the hydrocracking being stronger. As a result, the selectivity, which is the ratio between the iso-alkanes produced and the total conversion, go through a maximum. For the lower alkanes, this maximum is situated, depending on the catalyst, in the range situated between 20 and 30 bar, where the commercial processes operate.

This is illustrated by the data of Table 11.4 obtained on a bifunctional catalyst at 316°C and a molar ratio $H_2/C_6H_{14} = 4$ [25].

Coke formation is strongly reduced by the increase of pressure, which favors the hydrogenation of the alkenes, before they can form precursors of coke.

Table 11.4 Effect of Pressure on the Selectivity in
n-Hexane Isomerization

	p (bar)		
	6.3	22	49
Total conversion, % mole	60.7	32.0	14.5
Conversion to *i*-C$_6$, % mole	49.8	31.3	13.1
Selectivity	0.82	0.98	0.91

Source: Ref. 25.

Since the rate of coke formation increases with the increasing molecular mass of the feed, fundamental large differences exist between the operating pressure used in the processing of lube oils compared to the processing of light alkanes.

With the lower alkanes, the rate of coke formation is relatively low and does not influence the selection of the working pressure.

In the hydroisomerization of oil fractions, coke formation occurs in parallel to the desired reaction and is a determining factor in the selection of the operating parameters. Its reduction implies operating at pressures of the order of 100–110 bar.

11.5.3 The H$_2$/Feed Ratio

The molar ratio H$_2$/feed does not influence significantly the isomerization reaction, but prevents in an important way the deactivation of the catalyst by decreasing the coke deposits. Commercial units operate at molar ratios comprised between 2 and 6 for the lower alkanes and 8–20 for oils [26].

It is to be observed that the increase of the molar ratio requires the increase of the hydrogen recycle, which influences the economic parameters of the process. Process economics will allow finding the balance between increasing the operating pressure and the molar ratio hydrogen/feed for obtaining the desired hydrogen partial pressure that ensures a long catalyst life

11.5.4 The Feed

The hydroisomerization of two completely different feeds: the lower alkanes and the oil fractions, must be examined separately.

11.5.4.1 Lower Alkanes

For this feedstock, increasing the number of carbon atoms in the molecule increases the reaction rate and displaces the equilibrium toward the formation of iso-structures.

The behavior of alkanes with a similar number of carbon atoms is sufficiently close to allow their isomerization as a mixture. This is the case of the isomerization of mixtures of C$_5$-C$_6$ alkanes.

Since the ions with a higher molecular mass are formed with a higher rate, reactions of the type:

$$i\text{-}C_6H_{13}^+ + n\text{-}C_5H_{12} \rightarrow i\text{-}C_6H_{14} + n\text{-}C_3H_{11}^+$$

lead to an increase of the overall isomerization rate, favoring the isomerization of the alkanes of lower molecular mass.

The rates of the secondary reactions of hydrocracking and coking increase with the molecular mass, which require the selection of adequate temperatures, pressures, and hydrogen/feed molar ratios [42].

11.5.4.2 The Oil Fractions

The behavior of the oil fractions depends on their mean molecular mass and on their chemical structure, as will be discussed in conjunction with the industrial implementation of the respective processes. Despite the careful selection of the processing conditions, decomposition reactions take place generally in parallel with the hydroisomerization, which led to the occasional use of the wrong term of "hydrocracking," instead of hydroisomerization.

11.5.5 The Catalyst

In the development of alkane isomerization processes, four catalyst generations were developed and used in succession. The first two ($AlCl_3$ and metal on alumosilica) are no longer used.

For the catalysts currently in use, Pt/chlorinated Al_2O_3 (the third generation) and Pt/zeolite (the fourth generation), the ratio between the acid and the metallic sites have a determining role in the performance of the process. From this point of view the catalysts of Pt/chlorinated Al_2O_3 make it possible to modify the acidity by varying the amount of the chlorinating agent.

In exchange, the zeolites have the advantage of a well-defined structure of the pores, which allows one to select the zeolite with the structure best suited for the process.

An especially difficult problem is the selection of the pores structure of the zeolite used for the hydroisomerization of the oil fractions. In this case, zeolites with a bimodal distribution of the pore sizes generally are used.

The platinum amount deposited on the support increases on average with the molecular mass of the feed, being the highest in the hydroisomerization of oils, where the danger of coke formation is more important.

11.6 INDUSTRIAL HYDROISOMERIZATION OF LOWER ALKANES

An overview of the four generations of isomerization processes for the lower alkanes is given in Table 11.5 [18]. Only the last two are used currently and will be examined below.

11.6.1 Processes Using Pt on Alumina Catalysts

These are processes of the third generation, that use a strongly acidic support. Its strong acidity is maintained by means of introduction in the feed of very small amounts of organic chlorides. The catalyst being very active, the operating temperature is situated in the range 110–180°C, which favors the formation of iso-structures and in the case of C_5-C_6 fractions, of high octane numbers in single-pass operation.

The preparation of such catalysts may be illustrated by the method used by British Petroleum [27]. It consists of the treatment of the Pt/Al_2O_3 catalyst with

Table 11.5 Processes for the Isomerization of Lower Alkanes

Generation	Licensor	Feed	Phase	t (°C)	p (bar)	Catalyst
I	Shell	C_4	Gas	95–150	—	$AlCl_3$/bauxite/HCl
	UOP	C_4	Liquid	—	—	$AlCl_3$/HCl
	Standard Oil	C_5/C_6	Liquid	80–100	—	$AlCl_3$HCl
	Shell	$C_4, C_5/C_6$	Liquid	—	—	$AlCl_3/SBF_3$/HCl
II	UOP: Butamer	C_4	Gas	375	—	Pt/support
	Kellogg: Iso-Kel	C_5/C_6	Gas	400	20–40	Group (VIII)/support
	Pur Oil: Isomerate	C_5/C_6	Gas	420	50	Not noble metal/support
	UOP: Penex HT	C_5/C_6	Gas	400	20–70	Pt/support
	Linde	C_5/C_6	Gas	320	30	Pt/Si-Al
	Atlantic Refining Pentafining	C_5/C_6	Gas	450	50	Pt/Si-Al
III	UOP: Penex BT	$C_4, C_5/C_6$	Gas	110–180	20–70	Pt/Al_2O_3; $AlCl_3$
	BP	$C_4, C_5/C_6$	Gas	110–180	10–25	Pt/Al_2O_3; CCl_4
	IFP	$C_4, C_5/C_6$	Gas	110–180	25–50	Pt/Al_2O_3, AlR_xCl_y
	UOP: Butamer	C_4	Gas	110–180	20–30	Pt/Al_2O_3; AlC_3
IV	Shell: Hysomer	C_5/C_6	Gas	230–300	30	Pt/H-mordenite
	Mobil	C_6	Gas	315	20	H-mordenite (Pt, Pd)
	UOP	C_6	Gas	150		H-mordenite $+PtRe/AL_2O_3$
	Sun Oil	C_5/C_6	Gas	325	30	PtHY
	Norton	C_5	Gas	250	30	Pd/mordenite
	IFP	C_5/C_6	Gas	240–260	15–30	Pt/H-mordenite

CCl₄ at high temperatures. The reaction is very exothermal and measures are taken to prevent a strong increase of the temperature. As a result of the treatment, about 14% Cl_2 remain fixed chemically on the surface of the support, but its porous structure is not affected.

11.6.1.1 Isomerization of *n*-Butane

The operation of the first unit of *n*-butane isomerization on a bifunctional catalyst with Pt was started in 1941 by UOP, under the name BUTAMER, a name kept by UOP for all the subsequent variations. In 1986, 39 Butamer plants were in operation, with capacities of 160–2000 m³/day [28,29].

In its recent embodiment using Pt on chlorinated Al_2O_3, the process was first built in 1959 also by UOP. Subsequently, similar plants were built by British Petroleum [27] and by the Petroleum French Institute [30] as a result of development of their own catalysts.

In the process a small amount of hydrogen is used, together with the injection of small amounts of organic chlorides, in order to maintain the acidity of the support. The butane is separated from isobutane by fractionation and recycled.

The feed is submitted to desulfurization and drying before entering the reactor. Moreover, if the unit is coupled with an alkylation one with hydrofluoric acid, the *n*-butane that comes back to be isomerized must be contacted with molecular sieves, in order to retain the traces of HF, which are very noxious for the isomerization catalyst.

In the older variation of the process, two reactors were used in series with recycling of hydrogen. In the newer process. Butamer HOT (Hydrogen Once Through), one reactor without hydrogen recycling is used. Figure 11.9. This modification, which eliminates the high pressure separator and the recycling compressor for hydrogen and decreases the investments and the exploitation costs by about 20%, was the result of the improvements made to the catalyst together with the decrease of the hydrogen amount introduced in the process. Thus, at the inlet in the stabilizer, the hydrogen represents only 30% of that corresponding to the earlier process.

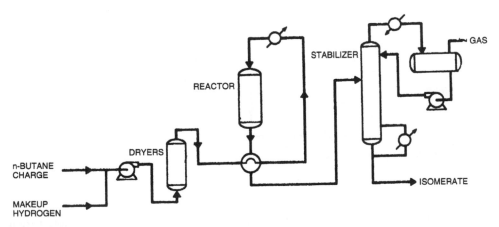

Figure 11.9 Flow diagram of UOP - Butamer HOT process.

The *n*-butane and *i*-butane in the reactor effluent are separated by fractionation. The *n*-butane is recycled in the reactor. The traces of HCl are neutralized.

11.6.1.2 Isomerization of C_5-C_6 Alkanes

Processes for the isomerization of the C_5–C_6 fractions on a catalyst of Pt on chlorinated Al_2O_3 were developed in 1965 by British Petroleum [27], in 1969 by UOP
under the name of "PENEX" [31], and later by the French Institute of Petroleum
[30].

The operating conditions of the process of the French Institute of Petroleum
are: temperature of 130–180°C, pressure of 20 bar, the volume space velocity $2\,h^{-1}$,
molar ratio H_2/hydrocarbons $= 4$.

As in the case of *n*-C_4 isomerization, the feed must be desulfurized and dried by
contacting with molecular sieves. The hydrogen also is dried before entering the
system. For both processes (*n*-C_4 and C/C_6) it is to be remarked that the presence
of sulfur decreases the activity of the catalyst, which however is recovered by resum-
ing operation with clean feed. The poisoning produced by the presence of water is
irreversible. Small amounts of water can significantly shorten the life of the catalyst.

The process flow diagram of the process of British Petroleum is given in Figure
11.10 and the typical data on the single-pass isomerization on catalyst of Pt/Al_2O_3
chlorinated, in Table 11.6.

The Penex (UOP) process allows a content of up to 10% C_7 in the feed, with a
minimal deterioration of the performances. Also, a content of up to 4% benzene and

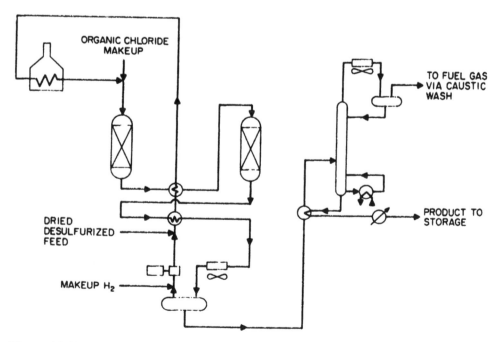

Figure 11.10 Flow diagram of the British Petroleum C_5 or C_5/C_6 isomerization. (From
Ref. 27.)

Table 11.6 BP Isomerization Process: Typical Commercial Data from C_5 and C_5-C_6 Single-Pass Isomerization with Pt on Chlorinated Al_2O_3 Catalyst

Unit	Industria Raffinazione Oil Minerali, Venice		Erdol Raffinerie Mannheim	
	Feed	Product	Feed	Product
Component, wt %				
C_4 and lighter	0.2	1.1	—	—
isopentane	24.6	39.7	2.6	77.6
n-pentane	26.8	11.8	95.9	21.2
cyclopentane	1.7	1.3	1.5	0.4
2,2-dimethylbutane	1.0	16.0		0.7
2,3-dimethylbutane	2.9	4.4		0.1
2-methylpentane	15.0	12.3	—	—
3-methylpentane	11.3	6.8	—	—
n-hexane	13.1	4.2	—	—
benzene	1.7			
methylcyclopentane	1.7	1.3	—	—
cyclohexane	...	1.1	—	—
Product data				
yield, wt %	...	98.5	...	99.5
sp. gr, 15/15°C	0.650	0.644	0.632	0.627
RON, clear	72.3	83.5	63.0	86.1
MON, clear	69.7	81.4	62.4	84.2
Reid, vapor pressure, lb/in²	12.0	13.9	14.6	18.0

Source: Ref. 27.

several percentage points of alkenes are acceptable in the feed and will not lead to formation of coke deposits on the catalyst.

The process flow diagram of a classical Penex process is given in Figure 11.11.

In the modern Penex-HOT process, the recycling of hydrogen is eliminated as in the Butamer HOT process (11.9). The high pressure separator is also eliminated. The Penex HOT was the first developed in 1987. On basis of the results obtained, the Butamer HOT process was developed [32,33].

Two separation systems were developed for the recovery of the isomerization product from the hydrocarbons which must be recycled:

Separation by fractionation
Separation by means of the molecular sieves Molex, which was implemented in several variations, presented in Figure 11.12 [31].

The various schemes presented in this figure lead to the following values for the clear octane number F_1 of the final isomerizate:

Feed	69
1. Without recycling	83
2. Recycling of n-C_5	86
3. Recycling of n-C_5 + n-C_6	89
4. Recycling of n-C_5 + n-C_6 + 2 and 3MeC$_3$	92

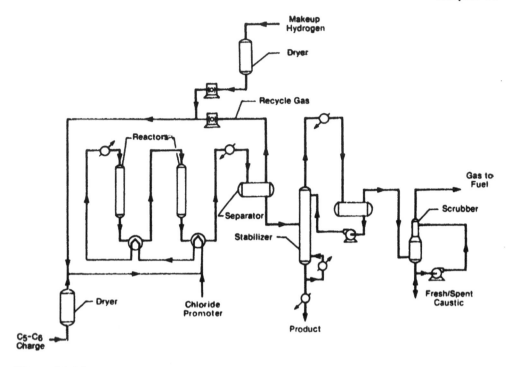

Figure 11.11   Process flow diagram of UOP - Penex unit.

The separation by fractionation is more complex, and by recycling the methyl-pentanes makes it possible to obtain a larger increase of the octane number than is possible with the method using molecular sieves.

Various original recycling schemes were suggested by the French Institute of Petroleum under the name of Ipsorb (separation with molecular sieves) and Hexorb (fractionation + molecular sieves) [30], as well as by British Petroleum [27].

11.6.2   Process Using Pt Supported on Zeolites

The catalysts of this type have a weaker acid function and therefore require a higher reaction temperature than the catalysts of chlorinated alumina. In exchange, they are less sensitive to poisons and do not require the neutralization of the reactor effluent.

Initially, the process was developed in several versions for the isomerization of the C_5/C_6 fractions. Later-on, Pt on H-zeolite catalyst was applied also for the isomerization of the C_4 fraction [33].

The first unit of this type was built in1970 by Shell at Spezia in Italy, under the name of Hysomer (license of Union Carbide Corporation) [34].

The catalyst used by Shell—Pt deposited on a H-mordenite—tolerates 35 ppm sulfur and 10–20 ppm water in the feed and is not poisoned by aromatic hydro-carbons, alkenes, or cycloalkanes.

UOP uses catalysts similar to that of Shell or variations thereof, whereas Procatalyse and the French Institute of Petroleum developed the catalyst I.S.632, which is much more tolerant to sulfur and does not require the drying of the feed and

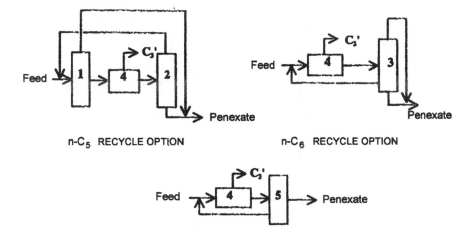

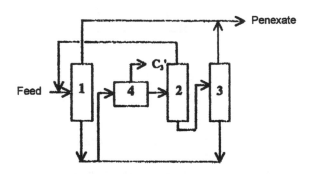

n-C$_5$/n-C$_6$+ MP RECYCLE VIA FRACTIONATION OPTION

Figure 11.12 Four systems for separation of isomerization products. 1 - isopentane removal; 2 - depentanizer; 3 - isohexanes separation; 4 - Penex; 5 - Molex.

hydrogen by means of molecular sieves [35]. A similar process, which tolerates up to 300 ppm sulfur, heavier hydrocarbons up to C$_9$, and does not require a drying by means of the molecular sieves, was commercialized under the name Isosiv, by Union Carbide [36].

The process named Hystomer is implemented either in a once-through in the reactor version, or by recycling with the separation of the products by adsorption on molecular sieves. In the latter case it is called *Total Isomerization Process* (TIP) [37,38]. It is presented in this form in Figure 11.13.

The operation conditions for this process are: temperature about 260°C, pressure 21–35 bar, molar ratio H$_2$/H$_c$ = 1–4, feed space velocity 1–3 hours^{-1} [34].

Typical results for the isomerization of a C$_5$/C$_6$ fraction, in the once through and recycle (TIP) versions are given in Table 11.7.

The catalyst is regenerated without removal from the reactor, at intervals of 2–2-$\frac{1}{2}$ years. The replacing of the catalyst becomes necessary after about 10 years [34].

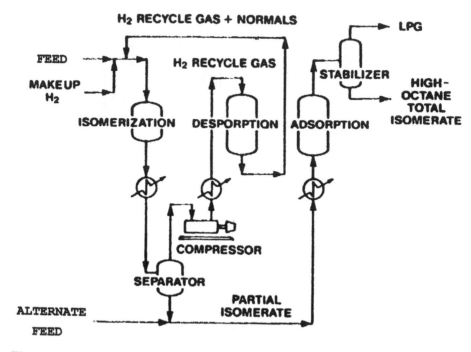

Figure 11.13   TIP isomerization of the C_5/C_6 fraction.

The process using catalysts of Pt on zeolite use temperatures of the order of those practiced in hydrofining. For feeds requiring a preliminary desulfurization, one may place a hydrofining reactor upstream of the isomerization reactors. On the basis of this concept the SafeCat process was developed, whose schematic process flow diagram is shown in Figure 11.14 [39].

Table 11.7   Typical C_5/C_6 Isomerization Results

	Feed	Hystomer	TIP
Density	0.6388	0.6360	0.6330
Distil. ASTM, i, °C	33	—	—
Distil. ASTM, f, °C	67	—	—
F_1 clear octane	73.2	82.1	90.7
Sensibility clear	—	2.0	2.0
C_5^+	—	98.0	97.0
Composition, wt %			
C_4H_{10}	0.7	1.8	2.8
i-C_5H_{12}	29.3	49.6	72.0
n-C_5H_{12}	44.6	25.1	2.0
2,2-dimethylbutane	0.6	5.0	5.5
2,3-dimethylbutane	1.8	2.2	2.5
methylpentanes	13.9	11.3	13.4
n-C_6H_{14}	6.7	2.9	<0.1
cyclo C_5/C_6	2.4	2.1	1.8

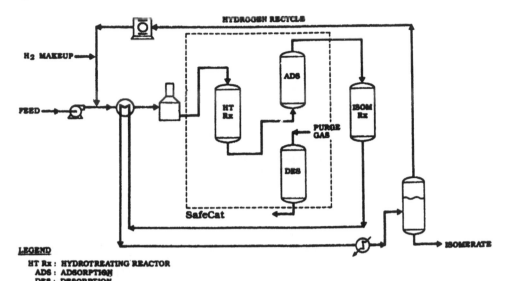

Figure 11.14   SafeCat process for isomerization with Pt on zeolite.

11.6.3 Yields and Product Quality

Economic data for comparing several typical isomerization processes are given in Table 11.8.

The utilities consumptions of various processes are shown.

Since the variation of the investment with plant capacity is not linear, the reference plant capacities are indicated.

Some more data concerning the investment for recent units are:

Lummus C_5/C_6 isomerization unit using Akzo Nobel's Pt on chlorinated alumina catalyst, giving a stabilized isomerizate 84-85 RONC, 3800–9400 $/m^3 feed, basis 1998 [43].

IFP (French Institute of Petroleum) C_5/C_6 isomerization unit, using Pt on chlorinated alumina or zeolite catalyst, capacity 800 m^3/day, 15,000 $/m^3 [44].

For SafeCat, which includes the hydrodesulfurization within the scheme of a hydroisomerization on Pt/zeolite, the total investment represents about 90% and the operation costs about 82% of those corresponding to the two separate plants [33].

In view of the restrictions imposed by the Clean Air Act on the content of aromatic hydrocarbons, especially benzene in gasoline, it is expected that iso-merization processes, especially of the C_5-C_6 fractions, will find increasing appli-cation. This will lead to an increase of the proportion of isomerizate in the gasoline pool, beyond the values reached in 1995, of 11.6% in the U.S. and 5.0% in Europe.

674   **Chapter 11**

Table 11.8   Several Representative Isomerization Economic Data

| | | | Catalyst Pt on chlorinated Al$_2$O$_3$ | | | Pt on H zeolite UCC (Shell) | |
| | British Petroleum[a] | | UOP[b] | | | | |
	C$_4$	C$_5$/C$_6$	Penex HOT C$_6$/C$_6$	Penex Molex C$_5$/C$_6$	[31] Hystomer C$_5$/C$_6$	[33] TIP C$_5$/C$_6$
	Consumptions on M^3 feed					
Electricity, kW	4.9	7.0	7.2	13.0	10.5	25.0
Fuel, 10^6 J	52	50	—	—	192	700
Steam 12–14 bar, kg	102	105	210	370	80	100
Steam, low pressure, kg	—	—	—	157	—	—
Water, m^3	Unspecified		1.8	2.6	2.0	4.0
H$_2$, m^3	Unspecified		Unspecified		15	20–30
Installation and license	1500 £	1600 £	4500 $	8500 $	3140 $	11,500 $
Catalyst and adsorbent		550£	800 $	1900 $		2300 $
Reference unit capacity	800	800	1600	1600	1270	1270
Reference Year	March 1983	January 1983	1985	1985	1983	1983/84

[a] Source: Ref. 24.
[b] Source: Ref. 28.

11.7 HYDROISOMERIZATION OF LUBE OILS AND MEDIUM FRACTIONS

The purpose of the hydroisomerization of lube oils is to decrease the pour point, the reason for which the process has also the name *catalytic dewaxing*. Diesel and turbine fuels are often submitted to the process, for the same purpose of decreasing the freezing point.

The distillates and the light oils that contain *n*-alkanes are processed for the isomerizations of the hydrocarbons chains on a catalyst of Pt on zeolite. Since the high melting points of the medium- and high-molecular mass *n*-alkanes decreases significantly with the isomerization of the chain, the products of the process have much lower melting points than the feed. The term "dewaxing" is therefore misleading, since the process does not remove high melting point wax but converts them to hydrocarbons of lower melting temperature.

It results that the gas oils and the light oils, which are rich in *n*-alkanes, have to be submitted to hydroisomerization, but not the medium, heavy, and residual oils containing hydrocarbons of nonlinear structures of carbon atoms, which have high melting temperatures. The delimitation between the fractions that may be submitted to hydroisomerization, and heavier fractions is therefore exactly similar to that for applying traditional processes for the production of paraffin wax by means of cooling and pressing.

Hydroisomerization produces transformer oils, oils for refrigeration machines, and other qualities of light oils.

The medium and heavy oils and the production of motor oils respectively require much more severe hydrocracking processes such as hydrogenation of the aromatic cycles and the opening of the naphthenic cycles. This leads to the oils having viscosity indexes of the order of 120. These processes are discussed in Section 12.9.

11.7.1 Process Technology

The catalyst used in the process is Pt on H mordenite [4] and the reaction conditions are:

Temperature	290–400°C
Partial pressure H_2	20–100 bar
Volume space velocity	0.5–5 hours^{-1}
Molar ratio H_2/feed	8–20

In these conditions the oil fractions undergo decomposition reactions in successive steps:

$$oil \rightleftharpoons isomerized\ oil \rightarrow hydrocracking\ products$$

The yield of hydroisomerized oil is between 70 and 85%, the rest being lighter products.

The use of catalysts of the type of Pt on chlorinated Al_2O_3, operating at lower temperatures, would have advantages. However, no commercial unit seems to use such a catalyst, probably due to its higher sensitivity to impurities in the feed.

A schematic process flow diagram of a hydroisomerization unit for oils is given in Figure 11.15.

Since the reaction is exothermal, temperature is controlled by cooling with hydrogen injections between the catalyst beds.

Actually, an additional hydrofining or hydrotreating reactor is coupled in series with the isomerization reactor [40]. In this manner, color stability and stability to oxidation is conferred to the produced oil.

11.7.2 Products

The data of Table 11.9 are examples of hydroisomerization of some heavy gas oils, with the products yields.

Table 11.10, gives the characteristics of an oil for refrigerating machines obtained by hydroisomerization.

Transformer oils require a very low freezing temperature and a very high stability to oxidation. These properties may be obtained either by hydrotreating, or by selective solvent refining, which may be implemented before or after the hydroisomerization. Both options, compared with the specifications of the International Electrotechnical Commission (IEC) are presented in Table 11.11.

11.7.3 Process Economics

The specific consumptions and the investments for a hydrodewaxing unit and one coupled with an additional hydrotreating reactor are given in Table 11.12.

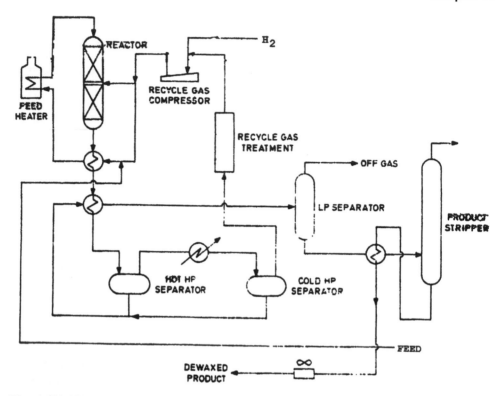

Figure 11.15 Flow diagram for hydroisomerization of lube oils.

Table 11.9 Hydroisomerization of Heavy Gas Oils

	I	II	III
Feed characteristics			
Pour point, °C	+15.5	+26.7	+23.9
Sulfur, wt %	1.92	1.73	0.43
Viscosity, cSt at 50°C	7.96	12.37	—
Paraffin, wt %	11.0	16.5	20.0
Distillation			
10% wt, °C	330	327	286
50% wt, °C	380	408	349
90% wt, °C	411	457	424
Products			
$C_1 + C_2$	0.1	0.4	0.1
C_3	2.5	5.2	3.8
C_4	6.4	9.4	11.0
C_5^+	8.2	10.4	16.3
Hydroisomerized gas oil	82.8	74.6	68.8
Gas oil characteristics			
Pour point, °C	−20	−20	−23
Viscosity, cSt at 50°C	9.04	16.35	—

Table 11.10 Characteristics of a
Refrigeration Oil Obtained by
Hydroisomerization

Kinematic viscosity, cSt at	
20°C	59.6
38°C	24.8
99°C	4.2
Pour point, °C	−40
Cloud point, °C	− 51
Flammability, °C	196
Acidity, mg KOH/g	0.01
Ashes, calcined at 775°C, wt %	<0.01
Freon insolubles at −40°C	0.01

Table 11.11 Comparison of Characteristics of Transformer Oils

	Solvent refining + Hydroisomerization	Hydroisomerization + Solvent refining	I.E.C specs for 2nd class oil
Density, at 20°C	0.850	0.850	max 0.895
Viscosity, cSt at:			
20°C	18.3	15.75	max 25
−30°C	550	444	max 1800
Pour point, °C	−45	−51	max −45
Flammability, °C	168	146	min 130
Acidity, mg KOH/g	0.01	0.01	max 0.03
Corrosivity sulfur	absent	absent	absent
Loss tangent at 90°C	3.5×10^{-4}	1.7×10^{-3}	max 5×10^{-3}
Oxidation test			
Insolubles, wt %	0.03	0.06	max 0.10
Acidity after oxidation, mg KOH/g	0.15	0.35	max 0.40

Source: Ref. 4.

Table 11.12 Economic Data for Lube Oil Hydroisomerization

	British Petroleum, without hydrotreating[a]	Texaco-Mobil, with hydrotreating reactor[b]
Utilities per m³ feed		
Electricity, kW	10.3	19
fuel, 10^6 J	200	265
steam, kg	60	88
water, m³	1.15	5.3
hydrogen, m³	unspecified	18–45
Investment	—	110 \$/m³
(basis 1992, 400 m³/day)		

[a] *Source*: Ref. 4.
[b] *Source*: Refs. 37, 45.

REFERENCES

1. S Ridgway, Ind Eng Chem 51: 1023, 1959.
2. GM Kramer, J Org Chem 40: 298, 1975.
3. CD Nenitescu, A Drăgan, Am Chem Ber. 66B: 1892, 1933.
4. JJ McKetta, Petroleum Processing Handbook. New York: Marcel Dekker Inc., 1992.
5. A Hennico, A Billon, PH Bigeard, JP Peries, Revue de l'Institut Français du Pétrole 48(2): 127, March–April 1993.
6. C Travers, C_5-C_6 Isomerization and Benzene Reduction in Gasoline with Advanced Isomerization Catalysts, OAPEC-IFP Conference, July 5–7, 1994, Report no. 4, IFP Rueil-Malmaison, France.
7. AL Petrov, Himia alcanov, Nauka, Moscow, 1974.
8. JE Germain, Catalytic Conversion of Hydrocarbons. London: Academic Press, London and New York, 1969.
9. M Giannetto, Thesis, University of Poitiers, 1985.
10. M Gisnet, F Alvarez, G Giannetto, G Perot, Catal. Today 1: 415, 1987.
11. R Domnesteanu, S Raseev, Revista de Chimie 21(1), 482; (8,II), 557; (9): 1970.
12. S Raseev, I Buricatu, R Domnesteanu, Revista de Chimie 21(6): 361, 1970.
13. R Domnesteanu, S Raseev, Bul Inst Petrol, Gaze si Geol, Bucharest 20:143, 1973.
14. R Domnesteanu, S Raseev, Bul Inst Petrol, Gaze si Geol, Bucharest 18:67, 1971.
15. R Domnesteanu, PhD Thesis, Institutul de Petrol Gaze si Geologie, Bucharest, 1973.
16. R Domnesteanu, S Raseev, Bul Inst Petrol, Gaze si Geol, Bucharest, 19:187, 1972.
17. JO Hernández, PhD Thesis, Institutul de Petrol, Gaze si Geologie, Bucharest, 1972.
18. M Belloum, C Travers, JP Bournoville, Revue de l'Institut Français du Pétrole 46(1): 89–108, 1991.
19. J Weitkamp, Prep Div Petrol Chem, ACS Monograph 20: 489, 1975.
20. JM Zhu, GX Huang, CL Li, FY Liu, Cui hmaxue bao (J Catal) 6: 44, 1985.
21. F Ribeiro, C Marcilly, M Guisnet, J Catal 78: 267, 1982; 78: 275, 1982.
22. V Fouché V, Thesis, University of Poitiers, France, 1989.
23. MA Baltanas, KK Van Raemdonek, GF Froment, SR Mohedas, Ind Eng Chem 28: 899, July, 1989.
24. GE Giannetto, FB Alvarez, MR Guisnet, Ind Eng Chem 27: 1174, July 1988.
25. RZ Magaril, Teoreticeskie osnovy himiceskih protzesov pererabotki nefti, Himia, Moscow, 1976.
26. JJ McKetta, Petroleum Processing Handbook, New York: Marcel Dekker Inc., 1992.
27. P Greenough, JRK Rolfe, BP Isomerization Process, In: RA Meyers ed. Petroleum Refining Process, New York: McGraw-Hill, 1986.
28. D Rosati, UOP Butamer Process In: RA Meyers ed. Handbook of Petroleum Refining Processes, New York: McGraw-Hill, 1986.
29. Hydrocarbon Processing 67: 55, April 1988.
30. JL Nocca, A Forestière, J Cosyns, Revue de l'Institut Français du Pétrole 49(5): 461, 1994.
31. JA Weiszmann, UOP Penex Process, In: RA Meyers ed. Handbook of Petroleum Refining Processes, New York: McGraw-Hill, 1986.
32. S Raghuram, RS Haizmann, DR Lowry, Oil and Gas J. 88: 66, Dec. 3, 1990.
33. ME Reno, RS Haizmann, BH Johnson, PP Piotrowski, AS Zarchy, Hydrocarbon Technology International 1990/1991, P Harrison ed. London: Sterling Publ Internat, p. 73, 1991.
34. NA Cusher, UCC(Shell) Hystomer Process, In: RA Meyers ed. Handbook of Petroleum Refining Processes, New York, McGraw-Hill, 1986.
35. A Hennico, JP Carion, Hydrocarbon Technology International 1990/91, P Harrison ed., London: Sterling Publ. Internat., p. 68.

36. MF Symoniak, Hydrocarbon Processing 59: 110, May 1980.
37. MF Symoniak, TC Holocombe, Hydrocarbon Processing 62: 62, May 1983.
38. NA Cusher, UCC Total Isomerization Process, In: RA Meyers ed. Handbook of Petroleum Refining Processes, New York, McGraw-Hill, 1986.
39. Hydrocarbon Processing 61: 171, Sept. 1982.
40. Hydrocarbon Processing 73:, 140, Nov. 1994.
41. O Clause, L Mank, G Martino, JP Frank, Trends in Catalytic Reforming and Paraffin Isomerization, Fifteenth World Petroleum Congress, vol. 2, John Wiley and Sons, 1997, pp. 695–703
42. JF Kriz, TD Pope, M Stănciulescu, J Mariner, Catalyst for Isomerization of C_7 Paraffins, Ind Eng Chem 37(12): 4560–4569, 1998.
43. Hydrocarbon Processing 77: 98, Nov. 1998.
44. Hydrocarbon Processing 77: 100, Nov. 1998.
45. Hydrocarbon Processing 71: 195, Nov. 1992.
46. Hydrocarbon Processing, 79: 131, Nov. 2000.
47. S Tiong, Acid Catalysed Cracking of Paraffinic Hydrocarbons. Part 3. Evidence for the Protonated Cyclopropane Mechanism from Hydrocracking/Hydroisomerisation Experiments. Ind Eng Chem Res 32: 403–408, 1993.
48. M Steijns, GF Froment, Hydroizomerisation and Hydrocracking. Kinetic Analysis of Rate Data for n-Decane and n-Dodecane. Ind Eng Chem Prod Res Dev 20: 880–888, 1981.
49. MA Baltanes, H Vensins, GF Froment, Hydroizomerisation and Hydrocracking. 5. Analysis of Rate Data for n-Octane. Ind Eng Chem Res Dev 22: 531–539, 1983.

12

Hydrocracking

A simplified description of hydrocracking is the decomposition of hydrocarbons on the acid sites of the catalyst's support, followed by hydrogenation of the cracking products on the metallic sites. The first stage of this process is analogous to catalytic cracking.

In reality, the process of hydrocracking is a lot more complex, because it also includes hydrogenation of aromatic hydrocarbons, opening of naphthenic rings, and isomerization. The activity of the two functions of the catalyst and the operating conditions of the process determine the conversion of these various reactions.

Basically, the aim of hydrocracking as well as other cracking processes, is the transformation of the heavy fractions of crude oil into light fractions. The use of this process is determined by the high quality of some of the products obtained, such as the jet fuel and the lubricating oils of high viscosity index. It is more expensive than catalytic cracking due to the high price of hydrogen and operation at high pressure. In addition, the role of hydrocracking increased due to the new requirements for gasoline, imposed in the U.S., by the 1990 Clean Air Act [1], now applied also in other countries. Hydrocracking gasolines have a low content of aromatic hydrocarbons and a limited content of volatile hydrocarbons.

The hydrocracking of gas oil also results in good quality because it has a high cetane number and low melting point. In the case of simultaneous separation of jet fuel, the gas oil would be lacking the light fractions that make possible their efficient use in combination with gas oils from other sources.

All the above factors contributed to the spectacular development of the hydrocracking process. From 1990 to 1995, the world capacity for hydrocracking increased from 300,000 to 400,000 m^3/day (130 million m^3/year), with another 46 million m^3/year under construction. The allocation of these quantities to the various regions of the world is graphically illustrated in Figure 12.1.*

---

* This is based on a report presented to the annual session of the American Institute of Chemical Engineers, November 15, 1994, San Francisco.

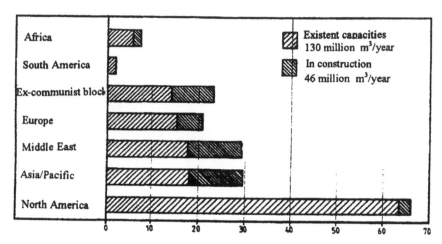

Figure 12.1 The distribution of hydrocracking plants in existence and under construction, at the end of 1994.

The feed most used for hydrocracking is vacuum distillate. There are two main options to accomplish this process:

> Using a two step reaction system, where gasoline is the main product (preferred in U.S.)
>
> Using a one step reaction system, where gas oil is the main product (preferred in Europe and Asia)

Apart from these two main systems, there are processes that perform direct hydrocracking of residues.

A special issue is the production of oils of very high viscosity index by hydrocracking that are of a higher quality than those produced by the conventional methods of solvent refining, dewaxing, and additivation.

Even though there are significant differences between these processes, their basic theory is similar and is presented for all together, with specifications where necessary.

12.1 THERMODYNAMICS OF HYDROCRACKING

As shown above, the process of hydrocracking consists of different types of reactions that are considered to be main reactions. This requires the analysis of the corresponding thermodynamic equilibria.

The following types of reaction will be analyzed, taking into account the limited availability of thermodynamic constants:

> Hydroisomerization of alkanes
> Hydrocracking of alkanes
> Partial or total hydrogenation of aromatic structures
> Hydrogenolysis of naphthenic rings.

12.1.1 Hydroisomerization of Alkanes

The previous chapter examined the isomerization equilibrium for the C_4-C_7 alkanes. For the C_8-C_{10} alkanes, that have thermodynamic data [2] available, the equilibrium compositions may be computed for a given temperature by writing the system of independent reactions and expressions for the equilibrium constants:

$$\left. \begin{array}{ll} N \rightleftharpoons I_1 & K_1 = \dfrac{pI_1}{p_N} \\[3mm] N \rightleftharpoons I_2 & K_2 = \dfrac{pI_2}{p_N} \\[3mm] \cdots\cdots & \cdots\cdots \\[3mm] N \rightleftharpoons I_i & K_i = \dfrac{pI_i}{p_N} \\[3mm] \cdots\cdots & \cdots\cdots \\[3mm] N \rightleftharpoons I_m & K_i = \dfrac{pI_m}{p_N} \end{array} \right\} \tag{12.1}$$

After summing up, it results:

$$\sum_1^m K_i = \frac{\sum_1^m pI_i}{p_N}$$

from which:

$$p_N = \frac{1}{1 + \sum_1^m K_i} \tag{12.2}$$

Knowing the values for the equilibrium constants, Eq. (12.2) allows the calculation of p_N, and by substituting it in Eq. (12.1) one can determine the partial pressures of all isomers.

Higher alkanes, having a high number of isomers (18 for C_8, 35 for C_9, and 75 for C_{10}), require a computer in order to perform the calculation.

In general terms, the proportion of normal alkanes at equilibrium is low and decreases with the number of carbon atoms in the molecule. Thus, as shown by the data given in the previous chapter (Figures 11.1–11.4) at 250°C the proportion of straight chain alkanes is 57% for C_4, 18% for C_6 and 10% for C_7. At the hydrocracking temperature it is 14% for C_7. In conclusion, for higher alkanes the proportion of *n*-alkanes at equilibrium does not exceed 10%.

12.1.2 Hydrocracking of Alkanes

The hydrocracking of alkanes is a stepwise process, made of two types of reactions that take place on catalytic sites of various types and can be illustrated as follows.

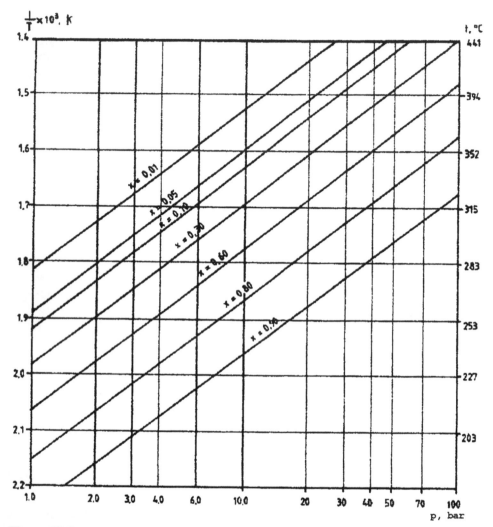

Figure 12.2   The thermodynamic equilibrium for the hydrogenation of tetrahydro-naphtha-lene to decahydro-naphthalene.

$$C-C-C-\overset{\overset{\displaystyle C}{|}}{C}-C\!\!\not\!\!-C-C-C-C \quad \overset{Ac}{\rightleftharpoons} \quad C-C-C-\overset{\overset{\displaystyle C}{|}}{C}=C \ + C_4H_8 \qquad \text{(a)}$$

$$C-C-C-\overset{\overset{\displaystyle C}{|}}{C}=C + H_2 \quad \overset{Me}{\rightleftharpoons} \quad C-C-C-\overset{\overset{\displaystyle C}{|}}{C}-C \qquad\qquad\qquad \text{(b)}$$

Where Ac and Me refer to the acidic and metallic sites of the catalyst, respectively.

The selection of an isoalkane for this representation is determined by the fact that, in general, the isomerization precedes the cracking on the acid sites.

It is known that for systems of successive reactions, such as the system (a)–(b), the thermodynamic calculation using a single overall reaction can lead to errors due to:

Intermediate substances modifying the overall equilibrium

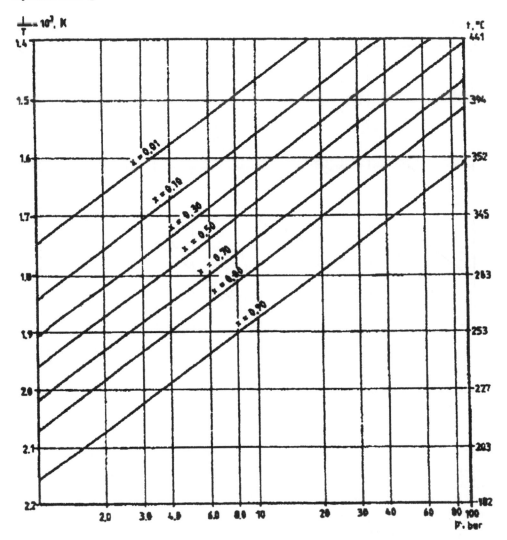

Figure 12.3 Thermodynamic equilibrium for the hydrogenation of butyl-benzene.

Large differences in the orders of magnitude of the two equilibrium constants

Therefore, if $(K_p)_a \ll (K_p)_b$ the conversion at equilibrium will be smaller than that calculated for the overall reaction using the equilibrium constant $K_p = (K_p)_a \cdot (K_p)_b$.

In order to avoid errors, the calculation in such cases is performed by considering the equilibrium for both reactions. Therefore, system (a)–(b) can be written:

$$(K_p)_a = \frac{(x-y)x}{1-x} \cdot \frac{p}{2+x-y} \tag{12.3}$$

$$(K_p)_b = \frac{y}{(x-y)(1-y)} \cdot \left[\frac{p}{2+x-y}\right]^{-1} \tag{12.4}$$

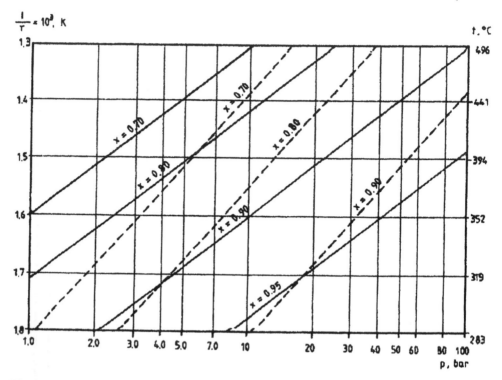

$\frac{1}{T} \times 10^3$, K

t, °C

Figure 12.4 Thermodynamic equilibrium of hydrodecyclization: — amyl-cyclopentane, --- butyl-cyclohexane.

where: x is the conversion corresponding to reaction (a), y is the conversion corresponding to reaction (b), and the initial composition is l mole isodecane and l mole hydrogen.

By solving the system of Eqs. (12.3) and (12.4), the correct values for the conversions x and y are obtained.

Both equilibrium constants—$(K_p)_a$ and $(K_p)_b$—have the same order of magnitude and values of about 10^2 at temperatures of 390–430°C, at which hydrocracking takes place. Thus, the calculations indicate that the hydrocracking of alkanes can take place without thermodynamic limitations.

The conclusion holds true for both higher and lower alkanes.

12.1.3 Aromatics Hydrogenation

The equilibrium for hydrogenating naphthalene to tetrahydronaphthalene was presented in Figure 10.4.

For the hydrogenation of tetrahydronaphthalene to decahydronaphthalene, the variation of the heat of formation and the entropy during the reaction have the following values:

$$\left(\Delta H_{298}^0\right)_r = -46,980\,\text{cal/mol} = -196,696\,\text{J/mol}$$

$$\left(\Delta S_{298}^0\right)_r = -92.553\,\text{cal/mol} \cdot \text{K} = -387.5\,\text{J/mol} \cdot \text{K}$$

Using the method described in Section 1.2, the equilibrium is plotted in Figure 12.2.

Using the following reaction for the analysis of the equilibrium of hydrogenation of the benzene ring

the following values are obtained for the variation of the heat of formation and the entropy during the reaction:

$$\left(\Delta H_{700}^{0}\right)_r = -50{,}230 \text{ cal/mol} = -210{,}303 \text{ J/mol}$$

$$\left(\Delta S_{700}^{0}\right)_r = -95.316 \text{ cal/(mol} \cdot \text{K)} = -399 \text{ J/mol} \cdot \text{K}$$

These values were used for plotting the graph for the equilibrium as shown in Figure 12.3.

The graphical representation in Figures 10.4, 12.2, and 12.3 infer that in the case of hydrocracking at temperatures of 390–430°C and pressures of 40–60 bar, the hydrogenation of naphthalenes to tetrahydronaphthalenes can take place with conversions of 60–70%, while the hydrogenation of the latter may reach 10–30%. For the case of alkylbenzenes, the equilibrium conversion may be of the order of 30–60%.

Therefore, one may deduce that hydrocarbons in the naphthalene series will hydrogenate to tetrahydronaphthalene in the process conditions of hydrocracking and following the opening of the saturated cycle thus formed, they will further be partially hydrogenated to alkylcyclohexanes.

The thermodynamic constants necessary for the calculation for other aromatic polycyclic hydrocarbons are not available.

The above calculations as well as those in the following section, use the thermodynamic constants published by DR Stull, EF Westrum Jr., and GC Sinke [2].

12.1.4 Hydrogenolysis of Naphthenic Rings

Taking into consideration the hydrocarbons present in crude oil and in the products, as well as the thermodynamic data available, it is of interest to examine the hydrogenolysis of rings with 5 and 6 carbon atoms that occur in the hydrogenolysis of the saturated ring of tetrahydronaphthalene and of one of the rings of decahydronaphthalene.

In addition, the partial hydrogenation of alkyl-naphthalenes followed by the hydrogenolysis of the saturated cycle can be examined.

The representative reaction for alkyl-cyclopentanes is:

The thermodynamic data for this reaction are:

$$\left(\Delta H^0_{700}\right)_r = -15{,}370 \, \text{cal/mol} = -64{,}351 \, \text{J/mol}$$

$$\left(\Delta S^0_{700}\right)_r = -19.972 \, \text{cal/mol} \cdot \text{K} = -83.62 \, \text{J/mol} \cdot \text{K}$$

These data lead to the lines of constant conversion plotted in Figure 12.4. The representative reaction chosen for alkyl-cyclohexanes is:

$$\text{(cyclohexane)}-C_4H_9 \; + \; H_2 \; \rightleftharpoons \; C-C-C-\overset{\overset{C}{|}}{C}-C-C-C-C$$

with the following thermodynamic data:

$$\left(\Delta H^0_{700}\right)_r = -10{,}600 \, \text{cal/mol} = -44{,}380 \, \text{J/mol}$$

$$\left(\Delta S^0_{700}\right)_r = -14.592 \, \text{cal/mol} \cdot \text{K} = -61.094 \, \text{J/mol} \cdot \text{K}$$

Based on these data, Figure 12.4 represents the lines of constant conversion for the equilibrium of hydrocracking of the ring of butyl-cyclohexane.

Note that for both cycloalkanes the breaking of the cycle occurred in β position relative to the tertiary carbon in the molecule, and led to a carbenium ion.

The values of the variation of the heat of formation and of the entropy during the hydrocracking of one ring of *cis*-decahydronaphthalene, to form butyl-cyclohexane are:

$$\left(\Delta H^0_{700}\right)_r = -9{,}690 \, \text{cal/mol} = -40{,}570 \, \text{J/mol}$$

$$\left(\Delta S^0_{700}\right)_r = -9.972 \, \text{cal/mol} \cdot \text{K} = -41.75 \, \text{J/mol} \cdot \text{K}$$

These data correspond to the lines of constant composition shown in Figure 12.5.

The thermodynamic data for hydrocracking in the saturated ring of tetra-hydronaphthalene to butyl-benzene are:

$$\left(\Delta H^0_{298}\right)_r = -9{,}990 \, \text{cal/mol} = -41{,}450 \, \text{J/mol}$$

$$\left(\Delta S^0_{298}\right)_r = -15.37 \, \text{cal/mol} \cdot \text{K} = -64.35 \, \text{J/mol} \cdot \text{K}$$

The corresponding equilibrium is given in Figure 12.5.

It is also useful to examine the direct reaction of hydrogenation-decyclization of alkyl-naphthalene to alkyl-benzene. The following reaction is considered to be representative:

$$\text{(naphthalene)}-C_5H_{11} \; + \; 3H_2 \; \rightleftharpoons \; \text{(benzene)}-C_9H_{19}$$

The thermodynamic data corresponding to this reaction are:

$$\left(\Delta H^0_{700}\right)_r = -41{,}530 \, \text{cal/mol} = -173{,}878 \, \text{J/mol}$$

$$\left(\Delta S^0_{700}\right)_r = -78.086 \, \text{cal/mol} \cdot \text{K} = -326.93 \, \text{J/mol} \cdot \text{K}$$

These data lead to the graphic representation shown in Figure 12.6.

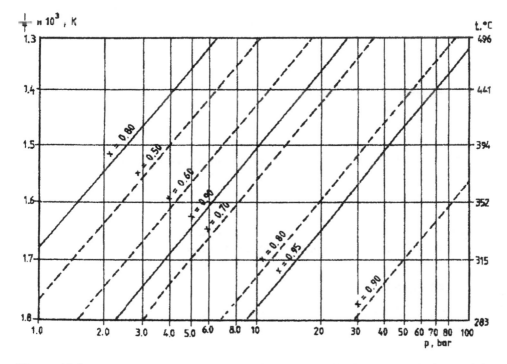

Figure 12.5 Thermodynamic equilibrium of dehydro-cyclization: — decahydro-naphthalene to butyl-cyclohexane, --- tetrahydro-naphthalene to butyl-benzene.

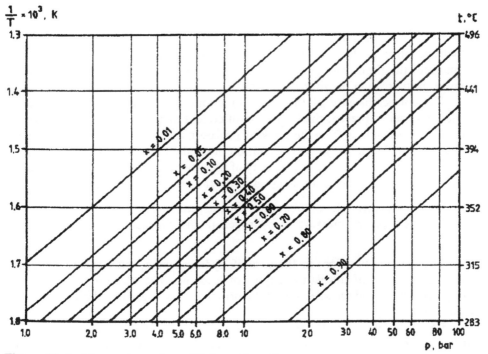

Figure 12.6 Thermodynamic equilibrium of reaction

The graphical representations in Figures 12.4–12.6 show that during hydro-cracking, reactions that open the naphthenic rings are possible and may occur as follows:

For monocyclic cycloalkanes with a cycle of 5 or 6 carbon atoms the possible conversions are up to 90–95%.

For tetrahydronaphthalene the conversion may reach 85%.

For decahydronaphthalene the process of opening one of the rings may reach conversions of approximately 95%.

For the hydro-dehydro-cyclization of alkyl-naphthalene to form alkyl-benzenes directly without going through the stage of alkyl-tetrahydro-naphthalenes, the conversions at equilibrium are of 70–80%.

In conclusion: for feeds made of aromatic hydrocarbons, the hydrogenolysis reactions of a naphthenic ring formed by a previous hydrogenation of an aromatic ring, might have the predominant role in the processes of hydrocracking.

12.2 CATALYSTS

The bifunctional hydrocracking catalysts have to promote the following main reactions to the required extents:

Cracking, according to mechanisms similar to those prevalent in catalytic cracking

Hydrogenation of the unsaturated hydrocarbons obtained from the cracking reactions

Hydrogenation of the aromatic rings

Hydrogenolysis of the naphthenic structures

In the first reaction, the metallic sites contribute to the initiation reactions of dehydrogenation to alkenes, which are later transformed into ions on acid sites. The second and the third reactions take place on the metallic sites. Both types of sites seem to participate in the fourth reaction.

The correct balance of these two functions, acid and metallic, is especially needed in order to prevent the formation of coke. For this purpose, the hydrogena-tion of alkenes resulting from the cracking process must take place over a short period of time, before they undergo polymerization or condensation to coke precursors. It is important to note that a balanced activity of these two functions does not rely only on the characteristics of the catalysis but also on the process conditions used:

High temperature favors reactions on the acid sites

Increased partial pressure of hydrogen favors hydrogenation on the metallic sites.

The metallic function, in the case of hydrocracking, must ensure the appro-priate performance of at least three types of reactions. The selection of the metal (or groups of metals) and the selection of the promoters used, are based on extended, in-depth studies.

Early studies under Raseev [3] led to the following conclusions: NiMo and NiW are the most appropriate as metal couples among the base metals. Among the noble metals, Pt and Pd are preferred.

Catalysts made of NiMo and NiW are used in plants with a single stage of hydrocracking. They are always used in the first stage in units with two stages. Pt and Pd are used exclusively in the second stage of systems with two stages.

In some cases, a bed of catalyst for hydrotreatment is placed at the upper part of the first reactor, with the purpose of performing the hydrogenolysis of the sulfur components and nitrogen components contained in the feed. This catalyst bed helps the partial hydrogenation of aromatic rings before they come in contact with the actual hydrocracking catalyst. This practice does not require the preliminary hydro-fining of the feed and is limited by the concentrations of H_2S and NH_3 resulting from the hydrogenolysis reactions, which are tolerated by the hydrofining catalysts. This matter is discussed further below.

The purpose of using promoters is to maximize the catalyst stability and efficiency with respect to specific reactions. The efficiency of the catalyst is not only dependent on the proportions of metals and the nature of the promoters but also on their appropriate dispersion on the support [4]. A good dispersion is also important for catalysts containing noble metals such as Pt and Pd.

In some cases, the selection of a metallic function is determined by the partial pressure of H_2S in the reaction zone. For example, Pd tolerates a very low partial pressure of H_2S, NiMo does not have any limitation, while NiW lies somewhere between them.

Initially, amorphous aluminosilica was used as support, which had been used many years earlier as a catalyst as for catalytic cracking.

The superiority of Y zeolites to amorphous catalysts (as demonstrated in the catalytic cracking process) led to their use as support in hydrocracking, as well.

A catalyst of this kind, NiW on Y zeolite, was commercialized under the name HTY by Shell and has been used in its refineries since 1983 [5]. This catalyst was then improved and commercialized under the name of S-753.

The zeolite-based catalysts show low coke formation, good performance stability, and longer cycle length between two regenerations, but they also require lower temperatures for achieving any given conversion (see Figure 12.7). The disadvantages of using zeolite catalysts are that they favor higher conversions to

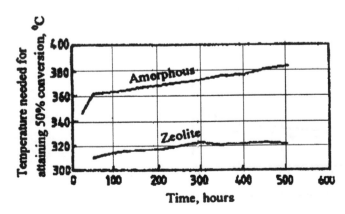

Figure 12.7 Comparison between hydrocracking catalysts based on amorphous support and on Y zeolite.

gasoline and gas and lower conversions to kerosene and gas oil. Due to this effect, some refiners prefer to use the catalysts based on amorphous support.

Further studies [5–7] proved that the excessive hydrocracking on zeolite catalysts was the consequence of having an excessive number of acid sites; the molecules that underwent an initial decomposition were readsorbed by other acidic sites and cracked to gasoline and gas.

The improvement of this situation led to the development of a second generation of zeolite catalysts for hydrocracking. They were characterized by a lower number of acid sites on the Y zeolite support obtained by the partial removal of aluminum from the zeolite lattice. At the same time, the composition of the zeolite support was modified. In some cases (Shell catalyst S-703) the reduction of the crystallite size is associated with the increase in volume fraction of the mesopores. This led to an increased selectivity (expressed by the conversion to kerosene) and stability (as seen in Figure 12.8) [5].

As an example, the transformation of the pore structure for the catalyst IFP HYC 642, is shown in Figure 12.9. This, together with the reduction in the number of acid sites led to the increase in selectivity [6, 7].

Another improvement mentioned by Shell is the use of two beds of catalyst within the reactor of the single stage system, or in the first reactor for systems with two reactors.

A pretreatment catalyst (such as C-424) may be introduced in the upper part of the reactor and a catalyst for hydrocracking is introduced in the lower part. In this way, a separate hydrofining step is no longer needed. This operation was made possible by the development of a catalyst not sensitive to the poisonous effect of nitrogen. One of the catalysts having this feature is known as Z-713 [4].

Z-713 has a density of 821 kg/cm^3, a specific surface area of 244 m^3/g, pore volume of 0.44 cm^3/g and an average pore diameter of 10.2 nm.

In parallel, the second stage catalyst was improved by the use of a support made of a mixture of amorphous aluminosilica and Y zeolite (such as the catalyst Z-603). Improved performances were obtained in this way, especially for the opening of naphthenic rings.

All of these improvements have a positive impact on cycle length and on the product's quality.

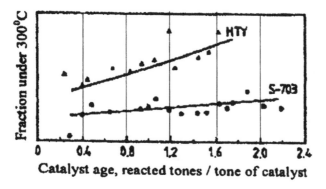

Figure 12.8 The increase in stability of second generation hydrocracking catalysts based on zeolitic support.

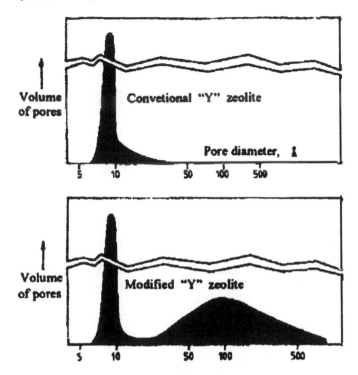

Figure 12.9 Modification of pore structure of standard zeolitic catalyst used for the preparation of the IFP HYC 642 catalyst.

The improvement of support is one of the major tasks in the development of catalysts for hydrocracking [8], and it is covered by a number of patents [9–23].

By 1990, UOP had also developed catalysts on amorphous support with improved performances under the name DHC-8, and subsequently the improved zeolite catalyst DHC-100. The zeolite catalyst has a high stability and a working temperature approximately 10°C lower than that of DHC-8 [24].

By moderating of the acidic function of the DHC-100 catalyst an almost equivalent production in medium distillates as with DHC-8 is obtained. The hydrocracking of a vacuum distillate produced the following conversions to products in volume %:

	DHC-8	DHC-100
Gasoline	18	24–27*
Jet fuel	48	48–49
Diesel gas oil	45	34–37

IFP is using a promoted NiMo catalyst on an alumina support HR-360 in the first stage, for hydrotreating, followed by the reactor containing the catalyst for hydrocracking HYC-642 based on zeolite. The zeolite catalyst HYC-642 generates amounts of distillates that are comparable to those generated by amorphous cata-

---

*The two values correspond to the beginning and the end of the cycle

lysts. As already mentioned (see Figure 12.9), this catalyst has a bimodal distribution of pores [6].

More details concerning the formulation of hydrocracking catalysts and their manufacture, activation, characterization and testing, as well as deactivation and reactivation in operation, are given in the excellent monograph of Scherzer and Gruia [106]

12.3 REACTION MECHANISMS

12.3.1 Hydroisomerization and Hydrocracking

Hydroisomerization and hydrocracking are linked together through their reaction mechanism, which refers to the alkanes and alkyl chains bound to the naphthenic or aromatic rings. The mechanism for the reaction of alkanes was analyzed in the previous chapter.

Numerous studies [25–31] clarified the mechanism of hydrocracking of alkanes, which today is universally accepted [8]. This mechanism is characterized by the following main steps:

- Dehydrogenation on metallic sites
- The adsorption of alkenes on acid sites with formation of carbenium ions
- Isomerization and cracking on the acid sites
- Hydrogenation of the alkenes resulting from isomerization and cracking, on the metallic sites

This mechanism is shown in Figure 12.10 for n-$C_{14}H_{10}$, including the steps of adsorption on two types of sites.

The reaction scheme explains the experimentally observed fact that in the reaction products, the ratios i-C_4H_{10}/n-C_4H_{10} and i-C_5H_{12}/n-C_5H_{12} are higher than those corresponding to the thermodynamic equilibrium. The scheme also shows that the process of hydrocracking, produces mainly iso-structures. The straight chain (normal) structures result from isomerization that does not reach equilibrium.

A similar reaction mechanism was developed for n-decane [8]. The hydrocracking of n-heptane [32], n-octane [33] and that of the higher alkanes was also studied [34]. The mechanism of hydrocracking for n-pentane [35] was presented in the previous chapter.

It is important to observe that the universally accepted initiation mechanism through dehydrogenation on metallic sites, followed by the adsorption of alkenes on Bronsted type sites (with the formation of carbenium ions) does not exclude the parallel formation of carbenium ions through the direct adsorption of alkanes on Lewis type sites of zeolite. Under normal conditions, it is likely for this reaction to be much slower than ion formation through the adsorption of alkenes on Bronsted sites and may be neglected. It is also possible that in other conditions, for example at high partial pressure of hydrogen—which diminishes the dehydrogenation reactions—the reaction on Lewis sites becomes the main source of carbenium ions.

There are two possibilities for the mechanism of reactions on metallic sites: the adsorption of alkenes and hydrogen occur on the same active site or on adjacent pairs [36,37].

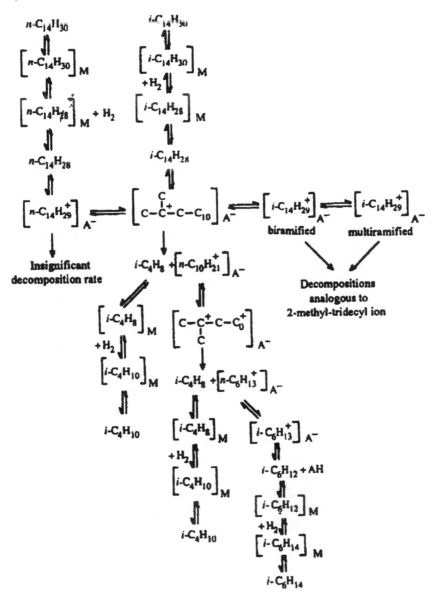

Figure 12.10 The mechanism of hydrocracking of tetra-decane.

A study of Raseev and El Kharashi was targeted at clarifying this issue. The study analyzed the experimental data obtained in the hydrocracking of a fraction of gasoline at 90–160°C, on an industrial catalyst of NiW supported on aluminosilica. A plug-flow, isothermal reactor was used at temperatures of 380, 420, 450°C, pressures of 60 and 80 bar, H_2/feed molar ratios of 5 and 10 and at a weight hourly space velocity in the range 1–7.5g/g cat. $\cdot$ hour^{-1} [38]. The analysis of the experimental data led to the conclusion that "single-site" adsorption is more plausible [39].

The reactions on acidic sites may follow two mechanisms. One involves the transfer of alkyl groups [8] and the other, the isomerization by means of ions with a cyclo-propyl structure [40]. The decomposition by cleavage in β position is not different from that operating in catalytic cracking. The only exception is that the hydrogenation to alkanes intervenes before a complete decomposition takes place, thus interrupting the cracking.

If the hydrogenating function of the catalyst is reduced compared to the acid function, and the partial pressure of hydrogen is low, the reactions lead to advanced decomposition of the feed. Most of the isomers have cracked and the ratio of isomerization products to cracking products will be small.

The reverse situation will occur in the case of a catalyst having a high ratio of metallic to acid functions respectively in process conditions that favor the hydrogenation reactions. In this case, the reaction products will be predominantly branched isomers and the cracking reactions might be reduced to a minimum. In this analysis it is important to take into account that the formation of branched isomers is successive. The isomers with only one branching are formed initially, followed by those with two or more branches.

The sequential formation of various intermediates makes it possible to select the conditions that are best suited to the purpose of designing the desired product. This represents one of the major advantages of hydrocracking over catalytic cracking.

The alkyl chains bound to the aromatic rings or naphthenic rings will undergo reactions of isomerization and decomposition based on similar mechanisms.

12.3.2 Hydrogenation of Aromatics and Hydrocracking of Naphthenic Rings

Only metallic sites participate in the hydrogenation of aromatic rings. Therefore the mechanism of reactions in this case is basically identical to that presented in Chapter 10. Nonessential differences might occur related to the specific properties of the metals that constitute the active sites.

The mechanism of hydrodecyclization (hydrogenolysis of naphthenic rings) in which both types of sites participate was much less studied. The participation of the metallic sites is obvious at least in the final stage, since it determines the saturated character of the final products.

The participation of the acid sites was demonstrated by the strong inhibiting effect of n-butylamine on the hydrogenation reaction followed by hydrodecyclization of 2-methyl-naphthalene [41]. n-Butylamine decomposes instantly to NH_3 during the reaction.

The hydrogenolysis of naphthenic rings generates carbenium ions,

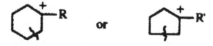

followed by the formation of iso-alkanes through cleavage in β-position.

As for alkanes, the carbenium ions are presumably formed by the participation of the metallic function through the intermediate formation of unsaturated hydrocarbons.

The isomerization might continue even after the opening of the ring and will generate isomers with several side chains. Reactions of decomposition might follow the ring opening whenever the hydrogenation function of the catalyst and the reaction conditions do not preferentially promote hydrogenation reactions.

For saturated polycyclic hydrocarbons, the reactions will take place similarly through the successive hydrogenolysis of cycles.

Experiments consisting of the hydrogenation of 1-methyl-naphthalene followed by the hydrodecyclization of the formed tetralines [41], are presented in Figure 12.11, showing the successive reactions.

The alkyl-indanes found among the products of reaction are considered by other authors [42,43] to be intermediate products when alkyl-benzenes are formed from methyl-tetralines. Our opinion is that their formation from methyl-tetraline is a reaction of isomerization that takes place in parallel to the reaction of hydrogenolysis of the rings.

The relevant part of this study is not the difference of opinion regarding the formation of alkyl-indanes, but rather the experimental proof that the dehydrocyclization reaction takes place by means of breaking the bond between a carbon atom and an aromatic ring.

It is important to note that the hydrogenation of the aromatic rings and the dehydrocyclization are the most important factors in assuring the superior quality of products obtained in hydrocracking. It is also essential to find the operating conditions under which these reactions (as well as isomerization) predominate, in order to obtain oils with a high viscosity index [44].

Polyaromatic hydrocarbons tend to accumulate in the recycle stream and special measures are needed in order to eliminate them, especially when high recycle rates are used, as when hydrocracking heavy feeds such as vacuum distillates or residues [8]. This can be explained by the fact that the reactions take place in a mixed vapor–liquid phase. Such conditions appear to prevent the heavy aromatics from reaching the active metallic sites in order to be hydrogenated.

The lack of interaction between these hydrocarbons and catalyst sites might be explained either through their high molecular volume or through the fact that they

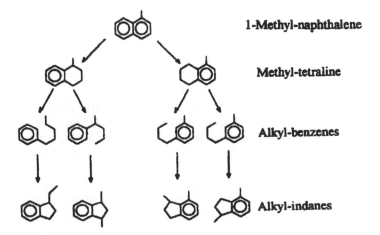

Figure 12.11 The hydrogenation of 1-methylnaphthalene followed by hydrogenolysis of tetralines to alkyl-benzenes and isomerization to alkyl-indanes.

do not succeed in diffusing from the liquid film to the surface of catalyst particles. At the reaction temperatures practiced in this process and without a contact with the active sites, the aromatic hydrocarbons undergo condensation and generate poly-aromatic structures.

12.4 KINETICS OF HYDROCRACKING

12.4.1 Hydrocracking of Individual Hydrocarbons

Early studies attempted to describe hydrocracking by using kinetic equations corresponding to a pseudohomogeneous process. Voorhies [45] expressed the hydrocracking of n-hexane by the following reaction:

$$[n\text{-}C_6H_{14} \rightleftharpoons i\text{-}C_6H_{14}] + H_2 \xrightarrow{k} \text{Hydrocracking products}$$

Considering the high excess of hydrogen and consequently its almost constant concentration, he obtained the following equation for a first order pseudohomogeneous process:

$$-\ln(1-x) = k\tau \tag{12.5}$$

Since the results obtained did not agree with the experimental data, Voorhies proposed an equation for a heterogeneous system where the Langmuir adsorption is accepted and the adsorption of hydrogen is predominant:

$$k = \frac{k_s b_{C_6H_{14}} \cdot p_{C_6H_{14}}}{(1 + b_{H_2} \cdot p_{H_2})^2} \tag{12.5'}$$

where

k = apparent rate constant of overall reaction
k_s = note constant of surface reaction
b = adsorption coefficients of substances indicated by subscripts

The exponent of the denominator for the above equation resulted from the analysis of experimental data obtained in a continuous reactor. Later studies did not confirm this equation either.

Several studies of the hydrocracking of n-alkanes correlated the hydrocracking with hydroisomerization by using a system of successive or successive-parallel equations [33,46–54]. Of these, the hydroisomerization-hydrocracking of n-octane [51], of n-decane, and of n-dodecane [33,46] will be reviewed here.

In all these studies, the experimental data were obtained in a Berty differential reactor*, using as catalyst 0.5% Pt on a US-Y zeolite, working at pressures of 5–100 bar, and temperatures of 180–240°C for n-octane and 130-250°C for both n-decane and n-dodecane [33,46]. The details of the hydrocracking of n-octane were published separately [51].

These studies considered not only the Langmuir-Hinshelwood adsorption model but also the Freundlich, Drachsel, and Dubinin isotherms of adsorption. The following set of reactions was adopted for the isomerization of n-octane:

---

*A rotating cage containing the catalyst.

I.	n-alkane (gas)	$\rightleftharpoons$	n-alkane (ads.)	
II.	n-alkane (ads.)	$\rightleftharpoons$	n-alkene (ads.) $+ H_2$	(metallic site)
III.1	n-alkene (ads.) $+ H^+$	$\rightleftharpoons$	R^+	
III.2	R^+	$\rightleftharpoons$	i-R^+	(acid site)
III.3	i-R^+	$\rightleftharpoons$	i-alkene (ads.)	
IV.	i-alkene (ads.) $+ H_2$	$\rightleftharpoons$	isoalkane (ads.)	(metallic site)
V.	i-alkane (ads.)	$\rightleftharpoons$	i-alkane (gas)	

It should be noted that the authors used the same assumptions as Hosten and Froment in 1971. They considered that on zeolites, a physical adsorption takes place as the reactant enters the pores; therefore it is irrelevant to consider the transfer of reactant molecules from one site to another by adsorption and desorption.

It was also considered that in the above reactions, step III is the most important. This is justified by the fact that for low reaction temperatures and for very active metallic sites (Pt), the reactions on the acidic sites are rate determining.

Thirty models were considered based on various assumptions regarding the isotherms of adsorption and the mechanism of reaction. The analysis of the experimental data, led to the selection of the following equation that best fitted the results on the isomerization of n-octane:

$$r = \frac{k\left(p_A - \frac{p_B}{k}\right)}{p_{H_2}[1 + b(p_A + p_B)]} \tag{12.6}$$

The correlation between the reactions of isomerization and hydrocracking was analyzed by considering the following plausible models in which the successive character of reactions of isomerization was proven in previous studies (see Section 12.3.1). In the following six reaction schemes, MB denotes the isomers that have one branch, MTB the isomers with several branches, and CR the hydrocracking products:

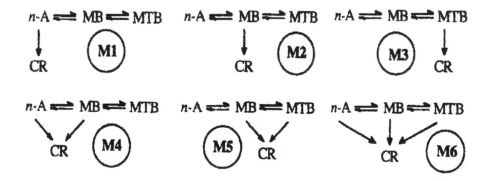

A seventh model, similar to M3, was disregarded because it was based on a completely unrealistic assumption that isomerization reactions are irreversible.

The processing of experimental data led to the conclusion that the only appropriate model is M3. The studies on n-decane and n-dodecane reached the same conclusions on the model that correlates the isomerization and the hydrocracking reactions [46].

The various possible forms for the adsorption term, i.e. the denominator, in the expressions for the reaction rate written for the M3 model, were then studied. The best agreement was obtained with the following expression:

$$p_{H_2}[1 + b(p_A + p_{MB} + p_{MTB} + vp_{CR}]\tag{12.7}$$

As in Eq. (12.6), here also the adsorption coefficients for all the participant hydrocarbons were considered to be equal. In conclusion, the system of kinetic equations for n-octane is identical with that previously developed for n-decane and n-dodecane [46].

$$r_{MB} = \frac{A_{MB} \cdot e^{-E/RT} \cdot (p_A - p_{MB}/K_1) - A_{MBT} \cdot e^{-E/RT}(p_{MB} - p_{MBT}/K_2)}{p_{H_2}[1 + b(p_A + p_{MB} + p_{MBT})]}$$

$$\tag{12.8}$$

$$r_{MBT} = \frac{A_{MBT} \cdot e^{-E/RT} \cdot (p_{MB} - p_{MBT}/K_{21}) - A_{CR} \cdot e^{-E/RT}p_{MBT}}{p_{H_2}[1 + b(p_A + p_{MB} + p_{MBT})]}\tag{12.9}$$

$$r_{CR} = \frac{A_{CR} \cdot e^{-E/RT} \cdot p_{MBT}}{p_{H_2}[1 + b(p_A + p_{MB} + p_{MBT})]}\tag{12.10}$$

In all reactions, the same value for the energy of activation, E, and the same value also for the coefficient of adsorption b, are used. These assumptions simplify significantly the form of the equations and the kinetic computations.

The equilibrium constant K_1 is for the isomerization of n-alkanes to isomers with one side branch, and the equilibrium constant K_2 corresponds to the isomerization of n-alkanes to isomers with more branches. Since in isomerization reactions the equilibrium constants are not significantly influenced by temperature, constant values were assumed for them.

The kinetic constants appearing in the Eqs. (12.8)–(12.10) are presented in Table 12.1 for each of the three hydrocarbons studied.

Table 12.1 Kinetic Constants for Eq. (12.8), (12.9), and (12.10)

Constants	n-Octane	n-Decane	n-Dodecane
A_{MB}, kmol/h · kg	$(4.36 \pm 0.85) \times 10^{14}$	$(1.8 \pm 0.3) \times 10^{15}$	$(4.0 \pm 1.0) \times 10^{16}$
A_{MTB}, kmol/h · kg	$(2.88 \pm 0.63) \times 10^{14}$	$(1.9 \pm 0.3) \times 10^{15}$	$(8.7 \pm 2.1) \times 10^{16}$
A_{CR}, kmol/h · kg	$(2.55 \pm 0.55) \times 10^{14}$	$(1.4 \pm 0.3) \times 10^{15}$	$(5.0 \pm 1.4) \times 10^{16}$
E, kJ/kmol	$(136.88 \pm 1.75) \times 10^3$	$(139.00 \pm 2.5) \times 10^3$	$(149.88 \pm 5.02) \times 10^3$
b, bar^{-1}	8.52 ± 0.51	13.9 ± 2.7	13.8 ± 3.9
K_1, bar^{-1}	5.21	7.6	7.6
K_2, bar^{-1}	0.952	2.2	2.2
Reaction rate constants at temperature of 200°C, $k_i = A_i \cdot e^{-E/RT}$			
k_{MB}, kmol/h · kg	0.338	0.820	1.146
k_{MBT}, kmol/h · kg	0.223	0.866	2.493
k_{CR}, kmol/h · kg	0.198	0.638	1.433

Source: Refs. 33, 46.

The modeling of the hydrocracking of cyclic hydrocarbons appears to be more difficult. The hydrocracking of methyl-naphthalene exemplifies the complexity of such a process [41].

The results of this study led to the chart presented in Figure 12.12, which also shows the apparent reaction rate constants and the apparent energies of activation that were measured. It should be noted that no mention is made of the reactions of hydrocracking of the products resulting from hydrogenation and ring opening.

Kinetic constants for the reactions indicated in the chart of Figure 12.12 are:

Reaction	k. hour^{-1}	E,kJ/kmol	Reaction	k. hour^{-1}	E,kJ/kmol
1	0.06	276	5	0.08	113
2	0.02	126	6	0.04	159
3	100.0	117	7	0.56	63
4	0.12	138			

The composition of the feed used for this experiment simulated the light recycle oil and contained: 40% 1-methyl-naphthalene, 57.75% n-heptane, 2% di-methyl-sulfide, 0.25% n-butyl-amine. Di-methyl-sulfide and n-butyl-amine are quickly split into H_2S and NH_3 in amounts similar to those in the real feed (1.4% S and 480 ppm N).

The catalyst was sulfated prior to use. It contained 2.7% NiO, 16.5% MoO_3, and 6.5% P_2O_5 on γ-alumina, and had a specific surface of 170 m^2/g and a bulk density of 3.4 g/cm^3.

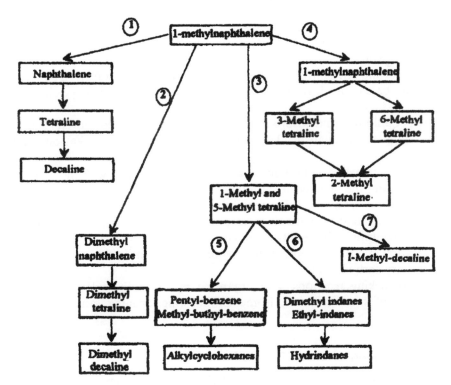

Figure 12.12 Chart for 1-methyl-naphthalene hydrocracking at 400°C. (From Ref. 41.)

An integral reactor was used in the following conditions: temperature 400°C, pressure 40 bar, H$_2$/feed ratio 8.3 mol/mol.

For the hydrogenation, the data was processed by accepting first order kinetics with respect to the hydrocarbon and to the hydrogen and by using the Langmuir-Hinshelwood kinetic equations.

The adsorption of NH$_3$ and H$_2$S was considered to dominate that of hydrocarbons and for this reason, the adsorption of the latter was ignored in the denominator of the equation.

As a result, the rate of hydrogenation for 1-methyl-naphthalene was expressed as follows:

$$r = \frac{k_S \cdot b_{1MN}\left[p_{1MN} - \dfrac{1}{K} \cdot \dfrac{p_{1MT}}{p_{H_2}^2}\right]}{1 + b_z p_z} \tag{12.11}$$

where K is the equilibrium constant. The subscripts: 1MN refers to 1-methyl-naphthalene, 1MT to 1-methyl-tetraline and z refers to NH$_3$ and H$_2$S. A similar equation was written for the formation of decaline through hydrogenation.

The analysis of experimental data using these two equations allowed the calculation of kinetic constants for reactions 3 and 7 (see Figure 12.12). The remaining constants were obtained on the basis of additional kinetic computations or by using graphical methods.

Most of the hydrocracked products are 1- and 5-methyl-tetraline that are produced from reaction 3, which is also rate-controlling for the process.

In a 1997 work [101], kinetic studies were used for determining the relationship between molecular structure and reactivity during the hydrocracking of model polynuclear aromatic hydrocarbons.

Naphthalene and phenantrene were reacted over a presulfided NiW/USY zeolite catalyst in an batch autoclave at $p_{H2} = 68.1$ bar and $t = 350$°C in a cyclohexane solvent.

The rate, equilibrium and adsorption parameter were estimated through fitting the kinetics data to a dual-site Langmuir-Hinshelwood-Hougen-Watson rate equation.

Adsorption parameters on both metal and acid sites increased with the number of aromatic rings and the number of saturated carbons.

The rate parameters showed that for a given total number of aromatic rings, hydrogenations at terminal aromatic rings were favored over hydrogenation of internal rings, which confirmed the results of other previous studies.

Isomerization and ring openings were favored at positions in α to the aromatic ring or the tertiary carbon atom.

The hydrocracking reactions were organized by the authors [101] into the reaction families of hydrogenation, isomerization, ring opening, and dealkylation.

12.4.2 Hydrocracking of Crude Distillation Cuts

Raseev and Kharashi [38] studied the hydrocracking of petroleum fractions in an integral reactor (see Figure 12.13). The employed catalyst was UOP (NiW/SiAl)

1 H_2 cylinder; 2 buffer vessel; 3 H_2 purifier;
4 H_2 dryer; 5 H_2 flowmeter; 6 Burette; 7. liquid
pump; 8 pump motor; 9 preheater; 10 reactor;
11 condensers; 12 high pressure separator;
13 cooling mantle; 14 collector, 15 gas flowmeter;
16 gasometer; 17 psanometer; 18 thermocouple;

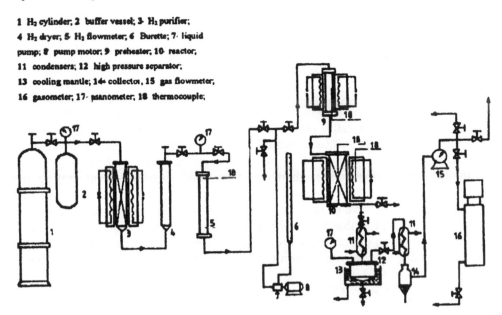

Figure 12.13   Laboratory unit for studying hydrocracking.

having a content of 24.3% WO_3, 8.64% NiO, and 0.236 Fe_2O_3, with a specific
surface area of 209 m^2/g and a density of 3.17 g/cm^3.

12.4.2.1   The Influence of Diffusion

In the hydrocracking of naphthas [55,56] and gas oil fractions [57], the influence of
the diffusion steps was assessed and operating conditions were selected in order to
eliminate its effect on the kinetics.

The established procedures of varying the linear velocity through the layer of
catalyst while keeping the other work parameters constant, allowed one to determine
the conditions in which the process is not influenced by external diffusion.

The results indicate that external diffusion does not influence the process at
Re > 6 for both feeds. The Reynolds number is based on the equivalent diameter of
the catalyst particle.

Five particle sizes (between 0.4 mm and 3.9 mm) were used under conditions of
Re > 6, at constant temperature, pressure, and molar ratio, varying only the
volumetric flowrate. From the results, the Thiele effectiveness factors were computed.

The results obtained for various values of conversion showed that internal
diffusion influences the overall reaction rate for catalysts with dimensions larger
than 0.4 mm for naphthas and higher than 0.35 mm for gas oils.

12.4.2.2   The Kinetics of Naphtha Hydrocracking

The kinetics of hydrocracking was studied on a straight-run naphtha obtained from
a Romanian crude oil [55]. The feed had distillation limits of 90–160°C and the
following composition:

n-alkanes	42%
i-alkanes	25%
Cycloalkanes	20%
Aromatics	13%

The experimental conditions were similar to those used in commercial operation: temperatures of 380, 420, and 450°C, pressures of 60 and 80 bar, and H_2/naphtha molar ratios of 5 and 10. The space velocity was varied between 1–7.5 g/g cat · hour. A sufficient number of experiments was performed in which only the space velocity was varied (in conditions in which diffusion had no influence) so that a finite differences method could be used, instead of integrating the differential equations [56].

The initial objective of the study was to develop a pseudo-homogeneous rate equation to be used in the design of the hydrocracking of petroleum fractions. A differential equation of *m*th order for a pseudo-homogeneous process was used for this purpose allowing the application of the method of finite differences as follows:

$$\frac{\Delta x}{\Delta(G_c/n_{oM})} = k(1 - \bar{x})^m \tag{12.12}$$

where:

x represents the conversion
G_c the quantity of catalyst in g
n_{oM} the feed rate in moles/hour.

Attempts to fit the experimental data with this equation led to unrealistic values for the exponent *m*, ranging between 6 and 10. The conclusion is that hydrocracking cannot be modeled as a pseudo-homogeneous process.

The hydrocracking reaction for a heterogeneous process may be written as follows:

$$M + (v - 1)H_2 \rightarrow vG$$

where M refers to the feed molecules, G to the reaction products and v is the stoichiometric coefficient.

Using the Langmuir adsorption law and considering that the hydrogenation of the reaction products requires a number v of available active sites, and assuming that hydrogen is adsorbed on the same sites as the hydrocarbons, the reaction rate equation will be:

$$r = \frac{k_S b_M P_M b_H^{(v-1)} p_H^{(v-1)}}{(1 + b_M p_M + b_H p_H + \sum b_G p_G)^v} \tag{12.13}$$

Considering that all the adsorption coefficients for the reaction products b_G are equal, the Σ sign can be omitted in the denominator. Using the following expressions for the partial pressures:

$$p_M = p\frac{1 - x}{1 + a} \qquad p_G = p\frac{vx}{1 + a}$$

where *a* is the molar ratio hydrogen/hydrocarbons and x is the conversion and using the method of finite differences, the result is:

$$\frac{\Delta x}{\Delta(G_c/n_{oM})} = \frac{k(1 - \bar{x})}{\left[1 + b_M'(1 - \bar{x}) + b_G'\bar{x}\right]^{\nu}}$$

(12.14)

where

$$k = \frac{k_S b_M b_H^{(\nu-1)} \cdot p_H^{(\nu-1)}}{(1 + b_H p_H)^{\nu}}$$

(12.15)

$$b_M' = \frac{b_M}{1 + b_H p_H} \cdot \frac{p}{1 + a}$$

(12.16)

$$b_G' = \frac{\nu b_G}{1 + b_H p_H} \cdot \frac{p}{1 + a}$$

(12.17)

In order to process the experimental data by using the least squares method, the relationship (12–14) was modified to:

$$\left[\frac{(1 - \bar{x})}{\Delta\bar{x}/\Delta\left(\dfrac{G_c}{n_{OM}}\right)}\right]^{1/\nu} = \frac{1 + b_G'}{k^{1/\nu}} + \frac{b_M' - b_G'}{k^{1/\nu}}(1 - \bar{x})$$

(12.18)

finally yielding the following values for the kinetic constants:

Temperature, °C	$(1 + b_G')/k^{1/\nu}$	$(b_M' - b_G')/k^{1/\nu}$
400	20.37	−19.88
450	13.31	−12.51

Because $b_G' \gg b_M$, Eq. (12.14) may be reduced to:

$$\frac{\Delta x}{\Delta\left(\dfrac{G_{\text{cat}}}{n_{OM}}\right)} = \frac{k(1 - \bar{x})}{\left(1 + b_G'\bar{x}\right)^{\nu}}$$

(12.19)

In all the experiments of this study involving the hydrocracking of naphtha, the stoichiometric coefficient had a constant value of 2.8. Obviously it will have different values for other feeds or conversions.

The strong adsorption of the reaction products indicated by the values obtained, suggests that at 400–450°C the rate-controlling step on NiW catalyst is the hydrogenation of alkenes resulting from hydrocracking. It is also true that on Pt catalysts at 150–250°C the rate-controlling step is the cracking on acid sites [33, 48]. These two observations are not contradictory because an increase in temperature favors cracking and prevents hydrogenation, leading to a change in the rate-controlling step.

In order to verify the hypothesis regarding the rate-controlling step, the cracking on Bronsted or Lewis (less probable) acid sites and the dehydrogenation/hydrogenation on metallic sites [39] were considered.

Since the reactions of dehydrogenation/hydrogenation take place on the same sites, these sites will be distributed between the two reactions based on their relative rates. Therefore, the reactions on metallic will become rate-controlling only when all of them are much lower than the reactions rate on acid sites. This case is also considered.

Controlling step—cracking of alkenes on Bronsted-type σ_A centers. In this section subscript o (as in M_O) refers to olefin (albene). The mechanism can be described by the following reactions:

$$
\left.
\begin{aligned}
M_O + \sigma_A H &\longrightarrow \sigma_A[M_O H^+] \\
\sigma_A[M_O H^+] &\longrightarrow (v-1)G_O + \sigma_A[G_O H^+] \\
\sigma_A[G_O H^+] + M_O &\longrightarrow \sigma_A[M_O H^+] + G_O
\end{aligned}
\right\}
\tag{a}
$$

The overall reaction is:

$$
M_O + \sigma_A H \xrightarrow{k_{SA}} v G_O + \sigma_A H
\tag{a'}
$$

Knowing that the resulting alkenes are not adsorbed on acidic sites and since only one site is involved in the reaction, the reaction rate is be given by the following equation:

$$
r = \frac{k_{SA} b_{M_O} p_{M_O}}{1 + b_{M_O} p_{M_O} + b_{G_O} p_{G_O}}
\tag{12.20}
$$

The dehydrogenation and hydrogenation are at equilibrium, therefore:

$$
\frac{p_{M_o} \cdot p_H}{p_M} = K_D \qquad \frac{p_G}{p_{G_o} \cdot p_H} = K_H
\tag{12.21}
$$

By substitution in Eq. (12.20), it becomes:

$$
r = \frac{k_{SA} b_{M_O} \dfrac{K_D}{p_H} p_M}{1 + \dfrac{b_{M_o} \cdot K_D}{p_H} p_M + \dfrac{b_{G_o}}{K_H p_H} p_G}
\tag{12.22}
$$

Controlling step—dehydrogenation and hydrogenation on the metallic sites σ_M. If a "single-site" absorption is accepted, the mechanism can be described by the following reactions:

$$
\left.
\begin{aligned}
M + \sigma_M &\longrightarrow M_O + H_2 \sigma_M \\
M_O + \sigma_A &\rightleftharpoons v G_O + \sigma_A \\
(v-1)H_2 + (v-1)\sigma_M &\longrightarrow (v-1)H_2 \sigma_M \\
v G_O + v H_2 \sigma_M &\longrightarrow v \sigma_M + G
\end{aligned}
\right\}
\tag{b}
$$

he overall reaction on the metallic sites is:

$$
M + (v-1)H_2 + \sigma_M \xrightarrow{k_{SM}} v G + \sigma_M
\tag{b'}
$$

The final result is identical to Eq. (12.13).

Controlling step—cracking of alkanes on Lewis sites. The reaction mechanism is:

$$
\left.
\begin{aligned}
M_O + \sigma_A &\longrightarrow \sigma_A H[M_O H^+] \\
\sigma_A H[M_O H^+] &\longrightarrow (v-1)G_O + \sigma_A H[G_O H^+] \\
\sigma_A H[G_O H^+] + M &\longrightarrow G + \sigma_A H[M_O H^+]
\end{aligned}
\right\}
\tag{c}
$$

The overall reaction is:

$$M_O + \sigma_A \longrightarrow (\nu - 1)G_O + G_1 + \sigma_A \qquad \text{(c')}$$

Taking into account also the hydrogenation of the alkenes G_o, i.e., the reactions (12.21) reaching equilibrium, the reaction rate equation become:

$$r = \frac{k_{SA} b_M p_M}{1 + b_M p_M + \left(\dfrac{b_{G_o}}{K_H p_H} + b_G\right) p_G} \qquad (12.23)$$

The final equations (12.22), (12.13), and (12.23) were reduced to more simple expressions by using several overall kinetic constants defined in Table 12.2.

The processing of the experimental data by means of the method of finite differences led to negative values of the kinetic constants for mechanism (c), which resulted in its being discarded.

The selection between mechanisms (a) and (b) was done following the integration of the rate equations.

For mechanism (a) the integral was:

$$\frac{G_c}{n_{OM}} = \frac{1}{k}\{-\ln(1 - x) + K_B[-x - \ln(1 - x)]\} \qquad (12.24)$$

For mechanism (b) and for $\nu = 3$ (close to the experimental value of 2.8):

$$\frac{G_c}{n_{OM}} = \frac{1}{k}\left\{-\ln(1 - x) + 3K_B[-x - \ln(1 - x)] + 3K_B^2\left[-\frac{x^2}{2} - x - \ln(1 - x)\right]\right.$$

$$\left. + K_B^3\left[-\frac{x^3}{3} - \frac{x^2}{2} - x - \ln(1 - x)\right]\right\} \qquad (12.25)$$

The processing of experimental data using these equations led to the values of Table 12.3 for the rate and equilibrium constants.

The variance values indicate that for hydrocracking on NiW catalysts, the reactions on the metallic sites are rate-controlling at temperature levels similar to those practiced in the commercial process.

Figure 12.14 shows the comparison between the conversion curves calculated using Eq. (12.25) and the experimental values obtained by Raseev and El Kharashi [56]. There is good agreement between the experimental and the calculated data.

The apparent energy of activation of the reaction on the catalyst surface for mechanism (b) was calculated dividing the equations for k and for K_B from Table 12.2. This gives:

$$\frac{k}{K_B^\nu} = k_{SM} \frac{b_M (b_H p_H)^{\nu - 1}}{(\nu b_G - b_M)^\nu} \cdot \left(\frac{p}{1 + a}\right)^{1 - \nu} \qquad (12.26)$$

Taking into account the large excess of hydrogen, one may write:

$$p_H = p \frac{a}{1 + a}$$

and using the notation:

$$b_G = n b_M$$

Table 12.2 Final Algebraic Forms of Eqs. (12.22), (12.13) and (12.23) and Their Expressions for the Kinetic Constants

Rate controlling step	Rate Eq.	Final algebraic form	Expression for kinetic constants	
			k	k_a
Cracking on Bronsted acid sites, mechanism (a)	12.22	$r = \dfrac{k(1-x)}{1+K_B x}$	$\dfrac{k_{SA} b_{MO} K_D/p_H}{(1+a)/p + b_{M_O} K_D/p_H}$	$\dfrac{\nu b_{G_O}/K_H p_H - b_{M_O} \cdot K_D/p_H}{(1+a)/p + b_{M_O} \cdot K_D/p_H}$
Single site reactions on metallic sites, mechanism (b)	12.13	$r = \dfrac{k(1-x)}{(1+K_B x)^\nu}$	$\dfrac{k_{SM} b_M (b_H p_H)^{\nu-1} p/(1+a)}{[1+b_H p_H + b_M \cdot p/(1+a)]^\nu}$	$\dfrac{(\nu b_G - b_M)p}{[1+b_H p_H + b_M \cdot p/(1+a)](1+a)}$
Cracking on Lewis acid sites, mechanism (c)	12.23	$r = \dfrac{k(1-x)}{1+K_B x}$	$\dfrac{k_{SA} \cdot b_M}{(1+a)/p + b_M}$	$\dfrac{\nu(b_{G_o}/K_H p_H + b_G) - b_M}{(1+a)/p + b_M}$

Table 12.3 Kinetic Constants for the Two Rate-Controlling Steps

Rate-controlling step	t, °C	k, mol/g cat · h	K_B	Variance
Reaction on metallic sites,	400	0.167	4.99	0.246
Equation (12.20)	450	0.203	3.14	0.354
Reaction on acid sites,	400	0.133	15.6	0.634
Equation (12.19)	450	0.233	10.1	0.830

Eq. (12.26) becomes:

$$\frac{k}{K_B^v} = k_{\text{SM}} \left(\frac{b_{\text{H}}}{b_M}\right)^{v-1} \cdot \frac{a^{v-1}}{(nv-1)^v} \tag{12.27}$$

while the equation for K_B becomes:

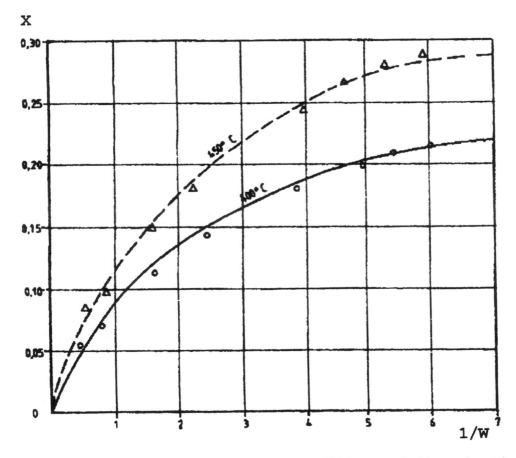

Figure 12.14 Conversion curves calculated using Eq. 12.25 compared with experimental values: ○ – for 400°C, △ – for 450°C.

$$K_B = \frac{n\nu - 1}{(1 + a)/p \cdot b_M + a\dfrac{b_H}{b_M} + 1} \tag{12.28}$$

Considering that in hydrocracking, the hydrogen to feed molar ratio ranges from 5 to 10, while the pressures are of the order of 60–80 bar or above, the first and last term of the denominator of Eq. (12.28) may be neglected, which results in:

$$\frac{b_H}{b_M} = \frac{n\nu - 1}{aK_B}$$

Substituting this result in (12.27), and regrouping and simplifying, we obtain:

$$k_{SM} = \frac{k}{K_B}(n\nu - 1) \tag{12.29}$$

Since n and ν do not vary significantly with temperature, by writing Eq. (12.29) for two temperatures and dividing results in:

$$\frac{(k_{SM})_{T_1}}{(k_{SM})_{T_2}} = \frac{(k_1)_{T_1} \cdot (K_B)_{T_2}}{(k_1)_{T_2} \cdot (K_B)_{T_1}} \tag{12.30}$$

Using the values obtained previously for the constants k and K_B at temperatures of 400 and 450°C, we obtain:

The apparent energy of activation: $E_k = 15.83\,\text{kJ/mol}$
The apparent overall heat of adsorption: $Q_{K_B} = 37.51\,\text{kJ/mol}$
The energy of activation of the reaction on surface, $E'_{k_{SM}} = 53.17\,\text{kJ/mol}$

12.4.2.3 The Kinetics of Hydrocracking of Gas Oil

A study was performed on the hydrocracking of a straight-run gas oil with the following characteristics [57]:

Density	0.8546
Distillation range	250–375°C
Molecular mass (cryoscopic)	250
nDM analysis	$C_A = 10.38\%$, $C_N = 31.82\%$, $C_P = 57.60\%$
FIA analysis	Aromatic hydrocarbons = 11.43%, Saturated hydrocarbons = 88.57%

The same NiW catalyst and a similar flow unit were used as shown in Figure 12.13. The catalyst particles were 0.35 mm in diameter in order to avoid the influence of internal diffusion.

Experiments were performed at temperatures of 410, 430 and 450°C, pressures of 50–80 bar, and hydrogen/feed molar ratios of 3.25 and 10.45 mol/mol. The space velocity was varied between very large limits using the technique of dilution of catalyst with inert matter.

As in the previous study, the attempts to process experimental data using kinetic equations assuming a pseudo-homogeneous reaction system (equations of type 12.12), resulted in apparently meaningless order of reaction.

The Langmuir law was assumed as valid and the following cases were considered: (1) the adsorption of hydrocarbons and hydrogen on a single active site, (2) on two different sites, and (3) the absorption of hydrogen with dissociation.

The various assumptions on the adsorption of the substances present in the reaction zone were checked by means of both the differential and integral methods. The best fit was obtained for the adsorption on two sites, with a dominance of the absorption of the reaction products. This agrees with the conclusions from the study regarding the hydrocracking of naphtha.

Consequently, Eq. (12.19) is valid also for hydrocracking of gas oil. In this case, the stoichiometric coefficient ν has the value of 2.0.

The energy of activation and the heat of adsorption for the constants of Eq. (12.19) have the values at temperature levels of 410–430°C found in Table 12.4.

12.4.2.4 The Kinetics of Hydrocracking of Heavy Fractions

The hydrocracking of heavier fractions such as gas oil from coking or heavy gas oil resulting from hydrocracking was studied by Yui and Sanford [58] using an industrial catalyst $NiMo/Al_2O_3$ containing 4 wt %. NiO and 20 wt %. MoO_3, with a pore volume of 0.3 cm^3/g and a specific surface area of 160 m^2/g.

The reactor used had biphasic flow over a fixed bed of catalyst (trickle-bed reactor).

The conversions were expressed in terms of the amount of gasoline with an end point of 185°C and an amount of light distillate with distillation range 185–343°C. The feed had an initial boiling temperature of 343°C.

Working temperatures were in the range of 360–400°C, pressures of 70–95 bar, and space velocities between 0.7 and 1.5 $l/l \times$ h. The molar ratio hydrogen/feed was maintained at a constant value of 600 Nm^3/m^3. The reactor was operated in practically isothermal conditions.

The following kinetic scheme was considered for processing the experimental data:

$$A \xrightarrow{k_1} B \xrightarrow{k_3} C$$
$$\underset{k_2}{\underrightarrow{\qquad}}$$

All reactions in this reaction scheme were considered to be of first order and pseudohomogeneous.

The assumption that the overall process is successive–parallel according to the above scheme was not satisfactory. Therefore the experimental data were treated assuming the process to occur according to two mechanisms: reactions in parallel ($k_3 = 0$) and successive reactions ($k_2 = 0$).

Table 12.4 Kinetic Constants Corresponding to the Hydrocracking of Gas Oil Eq. (12.19)

Pressure, bar	H_2/hydrocarbon mol/mol	E, kJ/mol	Q_{ads}, kJ/mol
80	10.5	42.3	35.6
80	3.3	58.6	37.3
50	10.5	62.8	56.6

By using the following equations a 1.5 reaction order was obtained for the desulfurization and a first order for the reactions of denitrification:

$$\frac{1}{S_i^{0.5}} - \frac{1}{S_f^{0.5}} = \frac{k_s p_{H_2}^\beta}{2 \text{LHSV}^\alpha}$$

$$\ln \frac{N_f}{N_i} = \frac{k_N p_{H_2}^\beta}{\text{LHSV}^\alpha}$$

(12.31)

where:

S_i and S_f represent the initial and final sulfur contents in wt %
N_i and N_f represent the initial and final nitrogen contents in wt %
p_{H_2} partial pressure of hydrogen in MPa
LHSV feed space velocity in vol liquid/vol catalyst hour
k_S, k_N, α and β are kinetic constants.

The following equations were used for hydrocracking:

For the parallel process.

$$C_A = C_{A_0} \exp\left[\frac{-(k_1 + k_2) p_{H_2}^\beta}{\text{LHSV}^\alpha}\right]$$

(12.32)

$$C_B = C_{B_0} + \frac{k_1}{k_1 + k_2} C_{A_0} \left\{ 1 - \exp\left[-(k_1 + k_2) p_{H_2}^\beta / \text{LHSV}^\alpha\right]\right\}$$

(12.33)

For the consecutive process.

$$C_A = C_{A_0} \exp\left(-k_1 p_{H_2}^\beta / \text{LHSV}^\alpha\right)$$

(12.34)

$$C_B = C_{B_0} \exp\left(\frac{k_3 p_{H_2}^\beta}{\text{LHSV}^\alpha}\right) + \frac{k_1}{k_3 - k_1} C_{A_0} \left[\exp\left(\frac{k_1 p_{H_2}^\beta}{\text{LHSV}^\alpha}\right) - \exp\left(-\frac{k_3 p_{H_2}^\beta}{\text{LHSV}^\alpha}\right)\right]$$

(12.35)

where

C_A and C_B are the percentages of heavy and light gas oil
C_{A_0} and C_{B_0} are the corresponding percentages in the feed
C_C and C_{C_0} are the differences of the percentages of naphtha

The other notations are the same as in Eqs. (12.31).

The variation of rate constants k_1, k_2, and k_3 with temperature is expressed by equations of the Arrhenius type:

$$k_i = k_{i0} \exp(-E_i / RT)$$

(12.36)

The kinetic constants used in Eqs. (12.31)–(12.36) are presented in Table 12.5 for the two raw materials studied.

As expected, the rate of hydrocracking of gas oil from coking was found to be higher to that of the gas oil produced in hydrocracking.

The activation energy for the formation of naphtha is larger than that for light diesel oil, which implies that the gasoline/light gas oil ratio increases with temperature.

GF Froment et al. [34,59,60] based their research on expanding the system of kinetic calculation based on the elementary steps of reactions. The principle behind their method is outlined in their article on synthesis [61].

The studies on the hydrocracking of n-octane show that even without differentiating between the four octane isomers with two or more side branches and their different rates of hydrocracking, the number of kinetic parameters that require experimental determination reaches 7.

Even though this type of simplification appears in the analysis of hydrocracking of crude oil fractions, insurmountable problems prevent obtaining exact solutions, for at least three reasons: (1) the extremely high number of participating hydrocarbons, (2) the difficulties in determining the kinetic constant for all these hydrocarbons requiring their availability in pure state, and (3) the difficulties in obtaining the precise analytic composition of the reaction product.

This creates a situation similar to the study of the kinetics of pyrolysis of a fraction of crude oil (Section 5.1.4), where it is possible to make simplifications and generalizations of the reaction mechanism, eventually leading to the correct treatment of the kinetics of the pyrolysis (the so-called "mechanistic method").

A similar approach was taken by GF Froment. The analysis of the single events of the mechanism of reaction allowed generalizations and simplifications, which led to a fundamentally theoretical approach to the hydrocracking of the crude oil fractions.

For n-octane, 383 events were identified: 86 charge isomerizations, 24 isomerizations through the migration of methyl group, 96 chain isomerizations via the cyclopropyl ion, as intermediate [62], 15 bond cleavages in β position, 75 protonations, and 85 deprotonations. Further simplifications and generalizations valid for

Table 12.5 Kinetic Constants

	Cooking gas oil		Hydrocracking gas oil	
	k_0	E/R, K	k_0	E/R, K
	Mild hydrocracking ($\alpha = 0.5$, $\beta = 0.4$)			
	(a) Parallel process ($k_3 = 0$)			
$k_1 + k_2$	8.754×10^4	8932	4.274×10^4	8674
k_1	8.544×10^3	7558	3.775×10^3	7208
k_2	1.780×10^8	14,987	6.847×10^8	16,189
	(b) Successive process ($k_2 = 0$)			
k_1	8.754×10^4	8932	4.274×10^4	8675
k_3	8.206×10^7	13,566	2.711×10^5	10,295
	Hydrodesulfurization			
k_S	9.375×10^9	15,756	8.283×10^{10}	15,848
	($\alpha = 1.0$, $\beta = 1.1$)		($\alpha = 1.0$, $\beta = 0.6$)	
	Hydrodenitration			
k_N	1.524×10^6	11,747	1.485×10^5	9108
	($\alpha = 1.0$, $\beta = 1.8$)		($\alpha = 1.0$, $\beta = 1.3$)	

all alkane hydrocarbons finally produced a method that enables computing the yields from the hydrocracking of crude oil fractions. In addition, the method provides the possibility of obtaining the kinetic constants for the hydrocracking of alkanes C_4–C_{10}, for which the identification and measurement of all reaction products poses no problems.

The kinetic model used in this study was developed by Froment in previous studies [33,46] and is represented by the reaction scheme:

$$A \rightleftharpoons MB \rightleftharpoons MTB \rightarrow CR$$

The types of reaction and adsorption phenomena comprised in this global mechanism are shown in Table 12.6. The bimolecular reactions of carbenium ion and the alkylation reactions of alkenes were not included in this presentation and the related calculations.

The number of rate constants corresponding to the elementary steps (single events) shown above was reduced in Froment's work [61], based on the following considerations and approximations:

1. Methyl ions and the primary ions, having a large energy of formation, were not considered when setting up the reaction system.
2. The type of carbenium ion (secondary or tertiary) determines its activity in the elementary steps of isomerization, rather than the number of carbon atoms. Accepting this approximation reduces the number of charges to 4 for isomerization, reduces the number of migrations of the methyl group to 5, and reduces the number of isomerizations through the cyclopropyl ion to 6.
3. The rate constant of protonation is independent of the nature of the alkene. This is justified by the structure of the activated complex and is similar to that of alkenes with an incompletely broken double bond [34].
4. The rate constant of deprotonation is independent of the number of atoms of carbon for C_5 and *higher homologues*.
5. The rate constant of cracking through cleavage in β position is independent of the nature of the alkene formed. The number of rate constants for cleavages in β position is therefore reduced to 4 events: the breaking takes place between: (1) two secondary carbons, (2) a secondary and a tertiary, (3) a tertiary and a secondary, (4) a tertiary and a tertiary.

In addition, the number of rate constants of elementary reactions was reduced based on the thermodynamic considerations applied to the reactions of deprotonation and isomerization [34,61].

By means of these procedures, the number of elementary rate constants was reduced overall to 17: protonation – 2, deprotonation – 2, charge transfer – 3, methyl group migration – 3, isomerization through cyclopropyl ion mechanism – 3, bond breaking in β position – 4. These constants are independent of the number of carbon atoms and are valid for all alkanes.

It must be remembered that the rate constants of elementary reactions refer to a single "event" and they have to be multiplied by the number of equivalent "events." Consider the migration of a methyl group of a carbenium ion as an example:

In the reaction from left to right, only one methyl group can be affected (single event possible). In the reverse reaction two methyl groups (two events possible) can be affected.

Table 12.6 Reactions in the Process of Hydroisomerization and Hydrocracking of Alkanes on Bifunctional Catalysts

Type of reaction	Reaction description
Adsorption of alkane in zeolite pores	$P(g) + \overset{EQ}{\rightleftharpoons} P(\text{ads.})$
Dehydrogenation of alkane on metallic sites	$P(\text{ads.}) \underset{k_{\text{H}}}{\overset{k_{\text{DH}}}{\rightleftharpoons}} O(\text{ads.}) + H_2$
Protonation of alkene on acid sites	$O(\text{ads.}) + H^+ \underset{k_{\text{mO}}}{\overset{k_{\text{Om}}}{\rightleftharpoons}} R_m^+$
Isomerization of R_m^+ on acid sites: charge transfer	$R_m^+ \underset{k_{qm}}{\overset{k_{mq}}{\rightleftharpoons}} R_q^+$
CH$_3$ group migration	$R_m^+ \underset{k_{rm}}{\overset{k_{mr}}{\rightleftharpoons}} R_r^+$
through intermediate cyclopropyl ions	$R_m^+ \underset{k_{nm}}{\overset{k_{mn}}{\rightleftharpoons}} R_u^+$
Cracking on R_m^+ on acid sites	$R_m^+ \overset{k_{cr}}{\rightleftharpoons} O''(\text{ads.}) + R_v^+$
Deprotonation of R_m^+ from acid sites: $h = m, g, I, u, v$; $O'' = $ isoalkene $(i - O)$ or resulted from cracking (O')	$R_h^+ \overset{k_{De}}{\rightleftharpoons} O''(\text{ads.}) + H^+$
Hydrogenation of isoalkenes on metallic sites	$i\text{-}O(\text{ads.}) + H_2 \overset{EQ}{\rightleftharpoons} LP(\text{ads.})$
Hydrogenation of alkenes resulted from cracking	$O'(\text{ads.}) + H_2 \overset{EQ}{\rightleftharpoons} P''(\text{ads.})$
Desorption of isoalkanes	$i\text{-}P(\text{ads.}) \overset{EQ}{\rightleftharpoons} i\text{-}P(g) + \text{zeolite}$
Desorption of alkanes resulted from cracking	$P''(\text{ads.}) \overset{EQ}{\rightleftharpoons} P''(g) + \text{zeolite}$

EQ = the reaction approaches equilibrium.
Source: Ref. 34.

The number of events can be determined on the basis of the structure of components or the symmetry changes when forming the activated complex [34].

The mechanism of reaction shown in Table 12.6 allows one to derive the reaction rate equations.

The rate of formation of alkanes can be written as follows:

$$r_p = k_H \theta_O \cdot p_{H_2} - k_{DH} \theta_p \tag{12.37}$$

where θ indicates surface concentrations.

The reactions of protonation of alkenes formed on the acidic sites (third reaction in Table 12.6) may be written as:

$$\left. \begin{array}{l} \theta_O + \theta_{H^+} \overset{k_{OS}}{\rightleftharpoons} \theta_{R^+ \text{sec}} \\[2ex] \theta_O + \theta_{H^+} \overset{k_{OT}}{\rightleftharpoons} \theta_{R^+ \text{tert}} \end{array} \right\} \tag{12.38}$$

where the only necessary distinction is that the ion formed is secondary or tertiary. In the reactions (12.38) the adsorption is competitive and it excludes the alkenes with five or less carbon atoms [34].

The radicals formed will decompose by bond cleavage in β position, as described by the general relationship:

$$\theta_{R_m^+} \overset{k_{CR}}{\rightleftharpoons} \theta_O + \theta_{R_i^+} \tag{12.39}$$

or they may deprotonize:

$$\theta_{R_m^+} \overset{k_{DE}}{\rightleftharpoons} \theta_0 + \theta_{H^+} \tag{12.40}$$

The rate of formation of a molecule of alkene may be written as:

$$r_O = k_{DH} \theta_p - k_H \theta_O p_{H_2} + k_{De} \theta_{R_{\text{sec}}^+} + k_{De} \theta_{R_{\text{tert}}^+} - (k_{OS} + k_{OT}) \theta_O \theta_{H^+} + k_{cr} \theta_{R_m^+} = 0 \tag{12.41}$$

Equalizing to zero results in the application of the theorem of stationary states. Combining with (12.37) results in:

$$r_p = k_{De} \theta_{R_{\text{sec}}^+} + k_{De} \theta_{R_{\text{tert}}^+} - (k_{OS} + k_{OT}) \theta_O \theta_{H^+} + k_{cr} \theta_{R_m^+} = 0 \tag{12.42}$$

The relative concentrations on the surface of R_{sec}^+ and R_m^+ ions can be determined using the expressions (12.38) and (12.39) and applying the theorem of stationary states, giving:

$$\theta_{R_{\text{sec}}^+} = \frac{k_{OS} \theta_0 \theta_{H^+}}{k_{cr}}$$

$$\theta_{R_{\text{tert}}^+} = \frac{k_{OT} \theta_0 \theta_{H^+}}{k_{cr}} \tag{12.43}$$

The relative concentration on the surface of alkenes θ_O can be expressed based on the equilibrium of the initial reaction of dehydrogenation:

$$K_D = \frac{\theta_O \cdot p_{H_2}}{\theta_p} \tag{12.44}$$

θ_{R+m}, which represents the total relative concentration on the surface of ions of carbenium, can be obtained through the application of the theorem of stationary states to the group of ions*:

$$r_{R_m^+} = (k_{OS} + k_{OT})\theta_O\theta_{H^+} + k_{cr}\theta_{R_m^+} - (k_{De} + k_{cr})\theta_{k_m^+} = 0 \tag{12.45}$$

The relative concentration on the surface of H^+ ions can be determined by the difference:

$$\theta_{H^+} = 1 - \sum \theta_{R_m^+} \tag{12.46}$$

The substitution of Eqs. (12.43)–(12.46) into (12.42) gives the rate of formation of alkanes as a function of the fraction of the surface θ_p occupied by them. This fraction of the surface can be expressed through the Langmuir adsorption:

$$\theta_p = \frac{b_p p_p}{1 + \sum b_i p_i} \tag{12.47}$$

In all these equations the relative concentrations on the surface, θ_i represent the concentration on the surface of the substance i relative to the saturation concentration of the acid sites θ_{sat}, or equivalently, the fraction of acid sites occupied by the substance i.

The use of the relative concentrations implies the addition of the saturation concentration θ_{sat} to the rate constants since the reaction rates r are expressed in kmol/kg catalyst hour:

$$k_i = k_i'\theta_{sat} \tag{12.48}$$

where:

k_i are the rate constants from the above equations

k_i' represents the product of elementary rate constants and the number of equivalent events

θ_{sat} is the total concentration of acid sites expressed in kmol adsorbed/kg catalyst

In this case, all rate constants will have the dimensions kmol/kg catalyst hour, except the rate constant k_H, which will be measured in kmol/kg catalyst hour bar.

The authors propose the extension of this system to other classes of hydrocarbons [61] and to take into account the sterical restrictions that might appear for some hydrocarbons when adsorbed on zeolites.

12.5 EFFECT OF PROCESS PARAMETERS

The analysis of the influence of the process parameters on hydrocracking has characteristics that are different from those of the processes examined in previous chapters. Although two types of catalytic sites intervene in the process just as in hydroisomerization, they promote different types of reactions in a feed of a much more complex composition. Therefore, it is necessary to examine the influence of the process parameters not only from the point of view of the catalyzed reaction

---

*The reactions of isomerization were not taken into account because they do not modify the final result.

but also for the change in structure taking place in the different types of hydrocarbons.

It is necessary to bear in mind also that hydrocracking involves a variety of processes that depend on the final products being sought. Such products can be jet fuel or diesel fuels with a high cetanic number and low freezing temperatures or oils with high viscosity index.

As will be indicated when specific commercial processes will be presented, in some cases additional factors need to be considered in order to assess the effect of process parameters.

12.5.1 Temperature

The increase in temperature accelerates the cracking reaction on acid sites and displaces the equilibrium of hydrogenation reactions towards dehydrogenation. As a consequence, above a certain temperature level the metallic sites become incapable of hydrogenating the alkenes produced by the cracking of acidic sites. Instead, the alkenes polymerize and form coke-like deposits, as during catalytic cracking.

The temperature level at which such phenomena can take place depends on the nature of the feed (variable cracking rate on the acidic sites). It also depends on the activity of the metallic sites, the pressure level, and the hydrogen/hydrocarbon molar ratio. All three influence the efficiency of the hydrogenation reaction. The upper limit of temperature depends therefore on the feed, on the catalyst, and on the process parameters and is situated in the range of 370–450°C.

The increase in temperature leads to a very significant increase in the conversion, especially for the highly active zeolite catalysts. Figure 12.15 presents the variation of conversion with temperature and with space velocity for two catalysts of AKZO-Chemie, KF742 used in the first stage of reaction and KF1010 used in the second stage. The examination of both plots indicates that a temperature increase of 20°C (i.e., from 400 to 420°C) induces an increase in conversion of $\sim 20\%$ for the first catalyst and of $\sim 27\%$ for the second. According to the second plot, at a temperature 20°C higher, the same conversion may be obtained at a three times higher space velocity.

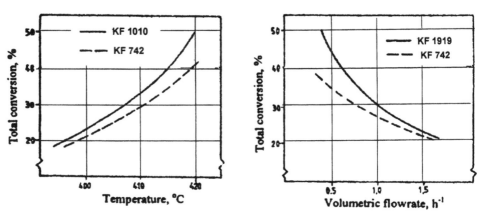

Figure 12.15 The influence of temperature and space velocity on conversion. (From Ref. 63.)

A similar conclusion was drawn when examining Figure 12.16 based on own data obtained [57] for hydrocracking of gas oil on a NiW/alumosilicate catalyst.

The increase in space velocity that can be gained through the increase in reaction temperature is very effective in reducing reactor size and investment costs. However, this increase is limited by the intensification of coke deposition on the catalyst, which shortens the length of the cycle, as well as by the quality of the products obtained.

For the alkanes and alkyl chains, the role of metallic sites is limited to their dehydrogenation to alkenes, which initiates cracking on the acidic sites, and also to the hydrogenation of the alkenes contained in the final product.

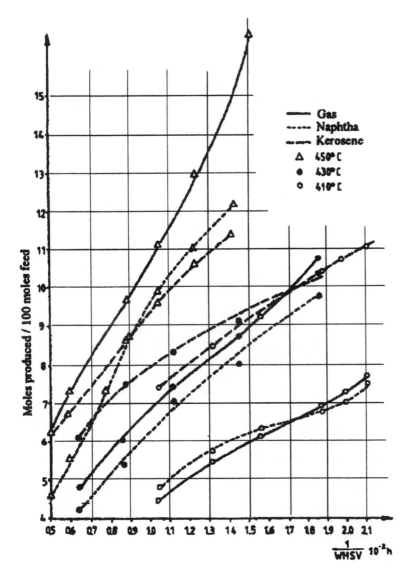

Figure 12.16 The influence of temperature on the conversion of a straight-run gas oil fraction. (From Ref. 57.)

In the case of aromatic structures the process is much more complex. These structures must be first hydrogenated and dehydrocyclized before they can be cracked on the acidic sites.

Thus, for alkanes, the cracking is preceded by a dehydrogenation favored by high temperatures, while for the aromatic structures it is preceded by the hydrogenation, which is favored by lower temperatures.

For this reason, the use of too high temperatures limits the hydrocracking of aromatic structures, which in most cases will influence the composition of the final products unfavorably. Thus, the presence of aromatic structures decreases the viscosity index of lubricating oils, the cetane number of the gas oils, and the height of the smokeless flame for jet fuels. All of these characteristics are essential for the quality of the above mentioned products.

The restrictions for protecting the environment limit the content of aromatic hydrocarbons in gasoline. Consequently, the high octane components with low content of aromatic hydrocarbons obtained in hydrocracking become very valuable.

In conclusion, for each hydrocracking process, the temperature should not exceed a certain maximum determined by the required quality of the target product.

12.5.2 Pressure

The pressure influences significantly the equilibrium of the dehydrogenation—hydrogenation reactions that take place on the metallic sites. Increasing pressure, also favors absorption on acid sites; maintaining the alkenes in adsorbed state enhances the formation of coke.

Higher pressure favors the hydrogenation of the alkenes produced on the acidic sites. Therefore, an increase in pressure releases these sites and hence, increases the overall rate of reaction. This effect is illustrated in Figure 12.17. However, a decrease in the conversion to gases can be observed, and it is probably due to the successive character of the reactions.

The increase in conversion is not monotonous and shows a maximum at pressures of 125–150 bar. The decrease in conversion at higher pressures may be explained by the retarding effect of a higher partial pressure of hydrogen on the reaction of dehydrogenation, which initiates the cracking on the acidic sites.

When heavy fractions are used as feed, a system of gas–liquid film–catalyst is produced in the reactor. Higher pressure increases the concentration of hydrogen dissolved in the liquid film and so it contributes to a more efficient reaction.

The increase in pressure which, for a given molar ratio H_2/feed corresponds to an increase in the partial pressure of hydrogen, will produce an increase of the conversion of aromatic structures to saturated products. Consequently, the quality will increase for the principal products of the process: jet fuel, diesel fuel, and oils with very high viscosity index. In the latter case, the aromatic structures are mostly bi-, tri-cyclic, or even naphthenic–aromatic structures. They are harder to hydrogenate and the process requires higher pressures than in the case of lighter distillates.

In addition, the rise in pressure leads to an extended cycle length between two regenerations of the catalyst, owing to a more active hydrogenation of the alkenes that are coke precursors.

Figure 12.18 illustrates the effect of pressure on the hydrogenation of aromatics and extending the cycle length of the catalyst.

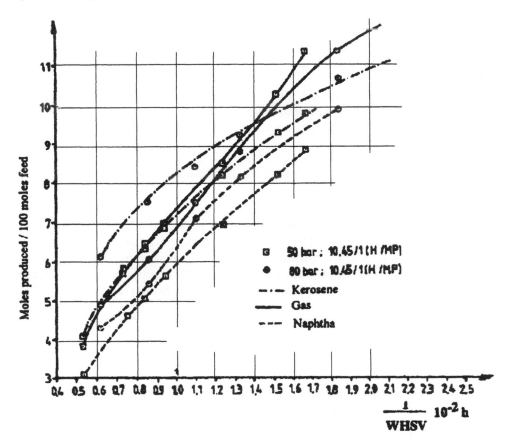

Figure 12.17 The influence of pressure on conversion to products at 450°C for a straight-run gas oil. (From Ref. 57.)

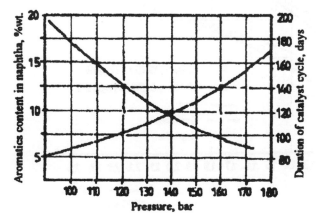

Figure 12.18 The effect of pressure on the conversion of aromatic hydrocarbons and on the cycle length (feed: vacuum distillate + residual oil; catalyst NiW on alumosilicate; $t = 450°C$).

12.5.3 Hydrogen to Feed Ratio

The increase in the molar ratio hydrogen/hydrocarbons leads to an increase in the partial pressure of hydrogen, the effects of which were examined in the previous section.

The distinguishing aspect is that higher pressures will cause the components present in the vapor phase to spend a longer time in the reaction zone, while the increase in the molar ratio diminishes the time spent by all reaction participants in the reaction zone. Hence, the effect on the hydrocracking of a gas oil fraction will be an increase in the conversion to kerosene and a decrease in the conversion to gases as shown in Figure 12.19 [57].

The situation is somewhat different in the case of heavy feedstocks which remain in liquid phase in the reaction zone. In this case, in the absence of hydrocarbons in the vapor phase, the increase in the molar ratio will not increase the partial pressure of hydrogen. The effect of the molar ratio shows that only after hydrocracking reactions, are hydrocarbons produced that remain in vapor phase under the conditions of temperature and pressure of the reaction zone. This explains why in some systems, hydrogen is introduced only partially at the start of the

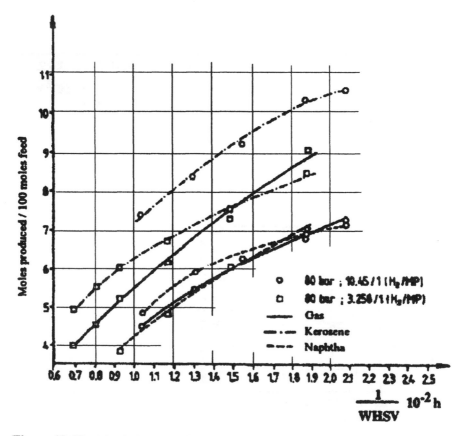

Figure 12.19 The influence of the molar ratio hydrogen/feed on hydrocracking of a straight-run gas oil at 410°C. (From Ref. 57.)

process, the rest being introduced later in the first reactor or, in the systems with two reactors*, between the first and the second reactor.

The increase in partial pressure of the hydrogen through the rise in pressure in the system or through the increase in the molar ratio has economic consequences. The increase in pressure entails higher investment for the reaction system and higher processing costs, related to the compression of hydrogen. The increase in molar ratio is linked to an increase in the reactor volume, and also its cost, more specifically expenditures related to hydrogen recycling.

From a process point of view, there is a certain range of pressure–molar ratio conditions that ensure a suitable partial pressure of hydrogen. Within this range, the choice of the operating conditions must be made for each case individually, depending on the local conditions and the economic constraints.

12.5.4 The Catalyst

From a process point of view, the ratio between the number of metallic and acidic sites is the most important factor concerning the catalyst. A study carried out on n-heptane by Gianetto et al. [64–67] led to the results presented in Table 12.7, describing the influence of this ratio on the main parameters of the process.

The most interesting aspect of this study is the way in which the reaction scheme changes, depending on the n_H/n_A ratio and the fact that this ratio must have a relatively high value for the hydrocracking process to produce satisfactory results.

Other issues regarding the applied types of catalysts, their evolution, structure, and their choice depending on the type of process and feed were discussed in Section 12.2.

Table 12.7 Influence of the Ratio Between the Number of Metallic and Acidic Sites of the Catalyst, n_H/n_A, on the Outcome of Hydrocracking

n_H/n_A	Small	Moderate	High
Activity of acid site	small	moderate	high
Stability	low	maximum	maximum
		Selectivity	
Reaction scheme	$nC_7 \rightleftarrows \begin{array}{l} M \\ B \\ C \end{array}$	$nC_7 \rightleftarrows M + B \rightarrow C$	$nC_7 \rightleftarrows M \rightleftarrows B \rightarrow C$
I/C	reduced	average	high
M/B	reduced	moderate	high
i/n	high	moderate	reduced

Note: I - isomerization products, C - cracking products, M - isomers with only one lateral group, B - isomers with several lateral groups, i/n - iso/normal hydrocarbons ratio in cracking products.

---

*The hydrocracking being exothermal, the introduced hydrogen has in many cases also the role of cooling agent.

Taking into account the influence of internal diffusion [57], the catalyst granules are mostly cylindrical or of specific shapes that shorten the access to the center of the particle.

12.5.5 Feedstock

The flexibility of hydrocracking allows the processing of varied types of feeds. However, for each feed the severity of the process and the amount of hydrogen consumed have to be set accordingly.

A higher content of aromatic hydrocarbons requires higher pressures and higher hydrogen/feed ratio, the lowest possible temperatures, and a higher hydrogen consumption than a nonaromatic feed in order to obtain the same final products.

The higher the difference between the molecular mass (or the average boiling temperature) of the feed and of the final products, the higher will be the consumption of hydrogen and the severity of the process.

When assessing the consumption of hydrogen of a commercial unit it is necessary to differentiate between the quantity of the reacting hydrogen that influences the modification of the structure and composition of the feed, and the quantities of hydrogen that is lost in streams from which it is not economically recoverable.

An approximate evaluation of the hydrogen bound by means of chemical reactions can be obtained with the aid of Figure 12.20 [63], which allows the estimation of hydrogen content of various products.

A more accurate calculation can be made using Figure 12.21 [68]. This helps determine the content of hydrogen more accurately as a function of the average boiling point temperature and of the Characterization Factor.

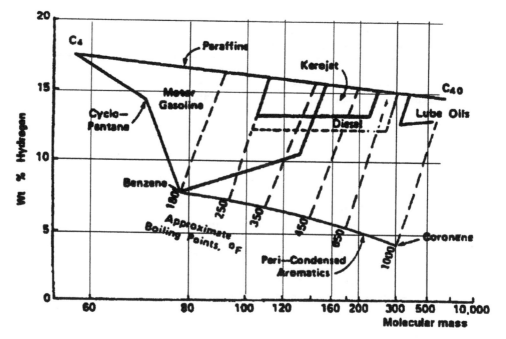

Figure 12.20 Hydrogen content related to the molecular mass. (From Ref. 63.)

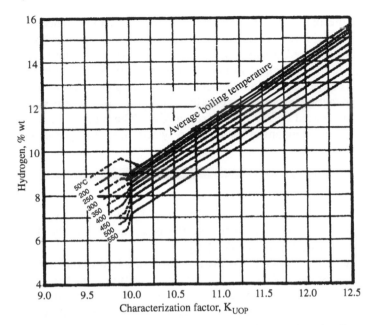

Figure 12.21   The hydrogen content of a fraction of crude oil as a function of the average boiling point and the Characterization Factor. (From Ref. 68.)

When evaluating the required quantity of hydrogen it is also necessary to take into account the hydrogen used for removing the heteroatoms. These quantities can be estimated using the following data [69]:

6–7 Nm^3 H_2/m^3 for removing 1% S
20–21 Nm^3 H_2/m^3 for removing 1% N
12–13 Nm^3 H_2/m^3 for removing 1% O

12.6 COMMERCIAL HYDROCRACKING OF DISTILLATES

Initially, the process was used for hydrocracking straight-run gas oil fractions. Subsequently it was used for processing vacuum distillates, which also today constitutes the main feed.

Presently, also other fractions resulting from the thermocatalytic processes were added to the hydrocracking feed, e.g., gas oil fractions resulting from coking, viscosity breaking and sometimes from catalytic cracking, soft waxes, and extracts from the production of lubricating oils. It is necessary to consider the large amount of hydrogen needed by fractions with a pronounced aromatic character whenever additions to the feed are considered, such as: gas oil fractions from catalytic cracking and from fluidized-bed coking and especially the aromatic extracts from solvent extraction of oils. A calculation of the expenses related to high hydrogen consumption might show that the use of such feeds is not justified economically.

Recently [1], UOP proposed a process for hydrocracking of heavy naphthas called *I. Forming*, with the purpose of producing especially *i*-butane for alkylation or for the production of MTBE after hydrogenation.

Various hydrocracking plants differ not only in the terms of their concept of the reaction system, but also with respect to the degree to which they accomplish the preliminary hydrofining of the feed. From this point of view, there are units that have a dedicated reactor for hydrofining, situated before the one used for hydrocracking, while other units have an additional layer of hydrofining catalyst placed in the upper part of the hydrocracking reactor. Finally, there are cases where standalone hydrofining units are placed before the hydrocracking. The manner in which this problem is dealt with is independent of the reaction system used for hydrocracking.

12.6.1   Mild Hydrocracking

The process of hydrocracking at medium pressures, known also as *Mild Hydrocracking*, appeared in 1980 and made use of the equipment available from the hydrodesulfurization of gas oil fractions [70,102].

This idea was spurred on by the fact that the flowsheet of a hydrocracking unit of this type is practically identical to that of a hydrodesulfurization unit [63]. In the first place it seemed necessary only to replace the catalyst and to adjust the fractionating tower for collecting an additional side fraction. These modifications are indicated with dotted lines on the flow diagram in Figure 12.22.

Nevertheless if hydrodesulfurization plants are to be used for hydrocracking, the following areas require revamping:

1.   *The reactor.* Hydrocracking is highly exothermic; therefore interval cooling through hydrogen injections between the catalyst beds is necessary. If the hydrodesulfurization or hydrotreating plant does not have such devices, they have to be installed.

2.   *Wash water*, introduced before the high pressure separator, is necessary in order to dissolve the ammonium salts resulting from denitration. Otherwise, these

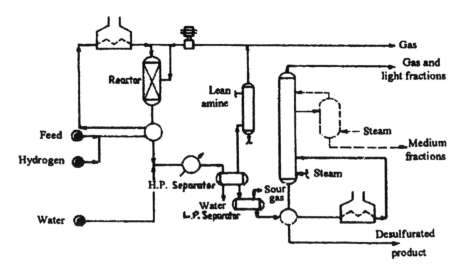

Figure 12.22   Flow diagram of a unit for the hydrodesulfuration of gas oil and the modifications (indicated by dotted line) required for converting to a middle hydrocracking unit.

salts will form deposits on the heat exchanger tubes, which will reduce the heat transfer, increasing the pressure drops and eventually plugging these tubes. If the hydrodesulfurization unit does not have the necessary devices for handling the wash water, it should be added.

3. *The fractionating tower* in a hydrodesulfurization unit has only the role of stripping the final product. In the case of hydrocracking, this tower with the equipment following it must ensure the recovery of the main products (jet fuel and gas oil) from the process. Figure 12.22 presents only the simplest case—the separation of a single product. This tower will be revamped or replaced with one designed adequately.

4. *Hydrogen consumption.* The significant increase in hydrogen consumption requires additional means for its production and compression.

In spite all of these alterations, the final cost is significantly lower than for building a new plant.

The comparison between process parameters in hydrodesulfurization of vacuum distillates to mild hydrocracking and at moderate and high pressures can be found in Table 12.8.

This table shows that mild hydrocracking operates at lower space velocity and at higher temperature than hydrodesulfurization. These conditions and the presence of acidic sites increase the danger of coking of the catalyst, thus shortening the length of the cycles.

The modifying of hydrodesulfurization units requires great care in the selection of the catalyst. Two systems were applied for this purpose:

1. The system with two catalysts, one destined for hydrofining and more importantly for denitration and the other destined for hydrocracking. For example: the catalysts HND-80 and NiW/A1-SC-R2x from American Cyanamid [71], the NiMo/zeolite catalyst from Katalystics [72] or KF 165 with KF 742 and KF 1011 with KF 1012 from Akzo.

2. Systems with a single catalyst, which fulfils both functions, provided the contents in N and S are not too high. Such catalysts are: UOP-MHC 10 [73], Shell S-

Table 12.8 Comparison Between Process Parameters for Hydrodesulfurization and Hydrocracking

	Hydrodesulfurization		Hydrocracking	
	Mild distillates	Vacuum distillate	Moderate	At high pressure
Pressure, bar	27.5–55	34.5	<100	>100
Volumetric flowrate, h^{-1}	2–4	1–2	0.4–1	0.4–1
Reactor average temperature, °C	315–370	360–415	385–425	315–400
H_2/feed ratio, m^3/m^3	85–200	170–340	500–850	675–1350
Hydrogen consumption, m^3/m^3	15–50	50–85	65–170	250–600
Conversions, %				
to naphtha	1	1	5–15	100
to mid products	—	10–20	20–50	—

454 (NiW) and S-444 (NiMo) [74], Leuna Werke K-8205 (NiMo) and Akzo KF 1014 [75].

The French Institute of Petroleum developed a system of two catalysts for hydrocracking, HR360 (NiMo) intended for a deep denitration and desulfurization of the feed and HYC-642 which is a zeolitic catalyst for hydrocracking with a reduced acidic function, to avoid overcracking [6,7].

The successes achieved through modifying existing units for gas oil hydro-desulfuration or hydrotreating [102] made it possible to start building units specifically for the mild hydrocracking process. The process flow diagram for a unit that produces both jet fuel and diesel fuel is shown in Figure 12.23. There are also flowsheets in which the diesel fuel is obtained from the main fractionation tower, thus eliminating the vacuum system, as well as other flowsheets in which only jet fuel is obtained. The fraction from the bottom of the column is entirely recycled [63].

The versatility of the hydrocracking units results in part from the possibility of varying the proportion of jet fuel to Diesel fuel simply by changing the temperature of separation between the two fractions (see Figure 12.24).

Recycling allows the complete conversion of feed to lighter products [76], as shown in Figure 12.23. However, it presents some specific difficulties in the case of mild hydrocracking units.

The hydrogenating activity in these units is less intense than in those operating at higher pressure ($\geq$ 100 bar); the condensed polycyclic aromatics will become concentrated in the recycle. When the solubility limit is exceeded, they form a separate phase, causing other similar components as well as heavy paraffins to separate. These form deposits, which decrease the heat exchange and can plug the lines.

To prevent such phenomena, the following measures are applied, as shown in Figure 12.25 [24]:

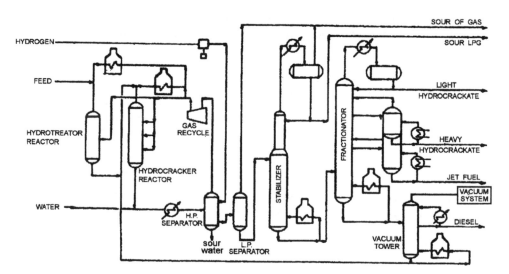

Figure 12.23 Hydrocracking unit with a single stage of reaction (hydrocracking at moderate or high pressure).

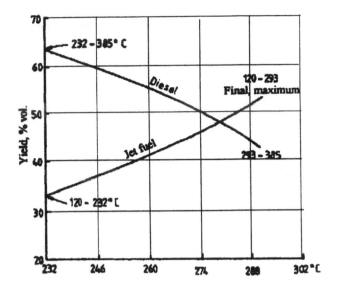

Figure 12.24 Modifying the proportion jet/diesel fuel by changing the separation temperature between the two fractions (Chevron Research Co.).

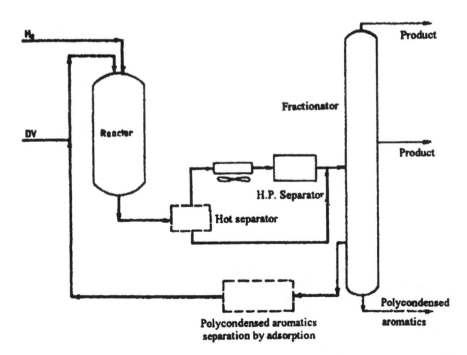

Figure 12.25 Device for eliminating polycondensed aromatics

Separation of polycondensed aromatics from the bottom of the fractionating
tower

Their separation through adsorption from the recycle flux

Introduction of a hot separator or modifying the existing flowsheet so that the
polycondensed aromatics bypass the cooling system

The process conditions of this type of unit can vary, depending on the design
specifications; some typical data are outlined in Table 12.9.

12.6.2 High Pressure Processes

Hydrocracking at high pressure ($\geq$100 bar) can be accomplished in two ways: with
one or two reaction stages [6,7,63,82,83].

The flow diagram of a unit with one stage of reaction is the same as that given
in Figure 12.23, while the flow diagram of a unit with two stages of reaction is shown
in Figure 12.26.

Although the hydrocracking units operating at pressures above 100 bar represent higher investments, they offer the advantage of permitting a higher conversion
of the aromatic hydrocarbons. This can be important when the products have high
quality specifications or in the case of processing feeds with a higher content of
aromatic hydrocarbons, especially polycyclic.

In two-stage units, the first stage operates without recycling, up to a conversion
of approximately 50% and employs a catalyst that is rather insensitive to poisoning.
As an example, the I.F.P. process with two reaction stages allows the conversion of
raw materials containing up to 2500 ppm nitrogen [6].

The second stage operates with recycling and since most of the poisons in the
feed have been removed, it can use catalysts with noble metals such as Pt or Pd.

Table 12.9 Operating Parameters of the Various Hydrocracking Plants

Licence	Source	Press., bar	Temp., °C	Vol. hourly space velocity, h^{-1}	Contact ratio H_2/feed, M^3/m^3	Yield to medium fractions, %
I.F.P. and BASF (Germany)	[77]	>31	$\geq$414	$\leq$2	>250	$\leq$40
Lummus Crest Corp (Germany)	[77]	42–84	400–450	0.4–1.2	—	25–48
Chevron Research (U.S.)	[77]	42–70	—	0.7–1.3	—	20–40
UOP (U.S.)	[78,79]	46–75	370–454	1–2	267	20–50
"Mobil Oil" (U.S.)	[71]	100	385–425	0.4–1	500–850	30–50
"Exxon" (U.S.)	[80]	54–70	—	—	—	20–35
Linde AG (Germany)	[81]	80–100	360–400	—	—	20–70
"Leuna Werke" (Germany)	[78]	75	360	1.5	—	15
Unocal (U.S.)	[74]	52	—	1	500	25

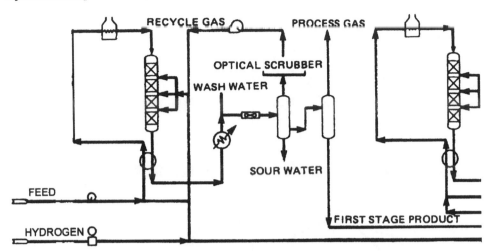

Figure 12.26 The flow diagram of a hydrocracking unit with two reaction stages. 1 - the first stage reactor. 2 - The second stage reactor.

This ensures lower temperatures in the reactor, creating thermodynamic conditions that are more favorable to the advanced hydrogenation of the aromatic hydro-carbons.

The two-stage system is used for improving product quality as well as for large capacity units (over 5500 m^3/day), when just one single reaction zone has insufficient capacity. In such cases, it is more economical to build one unit with two stages of reaction capable of ensuring the necessary processing capacity, rather than two one-stage units.

12.6.3 Yields and Product Quality

The industrial units already built are highly diverse. It is hard to generalize about the quality and yields of products obtained given the wide range of possibilities offered by the process with respect to the operating conditions and the variety of useable feeds [103]. Therefore, the following data have only orientation value.

Figure 12.27 [63] presents this type of general data. Some discrepancies that appear in this diagram result from the fact that the feed had a content of 13% heavy gas oil, which required a correction to this set of results.

A typical example is a feed consisting of a mixture of 70% heavy vacuum gas oil and 30% catalytic cracking gas oil, with the characteristics:

Sulfur wt %	2.35
Nitrogen, ppm	1.400
ASTM, °C initial	150
ASTM, °C 50 wt %	433
ASTM, °C final	605
C_7-Asphalthenes	< 100

When this feed is processed in a one-stage reaction hydrocracking system, the following product yields are obtained [7]:

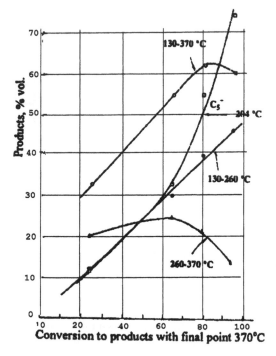

Figure 12.27 Products yields vs. conversion.

$H_2S + HN_3$	2.36%
$C_1 + C_2$	0.27%
$C_3 + C_4$	2.33%
Light gasoline	8.20%
Heavy naphtha	10.19%
Jet fuel	39.60%
Diesel fuel	38.15%
Residue	2.0%

Some characteristics of the main products are presented below:

	Jet fuel	Diesel fuel
Density	0.795	0.825
Sulfur, ppm	< 5	< 5
Nitrogen, ppm	< 5	< 5
Aromatics, % vol	10	12
Melting point, °C	< −60	< −15
Smoke point, mm	25	—
Cetane number	–	60

Both light gasoline and heavy naphtha contained less than 1 ppm sulfur and nitrogen. The light gasoline had an octane number of 80 Research.

The reactor pressure plays a major role in determining the quality of products. For a constant conversion of 70% and reactor pressure ranges between 30 and 100 bar, the product quality is given in Table 12.10 [72]. This table shows that the effect of pressure becomes significant especially when it exceeds 50 bar.

Table 12.10   Quality of Products as a Function of
Reactor Pressure for a Conversion of 70%

Products and specifications	Pressure, bar			
	30	50	70	100
Naphtha				
density, kg/m^3	780	750	750	710
S, ppm	300	150	50	5
N, ppm	50	5	2	—
aromatics, % vol.	30	25		3–5
Jet fuel				
density, kg/m^3	850	830	810	780
smoke point, mm	12	14	17	25–30
Light gas oil				
density, kg/m^3	880	880	850	800
S, ppm	200	100	50	—
N, ppm	100	50	10	—
cetane number	37–42	40–45	45–55	55–60
Hydrofined heavy fraction				
density, kg/m^3	900	833	880	870
S, ppm	500	300	70	20
N, ppm	500	200	15	5
Conradson carbon, %	0.10	0.07	0.07	0.00

A method to calculate the conversion and other important operating indicators was recently proposed [104] for two-stage high-pressure hydrocracking. The method allows the evaluation of the conversion for each of the stages.

12.6.4   Process Economics

Results of a thorough analysis of the economic aspects of the various hydrocracking processes were reported by the French Institute of Petroleum [6].

Calculations were performed for a heavy distillate resulting from a Middle Eastern crude oil, with the following main characteristics:

Density at 15°C	0.932
Sulfur wt %	2.95
Nitrogen, ppm	840
Viscosity at 100°C, cSt	11.1
ASTM distillation, °C	
5 wt %	405
50 wt %	485
95 wt %	565

Four types of units and operation modes were considered:

1. One reaction step without recycling, 90% conversion.
2. One reaction step with recycling, until the liquid fraction was completely processed, 70% conversion per pass.

3. Two reaction steps, with total conversion. In each reaction step, the conversion per pass was 50%, in order to favor the production of middle distillates.

4. Two reaction steps with an overall conversion of 85%. The conversion was 50% in the first reaction step and 70% in the second reaction step.

In all cases mentioned the plant capacity was 4770 m^3/day (30,000 bbl/day). The reactor pressure was 152 bar, and the length of the cycle was 3 years. All other operating parameters, yields, and main characteristics of the products are given in Table 12.11.

It should be noted that the reactors represented 30–40% of the total investment. The cost of hydrogen represents between 70% and 76% of the operation cost.

The investments are similar to those indicated by other constructing companies. For example, Chevron Research and Technology Co. announced a cost of $2800/bbl day [84,103] for a unit of the same capacity built in the U.S., and Unocal Corp. and UOP announced costs in the range of $2000–4000/bbl day without indicating the plant capacity [85].

Table 12.11 Operating Parameters, Yields, Products, Characteristics, and Economic Data

	Options			
	A	B	C	D
Process parameters				
catalyst volume, m^3	360	430	370	350
compressor recycling, m^3/h	280,000	280,000	470,000	410,000
reactor I capacity, m^3/day	4770	6840	4770	4770
reactor II capacity, m^3/day	—	—	4770	2380
distillation capacity, m^3/day	4770	6840	9540	7150
conversion in reactor I, % vol	80	70	50	50
conversion in reactor II, % vol	—	—	50	70
Yields, % wt				
$NH_3 + H_2S$	3.24	3.24	3.24	3.24
$C_1 + C_2$	0.40	0.40	0.30	0.25
$C_3 + C_4$	3.00	2.50	2.21	1.55
gasoline	9.00	8.00	7.50	4.50
naphtha	11.00	12.00	9.50	7.50
kerosene	38.50	39.00	33.70	30.00
Diesel	27.41	34.41	46.05	36.81
residue	10.00	3.00	0.00	18.50
Total	102.50	102.55	102.55	102.55
H_2 chemically bound, wt %	2.55	2.55	2.40	2.35
H_2 dissolved	0.20	0.28	0.40	0.29
Total	2.75	2.83	2.90	2.64
Quality of products				
Gasoline				
density at 15°C	0.665	0.665	0.666	0.670
RON, unethylated	80	80	80	79
Naphtha				
density at 15°C	0.743	0.745	0.741	0.741

Table 12.11 Continued

	Options			
	A	B	C	D
P/N/A, wt %	40/56/4	42/54/4	39/56/5	38/55/7
Kerosene				
density at 15°C	0.800	0.798	0.802	0.803
smoke point, mm	< 25	> 25	> 25	> 25
flammability, °C	> 40	> 40	> 40	> 40
freezing point, °C	< −60	< −60	< −60	< −60
Gas oil				
density at 15°C	0.826	0.828	0.837	0.832
freezing point, °C	−12	−12	−12	−12
cetane number	62	62	59	59
aromatics, wt %	< 10	< 10	< 10	< 10
sulfur, ppm	< 20	< 20	< 20	< 20
Residue				
density at 15°C	0.830	0.832		0.835
viscosity at 100°C, cSt	5.0	5.3		5.5
V.I. after dewaxing	125–130	120–125		120
Investment, U.S.$ × 10^6	105	124	147	122
Catalysts, U.S.$ × 10^6/year	0.83	1.07	0.83	0.78
Utilities, U.S.$ × 10^6/year	10.56	12.26	15.32	14.29
Hydrogen, U.S.$ × 10^6/year	36.08	37.14	38.06	36.64
Operating cost, U.S.$/bbl	6.4	7.0	7.8	6.9
Operating cost, U.S.$/m^3	40.25	44.03	49.05	43.4
Investment, U.S.$/bbl × day	3500	4130	4900	4070

12.7 RESIDUE HYDROCRACKING

The presently used systems for hydrocracking residues in ebullated bed reactors are a development of Cities Service and HRI. They were commercialized almost at the same time by HRI, as a process called H-Oil and by Lummus and Cities Service, as a process called LC-Fining [86–92].

As of end of 2002, worldwide there are seven commercial H-Oil units (presently licensed by Axens) and four LC-Fining units with three more in design and construction (presently licensed jointly by Chevron and Lummus) [92a].

The LC-Fining process flow diagram is given in Figure 12.28 and the H-Oil process flow diagram in Figure 12.29. The main difference between these two processes is in the reactors.

The LC-Fining reactor is a vertical cylinder 4.5 m diameter and 30 m height, which corresponds to a volume of about 250 m^3. The thickness of the walls is more than 30 cm and its weight exceeds 1000 tons. The details of the reactor are given in Figure 12.30. The liquid is recycled through the reaction zone with the help of a pump placed inside the reactor. Liquid feed and hydrogen makeup are introduced in the bottom of the reactor. The recycle flow maintains the catalyst in a moving state, expanded to a certain level, and rigorously controlled with the aid of density detec-

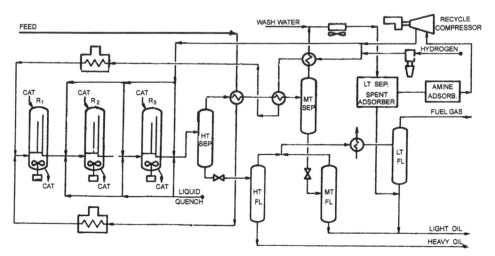

Figure 12.28 LC-Fining process flow for hydrocracking of residues in ebullated bed. SEP – separator; FL – vaporiser; HT, MT, LT – high, medium and low temperature, respectively.

tors. These detectors determine the flowrate of the recycle pump. The reactor operates as a well-stirred tank reactor. Consequently, in order to obtain the desired conversion the unit has three reactors through which the feed flows in series and the hydrogen is introduced separately. Since the process is exothermic, cooling between the reaction steps is provided by controlling the temperature of hydrogen. If necessary, additional cooling liquid is used.

The processing capacity of such a group of three reactors is about one million tons per year.

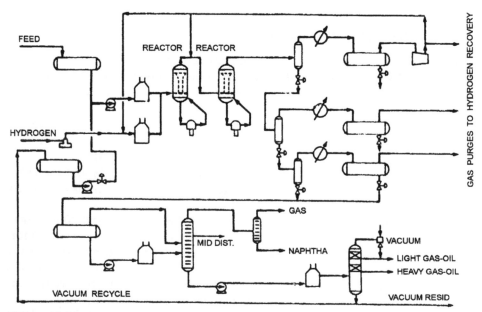

Figure 12.29 H-Oil unit for hydrocracking of residues in ebullated bed.

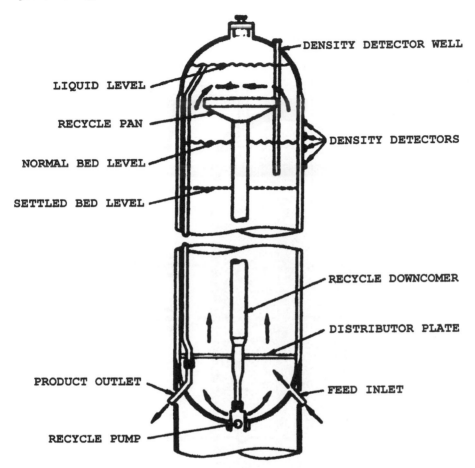

LIQUID LEVEL
RECYCLE PAN
NORMAL BED LEVEL
SETTLED BED LEVEL

DENSITY DETECTOR WELL
DENSITY DETECTORS
RECYCLE DOWNCOMER
DISTRIBUTOR PLATE

PRODUCT OUTLET
RECYCLE PUMP

FEED INLET

Figure 12.30 The reactor of a LC Fining unit for hydrocracking of residues.

If the removal of metals and sulfur from the residue is one of the objectives of the process, then two catalysts are used. The catalyst for metal removal is introduced in the first reactor and a gradual replacement is performed during operation by injecting fresh catalyst and extracting used catalyst (the dedicated lines are not shown in Figure 12.28). The fresh catalyst for desulfurization is introduced in the third reactor, where it gradually passes to the second reactor and finally is sent for reconditioning [86].

The intention to replace this catalytic system was expressed as early as 1987 [86]. Later publications (1990–1994) concerned with obtaining conversions of 95% mentioned the replacement of the catalyst used in the second and the third steps with a hydrocracking catalyst.

The details of the reactor used for the H-Oil plant are shown in Figure 12.31. The main difference between this reactor and the one shown in Figure 12.30 is the placement of the recycle pump outside the reactor. The inside wall of the reactor is protected by a 15 cm thick refractive layer.

In both types of reactors, the intense mixing of the catalyst leads to its attrition and consequently to a relatively high consumption of catalyst.

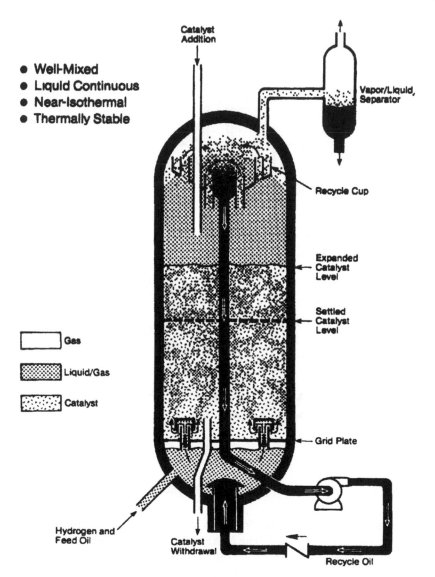

- Well-Mixed
- Liquid Continuous
- Near-Isothermal
- Thermally Stable

Gas

Liquid/Gas

Catalyst

Figure 12.31   The reactor of a H-Oil unit for hydrocracking of residues.

Both processes were exploited at moderate conversions of 55% to 60% when they appeared in 1985–1987. Conversions of 75% to 95% became feasible around 1990, probably due to the use of catalysts with zeolitic support. The only indication published by UOP [89] refers to a NiMo catalyst without clear specification as to the nature of the catalytic support.

The technological conditions of the two processes are given in Table 12.12. The yields of the LC-Fining process are illustrated by data [87] related to the processing of a vacuum residue prepared from a mixture of light and heavy crude oil from the Middle East (see Table 12.13). The yields of the H-Oil process is illustrated by similar data [90] related to the processing of a vacuum residue obtained from a medium crude oil from the Middle East (see Table 12.14).

Table 12.12 Operating Conditions for Residues Hydrocracking

	LC Fining [87]	H-Oil [90]
Reactor temperature, °C	385–450	410–450
Reactor pressure, bar	95–240	—
Hydrogen partial pressure, bar	70–185	70–170
Space velocity, h^{-1}	0.1–0.6	0.1–0.9
Conversion, %	40–97	40–95
Desulfurization, %	60–90	84–91
Demetallation, %	50–98	65–90
Reduction in Conradson carbon, %	35–80	63–82

The economic data show an investment of U.S.$4100–U.S.$6200 per barrel day for the LC-Fining process and U.S.$,3900–U.S.$5,300 per barrel day for the H-Oil process. These values are somewhat higher than for the hydrocracking of distillates.

The utilities consumptions can be estimated from the following data reported per volume (m^3) processed (see Table 12.15).

12.8 PROCESSES USING SLURRY PHASE REACTORS

A process from this category is *CANMET* built by Partec Lavalin, Petro-Canada, and Unocal for the hydrogenation of oil extracted from tar sands bitumen. The technology was then extended to the processing of heavy residues [93–95].

The catalyst (or "additive") is introduced into the feedstock as a suspension, which is not recovered and remains in the final residue. A low value mixture of carbon and iron [94] is used as catalyst and added to the feedstock in the proportion of approximately 10–12 kg/m^3.

The process flow diagram, including the hydrotreatment of distillates, is shown in Figure 12.32 [95]. This system is used for the processing of both atmospheric and vacuum residues, with 50 to 100% vol distilled above 520°C and with a sulfur content of up to 4–6% and of nitrogen up to 0.9%. The result of this process is a 90% vol distillate, the rest being a residue (pitch), having a softening point of 200°C, a viscosity of 400 cSt at 260°C and 10% ash, calculated without including the additive. The only use for this residue is to be crushed and burned, preferably in a

Table 12.13 LC-Fining Yields

Conversion (<550°C), % vol	60	75	95
Yields vol %			
C_4	2.35	3.57	5.53
C_5–175°C	12.60	18.25	23.86
175–370°C	30.62	42.65	64.81
370–550°C	21.46	19.32	11.92
> 550°C	40.00	25.00	5.00

Feed characteristics: density at 15°C = 1.040, S = 4.98%,
Ni = 39 ppm, V = 128 ppm.

Table 12.14 H-Oil Yields

Conversion (< 550°C), %vol	60	95
Yields		
H$_2$S + NH$_3$ wt%	5.6	5.1
C$_3$ – C$_4$ wt %	3.1	6.7
C$_4$ – 220°C vol %	17.6	23.8
204 – 370° vol %	22.1	36.5
370 – 565°C vol %	34.0	37.1
> 565°C vol %	33.2	9.5
Hydrogen chemically bound Nm^3/m^3	6.35	8.37

Feed characteristics: density at 15°C = 1.0374, S = 5.4%, Ni + V = 128 ppm.

fluidized bed, with the addition of lime, which binds approximately 90% of the contained sulfur as CaSO$_4$.

This process does not appear to find much interest, especially because of the difficulties encountered during the use of the pitch residue. Other processes such as coking are preferred for processing very heavy residues.

The process can hardly be considered to be a hydrocracking process as defined in this chapter, since there is very little catalytic hydrogenation and a lack of participation of the catalyst to the cracking reactions. Consequently, assigning the name of hydrocracking to this process by its authors is debatable.

The hydrocracking of heavy crudes and residual oils in slurry phase reactors was proposed by VEBA OEL [105]. Conversions of the residue of up to 95% are obtained at temperatures between 440 and 500°C. The distillates are sent to a fixed bed catalytic reactor, which operates at the same pressure as the first reactor.

Feedstock:
gravity, °API	−3 to 14
sulfur, wt%	0.7 to 7
metals (Ni, V), ppm	up to 2,180
asphaltenes, wt%	2 to 80

Yields:
naphtha < 180°C, wt%	15–30
middle distillates, wt%	35–40
vacuum gas oil, > 350 °C, wt%	16–50

Table 12.15 Utilities Consumption for Residue Hydrocracking

	LC-Fining, conversion			
	60	75	90	H-Oil
Fuel, 10^3 kJ	417	463	588	530–760
Electricity, kWh	87.4	103.8	144	56–107
Cooling water, m^3	3.91	3.92	5.95	2.4–5.7
Steam (produced), kg	197	277	614	—
Cost of catalyst, U.S.$	—	—	—	3.15–12.6

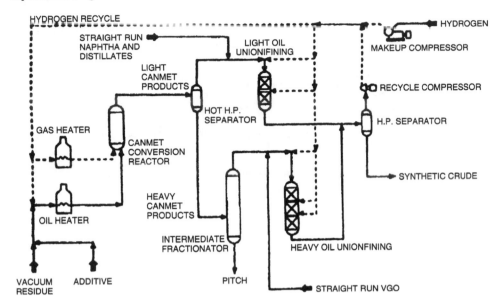

Figure 12.32   Process flow diagram of the CANMET process. (From Ref. 95.)

The products qualities are

naphtha: sulfur < 5 ppm
nitrogen: < 5 ppm
kerosene smoke point > 20 mm, cloud point < −50°C
Diesel: Sulfur < 50 ppm, cetane no. > 45
Vac. gas oil: sulfur < 150 ppm, CCR < 0.1 wt % metals < 1 ppm
Economics: Plant capacity 23,000 bpsd. Investment US MM$190 (USGC 1st Q, 1994).

Utilities:

fuel oil, MW	12
power, MW	17
steam, tph	−34
water, cooling, m^3/h	2,000

12.9   PRODUCTION OF HIGH GRADE OILS BY HYDROCRACKING

The potential ability of hydrocracking processes to hydrogenate aromatic rings of polynuclear hydrocarbons opened new perspectives for manufacturing lubricating oils with high viscosity index.

Hydrocracking has two major advantages when compared to the traditional methods used for the elimination of aromatic hydrocarbons such as solvent extraction:

1.   The possibility to produce lubricating oils with a viscosity index of 115–140, unattainable by the classical methods.
2.   Higher yields compared to those obtained by solvent extraction

Certain reactions, specific to hydrocracking illustrate the production of hydro-carbons that have a low freezing point and a high viscosity index:

These examples indicate that the catalyst and the process conditions used in hydrocracking must ensure a deep hydrogenation in order to obtain oils with a viscosity index in the range 115 to 140 and freezing point in the range -15 to $-20°C$. The hydrogenation of the aromatic structures needs to be almost complete, whereas only a limited acid function is required, just enough to catalyze the isomerization of the alkanes and alkyl chains, preferably without causing their cracking. Careful balancing of the two functions will have as effect the advanced ring opening of the polycyclic naphthenic hydrocarbons. Consequently, the operating temperatures need to be relatively low in order to prevent cracking reactions. The pressure needs to be high in order to ensure the almost complete hydrogenation of the aromatic structures.

These process conditions correspond to those used in high-pressure hydro-cracking units. It must be noted that for units that process heavy vacuum distillates, the product from the bottom of the fractionating column is actually an oil with viscosity index of more than 120. Such is the case of the options A, B, and D given in Section 12.6.4 and Table 12.11.

It is remarkable that seminal studies leading to the production of oils with very high viscosity index by means of hydrocracking were accomplished by A Nastasi and coworkers, in Ploiesti, Romania, as early as 1966 [96] starting from naphthenic oils [97]. The conclusions they reached agree entirely with the current state of the art for this process.

The production of oils having very high viscosity index by hydrocracking is not a matter of selecting process flow diagrams or units that are different from those already described in Section 12.6.2. Instead, it is a matter of selecting the process conditions, as well as of the appropriate conversion. Above all, the issue is that of selecting one or more catalysts that will ensure the optimum yield of the residual product obtained from fractionation. Therefore there are two main options: pro-cesses with one or with two reaction stages.

The processes with one reaction stage use a bifunctional catalyst on amorphous zeolite support with a low acidity. In order to achieve the required degree of hydrogenation, they operate at temperatures above 400°C, at low space velocities and high pressures.

Such process conditions correspond to conversions of 80–85% and they can be applied only to relatively light feedstocks. Consequently, oils with high viscosity index resulting from this process have viscosity in the range 3.5–4.0 cSt at 100°C. This corresponds to spindle oil, which is of only limited interest [98].

When processing heavy feedstocks, conversions above 70% are required in order to obtain the desired viscosity index. Under these conditions, the stability of the catalysts is limited. The cycle length is shorter than one year and the viscosity index decreases significantly along the cycle.

Catalysts on conventional zeolite Y require temperatures that are 35–45°C lower [10], thus solving the problem of the length of the cycle and making it possible to process heavy feedstocks. However, the yields of oil are much lower.

For this reason process designers chose either to use two catalysts placed in two sections of the same reactor, as in the Unicracking process of UOP [99] or with the two catalysts situated in two reactors through which the feed circulates successively, as in the DUAL process of IFP [10,98].

The Unicracking system uses a bifunctional catalyst on amorphous support and a catalyst on crystalline support. No detailed information is available regarding the metallic functions of the two catalysts. It is assumed that the two catalysts are placed in two different zones with intermediate injection of cold recycle gases, the feed coming into contact first with the catalyst on the amorphous support and then with the one on zeolite Y support.

Such an arrangement is justified by the fact that the amorphous support has pores of larger diameter. Thus, it easily catalyzes the hydrogenolysis of heteroatomic compounds, as well as the hydrogenation of polyaromatic hydrocarbons and the decyclization of poly-cyclic naphthenes with condensed rings, especially when heavy fractions are processed.

The second has the role of accomplishing the hydrogenation of the aromatic hydrocarbons as well as the reaction of isomerization and hydrocracking. It is placed in the same reactor. Therefore the second catalyst must tolerate the quantities of NH_3 and H_2S resulting from the hydrogenolysis of the components with nitrogen and sulfur that takes place on the first catalyst.

Table 12.16 shows the results obtained in hydrocracking in such a system at two levels of conversion of a mixture of vacuum distillate and residue from a propane deasphalting, while Figure 12.33 shows the variation of the viscosity and of the viscosity index with conversion.

The quantities of NH_3 and H_2S resulting from the process, i.e. 0.10 wt % and 2.59 wt % respectively, confirm the conclusion drawn above regarding the tolerance of the second catalyst to sulfur and nitrogen.

The dual system developed by IFP uses two reactors. A catalyst based on amorphous support is used in the first reactor to bring the nitrogen level to 10 ppm to desulfurize the feed and to hydrogenate the polycyclic aromatics. This is a typical catalyst for hydrotreatment (HR).

In the second stage, the process uses a typical hydrocracking catalyst (HYC) on zeolite support. The zeolite structure was especially developed in order to reduce the cracking reactions and to promote a good yield of medium fractions similar to the amorphous catalysts.

Table 12.16 Performance of the UOP Unicracking Process

	Low conversion	High conversion
Hydrocracking yields, wt %		
NH$_3$	0.10	0.10
H$_2$S	3.59	2.59
C$_1$-C$_4$	0.99	1.49
C$_5$	0.40	0.70
C$_6$	0.59	0.70
C$_7$–177°C	2.97	5.06
177–315°C	14.85	23.83
315°C+ (oil)	79.21	67.52
Total	101.70	101.99
Hydrogen chemically bound, Nm^3/m^3	4.64	5.4
Characteristics of 315°C+oil		
density	0.8524	0.8519
distillation, °C		
initial	315	315
50%	465	455
final	515	510
viscosity at 38°C, cSt	34.6	26.9
V.I. not dewaxed	100	120
V.I. dewaxed	95	115

Source: Ref. 99.

    Due to its pronounced cracking activity, the catalyst used in the second stage allows working at temperatures 35–45°C lower than those in the first stage (where temperatures are above 400°C). This allows completing the hydrogenation of the aromatic rings.

    The performance of a hydrocracking unit using this system is presented in Table 12.17.

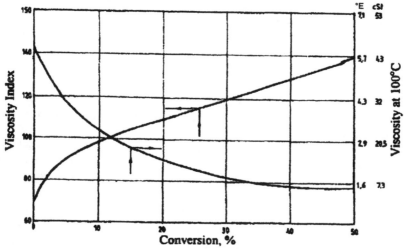

Figure 12.33 Variation of the viscosity index and of viscosity with conversion in a UOP Unicracking unit. (From Ref. 99.)

Economic data regarding hydrocracking for the purpose of producing oils with high viscosity index were published by Chevron for its own hydrocracking process called "Isocracking", which uses a single reaction zone [100].

The investment is based on a processing capacity of 800 m^3/day. The feed is a residue from Arabian light crude oil, with a density of 0.9254 and a viscosity of 9°E (68 cSt) at 38°C. A moderate severity was assumed for the process conditions.

Under these conditions, the investment represents U.S.$4100/bbl day, the reported utilities per m^3 processed being:

Fuel, 10^3 kJ	664
Electricity, kWh	18.9
Steam of 3.5 bar produced, kg	28.5
Steam of 10.5 bar consumed, kg	77.0
Cooling water, m^3	3.57

Table 12.17 The Performance of a Hydrocracking Unit Using the IFP Dual Catalytic System

	Feed				
	Vacuum distillate		Deasphalted (propane)		
Analysis of feed					
density	0.926		0.925		
viscosity at 50°C	46.1		884		
viscosity at 100°C	8.4		41.6		
sulfur, wt %	2.6		2.0		
nitrogen, ppm	1500		650		
Conradson carbon, wt %	0.3		1.5		
asphaltenes, ppm	< 100		< 200		
distillation ASTM, °C					
initial	310		463		
50%	451		586		
final	580		—		
Severity	I	II	I	II	III
Yields, wt %					
H$_2$S + NH$_3$	2.9	2.9	2.2	2.2	2.2
C$_1$-C$_4$	1.3	2.5	1.0	1.7	2.7
gasoline	4.4	8.3	2.4	5.0	11.1
naphtha	7.5	12.4	4.0	8.2	14.2
jet fuel	21.5	34.0	15.9	24.3	32.7
gas oil	25.6	24.3	17.0	19.4	21.5
residue (oil)	39.0	18.0	59.0	41.0	18.0
Characteristics of oil after dewaxing					
Yield, % of feed	30.0	13.5	45.5	30.2	13.5
density	0.854	0.833	0.866	0.860	0.855
viscosity at 100°C	5.5	4.5	13.8	12.7	11.5
viscosity index	119	127	112	116	117
freezing point, °C	−18	−18	−18	−18	−18
volatility, NOACK, wt %	< 20	< 20	—	—	—

Source: Ref. 44.

REFERENCES

1. RJ Schmidt, PL Bogdan, NL Giilsohof, Chemtech 41: Feb. 1993.
2. DR Stull, EF Westrum Jr, GC Sinke, The Chemical Thermodynamics of Organic Compounds. Malabar, Florida: Robert E. Krieger Publ. Co., 1987.
3. E Welther, S Feyer-Ionescu, C Ionescu, E Turjanski-Ghionea, Hidrofinare, Hidrocracare. S Raseev, ed. Editura Tehnica, Bucuresti, 1967.
4. T Huizinga, JMH Theunisen, H Minderhound, R Van Veen, Oil and Gas Journal 93: 40, 26 June 1995.
5. A Hock, T Hutzings, AA Escner, JE Maxwell. W Stork, FJ Van der Meerakker, O Sy, Oil and Gas Journal 89: 77, 22 Apr. 1991.
6. PH Bigeard, P Marion, A Espinosa, Revue de l'Institut Français du Petrole 49: 529–539, 5 Sept. 1994.
7. A Hennico, A Billan, PH Bigeard, Hydrocarbon Technology International 1992. P Harrison ed. London: Sterling Publ. Intern. Ltd., p. 18.
8. S Mignard, Past, present and future of hydrocracking catalysts, OAPEC-IFP Joint Workshop: Catalysts in the Downstream Operations, 5–7 July 1994, Raport 10 Institut Francais du Petrole, Ruel-Malmaison, France.
9. Hickey, Mobil Oil Corp., U.S. Patent 4,812,223.
10. JW Ward, Union Oil Company of California, U.S. Patent 5,228,979.
11. SJ Miller, Chevron R. and Tech. Company, U.S. Patent 5,139,647.
12. Cody, Exxon, U.S. Patent 4,906,599 (silylated NH_3 zeolite).
13. Flaningen, Union Carbide, U.S. Patent 4,759,919.
14. Kirker, Mobil Oil Corp. EP 284278 (pillared magadite).
15. Gortsema, Union Carbide, WO 8,806,614.
16. Hamilton, Shell International, U.S. Patent 4,640,764.
17. TF Degnan Jr, Mobil Oil Corp. U.S. Patent 5,183,557.
18. MR Apelian, Mobil Oil Corp. U.S. Patent 5,264,116.
19. Verduijn, Exxon EP 288,293 (ZK5 zeolite).
20. Vaughan, Exxon, EP 315,461 (ECR30 zeolite).
21. SI Zones, Chevron Research Company, U.S. Patent 5,007,997.
22. JL Casci, Imperial Chemical Industries, U.K. Patent 8,829,923.
23. Raatz, Institut Francais du Petrole, EP 258,127.
24. B Schaefer Pedersen, GR Ecurall, AJ Grula, MJ Humbach, DA Kauff, PJ Meurling, Middle-Distillate Hydrocracking, UOP, 1990.
25. HL Coonrach, WE Garwood, Ind Eng Chem Proc Des Develop 38(1), 1964.
26. RF Sullivan, J of Catalysis, 3(2): 183, 1964.
27. GE Langlois, RF Sullivan, Am Chem Soc Di Petrol 14(4): 8, 1969.
28. LP Masologites, LR Beckbarger, Oil and Gas J. 71(47): 49, 1973.
29. HF Shultz, JH Weitkamp, Ind Eng Chem Proc Res Develop 11(1): 46, 1972.
30. JH Weitkamp, HF Shulz, J. of Catalysis 29: 361, 1973.
31. JM Beelin, V Ponek, J of Catalysis 28: 396, 1973.
32. MJ Vàzguez, A Escardino, A Ancejo, Ind Eng Chem 27: 2029, 1988.
33. MA Baltanas, H Vansina, GF Froment, Ind Eng Chem Prod Res Dev 22: 531, 1983.
34. MA Baltanas, KK Van Raemdonck, GF Froment, SR Mohedas, Ind Eng Chem Prod Res Dev 28: 899–950, 1989.
35. JO Hernades, PhD dissertation, directed by S. Raseev, Institutul de Petrol, Gaze si Geologie, Bucharest Romania, 1972.
36. MR Gray, Upgrading Petroleum Residues and Heavy Oils, New York: Marcel Dekker Inc., 1994, Chapter 5.
37. BC Gates, Catalytic Chemistry, New York: John Wiley & Sons, Inc. 1991.
38. SD Raseev, S El Kharashi S, Revista de Chimie 29: 501, 615, 1978.

39. SD Raseev, S El Kharashi, Revista de Chimie 29: 1140, 1978.
40. TS Sie, Ind Eng Chem Res 32: 403, 1993.
41. M Bouchy, P Dufesne, S Kaszelan, Ind Eng Chem Res 31: 2661, 1992.
42. SA Qader, GR Hill, Am Chem Soc Div Fuel Chem Prep 16: 93, 1972.
43. SK Wuu, WS Hatcher Jr, Zeolite catalyst for hydrocracking polynuclear aromatic phenantren kinetics AIChE Spring National Meeting, Houston, 1983.
44. A Hennico, A Billon, PH Bigeard, JP Peries, Revue de l'Institut Francais du Petrole, 48: 2, March–April 1993.
45. A Voorhies, WJ Hatcher, Ind Eng Chem Prod Res Dev 8: 361, Dec. 1969.
46. M Steijns, GF Froment, Ind Eng Chem Prod Res Dev 20: 680, 1981.
47. M Steijns, GF Froment, PA Jacobs, J Uyteertooven, J Weltkamp, Erdöl, Kohle Erdgas Petrochem 31: 581, 1978.
48. M Steijns, GF Froment, PA Jacobs, J Uyterooven, J Weltkamp, Ind Eng Chem Prod Res Develop 20: 654, 1981.
49. J Wietkamp, PA Jacobs, Avard Symposium on Fundamentals of Catalysts and Thermal Reactions. 181st National Meeting of the American Chem Society, Atlanta, GA, 29 March–3 April 1981.
50. J Weitkamp, Ind Eng Chem Prod Res Dev 21: 550, 1982.
51. H Vansina, MA Baltanas, GF Froment, Ind Eng Chem Prod Res Dev 22: 526, 1983.
52. PA Jacobs, JA Martens, J Weitkamp, HK Beyer, Faraday Discuss Chem Soc 72: 123, 1981.
53. JA Martens, PA Jacobs, T Attempts, Appl Catal 20: 283, 1986b.
54. JH Martens, J Weitkamp, PA Jacobs, Catalysis by Acids and Bases. In B Imelik et al., eds., Studies in Surface Science and Cataltsis, vol. 20, Amsterdam: Elsevier, 1985.
55. S El Kharashi, PhD dissertation, directed by S Raseev, Institutul de Petrol si Gaze, Ploiesti, Romania, 1979.
56. S Raseev, S El Kharashi, Revista de Chimie 29: 815, 1978.
57. H Hassoun, PhD dissertation, directed by S Raseev, Institutul de Petrol si Gaze Ploiesti, Romania, 1978.
58. SM Yui, EC Stanford Ind Eng Chem Res. 28: 1278–1294, 1989.
59. MA Baltanas, GF Froment, Comp Chem Eng 9: 71, 1983.
60. KK Van Raemdonek, GF Froment, AIChE meeting. San Francisco 5–10 Nov., 1989.
61. GF Froment, Revue de l'Institut Français du Petrole, 46 (4): 491–502, July–Aug. 1991.
62. DM Brower, H Hogeveen, Prog Phys Org Chem 9: 179, 1972.
63. GE Weismantel, Hydrocracking. In: JJ McKetta, ed. Petroleum Processing Handbook. New York: Marcel Dekker Inc., 1992.
64. G Gianetto, Thesis, Poitier, France, 1985.
65. G Gianetto, G Perot, M Guisnet, Ind Eng Chem Prod Res Dev 25: 481, 1986.
66. G Gianetto, F Alvarez, FR Ribeiro, G Perat, M Guisinet, Guidelines for Mastering the Properties of Molecular Sieves, NATO ASI series, BD Barthoment, ed. vol. 221, 1990, p. 355.
67. M Guisnet, F Alvarez, G Gianetto, G Perot, Catalysis Today 1: 415, 1987.
68. P Wuithuier, Raffinage et Génie Chimique, vol. 1 , Technip, Paris, France, 1972.
69. WL Nelson, Oil and Gas Journal, 68: 138, 28 Dec., 1970.
70. TN Kalnes, PR Lamb, DG Tajbl, DR Pegg, Mild Hydrocracking. A Low Cost Rule for More Distilate. NPRA Annual Meeting, 25–27 March, 1984.
71. WR Derr, WJ Tracy, Energy Progress 6(1): 15, 1986.
72. EJ Aitken, RB Miller, The 6th Symposion on Fluid Catalytic Cracking, Budapest, 1–4 June 1987, p. 20.
73. N Basta, Chem Eng 93(1): 32, 1986.
74. RA Gorbetti, Oil and Gas Journal 85(40): 57, 1987.
75. HH Frü, HH Lovik, Erdôl und Kohle 40(8): 346, 1987.

76. SD Light, RV Bertram, JW Ward, Hydrocarbon Processing 60: 93, May 1981.
77. Anonymous, Hydrocarbon Processing 65(9), 1986.
78. LV Gankina, MJ Koni, Himia i technologia topliv i masel, No. 6, 1989, p. 42.
79. EM Hibbs, DA Kauff, MJ Humbach, E Yuh, Alternative Hydrocracking Application, UOP 1990.
80. Oil and Gas Journal, 82(16): 65, 1984.
81. S Goetzman, W Kreuter, HJ Wernicka, Hydrocarbon Processing 58(6): 109, 1979.
82. KT Powell, Oil and Gas Journal 61, 9 Jan., 1989.
83. J Mochida, XZ Zhao, K Sacanichi, Ind Eng Chem 29: 334, March 1990.
84. Anonymous, Hydrocarbon Processing, 73: 122, Nov. 1994.
85. Anonymous. Hydrocarbon Processing 73: 128, Nov. 1994.
86. RE Boening, NK McDaniel, RD Petersen, RP Van Driesen, Hydrocarbon Processing 66: 59, Sept. 1987.
87. Anon., Hydrocarbon Processing 69: 96, Nov. 1990; 71: 176, Nov. 1992; 73: 117, Nov. 1994; 77: 82, Nov. 1998; 79: 114, Nov. 2000.
88. RM Beaton et al., Oil and Gas Journal 84: 47, 7 July 1986.
89. TG Tasker, LJ Wisdom, Hydrocarbon Technology International 1988. P Harrison, ed. London: Sterling Publ. 1988.
90. Anon., Hydrocarbon Processing 71: 177, Nov. 1994; 73: 86, Nov. 1998; 77: 86, Nov. 1998.
91. RP Van Drisen, LL Fornoff, Hydrocarbon Processing 64: 91, Sept. 1985.
92. JF Le Page, SG Chatila, M Davidson, Raffinage et conversion des produits lourdes du petrole, Technip Paris, 1990, p. 157.
92a. Anon., Hydrocarbon Processing 81, p. 115, 118 Nov. 2002.
93. Anon., Hydrocarbon Processing, 61: 140, Sept. 1982.
94. Anon., Hydrocarbon Processing, 63: 95, Sept. 1984.
95. M Skripek, P Robinson, B Pruden, G Muir, In: P.Harrison ed., Hydrocarbon Technology International 1994. London: Sterling Publ. Ltd., 1994.
96. A Nastasi, J Zarna, D Grigoriu, Petrol si Gaze 17(10), 1966; 21(7), 1970; 22(9), 1971.
97. A Nastasi, PhD Dissertation, S Raseev advisor, Institutul de Petrol si Gaze Ploiesti, 1976.
98. A Hennico, A Billon, PH Bigeard, JP Peries, In: P Harrison, ed. Hydrocarbon Technology International, London: Sterling Publ Ltd., 1994, p. 25.
99. E Honde, P Harrison ed. Hydrocarbon Technology International London: Sterling Publ., 1993, p. 19.
100. Anon., Hydrocarbon Processing 69: 190, Nov. 1992.
101. SC Korre, MT Kleen, RJ Quan, Hydrocracking of Polynuclear Aromatic Hydrocarbons. Development of Rate Laws Through Inhibition Studies. Ind Eng Chem Res. 36: 2041–2050, 1997.
102. N Mawondza, A Rouquier, In: P Harrison ed. Converting a hydrotreater to a low pressure hydrocracker. Hydrocarbon Technology International, London: Sterling Publ. 1994.
103. Anon., Hydrocarbon Processing 75: 84, Nov. 1998.
104. A Neg, Assess hydrocracker performance using a short-cut method. Hydrocarbon Processing 74: 79–84, March 1997.
105. Hydrocarbon Processing 75: 88, Nov. 1998; 77: 118, Nov. 2000.
106. J Scherzer, AJ Gruia, Hydrocracking Science and Technology, New York: Marcel Dekker Inc., 1996.

13

Catalytic Reforming

The catalytic reforming process consists of a number of reactions which take place on bifunctional catalysts for converting the hydrocarbons contained in naphtha fractions to monocyclic aromatics.

Naphthenes with six carbon atom rings are subjected to dehydrogenation. Naphthenes with five carbon atom rings are subjected to isomerization followed by dehydrogenation, usually called dehydroisomerization. The alkanes go through cyclization followed by dehydrogenation, usually called dehydrocyclization. Simultaneously, the hydrocarbons and especially the alkanes undergo parallel, competing reactions of isomerization and hydrocracking with conversions sometimes comparable to the reactions producing aromatics.

There are two ways in which catalytic reforming may be used. One option is to process the heavy fractions of straight run naphthas in order to increase their octane rating by 40–50 units. The other way is to process a narrow fraction of gasoline such as C_6-C_8 or C_7-C_8. From the obtained reformate (called in this case BTX) are then separated the aromatic hydrocarbons (mainly benzene, toluene and xylenes), for the petrochemical industry. This second process is also called aromatization.

Both processing options are performed in the same units, working under similar operating conditions. The presentation that follows will be referring to both options at the same time.

The final part of the chapter (Section 13.9) will present the catalytic processes used for converting hydrocarbons obtained from the aromatization, in order to increase the production of those hydrocarbons that present a higher interest for the petrochemical industry: hydrodisproportionation or dealkylation of toluene and isomerization of xylenes.

In contrast with this classical image of catalytic reforming, a new process has been developed*, the feed of which is the propane-butane fraction. An extra step,

---

* First time communicated at the "American Institute of Chemical Engineers Summer National Meeting", Denver, Colorado 21–24 August 1988 [1].

dehydropolymerization, is added before cyclization and aromatization. This new process will be presented at the end of the chapter, but it is too early to estimate its impact on global processing.*

13.1 THERMODYNAMICS

As mentioned above, catalytic reforming consists of reactions of dehydrogenation, dehydroisomerization, and dehydrocyclization leading to the formation of aromatic hydrocarbons. The thermodynamics of other concurrent reactions, mainly involving alkanes, of isomerization and hydrocracking were examined in previous chapters.

Specific to catalytic reforming processes is the introduction of hydrogen together with hydrocarbon feed into the reactor system. Molecular ratios of 2 to 5 hydrogen/hydrocarbons are used in order to decrease the rate of coking of the catalyst. Since hydrogen is a product of the aromatization reactions, its presence in the feed to the reactor displaces the thermodynamic equilibrium of the reactions. This effect is accounted for in the following calculations and discussions.

The equilibrium calculations were performed for pressures between 2 and 40 bar in order to compare both older processes working at pressures of around 20–30 bar and newer processes working at much lower pressures.

13.1.1 Dehydrogenation of Cyclohexanes

The variations in heat of formation and entropy for the most important reactions of catalytic reforming were calculated based on the thermodynamic constants published by Stull et al. [2] (see Table 13.1).

Using the calculation method developed by us and presented in Section.1.2, the equilibrium of dehydrogenation of the cyclohexanes is shown in Figure 13.1. The hydrocarbons having similar values for the equilibrium conversion are grouped together.

The data of Figure 13.1 refer to stoichiometric conditions. However, if an excess of hydrogen is present in the system, further calculations will be necessary.

The equilibrium constant for the dehydrogenation of cycloalkanes is given by:

$$K_p = \frac{x(3x)^3}{1-x}\left(\frac{p}{1+3x}\right)^3 \tag{13.1}$$

where x is the conversion at equilibrium.

For an excess of n moles of hydrogen per mole of cycloalkane, the equilibrium constant becomes:

$$K_p = \frac{x'(n+3x')^3}{1-x'}\left(\frac{p}{1+3x'+n}\right)^3 \tag{13.2}$$

where x' is the equilibrium conversion in the presence of the hydrogen excess.

----

* So far this process is known by its trade names, such as "Cyclar" of BP-UOP or "Aroforming" of IFP. Names such as "dehydropolymerization of lower alkanes" and "catalytic poly-reforming" are our suggestions.

Table 13.1 Thermodynamic Data for Cyclohexanes Dehydrogenation

Reaction		$(\Delta H^0_{800})_r$, kcal/mol	$(\Delta S^0_{800})_r$, cal/mol·grd
cyclohexane ⇌ benzene + 3H$_2$		52.7	94.744
methylcyclohexane ⇌ toluene + 3H$_2$		51.75	94.894
ethylcyclohexane (C–C) ⇌ ethylbenzene (C–C) + 3H$_2$		50.86	94.824
propylcyclohexane (C$_3$) ⇌ propylbenzene (C$_3$) + 3H$_2$		50.57	95.254
pentylcyclohexane (C$_5$) ⇌ pentylbenzene (C$_5$) + 3H$_2$		50.14	95.324
dimethylcyclohexane ⇌ xylene + 3H$_2$	*cis*	48.39	95.184
	trans	49.95	95.404
dimethylcyclohexane ⇌ xylene + 3H$_2$	*cis*	50.55	96.424
	trans	48.76	95.304
dimethylcyclohexane ⇌ xylene + 3H$_2$	*cis*	48.79	95.114
	trans	50.37	95.894

Source: Ref. 2.

Since at any given temperature the two equilibrium constants are equal, Eq. (13.1) and (13.2) can be equalized and, denoting

$$x' = ax \qquad (13.3)$$

gives

$$\frac{x(3x)^3}{1-x}\left(\frac{1}{1+3x}\right)^3 = \frac{ax(n+3ax)^3}{1-ax}\left(\frac{1}{1+3ax+n}\right)^3$$

Finally, this expression becomes:

$$\left(\frac{1}{3ax+n}+1\right)^3 = \left(\frac{1}{3x}+1\right)^3\frac{a(1-x)}{1-ax} \qquad (13.4)$$

Equation (13.4) allows one to calculate the correlation between the values of a and n for any value of the conversion x. As a result, the variation of a vs. n may be plotted for various stoichiometric conditions x, as shown in Figure 13.2.

Figure 13.2 gives the value of a, for the equilibrium conversion x in stoichiometric conditions, (as read from the plots of Figure 13.1), at a given temperature and

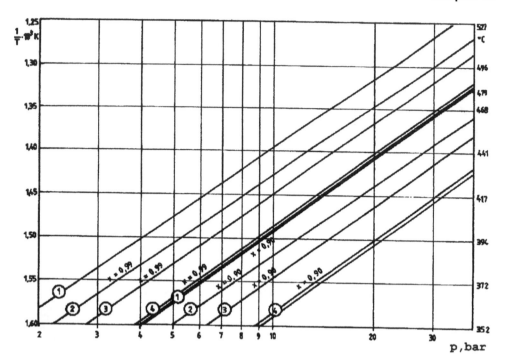

Figure 13.1 The thermodynamic equilibrium of dehydrogenation of cyclohexanes in stoichiometric conditions. 1. $\bigcirc \rightleftharpoons \bigcirc + 3H_2$ 2. $\bigcirc^C \rightleftharpoons \bigcirc^C + 3H_2$ 3. $\bigcirc^{C-C}_{C-C} \rightleftharpoons \bigcirc^{C-C}_{C-C} + 3H_2$ 4. C_8 alkylcyclohexanes $\leftrightarrow$ xylenes $+ 3H_2$, C_9-C_{11} alkylcyclohexanes$\leftrightarrow$ alkylbenzenes $+ 3H_2$.

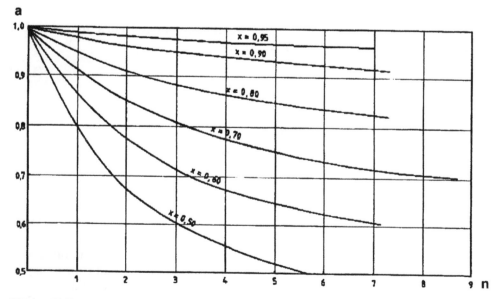

Figure 13.2 The influence of hydrogen excess upon the equilibrium of dehydrogenation and dehydroisomerization of cycloalkanes.

pressure and an excess of n moles of hydrogen. The equilibrium conversion in the presence of an excess of n moles of hydrogen is then calculated using Eq. (13.3).

Example. Calculate the cyclohexane-benzene equilibrium at 468°C, 33 bar, and an excess of 5 mol hydrogen per mol feed.

    ANSWER: In stoichiometric condition, Figure 13.1 gives $x = 0.90$.

    For $x = 0.90$ and $n = 5$, Figure 13.2 gives $a = 0.93$.

    According to Eq. (13.3), the equilibrium conversion with excess of hydrogen, x', will be:

$$x' = a \cdot x = 0.93 \times 0.90 = 0.837$$

13.1.2 Dehydroisomerization of Alkylcyclopentanes

Variations of the heats of formation and of the reaction entropies for dehydroisomerization of some representative alkylcyclopentanes were calculated in a similar manner (see Table 13.2).

    Using the same methodology as in the case of the cyclohexanes, the equilibrium for stoichiometric conditions is shown in Figure 13.3. Due to the fact that the stoichiometry of the reaction is identical to that for cyclohexanes, the expression in Eq. (13.4) and consequently the graph in Figure 13.2 for alkylcyclopentanes are

Table 13.2 Thermodynamic Data for Dehydro-isomerization of Alkylcyclopentanes

Reaction		$(\Delta H^0_{800})_r$, kcal/mol	$(\Delta S^0_{800})_r$, cal/mol · deg
(structure) $\rightleftharpoons$ (structure) $+ 3H_2$		49.24	85.414
(structure) $\rightleftharpoons$ (structure) $+ 3H_2$	cis	46.16	90.684
	trans	48.18	90.524
(structure) $\rightleftharpoons$ (structure) $+ 3H_2$	cis	47.98	90.524
	trans	47.44	90.524
(structure) $\rightleftharpoons$ (structure) $+ 3H_2$		46.20	88.344
(structure) $\rightleftharpoons$ (structure) $+ 3H_2$		46.39	88.634
(structure) $\rightleftharpoons$ (structure) $+ 3H_2$		46.02	89.034
(structure) $\rightleftharpoons$ (structure) $+ 3H_2$		45.78	86.994

Source: Ref. 2.

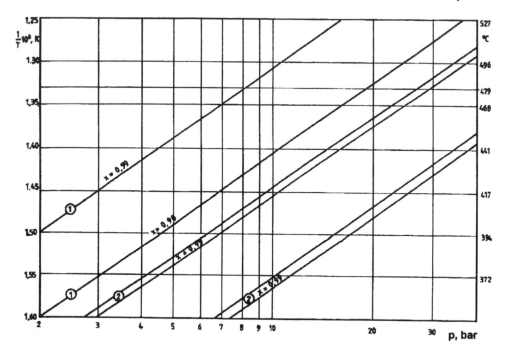

Figure 13.3 Thermodynamic equilibrium of dehydroisomerization of alkylcyclopentanes in stoichiometric conditions:

1. $\text{alkylcyclopentane} \rightleftharpoons \text{benzene} + 3H_2$

2. C_7–C_{11} alkylcyclopentanes $\rightleftharpoons$ alkylbenzenes + $3H_2$

also the same. Thus, the equilibrium in the presence of hydrogen excess can be calculated in a similar way.

13.1.3 Dehydrocyclization of Alkanes

Variations in the heat of formation and in the entropy for the reaction of dehydrocyclization were calculated for several typical hydrocarbons using the same source of thermodynamic data [2]. The results are given in Table 13.3.

In the calculations for the conversion of octanes, only the methylheptanes were taken into account, since dimethylhexanes are found in much smaller quantities in the straight-run gasoline (see Section 5.1.4).

The equilibrium conversion for dehydrocyclization under stoichiometric conditions is plotted in Figure 13.4.

Following the same logic used in obtaining Eq. (13.4), the equivalent expression for dehydrocyclization is:

$$\left(\frac{1}{4ax+n}+1\right)^4 = \left(\frac{1}{4x}+1\right)^4 \frac{a(1-x)}{1-ax} \tag{13.5}$$

Based on the results of Eq. (13.5), Figure 13.5 allows the calculation of the effect of the hydrogen excess on the thermodynamic equilibrium of the dehydrocyclization reactions, where a was defined by the same Eq. (13.3).

Table 13.3 Thermodynamic Data for the Dehydrocyclization of Alkanes

Reaction	$(\Delta H^0_{800})_r$, kcal/mol	$(\Delta S^0_{800})_r$, cal/mol · deg
$n\text{-}C_6 \rightleftharpoons$ (benzene) $+ 4H_2$	63.77	105.022
$n\text{-}C_7 \rightleftharpoons$ (methylbenzene) $+ 4H_2$	60.85	107.872
$2\text{-methyl-}C_6 \rightleftharpoons$ (methylbenzene) $+ 4H_2$	62.56	109.762
$3\text{-methyl-}C_6 \rightleftharpoons$ (methylbenzene) $+ 4H_2$	61.93	108.772
$n\text{-}C_8 \rightleftharpoons$ (C–C benzene) $+ 4H_2$	60.96	108.192
$n\text{-}C_8 \rightleftharpoons$ (dimethylbenzene) $+ 4H_2$	58.31	106.342
$2\text{-methyl-}C_7 \rightleftharpoons$ (dimethylbenzene) $+ 4H_2$	59.24	109.522
$3\text{-methyl-}C_7 \rightleftharpoons$ (C–C benzene) $+ 4H_2$	61.96	109.122
$3\text{-methyl-}C_7 \rightleftharpoons$ (dimethylbenzene) $+ 4H_2$	59.31	107.272
$3\text{-methyl-}C_7 \rightleftharpoons$ (dimethylbenzene) $+ 4H_2$	58.66	106.142
$4\text{-methyl-}C_7 \rightleftharpoons$ (dimethylbenzene) $+ 4H_2$	58.53	108.412
$n\text{-}C_9 \rightleftharpoons$ (C3 benzene) $+ 4H_2$	60.68	108.592

Source: Ref. 2.

13.1.4 Isomerization and Hydrocracking

Isomerization and hydrocracking reactions compete with aromatization. Consequently, the degree of influence depends on their relative rates.

The rate of aromatization of alkanes is the slowest. Therefore their overall conversion to aromatics is influenced to a large extent by the reactions of isomerization and hydrocracking. The rate of aromatization of the alkylcyclopentanes is much higher. Therefore they are less affected by such reactions. Finally, for alkylcyclohexanes, the rate of aromatization is very fast, so that the reactions of isomerization and hydrocracking may be completely ignored. The isomerization of alkanes has been examined in Chapter 11. The composition of various heptane isomers may be taken as representative of the higher alkanes. According to Figure 11.4, at the temperatures

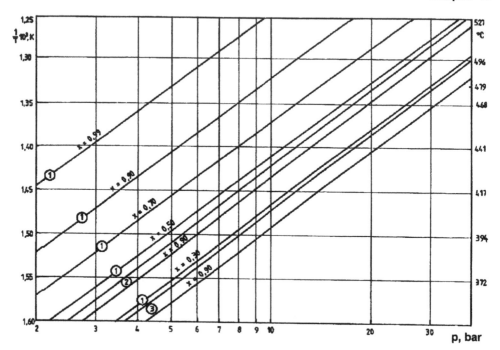

Figure 13.4 The thermodynamic equilibrium for dehydrocyclization of alkanes in stoichiometric conditions.

1. $n\text{-}C_6 \rightleftharpoons$ ⬡ $+ 4H_2$

2. $C_7 \rightleftharpoons$ ⬡^{C} $+ 4H_2$; $n\text{-}C_8 \rightleftharpoons$ ⬡$^{C-C}$ $+ 4H_2$

3. $C_8 \rightleftharpoons o,m,p\text{-xylenes}$; $C_9\text{-}C_{10} \rightleftharpoons$ ⬡^{R} $+ 4H_2$

practiced in catalytic reforming, the equilibrium corresponds to 14% *n*-heptane, 40% 2- and 3-methylheptanes, the rest being dimethylpentanes and trimethylbutanes. In conclusion, isomerization diverts a high proportion of alkanes towards molecular structures that can no longer undergo dehydrocyclization.

Furthermore, the reactions of hydrocracking, undergone mainly by the iso-alkanes, lower the conversion to aromatics even more.

The thermodynamic equilibrium of the overall reactions is influenced also by the fact that the hydrocracking reactions are highly exothermic while the reactions of aromatization are all highly endothermic. As a result, the hydrocracking reactions lead to an increase of the outlet temperature and therefore an increase in the conversion to aromatic hydrocarbons, especially in the last reactor of the unit.

The isomerization equilibrium of the alkylaromatic hydrocarbons produced in the process becomes important, especially for the xylenes, the proportions of which almost always correspond to the thermodynamic equilibrium of their isomerization. Since the equilibrium composition corresponds to approx. 20% ortho-, 20% para-, and 60% meta-xylene, the isomerization of meta-xylene becomes an important commercial process. It will be presented separately, at the end of this chapter (Section 13.9.5).

13.1.5 Conclusions

The calculations and results obtained in the previous sections lead to the following conclusions concerning the thermodynamic limitations to the process of catalytic reforming.

The temperatures do not change significantly with the process type and range between 410°C and 420°C for the exit from the first reactor and about 480–500°C for the exit from the last reactor.

In contrast, the pressure in the system varies more significantly. Older units were operated at a pressure between 18 and 30 bar while the units built after 1980 operate at pressures between 3.5 and 7 bar.

The molar ratio hydrogen/hydrocarbon varies between 8 and 10 for the older units and was lowered to 2 to 5 for the more modern ones.

Considering an effluent temperature of 480°C from the last reactor of the unit and a molar ratio hydrogen/hydrocarbon of 4, the equilibrium conversions were calculated comparatively for two levels of pressure, 20 bar and 2.5 bar. The following conclusions may be drawn:

1. The equilibrium conversions of cyclohexane and alkylcyclohexanes will be higher than 99%, both at lower pressures of 3.5–7.8 bar and at higher pressures of 20 bar. The conversion will drop to ∼ 90% for cyclohexane and to ∼ 95% for methylcyclohexane at pressures of 30–35 bar.
2. The equilibrium conversions of alkylcyclopentanes are somewhat less favorable. The equilibrium conversion of methylcyclopentane at a pressure of 20 bar reaches only 86% but increases to over 99% at pressures of 3.5–7 bar. The equilibrium conversion of higher alkylcyclopentanes will be almost quantitative at both low pressures and pressures of around 20 bar.
3. The equilibrium conversions of alkanes to aromatics are the least favored. For *n*-hexane, the conversion will reach only 30% at a pressure of 20 bar, decreasing to 10% at a pressure of 30 bar. The equilibrium conversion of *n*-hexane would reach 85–90% only at lower pressures of 3.5–7 bar. The conversion of heptanes and octanes to ethylbenzene will correspond to about 85% at a pressure of 20 bar, becoming almost complete at pressures of 3.5–7 bar. The conversions to xylenes and higher hydrocarbons will reach 85% at a pressure of 40 bar, increasing to 92–95% at 20 bar and becoming almost complete in low pressure processes.

It should be noted here that the graphs presented in Figures 13.1–13.5 allow an estimation of the thermodynamic limits to conversion, at the exits of any of the three (or four) reactors of the commercial catalytic reforming units.

13.2 THE CATALYSTS

The first catalyst used for industrial catalytic reforming consisted of 9% molybdenum oxide on alumina gel. The plant started in 1940 under the name of "Hydroforming."

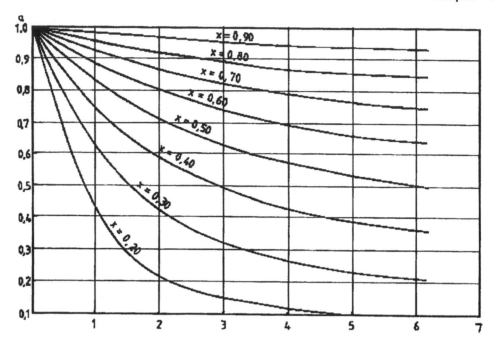

Figure 13.5   The effect of the hydrogen excess on the dehydrocyclization of alkanes.

The use of platinum instead of molybdenum oxide was patented by V Haensel of UOP in 1949 and the first plant using such a catalyst started working in the same year under the name of "Platforming."

The overwhelming advantages of platinum when compared to molybdenum oxide led to a complete replacement of the latter.

The catalyst of platinum on γ-alumina support underwent various improvements, both with respect to the support and by promoting the platinum with other metals such as iridium, palladium, tin, and rhenium.

Catalytic reforming catalysts patented by different manufacturers were reviewed by Aalund [3] and presented in the monograph of Little [4].

The dual function character of the catalysts for catalytic reforming is provided by the acid centers of the support, which catalyze the reactions of isomerization and hydrocracking, as well as by the metallic centers—platinum associated with other metals dispersed on the support—which catalyze the dehydrogenation reactions. A more detailed analysis of the reaction mechanisms (see Section 13.3) has to consider the formation of complex metal–acid centers. In order to achieve maximum efficiency for the process, a balance must be found between the acidic and dehydrogenating functions of the catalyst.

The most frequently used support is γ-alumina (γ-Al$_2$O$_3$), and its appropriate acidic level is achieved through a treatment with HCl or sometimes with HF. Hydrochloric acid is preferred, because the final acidity is easier to control. Sometimes CCl$_4$ or organic chlorides are used instead of HCl.

In the catalytic reforming process, traces of water tend to eliminate the hydrochloric acid fixed on the support; therefore a certain amount of hydrochloric acid is continuously injected into the reactors in order to maintain the required acidity level of the catalyst.

In the past [5], amorphous aluminosilica was used as support but alkaline substances needed to be added to compensate for its strong acidity. Since the required acidity was quite difficult to achieve, the use of amorphous aluminosilica was abandoned.

The progress achieved in the synthesis of zeolites led to attempts to incorporate them in the preparation of the alumina support, erionite [6] in particular. In laboratory studies, this led to an increase in octane number by 3 to 7 units [7].

The activity of erionite was explained in some studies by a bifunctional mechanism of cyclization [8], or in other studies by cyclization on the external surface of the erionite crystals [9,10].

Such catalysts were prepared by mixing bohemite with 15–35 wt % H-erionite, followed by an impregnation with a solution of H_2PtCl_6 [6].

The publications concerning laboratory results do not provide sufficient information to decide to include zeolites in the preparation of the supports for catalytic reforming catalysts. However, the multiple possibilities offered by zeolytes and the advances in their technology may increase their use in preparing more efficient supports for bifunctional catalysts.

Currently, γ-alumina is used almost exclusively as support for commercial reforming catalysts. Their preparation involves strict, proprietary rules that ensure the reproducibility of their activity and porosity.

The two main preparation methods are only mentioned here, since a detailed description of the preparation of catalytic reforming catalysts is the subject of specialized publications [188].

> One method prepares the alumina of the desired structure and porosity. After being washed, dried, and formatted, it is impregnated with the solution containing the metallic compound. The treatment with HCl that follows, sets the desired acidity. One option is to use chloroplatinic acid, which accounts also for the final acid treatment.
>
> The other method is coprecipitation by mixing the solutions containing the soluble alumina precursor with that of the platinum compound.

The first method is preferred for reforming catalysts because it allows a more precise adjustment of the characteristics of the final catalyst. Furthermore, this seems to be the only suitable method when preparing bi- and polymetallic catalysts, allowing a more precise dosage of the promoters.

At the beginning the only metal used was platinum. To improve its dispersion on the surface, a treatment with 0.1–0.5 wt % S was performed, usually within the reactors, just after loading the unit with a new batch of catalyst [4].

The content of platinum in catalyst varies between 0.2 and 0.6 wt %, usually being between 0.3–0.35 wt %. The content rarely exceeds this range.

The content of chlorine usually varies between 0.8–1.3 wt %.

An overview of the main characteristics of some bi- and polymetallic catalysts is given in Table 13.4 where data of six typical U.S. patents are presented.

It should be emphasized that the exact role of various metals added to platinum is not clear. Existing publications tend to justify their presence mainly by their

Table 13.4 Composition of Some Catalytic Reforming Catalysts, Patented in the U.S.

Patent Components	U.S. pat. 2,752,289 Range	U.S. pat. 2,752,289 Typical example	U.S. pat. 3,415,737 Range	U.S. pat. 3,415,737 Typical example	U.S. pat. 4,210,524 Range	U.S. pat. 4,210,524 Typical example
Platinum	0.01–1	0.3	0.2–1	0.7	0.05–1	0.375
Rhenium	—	—	0.1–2	0.7	0.05–1	0.375
Chloride	—	0.45	0.1–3	not specified	0.5–1.5	1.0
Fluoride	0.1–3	0.3	—	—	—	—

Patent Components	U.S. pat. 4,312,788 Range	U.S. pat. 4,312,788 Typical example	U.S. pat. 3,487,009 Range	U.S. pat. 3,487,009 Typical example	U.S. pat. 4,379,076 Range	U.S. pat. 4,379,076 Typical example
Platinum	0.05–1	0.375	0.01–1.2	0.3	0.2–0.6	0.3
Rhenium	0.05–1	0.375	0.01–0.2	0.03	—	—
Germanium	0.01	0.05	—	—	—	—
Iridium	—	—	0.01–0.1	0.025	0.2–0.6	0.3
Copper	—	—	—	—	0.025–0.08	0.05
Selenium	—	—	—	—	0.01–1	0.04
Chlorides or fluorides	0.5–1.5	1.0	0.1–2	not specified	0.7–1.2	0.9

effect on the activity, selectivity, and stability of the catalysts. Nevertheless, the published data allow some interesting assumptions and conclusions to be drawn.

The addition of metals of the platinum group such as palladium or iridium has the effect of augmenting the reactions promoted by platinum. The addition of a certain amount of iridium proves to be more efficient than increasing the content of platinum by the same amount, as shown in Figure 13.6.

Research on palladium was carried out mainly in Russia, with the purpose of replacing platinum, while research on iridium at the Institut Francais du Petrole led to the preparation of highly efficient bimetallic catalysts such as RG422 and RG423, as well as polymetallic catalysts with platinum, iridium, and rhenium.

There are several ways of expressing the behavior in time, i.e., the performance stability of a catalytic reforming catalyst. Usually, the stability is expressed by the decrease in time of the octane rating of the reformate. Sometimes it is expressed by the increase of temperature required for maintaining a constant activity or also, by the decrease of the selectivity, as expressed by the decrease in the volume of the liquid phase and of the hydrogen obtained.

Figure 13.7 [11] shows a comparison of the stability of a nonpromoted catalyst and of a catalyst promoted with rhenium.

The effect of iridium is explained by its higher hydrogenating activity compared to that of platinum. This prevents the formation of coke deposits on the active surface of the catalyst. This theory is supported by experimental results on platinum catalysts promoted with other metals, as shown in Figures 13.8 and 13.9 [12].

Figure 13.8 shows the activity of hydrogenating benzene to cyclohexane as a function of the ratio of the promoters (iridium, rhenium, and germanium) to plati-

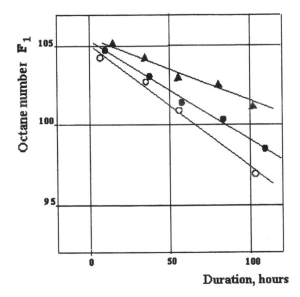

Figure 13.6   The effect of platinum and iridium concentrations on catalyst aging: ○ - 0.035 % Pt; ● - 0.60 % Pt; ▲ - 0.35 % Pt + Ir. Test conditions: Feed ASTM 95–200°C; PONA 50/-/42/8 % vol; pressure 10 bar; temperature 500°C; $H_2/H_C = 4$ mol/mol. (From Ref. 12.)

num. The activity of iridium increases continuously with the Ir/Pt ratio, that of rhenium reaches a maximum at about Re/Pt = 1.4, while the activity of germanium decreases continuously.

Similar results were obtained by Stoica and Raseev [13] when studying the promotion by rhenium of an industrial catalyst containing 0.35 wt % platinum. The maximum activity was obtained for 0.6 wt % Re, which corresponds to a ratio Re/Pt = 1.7.

Figure 13.9 presents a comparison of the rate of coke formation for four catalysts developed by the French Institute of Petroleum. The results show that when compared with a classic catalyst containing 0.6 wt % Pt, coke formation is about 40% lower on a catalyst containing 0.6 wt % Pt and 0.6 wt % Re, it is by about 67% lower on a catalyst containing 0.6 wt % Pt and 0.08 wt % Ir, while it is higher on a catalyst containing 0.6 wt % Pt and 0.22 wt % Ge. Both graphs in Figure 13.8 and 13.9 show the advantages of rhenium, which is the most frequently used promoter in the preparation of catalysts for catalytic reforming. Rhenium is also added to catalysts based on platinum and iridium.

The effect of rhenium on reducing the rate of catalyst aging is shown in Figure 13.10 [4], where a very efficient, multipromoted catalyst KX-130 developed by Exxon Research and Engineering Co. is also given for comparison.

A similar effect of reducing the formation of coke is observed for tin [14–16]. Furthermore, it decreases the amount of gas formed by decreasing the acidity of the acid centers.

The catalysts containing Sn were prepared in Russia at Riazan under the name SPR-2, as spheres of 1.5–1.8 mm diameter or as PR-42, as extrudates. Both catalysts were prepared in two forms:

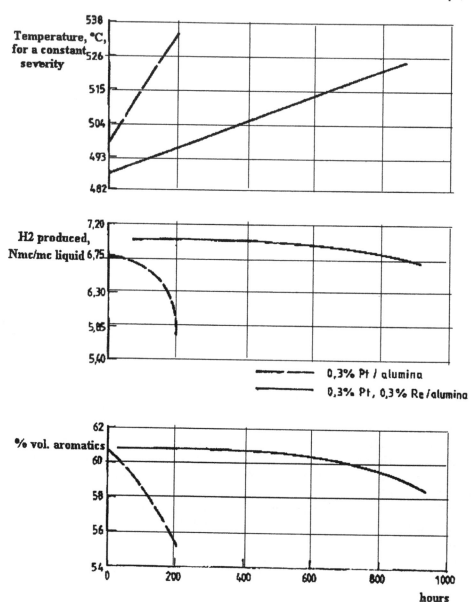

Figure 13.7 Effect of promoters on catalyst stability. (From Ref. 11.)

A: containing 0.38–0.40 wt % Pt, 0.25 wt % Sn, and 1.3 wt % Cl_2
B: containing 0.6 wt % Pt, 0.40 wt % Sn, and 1.3 wt % Cl_2

In the preliminary experiments the catalyst PR-42B gave the best results. It was successfully tested in an industrial unit [14] and it gave a better performance than the catalyst KR-110K (platinum-rhenium on alumina). The spherical catalyst SPR-2B also gave good results, comparable to those obtained using multimetallic catalysts [17].

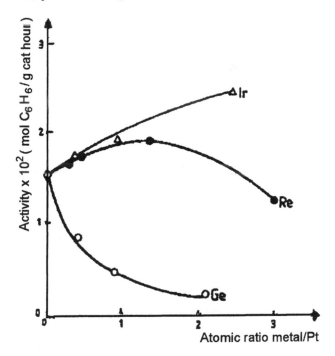

Figure 13.8 Hydrogenation activity vs. metal/platinum ratio for iridium, rhenium, and germanium. Benzene + 3H$_2$ → cyclohexane; $P = 1$bar; $t = 500°$C; space velocity 20 g/g/h; H$_2$/H$_C$ = 20 mol/mol. (From Ref. 12.)

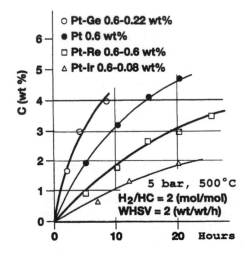

Figure 13.9 The dynamics of coke formation for various bimetallic catalysts. n-C$_7$H$_1$6 → toluene + gases + coke; $p = 5$ bar; $t = 500°$C; space velocity = 29 g/g/h; H$_2$/H$_c$ = 2 mol/mol. (From Ref. 12.)

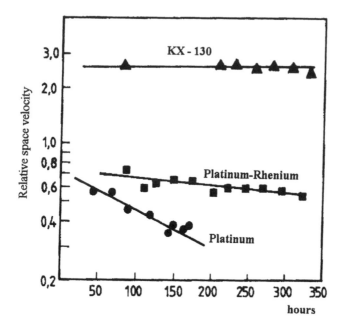

Figure 13.10   The time variation of the performance of several catalysts. (Feed: alkane naphtha, $p = 10.5$ bar, average temperature 499°C, F1 octane number of the reformate 102.5.)

Studies performed in Russia, investigated the promotion by cadmium as well as the general influence of the method of preparation on catalyst performance [19]. The studies examined promotion by 0.2–0.3 wt % Cd on the platinum-rhenium catalysts [18]. Although the higher relative volatility of cadmium leads to its elimination from the system, its effect seems to persist [20,21]. This could be explained by its influence on the dispersion of Re-Pt, which is maintained even after cadmium had left the catalyst by volatilization.

The dispersion of platinum and promoting metals in crystallites form plays a very significant role in the activity of reforming catalysts [4]. In the case of platinum-rhenium catalysts, the platinum crystallites may have dimensions between 8 and 100 Å [22]. The size of crystallites in a fresh catalyst should be smaller than 35 Å, preferably below 24 Å. Current trends are to achieve even smaller dimensions (nanometer sizes) for the dispersed noble metals particles on the support surface. Catalysts containing the noble metals as nanoparticles, will have higher activity (larger number of metallic sites) and lower costs (lower loading of noble metals) than the traditional ones.

One of the main roles of the promoters is to prevent the agglomeration of crystallites, a phenomenon known as "platinum sintering."

The dispersion effect can be observed in Figure 13.11, which shows the ratio between the amount of accessible platinum and the total amount of platinum for three catalysts: a nonpromoted catalyst, a Pt Ge/Al_2O_3 catalyst, and a Pt Ir/Al_2O_3 catalyst. The success of germanium as a promoter in maintaining the accessibility of platinum in time is very clear, justifying its use in catalyst preparation.

Additional details concerning the structure, characterization and testing of the reforming catalysts are given in the monographs edited by Antos et al. [189–191].

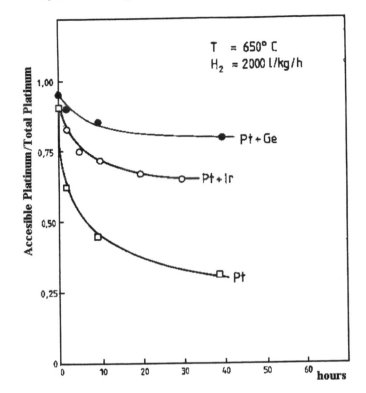

Figure 13.11   The effect of promoters on platinum efficiency. (From Ref. 12.)

13.3   REACTION MECHANISMS

Catalytic reforming was shown to consist of a number of reactions catalyzed by the two functions of the catalyst.

Figure 13.12A gives a schematic of the possible reactions for a feed of hexanes, while Figure 13.12B gives a similar sketch for heptanes. This reaction mechanism was analyzed by Raseev and Ionescu [23–25].

The sketches of Figures 13.12A and 13.12B require some explanations.

The precise mechanism of dehydrogenation of the six atom rings on Pt catalysts is still not completely understood [26]. Thus it was not possible to either confirm or deny the existence of the hexa-diene intermediate, already identified in the dehydrogenation on molybdenum oxide catalysts. For this reason such a step was not included in Figures 13.12 A and B.

The identification of the presence of cyclohexene is insufficient to confirm the dehydrogenation in stages or to reject the opposite theory, which considers the mechanism by which the molecule is adsorbed "parallel" to or "flat" on the catalyst surface, simultaneously on several centers—the "multiplet" theory of Balandin [27]. Cyclohexene may result from a parallel reaction occurring on pairs of sites, in places where the number of adjacent sites does not allow the adsorption on the multiplet. This aspect was mentioned by us earlier [23]. Consequently, the two mechanisms may take place in parallel, with the adsorption on multiplets having a dominant role. This latter mechanism is the only one accepted by various authors [28].

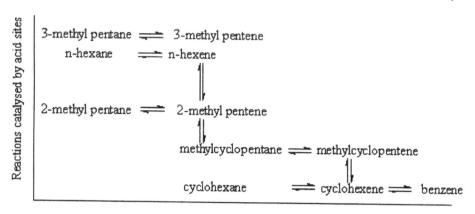

Figure 13.12A   Reactions of hexanes on bifunctional catalysts (hydrocracking reactions are not included).

The direct dehydrogenation of alkylcyclopentanes to aromatic hydrocarbons by enlargement of the cycle catalyzed by platinum was considered possible [29] even without the intervention of the acidic sites. In order to check this assumption, we prepared a catalyst of platinum on activated carbon [30]. Although the catalyst did not have any acidity, cyclohexene and benzene still formed from methyl-cyclopentane at 400°C in a batch reactor. The conversions however were two orders of magnitude lower than for a typical Sinclair Baker RD150 catalyst with 0.35 wt % Pt in the same conditions. The conclusion is that the presence of the acid sites is essential in the aromatization of methylcyclopentane.

The isomerization of cycloalkanes with five and six atoms in the ring follows the reactions suggested in Figures 13.12A and 13.12B. The presence of cyclohexene

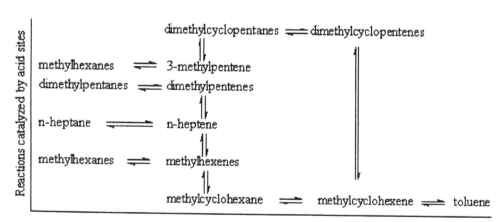

Figure 13.12B   Reactions of heptanes on bifunctional catalysts (hydrocracking reactions are not included).

and of methylcyclopentene has been identified in the study on methylcyclopentane aromatization [30].

It should be mentioned that the opening of the ring in methylcyclopentane leads to the formation of 2-methylpentene in agreement with the predominant formation of tertiary ions on acidic sites. The reaction is then followed by an isomerization to 3-methylpentene and n-pentene [30].

Our study confirmed the sequential character of the reactions. The ratio of 2-methylpentene to 3-methylpentene in the reaction product was almost two times larger than given by the thermodynamic equilibrium.

The ring-opening reaction is followed by reactions of hydrocracking of the iso-alkenes and also of the iso-alkanes.

The cyclization mechanism was examined in more detail by Raseev and Stoica within a study on the conversion of n-heptane on bifunctional catalysts in the unit [13] shown in Figure 13.13. The catalyst was diluted with inert material and the unit could be operated up to space velocities of 1000 g/g.hour.

At these high space velocities, *cis*- and *trans*-heptene-2, *cis*- and *trans*-heptene-3, and isoheptene were identified. No heptene-1 was found. Figure 13.14 shows the variation of total heptenes, methylcyclohexane, methylcyclopentane, and toluene with the weight hourly space velocity, while Figure 13.15 shows the variation of individual heptenes with the WHSV. The latter figure shows that the alkenes reach a maximum before methylcyclohexane, while toluene increases continuously. This fact supports the successive transformation

n-heptane $\leftrightarrow$ heptenes $\leftrightarrow$ methylcyclohexane $\leftrightarrow$ toluene

as shown also in Figure 13.12B.

The formation of intermediate alkenes was signaled by other authors [26,31] as well.

In most of the published studies it was observed that the 5-carbon atoms rings are formed in parallel with the 6-carbon rings. Their ratio depends on the characteristics of the catalyst.

With the exception of the methylcyclopentane-cyclohexane system, the probability of formation of five- or six-carbon atom rings is almost the same. This is reflected by the following equilibrium compositions calculated from thermodynamic data at a temperature of 480°C:

0.398   0.602

0.461   0.539

0.378   0.632

In the system methylcyclopentane-cyclohexane, the equilibrium is displaced towards the formation of methylcyclopentane; the equilibrium conversion at 480°C

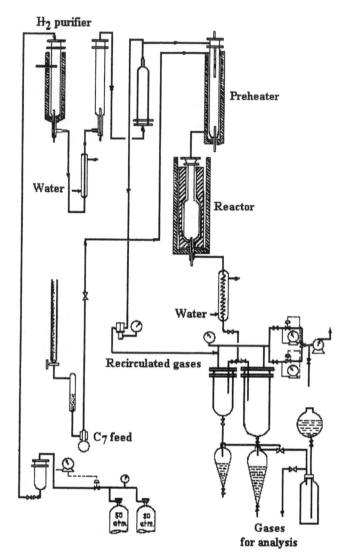

Figure 13.13 Continuous bench unit for the study of catalytic reforming.

corresponds to only 0.084 cyclohexane. Furthermore, the equilibrium of dehydro-genation of cyclohexane to benzene is less favorable than for the higher homologs (see Figure 13.1). These two effects explain why it is difficult to obtain high yields of benzene when subjecting to catalytic reforming naphthas rich in alkanes.

The type of ring resulting from the cyclization step depends on the manner in which the hydrocarbon is adsorbed on the active sites of the catalyst. Several assumptions were developed on this subject, ranging between that involving the positioning of the molecule of alkane between the platinum atoms (Kazanski [32]) and the adsorption of the atoms at the two ends of the carbon chain on two adjacent sites [13]. It is difficult to select from among these hypotheses. From the energy point of view, for hydrocarbons with seven or more carbon atoms, both types of rings have equal chance. This agrees with the experimental results and with most current opinions.

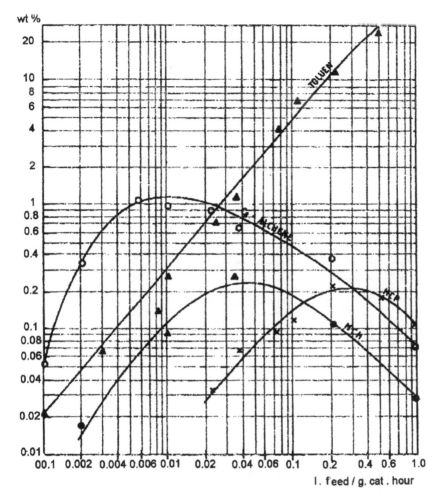

Figure 13.14 The formation of heptenes, methylcyclohexane, methylcyclopentane and toluene on a UOP-R11 catalyst in a continuous unit. $p = 10$ bar, $t = 500°C$, $H_2/C_7H_{16} = 6.5$. (From Ref. 13.)

The other reactions taking place on bifunctional catalysts, i.e., isomerization and hydrocracking, were already presented in detail in Chapters 11 and 12.

The formation of benzene from C_6 hydrocarbons, as a result of demethylation and hydrocracking, is of special interest. The sequence of reactions may be deduced from Figure 13.15 [13]. The position of the curves suggests that n-hexane first converts to cyclohexane, which then dehydrogenates to benzene.

The smaller value and the position of the maximum yield of methylcyclopentane at higher values of the space velocities than the maximum of cyclohexane suggests that methylcyclopentane appears from the isomerization of cyclohexane, rather than by cyclization of n-hexane.

At the beginning of this chapter we suggested the names "catalytic polyforming" and "catalytic dehydropolyaromatization" for the process with commercial

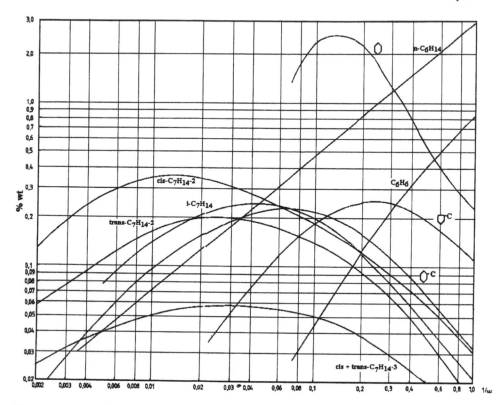

Figure 13.15   The conversion of *n*-heptane to C_6-hydrocarbons on a Pt/Re. Catalyst at $p =$ 10 bar, $t = 500°C$, H_2/hydrocarbon $= 1.5$ mol/mol/ (From Ref. 13.)

names "Cyclar" and "Aroforming." This process involves the reactions of dehydrogenation of C_3-C_5 alkanes on metal sites, followed by the dimerization of the formed alkenes on the acid sites, and subsequent cyclization and dehydrogenation to aromatic hydrocarbons. The reaction schematic developed by us is shown in Figure 13.16.

The process uses catalysts based on zeolite support, which do not favor the formation of polycyclic hydrocarbons and implicitly the formation of coke.

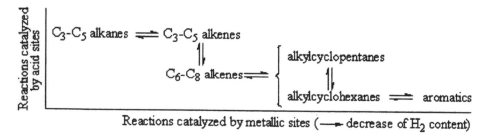

Figure 13.16   The reaction scheme of the aromatization process of C_3–C_5 alkanes on bifunctional catalysts.

However, since the operating conditions are more severe than in classic reforming, the catalyst requires continuous regeneration, either by continuously passing the catalyst through a regenerator (Cyclar) [33,34], or via the cyclic operation of the reactors (Aroforming) [35].

Details on the commercial implementation of these processes are given in Section 13.10.

13.4 THE KINETICS OF CATALYTIC REFORMING

13.4.1 The Influence of Diffusion Phenomena

The influence of external diffusion was studied by Raseev and Stoica [13] in a continuous bench unit by varying the linear velocity of the reactants through the catalyst bed at constant space velocity, shape, and size of the catalyst particles and constant parameters of the process. A plug flow through the catalyst was ensured. It has been found that external diffusion influences the process only at Reynolds numbers lower than 4. These were calculated for the diameter of the catalyst granules, at constant conditions of 500°C, 25 bar, molar ratio H_2/H_C of 6.5, and space velocity of 16.5 g n-heptane/g catalyst × hour.

Since in commercial reactors the Reynolds number is always above this value, the external diffusion has no influence on the yields obtained in commercial conditions.

Diffusion through the pores of catalysts depends very much on the pore structure of the particular catalysts, which makes it difficult to draw general conclusions. Previously published data [36–38] have shown a significant influence of the diameter of the catalyst granules on the overall rate of the process. The rate decreases with the increase in the average diameter of the catalyst, which demonstrates the influence of internal diffusion.

Experiments carried out using a Sinclair-Baker RD150 and UOP R11 catalysts of various dimensions led to the conclusion that internal diffusion begins to influence the overall reaction rate at granule diameters larger than 0.4–0.63 mm [13]. The process conditions were: temperature of 500°C, pressure of 25 bar, H_2/H_C molar ratio of 6.5, and WHSV of 50 g/g catalyst × hour.

Since the size of the catalyst particles in commercial reactors is larger than these values, one may conclude that internal diffusion will reduce the overall reaction rate of industrial processes. Naturally, the effect of internal diffusion may be decreased or even completely eliminated by using a more porous structure for alumina support. This explains the evolution of industrial catalysts from having a total pore volume of 0.48 cm³/g catalyst in 1975–1978 to a value of 0.6–0.7 cm³/g catalyst at the present time.

Usually, reforming catalysts are prepared as extruded cylinders with diameters of about 1.5 mm (or sometimes 3 mm) or as spheres with diameters of 2–3 mm.

The use of zeolite support with bimodal pore distribution and macrocirculation pores can reduce or completely eliminate the slowing effect of the internal diffusion on the reaction rate.

The decrease of the diameter of the catalyst particles lowers the effect of internal diffusion and increases the pressure drop in the reactor (see Figure 13.17). This increases the cost of compressing the recycle hydrogen-rich gases. The optimum

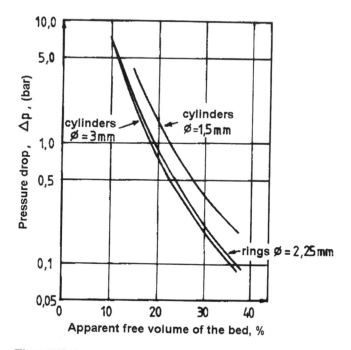

Figure 13.17 The effect of size of catalyst particles on the pressure drop. Operating conditions: $p = 30$ bar, $t = 380°C$, H_2/feed ratio $= 50$ Nm^3/m^3, feed flowrate 64 m^3/hour, catalyst bed diameter 1.5 m. (From Ref. 4.)

shape and size of the catalyst granules are determined as a function of the specific process operating conditions.

13.4.2 The Reaction Kinetics

The first kinetic models for catalytic reforming were proposed by Smith [39] and by Krane et al. [40] and were reviewed by Raseev et al. [23–25].

Smith [39] expressed the rate of reaction by the equations:

$$r = \frac{dN_i}{dV_R} \tag{13.6}$$

where N_i is the molar fraction of compound i transformed per mole of raw material and V_R is the ratio kg of catalyst per mole of feed per hour.

The equations were written based on the stoichiometry of the reaction. The rate constants for the direct reactions are given as follows:

- For dehydrogenation of cycloalkanes (the author did not differentiate between hydrocarbons with five and six atoms rings):

$$k_1 = 2.205 \cdot e^{23.21 - \frac{19,300}{T}} \tag{13.7}$$

- For dehydrocyclization of cycloalkanes:

$$k_2 = 2.205 \cdot e^{35.98 - \frac{33,100}{T}} \tag{13.8}$$

- For hydrocracking of alkanes and cycloalkanes:

$$k_3 = 2.205 \cdot e^{42.97 - \frac{34,600}{T}} \tag{13.9}$$

Krane et al. [40] adopted a more complex kinetic model, taking into account the reactions of n-heptane as typical for the whole catalytic reforming process:

$$\begin{array}{ccc} & n\text{-}C_7H_{16} & \\ \text{products of} \swarrow & \Updownarrow & \searrow \\ \text{hydrocracking} & i\text{-}C_7H_{16} & \end{array} \quad \text{cycloalkanes} \rightleftharpoons \text{toluene}$$

Neither this model nor that of Smith differentiate between cycloalkanes with rings of five and six carbon atoms.

The authors developed five empirical differential equations based on the above reactions.

For naphtha fractions, Krane et al. [40] proposed a kinetic equation for first-order homogeneous processes for all transformations taking place in the process. The general form of the equation is:

$$\frac{dN_i}{d\left(\frac{A_c}{w}\right)} = -k_i N_i \tag{13.10}$$

where A_c is the activity of the catalyst for a particular reaction and w is the space velocity.

The values of the rate constant recommended by the authors [40] are given in Table 13.5

The 53 differential equations written as Eq. 13.10 were then grouped into twenty differential equations, one for each type of hydrocarbon. The equation for the conversion of n-heptane becomes:

$$\frac{dP_7}{d\left(\frac{A_c}{w}\right)} = 0.0109 P_{10} + 0.0039 P_9 + 0.0019 P_8 + 0.002 N_7 + 0.0016 A_7 - 0.0122 P_7$$

where: P = paraffins (alkanes), N = naphthenes (cycloalkanes), A = aromatics. The subscripts 7, 8, 9, and 10 refer to the number of carbon atoms in the respective molecules.

By solving this system of differential equations, the time dependence of the composition of the reaction mixture may be calculated.

Although the method published by Krane was one of the first calculation methods, it is still used today for model development. The models are used to improve the performance of commercial catalytic reforming units [41,42].

The approaches of Smith and Krane do not take into account that the catalytic reforming process takes place in adiabatic conditions. Therefore, the equations have to be solved simultaneously with those describing the evolution of temperature (deduced from heat balances) along the reactor length.

Burnett [43,44] improved the approach. He started from the following reaction scheme:

Table 13.5 Reaction Rate Constants for the Calculation of Catalytic Reforming Kinetics by the Method of Krane et al.

Reaction	$k \cdot 10^2$	Reaction	$k \cdot 10^2$
$P_{10} \rightarrow N_{10}$	2.54	$A_{10} \rightarrow A_7$	0.00
$P_{10} \rightarrow P_9 + P_1$	0.49	$P_9 \rightarrow N_9$	1.81
$P_{10} \rightarrow P_8 + P_2$	0.63	$P_9 \rightarrow P_8 + P_1$	0.30
$P_{10} \rightarrow P_7 + P_3$	1.09	$P_9 \rightarrow P_7 + P_2$	0.39
$P_{10} \rightarrow P_6 + P_4$	0.89	$P_9 \rightarrow P_6 + P_3$	0.68
$P_{10} \rightarrow 2P_5$	1.24	$P_9 \rightarrow P_5 + P_4$	0.55
$N_{10} \rightarrow N_9$	0.54	$N_9 \rightarrow P_9$	0.54
$N_{10} \rightarrow P_{10}$	1.34	$N_9 \rightarrow N_8$	1.27
$N_{10} \rightarrow N_8$	1.34	$N_9 \rightarrow N_7$	1.27
$N_{10} \rightarrow N_7$	0.80	$N_9 \rightarrow A_9$	24.50
$N_{10} \rightarrow A_{10}$	24.50	$A_9 \rightarrow P_9$	0.16
$A_{10} \rightarrow P_{10}$	0.16	$A_9 \rightarrow A_8$	0.05
$A_{10} \rightarrow A_9$	0.06	$A_9 \rightarrow A_7$	0.05
$A_{10} \rightarrow A_8$	0.06	$P_8 \rightarrow N_8$	1.33
$P_8 \rightarrow P_7 + P_1$	0.19	$N_7 \rightarrow P_7$	0.20
$P_8 \rightarrow P_6 + P_2$	0.25	$N_7 \rightarrow A_7$	9.03
$P_8 \rightarrow P_5 + P_3$	0.43	$A_7 \rightarrow P_7$	0.16
$P_8 \rightarrow 2P_4$	0.35	$P_6 \rightarrow N_6$	0.00
$N_8 \rightarrow P_8$	0.47	$P_6 \rightarrow P_5 + P_1$	0.14
$N_8 \rightarrow N_7$	0.09	$P_6 \rightarrow P_4 + P_2$	0.18
$N_8 \rightarrow A_8$	21.50	$P_6 \rightarrow 2P_3$	0.27
$A_8 \rightarrow P_8$	0.16	$N_6 \rightarrow P_6$	1.48
$A_8 \rightarrow A_7$	0.01	$N_6 \rightarrow A_6$	4.02
$P_7 \rightarrow N_7$	0.58	$A_6 \rightarrow N_6$	0.45
$P_7 \rightarrow P_6 + P_1$	0.14	$P_5 \rightarrow P_4 + P_1$	0.12
$P_7 \rightarrow P_5 + P_2$	0.18	$P_5 \rightarrow P_3 + P_2$	0.15
$P_7 \rightarrow P_4 + P_3$	0.39		

A – aromatics, N – cycloalkanes, P – alkanes; the subscripts represent the number of carbon atoms in molecule.
Source: Ref. 40.

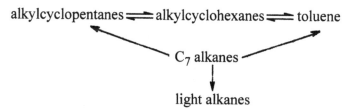

This scheme is debatable because of the well-known fact that alkanes do not convert directly to aromatic hydrocarbons. However, the scheme was used by the author for computer modeling, leading to plausible results.

Jorov and Pancenkov [45–47] used a simplified model, similar to that used by Smith:

$$\text{hydrocracking products} \leftarrow P \rightleftharpoons N \rightleftharpoons A$$

Again, they did not differentiate between the behavior of alkylcycloalkanes with five and with six carbon atoms in the ring, or between the behavior of normal- and iso-alkanes. Instead, they introduced several approximations as well as some "coefficients of adjustment" in order to fit the calculations to industrial data. This gives an empirical character to the whole approach. The results were used by the authors for developing recommendations on the distribution of catalyst among the reactors and for the optimum temperature profile.

Henningen and Bundgard-Nielson [48] developed a more complex model, that takes into account differences in the behavior of cycloalkanes with five and with six carbon atoms in the ring. It also considers differences in the behavior of normal- and iso-alkanes. The flowchart of the model is:

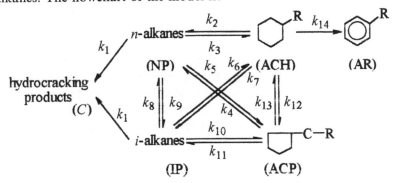

where:

NP = normal paraffins (n-alkanes)
IP = iso-paraffins (i-alkanes)
ACH = alkyl-cyclohexanes
ACP = alkyl-cyclopentanes
AR = aromatics

Considering the system as pseudohomogeneous, the differential equations are:

$$\frac{dP_c}{d\tau} = -k_1(P_{NP} + P_{IP}) \tag{13.11}$$

$$\frac{dP_{NP}}{d\tau} = -(k_1 + k_3 + k_5 + k_9)P_{NP} + k_2 P_{ACH} + k_4 P_{ACP} + k_8 P_{IP} \tag{13.12}$$

$$\frac{dP_{IP}}{d\tau} = (k_1 + k_6 + k_8 + k_{10})P_{IP} + k_7 P_{ACH} + k_9 P_{NP} + k_{11} P_{ACP} \tag{13.13}$$

$$\frac{dP_{ACH}}{d\tau} = -(k_2 + k_7 + k_{12} + k_{14})P_{ACH} + k_3 P_{NP} + k_6 P_{IP} + k_{13} P_{ACP} \tag{13.14}$$

$$\frac{dP_{ACP}}{d\tau} = -(k_4 + k_{11} + k_{13})P_{ACP} + k_5 P_{NP} + k_{10} P_{IP} + k_{12} P_{ACH} \tag{13.15}$$

$$\frac{dP_{AR}}{d\tau} = k_{14} P_{ACH} \tag{13.16}$$

A heat balance equation was added to the system of equations. This was a considerable improvement on previous models that treated catalytic reforming as an isothermal system:

$$\frac{dt}{d\tau} = -\frac{1}{C_p(n+1)}(k_1 P_{\mathrm{NP}}\Delta H_{\mathrm{NP}\to\mathrm{C}} + k_1 P_{\mathrm{IP}}\Delta H_{\mathrm{IP}\to\mathrm{C}} + k_2 P_{\mathrm{ACH}}\Delta H_{\mathrm{ACH}\to\mathrm{NP}}$$

$$+ k_3 P_{\mathrm{NP}}\Delta H_{\mathrm{NP}\to\mathrm{ACH}} + k_4 P_{\mathrm{ACP}}\Delta H_{\mathrm{ACP}\to\mathrm{NP}} + k_5 P_{\mathrm{NP}}\Delta H_{\mathrm{NP}\to\mathrm{ACP}}$$

$$+ k_6 P_{\mathrm{IP}}\Delta H_{\mathrm{IP}\to\mathrm{ACH}} + k_7 P_{\mathrm{ACH}}\Delta H_{\mathrm{ACH}\to\mathrm{IP}} + k_8 P_{\mathrm{IP}}\Delta H_{\mathrm{IP}\to\mathrm{NP}}$$

$$+ k_9 P_{\mathrm{NP}}\Delta H_{\mathrm{NP}\to\mathrm{IP}} + k_{10} P_{\mathrm{IP}}\Delta H_{\mathrm{IP}\to\mathrm{ACP}} + k_{11} P_{\mathrm{ACP}}\Delta H_{\mathrm{ACP}\to\mathrm{IP}}$$

$$+ k_{12} P_{\mathrm{ACH}}\Delta H_{\mathrm{ACH}\to\mathrm{ACP}} + k_{13} P_{\mathrm{ACP}}\Delta H_{\mathrm{ACP}\to\mathrm{ACH}}$$

$$+ k_{14} P_{\mathrm{ACH}}\Delta H_{\mathrm{ACH}\to\mathrm{AR}}) \tag{13.17}$$

In the above equations: τ is the reaction time, p_i the partial pressure of the component i, C_p the heat capacity, n the hydrogen/hydrocarbon ratio, k_i the reaction rate constant expressed using the Arrhenius equation, $\Delta H_{i\to j}$ the heat of reaction for the conversion i to j.

The values for the pre-exponential factors, the activation energies, and the heat of reaction recommended by the authors for the above model are given in Table 13.6.

A major improvement brought by the model developed by Henningen was the distinction made between cycloalkanes with five and with six carbon atoms in the cycle, and above all, the addition of the heat balance equation. Considering the process as nonisothermal during calculation brought the simulation results closer to reality.

Henningen's model [48] was compared with some simplified models [13] that also proved satisfactory for the conditions used.

Jenkins and Stephens [49] tried to improve Krane's method [40]. Krane's rate constants are indeed valid only for a pressure of 21 bar (300 psig). However, the system pressure influences in different ways the various types of reaction taking place during catalytic reforming.

Table 13.6 Values for the Parameters of Eqs. 13.11–13.17: Rate Constants, Pre-exponential Factors, Activations Energies, Heat of Reaction

Reaction	Relative rate constant, $t = 500°C$, $p_{\mathrm{H2}} = 30\,\mathrm{bar}$	Pre-exponential factor A (natural logarithm, ln A)	Activation energy, E, kcal/mol	Heat of reaction, ΔH, kcal/mol
NP → C	$k_1 = 0.10$	30.5	55.0	−12.0
NP → ACH	$k_3 = 0.04$	23.1	45.0	+10.0
NP → IP	$k_9 = 0.40$	21.9	40.0	0.0
NP → ACP	$k_5 = 0.04$	23.1	45.0	+10.0
IP → C	$k_1 = 0.10$	30.5	55.0	−12.0
IP → NP	$k_8 = 0.04$	19.6	40.0	0.0
IP → ACH	$k_6 = 0.04$	23.1	45.0	+10.0
IP → ACP	$k_{10} = 0.04$	23.1	45.0	+10.0
ACH → NP	$k_2 = 0.10$	24.2	45.0	−10.0
ACH → IP	$k_7 = 1.00$	26.5	45.0	−10.0
ACH → ACP	$k_{12} = 2.00$	23.5	40.0	0.0
ACH → AR	$k_{14} = 50.00$	20.4	30.0	+50.0
ACH → NP	$k_4 = 0.01$	21.6	45.0	−10.0
ACH → IP	$k_{11} = 0.10$	24.2	45.0	−10.0
ACH → ACP	$k_{13} = 0.20$	21.2	40.0	0.0

Consequently, Jenkins and Stephens proposed the introduction of exponential terms for pressure in order to correct the rate constants suggested by Krane:

$$k_i = k_i^0 \left(\frac{p}{300}\right)^{\alpha_i} \tag{13.18}$$

where k_i is the corrected rate constant, k_i^0 is the original rate constant given by Krane at the pressure of 300 psig, p is the system pressure in psig, and α is the correction coefficient. If the pressure is expressed in bars, the ratio within the brackets becomes $(p/21)$.

The values for the exponent α suggested by Jenkins for different reaction types are:

Isomerization	$\alpha = 0.37$
Dehydrocyclization	$\alpha = -0.70$
Hydrocracking	$\alpha = 0.433$
Hydrodealkylation	$\alpha = 0.50$
Dehydrogenation	$\alpha = 0.00$

heir values were calculated using the rate of reaction at two pressure levels. The zero value for the dehydrogenation of cycloalkanes indicates that at the almost complete conversion obtained at the process temperature, the effect of pressure is negligible.

Ancheyta-Juàrez and Anguilar-Rodriguez [41] of the Mexican Petroleum Institute generalized Krane's model further, adding a correction for temperature to Jenkins' pressure corrections. Thus, Eq. 13.18 became:

$$k_i = k_i^0 \left(\frac{p}{21}\right)^{\alpha} \exp\left[\frac{E_i}{R}\left(\frac{1}{T^0} - \frac{1}{T}\right)\right] \tag{13.19}$$

where T^0 is the reference temperature and T is the temperature corresponding to the constant k_i. The values for the activation energies E_i of various reactions were taken from Henningen (see Table 13.6).

The reference temperature corresponds to the temperature at which Krane calculated the rate constants given in Table 13.5. However, Krane [40] mentioned that for C_5-C_7 hydrocarbons the rate constants were obtained under isothermal conditions at 500°C, while for the C_8-C_{10} hydrocarbons the conditions were adiabatic in the temperature interval of 515–470°C. In that case, the reference temperature should be taken as the temperature equivalent to the adiabatic mean rate, Eqs. (2.163) or (2.164). For the present case, it is difficult and of no practical value to calculate the reference temperature with a high degree of precision for each reaction. Indeed, since the average rate depends on the activation energy of each reaction under consideration, the error using an average temperature is lower than the experimental error associated with the evaluation of the rate constants.

In general, the temperature equivalent to the adiabatic mean rate, t_{EVMA}, is somewhat higher than the arithmetic average for a specified interval. Hence, the same value of 500°C could be accepted for C_8-C_{10} hydrocarbons as for the C_6-C_7 hydrocarbons.

The model of Krane [40], improved by Jenkins [49], Henningen [48], and Ancheyta [41] was used by AguilarRodriguez and Ancheyta-Juàrez [42] for computer modeling of the process of catalytic reforming. The results were in remarkable agreement with the experimental effluent compositions and the dynamics of the reaction temperatures.

In all the above reported studies, the reactions were considered to be pseudo-homogeneous, which was reflected in the expressions used for the rate equations. The developed reaction models were shown to be satisfactory for reactor design and for optimization purposes.

The kinetics of catalytic reforming was treated as a heterogeneous catalytic process in 1967 by Raseev and Ionescu [51] within a study of the catalytic reforming of C_6-hydrocarbons in a bench unit with a plug flow, isothermal reactor connected to a gas chromatograph.

The reactions considered were:

light hydrocarbons

The reaction rates for the five participants were expressed as a function of the fraction of catalyst surface occupied by reactants without distinguishing between the two types: acid and metallic sites:

$$r_{CP} = -k_{-1}\theta_{CP} + k_1\theta_{CH}$$

$$r_A = k_2\theta_{CH} - k_{-2}\theta_A P_{H_2}^3$$

$$r_{C_6} = k_{-3}\theta_{CH}P_{H_2} - k_{-3}\theta_{C_6} - k_4\theta_{C_6}P_{H_2} \qquad (13.20)$$

$$r_{HC} = k_4\theta_{C_6}P_{H_2}$$

$$r_{CH} = k_{-1}\theta_{CP} + k_{-2}\theta_A P_{H_2}^3 + k_3\theta_{C_6} - (k_1 + k_2 + k_{-3}p)\theta_{CH}$$

At the relatively high pressures used in the experiments, (15 and 25 bar), the fraction of the surface occupied by the reactants was expressed by the simplified equations:

$$\theta_i = \frac{b_i p_i}{\sum b_i p_i} \qquad (13.21)$$

where b_i are the adsorption coefficients.

Despite such simplifications, the limited number of experimental points that were generated proved insufficient for evaluating the parameters of the kinetic model.

A more thorough treatment of the kinetics of catalytic reforming was performed by the research team of Martin and Froment for C_6-hydrocarbons [52] and by Van Trimpant, Martin, and Froment for C_7-hydrocarbons [53].

The authors used the following scheme for C_7-hydrocarbons without considering the hydrocracking reactions:

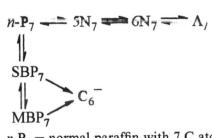

n-P$_7$ = normal paraffin with 7 C atoms (heptane), SBP$_7$ = single-branched C$_7$ paraffins (alkanes), MBP$_7$ = multi-branched C$_7$ alkanes, 5N$_7$ = C$_7$ Naphthenes (cyclo-alkanes) with 5 C atoms in the ring, 6N$_7$ = C$_7$ naphthene, with 6 C atoms in the ring.

The experimental data were obtained in an isothermal tubular reactor using a presulfided catalyst containing 0.593 wt % Pt and 0.67 wt % Cl$_2$ on γ-Al$_2$O$_3$. The feed was 99% n-heptane. The 996 experimental points obtained allowed the authors to verify their kinetic model and to determine the values of the reaction rate constants and the adsorption constants. The final equations were:

- For isomerization of alkanes

$$r_1^0 = \frac{Ae^{-E/RT}(p_A - p_B/K_{A-B})}{p_H \cdot \Gamma} \tag{13.22}$$

- For hydrocracking

$$r_2^0 = \frac{Ae^{-E/RT}p_A}{\Gamma} \tag{13.23}$$

- For cyclization

$$r_3^0 = \frac{Ae^{-E/RT}(p_A - p_B p_H/K_{A-B})}{p_H \cdot \Gamma} \tag{13.24}$$

- For isomerization of C$_5$ and C$_6$ cycloalkanes

$$r_4^0 = \frac{Ae^{-E/RT}(p_A - p_B/K_{A-B})}{p_H \cdot \Gamma} \tag{13.25}$$

- For dehydrogenation of methylcyclohexane

$$r_5^0 = \frac{Ae^{-E/RT}(p_A - p_B p_H^3/K_{A-B})}{(p_H \cdot \theta)^2} \tag{13.26}$$

Table 13.7 Pre-exponential Factors and Activation Energies

Equation	A	E, kJ/mol
(13.22)	1.83×10^6 kmol/kg cat. $\cdot$ h	87.75
(13.23)	1.43×10^{17} kmol/kg cat. $\cdot$ h $\cdot$ bar	256.4
(13.24)	2.48×10^{17} kmol/kg cat. $\cdot$ h	256.4
(13.25)	9.08×10^{17} kmol/kg cat. $\cdot$ h	256.4
(13.26)	1.47×10^4 kmol $\cdot$ bar/kg cat. $\cdot$ h	42.6

where:

> r_i^0 – reaction rates in kmol/kg catalyst hour in the absence of any coke deposits,
> p_A and p_B – the partial pressures of the initial and final components of the
> reactions, Γ and θ – adsorption factors.

The pre-exponential factors and the activation energies are given in Table 13.7.
The adsorption factors Γ and θ are given by the following equations:

The adsorption on acid centers—Γ:

$$\Gamma = \left(p_H + b_{C_6}P_{C_6} + b_{p_7}p_{p_7} + b_{N_7}p_{N_7} + b_T p_T\right)/p_H \tag{13.27}$$

where

$$b_{C_6} = 107\,\text{bar}^{-1}; \quad b_{p_7} = 21\,\text{bar}^{-1};$$

$$b_{N_7} = 659\,\text{bar}^{-1}; \quad b_T = 70.3\,\text{bar}^{-1}$$

The adsorption on metallic centers—θ

$$\theta = 1 + b_{MCH}p_{MCH} + Ae^{-\Delta H^0 RT}\left(p_{MCH}/p_H^2\right) \tag{13.28}$$

where

$$b_{MCH} = 0.27\,\text{bar}^{-1}; \quad A = 8.34 \cdot 10^9\,\text{bar}; \quad \Delta H^0 = 96.93\,\text{kJ/mol}$$

The authors considered the dehydrogenation of methylcyclohexane to toluene
to be the only reaction taking place by adsorption on the metallic centers. All other
reactions involved adsorption on the acid centers. The authors justified this separation by the fact that in isomerization, cyclization, and hydrocracking, the rate-
controlling steps are those involving the acid centers. This is in agreement with other
publications [54–56].

13.4.3 Catalyst Decay

The decay of catalyst activity was studied by many researchers [28,53,56–62]. Most
authors suggested various mathematical expressions that take into account the
decrease of the rate constants due to the formation of coke.
  Rabinovich et al. [28] proposed equations of the following type:

$$k_i = k_0 \frac{1}{1 + \alpha_i C_i} \tag{13.29}$$

where k_0 is the rate constant for the fresh catalyst, C_i is the coke content and α_i is a
deactivation constant.
  Equations similar to Eq. (13.29) have been written separately for the acid and
for the metallic sites of the catalyst. However, the authors did not give any values for
the constants in order to apply this method.
  Tanatarov et al. [57] suggested a deactivation equation of the following form:

$$k_i = k_0 e^{-k\tau} \tag{13.30}$$

Ostrovsky and Demanov [58] suggested the following form:

$$k_i = (1 - c/c_{max})^{\alpha_M/\alpha_K} \tag{13.31}$$

where c, c_{max} are the quantities of coke and α_M, α_K are the rate constants for the formation of coke on metallic and acid sites. The experimental values obtained by the authors gave a ratio of $\alpha_M/\alpha_K \cong 15$.

The kinetics of coke formation was also addressed in the paper on the catalytic reforming of C_6-C_7 hydrocarbons quoted above [53]. The following equation was developed for C_7-hydrocarbons:

$$R_c^0 = k_1 \frac{P_T \cdot P_{5N_7}}{p_H^2} + k_2 \frac{P_T \cdot P_{MCH}}{p_H^2} + k_3 \frac{p_{NP_7}}{p_H} \tag{13.32}$$

and for C_6-hydrocarbons:

$$R_c^0 = k_4 \left(\frac{p_{MCP}}{p_H^2}\right)^5 + k_5 \frac{p_{IH} \cdot p_B}{p_H^2} \tag{13.33}$$

In these equations, R_c^0 is the initial reaction rate for the fresh catalyst, expressed in kg coke /kg catalyst hour while p are the partial pressures. The following subscripts are used for hydrocarbons: T – toluene, $5N_7$ – cyclopentanes with seven carbon atoms, H – hydrogen, MCH – methylcyclohexane, NP_7 – normal-heptane, MCP – methyl-cyclopentane, IH – isohexanes, and B – benzene.

The values of the pre-exponential factors of the five rate constants and their corresponding activation energies are given below:

Rate constant	Pre-exponential factor, A, kg coke/kg catalyst hour	Activation energy, kJ/mol
k_1	7.25×10^5	58.64
k_2	1.00×10^6	58.64
k_3	1.01×10^2	58.64
k_4	3.18×10^{16}	205.7
k_5	3.77×10^2	58.94

Experimental data were used for selecting the following reactions for coke formation, which were incorporated in Eqs. (13.32) and (13.33):

For C_7 hydrocarbons

toluene + C_7 cyclopentadienes $\rightarrow C_{p1}$

toluene + C_7 methylcyclohexadienes $\rightarrow C_{p2}$

n-heptenes $\rightarrow C_{p3}$

where C_{pi} are the precursors of coke and the three reactions correspond to the three terms in Eq. 13.32.

For C_6 hydrocarbons

2-methylcyclopentadienes $\rightarrow C_{p4}$

benzene + methylpentadienes $\rightarrow C_{p5}$

where these reactions correspond to the two terms in Eq. (13.33).

In commercial plants, the space velocity is mostly constant. For a constant space velocity the ratio kg catalyst/kg feed per hour is constant, and it is given below:

$$\frac{dC_c}{d\tau} = R_c \tag{13.34}$$

where C_c is the content of coke on catalyst, in kg cok/kg catalyst and R_c is the rate of coke formation in kg coke/kg catalyst per hour.

The initial rates of various reactions given by Eqs 13.22–13.26 decrease in time due to the formation of coke on the catalyst. This deactivation is expressed by equations of the following form [53]:

$$r_i = r_i^0 \phi_i, \quad \text{where} \quad 0 \le \phi_i \le 1 \tag{13.35}$$

In a similar way, a decrease in the rate of coke formation [Eqs. (13.32) and (13.33)] may be expressed by:

$$R_c = R_c^0 \phi_c, \quad \text{where} \quad 0 \le \phi_c \le 1 \tag{13.36}$$

The deactivation functions are expressed by:

$$\phi_i = e^{-\alpha_i C_c} \tag{13.37}$$

$$\phi_c = e^{-\alpha_c C_c} \tag{13.38}$$

Equation (13.37) may be written for the deactivation of acidic sites and for the deactivation of metallic sites, replacing the subscript i with A or M, respectively.

Since the deactivation of acid sites leads to the deactivation of the catalytic reforming catalysts, the following expression is often used:

$$\phi_A = e^{-\alpha_A C_c} \tag{13.39}$$

The values for the α constants in the deactivation equations, in kg catalyst/kg coke, are:

$$\alpha_A = 14.95 \qquad \alpha_M = 16.79 \qquad \alpha_C = 24.18$$

After replacing ϕ from Eq. (13.35) in Eq. (13.38) and following integration, it results:

$$C_c = \frac{1}{\alpha_c} \ln(1 + \alpha_c R_c \tau) \tag{13.40}$$

This allows the calculation of the amount of coke formed on the catalyst in kg coke/kg catalyst, after a time on stream τ, operating the unit at a constant space velocity.

The value obtained for C_c may also be replaced in Eqs. (13.39) and (13.35) in order to calculate the decreased reaction rate due to the formed coke.

The amount of coke formed up to a certain distance z_j from the inlet of the isothermal reaction system or, the corresponding time τ_j, may be obtained by writing Eq. (13.40) in terms of the duration τ_j:

$$C_c(z_j, \tau_j) = \frac{1}{\alpha_c} \ln(1 + \alpha_c R_c \tau_j) \tag{13.40'}$$

Equations (13.40) and (13.40') may also be written as:

$$\alpha_c R_c \tau = e^{\alpha_c C_c} - 1$$

$$\alpha_c R_c \tau_j = e^{\alpha_c C_c(z_j, \tau_j)} - 1$$

Dividing and regrouping results in:

$$\frac{(e^{\alpha_c C_c} - 1)\tau_j}{\tau} = e^{\alpha_c C_c(z_j, \tau_j)} - 1 = \alpha_c R_c \tau_j$$

Substituting in Eq. (13.40) finally leads to:

$$C_c(z_j, \tau_j) = \frac{1}{\alpha_c} \ln\left(1 + \frac{e^{\alpha_c C_c} - 1}{\tau} \tau_j\right) \tag{13.41}$$

A later publication [59] studied the deactivation of reforming catalysts in processing naphtha fractions on catalysts of the Pt/Re supported on chlorinated γ-Al$_2$O$_3$. (The catalyst prepared by Leuna-Werke, Merseburg, Germany, contained 0.36% Pt, 0.28% Re and 1.3% Cl.)

The well-mixed microreactor contained 4.8 g catalyst and operated in iso-thermal conditions at a constant partial pressure of hydrogen. The experimental temperatures ranged from 753–773 K, the pressure ranged from 10–15 bar, and feed rates ranged from 80–160 cc n-heptane/hour.

The mathematical model was based on the following reactions:

$$n\text{-}C_7H_{16} \longrightarrow i\text{-}C_7H_{16} \tag{a}$$

$$i\text{-}C_7H_{16} \longrightarrow \text{(toluene)} + 4H_2 \tag{b}$$

$$\text{(toluene)} + H_2 \longrightarrow \text{(benzene)} + CH_4 \tag{c}$$

$$i\text{-}C_7H_{16} + H_2 \longrightarrow i\text{-}C_6H_{14} + CH_4 \tag{d}$$

$$n\text{-}C_7H_{16} + H_2 \longrightarrow n\text{-}C_6H_{14} + CH_4 \tag{e}$$

$$n\text{-}C_7H_{16} + H_2 \longrightarrow 2(C_1 \cdots C_5) \tag{f}$$

Assuming first order homogeneous reaction, the rates were expressed as:

$$r_i = k_i \phi_i p_j \tag{13.42}$$

where the deactivation function ϕ_i was defined by

$$\phi_i = k_i(\tau)/k_{i(\tau=0)} \tag{13.43}$$

and expresed via the exponential equation:

$$\phi_i = e^{-\alpha_i C} \tag{13.44}$$

This approach was similar to the one discussed above [53].

The processing of experimental data, summarized in Figure 13.18, led to the following expression for the rate of formation of coke:

$$\frac{dC}{d\tau} = (k_p p_p + k_A p_A) p_{H_2}^{n_1} C^{n_2} e^{-k_c C} \tag{13.45}$$

where:

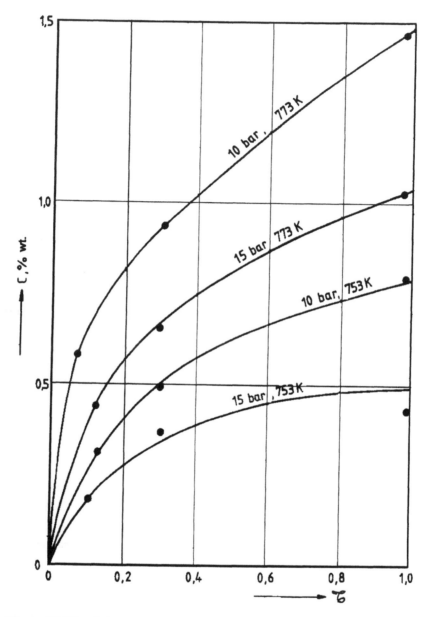

Figure 13.18 Coke formation vs. relative reaction time. The lines represent data calculated by using Eq. 13.45. (From Ref. 59.)

C is the coke contained in the catalyst in wt %
p is the partial pressures in MPa
subscripts:
P is for alkanes
A is for aromatics
H_2 is for hydrogen

exponents $n_1, n_2 < 0^*$

The three rate constants appearing in Eq. (13.45) are expressed by Arrhenius-type equations, their parameters being given in the table below:

Constant and dimensions	$\ln A$	E, kJ/mol
K_P, MPa$_1^{-(n+1)}$ h^{-1}	57.486	377.16
K_A, MPa$_1^{-(n+1)}$ h^{-1}	31.353	197.12
K_C, wt %$^{-1}$	-23.104	144.63

$1 MP_a = 10$ bar

The constant α_i in the exponent of Eq. (13.44) depends on the temperature and the type of reaction. Its value is given in Figure 13.19 for the six reactions (a to f)

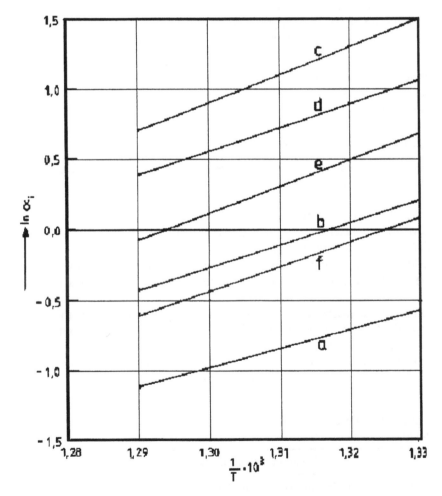

Figure 13.19 The deactivation constant α_i vs. temperature and reaction type. (From Ref. 59.)

---

* The value of exponent n_1 calculated by us from Figure 13.18 appears to be $n_1 \cong -2.5$.

used for the mathematical modeling. The lines of Figure 13.19 indicate that values of α_i and consequently the values of the deactivation function ϕ_i are reaction specific. This conclusion is different from that formulated by other authors [53], who accepted the same value for all reactions considered in the model.

13.5 THE EFFECT OF PROCESS PARAMETERS

Analyses of the effect of the process parameters should account for the presence of the two types of active sites that catalyze different types of reactions, which are sometime competing reactions.

13.5.1 Temperature

Values for the apparent activation energies for the main reactions taking place in catalytic reforming on platinum catalysts were selected and calculated [23–25] from published data [63–66]. They are given below:

	kJ/mol
Dehydrogenation of cycloalkanes	80.8
Cyclization of alkanes	157.2
Hydrocracking of alkanes	144.9
Hydrocracking of cycloalkanes	144.9
Hydrogenolysis of alkylcyclopentanes	146.5

These data reflect the fact that temperature has a higher effect on the reactions of cyclization and hydrocracking than on the reactions of dehydrogenation of cycloalkanes.

Consequently, increasing temperature will enhance the conversion of cyclo-alkanes by favoring the reactions of hydrocracking, while for alkanes it will enhance the conversion for the reactions of cyclization and dehydrogenation.

The following values of apparent activation energies for the conversion of n-heptane were derived from our research on some pure hydrocarbons [13] (see Table 13.8).

The values of the activation energies depend on pressure and on the actual catalyst as specified in the table above. This makes the comparison with other published data quite difficult.

This is the reason for which the influence of temperature on conversion is often represented by graphs where all other operating parameters are kept constant

Table 13.8. Activation Energies of n-Heptane Conversion on Different Catalysts

		Apparent activation energy, kJ/mol	
Catalyst	Pressure, bar	Overall conversion	Conversion to toluene
Sinclair-Baker			
RD-150	20	143.8	134.9
Pt + Re	10	166.3	152.7

[13,50,51]. Figures 13.20 and 13.21 show such plots for the catalytic reforming of *n*-heptane and methylcyclopentane on a Pt-Re catalyst based on the experimental results of Stoica and Raseev [13].

These graphs confirm the well-known fact that the conversion to aromatics increases with temperature, especially for hydrocarbons with seven or more carbons in the molecule. For hydrocarbons with six atoms, thermodynamic limitations influence the final conversion as shown above. As a result, higher temperatures are used in catalytic reforming, although they are limited by coke formation. Hydrocracking is also intensified by temperature increase, but to a lesser extent than aromatization and therefore it does not become a limiting factor.

Both cases (*n*-heptane and methylcyclopentane) confirm the formation of cycloalkanes as intermediates, their concentrations passing through a maximum as for any successive process.

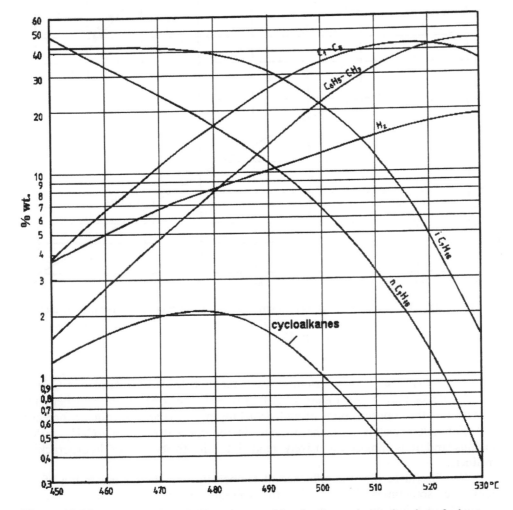

Figure 13.20   The variation of effluent composition for the catalytic reforming of *n*-heptane, versus temperature. Catalyst Pt-Re, $p = 10$ bar, $H_2/n\text{-}C_7 = 6.5$ mol/mol, space velocity = 2 g/ g catalyst × hour. (From Ref. 13.)

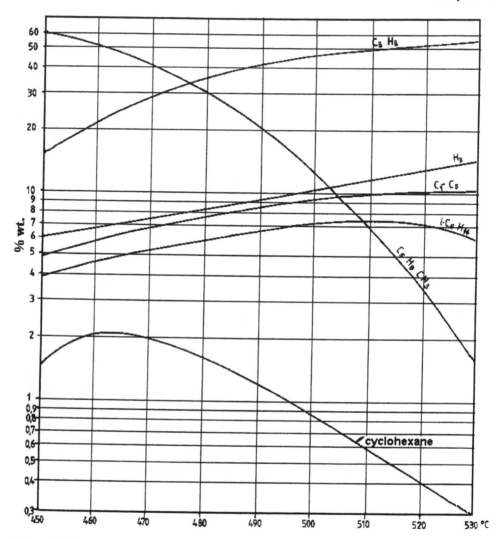

Figure 13.21 The variation of effluent composition for the catalytic reforming of methyl-cyclopentane, vs. temperature. Catalyst Pt-Re, $p = 10$ bar, $H_2/H_C = 6.5$ mol/mol, space velocity $= 2$ g/g catalyst · h (From Ref. 13.)

For n-heptane feeds, the isomerization reaction leads to a ratio of i-$C_7H_{16}/$ n-C_7H_{16} larger than 1 (see Figure 13.20). However, the concentration of heptanes decreases as the concentrations of hydrocracking products and of toluene increase.

For methylcyclohexane, the decrease in the concentration of i-C_6H_{14} is less obvious and it is almost exclusively related to the formation of hydrocracking products.

It is useful to plot the influence of temperature on the octane rating of the reformate, sometimes showing also the volume percent of the liquid fraction obtained. Such graphs are shown in Figures 13.22 and 13.23.

It was not possible to develop a general correlation expressing the influence of temperature [4] owing to the wide differences between the operating conditions and

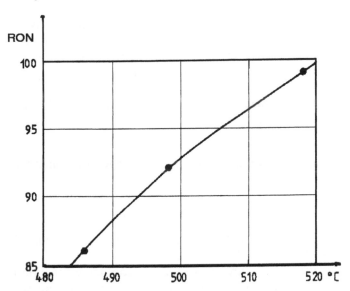

Figure 13.22   The variation in RON of the reformate with temperature. Feed composition: 76% vol cycloalkanes, 22% vol alkanes, 2% vol aromatics. Conditions: $p = 30$ bar, space velocity $= 1.5\,h^{-1}$. (From Ref. 4.)

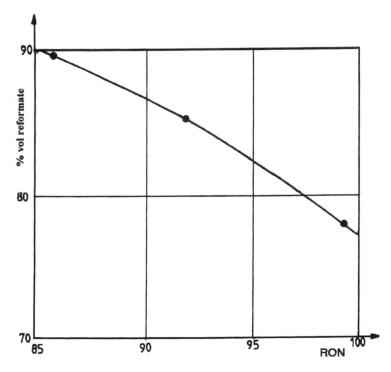

Figure 13.23   The variation of the reformate volume with the octane number. The feed composition and conditions are the same as in Figure 13.22. (From Ref. 4.)

the feed compositions used. The safest approach, is to use reaction models, the validity of which was confirmed by comparing with data from commercial units.

A rule of thumb for estimating the effect of temperature is that the octane number increases by one unit for an increase in the reaction temperature of approx. 2.8°C.

In industrial practice, two average temperatures have been defined, which make possible a quick comparison of the operating conditions for units using several adiabatic reactors:

> WAIT (Weighted Average Inlet Temperature) is the average of the inlet temperatures multiplied by the weight fraction of catalyst in each reactor.
>
> WABT (Weighted Average Bed Temperature) is calculated in the same way, using the average input and the output temperatures for each reactor.

Table 13.9 shows a set of experimental data from a commercial unit and the two average temperatures WAIT and WABT.*

The use of WAIT average temperature allows a more accurate estimation of the increase in octane rating as a function of the rise in temperature. An increase in WAIT temperature of 1.7–2.7°C is needed to increase the octane number by one unit for the octane number range RON 90 to 95. However, for the range 95 to 100, an increase in WAIT of 2.7–3.9°C is needed to increase the octane number by one unit.

The dynamics of the temperature in the reactors depends on the relative rate of various reactions and on their thermal effect.

All aromatization reactions are strongly endothermic and their relative rate decreases from alkylcyclohexanes to alkylcyclopentanes, to alkanes. The hydrocracking reactions are strongly exothermal and they take place in all the reactors of the unit, almost directly proportional to the amount of catalyst in the reactor.

The aromatization of alkylcyclohexanes will proceed to completion in the first reactor, while the aromatization of alkylcyclopentanes will be only partial. The conversion of the exothermic reactions of hydrocracking will represent only 10%

Table 13.9 Temperatures Profile in the Catalytic Reforming Reactors and the Corresponding Values for WAIT and WABT

Reactor	1	2	3	4	Total
Input temperature, °C	490	488	486	482	
Output temperature, °C	411	448	469	479	
Temperature drop, °C	79	40	17	3	139
Average temperature in reactor, °C	450.5	468	477.5	480.5	
Distribution of catalyst, wt %	11	17	17	55	100
WAIT, °C					485
WABT, °C					475

Source: Ref. 4.

---

* The example refers to a plant with a descending temperature profile in the reactors.

of the total conversion. The result is a strong temperature decrease in the first reactor (see Table 13.9).

The endothermic effect decreases gradually, as the rest of the alkylcycloalkanes are converted in the second and third reactors. At the same time, the exothermic effect increases due to the hydrocracking reactions, which are favored as the reaction mixture comes in contact with larger amounts of catalyst. Consequently, the temperature drop in the reactors will decrease when proceeding from the first reactor toward the last one.

The alkanes undergo aromatization more slowly, first passing through the cyclization reaction. Therefore their aromatization will take place mostly in the last reactor, which usually contains more than half of the catalyst of the whole unit. Under these conditions, the hydrocracking reactions will almost completely compensate for the reaction heat absorbed by the reactions of aromatization. Thus, the variation in temperature is almost insignificant.

13.5.2 Pressure

The influence of pressure on the catalytic reforming of *n*-heptane, using a Pt-Re catalyst is shown in Figure 13.24, based on the experimental results of Stoica and Raseev [13].

The conversions to toluene and to hydrocracking products are presented in the same graph, for comparison. The data were obtained under the same conditions, using a standard Pt catalyst. It can be seen that the conversions to toluene for a Pt-Re catalyst are much higher than those obtained with the Pt catalyst, for pressures up to 15 bar. At higher pressures there is little difference in performance, the standard Pt catalyst appearing to produce somewhat higher conversions.

Hydrocracking is faster on classic catalysts, at pressures below 15 bar, while at higher pressures the difference in rate is less important.

The above figure indicates that, in general, a higher pressure decreases the conversion to aromatic hydrocarbon, while the conversion to hydrocracking products remains almost constant.

The value of the operating pressure is selected such as to reduce the formation of coke and is a function of the catalyst and the reaction system used. The Pt-Re catalysts may be used at lower pressures since they can operate with higher concentrations of coke.

The reaction system may allow the regeneration of the catalyst at variable time intervals. These are also correlated with the system pressure (see Section 13.6).

The influence of pressure on catalytic reforming of methylcyclopentane is illustrated in Figure 13.25. It can be seen that the conversions remain almost constant at pressures exceeding approximately 15 bar.

Since the pressure in the unit decreases along the reaction system, average pressures were defined and used in various correlations. In general, the pressure in a catalytic reforming process is calculated as the arithmetic mean between inlet pressure in the first reactor and the outlet pressure of the last reactor. The pressure drop in older units, built before 1975, was approx. 1.5–2.0 bar, while in newer units it decreased to 0.7–1.0 bar.

The average pressure in older units built before 1975 used to be 20–35 bar (somewhat lower for units used for producing aromatic hydrocarbons from C_6-C_8

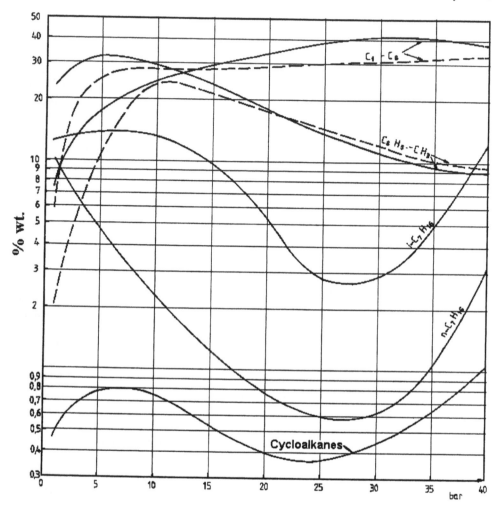

Figure 13.24 The variation of composition of the effluent from catalytic reforming of *n*-heptane, as function of pressure. — Pt-Re catalyst, ---- Pt catalyst (Sinclair-Becker); $t = 500°C$, $H_2/H_C = 6.5$ mol/mol, space velocity $= 2$ g/gcat $\cdot$ h.

fractions). After 1980 the system pressure was gradually decreased, reaching 3.5–7.0 bar in modern units.

The decrease of pressure leads to a decrease in the conversion of the hydro-cracking reactions and consequently to an increase in the yield of the liquid products, an increase in the conversion to aromatic hydrocarbons, and an increase in hydrogen production.

At a constant liquid yield, lower operating pressure leads to higher octane rating. This is reflected by the graph presented in Figure 13.26.

Pressure also influences the production of hydrogen; a decrease in pressure of 5 bar will increase the production of hydrogen by about 0.15–0.30 Nm^3/m^3 processed.

The main factor limiting the decrease in working pressure is the increase in the rate of deactivation of the catalyst due to the increase in the rate of coke formation.

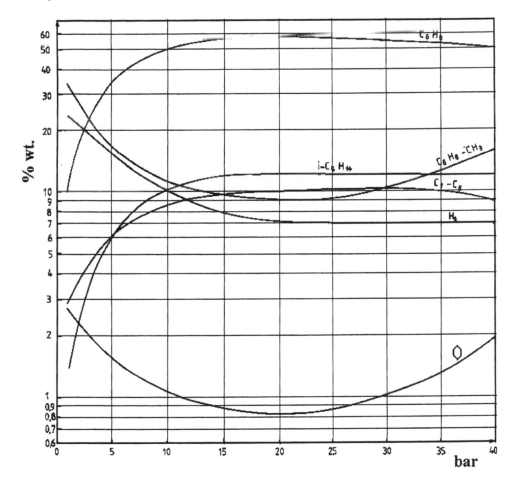

Figure 13.25 The variation of composition of the effluent from catalytic reforming of methyl-cyclopentane as a function of pressure. Pt-Re catalyst, $t = 500°C$, $H_2/H_C = 6.5$ mol/mol, space velocity $= 2\,g/gcat \cdot h$. (From Ref. 13.)

Expressed in terms of relative severity, this phenomenon is correlated with the working pressure and with the octane number of the reformate in Figure 13.27 [67].

The relative severity in Figure 13.27 is inversely proportional with the length of time of catalyst use. A decrease in pressure from 35 to 20 bar at an octane number of 96 RON requires doubling the relative severity (from 2 to about 4) and consequently to halving the length of time between two catalyst regenerations. An identical effect would be obtained by increasing the octane number from 92 RON to 96 RON at a constant pressure of 20 bar.

13.5.3 The Hydrogen/Hydrocarbon Molar Ratio

The influence of the hydrogen/hydrocarbon molar ratio on the dehydrocyclization of n-heptane on a Pt catalyst and on a Pt-Re catalyst is presented in Figure 13.28 [13].

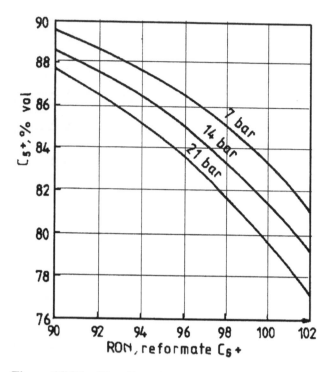

Figure 13.26 The effect of pressure on the liquid yield and RON.

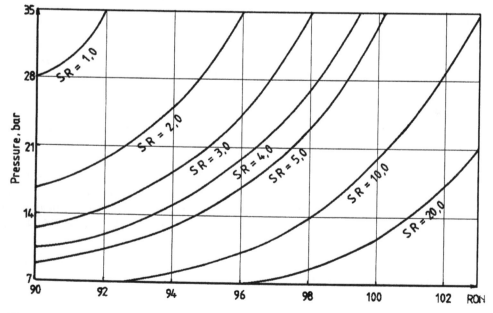

Figure 13.27 The relative severity (SR) as a function of pressure and RON of reformate. (From Ref. 67.)

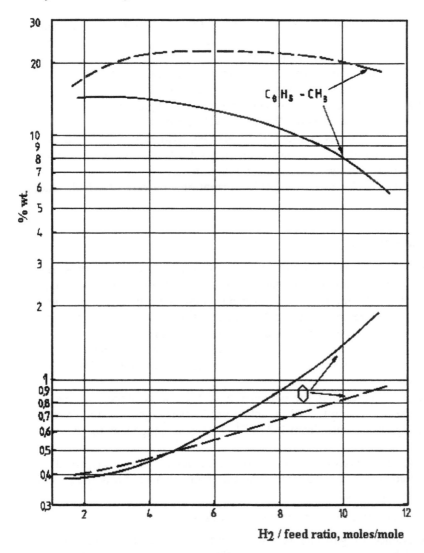

Figure 13.28 The effect of hydrogen/hydrocarbon molar ratio on the catalytic reforming of *n*-heptane. — Pt-Re catalyst, --- Pt catalyst; $p = 20$ bar, $t = 500°C$. (From Ref. 13.)

The molar ratio hydrogen/hydrocarbon as well as the system pressure have a direct effect on the formation of coke. Actually, the important factor is the partial pressure of hydrogen, which depends on the system pressure, and the concentration of hydrogen in the recycle gases.

The increase in the partial pressure of hydrogen causes an increase in the rate of hydrogenation of the coke precursors prior to their transformation into aromatic polycyclic hydrocarbons.

According to available information, the length of the catalyst cycle is reduced by about 20% for a decrease in the hydrogen/hydrocarbon molar ratio from 5 to 4 [68]. Other authors report that a decrease in the molar ratio from 8 to 4 corresponds to a decrease of the duration between regenerations by a factor of 1.75, while a reduction from 4 to 2 would decrease the duration by a 3.6 factor. Such information should be viewed with care since the effect of the molar ratio depends to a large extent also on the catalyst and on the feed.

Historically, units built in the 1950s used a molar ratio of 8/1 to 10/1, while starting from the 1980s, this ratio was reduced to 2/1 to 5/1.

Reducing the molar ratio has a direct economic consequence on the investment costs and on the operating costs for the gas recycle compressor.

There are some systems in which the hydrogen/hydrocarbon ratio in the first reactors is maintained at lower values in order to operate at higher values of the equilibrium conversion of the aromatization reactions. In the last reactor, the molar ratio is increased in order to prevent coke formation, the rate of which tends to increase towards the end of the reactor's path (see Section 13.6).

13.5.4 Space Velocity

The effect of the liquid space velocity derives from the kinetics of the process.

Owing to the large number of reactions taking place in catalytic reforming, a large number of kinetic models and approaches were developed. At this point, it seems useful to analyze the effect of the space velocity as such on the process performance rather than performing a detailed kinetic study.

Figure 13.29 [13] shows the effect of space velocity on the composition of the effluent from a catalytic reformer for n-heptane using two catalysts, a Pt-Re catalyst and a Pt catalyst.

The graph indicates that the conversion to toluene is higher for the Pt-Re catalyst while the conversion to hydrocracked products is higher for the Pt catalyst, over a range of space velocities from 1–2 $(h)^{-1}$. This is the range usually encountered in commercial units.

The curve of concentration of isoheptanes versus time (reciprocal space velocity expressed in $g/g \cdot cat. \cdot hour$) has a maximum that confirms the successive character of the reactions:

$$n\text{-}C_7H_{16} \rightleftharpoons i\text{-}C_7H_{16} \rightarrow \text{hydrocracking products}$$

Industrial experience of gasoline reforming indicates that decreasing the space velocity below 1 h^{-1} causes an increase in the side reactions, mainly hydrocracking. This leads to a decrease in the yield of the liquid product.

The octane rating decreases with increasing space velocity, as shown by the experimental data obtained in a pilot unit [4], operated at an inlet temperature of 520°C:

Volume space velocity, h^{-1}	1.2	2.0	3.0
RON, nonethylated C_5+	100.5	90.8	82.1

The two main targets, octane rating and yield, determine the selection of the space velocity within the range of 1.0 and 2.0 used in many commercial plants.

13.5.5 Feedstock Properties

The performance of catalytic reforming depends on the chemical composition of the feedstock, its distillation limits, and the content of heteroatoms that may lower the activity of the catalyst.

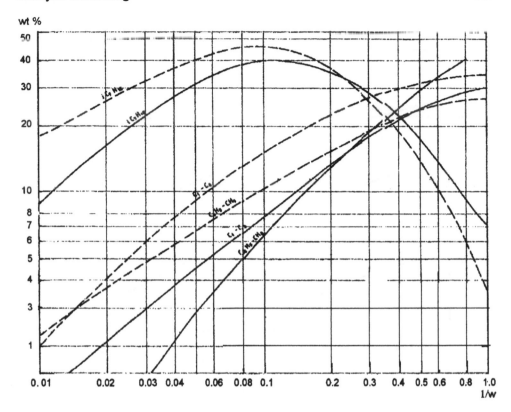

wt %

Figure 13.29 The variation of composition of effluent from catalytic reforming of n-heptane with the reciprocal of the space velocity. — Pt-Re catalyst, --- Pt catalyst; $p = 10$ bar, $t = 500°C$, $H_2/H_C = 6.5$ mol/mol. (From Ref. 13.)

Since cycloalkanes convert to aromatics easier than alkanes, it is important to know the types of hydrocarbons present in the feedstock as obtained from the PONA analysis.

In 1969, Allard of UOP proposed another method of characterization of the feed called the "N + 2A" index. This method, universally accepted, uses the sum of the volume percent of cycloalkanes (N) and twice the percent of aromatics (2A) in the feed. The author explained [70] that this index is an indirect estimate of the UOP characterization factor.

This agrees with the fact that, for the same molecular weight, the value of the characterization factor for cycloalkanes is by one unit lower and for aromatics by two units lower than for alkanes. Consequently, the lower the content of alkanes in the feedstock, the higher is the "N + 2A" index, and the overall performance (liquid yield and octane rating) of the catalytic reforming.

Figure 13.30 [67] shows the influence of the "N + 2A" index of the feed on the % volume reformate and on the RON of the gasoline.

It must be mentioned that n-alkanes produce more aromatics and less gases than iso-alkanes for the same molecular weight. This is a normal consequence of the reaction mechanism and it was confirmed experimentally for a Pt-Re catalyst, by adding n-alkanes to the feed [71].

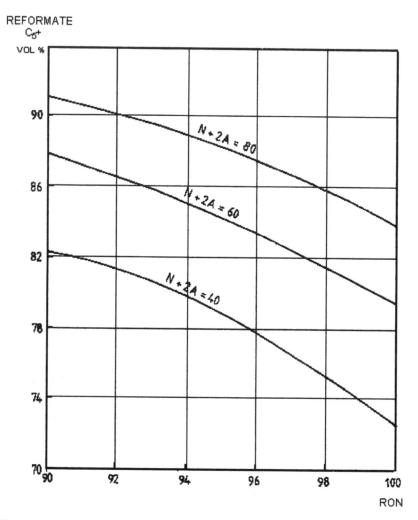

Figure 13.30 Correlation of the % vol reformate with RON, for several feed compositions expressed in terms of the N + 2A index. (From Ref. 67.)

The unsaturated hydrocarbons are not accepted in the feedstock for catalytic reforming because they significantly intensify the formation of coke. This is the reason why the heavy fractions of naphtha obtained from thermal or catalytic cracking have to be subjected to a strong hydrotreating before being used as catalytic reforming feed.

If a lower octane number is obtained, due to the use of feedstock with a lower "N + 2A" index, this may be compensated by an increase of the reaction temperature or by decreasing the liquid space velocity. This situation is illustrated in Figure 13.31. If the difference between the "N + 2A" index values of the two feedstocks is too large, the compensation by increasing the temperature is not sufficient. When processing feeds with lower "N + 2A" index values, a lower octane number of the reformate has to be accepted.

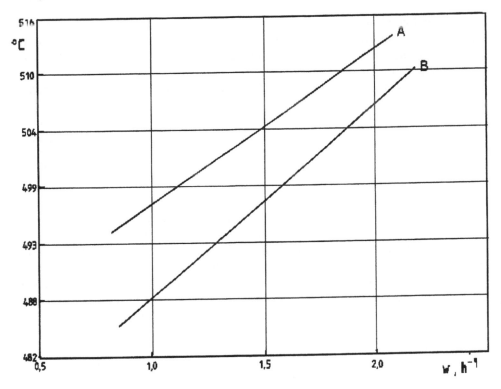

Figure 13.31   The relationship between the feed space velocity and temperature required for maintaining the constant RON = 95, for two feedstocks:

	Density	Boiling Range,°C	N + 2A
Feed A	0.7770	115–202	69
Feed B	0.7608	82–204	62

The correlation between the octane number of the reformate, the space velocity, and the temperature is characteristic to each feedstock. So far, no general correlations were developed on these parameters only.

The distillation range of the feedstock are determined as follows: the final boiling point in the crude oil distillation column and the initial boiling point in the fractionator following the hydrofining. In the latter, the hydrogenolysis of heteroatomic compounds leads to the formation of light hydrocarbons that should not be included in the feed of catalytic reforming.

Before 1980, the final boiling point of feedstock for catalytic reforming was limited to 180–185°C, in order to control the formation of coke on the catalyst. This was also supported by the fact that the reformate always has a higher end point than the liquid feed.

Today, the final boiling point is limited to 190–195°C, although there are circumstances [3] when a final limit as high as 204°C is accepted. The selection of this limit depends very much on the type of unit and its capabilities of regenerating the catalyst.

An end point above 204°C leads to a substantial increase in coke deposits and consequently to shorter cycles between catalyst regenerations. Industrial experience [72] indicates that an increase of the end point by 1°C in the interval 190–218°C leads to a shortening by approximately 2.5% of the length of time between two catalyst regenerations. This effect is smaller in the range 190–200°C but increases progressively over 200°C.

The initial boiling point is usually set somewhere between 90 and 95°C, to avoid including the C_6 fraction. One reason is that the unprocessed C_6 fraction already has a relatively high octane number. Also, aromatization is less efficient for this fraction.

When the C_6 cut is to be included in the feed, the initial boiling point is set between 65 and 70°C. This allows the inclusion of pentane in the catalytic reforming process as well, because it does not harm the catalyst. Instead, *n*-pentane isomerizes to isopentane which would not go through further transformations.

The actual distillation range of the feedstock, still within the limits discussed above, depends very much on the actual characteristics of the unit and the availability of feedstock. If the capacity of the plant is limited, it is advisable to decrease the final boiling point of the feed in order to increase the duration between two regenerations of the catalyst. This increases the actual production of the plant. The initial boiling point may be increased with the same purpose, but it is better not to go beyond a temperature of 90–95°C.

The following boiling points are set for the feedstock if the aim is to prepare aromatic hydrocarbons for petrochemistry:

60–85°C	for benzene
85–110°C	for toluene
105–145°C	for xylenes and ethyl-benzene

If the aim is to obtain the mixed fraction BTX, which would be separated after catalytic reforming, the feedstock boiling range has to be between 62–145 °C.

Table 13.10 [11] outlines the influence of feed source on the reforming results when the aim is to prepare aromatic hydrocarbons.

Heteroatoms contained in the feedstock have a damaging effect on catalytic reforming.

The action of sulfur is well known. It deactivates the metallic sites while nitrogen neutralizes the acid sites of the support. The action of nitrogen is the result of its conversion to ammonia during the process. This either neutralizes the acid sites directly, or it removes from the catalyst the chlorine, as volatile ammonium chloride [193].

In order to remove these effects, the sulfur and nitrogen contents are reduced to below 0.5 ppm through the prior hydrofining of the feedstock.

Thus, the hydrofining of the feedstock for catalytic reforming is recommended even if the naphthas under consideration here have sulfur and nitrogen content below 0.5 ppm, as in the case of Romanian crude oil.

There are two reasons that suggest such a precaution. The first is the sometimes unexpected variations in the characteristics of the crude oil with a higher sulfur content. The second reason is the extremely poisonous effect of some elements such as lead, arsenic, and copper, whose content in the feedstock is not under constant control.

Table 13.10 The Conversion to Aromatic Hydrocarbons of Several Feedstocks Using a Pt-Re Catalyst

	Source of heavy naphtha					
	Straight run		Hydrocracking of light catalytic cracking gas oil		Hydrocracking of vacuum gas oil	
Boiling range	C_6-C_8		C_6-C_8		C_6-C_8	
% vol. on crude oil	16.5		6.0		26.5	
N + 2A	34		91		52	
Process conditions						
pressure, bar	8.8	14.1	8.8	14.1	8.8	14.1
% conversion of C_7 alkanes	90	90	90	90	90	90
Yield, % vol						
benzene	5.6	5.4	5.8	5.6	6.8	6.6
toluene	16.5	15.5	27.3	25.4	21.2	20.1
C_8 aromatics	18.9	16.7	28.0	26.6	19.1	17.8
heavier aromatics	7.3	7.7	5.1	4.1	16.4	13.3
Total	48.3	45.3	66.2	61.7	63.5	57.8
% vol aromatics on crude oil	8.0	7.5	4.0	3.8	16.8	15.3
Production of hydrogen, Nm^3/m^3	6.62	4.97	10.97	9.27	8.51	7.25

There is a considerable decrease in the aromatics yield if the 0.5 ppm sulfur limit is exceeded, while the octane rating is less affected. The poisoning can be detected also through the considerable decrease in temperature gradient in the first reactor.

The effect of sulfur on catalysts of Pt-Re was analyzed in detail in a recent paper [73].

Poisoning with arsenic, lead, or copper also decreases the temperature gradient in the first reactor. This decrease is the result of poisoning of the metallic sites, which lowers the reaction rate of the dehydrogenation of cycloalkanes. As a result, these reactions continue in the second reactor, where the temperature gradient will increase. The graph of Figure 13.32 shows the variation over 75 days of operation of the temperature gradient in the reactors of a plant in which the feed to the first reactor contained 1000 ppm arsenic.

The chemical elements mentioned have a strong effect also on the hydrofining of the feed. The organometallic compounds with arsenic and lead are adsorbed by the catalysts of Co-Mo or Ni-Mo, causing their gradual poisoning.

Since this involves adsorption, the arsenic or lead concentration decreases along the catalyst bed. When saturation is reached, these elements pass through the catalyst bed. This means that the content of approx. 1 ppm acceptable in the reforming unit will be exceeded, and the hydrofining catalyst has to be replaced.

The arsenic and lead contents of a hydrofining catalyst at the moment of its replacement are given in the following table:

Position in the reactor	Arsenic, wt. %	Lead, ppm wt
Top (inlet)	0.49	297
Middle	0.26	111
Bottom (outlet)	0.15	49

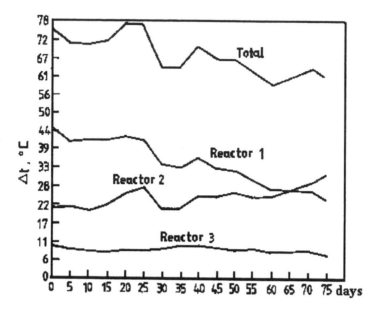

Figure 13.32 The effect of arsenic poisoning on the dynamics of temperature decrease in the reactors of an industrial unit. (From Ref. 4.)

The traces of water or oxygen contained in the feedstock cause the elimination of chlorine contained in the support of the reforming catalyst and lead to a decrease in its acidity. Since the chlorine content in the catalyst has to be maintained between the limits of 0.9–1.2% weight, the action of water has to be compensated by the addition of HCl or of organic chlorides.

If the chlorine content in the catalyst exceeds an upper limit (for some catalysts of 1.2%) the hydrocracking reactions become excessive and the reformate yield decreases.

The quantity of HCl in the recycle gases has to be maintained in the range of 105 ppm mol/mol to maintain optimum working conditions in the plant. This corresponds to 10–20 ppm mol per mol of water in the recycle gases. In order to maintain these values, the quantity of water in the feedstock has to be kept in the range of 3–4 ppm by weight.

13.5.6 The Catalyst

The state of the art for catalysts for catalytic reforming and a comparison between the different types of catalysts was examined in Section 13.2.

Since 1980, the tendency is to work under higher severity in order to increase the octane rating of the reformate. Consequently, the length of the cycle is reduced to 6 months. Also, based on extensive pilot studies, a trend to fine tune the catalysts used is observed so that different catalysts are used in each of the 3 or 4 reactors of the unit.

In practice, the performance of various catalysts is judged on basis of activity, selectivity, and stability.

The *activity* is determined by the temperature (WAIT or WABT*) required to obtain a specified octane rating of the reformate. The lower the temperature, the more active the catalyst is. Some authors prefer to compare the activity by determining the space velocity necessary to reach the same octane rating under constant temperature. The higher the space velocity, the more active the catalyst is.

The *selectivity* represents the yield in C_5^+ reformate produced having a specified octane rating.

The comparative selectivity of two catalysts is shown in Figure 13.33.

The *stability* is defined by the extent to which the activity and the selectivity are maintained in time. The graphs in Figures 13.6 and 13.10 given previously illustrate the behavior over time of some catalysts from the point of view of the preservation of their activity level. The graph in Figure 13.7 allows a broader definition of stability.

In general, the stability is assessed by considering that a catalyst must ensure a cycle of at least one year. It should allow processing more than 35 m³ of gasoline per kilogram of catalyst. The most reliable results for the evaluation of catalyst stability are obtained in pilot plant working under conditions that are severe enough for completely deactivating the catalyst during a cycle of 20–30 days.

Feedstock of pure hydrocarbons is most often used in experimental studies on the behavior of reforming catalysts, at least in the bench scale stages. This is also true for research linked to the optimization of new catalysts. Usually, *n*-heptane is used for behavioral studies, Valuable additional information is obtained from the use of cyclohexane and methylcyclopentane in addition to *n*-heptane, for the optimization

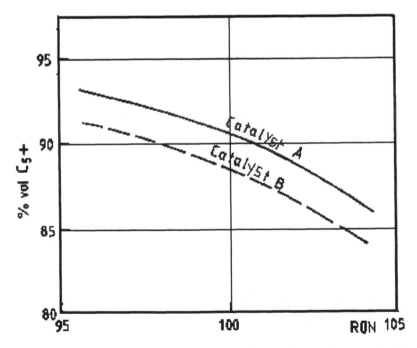

Figure 13.33   Relative selectivity of two catalysts; catalyst *A* is more selective than catalyst *B*.

* See Section 13.5.1.

of new catalysts. The analysis of reaction products is carried out by chromatographic and other adequate methods. Apart from the conversion to aromatic hydrocarbons, the composition of gases is also of interest. For example, large proportions of methane in the effluent generally indicate a low stability of the catalyst.

A similar technique was used by Raseev et al. [74] for the optimization of the ratio between the acidic and the metallic functions of some industrial catalysts. The modification of the ratio of these two functions was performed by treatment of the catalyst with nitrogen saturated with water vapor. The reduction in the acidic function thus obtained was compensated by increasing the temperature. This led to an increase in the percentage of aromatic hydrocarbons in the effluent, at the expense of hydrocracking products. The effect was stronger in the case of n-heptane and weaker in the case of methyl-cyclopentane.

Water present in the feed influences the loss of chlorine content of the support [75]. Considerable differences were noticed in the behavior of various catalysts with respect to the capacity of chlorine retention. This capacity appears to depend on the specific surface and the texture of the support.

The composition of catalysts can also influence the response to the poisoning. For example, the behavior of a catalyst with a high Rhenium content with respect to sulfur poisoning is compared to catalysts with a lower content in Figure 13.34 [76]. A catalyst with a ratio of $Re/Pt = 2.8$ had a much higher stability than one with $Re/Pt = 1$, when the feed had a sulfur content of 1 ppm weight. The stability increases considerably with a reduction in the sulfur content. However, for a sulfur content above 1 ppm, this catalyst has a lower stability than one with a ratio of $Re/Pt = 1$.

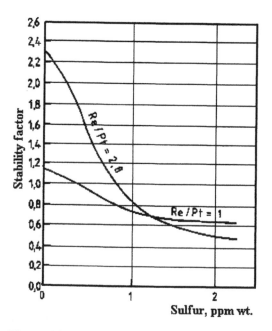

Figure 13.34 Relative stability of Pt/Re catalysts as function of the sulfur content in the feed.

In order to reduce the sulfur content below the limits obtained by hydrofining, a "guard bed" with an adsorbent which removes the traces of sulfur is introduced in the recycle stream of hydrogen. This approach was first proposed by Engelhard and is indicated in the flow diagram of a catalytic reforming unit (see Figure 13.35).

13.6 CATALYST REGENERATION

The regeneration of the reforming catalysts is carried out by burning the coke deposited on the surface. This is followed by procedures needed to bring the activity to the original level. These procedures have distinct steps depending on the metals contained: Pt, Pt + Re, Pt + Ir and on other specific characteristics of the catalyst. These are the subject of numerous patents and publications [77–90,195]. Catalyst suppliers provide their clients with precise instructions for catalyst regeneration.

The regeneration operation begins with the gradual reduction of the reactor temperature to 425–450°C. The feeding is interrupted, but the recycling of hydrogen continues until the temperature decreases to 370–400°C. In the next step, flammable gases are purged with nitrogen. The burning of coke is carried out in controlled conditions.

In cases that require the removal of the catalyst from the reactor for special reconditioning operations, its removal is carried out after the coke has been burnt. In the event that the catalyst has to be handled before regeneration, it must be remembered that it is pyrophoric. Therefore, the catalyst must be cooled down to 40–65°C and prevented from coming into contact with air, by blanketing with nitrogen.

The coke is burnt with air diluted with nitrogen under conditions in which the temperature of the catalyst does not exceed 450°C. To this end, during regeneration

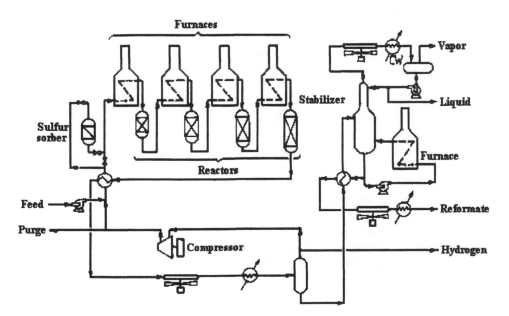

Figure 13.35 The inclusion of a sulfur guard bed (Engelhard) in a catalytic reforming unit. (From Ref. 76.)

the concentration of oxygen in the diluted air is kept in the range 0.2–0.5% vol. Initially, in order to initiate the controlled burning, the diluted air is fed with a low flowrate, which is adjusted as a function of the catalyst temperature.

The duration of the coke burning is approximately 36 hours for the regeneration of a 45 ton catalyst in a fixed bed reactor. This is the length of time is for the regeneration of the catalyst in the last reactor, which contains the largest amount of coke.

In the case of bimetallic catalysts, a chlorinating agent is injected in the first and sometimes also in the last reactor.

In order to avoid corrosion, a precise quantity of sodium carbonate or hydroxide is injected simultaneously with the burning of the coke into the outlet of the reactor.

With increasing severity, the frequency of the catalyst regeneration is increased. In order to shorten the duration of coke burning, a procedure at pressures higher than 28–30 bar is used for regenerating the catalyst in the high severity reforming (pressures of 5–10 bars). This is done without reducing the concentration of oxygen in the air fed to the regeneration.

After regeneration, spots in which the carbon was not burnt, having been bypassed by the regeneration air,* may remain in the last reactor. The removal of such deposits is carried out immediately after the burning of coke by increasing the temperature up to 480–510°C and gradually increasing the oxygen concentration up to 5–6% vol. The burning of coke is complete when the temperature of the burnt gases does not increase any more and the oxygen concentration in the effluent remains constant.

The effect of the high temperature, together with the sulfur and the coke, lead to the accumulation of metal or metals on the surface of the catalyst as crystallites with diameters of more than 10 Å, which considerably decrease in the activity of the catalyst.

The redispersion of platinum in standard platinum catalysts is performed by introducing it in the reactor of chlorinating agents at the same time as the diluted air containing 5–6% O_2, while maintaining a temperature of 480–510°C. This treatment lasts 2–5 hours.

For catalysts of $Pt + Re/Al_2O_3$ the operation is basically the same [77]; it differs considerably for catalysts of $Pt + Ir/Al_2O_3$ [78] owing to the susceptibility of iridium agglomerate in oxidizing conditions and to the difficulties in dispersing them. In fresh catalysts of $Pt + Ir/Al_2O_3$, the metallic sites tend to accumulate at temperatures that are 100°C lower than for Pt/Al_2O_3. Additionally, iridium oxide resists the action of chlorine, and the agglomerations of metallic iridium are covered by a layer of oxide. The experimental data for the agglomeration of metal in the case of catalysts of Pt/Al_2O_3, $Pt + Ir/Al_2O_3$, and Ir/Al_2O_3 are given in Figure 13.36. This graph shows that while at 600°C the surface of platinum does not decrease, the active surface of $Pt + Ir$ decreases to half its initial value, in less than one hour at 500°C. A very efficient procedure of redispersing the iridium [78] was developed on the basis of extended research. According to this

* This kind of situation took place at the Brazi Refinery (Romania) a few years after inauguration of the plant of catalytic reforming of heavy gasolines.

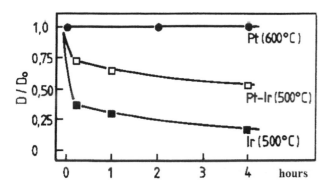

Figure 13.36 Sintering of metallic particles for selected catalysts: 0.3% Pt/Al_2O_3, 0.3% $Pt + 0.3\%$ Ir/Al_2O_3, 0.3% Ir/Al_2O_3. On the ordinate: D – metals dispersion following thermal treatment; D_0 – metals dispersion for the fresh catalyst. (From Ref. 82.)

procedure, after the reduction step the catalyst is treated with HCl up to the complete saturation of the Al_2O_3. Then, chlorine is introduced for redispersion of iridium. This procedure allows operating at higher regeneration temperatures and hence achieves a more complete elimination of coke.

In the case of the standard catalysts, the redispersion of the platinum is followed by a reduction with hydrogen at a temperature of 425–450°C. In the case of bimetallic catalysts, especially with iridium, the reduction precedes the redispersion [78,87,90].

Finally, in some cases [4] the sulfidation of the catalyst is practiced via the treatment of the catalyst with mercaptans, disulfides, or hydrogen sulfide. This method is used in order to balance the functions of the catalyst by attenuating the activity of the metallic sites, thus reducing the hydrocracking.

The concentration of the water remaining in the system following the regeneration step was brought down to below 50 ppm mols in the recycle hydrogen, prior to reaching the operating conditions.

The poisoning of the catalyst with As, Pb, Cu, or the deterioration by sintering are irreversible and require the replacement of the catalyst [193]. The old catalyst is sent for reconditioning to the supplier. The reconditioning consists of the recovery of platinum and the associated metal (such as rhenium) [194], which will be used for preparing new catalyst. Since only the catalyst from the top part of the first reactor is heavily poisoned with metals, it is sufficient to have this part only sent for reconditioning. The catalyst from the second reactor is moved to the first, and the fresh catalyst is added to the second reactor.

The catalyst suffers mechanical deteriorations, which will increase the pressure drops through the bed. To correct for this, the catalyst is removed from the reactors and is screened for removal of dust. This operation is more frequently required for the last reactor where larger amounts of coke are burned, causing a more extensive mechanical degradation of the catalyst.

All operations linked to the removal of the catalyst from the reactors must be carried out while respecting the safety measures mentioned above.

The catalyst regeneration described is basically the same also for systems with continuous circulation of the catalyst. Each step of the operation is carried out in a distinct zone in the regenerator of the units, as it will be described in the following section.

13.7 COMMERCIAL PROCESSES

The first commercial units of catalytic reforming were built practically simultaneously in the United States, in November 1940, under the name of Hydroforming [91], and in Germany, under the name of D.H.D. [23–25]. They used a catalyst of molybdenum or chromium oxide on alumina, discovered by Moldavschi and Kamusher in 1936 [92–94]. At the time, the process was used especially for producing toluene, which was necessary during World War II for the production of trinitrotoluene, by nitration. The catalysts of chromium and molybdenum oxide could not compete with platinum catalysts and were gradually abandoned, despite improvements achieved over the years, including the processes in moving bed or fluidized bed [23].

The first catalytic reforming unit using platinum catalyst on Al_2O_3 support was built by Dutch Refining in 1949.

The continuously improved performances were the result of modified process conditions with the corresponding modifications in the unit design and the use of bi- and polymetallic catalysts. As an example, Table 13.11 [95] presents the evolution of the UOP process "Platforming." The first two columns refer to units typical for the years 1950–1960, and have a stationary catalyst. The other two have circulating catalyst.

The high octane rating of the reformed gasoline, the efficiency of the process, and the broad availability of feedstock (naphtha) contributed to make Platforming in a relatively short period of time, the premier process for high octane gasoline. In addition, through continuous improvements, the process was capable of satisfing the increasing market requirements in the conditions of the gradual prohibition of the leaded gasoline.

Thus, as of January 1990, the worldwide Platforming capacity was 373.9 million tons/year (excluding the former Eastern Block), placing it in second place in the production of gasoline, after catalytic cracking with 539.5 million tons/year [96].

Table 13.11 The evolution of the "Platforming" Process of UOP

	1950	1960	1970	1990
Process conditions				
pressure, bar	35.1	21.1	8.8	3.5
flowrate, m^3/h	66.3	99.4	132.5	265
space velocity, h^{-1}	0.9	1.7	2.0	2.0
H_2/HC, mol/mol	7.0	6.0	2.5	2.5
interval between regenerations, months	12.9	12.5	—	—
regeneration flow rate, kg/h	—	—	300	2000
Yields and RON (gasoline)				
C_5+, % vol	80.8	81.9	83.1	82.9
total aromatics, % vol	38.0	45.0	53.7	61.6
H_2, Nm^3/m^3 feed	65.2	114.5	198.1	274.1
RON, unleaded gasoline	90	94	98	102

The evolution of reforming capacities from 1990–2010 is shown in Table 13.12 [196].

At the same time, the position of catalytic reforming increased strongly for the production of aromatic hydrocarbons, feedstocks for the petrochemical industry. It provides almost all the world's need for xylenes.

13.7.1 Fixed Bed Processes

13.7.1.1 Nonregenerative and Semiregenerative Systems

The nonregenerating system was used in the first type of catalytic reforming unit, the "Platforming" of UOP.

Operation at high pressure (above 35 bars) ensured a catalyst life of over 10 months when using as feed a naphtha with final boiling point between 192–205°C.

For this reason, the units are not provided with the means for regenerating the catalysts. The coked catalyst is replaced and sent off-site for regeneration.

The flow diagram of a unit of this type with three reaction steps is shown in Figure 13.37.

The advantages of operating at a higher severity, i.e., at lower pressure led to the conversion of the nonregenerative units to semiregenerative ones. The necessary equipment was added for the regeneration of the catalyst without its removal from the system. The unit was stopped for the duration of the catalyst regeneration.

The regeneration of the catalyst follows the procedures described in Section 13.6 and for the semiregenerative system provides a total duration of catalyst utilization of 7–10 years and even longer [4].

All catalytic reforming units typically involve 3 or 4 reaction steps with intermediate heating. This is a consequence of the endothermic effect of the aromatization and the exothermic effect of the hydrocracking reactions.

Dehydrogenation reactions of the alkylcyclohexanes take place in the first reactor almost to completion, while the reactions of dehydroisomerization of alkylcyclopentanes are only partially complete. The high rates of these reactions (the first in particular), and their strongly endothermic character, cause the temperature to drop by more than 50°C after contact with approximately 10–14% of the total

Table 13.12 Evolution of Reforming Capacity from 1990–2010 (thousands bpsd)

	1990	1995	1998	2000	2010
North America	4,487	4,278	4,333	4,380	4,700
Western Europe	2,017	2,268	2,386	2,420	2,500
Pacific Rim and South Asia	1,090	1,700	2,050	2,360	3,450
Eastern Europe and C.I.S.	na[a]	1,654	1,746	1,850	2,400
Latin America	na[a]	329	429	480	600
Middle East	311	640	790	850	1,100
Africa	530	404	429	470	690
Total	8,766	11,273	12,163	12,810	15,440

[a] Not available.
Source: DESP IFP.

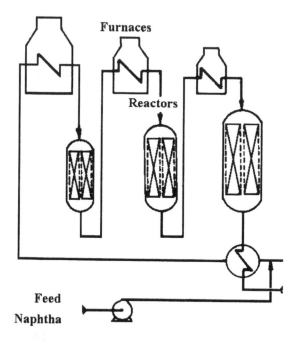

Figure 13.37 Process flow diagram of a fixed bed catalytic reformer.

quantity of catalyst in the system. The thermodynamic conditions reached at this point are unfavorable to aromatization and especially to dehydroisomerization. At the lowered temperature obtained, the reaction rate is low and would require reactor dimensions that would be uneconomical. Therefore, the reaction mixture is reheated in a furnace.

The dehydroisomerization reactions are completed in the intermediate reactor or reactors following the reheating. This is the stage where dehydrocyclization reactions take place and hydrocracking begins. The quantity of catalyst affected by these reactions represents approximately 35% of the total catalyst in the system, and the temperature decrease is lower than in the first step.

The reactions of dehydrocyclization and hydrocracking take place in the last reactor exclusively. The exothermic character of the latter reaction compensates almost fully the endothermic character of the former. This situation favors achieving a high equilibrium conversion of dehydrocyclization, which is strongly influenced by temperature in the range of 450–500°C (see Figure 13.4). The temperature drop in this last step is only a few degrees and is used as a criterion for the correct control of the process.

Two types of reactors are used in various industrial units. The reactor with radial flow described in Figure 13.38 is used most often. The annular space between the two walls may have the shapes "a" or "b" shown in the figure.

The radial reactors are of two types: with "hot walls" that have external insulation and with "cold walls" provided with internal insulation. The latter may be built from relatively inexpensive alloy steels, but fractures in the insulation would allow the access of hot vapors all the way to the metal and cause problems in operation [4].

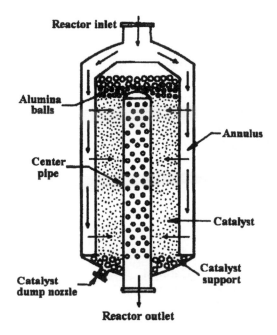

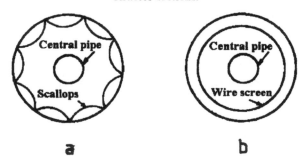

Figure 13.38 Radial flow reactor for catalytic reforming.

Axial flow reactors are operated with downflow, as indicated in Figure 13.39. Sometimes, they are spherical instead of cylindrical, the catalyst occupying a little more than the lower half of the sphere.

Since the pressure drop is usually larger in the cylindrical reactors with axial flow than in those with radial flow, the latter are usually preferred. The spherical reactors lead to smaller pressure drops than the cylindrical ones.

Plants of semiregenerative type were built by the engineering and construction companies Sinclair-Baker, Sinclair-Baker with Kellogg, under the names of "Catforming," "Savforming," etc. The units L-5, L-11, L-6, and L-6/11 built in Russia, are similar.

The following yields and economics are a typical example for a semi-regenerative unit [197]:

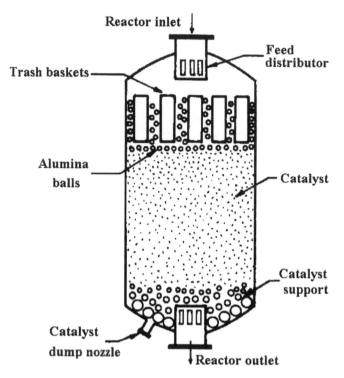

Figure 13.39 Axial flow reactor for catalytic reforming.

A feed with boiling range of 97–190°C, containing: paraffins 51.4%, naphthenes 41.5%, aromatics 7.1%,when processed at the average pressure of 14 bar produced 83.2% vol gasoline with 99.7 RONC.

Utilities per m^3 feed:

Fuel, kJ/kg (release)	1.83×10^6
Electricity, kwh	7.2
Cooling water, $\Delta t = 11°C$, m^3	0.82
Steam produced (12 bar), kg	45.0

A more detailed description of the working conditions and performances of these units were reviewed earlier by Raseev and Ionescu [23–25].

As of January 1984, 64% of the catalytic reforming units in the United States were of the semiregenerative type and only 1.2% of the nonregenerative type [4].

13.7.1.2 Regenerative Systems

The regenerative system includes an additional reactor, which is put on-stream to replace the reactor taken off-stream in order to undergo regeneration. This ensures the continuous running of the unit.

The operation of the unit can be cyclic, a mode of operation in which one of the reactors is constantly under regeneration, or in a swinging mode where the spare reactor (swing reactor), is coupled occasionally, when regeneration becomes necessary.

The regeneration of the catalyst without stopping the operation of the unit allows for a higher severity than that in the semiregenerative system. In general, this means a lower pressure in the reaction system, which leads to a higher octane rating at equal yield, or to a higher yield of reformate for a given octane rating. The second case is shown in Table 13.13 for two typical feeds [99]. The yield shown is greater by 2.5% for the cyclic mode at the same octane level.

The pressure in the cyclic reaction mode was 9–21 bars for a standard platinum catalyst, which otherwise, in the semiregenerative mode, requires a pressure of more than 35 bars. When bi- and polymetallic catalysts are used, the operating pressures in both systems are lower. However, the pressures in the two types of units remain significantly different.

The regeneration of the catalyst without stopping the unit allows the use of higher reactor inlet temperatures that may reach 550°C. Such operation can lead to octane ratings higher by approximately 5 units than the semiregenerative system.

The regeneration of the catalyst is carried out at short time intervals, on average 5–15 days. The reactor of the last step is regenerated more often. In many designs, the additional reactor occupies a central position on the platform of the industrial plant in order to shorten the length of the connecting lines.

The regenerative systems were built before the development of polymetallic catalysts and also before the process with circulating catalyst. At the present time, the interest for such units has somewhat decreased.

The first plant of this type was patented in 1953 and built in 1954 under the name of "Ultraforming" by Standard Oil of Indiana (Amoco). In 1956, the process "Powerforming" was patented and built by EXXON.

In January 1984, the regenerative system represented 22% of the total capacity of catalytic reforming units operated in the United States [4]. Compared with the semiregenerative units, their investment is about 10–15% higher.

Worldwide at the end of 1982, there were 39 "Ultraforming" units in use, with a total processing capacity of 84,000 m^3/day, while the "Powerforming" units had a total processing capacity of 220,000 m^3/day.

Table 13.13 Performance of Semiregenerative and Cyclic Catalytic Reforming Systems

Feed				
density	0.7511		0.7408	
paraffins, % vol.	57		61	
naphthenes, % vol.	30		30	
Reforming system	Semiregenerative	cyclic	semiregenerative	cyclic
Products				
H_2, % wt	2.5	2.7	2.8	3.0
C_1-C_4, % wt	12.8	9.5	16.0	13.6
C_5+, % vol	78.4	81.0	74.4	76.8
RON, clear	100	100	102	102

Source: Ref. 99.

13.7.1.3 Systems with Guard Reactor or Adsorber

The introduction of an additional reactor followed by a stripping column, before the classical reaction system, for the protection of the catalyst against compounds of S, N, and O, was patented by Houdry in 1951. The first unit of this type was built in 1953. The guard reactor used the same catalyst as the main reaction system.

The subsequent practice of hydrofining the feed to catalytic reforming eliminated the need for such a solution.

The purification of the feedstock within the unit before the reactors, is current practice. The distinction is that the purification is made by adsorption.

Such a system was developed by Chevron in the "Reniforming" process. The heavy naphtha passes through an adsorber before being mixed with recycle hydrogen. The performance of this process is presented in Table 13.14.

As suggested by the name of the process, it uses a Pt-Re catalyst that allows operating at low pressures.

Until 1990, Chevron had built 73 units of this type. The economics of the process, at the 1990 price level, based on a unit with a capacity of 3200 m^3/day are [97]:

Investment	9300 \$/m^3day
Catalyst	32 \$/m^3day
Utilities per m^3 processed:	
liquid fuel	0.05 m^3/m^3
electricity consumption	4.4 kWh/m^3
32 bar steam produced	18 kg/m^3
cooling water consumption	3.85 m^3/m^3

Table 13.14 Performance of "Reniforming" Process

Feedstock	Hydrofined Arabian naphtha		Hydrocracked naphtha
Boiling range, °C	93–166		93–199
Paraffins, % vol	68.6		32.6
Naphthenes, % vol	23.4		55.5
Aromatics, % vol	8.0		11.9
Sulfur, ppm	< 0.2		< 0.2
Nitrogen, ppm	< 0.5		< 0.5
Reactor outlet pressure, bar	6.3	14	14
Product Yields			
Hydrogen, Nm^3/m^3	6.8	5.4	6.3
C_1-C_3, Nm^3/m^3	0.72	1.64	0.72
C_5+, % vol	80.1	73.5	84.7
RON, clear	98	99	100
Paraffins, % vol	32.4	31.2	27.2
Naphthenes, % vol	1.1	0.9	2.6
Aromatics, % vol	66.5	67.9	69.9

13.7.1.4 The "Magnaforming" Process

The "Magnaforming" system, built by Engelhard Corp., is characterized by a small hydrogen/feed molar ratio in the first two reaction steps. This is followed by the injection of additional hydrogen-rich recycle gases, so that the ratio reaches normal values in the latter reaction steps.

The advantage of the system is that it favors the aromatization and the dehydrocyclization in the first two reaction stages as a result of the reduced H_2/hydrocarbon ratio that displaces the thermodynamic equilibrium in the desired direction. Since in the latter and especially the last reaction steps the danger of coke formation is higher, the H_2/hydrocarbon ratio is increased to prevent this occurrence.

The process flow diagram of a "Magnaforming" unit is given in Figure 13.40. The figure indicates also the alterations needed for obtaining such a unit by converting a classical unit. The details of these alterations and their respective costs are presented by McClung and Sopko [100].

In contrast to the flow scheme of Figure 13.40, the units of smaller capacity have one reactor instead of reactors 3 and 4 [101].

The process can be implemented in several ways using various types of catalyst. In some cases, a protective adsorber placed in the circuit of recycle gases is used for protecting catalysts that are very sensitive to sulfur (Engelhard E-600 bimetallic), as shown in Figure 13.35.

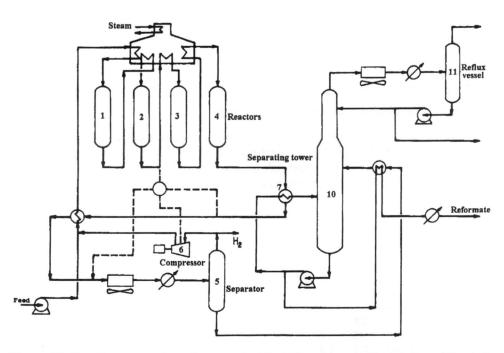

Figure 13.40 The process flow diagram of a Magnaforming unit. The portion of the unit added when revamping a standard unit is drawn with dotted line. 1,2,3,4 – reactors, 5 – separator, 6 – recycle compressor with split hydrogen recycle stream, 7 – heat exchanger (furnace in the original unit), 10 – fractionation tower.

Commercial units have been built in all three versions: nonregenerative, semi-regenerative, and cyclic (regenerative) operation.

The product yields depend on the pressure in the reaction system. Yields are given in Table 13.15 for a feed containing 55.0% vol alkanes, 34.4% cycloalkanes, and 10.6% aromatics (N + 2A = 55.6), for three pressures [101].

Typical operating conditions for the semiregenerative and cyclic modes are:

	Semiregenerative	Cyclic
Average pressure in reactors, bar	11.0	7.0
Weight space velocity, h^{-1}	2.19	4.42
Pressure drop in recycle loop, bar	4.5	4.5
H_2/H_C molar ratio in the least reactor	8.95	2.0

The investments for a semiregenerative unit, including the guard bed adsorber for sulfur, at price levels of November 1992 were \$6300–8800 per processed $m^3/$ day. The utilities per processed m^3 for a pressure of 8.75 bars in the separator are:

Fuel consumption, kJ	1.66×10^6
Electricity consumption, kWh	12.5
Cooling water, $\Delta t = 11°C$, m^3	1.0

Until November 1992, the number of operating "Magnaforming" units was 150, with a total processing capacity of 290,000 m^3/day.

13.7.2 Moving-Bed Systems

The greater efficiency of units with catalyst in moving bed, compared to those with fixed bed, triggered improvements at the commercial scale along two lines: (1) retrofitting existing fixed bed units in order to benefit at least partially from improved performances; (2) construction of new units with moving-bed catalyst.

13.7.2.1 The Modification of Fixed Bed Units

The various types of fixed-bed commercial units require specific modifications, in order to convert to the moving-bed technology.

The examination of the approach used for converting a typical semiregenerative unit is illustrative of the problems encountered when retrofitting other types of

Table 13.15 Yields for "Magnaforming" System

	Pressure, bar		
	24.5	17.5	10.5
H_2, % wt	2.5	2.8	3.1
C_1-C_3, % wt	8.5	6.2	4.0
i-C_4, % vol	3.0	2.2	1.4
n-C_4, % vol	4.1	3.0	2.0
C_5+, % vol	78.9	81.5	84.0
RON, clear	100	100	100

units. The process flow diagram of a semiregenerative unit with Pt-Re/Al$_2$O$_3$ catalyst is given in Figure 13.41. The performances before modification are given in the first two columns of Table 13.16.

The French Institute of Petroleum suggested two solutions for the modification of typical units for this type [102–104].

The first solution, named "Dualforming" is shown in Figure 13.42. It consists of placing an additional reaction section at the end of the existing system. The added system comprises a reheating furnace for the effluent and a reactor-regenerator system, with a moving bed of circulating catalyst. Thus, the reactor and the regenerator, both in the moving-bed system, are now integrated in the operation of the pre-existing reactors and work at the same pressure as the remainder of the system.

The revamping of this unit requires the replacement of the feed/effluent heat exchanger and the addition of the reactor-regenerator system. It is also necessary to add a reheating furnace and an extra compressor to compensate for the pressure drop due to the introduction of the additional reaction step.

These modifications allow the reduction of pressure in the pre-existing part of the unit. The pressure reduction is limited by the pressure drop through the system and the performance of the recycle compressors. In the actual revamping of a commercial unit the pressure was decreased from 26 to 15 bars.

The temperature in the conventional part of the unit is reduced so that the reduction in pressure does not shorten the life of the catalyst. The feed is passed through a sulfur adsorber that operates in liquid phase.

The ensuing reduction of the performance of the classical portion of the unit is fully compensated by the reaction step added. Moreover, the continuous regeneration of the catalyst in the added step makes it possible to operate at the high severity required for achieving a product with the desired octane rating.

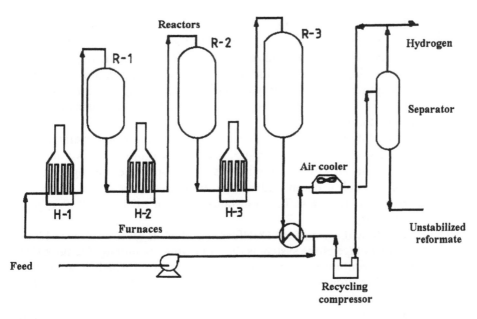

Figure 13.41 Process flow diagram of a classical fixed bed unit before revamping.

Table 13.16 Process Parameters for the Semiregenerative Systems, Revamped Units, and IFP Units with Integral Recycling

	Semiregenerative in: operating conditions		Dual Forming	Dual Forming plus	Octanizing	
	I[a]	II[a]				
Processing capacity, t/h	98.2	98.2	98.2	98.2	98.2	
Processing capacity, bbl/day	20,000	20,000	20,000	20,000	20,000	
Length of cycle, months	12	6	12/(cont)	12/(cont)	continuous	
Average reactor pressure, bar	26	26	15	20/5	7	3.5
Molar ratio, H_2/H_c	6	6	4.6	4.6	1.9	2.0
Yields:						
$\quad H_2$, % wt	1.6	1.7	2.4	2.6	3.1	3.8
$\quad C_5+$, % wt	78.6	76.3	81.9	83.8	87.4	88.0
$\quad$RON	95	100	100	100	100	102
Investment, millions US$	—	—	15.5	19.5	35.0	
Catalyst and utilities (excluding noble metals)	2.7[b]	2.7[b]	4.6	6.1	6.8[c]	
Profit from products, millions US$ per year	—	—	12.3	15.12	24.7	

[a] Higher temperature in operating condition II.
[b] Not including catalyst replacement.
[c] Utilities also include, per m^3 processed: fuel 1200 kJ, power 6 kWh, steam 80 kg, water 0.2 m^3, Nitrogen 0.6 Nm^3 [105].
Source: Refs. 103–105.

The performances of the "Dualforming" system are compared to the classic semiregenerative system in Table 13.16. The result is a 5.6% increase in the yield of the C_5^+ fraction and of 0.7 wt % yield of H_2, at unchanged octane rating.

The "Dualforming" system and its variants represent the cheapest solutions for revamping existing units. The possibility of implementing it depends however on the exact characteristics of the actual fixed bed unit and the space available for the placement of the additional equipment.

The second solution called "Dualforming Plus" can be applied to any unit with fixed-bed catalyst, since it does not require any modifications of it. All additional equipment can be installed without stopping the operation of the existing unit, which constitutes a considerable economic advantage. In addition, the added components may be located separately, at a sufficient distance from the existing unit, to adjust to local space availability.

The process flow diagram for "Dualforming Plus" is given in Figure 13.43.

The main advantage of the system is that following revamping, the pre-existing portion of the unit continues to function at the pressure at which it worked previously. The added reactor operates at a pressure of 5 bar, which ensures improved yields and octane number.

A comparison between the performances of a classical unit with "Dualforming" and "Dualforming Plus" units is given in Table 13.16 [103,104]. The "Dualforming

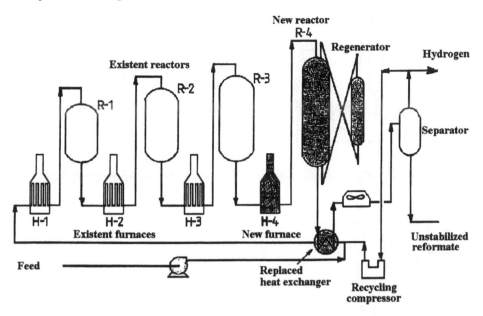

Figure 13.42 The Dualforming approach for modifying a typical fixed bed unit as shown in Figure 13.41. Crosshatched items represent equipment added for revamping. The additional recycle compressor is not shown. (From Ref. 103.)

Plus" system gives an additional yield increase of almost 2.0 wt % for the C_5^+ fraction and of 0.2 wt % for hydrogen, compared to "Dualforming."

The retrofitting implemented by UOP is known as "Hybrid CCP Platforming" and is somewhat similar to "Dualforming." The process flow diagram of such a unit, with the indication of the added equipment, is given in Figure 13.44 [106].

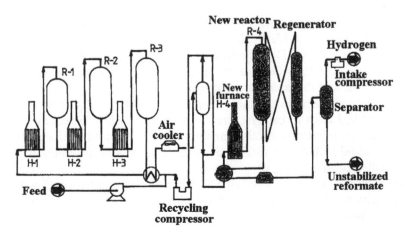

Figure 13.43 The Dualforming Plus approach of modifying an existing fixed bed reforming unit. Crosshatched items represent the added equipment. (From Ref. 103.)

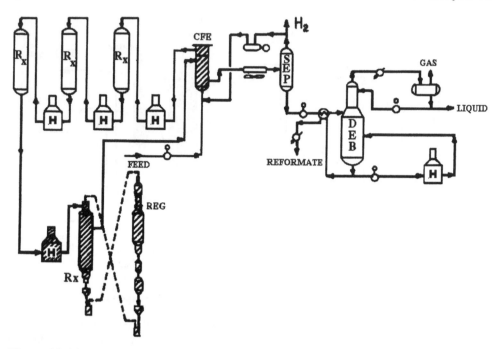

Figure 13.44 The UOP-Hybrid CCP Platforming process. Cross-hatched items represent the added equipment. Rx – reactors, H – furnaces, CFE – combined heat exchanger, SEP – separator, DEB – debutanizer, REG – regenerator. (From Ref. 105.)

The process parameters and the performance of this unit are given in Table 13.17. They are compared with those of units with integral circulation of the catalyst presented in the next section

13.7.2.2 Plants with Integral Circulation of the Catalyst

The French Institute of Petroleum developed a process with integral circulation of the catalyst named "Octanizing" as shown in Figure 13.45 [103]. The operating parameters are compared in Table 13.16 with those of the semiregenerative system and with the revamped processes, "Dualforming" and "Dualforming Plus."

This table indicates that the "Octanizing" unit offers the best performances both in terms of the yield and the production of hydrogen.

The economic data included in the table (September 1994, U.S.$) prove that the revamping of fixed-bed units to catalyst circulation is worthwhile. Selecting the best of the three solutions requires a detailed analysis of the situation in the particular refinery. This should include the market demand for increased production of reformate with specified octane rating.

Procatalyse offers the bimetallic catalyst CR201 in the form of spheres with high resistance to attrition. This is useful for systems with circulating catalyst such as Dualforming, Dualforming Plus, and Octanizing. For the production of aromatic hydrocarbons, Procalyse offers the catalysts AR401 and AR405.

The catalyst discharged at the bottom of the reactor (or regenerator) is conveyed by a stream of gas, through lift lines and is introduced at the top of the

Table 13.17 CCR-Platforming Units; Process Parameters and Performance

	Hybrid CCR	Full CCR	New CCR
Feed rate, m³/day	4000	4000	4000
Average pressure, bar	12.0	7.0	3.5
Catalyst	R-62/R-34	R-34	R-34
Molar ratio, H_2/H_c	4.6	1.9	2.5
Cycle length, months	12	continuous	continuous
Catalyst regeneration rate, kg/h	200	900	1360
Yields:			
hydrogen, Nm³/m³ feed	5.35	6.40	7.76
C_1-C_4, % wt	15.67	11.44	9.25
C_5+, % vol	74.34	77.54	77.55
Total aromatics, % vol	50.0	54.41	58.60
RON clear	101	101	103
Investments, millions US$	9.5	17	35
Utilities and catalyst, millions US$/year	1.7	4.5	7.2
Value of products, million US$/year	6.9	17.4	25.5
Investment recovery period, years	1.8	1.3	1.9

Source: Ref. 106.

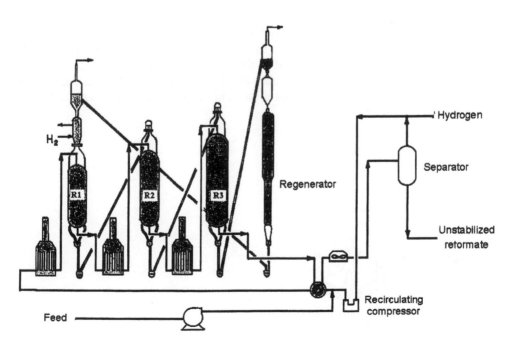

Figure 13.45 Schematic flow diagram of an Octanizing unit.

regenerator (or reactor). Devices such as the one presented in Figure 13.46 are installed at the bottom of the lift lines. The agent of transport is hydrogen between reactors and nitrogen between the reactor and the regenerator.

No valves are installed on the transport lines for the catalyst. The catalyst flow is controlled by the level of catalyst in the collector vessel situated above the reactor and is fine-tuned by modifying the flowrate of secondary hydrogen (as in Figure 13.46).

The catalyst regeneration system consists of four interdependent zones: primary combustion, final combustion, oxychlorination, and calcination.

The regeneration is monitored by a complex system of analyzers and controllers [102]. The regenerator can be isolated from the rest of the system through a system of four motorized valves. Two shutoff valves are placed on the feed lines for catalyst and for gas to the regenerator. Two are placed on the respective exit lines.

The system is completely automated and a computer records all the commands, alarms, messages, and operator interventions [102].

UOP developed a catalyst circulation system in two versions: "Full CCR Platforming" and "New CCR Platforming," which differ not so much through the process flowsheet, but through the average pressure in the reaction system. The former operates at 7 bar while the latter operates at 3.5 bar. The first system is capable of using some parts of the existing units [105].

The second system shows much better performance as a result of extensive process studies on the effect of pressure [107].

The flow diagram of these units is given in Figure 13.47 (the differences between the two versions lie in details not shown in the drawings). The process parameters are compared with those of the CCR Platformer Hybrid system in Table 13.17. Some economic parameters of each process are shown in the same table.

UOP uses the system of superposed reactors, which eliminates the need to lift the catalyst between reactors as shown in Figure 13.47. This eliminates the need to lift the catalyst between reactors, used in the I.F.P.'s "Octanizing" unit.

The performance of these units (as well as those built by I.F.P.) is superior to that of systems with fixed beds of catalyst. This explains the important effort of modernizing the latter.

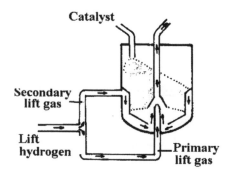

Figure 13.46   Catalyst lifting device.

13.8   ELEMENTS OF DESIGN AND MODELING

13.8.1   Effect of Thermodynamic Limitations

Thermodynamic limitations can intervene to varying degrees in the aromatization of hydrocarbons present in the feed to catalytic reformers. Their examination leads to the design of measures, which may be taken towards the (at least partial) elimination of these limitations.

The evaluation of the thermodynamic limitations may be done by means of calculation and graphical plotting of the equilibrium curves for the pressure conditions and the molar H_2/H_c ratio of the reaction system under consideration. It is evident that the thermodynamic limitations did not take place if the composition of the effluent from the last reactor corresponds to the equilibrium one.

The reaction conditions shown in the following table are typical for the various types of reaction systems discussed in the previous section.

		Pressure, bar	H_2/H_c ratio, mol/mol
Semiregenerative	Classic	25	6
Reconstructed	UOP, Hybrid CCR IFP DUALFORMING	12	4.6
Integral circulation	UOP Full CCR IFP Octanizing I	7	1.9
Integral circulation, new types	UOP New CCR IFP Octanizing II	3.5	2

For all units, the outlet temperatures from the reaction system range between 480–498°C, depending on the feed and the mode of operation.Figures 13.1 and 13.2 show that even under the least favorable conditions: $p = 25$ bar, $H_2/H_c = 6$,

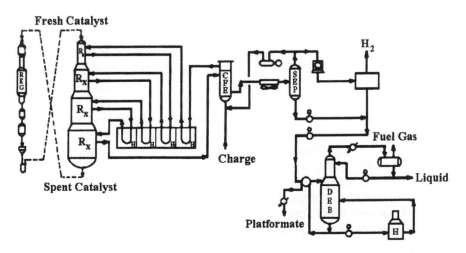

Figure 13.47   The flow diagram of the UOP CCP Platforming units. R_x – reactors, H – furnaces, *CFE* – combined heat exchanger, *SEP* – separator, *DEB* – debutanizer, *REG* – regenerator.

$t = 480°C$, the conversion can reach 99% for the C_7 cyclohexanes and over 99% for C_8^+. The equilibrium conversion for cyclohexane is approximately 92%, and above 99% for systems working at pressures equal or lower than 12 bar.

These data indicate that in catalytic reforming, there are practically no thermodynamic limitations for the conversion of alkylcyclohexanes to aromatic hydrocarbons.

The situation is somewhat different for alkylcyclopentanes. Even under the least favorable conditions corresponding to classical processes ($p = 25$ bar, $H_2/H_c = 6$, $t = 480°C$), the conversion of the C_7^+ hydrocarbons is of the order of 98% (see Figures 13.3 and 13.2). For methylcyclopentane, the equilibrium conversions under the same reaction conditions are of only 66%.[1]

However, revamped units working at pressures of 12 bars and H_2/H_c molar ratios of 4.5–5.0, achieve conversions of all alkylcycloalkanes in excess of 99%. In the case of methylcyclopentane, the conversion can reach approximately 97% and may exceed 99% if the operating pressure is below 8 bar.

The equilibrium conversions of alkanes are much lower. In order to analyze them it is necessary to differentiate between the C_6, C_7, and C_8^+ hydrocarbons, as well as to take into account the structure of the hydrocarbon feed and the structure of the products resulting from dehydrocyclization.

Table 13.3 shows the possible dehydrocyclization reactions and their respective thermodynamic constants for C_6-C_9 hydrocarbons. It takes into consideration the structure of alkanes and the mechanism of dehydrocyclization. The constants are used again in Table 13.18, which contains only the reactions that give the maximum and minimum conversions for the C_7 and C_8 hydrocarbons:

The equilibrium curves for the types of units under consideration were calculated on the basis of the above data and are plotted in Figures 13.48–13.50.

These curves were calculated using the equations and appropriate constants from Section 1.2. The effect of hydrogen excess (n) upon the equilibrium was determined by means of Eq. (13.5) and Figure 13.5.

The equilibrium curves are given in Figure 13.48 as plots of conversion vs. temperature, corresponding to the process conditions prevailing in the classic semiregenerative units: pressure of 25 bar, molar ratio $H_2/H_c = 6$. For n-hexane it is possible to observe an equilibrium conversion to benzene as low as 20–30% in the temperature interval of 490–500°C. For C_7 alkanes, the conversions are situated between 80–88% for the same temperature interval. For hydrocarbons C_8^+ they generally exceed 90%.

Figure 13.49 was derived for the process conditions of semiregenerative units that were upgraded by adding a reactor with circulating catalyst: $p = 12$ bar, and 4.6 H_2/H_c molar ratio. The equilibrium conversion is 80–88% for n-hexane in the temperature interval of 490–500°C and almost complete for C_7^+ alkanes.

Figure 13.50 was derived for two process conditions corresponding to units with integral circulation of the catalyst, i.e., 7 bar and H_2/H_c molar ratio of 1.9, respectively 3.5 bar and H_2/H_c molar ratio of 2.0. For all alkanes the equilibrium conversions to aromatic hydrocarbons are almost complete.

---

[1] For the older classic processes, the pressure was approximately 35 bar, and the molar ratio approximately 8. Under those conditions, the conversion of alkylcyclopentanes was of the order of 90%, and that of methylcyclopentane of approximately 60%.

Table 13.18 Thermodynamic Data for C_6–C_9 Alkanes Aromatization

Reaction (hydrogen omitted)	Conversion	$(\Delta H^0_{800})_r$, kcal/mol	$(\Delta S^0_{800})_r$, cal/mol · deg
n-C_6 ↔ benzene	—	63.77	105.022
n-C_7 ↔ toluene	maximum	60.85	107.872
3-methyl-C_6 ↔ toluene	minimum	61.93	108.772
3-methyl-C_7 ↔ ethyl-benzene	maximum	61.96	109.122
2-methyl-C_7 ↔ meta-xylene	minimum	59.24	109.522
Hypothetical mixture[a]	average	59.567	107.858
n-C_9 ↔ butyl-benzene	—	60.68	108.592

[a] Average thermodynamic values were used in the calculations.

In these calculations we do not take into account the reactions of isomerization that occur in parallel, and which could prevent some of the hydrocarbons from taking part in the aromatization reactions.

The delimitation of the zones corresponding to the equilibrium of alkanes C_7^+ in Figures 13.49 and 13.50 was set by taking into account the following two reactions which give minimum and maximum equilibrium conversions:

$$3\text{-methyl-hexane} \rightleftharpoons \bigcirc\!\!-\!\!C + 4H_2$$

$$2\text{-methyl-heptane} \rightleftharpoons C\!\!-\!\!\bigcirc\!\!-\!\!C + 4H_2$$

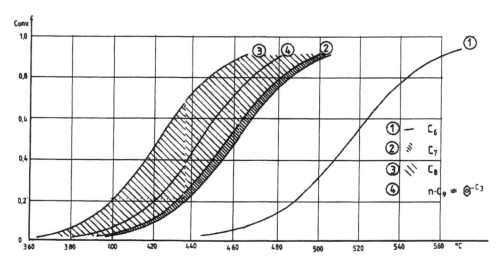

Figure 13.48 Thermodynamic equilibrium for dehydrocyclization of alkanes in the process conditions of the semiregenerative process. $p = 25$ bar, molar ratio $H_2/H_c = 6$.

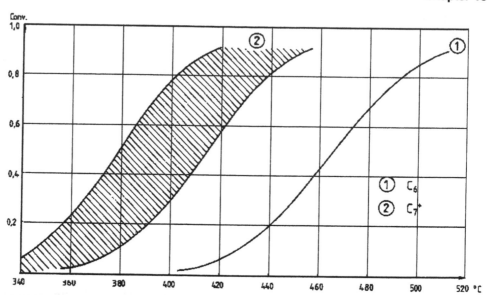

Figure 13.49. Thermodynamic equilibrium of dehydrocyclization of alkanes in the process conditions of classic units retrofitted by adding a reactor with circulating catalyst $p = 12$ bar, molar ratio $H_2/H_c = 4.6$.

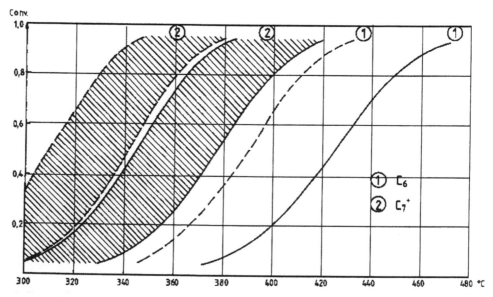

Figure 13.50 Thermodynamic equilibrium of dehydrocyclization of alkanes in units with moving bed, with continuous regeneration of catalyst. — regime I, $p = 7$ bar, molar ratio $H_2/H_c = 1.9$, ... regime II, $p = 3.5$ bar, molar ratio $H_2/H_c = 2.0$.

The direct input of heat in fixed bed reactors cannot be accomplished at the 500°C temperature level at which catalytic reforming occurs. This led to the use of adiabatic reactors. The strongly endothermic character of the reactions and the necessity for high outlet temperatures require the reheating of the reactant flow at an intermediate stage, i.e., the use of multizonal adiabatic reaction systems.

The multizonal adiabatic reactor systems are also used in exothermic processes, such as oligomerization, hydrotreating, or hydrocracking. This is due to the impossibility of letting the temperature exceed a certain level without intensifying undesirable secondary reactions (i.e., formation of higher polymers and of coke, thermal cracking, etc.). Under these circumstances, it is customary to perform intermediate cooling of the reaction mixture between the reaction zones.

Occasionally, a limiting factor of a thermodynamic nature intervenes and forces the decision to cool the reaction mixture between the reaction zones. In this case, the limiting factor is the maximum possible conversion under the given conditions. An example of such a situation is presented in Figure 13.51a.

As the system approaches equilibrium, the rate of the forward reaction decreases owing to the increasing rate of the reverse reaction. As result, owing to the limited residence time in real reactors, the conversion does not reach equilibrium and its value is under "kinetic control."

In a similar way, for endothermic reactions, such as those occurring in catalytic reforming, the decrease in temperature leads to a considerable decrease of the reaction rate. Since operating in such conditions would increase the reactor size and investment cost beyond acceptable values, it becomes necessary to heat the reaction mixture before it is introduced into the next reaction zone.

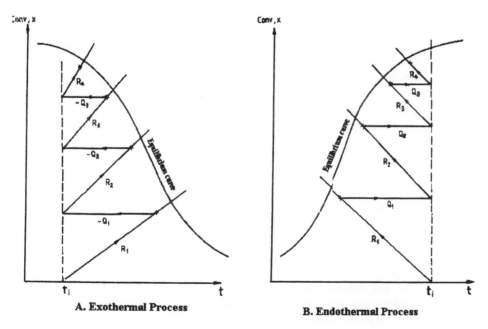

A. Exothermal Process

B. Endothermal Process

Figure 13.51 The use of multistage adiabatic reactor system for overcoming thermodynamic limitations. R_1–R_4 – successive reactors; Q_1–Q_3 – successive stages of intermediate heating or cooling. Kinetic reasons prevent the conversions to ever reach the equilibrium values.

The thermodynamic limitations intervene here since the equilibrium conversion decreases with decreasing temperature. Heating of the reaction mixture will compensate for this effect (see Figure 13.51b).

The chemical transformations taking place in catalytic reforming cannot be represented by a single equilibrium curve. However, the diagram of Figure 13.51b, combined with the equilibrium curves of Figures 13.48 to 13.50, prove for catalytic reforming the need to use multiple alternating reaction and reheating zones for the reaction mixture.

Figures 13.48–13.50 indicate that thermodynamic limitations are present in classical systems with 3–4 reactors with fixed catalyst beds (Figure 13.48) and for the conversion of C_6 alkanes to benzene, in systems retrofitted with an additional moving bed reactor (Figure 13.49). These limitations are less likely to occur in systems with integral circulation of the catalyst.

The application of graphical design methods, such as the one of Figure 13.51b to catalytic reforming, is not possible owing to the complex composition of the raw material, to the multitude of aromatization reactions, as well as to their interdependent character. Semiquantitative evaluations are useful, nevertheless. Indeed, they facilitate the understanding of the factors leading to the superior performances of modern systems with circulating catalyst.

Thus, for the classical semiregenerative processes, the actual conversion to benzene cannot go beyond 10%. It is limited to 60–65% when using an additional moving bed reactor and becomes almost complete in systems with integral circulation of the catalyst, especially for those with operating pressures of 3.5–5.0 bars.

The conversion of C_7 alkanes can reach values of approximately 65% in the classical, fixed bed system and is not complete in the other systems either.

The conversions of the C_8 and higher hydrocarbons can reach 90% for the semiregenerative process and are almost complete for the other systems.

In conclusion, the pressures and the molar ratios H_2/H_c are larger for the older, still operating classical units than for the more modern ones considered in the analyses undertaken here. The thermodynamic limitations are more important and the conversion of the C_6 and C_7 hydrocarbons is lower. The conversion may be calculated for each individual case, based on the methodology outlined above.

13.8.2 Kinetic Calculations and Process Modeling

Models that consider the whole range of phenomena taking place in the multibed adiabatic reaction system extend the kinetic models discussed in Section 13.4.2.

Several recent papers addressing this subject [42,108,109] are reviewed in the following paragraphs.

A detailed presentation of the numerical simulation of the system of reactors for catalytic reforming with fixed bed catalyst is given by Ferschneider and Mege [109]. The chemical transformations are analyzed as well as the thermal and, in more detail, the hydraulic phenomena.

The conversion of the C_8 hydrocarbons are considered representative. Their transformation are expressed by the following reactions and rate equations:

1. Dehydrogenation of cycloalkanes:

$$C_8H_{16} \rightleftharpoons C_8H_{10} + 3H_2$$

$$r_1 = \frac{k_1}{K_1}\left[K_1 \cdot p \cdot x_N - p^4 \cdot x_A \cdot x_{H_2}^3\right] \qquad (13.46)$$

2. Cyclization of alkanes:

$$C_8H_{18} \rightleftharpoons C_8H_{16} + H_2$$

$$r_2 = \frac{k_2}{K_2}\left[K_2 \cdot p^2 \cdot x_N x_{H_2} - p \cdot x_p\right] \qquad (13.47)$$

3. Hydrocracking of alkanes:

$$C_8H_{18} + \frac{5}{3}H_2 \rightarrow \frac{8}{15}C_1 + \frac{8}{15}C_2 + \frac{8}{15}C_3 + \frac{8}{15}C_4 + \frac{8}{15}C_5$$

$$r_3 = k_3 \cdot x_p \qquad (13.48)$$

4. Hydrocracking of cycloalkanes:

$$C_8H_{16} + \frac{8}{3}H_2 \rightarrow \frac{8}{15}C_1 + \frac{8}{15}C_2 + \frac{8}{15}C_3 + \frac{8}{15}C_4 + \frac{8}{15}C_5$$

$$r_4 = k_4 \cdot x_N \qquad (13.49)$$

5. Formation of coke:

$$3C_8H_{10} \rightarrow C_{24}H_{12} + 9H_2$$

$$r_5 = \frac{k_5}{p^3 x_{H_2}^4}\left[0.25 x_{C_5} + 4x_p + 4x_N + 8x_A\right] \qquad (13.50)$$

In Eqs. (13.46)–(13.50), p is the system pressure in Pa and x is the mole fractions. The equilibrium constants and the rate constants are expressed by equations of the form:

$$K_i = A \exp\left(B - \frac{C}{T}\right) \qquad (13.51)$$

$$k_i = D \exp\left(E - \frac{F}{T}\right) \qquad (13.52)$$

The numerical values of constants A–F, that appear in these equations are given in Tables 13.19 and 13.20.

Table 13.19 Thermodynamic Coefficients for Eq. (13.51)

K_i	A	B	C	K_i (773 K)
$K_1\, Pa^{+3}$	1015	46.15	25,581	4.68×10^{20}
$K_2\, Pa^{-1}$	10^{-5}	-7.12	-4444	2.54×10^{-6}

Table 13.20 Kinetic Coefficients for Equation (13.52)

K_i	D	E	F	k_i (773 K)
k_1 (mol/kg$_{cat}$/s/Pa)	0.2780×10^{-05}	23.21	19,306	4.75×10^{-07}
k_2 (mol/kg$_{cat}$/s/Pa)	0.2780×10^{-10}	35.98	33,111	2.93×10^{-14}
k_3 (mol/kg$_{cat}$/s/Pa)	0.2780×10^{-00}	42.97	34,611	4.57×10^{-02}
k_4 (mol/kg$_{cat}$/s/Pa)	0.2780×10^{-00}	42.97	34,611	4.57×10^{-02}
k_5 (mol/kg$_{cat}$/s/Pa)	0.2780×10^{-03}	24.90	20,000	1.05×10^{-04}
		$-12.5[C]$		

For calculating the temperature dynamics in the reactors, the authors use data of the American Institute for Physical Property Data Version no. 4.0, (Feb. 1989) [110]. In particular, the variations of the specific heats with temperature for the substances involved in the process are given by the relationship:

$$C_p = A + B \left[\frac{C}{T \sinh \dfrac{C}{T}} \right]^2 + D \left[\frac{E}{T \sinh \dfrac{E}{T}} \right]^2 \tag{13.53}$$

The heat of formation and coefficients in this relationship are given in Table 13.21.

For hydraulic calculations, an extruded catalyst was considered which had an equivalent diameter of 1.620 mm, a free volume fraction of 0.385, and a bulk density of 0.620.

The calculations were made for a system of three fixed bed reactors with radial flow of a conventional type, represented in Figure 13.52. The dimensions of the reactors 1 and 3 (in mm) can be found in Figure 13.53.

The operating conditions in reactors 1 and 3 are given in Table 13.22.

Table 13.21 Heat of Formation for Hydrocarbons Participating in the Catalytic Reforming Process and Coefficients for Eq. (13.53)

							Heat Capacity (J/mol · K)		
Comp.	ΔH_f^0 (J/mol)	$A \times 10^3$	$B \times 10^3$	C	$D \times 10^3$	E	C_p (723 K)	C_p (773 K)	C_p (823 K)
$C_8H_{18}P$	219,200	107.10	497.30	1.497	327.80	0.6328	373.8	388.6	402.5
$C_8H_{16}N$	180,000	111.20	465.28	−1.670	332.00	0.7934	355.3	370.8	385.1
$C_8H_{10}A$	17,320	75.80	339.24	1.496	224.70	−0.6759	262.3	273.0	282.8
H_2	0	27.61	95.60	2.466	3.76	0.5676	29.4	29.5	29.6
CH_4	−74,520	33.29	80.29	2.101	42.13	0.9951	59.4	62.0	64.6
C_2H_6	−83,820	35.65	135.20	1.430	61.80	0.6120	101.0	105.7	110.0
C_2H_8	−104,680	44.00	193.80	1.369	98.00	0.5830	145.3	151.8	157.8
C_4H_{10}	−125,790	71.34	243.00	1.630	150.33	−0.7304	190.4	198.2	205.5
C_5H_{12}	−146,760	88.05	301.10	1.650	182.00	−0.7476	235.4	245.2	254.3

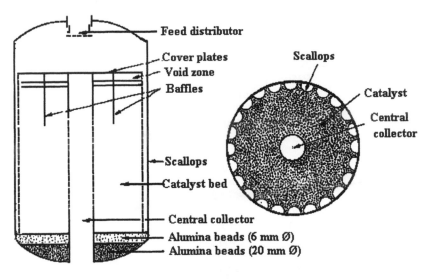

Figure 13.52   Conventional axial flow reactor.

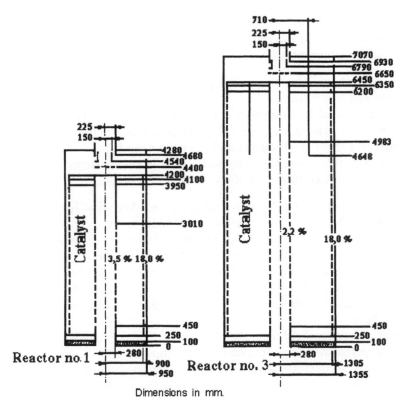

Dimensions in mm.

Figure 13.53   Dimensions of reactors 1 and 3.

Table 13.22 Technological Parameters of Reactors 1 and 3 in Figure 13.53

	Reactor 1	Reactor 2
Catalyst inventory, kg	5,380	20,336
Flowrate, kg/h	61,610	61,620
Space velocity, h^{-1}	5.8	1.5
Inlet temperature, °C	490	490
Outlet pressure, bar	15	14
Distributors		
inlet velocity, m/s	11.20	6.70
velocity at screen, m/s	1.10	0.48
Catalyst bed		
inlet radial velocity, m/s	0.50	0.22
outlet radial velocity, m/s	1.80	1.10
radial pressure drop, Pa	6,550	2,900
Collector		
velocity in holes, m/s	19.8	19.25
axial pressure drop, Pa	950	900
total pressure drop, Pa	7,500	3,800

The computer modeling of the reaction system was carried out on the basis of these data. The main conclusions are:

The flow is perfectly radial in the central part of the reactors and disturbed in their upper and lower ends.

Flow irregularities lead to differences in concentration and in thermal effects. In areas where flowrates are lower, the reactions reach higher conversion, which increases the thermal effects and the coke deposits.

The kinetic model may be improved by taking into account a larger number of hydrocarbons and elementary reactions.

The modeling carried out by Aguilar-Rodriguez and Ancheyta-Juárez of the Petroleum Institute of Mexico involves a model similar to that of Krane [40], expressed by 53 irreversible kinetic equations based on first-order reaction kinetics written for alkanes, cycloalkanes, and aromatic C_6-C_{10} hydrocarbons. At the same time, they take into account the thermal phenomena occurring in the systems of adiabatic reactors. The influence of hydraulic phenomena is not considered.

The modeled reaction system consists of three reactors with fixed bed of catalyst and axial flow.

The equation system is solved on a digital computer and the calculated effluent composition is compared with that obtained in an industrial unit. The agreement between the two is very satisfactory. A very good agreement is also obtained between the computed temperature profile and that measured in the actual reaction system.

The paper by Turpin [108] discusses in particular the efficiency of the elimination of the C_6 fraction from the catalytic reforming feed.

13.9   PRODUCTION OF AROMATICS

13.9.1   Separation of Components of the BTX Fraction

The benzene-toluene-xylene fraction, called currently BTX, may be separated by fractionation from the reformate or may be obtained as such by the catalytic reforming of a naphtha with distillation limits 62(65)–140°C. In the latter case, the pressure in the reaction system may be lower than in the catalytic reforming of a wider naphtha fraction. This favors the aromatization reactions without increasing the coke deposits.

Occasionally, the BTX fraction resulted from catalytic reforming may be combined with the fraction of C_6-C_8 aromatics obtained from the pyrolysis of middle distillates. Prior to this, the latter requires however a deep hydrotreating in order to hydrogenate the unsaturated hydrocarbons and for the hydrogenolysis of the sulfur, nitrogen, and oxygen compounds contained therein.

The content of aromatic hydrocarbons in the BTX fraction depends on the type of reforming process and may be illustrated by the following values:

Semiregenerative	1950	41%
	1960	48%
Catalyst circulation with continuous regeneration	1971	63%
The same at very low pressures	1988	71%

Typical yields and economics for low pressure BTX production are given in Table 13.23 [198].

Table 13.23   Aromatization Yields and Economics for IFP Process

Feed		Products	
TBP cut points, °C	80–150	Hydrogen	4.1%
Paraffins, wt %	57	C_5^+	87
Naphthenes, wt %	37	Benzene	8.5
Aromatics, wt %	6	Toluene	26.3
		Xylenes	26.1
		Total aromatics	74.3

Economics	
Feed rate	4000 m³/day
Investment, including initial catalyst inventory exclusive of noble metals	46 million US$
Typical utility requirements	
fuel, 10^6 kcal/h	76
steam, HP t/h (net export)	(17)
electricity, kWh/h	5.900
Catalyst operating cost, $/ton feed	0.5

Commercial plants: six in operation (March 1999); licenser: IFP.
Source: Ref. 198.

A typical separation scheme for the effluent of catalytic reforming is shown in Figure 13.54 [111]. Similar schemes are suggested also by others [112,113].

The whole amount of effluent enters a fractionating column at the top of which the C_6-C_7 fraction is separated.

This fraction, which also contains the unreacted alkanes and alkyl-cycloalkanes, is submitted to a solvent extraction, combined with an extractive distillation after which the benzene and the toluene are separated by fractionation.

The nonaromatic portion may be converted to benzene and toluene using a supplementary reforming capacity included in the existing reforming unit.

The C_8^+ fraction from the bottom of the column is fed to a second fractionating column, at the top of which a mixture of ethyl-benzene, meta- and para-xylene, with small amounts of ortho-xylene are separated. The rest of the o-xylene and the C_9^+ aromatics form the bottom product of the column.

The stream of xylenes goes to the recovery of the para-xylene. Before 1970 this was performed by crystallization; after this date the separation by adsorption on zeolites (Parex) was preferred, Figure 13.55.

The recovery of the para-xylene by crystallization [115] has yields of about 65%, while by adsorption (Parex system), yields of 97% are obtained. As shown in

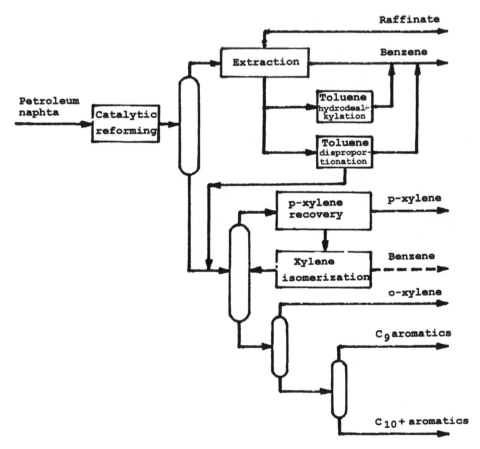

Figure 13.54   Typical BTX separation system. (From Ref. 111.)

Figure 13.55, replacing the crystallization by adsorption leads to the same processing capacity of the xylenes isomerization, to an increase by almost 50% of the recovered p-xylene [115].

Improvements brought to the Parex process of separation of p-xylene consist of among others, the use of new adsorbents. Thus, the replacement of the adsorbent ADS-3 by ADS-27 increased processing capacity by 60%, and the replacement of ADS-7 by ADS-27 decreases by 10% the investments and by 5% the consumptions [115].

These processing improvements are determined by the increasingly high purity required for the products. Thus, the average purity of commercial p-xylene, was of 99.2 wt % in 1970, reached 99.5% in 1980, and 99.7% in 1990. The new separation units are designed for producing 99.9% p-xylene.

The described separation system and similar ones recover the individual aromatic hydrocarbons contained in the reformate but do not allow the modification of their relative amounts. The amounts of the individual compounds in the BTX cut do not correspond to the current requirements of the market or to those predicted

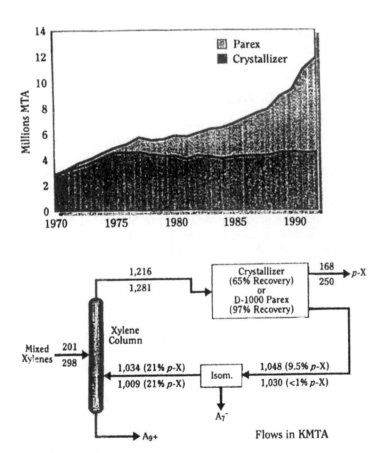

Figure 13.55 p-xylene recovery by crystallization vs. by the Parex process. Figures are flowrates in 1000 t/year. Upper row refers to crystallization; lower row refers to Parex process (with additional fresh feed). (From Ref. 115.)

for the future. This situation is corrected by submitting to further processing the aromatic hydrocarbons resulted from the catalytic reforming.

Processes were developed for the reciprocal conversion of the components of the BTX fraction in order to respond to market requirements. The following trends had to be accommodated:

1. The steadily increasing demands of p-xylene made it the main product in the processing of BTX. The yearly growth of the consumption of p-xylene, is determined mainly by the polyester fibers, between 1984 and 2000 the growth rate was 6.5–8.8%/year. World production in 2000 was 16.4 million tons.

2. Since the demand for toluene is less than its content in the produced BTX, dealkylation and disproportionation processes were developed for converting it in more valuable products. Often, toluene is added to the gasoline.

3. The low demand for meta-xylene determines its isomerization to more valuable isomers.

4. High demand for ethyl-benzene, mainly for dehydrogenation to styrene. The world consumption of styrene increases by about 4.5% per year [199].

5. The demand for benzene is determined mainly by its consumption in the chemical industry, which increases by about 4.9%/year. These increases are covered to a large extent by the benzene produced in the pyrolysis of liquid fractions and less by that produced in catalytic reforming.

13.9.2 Problems Concerning the C_6 Fraction

Limits on benzene content of gasoline, first imposed in the U.S., then to Western Europe and gradually extending to the rest of the world, prevent the introduction in gasoline of the benzene resulting from catalytic reforming.

The most simple solution is to increase the initial boiling point of the feed to reforming from 62 to 85°C. The C_6 hydrocarbons are in this way excluded from the feed. They may be isomerized in mixture with the C_5 fraction, to increase in this way the octane rating of the gasoline. The process is described in Section 11.6.

The presence in the C_6 fraction of some benzene does not preclude the use of this approach. Processes, such as UOP's Penex Plus, allow the direct isomerization of the C_5-C_6 fraction, which may contain up to 30% benzene [116].

In absence of an isomerization unit for the C_5-C_6 fraction, one may take the approach of submitting the C_6 fraction produced in the catalytic reforming to a nonselective alkylation with C_2-C_4 alkenes contained in the gases from the catalytic or thermal cracking. This approach warrants the inclusion of other fractions rich in benzene in the feed to the alkylation.

The nonselective alkylation is performed by two processes: with catalyst in fixed bed [117,118] or in fluidized bed [119].

In fluidized bed process using a ZSM-5 catalyst, the working pressure is comprised between 7 and 14 bar. The conversion of benzene in these conditions is 50-60%. The RON increases from 80 to 90. The process is slightly exothermal.

A variation of this process is to feed the unit with a C_6 fraction resulted from reforming, together with naphtha and gases from catalytic cracking. In this case, the alkylation of benzene is accompanied by the saturation of the alkenes from the catalytic cracking naphtha.

The units with fixed bed [118] seem to achieve a more complete conversions of benzene of the order 92–95%. When a C_3 fraction is used for alkylation, the product contains besides the isopropyl-benzene, also di-isopropyl-benzene and higher oligomers. Often, the molar ratio between isopropyl and di-isopropyl benzene is of 4/1.

The hydrogenation of benzene to cyclohexane is another option for the use of benzene. A unit of this type with a capacity of 150,000 t/year is operating at the Kawasaki refinery, in Japan.

In 1998, UOP had licensed three benzene hydrogenation units (Bensat) and three units for converting C_5-C_6 feedstocks (Penex-Plus) with a high benzene content. The conversion of benzene was complete [197].

13.9.3 Toluene Hydrodealkylation

Thermal hydrodealkylation of toluene was discussed in Section 4.3.4.

The process flow diagram of the unit is basically the same for both the thermal process and the catalytic one, as shown in Figure 13.56. The main difference between the two processes is the temperature level at which the reaction takes place: 700–750°C for the thermal process and 540°C for the catalytic one. The space velocity is higher in the catalytic process, which therefore, requires a smaller reactor. Also, the catalytic process is more selective; for example, it requires a lower consumption of hydrogen.

The catalysts contain nickel, molybdenum, or cobalt, supported on alumina or silica.

A more in-depth comparison between the thermal and the catalytic processes was reported by Hussein and Raseev. They also studied the effect of some added hydrocarbons on the rate of dealkylation [121].

The experiments were performed in a bench unit sketched in Figure 13.57 with a Houdry commercial dealkylation catalyst. The catalyst used was $Cr_2O_3/$

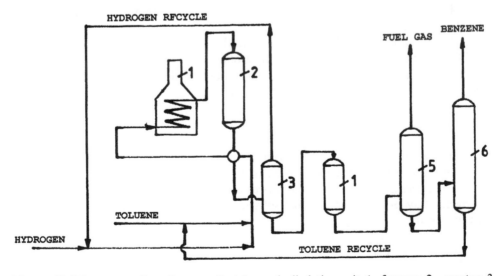

Figure 13.56 Process-flow diagram of a toluene dealkylation unit. 1– furnace, 2 – reactor, 3 – separator, 4 – benzene purification (by adsorption), 5,6 – fractionation columns.

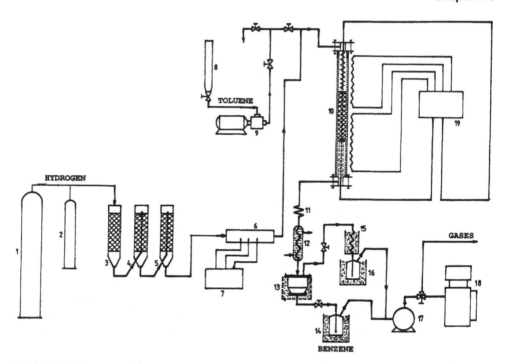

Figure 13.57 Toluene dealkylation laboratory unit. 1,2 – hydrogen, 3,4,5 – hydrogen purification, 6,7 – flowmeter, 8,9 – toluene feed system, 10 – reactor, 12,13,14,15,16 – liquid products separation, 17,18 – gases, 19 – temperature controller.

Al_2O_3 and had a specific surface of 84.3 m^2/g. For the thermal dealkylation study, the reactor was filled with quartz chips, having the same size as the catalyst.

The tests were carried out in conditions that eliminate the effect of mass transfer on the reaction kinetics.

The conversions of the thermal and catalytic dealkylations is compared in Fig. 13.58. In Figures 13.59 and 13.60, the effect of adding 1% n-heptane, cyclohexane, or decaline to the feeds to the thermal, and catalytic dealkylations. Each of these additives increases the reaction rate in both cases.

For the thermal process, this effect may be explained by the additional free radicals formed that promote the dealkylation of toluene. This hypothesis is confirmed by the identical effect of the three added hydrocarbons.

In the catalytic process, the mechanism seems to be more complex, since the effect of the three added hydrocarbons is different.

The experimental results were processed [121] using the equation suggested by Silsby and Sawler [122] and adopted subsequently by others [123,124]. The equation has the form:

$$\tau = k \cdot C_T \cdot C_{H_2}^{0.5} \tag{13.54}$$

where C are the concentrations, the rate is first order in toluene and of 0.5 order in hydrogen.

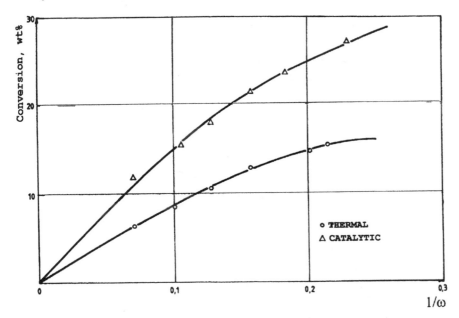

Figure 13.58 Comparison between thermal and catalytic toluene dealkylation. $p = 60$ bar, $t = 635°C$, $H_2/\text{Toluene} = 7.5$ mol/mol.

This equation was used by Hussein and Raseev [121] in the integral form:

$$k = \frac{-\ln(C_T/C_{T_0})}{\tau \cdot C_{H_2}^{0.5}} \tag{13.55}$$

The linearized experimental results for the thermal and catalytic dealkylations, using this equation are given in the graphs of Figures 13.61 and 13.62.

In all cases, excepting for the addition of 1% n-heptane to the catalytic dealkylation, the linearization is quite satisfactory.

The rate constants for the experimental conditions $p = 40$ and 60 bar, $t = 635°C$, $H_2/H_3 = 4.5$ and 7.5 mol/mol have the following values expressed in $(\text{kmol/m}^3)^{-0.5}\,\text{s}^{-1}$:

	Thermal	Catalytic
$p = 40$ bar, $H_2/H_c = 4.5$	0.601	—
$p = 60$ bar, $H_2/H_c = 7.5$	0.621	1.275
1% n-heptane	1.081	2.052
1% cyclohexane	1.008	3.153
1% decaline	1.121	2.831

These values confirm the higher rate of the catalytic process compared to the thermal process and the favorable effect of the additions of alkanes or cyclo-alkanes to the feed.

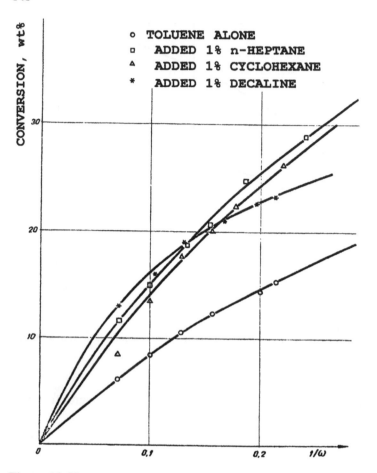

o TOLUENE ALONE
□ ADDED 1% n-HEPTANE
△ ADDED 1% CYCLOHEXANE
* ADDED 1% DECALINE

Figure 13.59 Effect of added hydrocarbons on thermal dealkylation of toluene. $p = 60$ bar, $t = 635°C$, $H_2/Toluene = 7.5$ mol/mol.

13.9.4 Toluene Disproportionation

In conditions of a decreasing market for benzene and a strongly increasing interest for p-xylene, the disproportionation of toluene is preferred over its dealkylation.

Besides, the increase of xylenes production, resulting by processing a mixture of toluene with C_9 aromatic hydrocarbons also produced in the catalytic reforming [125], greatly increased the interest in disproportionation.

13.9.4.1 Process Thermodynamics

Thermodynamic calculations show that the disproportionation reactions are influenced very little by the temperature. Thus, for the reaction:

$$C_6H_5 \cdot CH_3 \rightleftharpoons \frac{1}{2}C_6H_6 + \frac{1}{2}C_6H_4(CH_3)_2 \tag{a}$$

the equilibrium conversion ranges between 0.452 at the temperature of 400 K and 0.446 at the temperature of 800 K.

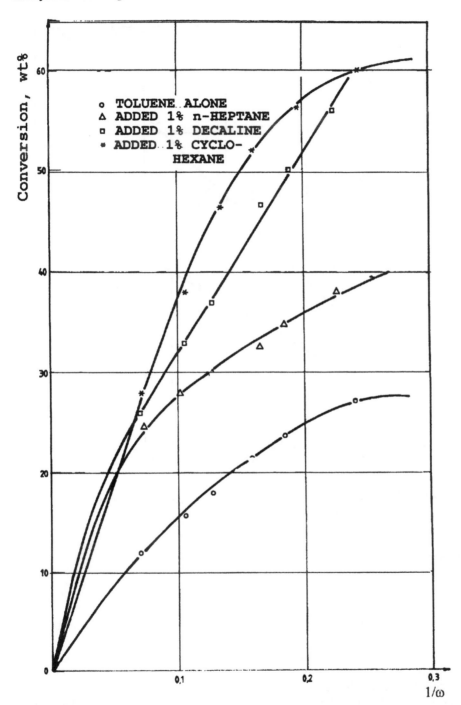

Figure 13.60 Effect of added hydrocarbons on catalytic dealkylation of toluene. $p = 60$ bar, $t = 635°C$, $H_2/$Toluene $= 7.5$ mol/mol.

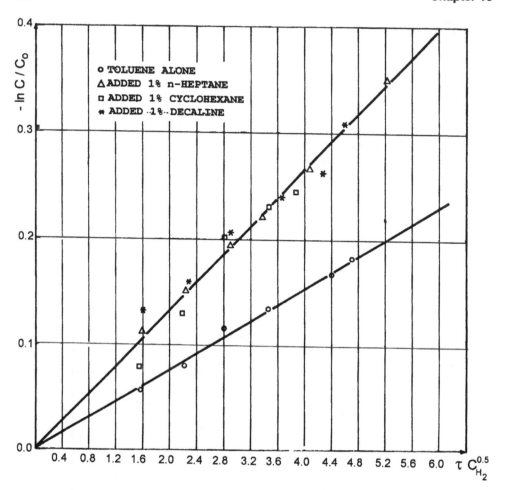

Figure 13.61   Thermal dealkylation of toluene. experimental data linearization, $p = 60$ bar, $t = 635°C$, $H_2/\text{Toluene} = 7.5$ mol/mol.

For the reaction:

$$C_6H_5 \cdot CH_3 + 1, 2, 4\,C_6H_3(CH_3)_3 \rightleftharpoons 2\,C_6H_4(CH_3)_2 \qquad \text{(b)}$$

the equilibrium conversion of toluene ranges between 0.308 at 500 K and 0.329 at 800 K.

In these and the following thermodynamic calculations, the C_8 hydrocarbon is represented by m-xylene (which is predominant in the reaction products). The pressure has no influence on the equilibrium conversion, since the number of moles in the reactions is constant. Since there are two possible reactions, (a) and (b), the conversion of the C_8 hydrocarbons may be influenced by changing the ratio between trimethyl-benzene and toluene in the feed.

Using for benzene, toluene, meta-xylene, and trimethyl-benzene the notations: B, T, MX and TMB, the equilibrium conditions for the initial state of 1 mol T and n moles TMB, for the reactions (a) and (b) will be:

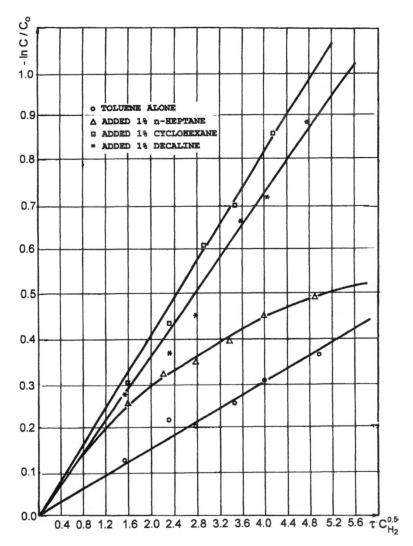

Figure 13.62 Catalytic dealkylation of toluene: experimental data linearization. $p = 60$ bar, $t = 635°C$, H_2/Toluene $= 7.5$ mol/mol.

$$\underset{(1-x-y)}{T} \overset{K_a}{\rightleftharpoons} \underset{(\frac{1}{2}x)}{\tfrac{1}{2}B} + \underset{(\frac{1}{2}x+2y)}{\tfrac{1}{2}MX} \tag{a}$$

$$\underset{(1-x-y)}{T} + \underset{(n-x)}{TMB} \overset{K_a}{\rightleftharpoons} \underset{(2y+\frac{1}{2}x)(b)}{2MX}$$

Here x and y are the number of toluene reacted according to reactions (a) and (b) respectively.

The expression for the equilibrium constant for reaction (a) becomes:

$$K_a = \frac{(\tfrac{1}{2}x + 2y)^{0.5} \cdot (\tfrac{1}{2}x)^{0.5}}{1 - x - y} \tag{13.56}$$

which, for ease of calculation, may be written as:

$$K_a^2 = \frac{(1/2x + 2y) \cdot 1/2x}{(1 - x - y)^2}$$

(13.57)

For reaction (b) we have:

$$K_b = \frac{(1/2x + 2y)^2}{(1 - x - y)/(n - y)}$$

(13.58)

For a temperature of 700 K, which is typical for this process, using published thermodynamic data [2] the following values result:

For reaction (a):

$$(\Delta H_{700}^0)_r = -115 \, \text{cal/mol} \qquad (\Delta S_{700}^0)_r = -1.945 \, \text{cal/mol} \cdot \text{deg}$$

$$(\Delta G_{700}^0)_r = 1246.5 \, \text{cal/mol} \qquad K_a = 0.408, \ K_a^2 = 0.166$$

For reaction (b):

$$(\Delta H_{700}^0)_r = -570 \, \text{cal/mol} \qquad (\Delta S_{700}^0)_r = -0.630 \, \text{cal/mol} \cdot \text{deg}$$

$$(\Delta G_{700}^0)_r = -129 \, \text{cal/mol} \qquad K_b = 1.097$$

In order to determine the effect of n on the equilibrium of the two reactions, the system of Eqs. (13.57) and (13.58) are solved for n and y, for arbitrary values of x, between 0 and 0.4494. The latter value corresponds to the equilibrium of reaction (a) for $y = 0$.

By substituting these values in Eq. (13.57), the y-values are obtained which, upon substitution in Eq. 13.58, make possible to calculate the term n.

The pairs of y and x values allow the calculation of the number of moles at equilibrium for the four hydrocarbons involved.

The dependency of the equilibrium composition on n is plotted in Figure 13.63.

The graph shows the strong increase of the conversion to xylenes as a result of the addition of C_9 aromatic hydrocarbons. Also, if the ratio $n \sim 0.8$ in the feed is exceeded, a strong increase of the unreacted C_9 results, which correspondingly increases the recycling cost. The graph of Figure 13.64 allows estimating on this basis the most economical operating conditions.

13.9.4.2 Disproportionating Catalysts

Both ZSM-5 zeolite, often modified by additions of magnesium, boron, or phosphorus [125–130], and Y zeolite, as such or promoted with transitional metals [131–133], have been reported for the disproportionation of toluene.

The ZSM-5 is used in the protonated form HZSM-5 and may be of two types: with SiO_2/Al_2O_3 ratio comprised between 30 and 75, e.g., the catalysts V-J4, Rikert, Mobil, and CSIRO; with a SiO_2/Al_2O_3 ratio comprised between 123 and 167, the catalysts VJ5 ($SiO_2/Al_2O_3 = 123$) and VJ2 ($SiO_2/Al_2O_3 = 167$). It is claimed that the shape selectivity, characteristic to the zeolites of ZSM type, leads to a concentration of p-xylene in the reaction products, larger than that corresponding to the thermodynamic equilibrium among the three xylenes [134]. This has been interpreted as meaning that owing to the shape selectivity, mostly p-xylene is produced, and it is subsequently isomerized to m- and o-xylene.

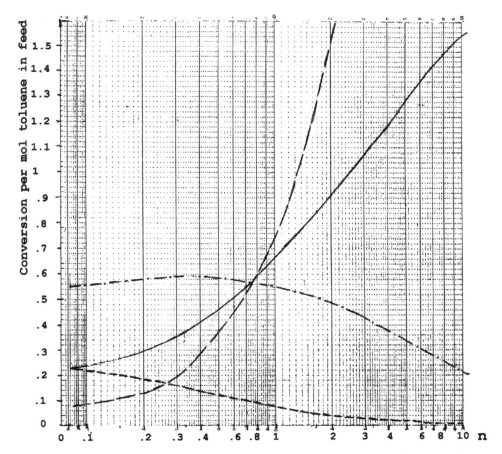

Figure 13.63 Equilibrium compositions for the disproportionation of mixtures of toluene-1,2,4-trimethylbenzene. — conversion to xylenes, --- conversion to benzene, – – – unreacted toluene. – – unreacted trimethylbenzene.

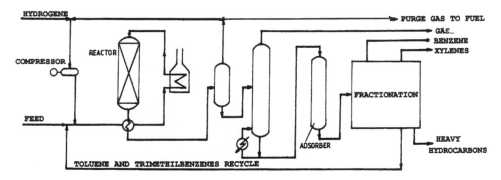

Figure 13.64 Process flow of a toluene disproportionation fixed bed unit.

Since *p*-xylene is in highest demand, among xylenes, maximizing its yield has important economic benefits. Therefore, dedicated units were developed for the isomerization of *m*-xylene and *o*-xylene, to *p*-xylene (see Section 13.9.5).

The effect of the SiO_2/Al_2O_3 ratio on the yields of the produced xylenes is illustrated in Table 13.24 [134].

The equilibrium calculations at 800 K (527°C), on the basis of the thermo-dynamic constants give the ratio of the *p/m/o*-xylenes (in %), as: 23.1/51.3/23.6. By comparing these data with experimental ones given in Table 13.24 it results that in the case of catalysts with relatively low SO_2/Al_2O_3 ratios, the disproportionation leads to the formation of *o*- and *p*-xylenes, the *m*-xylene being formed by isomerization. For catalysts with ratios of 120–170, the shape selectivity leads to the primary formation of *p*-xylene, which is then isomerized to the other two xylenes.

The catalysts of *Y* zeolite is not shape selective but the obtained selectivities may be influenced by constraints appearing in the transition state [135–137].

13.9.4.3 Reaction Mechanism and Kinetics

Two mechanisms have been suggested for the main disproportionation reaction:

1. Transfer of the alkyl group on alumosilica and zeolite catalysts [138,139]:

$$T + S \rightleftharpoons [T]S$$

$$[T]S + T \rightleftharpoons B + [X]S$$

$$[X]S \rightleftharpoons X + S$$

where:

S is the concentration of the active sites of the catalyst, the nature of which is not specified

B, T, and X are the concentrations of benzene, toluene, and xylenes.

2. A diphenylalkane intermediate [129,131,139]:

Table 13.24 Xylenes Isomerization. Product Composition for Different Catalysts

Catalyst	SiO₂/Al₂O₃ ratio	Na, % content	*p/m/o*-xylene, % product composition	
			502°C	532°C
VJ4	30	< 0.1	24/47/29	24/48/28
Rickert	40	~ 0.28	29/48/23	28/48/24
Mobil	70	~ 0.2	25/47/28	25/46/28
CSIRO	75	< 0.1	25/46/29	25/47/28
VJ5	123	< 0.1	44/34/22	45/36/19
VJ2	167	< 0.1	44/33/23	46/37/18

$$T + S^-[H^+] \rightleftharpoons S^-[T^+] + H_2$$

$$S^-[T^+] + T \rightleftharpoons S^-[D^+]$$

$$S^-[D^+] \rightleftharpoons B + S^-[X^+]$$

$$S^-[X^+] + H_2 \rightleftharpoons S^-[X^+] + X$$

where $[D^+]$ is the concentration of the diphenylalkane intermediate, which is adsorbed on the active sites. The other notations are as above.

Neither kinetic studies [133,140,141] nor mechanistic studies using radioactively tagged reactants [142] allow a definitive conclusion on which of the two mechanisms is correct [143]. Actually, it seems that the mechanisms may be different depending on the catalyst used.

Among the secondary reactions, the most important are those of coke formation and of toluene dealkylation. Both reactions are strongly influenced by temperature, which is different from the disproportionation.

In order to limit the extent of these reactions, the process is carried out at a temperature not higher than, approximately 500°C, which is sufficient to obtain a satisfactory reaction rate on the employed catalyst.

In the processes that use a fixed bed of catalyst, one practices the introduction of hydrogen in the reactor, which limits the coke formation and allows cycle lengths (catalyst life) of several years without the need to regenerate the catalyst [125]. The amount of hydrogen introduced is limited by the promotion of dealkylation reactions, which diminish the yield of xylenes. At higher partial pressures of the hydrogen, secondary hydrogenation reactions may also occur with the formation of saturated rings, which may crack to produce lower hydrocarbons.

All these undesired reactions are avoided in the moving bed systems, where no hydrogen is introduced in the reactor, and the continuous regeneration of the catalyst is carried out.

Since the disproportionation reactions are reversible, they lead not only to benzene and xylenes, but also to polyalkylated aromatic hydrocarbons, which are recycled to the process in order to increase the yield of the desired product.

The relative rate of the various reactions and thus the selectivity to various products are strongly influenced by the size of the catalyst crystals [134]. Numerous studies have examined the inter-crystal transport phenomena and the transport between the catalyst particles [129–131,133,137,140,144–150]. These studies are of general interest for understanding the effect of the mass transfer on the performance of the zeolite catalysts.

The effectiveness factors and the temperature gradients determined in these studies as well as the equations used are reported by Dooley et al. [143].

A number of 42 expressions for the reaction rates were formulated on basis of the two reaction mechanisms indicated above, in some cases modified by using additional assumptions [143]. The processing by nonlinear regression of the experimental data allowed the selection of four expressions of the reaction rates that fitted the data equally well.

The equation which gives the best agreement for the majority of the catalysts tested is:

$$r = \frac{kp_T^2}{1 + b_T p_T} \qquad (13.59)$$

This is a Langmuir-Hinshelwood equation, of second order in toluene and with a controlling adsorption of toluene. It results from the above mechanism I, transfer of the alkyl group, the rate-controlling step being the surface reaction of one adsorbed toluene molecule.

Equation (13.59) is also valid for the disproportionation of toluene on amorphous alumosilica catalysts [139].

Another study [140] that used a pentasil zeolite, with a ratio $Si/Al = 12.2$ and a SiO_2 content of 20 wt % deduced the equation:

$$r = \frac{kp_T^2}{(1 + b_T p_T)^2} \qquad (13.60)$$

corresponding to the adsorption of two reacting toluene molecules on two active sites.

Finally, the experimental data obtained on zeolites of the type $Cu/AlF_3/Y$ [131,144] are best fitted by an equation of the form:

$$r = \frac{kp_T}{1 + b_H p_H + b_T p_T} \qquad (13.61)$$

Concluding, the kinetics of the process depends to a large extent on the type of the catalyst used. The most probable mechanism involves the transfer of an alkyl group. Among the kinetic equations developed, Eq. (13.59) is preferred since it seems to be valid in the broadest range of conditions.

13.9.4.4 Fixed Bed Processes

The schematic flow diagram of this unit is given in Figure 13.64. The fractionation may be shared with the extraction of the aromatics contained in the reformate.

The feed to the unit may be toluene or a mixture of toluene and trimethyl-benzenes.

The conversions obtained in the process are close to those corresponding to the thermodynamic equilibrium among the xylenes, as shown for some typical cases in Table 13.25 [151].

When the feed is toluene, the amount of hydrogen introduced in the process is $0.85 \ Nm^3/m^3$ feed [125].

The produced benzene has 99.9% purity and a freezing point of $+5.45°C$. The typical composition of the C_8 fraction is: 23–25% p-xylene, 50–55% m-xylene, 23–25% o-xylene, with only 1–2% ethylbenzene.

An investment of US$9.8 million was calculated (January 1983 prices) for a UOP unit with a processing capacity of 335 m^3/day toluene.

The consumptions for the unit (excluding the purification by adsorption on clays and the fractionation) using electric motors for pumps and partial cooling by air heat exchangers are:

Processing capacity	335 m^3/day
Operation days/year	340
Electric power, kW	577

Table 13.25   Fixed Bed Disproportionation Results

Feed composition, wt %				
Toluene	100	70	50	—
Trimethylbenzenes	—	30	50	100

Products, wt %				
Benzene	41.6	29.6	20.4	5.0
Ethyl-benzene	2.4	2.3	2.2	2.1
o-xylene	11.7	13.0	14.5	13.5
m-xylene	28.8	31.8	35.7	33.1
p-xylene	12.8	14.1	15.9	14.7
Light hydrocarbons and C_{10}^+	2.7	9.2	11.3	31.6

Steam, t/hour	7.07
Produced condensate, t/hour	7.07
Fuel (net contribution), 10^6 kJ	10.25
Cooling water m^3/mm	0.72

The Trans Plus process of Mobil converts C_9^+ aromatics alone or with cofeeding toluene or benzene, to mixed xylenes [200]. The first commercial unit was started in Taiwan in 1997.

13.9.4.5 Moving Bed Processes

The process flow diagram of such a process is shown in Figure 13.65 [152].

The regeneration of the catalyst is performed continuously and the introduction and recycling of the hydrogen to the reactor is absent, which eliminates the corresponding investments.

The main characteristics of the moving bed systems were presented in Section 5.3.4. In the toluene disproportionation units, the flow of the reaction mixture is in down-flow, cocurrent with the catalyst.

One of the important process parameters is the catalyst/feed ratio. Tsai et al. [152], studied the effect of this ratio on thermal balance and the efficiency of the process based on the experience of the Taiwan unit.

A function on the composition of the feed, i.e. toluene, or a mixture of toluene with C_9 aromatics, the process yields vary between 55–84 wt % xylenes and 10–40 wt % benzene. Overall, the molar conversion to these products is 95% [153].

The data show that the yields are better than in the process with fixed bed. This is easy to explain by the absence of hydrogen feed to the reactor and the improved capability to influence the selectivity. The absence of hydrogen eliminates the secondary hydrogenation reactions that generate gaseous products.

The utilities consumption per ton of feed [153], is:

Fuel (adsorbed heat), 10^6 kJ	4.16–7.07
Power	20–35 kWh
Steam	100–200 kg
Cooling water	3.5–7.0 m^3

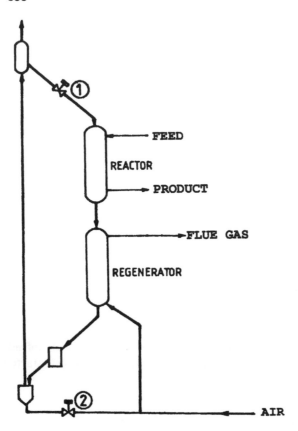

Figure 13.65   Process flow of a toluene moving-bed disproportionation unit: 1,2 – catalyst circulation control.

13.9.5   Xylenes isomerization

13.9.5.1 Process Thermodynamics

In the effluent from commercial processes, the relative amounts of the three xylenes are close to those corresponding to the thermodynamic equilibrium. It is of practical interest to calculate the effect of temperature on the equilibrium compositions.

The variation with the temperature of the equilibrium concentrations of the ortho- and para-xylene are plotted in Figure 13.66. The concentration of meta-xylene results by difference.*

The concentration of ethyl-benzene formed in the isomerization processes of xylenes does not reach, with rare exceptions, the values corresponding to the thermo-dynamic equilibrium. Therefore, it was not included in the plot of Figure 13.66. For cases when ethyl-benzene is of interest, the variation with temperature of the equilibrium concentrations in the system xylenes–ethyl-benzene are shown the plots of Figure 13.67.

---

* In the calculations the thermodynamic constants published by Stull, Westrum, and Sinke [2] were used.

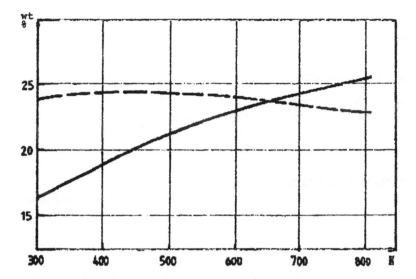

Figure 13.66 Composition of xylenes at thermodynamic equilibrium. — o-xylene, --- p-xylene, m-xylene by difference.

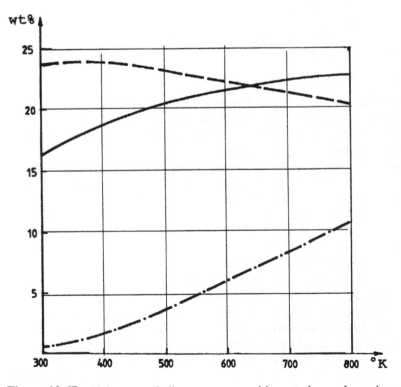

Figure 13.67 Xylenes + ethylbenzene composition at thermodynamic equilibrium. — o-xylene, --- p-xylene, – · – ethylbenzene, m-xylene by difference.

While the equilibrium concentrations of the xylenes do not vary much with temperature, that of ethyl-benzene increases from 0.54% at 300 K to 10.95% at 800 K.

13.9.5.2 Isomerization Catalysts

Initially, the catalysts used for the isomerization of the xylenes were of the Friedel-Crafts type, such as aluminum chloride [154–157], which were operated at temperatures of 0–30°C. They were followed by catalysts based on boron fluoride.

Greensfelder and Voge [158] were the first to use catalysts of alumosilica. They were followed by other researchers [159,160]. In the same period, the first uses of bifunctional catalysts with platinum were reported [161–163].

After the appearance of the zeolite catalysts, many studies were devoted to the isomerization of xylenes on these catalysts [164–173]. Both small-pore zeolites of the ZSM-5 type, with different Si/A ratios, and larger-pore Y-type zeolites and USY (ultrastable Y) type [174] were studied. Noble metals Pt or Pd, are deposited on the zeolite in order to reduce the secondary reactions, which lead finally to the formation of coke. Nickel was also tested in place of platinum [175], as well as the promotion of mordenites with Zr, Ge, Sn, or Mg with the purpose of increasing isomer conversion [176].

The zeolites by themselves without a noble metal catalyze the isomerization of xylenes, whereas the conversion of ethyl-benzene to xylenes requires the hydrogenation function of Pt or Pd [176]. The precious metal has the important role of allowing the use of the catalyst for long periods without regeneration. The amount of the supported platinum is generally 0.3–0.4 wt %.

It is to be remarked that the shape selectivity of some zeolites of the ZSM-5 type favors the formation of p-xylene.

13.9.5.3 Reaction Mechanism

The mechanism explaining the isomerization reactions is shown in Figure 13.68.

The isomerization of xylenes takes place on the acid sites without the participation of the metallic sites, by the migration of the methyl group along the ring of the adsorbed ionic species (see Figure 13.68a).

In the case of ethyl-benzene the mechanism is more complex. First, the adsorption on the metallic site and the partial hydrogenation of the aromatic ring takes place with the formation of a cyclene (cyclo-alkene), which is desorbed from the metallic site and readsorbed on the acidic site as an ion. The adsorbed ion is now isomerized via a 5-carbon atoms ring, which allows the reshaping of the ethyl group to form two methyl groups attached to an ionic 6-carbon atoms ring. By desorption, a dimethyl-cyclene is formed, which is readsorbed on the metallic site and dehydrogenated to give after desorption, o-xylene (see Figure 13.68b).

According to this mechanism, the isomerization of the xylenes takes place exclusively by the migration of the methyl group to the neighboring carbon atom, which excludes the direct isomerization o-xylene $\rightleftharpoons$ p-xylene. The absence of this isomerization was demonstrated by Olteanu and Raseev [177] at very high space velocities by feeding the experimental unit with pure o-xylene or pure p-xylene. In both cases, the exclusive formation of m-xylene was observed. The same results were obtained subsequently also by other researchers [165,169] on acid catalysts, including on zeolites [171].

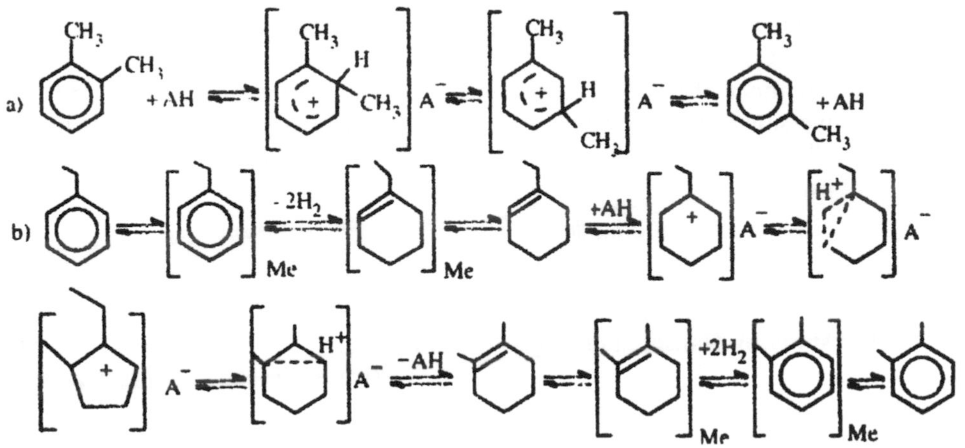

Figure 13.68 Reactions mechanism of C_8 aromatics isomerization. (a) xylenes isomerization, (b) ethylbenzene isomerization.

Some researches [178] don't eliminate completely the isomerization between o- and p-xylene but state that the rate of this reaction is very low. This conclusion was formulated on the basis of experimental data obtained at high conversions, where relatively large amounts of trimethylbenzene are also produced. In this case, the isomerization of o-xylene $\rightleftharpoons$ p-xylene may take place by the intermediary formation of 1,2,4-trimethylbenzene.

The participation of the dehydrogenation—hydrogenation steps in the isomerization of ethyl-benzene to xylenes is proven by the fact that the isomerization has an insignificant rate on catalysts without the metallic function [176]. Also, as found by Raseev and Olteanu [177,179,180], the isomerization of ethyl-benzene on bifunctional catalysts is enhanced by operating in conditions that favor hydrogenation, e.g., pressures of 12–40 bar and temperatures of 390–400°C .

The secondary reactions that lead to the decrease of the yields in xylenes are especially those of disproportionation. A pertinent mechanism that assumes the formation of an intermediary diphenyl-alkane, or better said, of the corresponding ion, is shown in Figure 13.69.

The intermediary ion may be formed on the active sites situated on the external surface of the catalyst and to a very small extent on the site within the pores, owing to the large size of the diphenyl-alkyl ion, compared to the diameter of the pores of the zeolite. This was proven by Benazzi et al. [176] by passivating the active sites

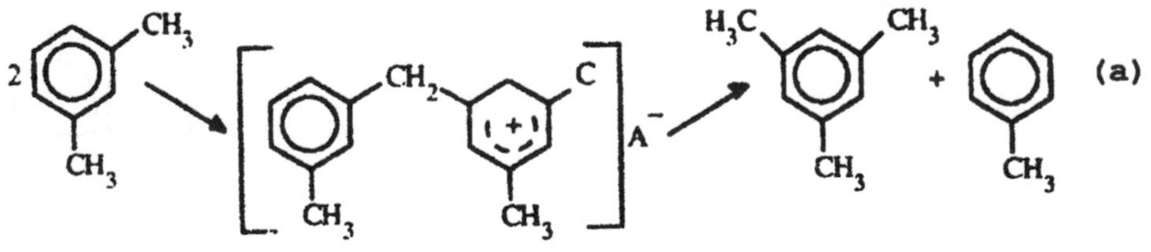

Figure 13.69 Formation of trimethylbenzenes.

situated on the exterior of the catalyst particles by means of organometallic compounds of Mg, Ge, Sn, and Zr.

In this manner, the rate of the secondary reaction (disproportion) was reduced
and the conversion to xylenes was increased. This effect was promoted mostly by the
organometallic compounds of zirconium and germanium [176].

13.9.5.4   Kinetics of Isomerization

The kinetics of xylene isomerization on alumosilica catalysts was studied by Silvester
and Prater [181], considering that the reactions were pseudohomogeneous of first
order. The study by Hanson and Engel [182] on the same catalyst, used also simplified kinetics but took into account the successive character of the reactions
(o-xylene $\rightleftharpoons$ m-xylene $\rightleftharpoons$ p-xylene).

In the studies of Raseev and Olteanu [177,179,180,183], the experimental unit
of Figure 13.70 was used [179].

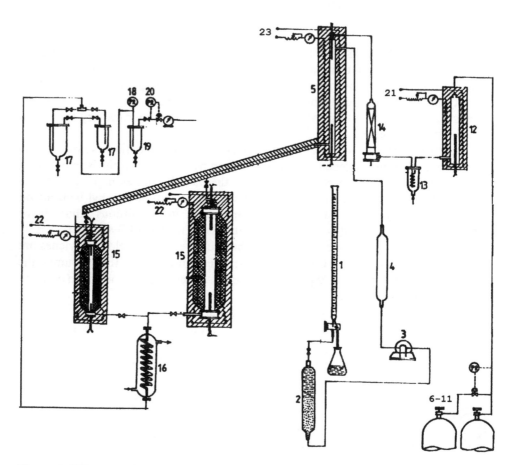

Figure 13.70 Experimental unit for studying the isomerization of xylenes. 1-4 –
Hydrocarbon feeding system, 5 – Evaporator, 6-11 – Hydrogen feeding system, 12 –
Removal of O_2 traces, 13,14 – Hydrogen drying, 15 – Reactors for different quantities of
catalyst, 16-19 – Products separation, 20 – Pressure regulator, 21-23 – measures and regulators.

Several types of industrial catalysts were used: Pt on SiAl, Pt on activated Al_2O_3, SiAl, W + Mo on SiAl. The bifunctional catalysts had a platinum content comprised between 0.35–0.40 wt % and a specific surface area of 170–360 m^3/g. The purities of the o-, m- and p-xylene were respectively of 99.7%, 96.0%, and 99.1%.

Preliminary evaluations determined the relative activities of the catalysts at temperatures of 430–520°C and pressure of 12 bar in the presence of hydrogen and at atmospheric pressure in the absence of hydrogen. On this basis, the platinum catalyst on alumosilica was used in this study.

The study examined the effect of the operating parameters on process performance [180]. For the molar ratios of hydrogen/feed usually used in commercial units, the optimum range of operating conditions is: temperatures 430–450°C, pressure 5–10 bar, space velocities of 2–3 g hydrocarbons/g cat. $\cdot$ x hour.

The temperatures above 470°C lead to increased disproportion reactions and to hydrodealkylation, while those below 430°C favor hydrogenation, which as it was mentioned above, increases the isomerizations in the xylenes–ethyl-benzene system.

The tests were carried out in conditions for which the diffusion effects had no effect on the reaction rates. In order to eliminate the influence of the internal diffusion, the catalyst granules had an average diameter of 0.56 mm.

The kinetics of the process were studied by Raseev and Olteanu [177,183] in the temperature range 430–470°C, at a pressure of 11.6 bar, and for constant H_2/hydrocarbon molar ratio. The following kinetic diagram was used:

$$o\text{-xylene} \underset{k'_{-1}}{\overset{k'_1}{\rightleftharpoons}} m\text{-xylene} \underset{k'_{-2}}{\overset{k'_2}{\rightleftharpoons}} p\text{-xylene} \tag{a}$$

The absence of direct o-xylene to p-xylene isomerization was proven by us earlier [177] (see Section 13.9.5.3).

Assuming a Langmuir adsorption law led to the formulation of the rate equations for the first and the last reaction as:

$$-r_{OX} = k'_1\theta_{OX} - k'_{-1}\theta_{MX} \tag{13.62}$$
$$r_{PX} = k'_2\theta_{MX} - k'_{-2}\theta_{PX}$$

By neglecting the adsorption of hydrogen on the acid sites on which the isomerization takes place:

$$\theta_{OX} + \theta_{MX} + \theta_{PX} = 1$$

The surface fractions occupied by the three xylene isomers on the acid sites can be expressed as:

$$\theta_i = \frac{b_i p_i}{1 + b_{OX}p_{OX} + b_{MX}p_{MX} + b_{PX}p_{PX}} \tag{13.63}$$

Considering that the three xylene isomers have the same adsorption coefficients, Eq. (13.63) becomes:

$$\theta_i = \frac{b_i p_i}{1 + b_i(p_{OX} + p_{MX} + p_{PX})} \tag{13.64}$$

If the secondary reactions are negligible, one may write:

$$p_{OX} + p_{MX} + p_{PX} = \text{const.}$$

and Eq. (13.64) becomes:

$$\theta_i = K \cdot p_i \tag{13.65}$$

where K is a constant which, when incorporated in the rate constants which become:

$$k_i = \frac{k_i' b_i}{1 + b_i(p_{OX} + p_{MX} + p_{PX})} \tag{13.66}$$

the system of rate Eq. (13.62) is thus reduced to a pseudohomogeneous one:

$$\left. \begin{aligned} \frac{dp_{OX}}{d\left[\dfrac{\pi W}{(1+n)F_0}\right]} &= k_{-1}p_{MX} - k_1 p_{OX} \\[2em] \frac{dp_{PX}}{d\left[\dfrac{\pi W}{(1+n)F_0}\right]} &= k_2 p_{MX} - k_{-2}p_{PX} \end{aligned} \right\} \tag{13.67}$$

where:

> π is the pressure in bar
>
> W — the catalyst amount in g
>
> n — the molar ratio H_2/hydrocarbon
>
> p — the partial pressures of the components
>
> F_0 — the feed rate in moles of hydrocarbon/second.

When high purity ortho- or para-xylene are fed at high space velocities of the order of 1000, the reverse reactions of formation of these hydrocarbons from meta-xylene may be neglected and the first right-hand term of Eqs. (13.67) may be neglected.

In these conditions, these equations were written in terms of finite differences, using the subscript i for the initial state:

$$\left. \begin{aligned} \frac{\dfrac{\Delta p_{OX}}{\pi W}}{(1+n)F_0} &= k_1\left(p_{OX_i} - \frac{\Delta p_{OX}}{2}\right) \\[2em] \frac{\dfrac{\Delta p_{PX}}{\pi W}}{(1+n)F_0} &= k_{-1}\left(p_{PX_i} - \frac{\Delta p_{PX}}{2}\right) \end{aligned} \right\} \tag{13.68}$$

This equation system allowed the calculation of the rate constants k_1 and k_2. The constants for the reverse reactions were then calculated using the equilibrium constants on a thermodynamic basis.

For Pt on alumosilica catalyst, [183] the values of the obtained rate constants are given in Table 13.26.

The activation energies are:

Table 13.26 Kinetic Constants for Isomerization of Xylenes

	430°C	470°C
k_1, mol/s · g cat. · bar	1.11×10^{-5}	2.37×10^{-5}
k_{-1}, mol/s · g cat. · bar	0.52×10^{-5}	1.14×10^{-5}
k_2, mol/s · g cat. · bar	1.3×10^{-5}	2.04×10^{-5}
k_{-2}, mol/s · g cat. · bar	1.52×10^{-5}	4.54×10^{-5}

$$E_1 = 82.48 \, \text{kJ/mol}$$

$$E_{-1} = 85.37 \, \text{kJ/mol}$$

$$E_2 = 64.35 \, \text{kJ/mol}$$

$$E_{-2} = 64.02 \, \text{kJ/mol}$$

Calculations using the kinetic constants thus obtained by us showed good agreement with the experimental data [183].

Robschlager and Christoffel [167] used a model that takes into account all possible reactions among these four C_8 hydrocarbons:

Other hydrocarbons   MX ⇌ OX   Other hydrocarbons   (b)

PX ⇌ EB

Hsu et al. [174] used several zeolite catalysts for which the activity increased in the order:

$$\text{Pt/mordenite(MN)} \ll \text{Pt/USY} < \text{Pd/ZSM5} < \text{Pt/ZSM5}$$

The kinetic studies used a laboratory unit with integral reactor, a Pt/ZSM5 catalyst, the feed was a mixture of ethyl-benzene/m-xylene in the ratio of 1/3 and a ratio H_2/hydrocarbon of 10 moles/mol.

The kinetic model of Robschlager and Christoffel was simplified somewhat, by considering that ethyl-benzene is formed exclusively from o-xylene, which agrees with what was said above about the reaction mechanism (Figure 13.69b). This model has the form:

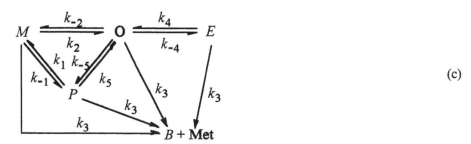

(c)

The model neglects the hydrogenation and the disproportionation reactions, considering as secondary reactions only the transformation of the C_8 aromatics to benzene and methane.

The process was assumed to be pseudohomogeneous, which led to the following system of kinetic equations:

$$\left.\begin{array}{l} \dfrac{dX_E}{d\left[\dfrac{W}{F_O(1+n)}\right]} = k_{-4}X_O - (k_3 + k_4)X_E - k_3 X_E \sum X_{C_8} \\[4ex] \dfrac{dX_p}{d\left[\dfrac{W}{F_O(1+n)}\right]} = k_{-5}X_O + k_{-1}X_M - (k_1 + k_3 + k_5)X_P - k_3 X_P \sum X_{C_8} \\[4ex] \dfrac{dX_M}{d\left[\dfrac{W}{F_O(1+n)}\right]} = k_1 X_P + k_{-2}X_O - (k_{-1} + k_2 + k_3)X_M - k_3 X_M \sum X_{C_8} \\[4ex] \dfrac{dX_O}{d\left[\dfrac{W}{F_O(1+n)}\right]} = k_2 X_M + k_4 X_E + k_5 X_P - (k_{-2} + k_{-4} + k_{-5} + k_3)X_O \\[3ex] \qquad\qquad\qquad - k_3 X_O \sum X_{C_8} \\[3ex] \dfrac{dX_B}{d\left[\dfrac{W}{F_O(1+n)}\right]} = k_3(1 - X_B) \sum X_{C_8} \\[4ex] \dfrac{dX_{Met}}{d\left[\dfrac{W}{F_O(1+n)}\right]} = k_3(1 - X_M) \sum X_{C_8} \end{array}\right\} \quad (13.69)$$

where:

X_i are the molar fractions of the species i,
W the amount of catalyst in g
F_O the feed in moles/hour
n the molar ratio H_2/hc
The subscripts E, P, M, O, B, Met in Eqs. (13.69) correspond to the notations of scheme (c) and have the meanings: ethyl-benzene, p-, m-, o-xylene, benzene and methane.

In Eqs. (13.69):

$$\Sigma X_{C_8} = X_E + X_P + X_M + X_O$$

The solution of this system of equations led to the values of the kinetic constants listed in Table 13.27 [174].

The values of the constants for the direct isomerization o-/p-xylene are small relative to the other isomerization constants.

Table 13.27 Rate Constants for Eqs. (13.69)

Temperature °C	$k (\times 10^3)$, mol/g cat. · h								
	k_1	k_1	k_2	k_2	k_3	k_4	k_4	k_5	k_5
300	145	64.6	33.6	79.9	1.56	4.16	1.04	7.80	8.25
350	302	135	61.1	138	7.17	13.5	4.06	32.8	33.3
400	513	231	88.5	192	24.4	31.1	10.9	104	101
E, kJ/mol	39.4	39.4	29.7	26.8	85.8	60.7	70.8	80.8	78.3
A, mol/g cat. · h	566	268	17.5	23.2	115.000	1590	3430	198,000	121,000

It is to be mentioned that the values of the kinetic constants obtained on the basis of experiments that used mixtures with constant ethyl-benzene/m-xylene ratios exclusively are less reliable than when this ratio is varied.

Cappellazzo et al. [184] studied the isomerization of the three xylenes with the purpose of identifying the existence of shape selectivity for p-xylene. The impetus to their study were the conclusions of Olson [185] that the diffusion coefficient of m-xylene in ZSM-5 is almost 10^3 times smaller than that of p-xylene. On such catalysts, the formed p-xylene can leave immediately the pores of the zeolite, whereas m- and o-xylene will be retained a much longer time, during which they may be submitted to isomerization reactions. The consequence of such a mechanism would be a concentration of p-xylene in the products, larger than that obtained on catalysts that have no shape selectivity.

The authors used a Mobil ZSM-5 isomerization catalyst in an integral reactor. The reaction temperatures ranged between 250 and 300°C and the pressure was 30 bar. In these conditions the reaction took place in liquid phase.

Two kinetic schemes were used: one is identical with scheme (a) given above, which does not allow the direct isomerization o-xylene/p-xylene and a second one, which allows such isomerization.

The experimental results were plotted in ternary diagrams and led to the conclusion that both kinetic schemes may be used in calculations.

Concerning the shape selectivity, it was concluded that by selecting adequate values for the temperature and the space velocity, the concentration of p-xylene in the effluent could be increased by approximately 2% over that corresponding to the thermodynamic equilibrium.

13.9.5.5 Effect of Process Parameters

By adequately modifying the operating conditions, the conversion in the system ethylbenzene/xylenes may be influenced towards maximizing the selectivity of the desired product.

Indeed, in the conditions of commercial operations the ratios between the produced xylenes are close to those for the equilibrium, whereas the conversion of ethyl-benzene to xylenes requires an intermediary hydrogenation step. Increasing the rate of this step requires an adequate selection of the operating conditions.

The influence of the temperature on the thermodynamic equilibrium was depicted in Figures 13.67 and 13.68. It results that the equilibrium among the xylenes is hardly influenced by the temperature, whereas the equilibrium conversion of ethyl-benzene increases strongly with increasing temperature.

The pressure and the molar ratio H_2/feed do not influence the ratios of the concentrations of the xylenes, but favor the reciprocal transformations ethyl-benzene-xylenes, and bring their concentrations closer to the values corresponding to thermodynamic equilibrium. This fact is a result of the favorable effect of the increase of the pressure and of the H_2/feed molar ratio on the formation of the hydrogenated intermediates.

In the studies of Raseev and Olteanu [180], ethyl-benzene of 99.7% purity was fed to the experimental unit, which was described previously (Figure 13.71), with a space velocity of 1.1 h^{-1}. The catalyst was of Pt on amorphous alumosilica. The results are summarized in Table 13.28.

From these data it results that the best conversions to xylenes (selectivity higher than 70%) are obtained at moderate pressures and at temperatures of the order of 390°C.

The commercial catalysts of platinum on zeolite, which are currently in commercial use, produce better performances than those shown in this table.

The presence of hydrogen and the elevated pressure maintained in the reactor favor the hydrogenation of the heavier aromatic hydrocarbons, which result from disproportioning reactions as well as from other coke generating compounds. For this reason a pressure of the order of 12 bar and the presence of hydrogen are necessary in order to ensure life cycles for the catalyst of several months or even years.

The catalyst factor may be analyzed from the point of view of the support and of the supported metal.

The zeolite support is much superior to the amorphous alumosilica. An important preoccupation is the selection of zeolitic supports of the ZSM-5 type, having a structure which should produce, by shape selectivity, reaction effluents having concentrations of para-xylene higher than those corresponding to the thermodynamic equilibrium.

Platinum is the metal of choice for these catalysts. Owing to its strong hydrogenating activity, Pt ensures operating cycles for the catalyst of many months and even of several years.

The chlorination of the catalyst is not necessary.

Table 13.28 Effect of Pressure and Temperature on Ethyl-Benzene Conversion

Pressure, bar	Temperature, °C	Conversion, wt %						
		Xylenes	TMB and higher	Benzene	Toluene	P + N	C_1-C_4	Total conversion
12	390	53.8	1.4	2.0	3.9	14	1.5	76.6
12	470	20.3	6.1	14.9	1.9	0.6	10.9	54.7
40	390	13.1	0.4	9.7	3.2	65.1	7.6	99.1
40	470	30.4	4.6	2.6	7.7	23.4	28.4	97.1

Source: Ref. 180.

13.9.5.6 Commercial Implementation of the Process

The process diagram of a unit for the isomerization of xylenes, including the products separation section, is depicted in Figure 13.71.

The operating conditions depend on the characteristics of the catalyst and on the interest in performing or not the conversion of the ethyl-benzene to xylenes. The usual operating conditions are situated in the range of temperatures comprised between 400–500°C, pressures of 10–20 bar, and molar ratios H_2/feed of 15:20.

Since the process has a low thermal effect, no reactor cooling by hydrogen injection is required.

The performance of an isomerization unit using a nonzeolitic catalyst (amorphous support) may be exemplified by the following data [184]:

Feed	ethyl-benzene	25.5 wt %
	p-xylene	14.0 wt %
	m-xylene	41.0 wt %
	o-xylene	19.5 wt %
Effluent	o-xylene	19.6 wt %
	p-xylene	70.6 wt %
	Total	90.2 wt %

Zeolite catalysts, the external surface of which have been passivated by zirconium and germanium, have better performances than the classical catalysts on amorphous support. The high isomerization efficiency of these catalysts leads to the improvement of the performances of the entire isomerization/separation complex.

Mobil Oil Corp. together with IFP (French Institute of Petroleum) proposed [201] the "Mobil High Activity Isomerization" process (MHAI), which assures a conversion to p-xylene 102% of the equilibrium and a 60–75% ethyl-benzene conversion.

The process economics are:

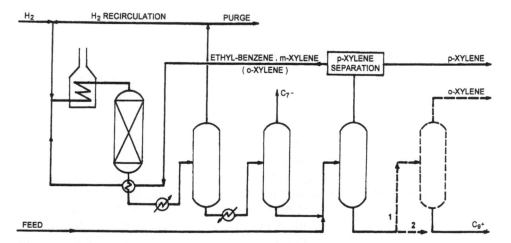

Figure 13.71 Process flow diagram for the isomerization of C_8 aromatics. When o-xylene is recycled, the last column (dotted lines) is missing.

Investment (1997)	57 $/tpy para-xylene
Utilities per metric ton of para-xylene produced	
electricity, kWh	47
fuel, 10^3 kcal/h	1300
cooling water (10°C rise) m^3/h	5
Catalyst fill lb/lb feed converted	3.0×10^{-6}
Maintenance per year as % of investment	2

UOP proposed [201] an isomerization unit (Parex-Isomar) with following yields:

Composition	Fresh feed wt units	Products wt units
Ethylbenzene	25.5	—
p-xylene	14.0	71.1
m-xylene	41.0	—
o-xylene	19.5	19.6

13.10 DEHYDROPOLYMERIZATION OF LOWER ALKANES

This process, called here "Catalytical polyaromatization" has the commercial names "Cyclar" (developed by UOP-BP) and "Aroforming" (developed by IFP). These processes were first commercialized only in 1988.

In this processes C_3-C_4 alkenes are converted to aromatic hydrocarbons on bifunctional catalysts.

Propane with *i*- and *n*-butane and, in some cases, also with pentane, are submitted to the following successive reactions: dehydrogenation to alkenes, dimerization of the alkenes, dehydrocyclization of the C_6-C_8 alkenes to yield finally, aromatic hydrocarbons. The reactions that take place in the case of propane may be written as:

$$2C_3H_8 \xrightleftharpoons{-2H_2} 2C_3H_6 \rightleftharpoons C_6H_{12} \xrightleftharpoons{-3H_2} \bigcirc$$

Similar reaction sequences may be written for *n*- and *i*-butane.

13.10.1 Process Thermodynamics

Since the process involves successive steps, the thermodynamic analysis must take into account the equilibrium of the three steps (reactions) in order to obtain correct results. Thus, if the reaction of the first step is thermodynamically restricted, the intermediary products will actually be absent, and the calculation of the overall equilibrium will give correct results. However, if the last step is thermodynamically restricted, or if the equilibrium constants of the reaction steps have orders of similar magnitude, the intermediate compounds will be present in the final product, and the calculation limited to the overall equilibrium will give wrong results. In such cases, all the reaction steps must be taken into consideration, or at least those that have less favorable equilibriums in order to obtain the correct composition of the effluent corresponding to the thermodynamic equilibrium of the process. For a process for which the equilibrium of the steps are unknown, it is useful to perform the analysis beginning from the last reaction step.

In the reaction condition for the dehydro-polyaromatization, the equilibriums of the last step—dehydrocyclization of the alkenes—is completely displaced to the right for all the hydrocarbons that participate. Thus, for C_6-C_8-alkenes, the equilibrium conversion is above 99% at atmospheric pressure and temperatures that exceed 230°C, or at 340°C and a pressure of 7 bar. In connection with this, it must be remarked that for the dehydrocyclization of alkenes the thermodynamic equilibriums are much more favorable than those of the alkanes represented in Figure 13.4.

Taking into account the last two steps, it results that here also, the equilibriums are completely displaced to the right. The equilibrium conversion of butenes to aromatic hydrocarbons exceeds 99% at temperatures above 220°C and at a pressure of 7 bar. For propene, the same conversion is obtained at temperatures of above 515°C and 7 bar, or at 365°C and atmospheric pressure.

The equilibrium of the first reaction, the dehydrogenation of the $C_3°$-C_4 alkanes, is the most disfavored and constitutes the thermodynamically limiting step of the process. In this situation, the composition at equilibrium may be correctly calculated by taking into consideration the overall process of chemical conversion.

Such a calculation leads to the equilibrium represented in Fig 13.72 for propane, Figure 13.73 for *n*- and *i*-butane and Figure 13.74 for mixtures of propane-butane.

From these graphs it results that from the thermodynamic point of view, the process could lead to conversions higher than 95% at atmospheric pressure and at temperatures above 569°C, a temperature much lower than the approximately 620°C at which the secondary reactions of thermal cracking become important [187].

The equilibrium conversions are diminished by pressure. Thus, at 7 bar and 560°C, the equilibrium conversion of propane is only 60% (Figure 13.73). It is 70% for mixtures of propane with butane (Figure 13.75).

13.10.2 Catalysts

The catalysts used in the process are base metals, such as oxides supported on zeolites or on zeolites included in an inert binder. The latter solution is used in the case of the moving bed processes [34] where the catalyst used is shaped as spheres, which have a good resistance to attrition.

13.10.3 Reaction mechanism and Kinetics

The mechanism of these reactions was pictured in Figure 13.16. It consists in the adsorption of the alkane on the metallic site, followed by dehydrogenation and desorption of the formed alkene. The alkene is adsorbed on an acidic sites of the zeolite where it dimerizes. The pore size and structure of the zeolite are the probable factors that prevent more extensive polymerization reactions. The formed C_6-C_8 alkenes are readsorbed on the metallic sites and are dehydrocyclized with the formation of the corresponding aromatic hydrocarbons.

Since no hydrogen is introduced in the process and the pressure is limited, the coke deposits are rather important. They need to be periodically burned, thus regenerating the catalyst. The burning process leads to the concomitant reoxidation of the

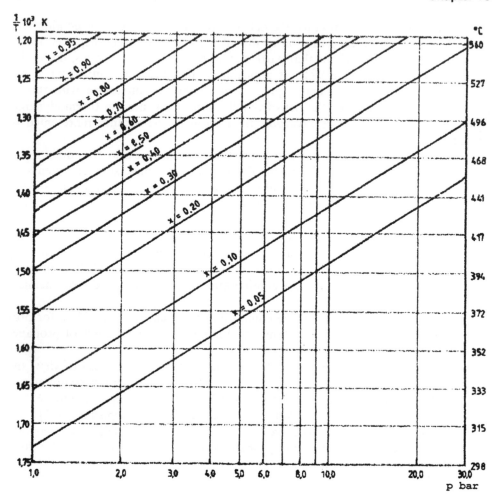

Figure 13.72   Reaction equilibrium for: $2\,C_3H_8 \rightleftharpoons C_6H_6 + 5\,H_2$.

catalyst. In operation, the oxides of the deposited metals are partly reduced by the hydrogen resulting from the process.

Other secondary reactions are those of cracking, which lead to the formation of methane and ethane and to the decrease of the yield of aromatic hydrocarbons. The activation energy of thermal cracking is higher than that of the catalytic reactions taking place in the process. The reactions of thermal cracking are favored by higher temperatures and determine the highest acceptable level of temperatures.

Concerning the kinetics of the process, the rate-determining step is the de-hydrogenation of the C_3-C_4 alkanes. The catalyst used should have good dehydrogenation activity.

No direct studies of the process of kinetics seem to be available. By taking into account that the rate determining step is the dehydrogenation, conclusions drawn from the analysis of the kinetics of the dehydrogenation of lower alkanes are useful and make easier the kinetic study of dehydro-polyaromatization.

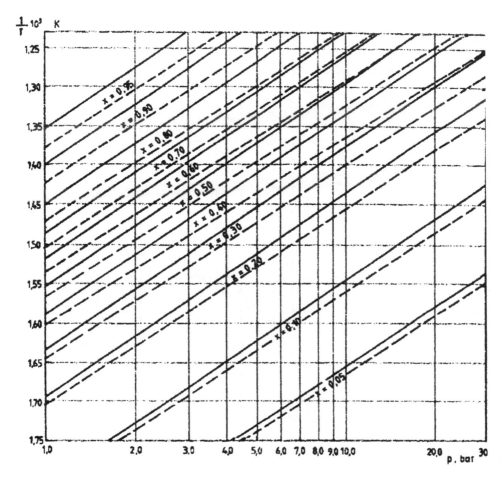

Figure 13.73 Equilibrium of reactions. — $2n\text{-}C_4H_{10} \rightleftharpoons m\text{-}C_6H_4(CH_3)_2 + 5H_2$, --- $2\,i\text{-}C_4H_{10} \rightleftharpoons m\text{-}C_6H_4(C_3)_2 + 5H_2$.

13.10.4 Effect of Process Parameters

Only few data are available on the operating conditions of these units, which are of recent vintage. The first experimental unit was put into operation in January 1990 [34] and served mainly to determine the optimal process conditions.

The tests were first carried out at pressures named "reduced" (probably close to the atmospheric) and then at pressures called "high" with the mention that in any case they did not exceed 7 bar [34].

The operation at low pressures is justified, as shown in Figures 13.73–13.75, by reasons of thermodynamic equilibrium, which at such pressures, favors the desired products. Since the process takes place in vapor phase, the "high" pressure, corresponds to a longer contact time and therefore to higher conversion. At constant conversion, higher pressures lead to the decrease of the reactor size and of the catalyst inventory, with the corresponding economic advantages.

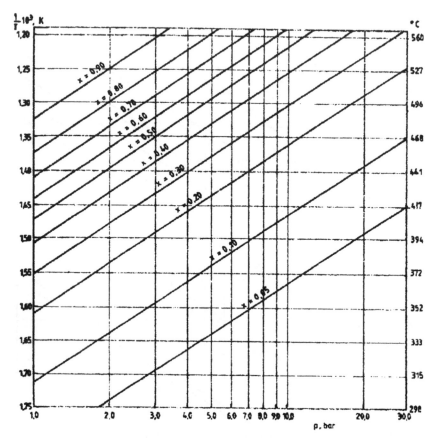

Figure 13.74   Reaction $C_3H_8 + C_4H_{10} \rightleftharpoons C_6H_5 \cdot CH_3 + 6H_2$.

The highest acceptable temperature is determined by the acceptable concentration of products from the secondary reactions of thermal cracking. The lowest acceptable temperature results from considerations of thermodynamics and reaction rate. The selection of the operating temperature is determined also by the required purity of the desired product, which decreases with the increase of temperature.

13.10.5 Commercial Implementation

The process knows two embodiments: one with moving bed of catalyst—"Cyclar" developed by BP and UOP [33, 34] and another, with fixed bed of catalyst "Aroforming," developed by IFP [35].

The continuous regeneration of the catalyst as practiced in the system with moving bed "Cyclar," allows a more severe operating regime with relatively intense deposition of coke. The process with fixed bed "Aroforming" needs a milder regime to ensure cycles of operation of 12 hours, between regenerations [35].

The strong endothermal character of the process causes important difficulties for its implementation with fixed bed of catalyst.

13.10.5.1 The "Cyclar" Process

The process flow schematic of the "Cyclar" unit is shown in Figure 13.75. The unit uses a reactor subdivided in several zones, aligned vertically, in one shell. The catalyst is fed to the top of the upper zone and flows by gravity through all zones in series, while forming in each a moving bed. The reactant flows downwards, cocurrent with the catalyst. Since the process is strongly endothermal, the reaction mixture is reheated between the reaction zones, by passage through dedicated furnaces. The catalyst leaving the last reaction zone is regenerated by burning of the deposited layer of coke. The regenerated catalyst is reintroduced in the upper part of the reactor in the top zone. The system is therefore similar to the catalytic cracking with moving bed, the difference being that it has several reaction zones with the intermediary reheating of the reaction mixture.

After leaving the reaction system, the products are cooled and sent to a gas–liquid separator. The separated gases are compressed and processed for the recovery of the hydrogen generated on the reactions and of the light hydrocarbon produced by the secondary reactions of thermal cracking. The aromatics are recovered from the liquid product.

The first unit of this type was built in the Grangemouth refinery in Scotland and started operation in January 1990.

The tests performed in this unit in two operating conditions (two pressure levels) led to the material balance in wt % given in Table 13.29. The feed was a mixture of 50% propane and 50% butane.

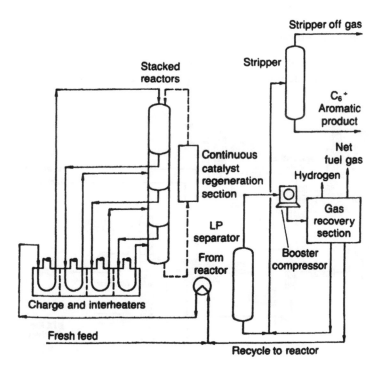

Figure 13.75 Cyclar process flow diagram..

Table 13.29 Typical Yields for
the Cyclar Process (wt %)

	Pressure	
	low	high
Hydrogen (purity 95%)	6.9	4.4
Gases	29.6	43.1
Benzene	15.6	9.9
Toluene	27.6	22.7
Xylenes	14.7	13.9
Aromatics C_9^+	5.6	6.0
Total	100.0	100.0

The quality of the obtained gasoline is:

	Fraction	
	C_6^+	C_7^+
The octane number research, unleaded	111.8	113.2
The octane number motor, unleaded	100.4	102.5

The economics of the process depend very much on the cost of the C_3-C_4 fractions compared to the cost of the aromatic hydrocarbons and besides, to a great extent, on the interest for the hydrogen produced in the process. For this reason, economic justification has to be established on a case-by-case basis.

13.10.5.2   The "Aroforming" Process

The process schematic of the "Aroforming" unit is shown in Figure 13.76 [35]. The reaction system consists of two identical reactors, operating alternately, with fixed

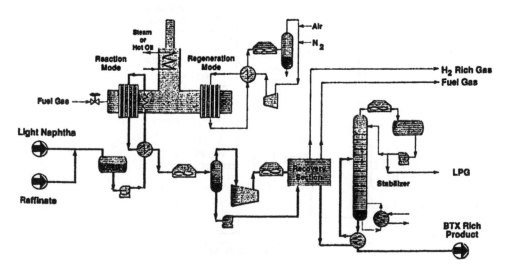

Figure 13.76   Aroforming process flow diagram.

bed of catalyst, placed in a system of parallel tubes. The tubes are located within a furnace. The reactors operate in 12 hours cycles. While one of the reactors is operated, the other is being regenerated.

The system may process not only propane and butanes but also light naphtha.

The process was developed based on studies which started in 1983 and was commercialized by a consortium of I.F.P. and Salutec of France, with the Institute of Technology RMIT of Melbourne (Australia). In 1990, burners with ceramic fibers, developed by Kinetic Technology International (KTI), were incorporated. These burners allow the direct heating of the reaction tubes placed in the furnace in a manner that allows the reaction to take place in practically isothermal conditions, which is an important advantage.

The material balances (wt %) for the processing of propane or butane are presented in Table 13.30.

The investments for a unit having the processing capacity of 800 m^3/day are estimated at US\$ 21 million for the processing of the C_3-C_4 fraction and at US\$16 million for a feed of light naphtha (1990 basis).

The utilities consumptions for the same unit are:

Fuel gases	kJ/h	58.5
Electricity	kWh/h	4000
Cooling water	m^3/h	20
Nitrogen	Nm^3/h	30

13.10.6 Comparison of the Two Commercial Processes

The technical approaches adopted in the two processes, i.e., adiabatic multizone reactor or isothermal reactor, may be compared by examining the change of conversion and of the temperature in the two reaction systems.

To this purpose, Figure 13.78 gives the equilibrium compositions for the dehydro-polyaromatization of butanes, propane and the mixtures butane-propane, as derived from the graphs of Figures 13.73, 13.74, and 13.75. The graph refers to the pressures of 1 and 7 bar, the latter being the highest pressure used in the analyzed processes.

For the adiabatic multizone process, by neglecting the secondary reactions, the decrease of the temperature as a function of the conversion may be expressed, by the equation:

Table 13.30 Typical Yields for the Aroforming Process

	Feed	
	C_3H_8	C_4H_{10}
H_2	6.0	5.6
Gases	30.8	29.5
Benzene	20.0	18.0
Toluene	26.1	27.1
Xylenes	11.0	14.0
Aromatics C_9^+	6.1	4.8

$$x \cdot (\Delta H_T)_r - \overline{C}_p \Delta t = 0 \qquad\qquad (13.71)$$

where:

$(\Delta H_T)_r$ – the thermal effect of the reactions,
$\overline{C}_p$ – the heat capacity of the reaction mixture for the average temperature in
    the zone
Δt – the temperature decrease in the considered zone
x – the conversion

According to this equation, the variation of Δt with conversion x is linear, and therefore in the graph of Figure 13.78 it is traced as a straight line. One of the points of this straight line corresponds to the conditions of inlet in the considered zone. A second point may be easily determined by using Eq. (13.71) for calculating the conversion for a temperature decrease conveniently selected. The obtained straight line is extended until it intersects the equilibrium curve.

The above calculation neglects the secondary reactions. An easy remedy is to take for $(\Delta H_T)_r$ the real value of the thermal effect of the process as calculated according to the methods given in Section 1.1.

The intersection point of the straight line for the temperature decrease with the equilibrium curve corresponds to attaining the equilibrium composition. Actually, this point will not be reached due to the limited residence time in the zone. The actual conversion attained at the exit from any zone is computed on basis of the kinetic model and represents the corresponding point on the straight line of temperature decrease traced in the graph.

For illustrating the procedure, the process of dehydro-polyaromatization of propane and a temperature of about 800 K is considered. Neglecting the secondary reactions and considering the overall reaction:

$$2C_3H_6 \rightleftharpoons C_6H_6 + 5H_2$$

the following thermodynamic values will be used [2]:

$$\left(\Delta H^0_{800}\right)_r = 317.01 \text{ kJ/mol}$$

$$\left(\bar{C}_{p_{800K}}\right)_{feed} = 310.49 \text{ J/mol} \cdot \text{K}$$

$$\left(\bar{C}_{p_{800K}}\right)_{products} = 336.83 \text{ J/mol} \cdot \text{K}$$

for a conversion of 10% in the first reaction zone:

$$\left(\overline{C}_{p800K}\right)_1 = 313.13 \text{ J/mol} \cdot \text{K}$$

Selecting arbitrarily $x = 0.2$, by using Eq. (13.71) it results:

$$\Delta t = 0.2 \frac{317,010}{313.13} = 202.5°C$$

which leads to the representation in the graph of Figure 13.77 of the straight line for the decrease in temperature in the first zone, for which the inlet temperature in the reactor is 640°C.

Assuming that in each zone the reaction reaches 90% of the respective equilibrium conversions, it is possible to plot the intermediary reheating steps and the

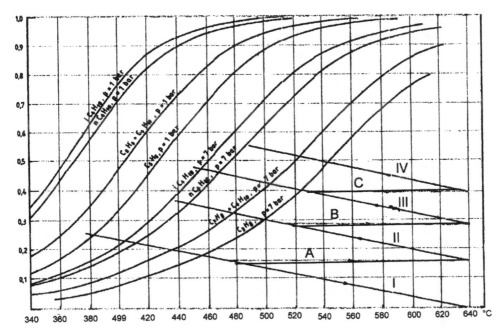

Figure 13.77 Graphical calculation of the adiabatic multizone system for the dehydropolydehydrocyclization of light alkanes. I-IV – Reaction zones, A-C – Intermediary heating.

conversions in zones II, III, and IV. Such a procedure is shown in Figure 13.77 for a pressure of 7 bar.

For zones II–IV, Eq. (13.71) was used in the form:

$$\Delta t_i = \Delta x_i \cdot \frac{(\Delta H_T)_r}{(\overline{C}_p)_i} \tag{13.72}$$

According to Figure 13.77, at the outlet from the 4th reaction zone the conversion is 48.5% and the temperature is 550°C.

The position of the equilibrium curves allows us to conclude that for the same inlet temperature:

The final conversion is higher at a lower pressure than at higher ones;
The conversion is higher for butanes or mixtures butanes + propane than for propane alone.

The temperature of the feed at the reactor inlet has a strong effect on the final conversion. Higher inlet temperature leads to higher conversions but has the disadvantage of emphasizing the secondary reactions.

The above approach for evaluating the conversion obtained in the adiabatic multizone reaction system and the calculation method presented in Figure 13.77 can be applied also to other processes, such as oligomerization or catalytic reforming, by using the reaction models and equilibrium parameters that are specific for the respective process.

The isothermal reaction system of "Aroforming," has the advantage that for a given inlet temperature, the conversion at equilibrium is much higher than for an adiabatic system. However, the cyclic alternation of conversion and regeneration periods obliges operating at lower temperature in order to decrease the rates of the coke-producing reactions. This decreases significantly the advantages of the isothermal system.

In view of this situation, a realistic comparative analysis of the isothermal and adiabatic systems has to be based on accurate data for the operating conditions of both processes.

REFERENCES

1. DC Martindale, PJ Kuchar, RK Olson, Aromatics from LPG using Cyclar process, American Institute of Chemical Engineers, Summer National Meeting, Denver, Colorado, 21–24 August, 1988.
2. RJ Stull, EF Westrum Jr., GC Simke, The Chemical Thermodynamics of Organic Compounds, Malabar, Florida: Robert E. Krieger Publ. Co., 1987.
3. LR Aalund, Oil and Gas J 82: 55, Oct. 8, 1984.
4. DM Little, Catalytic Reforming, Tulsa, Oklahoma: PennWell Publ. Co., 1985.
5. FG Ciapetta, U.S. Patent 2,550,531, 1951.
6. J Liers, J Mensinger, A Mösch, W Reschetilowski, Hydrocarbon Processing 72: 181, Aug. 1993.
7. RH Heck, NY Chen, Appl Catal A86: 83, 1992.
8. H Kalies, F Rossner, HG Karge, KH Steinberg, Stud Surf Sci Cat 69: 425, 1991.
9. J Dermietzel, W Jockisch, HD Neubauer, Chem Tehn. 42: 164, 1990.
10. J Dermietzel, G Strenge, J Bohm, HD Neubauer, Catal Today 3: 445, 1988.
11. EM Blue, GD Gould, CJ Egan, TR Huges, Production of Aromatic Hydrocarbons by Low-Pressure Rheniforming, Chevron Research Co., California, Oct. 21, 1975.
12. JP Frank, OAPEC-IFP Joint Workshop-Catalysis in the Downstream Operations, July 5-7, 1994, Report no. 3, Institut Français du Pètrole, Rueil-Malmaison, France.
13. A Stoica, PhD thesis, Sci. directed by S. Raseev, Institutul de Petrol-Ploiesti, Romania, 1976.
14. JN Kolomytshev, AS Belîi, VK Dupliakin, AJ Lugovskoi, PM Vascenco, SA Loghinov, VA Romashkin, Himia i Tehnologhia Topliv Masel 4(1), 1991.
15. VK Dupliakin, ed. Nanesenyie metaliceskie katalizatory prevrascenia uglevodorodov, pp. 79, Catalysis Institute of USSR Acad. Novosibirsk, 1978.
16. J Völter, G Lietz, M Uhlemann J Catalysis 68(1): 42, 1981.
17. AS Belyi, IN Kolomytshev, AP Fedorov, LP Vorontzova, VK Dupliakin, Himia i Tehnologhia Topliv Masel, 19(2), 1991.
18. VB Maryshev, IA Skipin, Himia i Tehnologhia Topliv Masel, 4(12), 1991.
19. BB Jarkov, VI Georgievski, BV Kpacii, SB Kogan, TM Klimenko., LA Saiko, Himia i Tehnologhia Topliv Masel, 10(1), 1991.
20. AF Mahov, NP Smirnov, GC Teliashev Neftepererabotka i Neftehimia, 25(8), 1976.
21. AF Mahov, NP Smirnov, RI Safin. Himia i Tehnologhia Topliv Masel, 17(2), 1984.
22. BC Gates, JR Katzer, GCA Schmitt, Chemistry of Catalytic Process, New York: McGraw Hill. 1979.
23. SD Raseev, CD Ionescu, Reformarea catalitică, Editura Tehnică, Bucharest, 1962.
24. SD Raseev, CD Ionescu, Reformowanie Katalityczne, Wydawnictwa Naukowo-Techniczne, Warszawa, 1965.
25. SD Raseev, CD Ionescu, Katalytisches Reformieren, VEB Deutscher Verlag für Grundstoffindustrie, Leipzig, 1966.
26. BC Gates, Catalytic Chemistry, John Wiley & Sons Inc., 1991.

27. AA Balandin, Multipletnaia teoria, Nauka, Moscow, 1971.
28. GB Rabinovitch, ME Levinter, MN Berkovitch, Optimizatzia protzesa katalicheskovo reforminga s tzeliu snijenia energopotreblenia, Ministry of Petrochemistry and Petroleum Processing, Petroleum Processing Series, 1985.
29. J Barron, G Maire, JM Müller, FG Gault, Journal of Catalysis 5: 428, 1966.
30. E Ghionea, PhD thesis, Institutul de Petrol si Gaze-Ploiesti, Romania, 1979.
31. GA Mills, H Heinemann, TH Milliken, AG Oblad, Ind Eng Chem. 45: 134, 1953.
32. BA Kazanski, AL Liberman, MJ Bratuev Dokladi Akad Nauk SSSR 61: 67, 1948.
33. PC Doolan, PR Pujado, Hydrocarbon Processing 68: 72, Sept. 1989.
34. CD Gosling, FP Wilcher, L Sullivan, RA Mountford, Hydrocarbon Processing 70: 69, Dec. 1991.
35. L Mank, A Minkkinen, R Shaddiek, Hydrocarbon Technology International-1992, P. Harrison, ed. London: Sterling Publ. Internat. Ltd., 1992.
36. PB Weisz, Science 123: 888, 1956.
37. SG Hindin, SW Weller et al. J Phys Chem 62: 224, 1958.
38. FG Ciapetta, RN Dobres, RW Baker. Catalysis, vol. VI, pp. 445, London: Reinhold Publ. Corp., 1958.
39. RB Smith, Chem Engineering Progress 55: 76, 1957.
40. HG Krane, BA Groth, LB Schulman, HJ Sinfeld, Fifth World Petroleum Congress-Section III, New York, June, 1959, pp. 39.
41. J Ancheyta-Juárez, E Aguilar-Rodriguez, Oil Gas J 92: 93, Jan. 31, 1994.
42. E Aguilar-Rodriguez, J Ancheyta-Juárez, Oil Gas J. 92: 80, July 25, 1994.
43. RL Burnett et al. Preprint Papers Gen. Div. Petrol. Chem., Detroit, Michigan, 1965.
44. RL Burnett et al. J Am Chem Soc 87(1): 17, 1995.
45. JM Jorov, GM Pancencov, Kinetica i Kataliz 6: 1092, 1965.
46. JM Jorov, GM Pancencov, Kinetica i Kataliz 8: 658, 1967.
47. JM Jorov et al. Himia i Tehnologhia Topliv Masel 11: 37, 1970.
48. H Henningen, M Bundgard-Nielson, British Chem Eng. 15: 1433, 1970.
49. JH Jenkins, TW Stephens, Hydrocarbon Processing 59: 163, Nov. 1980.
50. S Raseev, C Ionescu, Bul. Inst. Petrol Gaze, Geologie, Bucharest, vol. XV, 1967, pp. 89.
51. C Ionescu, S Raseev, Bul. Inst. Petrol Gaze, Geologie, Bucharest, vol. XIV, 1966, pp. 105.
52. GB Martin, GF Froment, Chem Eng Sci. 37: 759, 1982.
53. PA Van Trimpout, GB Martin, GF Froment, Ind Eng Chem Res 27: 51, 1988.
54. LH Hosten, GF Froment, Ind Eng Chem Process Des Dev 10: 280, 1971.
55. RP De Pauw, GF Froment, Chem Eng Sci 30: 789, 1975.
56. GB Marin, GF Froment, Chem Eng Sci 37: 759, 1982.
57. MA Tanatarov, GM Panchencov, MF Galiulin, Kataliticheskie reactzii prevrashcenia uglevodorodov, no. 6, Nauka, Kiev, 1981.
58. NM Ostrovski, JK Demanov, Himia i Tehnologhia Topliv Masel 2: 35, 1991.
59. B Schöder, C Salzer, F Turek, Ind Eng Chem Res 30(2): 326, 1991.
60. J Barbier, P Marecot P, J Catal 21: 102, 1986.
61. PH Schripper, KP Graziani, The 8th Proc Int Symp Chem React Eng, 1984.
62. MP Ramage, KP Graziani, FJ Kranbeck, Chem Eng Sci. 35: 41, 1980.
63. BA Kazanski, TF Bulanova, Doklady Akad Nauk SSSR 62: 83, 1948.
64. AA Balandin, Z Phys Chem B, 2: 289, 1929.
65. EV Backensto, RW Manuel, Oil Gas J 56(20): 131, 1958.
66. BN Aspel, GS Golov, VD Pohojev, Himia i Tehnologhia Topliv Masel 5(5): 1, 1960.
67. J D'Auria, WC Tieman, G Antos, Recent Platforming Catalyst Developments, Paper AM-80-49, NPRA Annual Meeting, 1980.
68. MD Edgar, Catalytic Reforming of Naphtha in Petroleum Refineries, In: Applied Industrial Catalysis, 1, New York: Academic Press Inc., 1983.

69. NS Figoli. Applied Catalysis 5: 19, Jan. 14, 1983.
70. NPRA Annual Meeting—1969, Questions and answers, 65, question no. 27.
71. JM Grau, RJ Verderone , CL Pieck, EL Jablonski, JM Parera, Ind Eng Chem 27: 1751, 1988.
72. NPRA Annual Meeting—1978, Questions and answers of E. Dean and L.R. Mains.
73. GM Bickle, JN Beltramini, DD Do, Ind Eng Chem 29: 1801, 1990.
74. S Raseev, C Ionescu, R Kramer, Bul Inst Petrol Geologie 19: 207, 1972; 20: 139, 1973.
75. L Verman, I Zîrnă, I Ghejan, D Grigoriu, T Filotti, D Iorga, S Raseev, Petrol si Gaze 19: 62, suppl. 1967.
76. RG McClung, Oil Gas J 88: 98, Oct. 8, 1990.
77. GJ Straguzzi, HR Aduriz, CE Gigola, J Catal 66: 171, 1980.
78. SC Fung, Chemtech 40, Jan. 1994.
79. MFL Johnson, SL Graff, U.S. Patent 3,781,219, 1973.
80. K Lorenz, W Petzerling, HD Neubauer, H Schnetter, H Franke, H Hergeth, K Becker, H Engelmann, U.S. Patent 150,986, 1979.
81. H Lieske, G Lietz, H Spindler, J Volter, J Catal 8: 81, 1983.
82. GB McVicker, JJ Ziemiak, Appl Catal 14: 229, 1985.
83. TF Garetto, CR Apesteguia, Appl Catal 20: 133, 1986.
84. DR Hogin, RC Morbeck, HR Sanders, U.S. Patent 2,914,444, 1959.
85. JH Sinfelt, Bimetallic Catalysts—Discoveries, Concepts and Applications, New York: John Wiley & Sons, 1983, p. 83.
86. YJ Huang, SC Fung, WE Gates, GB McVicker, J Catal 118: 192, 1989.
87. DJC Yates, WS Kmak, U.S. Patent 3,937,660, 1976; DJC Yates, U.S. Patent 3,981,823, 1976.
88. JD Paynter, RR Cecil, U.S. Patent 3,939,061, 1976.
89. DJC Yates, WS Kmak, U.S. Patent 4,172,817, 1979.
90. K Foger, D Hay, HJ Jaeger, J Catal 96: 154, 170, 1985.
91. RL Davidson, Petr Proc 10: 1170, 1955.
92. BL Moldavski, GD Kamusher, MB Kobylskaia, Doklady Akad Nauk SSSR, 1936, p. 343.
93. BL Moldavski, GD Kamusher, J Obs Himii 7: 131, 1937.
94. BL Moldavski, GD Kamusher, Sbornik Trud. TZIATIM, 4: 69, Gostoptehizdat, 1947.
95. RL Peer, RW Bennet, DE Felch, RG Kabza, Platforming Leading Octane Technology, UOP, 1990.
96. G Heinrich, M Valais, M Passot, B Chapotel, Mutations in World Refining Challenges and Answer, Proceedings of the Thirteenth World Petroleum Congress, 1982, Vol. 3, pp. 189, New York: J. Wiley & Sons.
97. Hydrocarbon Processing 69: 118, Nov. 1990.
98. Hydrocarbon Processing 61: 169, Sept. 1982.
99. Hydrocarbon Processing 61: 167, Sept. 1982.
100. RG McChing, JS Sopko, Hydrocarbon Processing 62: 80, Sept. 1983.
101. Hydrocarbon Processing 71: 118, Nov. 1992.
102. A Hennico, E Bourg, L Mank L, Hydrocarbon Technology International, 1988, P Harrison, ed., London: Sterling Publ., 1988.
103. G Stephens, P Travers, L Mank, Vers le XXI^{-e}siecle avec les Technologies de l'IFP, Institut Français du Pètrole, June, 1994.
104. G Stephens, P Travers, L Mank, Revue de l'Institut Français du Pètrole 49: 453, Sept.–Oct. 1994.
105. Hydrocarbon Processing 73, Nov. 1994.
106. MW Golem, NL Gilsdorf, DE Felch, In: P Harrison, ed., Hydrocarbon Technology International, 1989/1990, London: Sterling Publ. 1991.

107. RL Peer, RW Bennett, ST Bakas, Oil and Gas J 86: 52, May 30, 1988.
108. LE Turpin, Hydrocarbon Processing 71: 81, June, 1992.
109. G Ferschneider, P Mège, Revue de l'Institut Français du Pètrole 48: 711, Nov.–Dec. 1993.
110. The American Institute for Physical Property Data (DIPPR). The American Institute of Chemical Engineers, version no. 4.0, Feb. 1989.
111. JJ Jeanneret, CD Low, V Zukanskas, Hydrocarbon Processing 73: 43, June, 1994.
112. KW Rockett, UOP Scheme, presented at The UOP Symposium, Bucharest, January 1977/1978.
113. JR Mowry, UOP Isomar Process for Xylene Isomerization, Chapter 5.7, In: Handbook of Petroleum Refining Processes, RA Meyer, ed., New York: McGraw Hill, 1986.
114. Hydrocarbon Processing 62: 158, Nov. 1983.
115. RE Prada., SM Black, SH Herber, JJ Jeanneret, JA Johnson, In: P Harrison, ed., Hydrocarbon Technology International, 1994, London: Sterling Publishing Co.
116. Hydrocarbon Processing 71: 144, Nov. 1992; 73: 92, Nov. 1994.
117. Hydrocarbon Processing 73: 90, Nov. 1994.
118. BM Wood, ME Reno, GJ Thompson, Alkylate Aromatics in Gasoline via the UOP Alkymax Process, UOP, 1990.
119. Hydrocarbon Processing 73: 90, Nov. 1994.
120. A Nicoară, Hidrodezalchilarea, In: Ingineria prelucrării hidrocarburilor GC Suciu, RC Tunescu, coord. eds, Vol. 2, Editura Tehnică, Bucharest, p. 399.
121. H Hussein, PhD thesis, directed by Raseev. Institutul de Petrol, Gaze si Geologie, Bucharest, 1970.
122. RJ Silsby, EW Sawler, Journ Appl Chem 8(3): 347, 1956.
123. AH Weiss, Hydrocarbon Processing 41(6): 185, 1962.
124. S Shull, H Hixon, I.E.C. Proc Design Develop 5(2): 146, 1966.
125. JR Mowry, UOP Aromatics Transalkylation Tatoray Process, In: RA Meyers, ed. Handbook of Petroleum Refining Process, McGraw Hill, (1986.
126. WW Kaeding, LB Young, U.S. Patent 4,034,053, 1977.
127. WW Kaeding, U.S. Patent 4,067,920, 1978.
128. NY Chen, WW Kaeding, FG Dwyer, J Am Chem Soc 101: 6783, 1979.
129. WW Kaeding, C Chu, LB Young, SA Butter, J Catal 69: 392, 1981.
130. NR Mersham, J Chem Technol Biotechnol 37: 111, 1987.
131. LE Aneke, LA Gerritsen, J Eiler, R Trion, J Catal 59: 37, 1979a.
132. N Davidova, N Peshev, D Shopov, J Catal 58: 198, 1979.
133. JC Wu, A Leu, YH Ma, Zeolites 3: 118, 1983.
134. GV Bhaskar, DD Do, Ind Eng Chem Res 29(3): 355, 1990.
135. PB Weisz, Pure Appl Chem 52: 2091, 1980.
136. WC Haag, RM Lago, PB Weisz, Discuss Faraday Soc 72: 319, 1982.
137. DH Olson, WO Haag, Am Chem Soc Symp Ser 248: 275, 1984.
138. GW Pukanic, FE Massoth, J Catal 28: 308, 1978.
139. ML Poutsma, Mechanistic Considerations of Hydrocarbon Transformations Catalysed by Zeolites, In: JA Rebo ed., Zeolite Chemistry and Catalysis, American Chemical Society, Washington D.C., 1976, pp. 437.
140. P Beltrame, PL Beltrame, P Carniti, G Zuretti, G Leofanti, E Moretti, M Padovan, Zeolites, 7: 418, 1987.
141. JA Amelse, Catalysis 1987. JW Ward, ed., Elsevier, Amsterdam, 1988, p. 165.
142. A Streitweiser Jr, L Reif, J Am Chem Soc 82: 5003, 1960.
143. KM Dooley, SD Brignac, GL Price, Ind Eng Chem Res 29: 789, 1990.
144. LE Aneke, LA Gerristsen, PJ Van den Berg, WA de Long, J Catal 59: 26, 1979b.
145. JR Chang, PC Shen, YM Cheng, JC Wu, Appl Catal 33: 39, 1987.
146. NS Gnep, M Guisnet, Appl Catal 1: 329, 1981.

147. H Matsumoto, Y Morita, J Jap Petr Inst 8: 572, 1967.
148. VS Nayak, L Riekert, Appl Catal 23: 403, 1986.
149. G Schutz Ekloff, NI Jaeger, C Vladov, L Petrov, Appl Catal 33: 73, 1987.
150. T Yashima, H Moslehi, N Hara, Bul Jap Petr Inst 12: 106, 1970.
151. Hydrocarbon Processing 62: 83, Nov. 1983.
152. TC Tsai, HC Hu, JY Tsai, Oil Gas J 92: 115, June 13, 1994.
153. Hydrocarbon Processing 62: 159, Nov. 1983.
154. JE Norris, DJ Rubinstein, G Vaal, J Am Chem Soc 61, 1939.
155. GH Baddeley, D Voos, J Chem Soc 21(1–2): 100, 1952.
156. AP Caulay-Lien, J Am Chem Soc 74, 1952.
157. HC Brown, H Jungk, J Am Chem Soc 77, 1953.
158. BS Greensfelder, HH Voge, Ind Eng Chem 37, 1945.
159. ER Bordeker, EW Benkiser, J Am Chem Soc 76(13), 1956.
160. JG Mamedaliev, Doc Ak Nauk SSSR 106(6), 1956.
161. PM Pitts, JE Connor, LN Leum, Ind Eng Chem 5, 1955.
162. BG Ranby, JK Johnson, The Sixth World Petroleum Congress, Vol. IV, 28, 1963.
163. J Mullarkey, Chem News 35, 1969.
164. P Chutoransky, FG Dwyer, In: WM Meyer, JB Uytterhoven, eds., Molecular Sieve, Advances in Chemistry Series, no. 121, American Chemical Society, Washington D.C., 1973, p. 540.
165. A Cortes, A Corma, J Catal 51: 338, 1978.
166. A Cortes, A Corma, J Catal 57: 444, 1979.
167. KH Robschlager, EG Christoffel, Ind Eng Chem Prod Res Dev 18: 347, 1979.
168. M Nitta, PA Jacobs, In: B Imelik, C Naccacho, Y Ben Taarit, JC Vedrine, G Condurier, H Praliand eds., Catalysis by Zeolites, Amsterdam: Elsevier, 1980, p. 251.
169. A Corma, A Cortes, Ind Eng Chem Process Des Dev 19: 263, 1980.
170. LB Young, SA Butter, WW Kaeding, J Catal 76: 418, 1982.
171. DJ Collins, KJ Mulrooney, RJ Medina, BH Davis, J Catal 75: 291, 1982.
172. DJ Collins, RJ Medina, BH Davis, Canad J Chem Eng 61: 29, 1983.
173. RC Sosa, M Nitta, HK Beyer, PI Jacobs, Proceedings of Sixth International Zeolite Conference, London: Butterworths, 1984, p. 508.
174. YS Hsu, TY Lee, HC Hu, Ind Eng Chem Res 27: 942, 1988.
175. V Sreedharan, S Bhatia, Chem Eng J 36: 101, 1987.
176. E Benazzi, S De Tavernier, P Beccat, JF Joly, C Nedez, A Choplin, JM Basset, Chemtech 24(10): 13, 1994.
177. B Olteanu, PhD thesis, directed by S. Raseev, Institutul de Petrol; Gaze si Geologie, Bucharest, 1971.
178. IM Kolesnicov, PN Pisciolin, IN Frolova, PS Belov, SM Koutcherenco, IV Borisenco, Nauka, Kiev, 1981, p. 47.
179. S Raseev, B Olteanu, Revista de Chimie 28(2): 111, 1977.
180. S Raseev, B Olteanu, Revista de Chimie 28(4): 319, 1977.
181. AJ Silvestri, CD Prater, J Phys Chem 68(6): 328, 1977.
182. KL Hanson, AJ Engel, Chem Eng J 3: 260, 1967.
183. S Raseev, B Olteanu, Revista de Chimie 28(6): 521, 1977.
184. O Cappellazzo, G Cao, G Messina, M Mordibelli, Ind Eng Chem Res 30: 2280, 1991.
185. DH Olson, GT Kokotailo, SL Lawton, WM Meier, J Phys Chem 85: 2238, 1981.
186. Hydrocarbon Processing 62: 160, 1983.
187. H Bölt, H Zimmerman, In: P Harrison, ed., Hydrocarbon Technology International, 1992, London: Sterling Publ. Int.
188. JP Boitiaux, JM Devès, B Didillon, CR Marcilly, In: GJ Antos, AM Aitahi, JM Parera, eds., Catalytic Naphtha reforming. New York: Marcel Dekker Inc., 1995, pp. 79–111.

189. KR Murthy, N Sharma, N George, Structure and Performance of Reforming Catalysts, In: GJ Antos, AM Aitahi, JM Parera, eds., Catalytic Naphtha Reforming. New York: Marcel Dekker Inc., 1995, pp. 207–255.

190. BH Davis, GJ Antos, Characterization of Naphtha Reforming Catalysts, In: GJ Antos, AM Aitahi, JM Parera, eds., Catalytic Naphtha Reforming, New York: Marcel Dekker, 1995, pp. 118–180.

191. S Tung Sie, Evaluation of Catalyst for Catalytic Cracking, In: GJ Antos, AM Aitahi, JM Parera, eds., Catalytic Naphtha Reforming. New York: Marcel Dekker Inc., 1995, pp. 181–206.

192. P Marécot, J Barbièr, Deactivation by Coking, In: GJ Antos, AM Aitahi, JM Parera, eds., Catalytic Naphtha Reforming, New York: Marcel Dekker Inc., 1995, pp. 279–311.

193. JN Beltramini, Deactivation by Poisoning and Sintering, In: GJ Antos, AM Aitahi, JM Parera, eds., Catalytic Naphtha Reforming. New York: Marcel Dekker Inc., 1995, pp. 313–363.

194. TN Anghelidis, D Rosopoulau, V Tzitzios, Selective Rhenium Recovery from Spent Reforming Catalysts, Ind Eng Chem 38(5): 1830–1836, 1999.

195. JN Beltramini, Regeneration of Reforming Catalysts, In: GJ Antos, AM Aitahi, JM Parera, eds., Catalytic Naphtha Reforming, New York: Marcel Dekker Inc., 1995, pp. 365–394.

196. O Clause, L Mank, G Martino, JP Frank, Trends in Catalytic Reforming and Paraffin Isomerization, Fifteenth World Petroleum Congress, 1997, Vol. 2, John Wiley and Sons, Chichester, New York, Weinheim, Brisbane, Toronto, Singapore, 1998, pp. 695–703.

197. Hydrocarbon Processing 77: 60, March 1998.

198. Hydrocarbon Processing 78: 99, Nov. 1999.

199. W Weirauch, Benzene Demand to Grow 4.9%/year, Says New DeWitt Forecast, Hydrocarbon Processing 74: 27, Sept. 1995.

200. Hydrocarbon Processing 78: 122, March 1999.

201. Hydrocarbon Processing 78: 150, March 1999.

14

Process Combinations and Complex Processing Schemes

14.1 DEFINITION OF OBJECTIVES

14.1.1 Combination of Processes

The term "combination of processes" refers to permanent, direct links between two, or occasionally several, processes, whereby the stream of reactants flows successively through the respective units.

Most of the time such links refer to the pretreatment or purification of the raw materials before feeding them to the main processing unit. This role is played by the hydrofining of the feed before the hydroisomerization or the catalytic reforming on platinum catalysts, the elimination of mercaptans from the feed before oligomerization or alkylation, the drying of the feed to the alkylation, etc. In all these cases, the combination of the processes has a compulsory character, necessary for a successful implementation of the technology. Saturation of the naphthas produced in coking and in visbreaking processes is also required before their use as feed to catalytic reforming, otherwise the coke formation would become excessive.

There are also cases where the preliminary improvement of the reactant stream has not a compulsory character as above, but serves to improve the performance of the main process. The hydrofining of the feed to catalytic cracking has this character. In such cases, the decision whether to combine the processes or not is based on economic criteria.

The combination of processes may have as its purpose energy savings, the elimination of intermediary storage, or a common automation system for reducing supervisory personnel. Such is the case with atmospheric and vacuum distillation, commonly combined.

To extend such solutions by the complex automation of a larger number of processes has not found general acceptance, although it is realized sometimes in pyrolysis plants. The reason is that such a system is excessively rigid, and allows with difficulty, modifications to the feed or to the operating conditions. It is difficult to adapt the system to situations that occur in exploitation.

It results that the decision to combine processes is taken on the basis of the information reviewed in the previous chapters and does not require additional explanations.

14.1.2   Development of Complex Processing Flow Sheets

The development of a general flow sheet for the complex processing of a specified crude oil consists in defining the physical and chemical transformations to which it must be submitted and to produce in optimal economic conditions the end products with the required yields and qualities.

The problem is much more difficult than that of process combination, since it requires the comparative examination of the various possible options. It requires a broad process experience and knowledge of the various domains of crude oil processing.

It is obvious that for the development of a complex processing flow sheet it is necessary to have available the complete analysis of the crude oil or oils considered for processing and the detailed specification of the final products. The developers of the complete processing scheme will define also what properties will be analyzed for (see Section 14.4), in order to evaluate the performances of the process steps. When defining the final objectives of the complex process, i.e., the products specifications, the starting points are of course, the requirements of the market. However, just taking market requirements into consideration is not sufficient. The required quantities of the target products and especially their quality expectations change extremely fast. Thus, when specifying the final objectives, the evolution of the market must be taken into consideration [47]. To serve as orientation in this sense, the evolution of the product slate and that of products specifications is discussed in some detail in the following section.

On the basis of these initial elements—the characteristics of the crude oil and the quantities and specifications of the target products—the complex processing scheme must specify the processing technologies, including their characteristics and performances, the flowrates of the various streams and the production of the finished products by mixing and compounding, the supply with utilities (water and electrical energy), the purification of the process waste waters, the recycling of the cooling water, the transportation and the storage of the crude oil and of the products. The selection of the site for a new refinery site is often closely related to the manner in which these problems are solved. For this reason the justification for selecting each site must accompany the documentation concerning the recommended complex flow diagram. Also, the recommendation must address the measures to supply manual labor, qualified personnel, necessary training, and if possible, in similar units, the construction of housing and social facilities.

Since the recommended scheme is the result of a selection between various possible options, all options should contain the necessary elements for performing a correct comparison. Since the selection is based mainly on economic criteria, all schemes must be accompanied by the necessary data for investments and operating costs. The latter should also take into consideration the different ways the scheme could evolve.

If the number of options is large, it is often practical to perform successive evaluations of increasing detail in order to eliminate the uneconomical solutions as early as possible so as to perform exact calculations only in the framework of the one or two final options.

14.2 EVOLUTION OF THE RANGE AND SPECIFICATIONS OF PRODUCTS

Market demand determines the range of products and their specifications. The processing schemes of the crude are selected such as to supply the required products in economically competitive conditions.

The development of the petroleum refining industry provides numerous examples of advances in several techniques:

the development of thermal cracking as a result of the increased demand for gasoline

development and expansion of catalytic cracking as a result of the increased demand for higher octane numbers for gasoline

development and the expansion, for the same reason, of oligomerization, of the alkylation of isoalkanes and later, of the catalytic reforming

development of the hydrotreating of gas oils to produce increased amounts of Diesel fuel, with a high cetane number and low sulfur content

development of hydrocracking as a result of the increased qualitative requirements imposed on the products

Parallel to these developments, which represent technology advances, all refining processes were optimized to increase efficiency and profits.

The appearance of laws addressing the concern for public health and environmental protection represents a radical change away from economics of the driving forces leading to new process technologies. The new quality requirements imposed on the fuels as a result of these laws determined the development of new catalysts and processes.

A typical example is the reduction of the maximum sulfur content in fuel oil as result of the disastrous effects of acid rain on the vegetation and especially on forests. The Clean Air Act Amendment (CAA) of 1990 was the first step toward the protection of public health. It was followed by similar measures taken first in Western Europe and then also in other countries.

The evolution of the demand for the range of products and their quality specifications is a useful indicator of their foreseen development and related to it, of the directions for the development of processing technologies and complex processing schemes.

14.2.1 Evolution of Products Demand

The worldwide consumption of the main categories of products obtained by the processing of crude oil is given in Table 14.1.

The analysis of the trends of market demands [58] indicates the important part played by the products of catalytic processes.

Table 14.1 Worldwide Evolution of Main Petroleum Products

	1973 10^6 t/year	%	1980 10^6 t/year	%	1990 10^6 t/year	%	2000 10^6 t/year	%
Gasoline + C_3/C_4	789	28.2	883	29.2	1031	33.0	1150–1230	35.4
Kerosene + Gas oil	221	7.9	297	9.8	419	13.4	540–570	16.6
Fuel oil	582	20.8	594	19.6	641	20.5	630–670	19.4
Residual fuel	928	33.1	955	31.6	674	21.6	530–560	16.3
Petrochemistry	125	4.5	130	4.3	175	5.6	200–210	6.1
Other products	155	5.5	160	5.3	180	5.8	200–210	6.1
Total	2800		3025		3120		3250–3450	

The data show a strong decrease of the consumption of residual fuels, which in 2000 represented about 16% of the total processed crude, as compared to 33.1% for 1973. This tendency became stronger after the year 2000.

In the same time, there is a continuous increase of the consumption of motor fuel, especially of gas oils and jet fuel, which in 2000 represented 16.6% of processed crude oil, as compared to 7.35% in 1973. The gasoline demand increased by 50% during the period 1973–2000.

Petrochemical products, which in this statistic include the fuel used in the respective industrial complexes, know a moderate but continuous increase.

It is to be remarked that these important modifications take place in the conditions of a very slow increase of the total amount of the processed crude oil. The average growth for the period 1973–1980 was of only 1.66% per year, followed by a decrease to 0.31% per year for the period 1980-1990, as a result of the substantial increase of the price of the crude oil.

In the conditions of the increased crude price it became necessary to continuously reduce the amounts of the low-value products, which explains the indicated changes.

The indicated increase of the middle distillates at the expense of the heavy ones is shown in Figure 14.1 [1], for various areas of the world. Important differences are observed among various countries, even among developed countries (e.g., Japan compared to Western Europe). Nevertheless, the general trend, i.e., the decrease of the consumption of heavy fuels, and the substantial increase of the middle distillates is consistent. The data may be completed with some information concerning crude oil refining in the U.S. and Russia.

In the U.S. in 1996, the total consumption of crude oil and liquefied gases was 1028 million m^3, of which gasoline accounted for 43.8%, jet fuel for 8.6%, the Diesel fuel for 18.1%, residual fuel oil for 4.9%, and the rest of the products for 24.6% [55].

In Russia, in 1994, the total amount of fuels was 156.5 million m^3, of which gasoline accounted for 22.5%, gas oils for 33.8% and the residual fuel oil for 43.7%. In 2001, the fuels should reach 197 million m^3, the gasoline accounting for 25.1%, the gas oils for 36.4%, and the residual fuel for 38.5% [56].

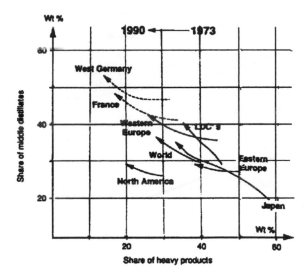

Figure 14.1 Evolution of demand for residual fuels and middle distillates from 1973–1990. (From Ref. 1.)

14.2.2 The Quality and Production of the Finished Products

14.2.2.1 Gasoline

Commercial gasoline is obtained by mixing together various components obtained from several processes. It is therefore important to estimate the properties of the mixture from the properties and relative amounts of the components. Such estimation poses no problems for additive properties, such as density, chemical composition, etc. Specific methods were developed for the evaluation of nonadditive properties, such as the octane rating [57].

The quality requirements for gasoline saw in the last 10–15 years a significant evolution, mainly as result of the regulations for the protection of the health of the population and of the environment.

The first measure taken was to restrict the use of leaded gasoline in order to prevent the poisoning by lead oxides that result from burning the tetraethyl-lead. Gradually, the trend to completely eliminate lead from gasoline appeared, so that in 1990, only 45% of the gasoline produced worldwide was ethylated [2].

The second measure, the fight against the emissions of carbon monoxide and especially of nitrogen oxides, led to the extension of the catalytic conversion of the exhaust gases [42]. This measure is compulsory in many countries, at least for new cars.

The catalytic converters use catalysts containing noble metals, especially platinum, which are poisoned by lead oxides. In order to avoid the loss of catalyst activity, high octane, unleaded gasoline was developed, which resulted in the gradual disappearance of leaded gasoline.

Figure 14.2 illustrates the evolution in the U.S., of the various types of automotive gasoline in the period 1980–1995 [2]. As shown, leaded premium gasoline disappeared in 1983, while the regular leaded was no longer sold from 1992 on. The

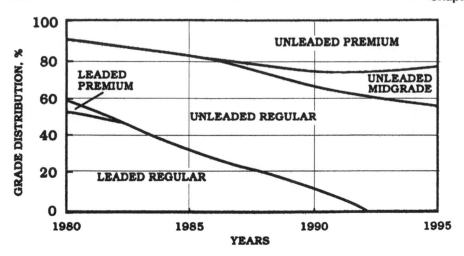

Figure 14.2. Trends in U.S. gasoline quality.

unleaded grade, intermediary between the premium and regular, was introduced in 1986.

In Europe, depending on the country, the consumption of unleaded gasoline in 1989 represented between 25% and 55% of the total consumption [2]. This value increased significantly in the following years.

The complete replacement of leaded gasoline imposes an important requirement on the refining industry and considerable investments. However, the time given by various countries to implement this measure depends on the capacity of the respective refining industry to adapt to the new specifications. New refineries must provide the technologies to ensure the exclusive production of unleaded gasoline with the required octane rating.

Another constraint that impacts gasoline production is the imposed limitations on the content in aromatic hydrocarbons (maximum limit) and oxygenated compounds (minimum limit).

The first measures in this sense were taken by the Clean Air Act Amendment (CAA) and they are followed by regulations issued by the Environmental Protection Agency (EPA). These measures refer to: the reduction of the content of benzene in gasoline (owing to its carcinogenic effect) to max 1 vol % and the reduction of other toxic substances: 1–3-butadiene, formaldehyde, acetaldehyde and polycyclic hydrocarbons, either present in the gasoline or resulted from combustion.

In the presence of nitrogen oxides (NO_x) and sunlight, hydrocarbon vapors lead to the formation of ozone at ground level, which may produce grave lung lesions. Such a reaction is undergone by lower alkenes, especially i-amylenes. This undesired reaction is controlled in two ways: by the reduction of the vapor tension of gasoline in order to reduce the amounts of hydrocarbon vapors released to the atmosphere, and by the compulsory introduction in the gasoline of oxygenated compounds.

Because of the difficulties for the refining industry to put into application immediately the required measures [44], two steps were provided.

Beginning January 1, 1995, gasoline had to satisfy the following conditions:

- Benzene, max. 1 vol %
- Oxygen as contained in oxygenated compounds, min 2.0 wt %
- Aromatic hydrocarbons max 25 vol %
- The reduction in each refinery of the toxic compounds by 15%
- The maintaining of the 1990 levels for sulfur and alkenes content, and of the temperature for ASTM 90% vol distillation
- The vapor tension in the summer:
  - max 0.56 bar (8 psi Rvp) in the North
  - max 0.50 bar (7.2 psi Rvp) in the South

EPA regulations provided also the equations for calculating the allowable concentrations of toxic compounds.

The more severe constraints that came into effect after May 1, 1997 and during the following years, were the result of a collaborative effort between the EPA and the refining industry, cars manufacturers, and governmental authorities. The further decrease, by 29% of the volatile substances content, by 22% of the toxic compounds, and by 6.8% of the NO_x emissions [4], and the decrease of aromatic hydrocarbons and alkenes were also imposed.

More severe limits than those valid for the rest of the U.S. were introduced in 1996 in California, by the California Air Resources Board (CARB), namely:

Oxygen, wt %	min 2.0
Benzene, vol %	max 1.0
Aromatic hydrocarbons, vol %	max 25.0
Sulfur, ppm	max 40
Alkenes, vol %	max 6
50% ASTM	max 99°C
90% ASTM	max 149°C
Vapor pressure	max 0.48 bar (7.1 psi Rvp)

In order to estimate the effort made by the refiners to comply with the new specifications it is interesting to compare these figures with those for the average quality gasoline in the U.S. in 1990:

Oxygen, wt %	0.0
Benzene, vol %	1.53
Aromatic hydrocarbons, vol %	32.0
Sulfur, ppm	339
Alkenes, vol %	9.2
50% ASTM	93°C
90% ASTM	149°C
Vapor pressure	0.6 bar (8.8 psi Rvp)

These measures imposed the changes in the composition of the gasoline given in Table 14.2 [43].

The World Health Organization (WHO) assigns a high importance to the content of benzene in gasoline, since it induces leukemia [43] and is working to eliminate benzene completely. Some benzene standards for ambient air are given in Table 14.3.

Table 14.2 Changes to the Gasoline Composition to Satisfy the California Air Resources Board (CARB)

Components, % vol	Gasoline composition	
	usual	CARB
Reformate	33.0	22.2
Light cat. gasoline	16.0	12.7
Heavy cat. gasoline	21.0	19.5
Alkylate	11.0	13.3
Light straight run	7.0	11.9
Hydrocrackate	7.5	9.3
Butane	4.5	0
MTBE	0	11.0
Total	100.0	100.0
Octane number $(F_1 + F_2)/2$	88.4	87.7

Measures similar to those taken in the U.S. have been taken or being considered by the European Community and other developed states. Besides the complete elimination of the lead in gasoline, they will reduce the content of benzene, of sulfur, and vapor pressure, and will include oxygenated compounds.

For the moment, the proposals formulated by the European Commission include the following:

Table 14.3 Benzene Ambient Air Standards

Source/organization	Recommendation: ambient level		Measurement criteria	Acceptable risk criteria
	Near-term objective	Long-term goal		
Department of Environment, United Kingdom	5 ppb	1 ppb	Running annual average	Nonobservable risk with safety factors
National Institute for Public Health and Environmental Protection, The Netherlands	3.1 ppb	1.6 ppb	Annual average	1 excess leukemia/ 10^8 population
U.S. Environmental Protection Agency	None	0.037 ppb[a]	Note (a)	1 excess leukemia/ 10^6 population
Health Canada	0.92 ppb[b]	0.009 ppb[c]	Not specified	$< 2 \times 10^{-4}$ of the $TD_{0.05}^d$ ($TD_{0.05} = 4.6 \times 10^3$ ppb)

[a] Not adopted as a criteria pollutant; regulations allow a certain number of noncompliances per year.
[b] Ambient level required for "medium" rather than "high" priority.
[c] Ambient level required for "low" priority.
[d] Toxic dose affecting 5% of the population.
Source: Ref. 53.

- The control of the aromatic hydrocarbon content
- The decrease of the content of
  - benzene from 5% to max 2%
  - sulfur from 0.05% to max 0.010%
  - lead from 0.013% to max 0.005%
- The decrease of the vapor pressure to max 60 kPa (8.7 psi Rvp)

These regulations are to a large extent inspired from those implemented in the U.S. They also take into account the specific conditions in the respective countries such as cars with higher compression ratios, which require higher octane numbers, the composition of the gasoline usually produced, and the capability of the local refining industry to adapt to new specifications. Also, the climate conditions specific to the various countries are taken into consideration.

All these trends, including that of gradually eliminating the use of leaded gasoline provide a general orientation on the direction of development of the refining industry. They are factors that must be taken into account in the development of general processing strategies of the new refineries and in the modernization of those that already exist [46].

The overall result is an increased part played by the alkylation and isomerization of C_5-C_6 fractions and to the decrease of the importance of catalytic reforming.

A new element, which will strongly influence the processing of the crude oil is the requirement that will be surely generalized, to introduce oxygenated compounds in the gasoline composition.

Following its introduction, the methyl-tert-butyl-ether (MTBE) was the oxygenated compound of choice for obtaining so-called "reformulated gasoline" (RFG). MTBE is produced by the reaction of methanol with isobutene in presence of an acid catalyst (e.g., an acid ion exchanger). Another ether, the tert-amyl-methyl-ether (TAME), has the advantage of a lower vapor pressure than MTBE, but also the benefit to use as raw material the iso-amylenes, which by themselves have the undesired property of having a very pronounced photochemical reactivity, leading to the formation of ozone.

The use of ethanol in place of methanol in the production of ethers leads to ethyl-tert-butyl-ether (ETBE) and of ethyl-tert-amyl-ether (ETAE).

The main characteristics of the mentioned ethers are given in Table 14.4 [4]. Their high octane numbers are remarkable, and was a determining factor among others, in their selection as components of the RFG.

Since MTBE contains 18.2 wt % oxygen for obtaining the content of 2.0–2.1 wt % O_2 required by the regulations, it is necessary to introduce in the gasoline about 11% MTBE. The corresponding figures are 13% for ETBE or TAME.

In these conditions, the availability of the amounts of isobutene and iso-amylene required for the production of the respective ethers became a problem of first importance for the refining industry, which had effects on the entire process [45]. This will be examined separately in Section 14.3.

High chemical stability and water solubility are the cause of a movement to ban the use of MTBE, which started in California. Accidental spillages and leaks from storage tanks of gasoline led to ground water contamination with MTBE that persisted long after the gasoline itself has dissipated. The sustained movement started by activist groups for eliminating the use of MTBE from gasoline is the

Table 14.4 Properties of Ethers Used in Gasoline
Preparation

Characteristics	MTBE	ETBE	TAME
F_1 blending octane number	118	118	114
F_2 blending octane number	102	102	98
Oxygen, wt %	18.2	15.7	15.7
Vapor pressure, bar	0.55	0.40	0.25

driving force for the enhanced use of other oxygenates (ethanol, *t*-butanol) in order to achieve the required oxygen content in the RFG.

14.2.2.2 Jet Fuel and Gas Oil

As shown in Table 14.1, during the last 30 years of the 20th century, the consumption of gas oil and jet fuel grew at an average of 3% per year. It is estimated that this rate of increase is the same to date.

The consumption of middle distillate fractions in the various regions of the world presents important differences [2]. Contrary to the U.S., the use of Diesel motors is widespread in Europe, resulting to a larger consumption of the middle distillate fractions. In 1970, these fractions represented 32% of the total processed crude oil and reached 41% in 1990. The increase was much more spectacular in Japan, where the change was from 17% in 1970 to 37% in 1990 [2].

With respect to the quality specifications, the restrictive measures address especially the sulfur content and that of aromatic hydrocarbons in Diesel fuel, as well as their cetane number.

Starting in 1993, the maximum content of sulfur accepted in the U.S. was reduced from 0.25 wt % to 0.05 wt % [5]. The same limitation was introduced in other countries, as follows: 1990 in Sweden, 1994 in Switzerland, Mexico, and Canada, in 1995 in Germany and Austria, and the rest of Europe in 1996, and Japan in 1997. Overall, this measure is a consequence of legislation in 1980 concerning the limitation of sulfur oxides in the atmosphere and is expected to have two effects on the construction and behavior of Diesel motors:

Increased efficiency of catalytic filters for discharged gases, in the Diesel motors, in order to reduce the particles emission
Decreased motor wear as a result of corrosions

The refining industry supported the implementation of these regulations.

The limits set by the EPA on content of aromatic hydrocarbons and the cetane number are 34 vol % aromatic hydrocarbons and a cetane number of min 40. In California, CARB provides the adoption of much more strict norms i.e., between 10 and 20 vol % for the content of aromatic hydrocarbons. These reductions will lead to lowering of the combustion temperature within the motor and to a decrease of the NO_x content in the exhaust gases. The refiners' objection is that the same result may be obtained by increasing the cetane number from 40 to 50 without changing the 34 vol % limit for the aromatic hydrocarbons, but requiring lower investments. Thus, it is estimated [5] that the 10 units increase of the cetane number would cost 2–2.5 cents/gallon, whereas the decrease of the content of aromatic hydrocarbons to a level of 20 vol % would cost 6–11 cents/gal, or to a level of 10 vol %, would cost 12–14 cents/gal.

In Western Europe beginning in 1996, specifications similar to those in the U.S. were implemented: the maximum content of sulfur was set at 0.05 wt %.

Later, the following measures with general character were set in Western Europe:

Cetane number increase	from 49 to min 51
Density reduction	from 0.860 to max 0.845
Distillation	95% at 350°C
Distillation for Italy, France, Spain and Portugal	95% at 370°C
Polyaromatic hydrocarbons	max 6%
Sulfur	max 0.035%

Moreover, special attention is paid to exhaust gas emission of fine particles having the sizes ranging between 0.01–2.0 μm. Such particles lead to respiratory diseases, besides being carcinogenic [54].

Some additional data on the characteristics required for Diesel fuels were given in Section 10.7.2.

Besides the measures related to the characteristics of Diesel fuels, a reduction of the pollution will be obtained by constructive means: the outfitting of the motors with catalytic converters for the exhaust gases and with ceramic filters to retain fine particles. To prevent their plugging, the retained particles are burned, which imposes the use of some additives.

The European Commission has specified the following emissions, per km traveled: 0.4 g NO_x and 0.04 g fine particles.

14.2.2.3 Residual Fuels

The trend to continuously reduce the share of residual fuels was illustrated by the data of Table 14.1. Processing schemes were suggested that completely eliminate the use of residual fractions: by deasphalting of the vacuum residue and gasification of the asphalt with production of steam and electric energy; or by hydrocracking vacuum residue at 90% conversion with recycling, and the partial oxidation of the remaining 10% with production of hydrogen [6].

The possible ways of reducing the quantity of residue are: (1) carbon rejection by submitting the residue to catalytic cracking, coking or visbreaking, (2) enriching with hydrogen by hydrocracking, (3) deasphalting with propane or butane. The deasphalted oil is used as feed for catalytic cracking, while the asphalt is used as such, or is submitted to coking or gasification [7,8].

Interesting data was published [9] concerning the various possible options, for processing light and heavy Arabian crudes, including mass balances and economic evaluations. The analytical information provided makes it possible to apply the evaluation procedure to the similar processing of other crude oils.

The limitations imposed on the sulfur content of the residual fuel oils are of special importance. This is illustrated by the fact that a content of 1% sulfur in the fuel oil results in flue gases containing 1700 mg SO_2 per Nm^3.

The problem of reducing the sulfur content in the residual fuels is made worse by the continuous increase of the sulfur contained in the crude oils extracted world-wide, as shown in Figure 14.3 [6].

The resulting stress on processing capabilities, has led the Italian National Electric Administration (ENEL), one of the greatest fuel consumers in the world (20 million t/year), starting in 1997, to limit to 0.25% sulfur the fuel they purchase.

The norms to be implemented in North America and Europe specify a content of maximum 250 mg SO_x/Nm^3 in flue gases, which corresponds to a sulfur content in the fuel of max 0.15%. Presently, the hydrotreating of residual fuels cannot produce such low sulfur concentrations. Even if it will become feasible, the economics of this operation seem noncompetitive. On the other hand, deasphalting followed by hydro-desulfurization of the deasphalted oil allows one to reach a content of 0.15% sulfur. The problem is transferred to the use of the produced asphalt by gasification or partial oxidation.

Another possible solution [10] is to combine the residue hydrodesulfurization with the elimination of SO_x from the flue gases, which seems to make good sense from the economic point of view.

In parallel with the limitation of the content of sulfur oxides (SO_x), it is fore-seen to limit the nitrogen oxides (NO_x) in the flue gases to 450 ml NO_x/Nm^3. This is achieved by using burners designed for this specification. When this is not possible, one resorts to treating the flue gases.

Besides sulfur and nitrogen, the content of metals, especially of vanadium, is being limited. Qualitative limits are set also for ensuring the complete combustion, thus avoiding the emission to the atmosphere of solid or liquid unburnt particles.

14.3 ADDITIONAL RESOURCES

Additional resources are the substances other than hydrocarbons which directly, or as result of processing, are part of the composition of the finished products. To this category belongs hydrogen, the internal resources of which, as ensured by catalytic reforming, are many times insufficient, and the ethers, which have to be added in important amounts to the RFG (14.2.2.1). The production of MTBE and of TAME require the internal resources of isobutene, and iso-amylene respectively.

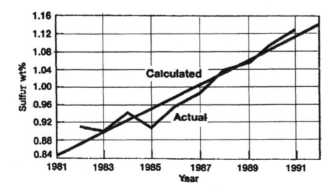

Figure 14.3.   Sulfur content in crude oils.

The technology for the production of ethers is not specific to petroleum refining. Its review is nevertheless useful since the ethers are closely related to the production of reformulated gasoline and therefore, to the overall refinery processing schemes.

14.3.1 Hydrogen

In the past, until 1970–1980, the high purity hydrogen resulting from the high pressure separator of the catalytic reforming units satisfied the requirements of the hydrofining of naphthas and of the middle distillates.

The subsequent development in crude oil processing introduced two new elements. The first, of a somewhat secondary importance, is the decrease of hydrogen purity as result of lower operating pressure in catalytic reforming units, compensated partly by the increase of the conversion and therefore of the amount of produced hydrogen. The second, which is the controlling factor, is the increase of hydrogen consumption, as a result of the introduction of processes that consume large amounts of hydrogen, such as the dearomatization by hydrotreating and hydrocracking and of the extension of hydrotreating to lubricating oils and residues.

Two solutions exist for ensuring the supply of the increased amounts of hydrogen required: (1) its production by traditional processes, (2) concentration and purification of the hydrogen resulting from various processes, accompanied wherever necessary by its compression [49].

The description of the traditional processes for the production of hydrogen by steam reforming of hydrocarbons or by their partial oxidation is beyond the scope of this work and can be found in the literature [11–16].

The use of the processes of the second category, depends on the units actually existing in the refinery and therefore on the general processing flow diagram. Thus, the task is to recover hydrogen contained in the low-pressure gases from the catalytic reforming of naphthas as well as of that contained in the pyrolysis gases or obtained from the dehydrogenation processes. Also, it is useful to concentrate the hydrogen contained in the recycling gases of the processes that consume hydrogen such as hydrofining, hydroisomerization, hydrotreating, and hydrocracking. Usually, the hydrogen recovered from these streams must be purified.

Three methods have been developed for the recovery of hydrogen from gases where its concentration is relatively low [17]: absorption at pressure, separation by deep cooling [48], and separation by membranes. The first two methods are well established as operations of chemical engineering [18] and do not require additional explanation. The third method, the separation through membranes of cellulose acetate [17], fibers [17], etc. is of more recent date. The separation by permeation through membranes is gaining ground over the other methods, since it requires much less energy. The method is based on the fact that hydrogen passes through the membrane much faster than gases with higher molecular mass, as shown in Figure 14.4.

The inclusion of the membranes separation system in the hydrogen recycle loop of a hydrocracking unit with a capacity of 2400 m^3/day and a hydrogen consumption of 0.178 Nm^3/m^3 product is depicted in Figure 14.5 [20].

Table 14.5 shows the flowrates and compositions for different streams as marked on the sketch of Figure 14.5.

Three cases have to be taken into account:

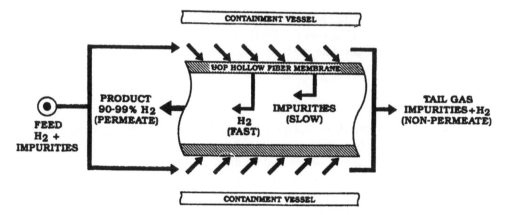

Figure 14.4. Membrane system for hydrogen separation.

Case 1. Base case: without separation through the membrane system (line 6 and the corresponding equipment).

Case 2. The purification through the membrane system treats 509,000 Nm³/day recycle gases and rejects 42,000 Nm³/day gases containing 34.1 vol % H_2 compared to 85,000 Nm³ containing 78 vol % H_2 for the base case. This recovery of hydrogen makes it possible to reduce by 53,000 Nm³/day the hydrogen makeup and to decrease the pressure in the hydrocracking system from 139 to 121 bar, while keeping unchanged the partial pressure of hydrogen. Thus, savings in the operating cost are obtained without modifying the hydrocracking capacity.

Case 3. In this case the increase of the purity of the hydrogen, as a result of the incorporation of the membrane system, is not accompanied by a decrease of the pressure. The higher partial pressure of hydrogen is used to increase by 10% the processing capacity of the hydrocracking unit.

Table 14.5 Flowrates Marked in Figure 14.5

Stream number	1	2	3	4	5	6	7
Case 1							
Flowrate, Nm³/day · 10³	1,007	4,460	3,537	3,453	85	0	0
Pressure, bar	24	139	114	114	144	—	—
Purity, % vol	97	82.1	78	78	78	—	—
Case 2							
Flowrate, Nm³/day · 10³	954	3,908	2,994	2,952	0	509	42
Pressure, bar	24	121	97	97	—	97	96
Purity, % vol	97	93.7	91.8	92.6	—	91.8	34.1
Case 3							
Flowrate, Nm³/day · 10³	1,056	4,701	3,693	3,645	0	340	48
Pressure, bar	24	139	114	114	—	114	113
Purity, % vol	97	90.4	87.8	88.5	—	87.8	36.5

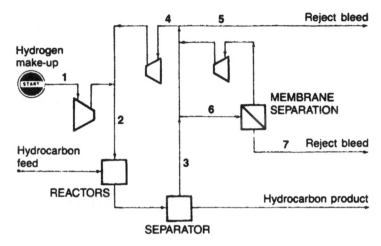

Figure 14.5. Incorporation of a hydrogen membrane separator in a hydrocracking plant. (From Ref. 20.)

In the cases 2 and 3, the use of membranes for separation of hydrogen improves the economics of the hydrocracking process.

14.3.2 Ethers as Gasoline Components

As indicated earlier, the addition of ethers must ensure, according to the current regulations in the advanced countries, a content of 2.0–2.1% by weight O_2 in the product gasoline, which corresponds to 11–13% added ethers.

The ethers are produced by the reaction of alcohols with the iso-alkenes [21] according to the reactions:

For methyl-tert-butyl-ether (MTBE), which presently is the most used:

$$(CH_3)_2C=CH_2 + CH_3OH \rightarrow (CH_3)_3C-O-CH_3$$

For the ethyl-tert-butyl-ether (ETBE):

$$(CH_3)_2C=CH_2 + C_2H_5OH \rightarrow (CH_3)_3C-O-C_2H_5$$

For the tert-amyl-methyl-ether (TAME):

$$CH_3 \cdot C_2H_5 \cdot C=CH_2 + CH_3OH \rightarrow CH_3 \cdot C_2H_5C-O-CH_3$$

Taking into account the large amounts of the ethers required for producing high grade gasoline, a first problem is the assurance of the supply of isobutene and iso-amylenes necessary for their production. It is to be noted that the production of ethers requires C_4 and C_5 fractions, with isobutene and iso-amylenes respectively, concentration comprised between 10% and 60%, provided they do not contain dienes, nitrogen, and sulfur compounds.

The production of ethers can use C_4 and C_5 fractions respectively, recovered from catalytic cracking, pyrolysis, and coking. The iso-alkenes are also produced by the catalytic dehydrogenation of the corresponding iso-alkanes.

It is to be remarked that the conversion of practically 100% isobutene, by using the C_4 fraction as feed for the ether production unit makes it easier to obtain high purity butene by fractionation. Indeed, butene will not be contaminated with the close-boiling isobutene, since the latter was completely converted to MTBE.

Among the mentioned sources of isobutene and iso-amylene, the respective fractions from the catalytic cracking are the ones preferred, both owing to the high concentration of iso structures, and of the relatively low content of contaminants. The C_4 fraction depleted of isobutene, resulting from the production of ethers, may be used directly as feed for the alkylation.

The most expensive route is to obtain the C_4 and C_5 iso-alkenes by the dehydrogenation of the corresponding iso-alkanes. The respective processes are covered by a large number of publications and reviews [22–27] and will not be addressed here.

Butacracking is a process which performs the thermal cracking of the C_4 fraction, with the purpose of producing isobutene for the synthesis of MTBE concomitantly with other alkenes [28].

Several licenses and variations of the process for producing MTBE, ETBE, and TAME were developed: UOP [21], Hüls and UOP [20], The Petroleum French Institute [30], CD Tech [31], Philips Petroleum Co [32], Snamprogetti, [33], Deutsche BP with EC Erdölchemie [52] and others.

In all these processes the ethers are produced by the reaction of methyl- or ethyl alcohol with isobutene or iso-amylene, at moderate temperatures in the presence of a catalyst, which usually is an acidic ion exchange resin. The reactions that take place were written above.

A typical catalyst, Amberlyst 35 produced by Rohm and Haas, is used by many producers including Lyondell, and has the following characteristics [34]:

Particle size	0.4–1.25 mm
Apparent density	0.8 g/cm^3
Specific surface area	44 m^2/g
Moisture content	56%
Concentration of the acid sites	1.9 meq/ml (5.6 meq/g)
Pore volume	0.35 cm^3/g
Mean pore diameter	300 Å

The working conditions recommended for this catalyst are:

Maximum temperature	140°C
Minimum height of the catalyst bed	0.61 m
Liquid hourly space velocity	1–5 h^{-1}

The process takes place in liquid phase.

The operating conditions for the synthesis of MTBE from the fraction of C_4 gases of a fluid catalytic cracking are [34]:

Content of isobutene in the gases	10–15%
Methanol/isobutene ratio	1.0–1.1
Reactor inlet temperature	73–82°C
Liquid space velocity	3–5 h^{-1}

Most processes use two reaction stages. In a first reactor with a fixed bed of catalyst, in the production of MTBE the conversion is 96–97%, for ETBE it is 90%, and only 70% for TAME. In a second step, the conversion reaches 99% for MTBE and 90% for TAME [29].

The typical flow diagram of an commercial unit is given in Figure 14.6 [31]. The two reaction steps are marked by 1 and 2 in the figure. In the CDTech process, the reactor of the second step (2) is also used as a fractionating column (catalytic distillation) in order to separate MTBE from the unreacted C_4 hydrocarbons and from the slight excess of methanol. The methanol is recovered in the column system 3, 4 and recycled to the reaction. The process allows the processing of the C_4 fractions with a content of 10–60% isobutene.

The investments for such a unit, in conditions 1996/U.S. Gulf Coast [51] is 25,000-38,000 $ per m^3 MTBE.

The utilities consumption per m^3 of MTBE [51]:

Power	7.5–11.3 kWh
LP steam	570–850 kg
Cooling water ($\Delta t = 10°C$)	38–55 m^3

14.4 INITIAL DATA FOR THE SELECTION OF REFINERY CONFIGURATION

The starting point in the development of a complex processing flow diagram and of variations thereof, is the quality of the available crude oil and the amounts and quality of the end products. These two elements have a prevailing role in the selection of the complex processing flow diagram.

The local conditions have a secondary role, the energy resources, availability of cooling water, shipping costs for products and crude oil, environmental issues, etc. These considerations determine to a large extent the selection of the site of the refinery influencing slightly the complex processing flow diagram.

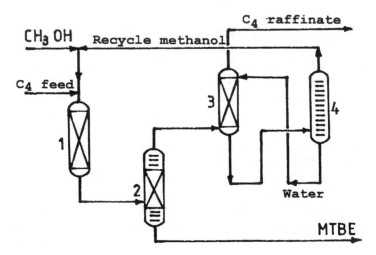

Figure 14.6. MTBE production flow diagram. (From Ref. 31.)

14.4.1   Availability and Composition of Crude Oil

The decision to build a refinery is from the beginning dependent on the availability of crude oil, whether coming from the internal resources of the respective country or from import.

At the start it is prudent to foresee possible changes in supply and quality that could intervene over time and therefore to design the refinery to be capable of processing a reasonably broad range of crude oil qualities. The changes of crude oil quality that could occur in time are due to the exploitation of new petroleum sources, situated generally at greater depths and often having a higher sulfur content, and to factors affecting import.

The degree to which the changed characteristics of crude oil influence processing activity depends also on the refinery profile. They may significantly influence the relative amounts of the straight run fractions obtained with corresponding consequences to the downstream processing units, or may influence these processes as a result of specific qualitative characteristic, such as contents of heteroatoms, of resins, of polycyclic hydrocarbons, etc. The characteristics of the processed crude oil have the strongest effect on the production of motor oils by means of the traditional processes: deasphalting, solvent refining, dewaxing, etc. For this use, the selection of the crude oils to be used as feed must be made with special care, on the basis of detailed analytical studies.

These difficulties are to a large extent attenuated by the use of modern technologies of catalytic hydrogen treating, described in Sections 10.6.3, 10.7.3, 11.7, and 12.9.

Besides the characterization by the usual methods, the preliminary analytical study of crude oil, or of the crude oil considered as raw material must generate also the data necessary for estimating the performances of the processes incorporated in the complex process flow diagram. Taking into account this relationship between the considered processes and the analytic study, the latter must take into account the analytical data provided by the potential suppliers of the process units. Usually the study comprises more analytical data than those which will be used in the final processing scheme. This situation is to be preferred to the opposite one, i.e., unavailability of the data needed for preparing a correct estimation.

The general characterization of the crude oil comprises the general analytical data, such as: viscosity, density, the contained impurities, sulfur content, nitrogen content, metals (especially Ni and V), etc., and the TBP atmospheric-vacuum distillation curve. The latter must be accompanied by the curves of the important properties such as density, viscosity and possibly other analytic data determined for narrow distillation fractions of 5–10°C. For some fractions, analytical data required by certain of the considered processes are also required, as for example the PONA analysis for the fraction corresponding to the feed to catalytic reforming or the content of Ni and V for the fractions that will be used as feed to the catalytic cracking.

One cannot foresee all the situations that might occur and all the analyses that could become necessary. They must be indicated in the analytical program for crude oil on the basis of the in-depth knowledge of the specific processes and of the required specifications of the products. The analytical characteristics of the raw materials that allow one to anticipate the product's characteristics were given in the previous chapters, separately for the various types of thermocatalytic processes of hydrocarbons and crude oil fractions.

On many occasions, the study of crude oil must involve tests in pilot plants in order to obtain a realistic measure of the performances to be obtained. Such an approach is very useful when preparing for the production of motor oils and sometimes for catalytic cracking and for hydrocracking.

14.4.2 Range and Specifications of the End Products

Besides the quality of the processed crude oil, the range and quality specifications of the end products constitute, even to a higher degree, a determining factor for the structure of the complex processing scheme.

Most refineries have a general "orientation" which may be for the production of automotive fuels and petrochemical products, or of automotive fuels and lubricating oils. The main products of most refineries are the automotive fuels. The petrochemical products and the lubricating oils each account for only a small percentage of the total amount of processed crude.

Within the complex processing scheme in the general orientation of automotive fuels, important differences may appear depending on the importance allocated to Diesel fuel, to gasoline, the specifications of these products (especially of gasoline), and on the importance of the residual fuel oils.

A clear picture of the range, types of products to be offered, and their quality specifications has to be developed before the complex processing flow diagram can be defined.

The requirements of the market for these products have to be understood as well as the anticipated trends concerning the range of products and their specifications.

In evaluating these trends, both the specifics of the local markets and the general trends in the worldwide market need to be considered (Section 14.2).

The correct evaluation of the specific refining objectives is very important and is tightly connected to the investments for the processing units that will produce the target products. A practical approach is a stepwise investment. In a first step the units are built that will provide the products presently required in terms of types and qualities, possibly with a small capacity reserve. The possibility to expand the capacity and/or the range of products is accommodated by reserving space and providing for the needed infrastructure for the next construction step. The overall flow sheet of the refining development should comprise two or even three implementation stages.

The quality specifications of the automotive fuel to be produced have to satisfy the requirements of the motors used in the developed countries.

In setting the quality specifications for the residual fuel oils, one has to take into account the conditions in the country or area where these fuels will be used. While producing fuel oils, with sulfur content, e.g., higher than the local specifications is unacceptable. Producing at specifications tighter than those required by the environmental regulations may not be justified economically. However, certain flexibility has to exist in the range of products and in their quality specifications.

14.4.3 Local Conditions and Site Selection

Local conditions influence to a significant extent the supply with utilities (water and energy), the cleaning up, and the evacuation of the waste streams, and the transportation of crude oil and refinery products. For these reasons they are controlling factors in selecting the site of the refinery. They also influence the execution of the process units with reference to the selection of the cooling systems, the preference given to steam or electricity to drive pumps, compressors, etc.

Local conditions have only a small effect on the overall processing flow diagram. This effect is manifested occasionally by the use of the residual fractions for the generation of steam and electric energy for covering their own needs. Situations may be economically beneficial where the residual fractions are consumed in their totality for generating steam and energy supplied to outside consumers or to the grid, after covering internal needs [6] ("cogeneration refinery"). These are still isolated cases to be considered in special conditions.

Most refiners find that the cost of energy purchased from the outside grid is lower than if produced internally. In special situations, such as isolated locations or unreliable supply, cogeneration may become attractive.

Refinery gases are the fuel of choice for operating the process plants and generating steam. The gases obtained after the separation of the liquefied gases, or imported natural gases, cost less than liquid fuel and ensure greater flexibility in operation.

Local regulations concerning environmental protection are an important factor within local conditions.

Air pollution may be prevented by implementing the measures indicated earlier for retaining the catalyst dust contained in the flue gases of the catalytic cracking plants and by practicing the accepted methods for limiting the SO_x and NO_x in the flue gases resulting from the burning of the fuel oils.

The evacuation of residual waters can be more problematic. First of all, there must be a rigorous separation of the process waters, i.e., those which have been in contact with petroleum products and therefore are contaminated, from the cooling water and recycled water. Even when they are partly evacuated and replaced with fresh water, the latter are, in most cases, not polluting. On the other side, the process waters contain various contaminants: tiny drops of petroleum products, sometimes as emulsions, phenols and cresols, nitrogen compounds, etc. Their purification consists of a series of stages of decontamination by decantation, followed by chemical and biological treatments. The technology and the purification methods depend on the nature of process plants from which they originate, on their flowrates, and on the prevailing purity specifications for the dischargeable water streams. In some cases, besides the reduction to a minimum of the volume of waste waters, it is more efficient to keep streams from different sources separate, depending on the contained contaminants, and have them treated individually.

In all cases, attention must be paid to the issue of the treatment of waste waters, even in the phases of site selection and of designing of the refinery, since it could have an important impact on the investment.

An important problem closely related to the selection of the site is the transportation of the crude oil and of the sellable products.

Taking into account the variety of the end products that must be distributed, it is logical to locate the refinery closer to the market for the products than to the source of the crude oil. Since one single product, the crude oil, has thus to be transported at large distances, relatively inexpensive means are used, such as pipelines or sea tankers. These economic advantages explain why most of the refineries are located in the countries that are large consumers of petroleum products (e.g., the U.S., Western Europe, etc.) and much less in the crude producing countries.

For these reasons, many of the largest refineries are located in harbors or in their vicinity, which allows the access and discharge of tankers of large capacity. Often, refineries are located at a certain distance from the harbor to which they are connected by dedicated pipeline systems. This is determined by strategic reasons, or in view of the large extent of land required by the refinery, by the price, if located in the immediate neighborhood of the harbor.

The site selection for a refinery must therefore consider a number of complex issues: the transportation of the crude oil, the distribution of the products, supply with water and energy, local regulations concerning the sewerage of the waste waters, and other specific problems, such as the availability of qualified work force. The detailed study will examine all the above issues and recommend the approach that will produce the optimal solution.

14.5 APPROACH FOR ESTABLISHING THE CONFIGURATION OF A MODERN REFINERY

The complex processing flow-diagram is developed in several stages:

- Selection of the options that should be taken into account
- Technical-economical evaluations for each option
- Estimation of impact of trends and future developments
- Selection of the final solution

All these stages are worked out by taking into account the general conditions and interrelations discussed in the previous sections.

14.5.1 Identification of Options

The options are a consequence of the properties of the available crude oil but especially of the types of products that will be produced and of their specifications. Often, there are more ways to satisfy these requirements. The selection is made on the basis of a complex of criteria, among which the economics are of primary importance: the value of the investment and the financing possibilities, the operating costs, the availability and the cost of licenses etc. Specific local conditions may also intervene in the selection.

It is useful to start by considering a broad range of options. In their selection one has to be aware of the latest technical progress achieved in processes, which in older embodiments were economically uninteresting. As an example, hydrocracking was for too long of a time too expensive to be considered as a competitive technology, which was perfectly true in its early years. However, when combined with catalytic cracking, propane deasphalting, and coking, and taking also into account the progress made in hydrogen production, hydrocracking has become quite feasible and is being increasingly included in the complex flow diagrams.

A general picture of the usefulness of the various processes that should be considered for inclusion in the scheme is given by their processing capacities in various regions of the world, as shown in Tables 14.6 and 14.7 [1]. When using such data, one should take into account the changes that have taken place since

Table 14.6 Distribution of Various Processes as of Jan. 1, 1990

	Cat. cracking	Hydrocracking	Visbreaking	Coking	Thermal cracking	Residue hydroconversion
North America	299.6	71.8	11.2	83.6	14.8	10.0
South America	62.9	3.1	23.5	9.5	5.2	1.0
Western Europe	100.3	16.2	69.6	14.2	17.3	1.5
Africa	9.3	1.8	4.1	1.0	0.2	—
Middle East	12.5	22.1	12.7	3.8	0.3	7.0
Japan	33.1	8.2	1.2	1.3	2.5	0.7
Asia and Pacific	21.8	11.8	16.1	5.9	4.7	5.3
Total	539.5	135.0	138.4	119.3	45.0	25.5

Capacities in millions t/year.
Source: Ref. 1.

Table 14.7   Gasoline Production Processes

	Cat. reforming	Isomerization	Alkylation	Oligomerization	MTBE
North America	187.3	16.1	43.8	4.8	3.5
South America	21.6	0.8	3.7	0.8	0.3
Western Europe	84.0	5.0	6.9	2.3	1.9
Africa	14.3	0.2	0.7	0.1	—
Middle East	20.4	0.1	0.7	—	0.5
Japan	23.2	—	0.8	0.4	0.2
Asia and Pacific	23.1	0.5	2.2	0.4	—
Total	373.9	22.7	58.8	8.8	6.4

Capacities in millions t/year.
Source: Ref. 1.

the date they were developed and the anticipated developments based on the current trends in technology and market.

When selecting among various licensors, it is recommended to take into consideration besides the technical merits, also the number and performance of the units built by each.

The number of possible options for processing the large variety of crude oils for producing multiple types and qualities of products is too large to allow an exhaustive treatment. For this reason, in the following, only some typical solutions and general guidelines will be presented as illustrations.

For ease of presentation, the solutions will be presented separately for the conversion of the residues, gas oil production, gasoline production, gas processing, and production of lubricating oils.

The desalting and the atmospheric and vacuum distillation of the crude oil belong in any processing flow diagram and therefore do not require a separate treatment.

14.5.1.1   Residue Conversion

The present trend of increasing the amount of automotive fuels derived from crude, led to processes for converting the residues to lighter fractions.

The first of such processes was the thermal cracking of the straight-run residue, today no longer used. It was followed, beginning in 1945–1950, by catalytic cracking. In that period, only distillates such as vacuum distillates were used as feed. The residue was submitted to visbreaking, coking, or deasphalting The resulted distillates, respectively the deasphalted oils, were included in the feed to the catalytic cracking.

These classic schemes, depicted in Figure 14.7, were applied for a long period of time. They are still in use where no new requirements for improved quality and types of the products supplied by the refineries appeared and no improvements in the catalytic cracking technology were implemented. Increasingly stringent requirements for the fuel oils refer to improved stability and lower contents of sulfur, nitrogen and metals, especially vanadium. For the Diesel fuels it refers to a high cetane number and new quality specifications for gasoline and the sulfur content for coke.

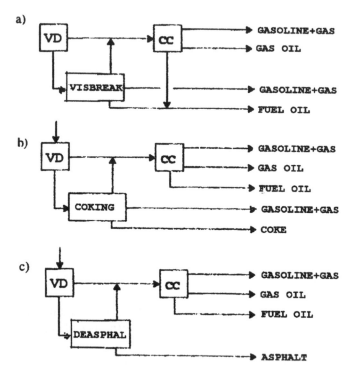

Figure 14.7. Classic schemes for processing of vacuum residue: (a). visbreaking, (b) coking, (c) propane deasphalting. B - naphtha, G - gases, M - gas oil, C - fuel oil.

The improvements brought to catalytic cracking led to the decrease of the amount of coke generated on the catalyst and made possible the catalytic cracking of residues. At the same time, new zeolitic catalysts, sensitive to poisoning by heavy metals (especially Ni and V) made it necessary to develop new methods for pretreating the feed.

These improvements are performed often by incorporating processes for the treatment with hydrogen of the various streams of the diagrams of Figure 14.7, while keeping catalytic cracking as the main process in the production of gasoline.

Thus, the vacuum distillate that feeds the catalytic cracking is submitted to hydrogen treating, a process that has usually the objective to achieve denitration and demetallation of the feed. To a similar treatment are submitted also the distillates produced in visbreaking and in coking, and in some cases, the deasphalted oil. In the case of fluid coking, the process must be more severe in view of the aromatic character of the distillate. The need to treat the deasphalted oil with hydrogen is determined by the operating conditions of the process and on the use for deasphalting of higher hydrocarbons, which increase the heteroatoms content in the deasphalted oil. The improved performance of catalytic cracking as a result of this treatment of the feed was analyzed in Section 7.1.

On rare occasions the vacuum residue is submitted to hydrocracking, which is generally more expensive. A comparison of such a solution to that of hydrotreating of the vacuum distillate alone [7] gives useful indications for the selection of such approaches.

A complete comparison, including the mass balances for six processing options, was performed [9] for a heavy Arabian crude oil. The following processing schemes of the atmospheric residue were considered:

1. Vacuum distillation with the catalytic cracking of the distillate, delayed coking of the distillation residue, and hydrofining of the resulted light products.
2. The same scheme with the hydrofining of the feed to catalytic cracking.
3. The same scheme where the delayed coking is replaced by flexicoking.
4. The C_3-C_5 deasphalting of the atmospheric residue followed by the hydrofining of the oils, including the feed to catalytic cracking.
5. Fluidized bed hydrocracking of the vacuum residue, with the hydrofining of the products. The feed to the catalytic cracking is the vacuum distillate with the distillate resulting from hydrocracking.
6. Hydrocracking of the atmospheric residue, with the catalytic cracking of the fractions with the boiling temperature above 375°C.
7. Hydrofining of the atmospheric residue followed by vacuum distillation, the catalytic cracking of the distillate, and coking of the residue.

It is remarkable that visbreaking is not taken into consideration in any of these schemes. The reason is the current lack of interest for heavy fuel oil, which has a low value on the market.

Occasionally, schemes are proposed, in which hydrocracking [6] by itself, or in combination with catalytic cracking produce only distillates: gas oil, jet fuel, gasoline, and gases. Such approaches seem to have a good future [1].

One of the currently used processing schemes for residue conversions including some possible options is shown in Figure 14.8.

14.5.1.2 Production of Gas Oil and Jet Fuel

Merchant gas oils may be obtained by mixing gas oil cuts from the following sources: straight-run gas oils, those of catalytic cracking and of hydrocracking, as well as the corresponding distillates from coking and visbreaking.

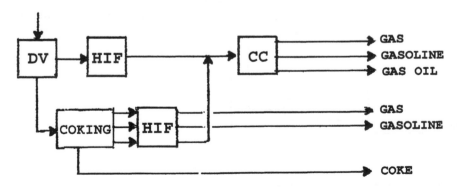

Figure 14.8. Residue conversion scheme. The coking may be replaced by visbreaking and the catalytic cracking (CC) by hydrocracking (DV - vacuum distillation, HIF - hydrofining).

In order to obtain Diesel fuel that satisfies the present requirements of stability and cetane number, coking and visbreaking gas oils are hydrofined to acquire a saturated character. Those from catalytic cracking require deep hydrotreating for the saturation of the polycyclic aromatics.

The hydrocracking gas oils have high cetane numbers and low freezing points; therefore they do not require any special treatments.

The straight run gas oils resulting from crude oils with high sulfur content, must be submitted to hydrofining in order to comply to the present norms for acceptable sulfur levels. If they have an aromatic character, these gas oils are submitted to hydrotreating.

The hydrocracking gas oils and those of catalytic cracking have low freezing points, and the latter have a depressive character on the mixed freezing point. In some cases, in order to reduce the freezing point it is necessary to lower the end point of the primary gas oil and to include in the gas oil a corresponding amount of kerosene. Additional information has been reported in some publications [5,17].

A general scheme for the production of good quality Diesel gas oil is shown in Figure 14.9. Selected straight-run and hydrocracking kerosene are the preferred raw materials for jet fuel. The latter is of exceptional quality and has a lower content of aromatic hydrocarbons than admitted by the jet fuel specifications. If used in the mixture with the straight run kerosene, hydrocracking kerosene may compensate for the somewhat higher content of aromatic hydrocarbon of the straight run fractions.

The straight run kerosene usually needs desulfurization.

If the content of aromatic hydrocarbon exceeds the specifications, the hydrofining is replaced by hydrotreating. The latter is preferred to solvent refining, which generally increases the freezing point.

14.5.1.3 Gas Processing

The processing of the hydrocarbon gases is often correlated with their use in petrochemical processes. The examination of these processes exceeds the scope of this

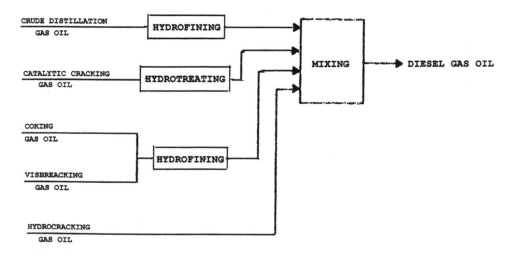

Figure 14.9.   Production of Diesel fuel.

book, so that present discussion is limited to the production of components of automotive fuels, and gasoline respectively.

Three main processes are presented: alkylation, oligomerization, and production of ethers.

The inclusion of ethers in the composition of high grade gasoline became compulsory in technically advanced countries and will with certainty be increasingly expanded as measures for environmental protection will be legislated in other countries. For these reasons, the available resources of isobutene and iso-amylenes are almost exclusively used for ethers production at the expense of petrochemical uses.

The iso-amylenes are obtained almost exclusively from catalytic cracking, while isobutene is also obtained from catalytic cracking and to a lesser extent from thermal cracking processes.

On occasions, when the available resources of isobutene and iso-amylene did not cover the requirements for ethers production, isobutene was produced by the dehydrogenation of isobutane produced by the isomerization of n-butane [28,32].

The selection between alkylation and oligomerization is based on consideration of the product quality, the cost of the process, and the available raw materials.

The unsaturated character of the oligomerization gasoline and the restrictions concerning the content of alkenes in the finished gasoline, which will probably become tighter in the future, make alkylation the preferred choice for new refineries or for the upgrading of existing ones. The only reason in favor of the oligomerization is lower investments and operation costs.

Starting from the same amount of alkenes, alkylation produces about double the amount of end product than oligomerization. The alkylate is saturated and has excellent octane rating.

Most of the isobutane used for alkylation is recovered from the catalytic cracking gases and from hydrocracking gases. If necessary, isobutane can be produced by the hydroisomerization of n-butane.

The C_3-C_4 alkenes are recovered from the catalytic cracking gases and from the gases produced in the processes of coking, visbreaking, and pyrolysis. Before being fed to the alkylation unit, the C_4 fraction passes through the MTBE unit, where the isobutene contained in it is converted to ether.

The alkylation process selected to be incorporated in the general refining scheme depends on the availability of i-butane (possibly i-pentane), n-butenes, propene, and possibly amylenes. The differences between the performance of the alkylation with sulfuric acid and that with hydrofluoric acid depend on the alkenes processed. Generally, if propene is included in the feed the alkylation with hydrofluoric acid performs better (see Section 9.2). The quality of the alkylate depends on the concentration of the various butanes in the feed.

14.5.1.4 Gasoline Production

The resources for finished gasoline production are the straight-run naphthas, those resulted from the processing of the heavier fractions (catalytic cracking, hydrocracking, coking, and visbreaking) and from gas processing (alkylate, oligomerization gasoline, and ethers).

In general, there are more possible options and the problems to be solved are more difficult than for other products, since there are several qualities of gasoline that are in demand in parallel and a strong interdependence between the components used. Also, the quality requirements for gasoline, both related to the performance of the motors and the restrictions related to environmental protection, vary much faster than for other products. Followed by the exclusion of lead compounds from the gasoline, the reduction of the contents in amylenes, benzene, or of aromatic hydrocarbons occurred within a short time period and forced the refining industry to make significant investments in order to comply with new quality requirements.

It follows that gasoline production, more than most other processes, has to take into account the direction of evolving requirements and to evaluate a larger number of options when deciding on modifications of the processing scheme.

Consensus regarding the evolution of gasoline quality makes the incorporation of some processes a definite obligation. Thus, catalytic reforming of straight-run naphtha ensures such a large increase of the octane number (about 40 units) that the process cannot be omitted from the general flow sheet. Also, the amount of gasoline presently required by the market cannot be supplied without the processes of catalytic cracking, hydrocracking, or the combination of their products.

Other processes are incorporated in the general scheme in order to comply with specific requirements. Thus, the limitations imposed on benzene content in gasoline determined the extension of the hydroisomerization to the C_5-C_6 fractions instead of including the C_6 fractions in the feed to catalytic reforming. Similar changes determined the expansion of the hydrocracking, which, different from the catalytic cracking, produces a saturated product having a high octane number.

The processes that use noble metals catalysts, especially platinum, require the preliminary hydrofining of their feeds. The corresponding units have to be included in the general processing scheme.

The naphtha fractions resulting from coking and visbreaking are usually hydrofined together with the straight run naphtha and included in the feed to catalytic reforming.

The processing of refinery gases produces alkylate, oligomerization gasoline, and ethers as components of the finished gasoline.

The large number of components, the demand for 2–4 types of gasoline, the high number of qualitative characteristic and their nonadditive character, increase considerably the number of the possible options for the general processing scheme and sometimes leads to difficulties in anticipating the characteristics of the produced gasoline.

The correct selection of the possible options for the production of the various types of gasoline and their correct evaluation is the most difficult part of the development of the general processing scheme. Several publications provide good pointers for solving this task [1-4,9,21,28,35].

14.5.1.5 Production of Lubricating Oils

Within the general processing scheme, the production of lubricating motor oils may pose special problems.

Their production has not changed significantly. It consists of submitting the vacuum distillates to the traditional sequence of operations: solvent refining followed by dewaxing and finishing. For vacuum residues, these operations are preceded by propane deasphalting.

Owing to the poor stability of furfural [36] and the toxicity of phenol, until 1982 in the U.S., 12 solvent extraction units were modified to replace the phenol by

N-methyl-2-pyrolidone (NMP). A large number of solvent refining units using NMP were put into operation since then [37]. Owing to the physical properties and the performance of this solvent, the refining with NMP proved in many instances to be superior to that using phenol or furfural [38,39].

In parallel, improvements were brought to the solvent recovery systems leading to energy economies of 40–50% over the traditional systems [37,39].

The dewaxing process with methyl-ethyl-ketone was improved with respect to the dilution technique, the implementation of two-step filtration, and other measures, which increased filter productivity by about 30% [39]. Also, improvements were made to the solvent recovery section.

The most important changes to the production of lubricating oils are the result of the introduction of hydrogen-treating processes.

Initially, hydrogen treating replaced refining with sulfuric acid followed by clay contacting, leading to improved yields, better color, and in some cases an improved stability.

Later on, hydrogen treatment was expanded and replaced solvent refining. By submitting the oils to hydrotreating, one achieves the hydrogenation of the poly-cyclic aromatic hydrocarbons and as a result, the increase of the viscosity index. Different from solvent refining, which in order to achieve good yields requires the use of selected crude oils, hydrotreating makes it possible to produce oils with high viscosity indexes from crude oils of low quality. The process is very efficient for the production of "multigrade" oils that have viscosity indexes of the order 120–130 [40,41]. However, hydrotreating is more expensive than solvent refining, both in terms of investment and of operating costs.

In order to lower the cost, a hybrid solution was applied in Geelong (Australia), in 1980. High viscosity index oils are produced by performing in a first step a mild solvent refining, followed by hydroconversion [37].

The early concerns about the higher costs of hydrotreating compared to sol-vent refining may not be warranted in view of the important progress achieved in the performance of the catalysts used in hydrotreating technology, as well as in the use of hydrogen recovered from diluted streams [48].

A one-step catalytic dewaxing was developed using platinum catalysts on mor-denite, and well-suited for the dewaxing of light oils [37]. The use of two reaction zones in the MLDV process uses in a first step a ZSM-5 zeolite catalyst followed by a hydrofining step (hydrogen consumption is 18–36 Nm^3/m^3 feed) and has better process economics than dewaxing [37].

Additional details on the processes for hydrogen treating of oil fractions are given in Section 11.7.3 and especially in Section 12.9.

The combination of the traditional technology for the production of lubricat-ing oils with the hydrogen treating processes allows development of several varia-tions of the general scheme, so as to optimize the process economics.

14.5.1.6 Coordination of Processing Systems

The general schemes discussed in the previous sections refer largely to the processing of a specific fraction. They must be correlated in order to obtain the overall scheme for the processing of a given crude oil in order to obtain the desired types of products and qualities. These general schemes make it possible to select the options that have the most competitive economics [9].

The correlation of the partial schemes leads to the increase of the number of options that should be evaluated, as discussed in the following sections.

14.5.2 Process Economics

In order to obtain specified types and properties of products from a given crude oil, or types of crude oil, the evaluation of the overall processing scheme will involve the material balances, the investments, the utilities, and the processing costs. All these evaluations are indispensable for comparing the economics of the various process scenarios developed.

The overall mass balance for the complex scheme is obtained by combining the mass balances of all participating processes. The quantitative and qualitative restrictions imposed by the project objective refer usually not to all, but only to a portion of the products and indicate the allowed variation from the specifications. For this reason, the final mass balances may show significant differences between the types and quality of the products anticipated to be obtained, even though all prescribed conditions were observed.

The evaluation of the balances and of the processing units included in the various scenarios may present difficulties and require the attentive analysis of all available data. Sometimes, a testing campaign in a pilot plant is necessary in order to obtain reliable mass balances.

Similarly, the published data concerning the investments and the consumption of utilities taken after an attentive analysis, constitute a first-pass comparison of the general schemes. The final decisions requires, however, evaluating the offers provided by the suppliers [49,50].

The consumptions of utilities, obtained as the sum of the consumptions in all processing units, lead to the overall consumptions corresponding to each of the schemes considered. On their basis one may perform a first-pass comparison among the values corresponding to each scheme. Similarly, it is possible to calculate the cost of labor and the investments. Thus, a first preliminary comparison results between the suggested schemes.

A more precise evaluation follows that takes into account the specific manner for solving the following problems: the supply with power, steam, and water, the management of the high- and low-pressure steam, of the condensed water, of the recycled water, etc. Special attention is paid at this stage to the waste waters, to the possibility of maintaining separately the cooling water and the process water, if this is justified by the contaminants present. In this analysis it is useful to design on the basis of the anticipated more severe conditions that will influence in the future the specifications of the water that may be discharged in the rivers or the seas or oceans.

In order to efficiently find the correct solutions to the problems concerning the supply and the management of utilities it is useful to resort to the services of specialists in the respective domains.

Hydrogen belongs to the utilities for which an overall balance must be established. If the hydrogen produced in catalytic reforming is not sufficient, various options for closing the balance must be evaluated and compared: recovery from diluted streams, or its production, indicating the considered technology [17,49]. For the selection of the most competitive method of generating the makeup hydrogen, the involvement of specialists is beneficial.

The accurate economic evaluation of each of the considered schemes requires an important effort, and in view of the large number of schemes generated by the number of options selected, it is recommended to proceed by trial-and-error.

Thus, a first-pass selection is made on the basis of evaluations that may contain reasonable assumptions. The remaining schemes are compared on the basis of a more accurate evaluation. Thus, two or three successive stages are required for selecting the optimal scheme.

This approach must be applied paying attention to the validity and consistency of the assumptions or approximations incorporated in the earlier steps. Indeed, if the accepted approximations were not correctly selected, schemes could be eliminated that are actually superior to the one selected.

14.5.3 Evaluation of Trends. Selection of Final Process Configuration

The selection among the several process schemes considered comprises a succession of increasingly precise comparative evaluations followed by the rejection of the less competitive solutions (trial-and-error), or of those that appear to be difficult to implement within the given practical conditions.

Eventually, only a small number of solutions remain. The selection among them on the exclusive basis of economic advantage is difficult.

In this situation it becomes necessary to consider the market and technical trends and the flexibility or ability of each of the process schemes to adapt to foreseen changes in the demand (capacity) and tightening of purity specifications for the products.

The capacity to adapt to future requirements concerning the evolution of the types and quality of products, comprises two aspects:

1. Capacity of the units to be built to adapt by modification of the process operating conditions
2. Incorporation in the general scheme of additional process units

Many times, the consideration of such additional factors may determine the selection between two general schemes, which are otherwise close, from the point of view of the instant economics. Consideration of foreseeable trends may determine the final selection of a process scheme, the economics of which are close to but at the present time are not the optimal.

Issues related to the financing of the project may also come into play. Thus we see, the realization of the project in stages of compatible process units selected from among those that are part of the overall process scheme. Such a stage-wise realization is often worth considering, since it also allows reevaluation of the trend factors in the course of the project execution.

REFERENCES

1. G Heinrich, M Valais, M Passot, B Chapotel, Mutations of World Refining, Proceedings of the Thirteenth World Petroleum Congress, Vol. 3, 189, New York: John Wiley & Sons, 1992.

2. RF Denny, NL Gilsdorf, FM Hibbs, EJ Houde, ME Reno, RP Silverman, The Refining Challenge for the 1990's, UOP, 1990.
3. RI Schmidt, PL Bogdan, NL Glisdorf, Chemtech 41: February 1993.
4. JL Nocca, A Forestière, J Cosyns, Revue de l'Institut Français du Pétrole 49: 461, 1994.
5. GJ Thompson, NL Gilsdorf, JK Gorawara, DA Kauff, Diesel Regulations, UOP, 1990.
6. GL Farina, M Fontanam, Hydrocarbon Processing 72: 52, Nov. 1993.
7. HD Sloan, Hydrocarbon Processing 70: 99, Nov. 1991.
8. JF Le Page, SG Chatila, M Davidson, Raffinage et conversion des produits lourds du pétrole, Technip. Paris, 1990.
9. HR Slewert, AH Koenig, TA Ring, Hydrocarbon Processing 64: 61, March 1985.
10. F Tamburrano, Hydrocarbon Processing 73: 70, Sept. 1994.
11. I Opris I, Producerea de hidrogen, In: G Suciu, R Tunescu, eds., Ingineria prelucrării hidrocarburilor, vol. 2, Bucharest: Editura Tehnică, 1974, p. 349.
12. Texaco introduces a new high purity hydrogen generation process: HyTEX, In: P Harrison, ed., Hydrocarbon Technology International 1992, London: Sterling Publishing International.
13. GW Bridger, W Wyrwas, Chem Process Eng. 38(9): 101, 1967.
14. G Berlioux, L'Ind Pétrole Eur 417: 17, 1971.
15. Kirk-Othmer, Encyclopedia of Chemical Technology, 2nd ed., vol. XI, New York: John Wiley & Sons Inc., 1966, pp. 338–379.
16. GJ Van der Berg, Chem Progress Eng 52(10): 49, 1971.
17. EC Haun, RF Anderson, DA Kauff, GQ Miller, J Stoeker, The Efficient Refinery Hydrogen Management in the 1990's, UOP, 1990.
18. Ingineria prelucrării hidrocarburilor, vol. 1-3, GC Suciu, ed., Bucharest: Editura Tehnică, 1983, 1985, 1987.
19. H Yamashiro, M Hirajo, WJ Schell, CF Maitland, Hydrocarbon Processing 64: 87, Feb. 1985.
20. JL Glazer, ME Schott, LA Staph, Hydrocarbon Processing 67: 61, Oct. 1988.
21. BV Vora, ST Bakas, RF Denny, CP Luebke, PR Prijado, RL Venson, Ethers for Gasoline Blending, UOP, 1990.
22. I Velea, G Ivănus, Monomeri de sinteză, vol. 1, Bucharest: Editura Tehnică, 1989.
23. RG Craig, DC Spence, Catalytic Dehydrogenation of Liquefied Petroleum Gas by the Hundry CATOFIN and CATADIENE Processes, In: RA Meyers, ed., Handbook of Petroleum Refining Processes, New York: McGraw Hill Book Co., 1986.
24.. T Hutson Jr, WC McCarthy, Philips STAR Process for Dehydrogenation and Dehydrocyclization of Refinery Paraffins, In: RA Meyers, ed., Handbook of Petroleum Refining Processes, New York: McGraw Hill Book Co., 1986.
25. FP Wilcher, CP Luebke, PR Pujado, Production of Light Olefins From LPG, In P Harrison, ed., Hydrocarbon Technology International 1992, London: Sterling Publishing Intern. Ltd., p. 93.
26. JJ McKetta, Petroleum Processing Handbook, New York: Marcel Dekker Inc., 1992, p. 544.
27. D Sanfilippo, Chemtech. 35: Aug. 1993.
28. JL Monfils, S Barendregt, SK Kapur, HM Woerde, Hydrocarbon Processing 71: 47, Feb. 1992.
29. Hydrocarbon Processing 73: 106, Nov. 1994.
30. Hydrocarbon Processing 73: 109, Nov. 1994.
31. Hydrocarbon Processing 73: 104, Nov. 1994.
32. Hydrocarbon Processing 63: 114, Apr. 1984.
33. Hydrocarbon Processing 71: 168, Nov. 1992.
34. Oil and Gas J. 68: Oct. 10.

35. NL Gilsdorf, AP Kelly, CA LeMerle, ME Reno, WC Tieman, Gasoline Processing for 1990, UOP.
36. F Richter, Hydrocarbon Processing 57(5): 1881, 1978.
37. DH Shaw, JD Bushnell, AD Reichle, RH Coulk, Recent Developments in Oil Refining, Proceedings of the Eleventh World Petroleum Congress, 1983, vol. 4, p. 345.
38. JD Bushnell, RJ Fiocco, Proc Div Refining Am Petrol Inst 59: 159, 1980.
39. AJ Segneira, MR McClure, CW Harrison, R Maxelon, Proc Div Refining Am Petrol Inst 59: 133, 1980.
40. R Hournac, Hydrocarbon Processing 60(1): 207, 1981.
41. JW Gleitsmann, Heat Eng 50(3): 45, 1981.
42. M Prigent, La Catalyse d'Epuration des Gazes d'Echappement Automobiles. Situation Actuelle et Perspectives, Revue de l'Institut Français du Pétrole 51(6): 829–842, Nov.–Dec. 1996.
43. W Weirauch, Benzene exposure risk is questioned, Hydrocarbon Processing 74: 23, Aug. 1995.
44. JE Johnson, FM Peterson, Watch-out: Here comes reformulated gasoline, Chemtech May 1991: 296–298, 1991.
45. J Cosyns, JL Nocca, Maximising Reformulated Gasoline Production, In: P Harrison, ed., Hydrocarbon Technology International 1993. London: Sterling Publishing Ltd., 1993.
46. JJ McKetta, Petroleum Processing Handbook, New York: Marcel Dekker, 1992, pp. 50–64.
47. A Donaud, Tomorrow's Engines and Fuels. A World Perspective on Transportation Trends for the Next 10 years, Hydrocarbon Processing 74: 55–61, Feb. 1995.
48. B Pacalowska, M Whysall, MV Narasimhan, Improve Hydrogen Recovery from Refinery Offgases, Hydrocarbon Processing 75: 55–59, Nov. 1996.
49. GH Shahani, LJ Garodz, KJ Murphy, WF Baade, P Sharma, Hydrogen and utility supply optimization, Hydrocarbon Processing 77(9): 143–150, Sept. 1998.
50. C Baudouin, JP Favennec, Marges et Perspectives du Raffinage, Revue de l'Institut Français du Pétrole 52(3): 337–347, May–June 1997.
51. Hydrocarbon Processing 77: 78, Nov. 1998.
52. K Rock, H Semerak, P Greenough, CP Hälsig, H Tschorn, Options for Selective Hydrogenation and Etherification of Refinery Feeds, In: P Harrison ed., Hydrocarbon Technology International 1994, London: Sterling Publishing Ltd., 1994.
53. Hydrocarbon Processing 74: 27, Nov. 1995.
54. F Brunngnell, P Liebert, S Le Vaillant, D Poivet, Diesel faut il l'interdire en ville? Science et Avenir May 1996: 36–42.
55. Hydrocarbon Processing 74: 27, Dec. 1995.
56. Hydrocarbon Processing 75: 35, April 1996.
57. CH Twu, JE Coon, Predict octane number using a generalized interaction method, Hydrocarbon Processing 75: 51, Feb. 1996.
58. M Absi-Halbi, A Stanislaus, H Quabazand, Trends in catalysis research to meet future refining needs. Hydrocarbon Processing 76: 45–54, Feb. 1997.

Appendix

Influence of the *n/i*-Alkanes Ratio in the Pyrolysis Feed on the Ethene/Propene Ratio in the Products

In order to evaluate the influence of the *n/i*-alkanes ratio in the feed on the ethene/propene ratio in the pyrolysis effluent, three representative hydrocarbons were selected: *n*-octane, 4-methyl-heptane and 2,5-dimethyl-hexane.

For these hydrocarbons, the product's composition was calculated using the F.O. Rice method at a temperature of 1100 K, which is typical for those used in pyrolysis.

For *n*-octane, the complete decomposition of the radicals formed by different substitution reactions gives:

$$\text{C-C-C-C-C-C-C-}\overset{\bullet}{\text{C}} \longrightarrow 4\,C_2H_4 + \overset{\bullet}{H} \tag{a}$$

$$\text{C-C-C-C-C-C-}\overset{\bullet}{\text{C}}\text{-C} \longrightarrow C_3H_6 + 2\,C_2H_4 + \overset{\bullet}{C}H_3 \tag{b}$$

$$\text{C-C-C-C-C-}\overset{\bullet}{\text{C}}\text{-C-C} \left\langle \begin{array}{l} C_7H_6 + \overset{\bullet}{C}H_3 \\ C_4H_8 + 2\,C_2H_4 + \overset{\bullet}{H} \end{array}\right. \begin{array}{l} \text{(c)} \\ \text{(d)} \end{array}$$

$$\text{C-C-C-C-}\overset{\bullet}{\text{C}}\text{-C-C-C} \left\langle \begin{array}{l} C_6H_{12} + C_2H_4 + \overset{\bullet}{H} \\ C_5H_{10} + C_2H_4 + \overset{\bullet}{C}H_3 \end{array}\right. \begin{array}{l} \text{(e)} \\ \text{(f)} \end{array}$$

The dissociation energies admitted by Rice for the bonds between hydrogen and the primary, secondary, and tertiary carbons atoms (92.0, 90.8, and 88.0 kcal/mol), give for 1100 K, according to Eq. (2.4):

$$\frac{r_{sec}}{r_{prim}} = 1.7 \qquad \frac{r_{tert}}{r_{prim}} = 5.7$$

Each of the two last reactions have two possible paths for the decomposition of the initial radicals (c,d and e,f). The relative rates of these may be calculated using the dissociation energies given in Table 2.4. It results:

$$\frac{r_d}{r_c} = e^{\frac{138,000-113,000}{1100 \cdot R}} = 15.6$$

$$\frac{r_f}{r_e} = e^{\frac{130,000-121,000}{1100 \cdot R}} = 2.5$$

The relative rates of the reactions (a)-(f) will be:

Reactions	Relative rate	Conversion per mol n-octane
(a)	$1 \times 6 = 6$	0.22
(b)	$1.7 \times 4 = 6.8$	0.26
(c)	$1.7 \times 4 \times \dfrac{1}{15.6} = 0.436$	0.02
(d)	$1.7 \times 4 \times \dfrac{14.6}{15.6} = 6.364$	0.24
(e)	$1.7 \times 4 \times \dfrac{1}{2.5} = 2.72$	0.10
(f)	$1.7 \times 4 \times \dfrac{1.5}{2.5} = 4.08$	0.16

Taking into account the number of molecules of ethene and propene formed in the reactions (a)-(f), it results that for 1 mol of reacted n-octane the following amounts of ethene and propene will be obtained:

0.22 4 + 0.26 2 + 0.24 2 + 0.10 + 0.16 = 2.14 mol ethene and
0.26 mol propene
The molar ratio ethene/propene = 8.23

For 4-methyl-heptane the different substitution reactions give:

$$C-C-C-\underset{\underset{C}{|}}{C}-C-C-\dot{C} \begin{cases} \longrightarrow C_2H_4 + C_5H_{10} + \dot{C}H_3 & \text{(a)} \\ \longrightarrow 2\,C_2H_4 + C_3H_6 + \dot{C}H_3 & \text{(b)} \end{cases}$$

$$C-C-C-\underset{\underset{C}{|}}{C}-C-\dot{C}-C \longrightarrow 2\,C_3H_6 + C_2H_4 + \dot{H} \qquad \text{(c)}$$

$$C-C-C-\underset{\underset{C}{|}}{C}-\dot{C}-C-C \begin{cases} \longrightarrow n\text{-}C_7H_{14} + \dot{C}H_3 & \text{(d)} \\ \longrightarrow C_5H_{10} + C_2H_4 + \dot{C}H_3 & \text{(e)} \\ \longrightarrow i\text{-}C_7H_{14} + \dot{C}H_3 & \text{(f)} \end{cases}$$

$$C-C-C-\underset{\underset{C}{|}}{\dot{C}}-C-C-C \longrightarrow C_6H_{12} + C_2H_4 + \dot{H} \qquad \text{(g)}$$

$$C-C-C-\underset{\underset{C}{|}}{C}-C-C-C \longrightarrow n\text{-}C_5H_{10} + C_2H_4 + \dot{C}H_3 \tag{h}$$

Estimating the relative rates of the reactions (a)/(b), (d)/(e), and (f)/(e) as in the pyrolysis of *n*-octane, it results:

Reactions	Relative rate	Conversion per mol 4-methyl-heptane
(a)	$1 \times 6 \times \frac{1}{3} = 2$	0.07
(b)	$1 \times 6 \times \frac{2}{3} = 4$	0.14
(c)	$1.7 \times 4 = 6.8$	0.24
(d)	$1.7 \times 4 \times \dfrac{1}{15.6} = 0.436$	0.02
(e)	$1.7 \times 4 \times \dfrac{13.6}{15.6} = 5.928$	0.21
(f)	$1.7 \times 4 \times \dfrac{1}{15.6} = 0.436$	0.02
(g)	$5.7 \times 1 = 5.7$	0.20
(h)	$1 \times 3 = 3$	0.10

The number of moles of ethene and propene formed from one mole of 4-methyl-heptane decomposed will be:

0.07 + 0.14 2 + 0.24 + 0.21 + 0.20 + 0.10 = 1.10 moles ethene and
0.14 + 2 0.24 = 0.62 moles of propene
The molar ratio ethene/propene = 1.77

For 2,5-dimethyl-hexane, a similar reasoning gives:

$$C-\underset{\underset{C}{|}}{C}-C-C-\underset{\underset{C}{|}}{C}-\dot{C} \begin{cases} \longrightarrow i\text{-}C_7H_{14} + \dot{C}H_3 & \text{(a)} \\ \longrightarrow 2\,C_3H_6 + C_2H_4 + \dot{H} & \text{(b)} \end{cases}$$

$$C-\underset{\underset{C}{|}}{C}-C-C-\underset{\underset{C}{|}}{\dot{C}}-C \longrightarrow C_4H_8 + C_3H_6 + \dot{C}H_3 \tag{c}$$

$$C-\underset{\underset{C}{|}}{C}-C-\dot{C}-\underset{\underset{C}{|}}{C}-C \begin{cases} \longrightarrow C_7H_{14} + \dot{C}H_3 & \text{(d)} \\ \longrightarrow C_5H_{10} + C_3H_6 + \dot{H} & \text{(e)} \end{cases}$$

In the reactions (b) and (c) it is considered that the propyl radical decomposes to propene and atomic hydrogen, since in this case the dissociation energy is only 167 kJ/mol. Analogously, in this example it was considered that the ethyl radical is decomposed completely to ethene and atomic hydrogen. Both of these considerations correspond to the temperature and pressure conditions that are specific to pyrolysis.

The relative rates of the reactions (a)-(e) will be:

Reactions	Relative rate	Conversion per mol 2,5-dimethyl-hexane
(a)	$1 \times 12 \times \dfrac{1}{15.6} = 0.77$	0.04
(b)	$1 \times 12 \times \dfrac{14.6}{15.6} = 11.23$	0.52
(c)	$1\ 5.7 = 5.7$	0.26
(d)	$1 \times 4 \times \dfrac{1}{15.6} = 0.26$	0.01
(e)	$1 \times 4 \times \dfrac{14.6}{15.6} = 3.74$	0.17

The number of moles of ethene and propene formed from a mol of 2,5-dimethyl-hexane will be:

0.52 moles ethene and
$0.52 + 2 + 0.26 + 0.17 = 1.47$ moles propene
The molar ratio ethene/propene $= 0.35$
Finally, it results:

Feed	C_2H_4/C_3H_8 ratio in the effluent
n-C_8H_{18}	8.23
2-$C_7H_{15}CH_3$	1.77
2,5-$C_6H_{12}(CH_3)_2$	0.35

Index

Milton Keynes UK
Ingram Content Group UK Ltd.
UKHW051850071024
449327UK00025B/1904